Lecture Notes in Electrical Engineering

Volume 122

Gary Lee (Ed.)

Advances in Automation and Robotics, Vol.1

Selected Papers from the 2011 International Conference on Automation and Robotics (ICAR 2011), Dubai, December 1–2, 2011

Springer

Prof. Gary Lee
Information Engineering Research Institute
100 Continental Dr
Newark, DE 19713
United States
E-mail: garrylee@ier-institute.org

ISBN 978-3-642-25552-6 e-ISBN 978-3-642-25553-3

DOI 10.1007/978-3-642-25553-3

Lecture Notes in Electrical Engineering ISSN 1876-1100

Library of Congress Control Number: 2011941700

Typeset by Scientific Publishing Services Pvt. Ltd., Chennai, India.

Printed on acid-free paper

9 8 7 6 5 4 3 2 1

springer.com

Preface

The international conference on Automation and Robotics-ICAR2011 is held during December 12–13, 2011 in Dubai, UAE. The proceedings of ICAR2011 have been published by Springer Lecture Notes in Electrical Engineering, which include about 163 excellent papers selected from more than 400 submitted paper(s).

The conference is intended to bring together the researchers and engineers/technologists working in different aspects of intelligent control systems and optimization, robotics and automation, signal processing, sensors, systems modeling and control, industrial engineering, production and management. In addition to the contributed papers, the conference committee has invited papers by active researchers from various countries in relevant topic areas covered at ICAR2011. Internationally known experts from several countries are invited to deliver plenary/Keynote lectures at ICAR2011.

Much of the credit of the success of the conference is due to topic coordinators who have devoted their expertise and experience in promoting and in general co-ordination of the activities for the organization and operation of the conference. The coordinators of various session topics have devoted a considerable time and energy in soliciting papers from relevant researchers for presentation at the conference. The Chairpersons of the different session played important role in conducting the proceedings of the session in a timely and efficient manner and the Committee expresses sincere appreciation for their involvement.

The reviewers of the manuscripts, those by tradition would remain anonymous, have been very helpful in efficiently reviewing the manuscripts, providing valuable comments well within the time allotted to them. The Editors express their sincere and grateful thanks to all reviewers.

The ICAR2011 International Committee is proud to dedicate this conference proceeding to the Honor Chair David Wang from IEEE Nanotechnology Council Cambodia Chapter Past Chair, Cambodia. And Mark Zhou, from Hong Kong Education Society, Hong Kong, was not only an outstanding researcher and visionary but was also a humble human being.

Finally, our sincere thanks to the supper from Information Engineering Research Institute (USA) for designing the conference web page and also spending countless days in preparing the final conference program in time for printing. Many thanks to Miss Han: ICAR2011 secretary for arranging to type a large number of the invitation letters and for establishing the vast ICAR2011 conference address data base. Sincere thanks to our organization committee for their several months' hard work in sorting out manuscripts from various authors. Also thanks to other staffs assisting in the various stage of the editorial work, especially the international committees of our conference.

Garry Lee
Information Engineering Research Institute, USA

Conference Committee

Honor Chair

Gerald Schaefer — Loughborough University, U.K.

General Chair

Garry Lee — Information Engineering Research Institute, USA
Xiang Li — IERI Circuit and System Society, USA

Publication Chair

Garry Lee — Information Engineering Research Institute, USA

Organizing Chair

Ming Ma — Singapore NUS ACM Chapter Chair, Singapore
Paris Kitsos — American Applied Sciences Research Institute, USA

International Committee

Yuanzhi Wang	Anqing Normal University, China
Xiaofang Zhu	Jiangxi Normal Univesity, China
Gerald Schaefer	Loughborough University, UK
Qihai Zhou	South Western University of Finance and Economics, China
Srinivas Aluru	ACM NUS Singapore Chapter, Singapore
Tatsuya Akutsu	ACM NUS Singapore Chapter, Singapore
Aijun An	National University of Singapore, Singapore
Minli Dai	Suzhou University, China
Yijin Wu	Asia Pacific Environmental Science Research Center, Hong Kong
Zhenghong Wu	East China Normal University, China
Biswanath Vokkarane	Society on Social Implications of Technology and Engineering
Nian Zhang	Suzhou University, China
Xiang Li	IERI Circuit and System Society, USA

ICAR 2011 Sponsored by

Information Engineering Research Institute, USA

IERI Circuits & Systems Society

Contents

Formal Description of 3D Semi-tunnel Component

Enke Hou[1], Zhihua Zhang[2,*], and Xiaoxia Luo[3]

[1] School of Geology and Environment, Xi'an University of Science and Technology, 710054 Xi'an, China
[2] School of Mathematics, Physics and Software Engineering, Lanzhou Jiaotong University, 730070 Lanzhou, China
[3] College of Computer Science and Technology, Xi'an University of Science and Technology, 710054 Xi'an, China
houek@163.com, zhzhihua99@163.com, luoxx3@163.com

Abstract. Formal description is an abstract mathematical description based on the mathematical model, and it is a theory foundation of three-dimensional (3D) modeling and system development. The semi-tunnel is component which constructs 3D tunnel model as research object, using Object-oriented approach and the collection operation for specific ways, by analyzing the construction elements of semi-tunnel, all modeling elements of semi-tunnel are strictly defined and made formal description by utilizing the theory of topology. Then, based on the formal description of modeling elements, the formal description of semi-tunnel is expressed, it is easier to design and program the semi-tunnel. According to the formal description of semi-tunnel, it also simplifies the construction of 3D tunnel network model, and facilitates the spatial analysis and query.

Keywords: Semi-tunnel, three-dimensional, formal description, Data model, Component.

1 Introduction

Tunnel is a 3D entity, and it plays a very important role in exploration and exploitation of mining resources. 3D modeling of tunnel is not only make mining digitalized, but also to guide the mine safe production, gas monitoring and other work. The research on data model of 3D tunnel directly determines the difficulty of modeling, and it is also the key issue of spatial analysis. So, it occupies a central position. A semi-tunnel model is component to construct 3D tunnel network. In order to make the semi-tunnel have more prominent advantages, such as update model, topological description and seamless integration of models etc, we need to define and describe from the perspective of topological, so that it is easier to construct the model and to do spatial analysis. Through formal description on semi-tunnel, the paper will directly derived the formal description of 3D tunnel network model.

In formal description, a lot of domestic and foreign scholars have focused on spatial topological relationships, for example, Wang L.P [1] and Deng N.D [2] had

* Contact author.

G. Lee (Ed.): Advances in Automation and Robotics, Vol. 1, LNEE 122, pp. 1–8.
springerlink.com

researched on formal description of topological relations; Zhang S.J[3] had formally described topological relations on spatial lines; Deng M [4] discussed topological relations between spatial surfaces; Yan H.W [5-7] established the model of spatial direction relations on the basis of Voronoi diagram, then described this model formally. In addition, Egenhofer [8-10] proposed 4 intersection model (4IM) and 9 intersection model (9IM) on the basis of point set topology to describe the topological relations. In expression of model, OO-Solid data model was formally described [11]. Although there are many formal description on data model in the literature, but it is less on the specific 3D solid model, and most of above researches focus only on the topological relationships, and do not consider the specific model, it will not apply into reality. Therefore, this paper will formally describe on semi-tunnel and its elements which the semi-tunnel have been constructed, meanwhile, take topological relationships into account between elements of semi-tunnel. Based on this component, the whole tunnel network will be expressed.

2 Semi-tunnel and Its Elements

2.1 Composition of Semi-tunnel

Take arch tunnel for example, tunnel will be split in half along the center line by vertical surface. We get a section between two adjacent nodes, and then the component of semi-tunnel model will be obtained, as shown in Fig.1.

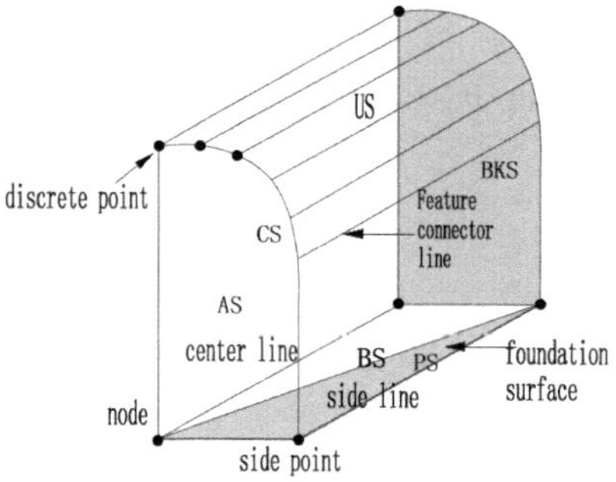

Fig. 1. Semi-tunnel and its elements

The semi-tunnel was put forward as a component model of 3D tunnel modeling there are many advantages, for example, when the model of intersections among some tunnels is constructed, it ensure that the intersection model has an internal connectivity; in addition, we can describe the network model of whole 3D tunnel by using this component model [12].

From Semi-tunnel component, there are point class (node, discrete points, and side point), line class (arc, tunnel center line, and feature connector line), and surface class (triangle, quadrilateral) which the semi-tunnel will be constructed by.

Through these basic geometric elements, component surfaces of semi-tunnel will be composed, there are bottom surface(*BS*), profile surface(*PS*), up surface(*US*), center surface(*CS*), front surface(*FS*) and back surface(*BKS*). Although the *FS*, *BKS* and *CS* does not need to build in practice, in order to facilitate the description, all of these auxiliary surface will be given, thus, a closed solid model is generated.

2.2 Elements of Semi-tunnel

The arch of semi-tunnel is discretized into five equal parts, that is the dome surface of semi-tunnel can be split into five interlocking surfaces, and there are US_1, US_2, … , US_5 from profile to center surface. The division of component surfaces will be conducive to modeling different surfaces and express surface objects alone, such as the bottom can represent tunnel floor. In addition, it can help to solve the correspondence between components, simplify other types of tunnels to construct, such as the construction of simple tunnel.

The geometric elements of Semi-tunnel include point class, line class and surface class, point class contains the nodes, side points, discrete points, among them, there are two nodes which expressed as Nd_1 and Nd_2, two side points which are represented by Sp_1 and Sp_2, 12 discrete points on arch which are represented by $Dp1_0$, $Dp1_1$, ..., $Dp1_5$(front surface) and $Dp2_0$, $Dp2_1$, ..., $Dp2_5$(back surface). Node is the start point and end point on the arc, in order to facilitate modeling, the concept of feature point is given, which is point on the front surface and back surface of semi-tunnel, and it is mainly used to build feature connector line. For example, side point and discrete point belong to the scope of feature points.

Line class includes arc, center line, side line, section outline and feature connector line. There are one arc, one center line, one side line and some feature connector lines in semi-tunnel component. The number of feature connector lines changes because of types of surface component, if we adopt triangle component, the connector lines will be more; on the contrary, it will be few as soon as the quadrilateral component is used. Assume that the feature connector lines define with counter-clockwise which the order begins from node on the center line, i.e. *FCL_BS*(feature connector line of *BS*), *FCL_PS*(feature connector line of *PS*), FCL_US_i(i=1,2…,11)(feature connector line of *US*, among them, the line connects between two discrete points on the AS and BKS, if it parallels to centerline, it will be named as parallel feature connector line, and represent by *FCL_PXX*; if it isn't parallel to centerline, it is diagonal feature connector line with *FCL_DJX* to represent.), *FCL*_CP(feature connector line of *CP*). The delineation of *FCLs* can easily express every component surfaces of semi-tunnel. Section outlines are mainly used to express *AS* and *BKS*. Take *AS* for example, There are Section outlines (represent by PMX_AS_i $i\in\{1,\ldots,8\}$)from beginning of node with counterclockwise order. Section polygon can be constructed by *PMS*, which can be used to describe the topological relationships between semi-tunnels, for example, the intersection of two adjacent semi-tunnels is a section polygon.

Foundation surface class has triangular component and quadrilateral component, because these two types are basic elements to build any surface. Similarly, all surfaces of semi-tunnel component can be expressed by foundation surface. Triangular component is represented by using *TFS*, and quadrilateral component is expressed with *QFS*. So, all component surfaces can be expressed with *TFS* or *QFS*. If we use *TFS*, there are *BS_TFS1* and *BS_TFS2* within *BS*, *BS*={*BS_TFS1*, *BS_TFS2*}, i.e. *BS* composes of *BS_TFS1* and *BS_TFS2*, the intersection set between *BS_TFS1* and *BS_TFS2* is a feature connector line. *PS* is similar to BS, $PS=\{PS\_TFS_1, PS\_TFS_2\}$. $US=\{US\_TFS_1, US\_TFS_2,..., US\_TFS_{10}\}$, there are 10 *TFSs* on the arch, any two adjacent US_TFS_i and US_TFS_{i+1} has intersection set: a feature connector line which is *FCL_PXX* or *FCL_DJX*. *CS*, *AS* and *BKS* needn't to construct in fact, but they have outlines. So, we adopt their contour polygons to describe.

According to above analysis, object-oriented 3D mathematical model of semi-tunnel can be simply formal described as follows:

$Semi\text{-}tunnel_i = <Surface_1, ..., Surface_n>$,
$Surface_i = < TFS_1, ..., TFS_n>$,
$TFS_i = <FCL_1, FCL_2, FCL_3>$,
$FCL_i = < Node_1, Node_2>$,

3 Formal Description of Modeling Elements

Some definitions will be given before formal description of elements.

Definition 1. Assuming that $P_0,P_1,\ldots,P_n$ are the point set within n-dimensional Euclidean space (*Rn*), According to topology theory, if $P_0,P_1,\ldots,P_n$ are linearly independent, n-simple(*Sn*) is the point set which is smallest convex set on *Rn*, then, $P_0,P_1,\ldots,P_n$ are vertexes of n-simplex, and express as $Sn=(P_0, P_1, ..., P_n)$[10]. If n=0, *Sn* represent a 0D simplex, and can describe a point object; if n=1, *Sn* represent a 1D simplex, and can describe a line object; if n=2, *Sn* represent a 2D simplex, and can describe a foundation surface, such as *TFS* or *QFS*; when n=3, *Sn* represent a 3D simplex, and can describe entity object, such as semi-tunnel component.

3.1 Description of Point

Definition 2. Assuming that *N* is the point set in 3D Euclidean space, then any two nodes p_i and p_j in this set must meet $P_i \cap P_j = \Phi$, $\bigcup_{i=0}^{n} P_i = N$, $P_i \in N\ (i \neq j)$.

There are node set, side point set, discrete point set and Feature Point Set in semi-tunnel. Feature Point Set includes *AS Feature Point Set* and *BKS Feature Point Set* which we set mainly to facilitate the subsequent description.

POINT SET={*Node Set, Side Point Set, Discrete Point Set* }={*AS Point Set, BKS Point Set* };
Node Set={ Nd_1, Nd_2};*Side Point Set* = { Sp_1, Sp_2};
Discrete Point Set = { $Dp1_0, Dp1_1,...,Dp1_5, Dp2_0,Dp2_1,...,Dp2_5$} = { $Dp1_i,Dp2_i| \ i \in$ {1,2…,5}};
AS Point Set={ $Nd_1,Sp_1,Dp1_i | i\in$ {0,1…,5}}; *BKS Point Set*={ $Nd_2,Sp_2,Dp2_i| \ i\in$ {0,1…,5}};

The geometric data of point is space coordinate, and express with <x,y,z> triple. Assuming that P is the point set of semi-tunnel on the 3D space, then any two points p_i and p_j are geometric independent, they must meet $p_i \cap p_j = \Phi(i \neq j)$.

$$point = \begin{vmatrix} (x, y, z) \\ p_i \cap p_j = \Phi, i \neq j \end{vmatrix} \tag{1}$$

Point type is distinguished by using an integer field, according to different field value, and it can express different types of point objects.

3.2 Formal Description of Line

Line class include Arc(*Al*), center line(*Cl*), side line(*Sl*), *FCL*, here the arc coincides with the center line. The paper adapts *TFS* to construct component surface, then *FCL* not only includes contour line of *AS* and *BKS*, but also contain connector line of *US*.

LINE SET={*Al* , *Cl* , *Sl*, *FCL SET*};
Al={(Nd_1, Nd_2)}; Cl={(Nd_1, Nd_2)};Sl={(Sp_1, Sp_2)};
FCL SET= {*FCL_PMX Set*, *FCL_PXX Set*, *FCL_DJX Set*};
FCL_PXX Set={*Al*, *Sl*, PXX_j| $j \in \{1,2...,6\}$ };*FCL_DJX Set*={ DJX_k,| $k \in \{1,2...,8\}$};
FCL_PMX Set={ *FCL_ASi*, *FCL_BKSi*};//contour lines on *AS* and *BKS*

FCL_PXX set is collection of parallel feature connector lines, FCL_DJX set is collection of diagonal feature connector lines. FCL_PMX set is collection of connector lines on AS and BKS.

Definition 3. Line set is collection of all line objects in 3D Euclidean space, any one $line_i$ must meet $linei \in Line(0 < i < n)$, and all line elements must satisfy $\bigcup_{i=0}^{n} line_i = Line$. If $line_i$ intersects with $line_j$, the intersection result must meet $line_i \cap line_j = P_k$ or $\Phi(i \neq j)$.

According to above definition, in semi-tunnel model, all types of lines are connected together by two points. Assumed that the IDs of AS and BKS are m and n, then $line = (p_{mi}, p_{nj})$, p_{mi} and p_{nj} are on *AS* and *BKS* respectively ($m \neq n, |i - j| \in \{0,1\}$). If p_{mi} and p_{nj} are on the same section, $m = n, |i - j| = 1$. When p_{mi} and p_{nj} are on the different section, $m \neq n$, if $|i - j| = 0$, (p_{mi}, p_{nj}) belongs to FCL_PXX set; If $|i - j| = 1$, (p_{mi}, p_{nj}) is in FCL_DJX set. i.e.

$$(p_{mi}, p_{nj}) \begin{cases} \in FCL\_PXX, m \neq n, |i - j| = 0 \\ \in FCL\_DJX, m \neq n, |i - j| = 1 \\ \in FCL\_PMX, m = n, |i - j| = 1 \end{cases} \tag{2}$$

Above expression, when $m \neq n$, $i = j = 1$, (p_{mi}, p_{nj}) is center line, p_{mi} and p_{nj} represent two nodes respectively. The formal description of line is as follows:

$$line = \begin{cases} (p_{mi}, p_{nj}) \begin{cases} m = n, |i-j| = 1, p_{mi} \in AS\ PointSet, p_{nj} \in AS\ PointSet \\ m \neq n, |i-j| \in \{0,1\}, p_{mi} \in AS\ PointSet, p_{nj} \in BKS\ PointSet \end{cases} \\ line_p \cap line_q \in \{point, \Phi\}, p \neq q \end{cases} \quad (3)$$

All line types can be characterized by an attribute field to distinguish. For example, $attr_i$ is name of attribute field, when $attr_i = 1$, the line represents center line; when $attr_i = 2$, the line is side line.

3.3 Formal Description of Surfaces

Surface class is component surfaces, and component surface are composed by TFS. Surface class and component surfaces represent as follows.

SURFACE SET={ *BS*, *PS*, *US*, *CS*, *AS*, *BKS*};
$BS=\{TFS\_BS_1, TFS\_BS_2\}$;$PS=\{TFS\_PS_1, TFS\_PS_2\}$;
$US=\{TFS\_US_1, TFS\_US_2, \ldots, TFS\_US_{10}\}$;$CS=\{TFS\_CS_1, TFS\_CS_2\}$;
AS and *BKS* are represented by outline polygon, i.e. $PMX\_AS=\{FCL\_AS_i | i\in\{1,2\ldots,8\}\}$, $PMX\_BKS=\{FCL\_BKS_j | j\in\{1,2,\ldots,8\}\}$.

According to analysis the triangular foundation surface, it can be represented by a unique triple array of vertices, namely $Tri = < p_{mi}, p_{mj}, p_{nk} >, i \neq j \neq k, m \neq n$. in the semi-tunnel model, there are two vertexes in triple which must be adjacent and on the same contour polygon of *AS* or *BKS*, then, $|i-j|=1$ or $|i-k|=1$ or $|k-j|=1$.

Definition 4. *F* is a collection of all *TFS* in the 3D Euclidean space, then, f_i within *F* must meet $\bigcup_{i=0}^{n} f_i = F$. If any two *TFSs* intersect, $s_i \cap s_j = Line_k$ or $\Phi (i \neq j)$.

Definition 5. *SS* is a set of all surface objects in 3D Euclidean space, s_i in *SS* must satisfy $\bigcup_{i=0}^{n} f_i = s_i$, $\bigcup_{i=0}^{n} si = SS$. If any two surfaces which are s_i and s_j intersect, $s_i \cap s_j = Line_k$ or $\Phi (i \neq j)$.

According to above definition, the formal description of TFS is following.

$$Tri = \begin{cases} < p_{mi}, p_{mj}, p_{nk} >, i \neq j \neq k, m \neq n, |i-j| = 1\ or |i-k| = 1\ or |j-k| = 1, i, j, k \in \{1,2,\ldots,8\} \\ Tri_i \cap Tri_j \in \{line, \Phi\}, i \neq j \end{cases} \quad (4)$$

With the formal description of *TFS*, description of component surfaces can be easily described by *TFS*. Among component surfaces, every surfaces of *BS*, *PS*, *CS* and *US* can be expressed by double adjacent *TFSs*.

$$surface = \begin{cases} Tri_i \cup Tri_j, i \neq j, |i-j| = 1 \\ Tri_i \cap Tri_j \begin{cases} \in \{FCL\ SET\}, |i-j| = 1 \\ \in \Phi, |i-j| \notin \{0,1\} \end{cases} \end{cases} \quad (5)$$

We can set a field to character the current component surface, and then differentiate each component surface through different value of the field. There are five surfaces in *US*, so it can be described as follows.

$$us = \begin{vmatrix} \bigcup_{k=1}^{n} surface_i = \bigcup_{k=1}^{n} (Tri_i \bigcup Tri_j), Tri_i, Tri_j \in \{US\} \\ surface_i \bigcap surface_j \begin{cases} \in \{FCL\ SET\}, \left|i-j\right| = 1 \\ \in \{\Phi\}, \left|i-j\right| \notin \{0,1\} \end{cases} \end{vmatrix} \tag{6}$$

3.4 The Description of Semi-tunnel Component Model

The description of component surfaces facilitates the description of semi-tunnel component model.

Definition 6. Every 3D object *Vol* can be described as $Vol = \bigcup_{i=0}^{n} vol_i$, vol_i is subset which the *Vol* is composed of, it must satisfy $\bigcup_{i=0}^{n} s_i = vol_i$, $vol_i \bigcap vol_j = s_k$ or $\Phi (i \neq j)$.

Semi-tunnel component model has six parts, according to Def.6, the description of semi-tunnel component is union with all component surfaces.

$$Semi\_tunnel = \begin{vmatrix} \bigcup_{i}^{n} surface_i ,\ surface_i \in \{SURFACE\ SET\} \\ surface_i \bigcap surface_j \begin{cases} \in \{FCL\ SET\}, | i-j | = 1 \\ \in \{\Phi\}, | i-j | \in \{0,1\} \end{cases} \end{vmatrix} . \tag{7}$$

We can conclude some from above illustration.

1) Any semi-tunnel can be represented by above basic elements; 2) Double adjacent semi-tunnels have intersection set that is contour polygon, which is *AS* of first semi-tunnel and *BKS* of last semi-tunnel; 3) Tunnel section is a special type of division surface, if and only if the two semi-tunnels can consist of a new semi-tunnel. *AS* of new semi-tunnel is *AS* of original semi-tunnel 1, and *BKS* of new semi-tunnel is *BKS* of original semi-tunnel 2; 4) Definitions 3 to 6 are four constraints to establish topological space in theory. According to definitions, we define the point, line, *TFS*, component surface, semi-tunnel and constraint relations between these objects. It is a foundation that 3D spatial analysis, query and 3D reconstruction must need, it also guarantees 3D objects constructed with correctness and uniqueness [13].

4 Conclusion

The paper uses the theory of topology to define the elements of semi-tunnel tunnel and formally describe every type of tunnel entity. It is the foundation that the semi-tunnel can be applied into practice as component of new data model. Meanwhile, it gives the theoretical basis for 3D tunnel modeling, and ensures 3D spatial analysis, query, dynamic update of model etc.

Acknowledgments. The work was supported by a grant provided by National Natural Science Foundation of China (Project No.40972204), and partially supported by a grant of the Education Department of Shaanxi Province (Project No.2010JK682). The authors are grateful to the editors and anonymous reviewers more enough for their self-giving works.

References

1. Wang, L.P., Pan, Y.H., Qiu, F.Y.: Formal Description of the Topological Spatial Relationship in GIS. J. of Zhejiang Univ. of Tech. 27, 168–172 (1999)
2. Deng, N.D., Hou, E.K., Zhang, Z.H.: Formal description and the3-D modeling topological relationships. J·Xi'an Univ. of Arch. & Tech(Natural Science Edition) 39, 873–877 (2007)
3. Zhang, S.J., Li, Y.S.: Formal Model Describing Topological Relations Between Spatial Linear Objects in GIS. J. of Geomatics Science and Technology 26, 292–295 (2009)
4. Deng, M., Liu, W.B., Feng, X.Z.: A Generic Model Describing Topological Relations among Area Objects in GIS. Acta Geodaetica et Cartographica Sinica 34, 85–90 (2005)
5. Yan, H.W., Guo, R.Z.: On Fundamental Problems of Directional Relationships. Acta Geodaetica et Cartographica Sinica 32, 42–46 (2003)
6. Yan, H.W., Guo, R.Z.: Research on Formal Description of Directional Relation Based on Voronoi Diagram (I). Science of Surveying and Mapping 27, 24–27 (2002)
7. Yan, H.W., Guo, R.Z.: Research on Formal Description of Directional Relation Based on Voronoi Diagram (II). Science of Surveying and Mapping 27, 4–7 (2002)
8. Egenhofer, M., Franzosa, R.: Point-set topological spatial relations. International Journal of Geographical Information Systems 5, 161–174 (1991)
9. Egenhofer, M.: Reasoning about Binary Topological Relations. In: Proceedings of the 2nd International Symposium on Advances in Spatial Databases, pp. 143–160. Zurich, Swiss (1991)
10. Egenhofer, M., Franzosa, R.: On the Equivalence of Topological Relations. International Journal of Geographic Information Systems 9, 133–152 (1995)
11. Deng, N.D., Hou, E.K.: Rectification and Formal Description of 3D OO-solid Model. Geomatics and Information Science of Wuhan University 34, 52–56 (2009)
12. Zhang, Z.H.: Research on Method of 3D Tunnel Network Modeling and Path Analysis. Ph.D, Xian University of Science and Technology, Xi'an (2010)
13. Shi, W.Z., Wu, L.X., Li, Q.Q., et al.: Model and Algorism of Three-dimensional Spatial Information System. Electronics Industry Press, Beijing (2007)

Application of Neural Networks to the Correction of a Stiffness Matrix in a Finite Element Model

Zhong Yang and Haifei Si

Institute of Mechanical and Electrical Engineering,
Jinling Institute of Technology
211169 Nanjing, China
yz@jit.edu.cn, sihaif@jit.edu.cn

Abstract. It is important how to find the sources of error in a finite element model and how to correct them because the model established by analysis could not correspond to the real structure completely. One practical method is to correct the stiffness matrix by a static test first, then correct other matrixes according to the corrected stiffness matrix. The problem with this method is that the result is often highly sensitive to differences in relative errors in static displacements. The neural networks were applied in this study and preliminary results were obtained. The new approach is able to get rid of the high sensitivity and can modify stiffness matrix well.

Keywords: neural networks, stiffness matrix, finite element model, correction.

1 Introduction

The back-propagation of errors algorithm is probably the most widely used neural network learning algorithm. Although this algorithm is slow, it minimizes mean squared error and is quite simple to implement, two characteristics that account for its popularity among neural network experimenters. Back-propagation is a form of highly supervised learning, which means that the desired output is required for every member of the input set. The basic conceptual procedure behind back-propagation is to give the network an input and allow it to generate an output. The desired output is then used with the generated output to produce a set of error values. These error values are then used to modify the weights of the network so that the error is decreased. The output error values must be propagated back into the network for the modification of the weights of the other layers of elements. The back-propagation algorithm derives its name from this procedure.

Each of the training set patterns represents a single point in the multidimensional input space of the network. Each input is mapped to a desired output value through the general procedure outlined above. Through repeated exposure to the training set, the network gradually accumulates the ability to map the input to the desired output. One drawback of this algorithm is that the network will eventually memorize the patterns presented to it. This is acceptable if the network will always receive its inputs from the training set, without distortions or noise in the inputs. Such a network is undesirable if any

G. Lee (Ed.): Advances in Automation and Robotics, Vol. 1, LNEE 122, pp. 9–16.
springerlink.com

generalization abilities are needed. If generalization capabilities are desired, care must be taken not to over train the network to the point of memorization. One way to avoid the problem of over training is to inject noise into the training set images before they are presented to the network during training. The single point in input space now becomes a region around that point, whose size is a function of the noise level. The greater the amount of noise added to the training set, the larger the region. The network learns to associate a set of similar inputs to a single output, increasing the generalization capabilities of the network.

Such an approach has been shown to be effective in many different back-propagation network applications. The only major drawback of such an approach is that the additional noise can dramatically increase the time it takes to generate a working network. The number of training iterations necessary for the network to converge to the desired error level increases as the noise level increases. If training time is not a concern, noise training sets are a fairly simple way to increase the usefulness of the network. In a conceptual sense, noise in the input set can be seen as analogous to damage in the network. Information is either being deleted or added in much the same way as if the network were faulty. This paper studies the relationship between noisy training sets and the fault tolerance of the back-propagation network. The fundamental question addressed here is whether training using noisy inputs increases the fault tolerance of the network in the same way as it increases the network's generalization abilities.

Neural networks, by nature of their use of large numbers of highly connected elements, are often claimed to be inherently fault tolerant. The distributed internal representation allows the removal of significant portions of the internal structure before the networks fail. In addition, neural networks seem to degrade gracefully, in that the performance gradually drops off as the number of faults increases. Fault tolerance and the graceful degradation are both useful properties of complex systems that may be subject to faults. This study seeks to determine if this fault tolerance can get rid of the sensitivity of the correction of a stiffness matrix in a finite element model.

2 Network Structure and Functions

The network used is a conventional back-propagation network using 3 layers of elements. The input layer is fully connected (every element of one layer connects to every element of the other layer) to the hidden layer, which is correspondingly fully connected to the output layer. Associated with each connection is a weight value that is analogous to the synapse strength in biological systems. The inputs of each element are combined by a weighted sum activation function, which is used with a nonlinear output function to generate that element's output. There are no feedback connections, so the network has no real dynamic behavior.

The functions animating the network are the same for all layers. The activation function is a simple weighted sum combination of the inputs:

$$y_i = w_{io} + \sum_{j=1}^{N} w_{ji} x_j \tag{1}$$

Where y_i is the activation of element i, x_j is the input from element j in the previous layer, w_{ji} is the connection weight from element j to i, and w_{io} is the bias weight. This is used in an output function of the form:

$$z_i = 1/(1+e^{-y_i}) \tag{2}$$

Where z_i is the output of element i. These 2 functions describe the operation of the network.

The power of the back-propagation network comes not from these equations, which animate the network during its normal operation, but from the learning algorithm that builds the set of weights from a randomized starting point. As discussed earlier, the back-propagation learning paradigm is based on an error vector generated from the desired and actual outputs of the network. This is used to determine a weight modification value for each weight, which is added to the current value according to:

$$w_{ij}(t+1) = w_{ij}(t) + \Delta w_{ij}(t) \tag{3}$$

Where Δw_{ij} is the weight modification value.

This value is determined through a fairly simple set of equations. The error values for the output layer are first determined by:

$$error_i = d_i - z_i \tag{4}$$

Where d_i is the desired output for element i. This is then used to determine a δ value for the element:

$$\delta_i = f(y_i) error_i \tag{5}$$

Where f is the derivative of the output function (shown in (2)) evaluated at the activation value (determined by (1)). This allows the weight modification value to be generated by:

$$\Delta w_{ij} = \eta \delta_j x_i \tag{6}$$

Where η is a learning parameter that determines the rate of learning.

Learning in the non-output layers is not so simple because there is no general way to produce an error vector for those elements. The back-propagation algorithm used a technique that estimates the hidden layer's errors based on the output layer's errors and the connection weights from the hidden to the output layer. This approach can be generalized to networks with an arbitrary number of layers.

The error value is the only part of the algorithm that is different for hidden layers. The error value for a hidden layer element is calculated by:

$$error_i = \sum_{k=1}^{M} \delta_k w_{ik} \tag{7}$$

Where the δ_k is the value determined for the output layer and the w_{ik} is the weight from hidden layer element i to output layer element k. The rest of the learning algorithm is unchanged (using (5) and (6)).

This approach can be continued back another layer by using the hidden layer value for the output layer values in (7).

Typically, this algorithm is allowed to continue until some desired error level is attained. The back-propagation algorithm can approximate the mapping of a compact set onto any other compact set to an arbitrary degree of accuracy.

3 Sensitivity in the Correction of a Stiffness Matrix

To correct the stiffness matrix in a finite element model by a static test needs 3 steps. First, establish the stiffness matrix of the structure by analysis; then, do a static test and measure or calculate the static displacements; third, modify the stiffness matrix by the static displacements and static force which we have known. In order to study this subject conveniently, we conduct a cantilever beam test to determine if the back-propagation neural network can get rid of the sensitivity in the correction. In fact, there is highly sensitive problem in the modification using conventional approaches. A back-propagation neural network is applied in this problem for the seek of using its ability about fault tolerance.

The model of the structure which we have studied is a cantilever beam shown in figure 1.

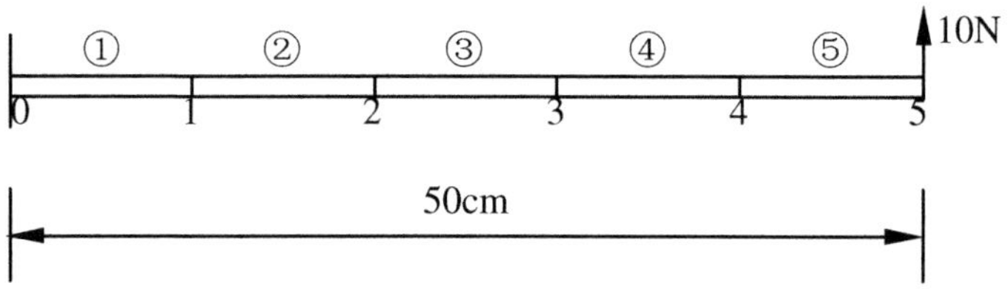

Fig. 1. Model of the structure to be studied

Where the mass(m) and the modulus of elasticity(E) of the cantilever beam are shown below:

$$m = 0.0624\text{kg} \tag{8}$$

$$E = 2.1\times10^{11}\text{N}/\text{m}^2 \tag{9}$$

$$EI = 4.48\text{Nm}^2 \tag{10}$$

By dividing the cantilever beam into 5 parts equally and adding a force upward shown in figure 1, we can get the finite element model.

The stiffness matrix of a structure is made up of the element stiffness matrixes. If we choose the element as a whole in the correction, we can get the following equation:

$$K = \sum_{i=1}^{5} \beta_i K_i \tag{11}$$

Where K is the stiffness matrix of the cantilever beam, K_i is the element stiffness matrix of the 5 elements and β_i is the coefficient to be corrected. It is not difficult to understand: $\beta_i = 1$ shows that the correspond element stiffness matrix K_i is correct; $\beta_i \neq 1$ shows that the correspond element stiffness matrix K_i has errors and it needs to be modified.

As a preparation, we give a simple and important equation:

$$KX = F \tag{12}$$

The conventional method is to reform the equation (11) and equation(12) into the forms:

$$\left(\sum_{i=1}^{5} \beta_i K_i\right) X = F \tag{13}$$

$$A\beta = F \tag{14}$$

Where $A \in R^{12\times5}$ is a coefficient matrix, $\beta \in R^{5\times1}$ is a coefficient vector and $F \in R^{12\times1}$ the force vector. We can get every coefficient by solving the equation (14).

However, this approach has a question that the result is often highly sensitive to differences in relative errors in static displacements and this question is difficult to be settled. Actually, there are always differences in relative errors in static displacements not only in test but also in engineering. Therefore, this approach is not a good one and it is not accepted in engineering. To find a new method to solve this problem is always a task among scientists in the field of vibration.

4 Application of Back-Propagation Neural Networks to the Correction of Stiffness Matrix

Back-propagation neural networks were chosen to study the modification of stiffness matrix in finite element model by static test. A back-propagation neural network program in computer language C was written and it has following abilities:

(1) To choose the topology of neural networks, including the number of hidden layers and the element numbers of input layer, output layer and every hidden layer.

(2) To adjust the connection weights and bias weights by training sets.

(3) To give the correspond outputs of new input sets. It can be used for classification tasks.

According to the structure to be studied shown in figure 1, the network uses 10 input elements, one for each displacement and 5 output elements, one for each coefficient. The number of hidden layers and the element numbers of every hidden layer were changed in computing in order to obtain the best result. When the network uses one hidden layer and 10 hidden elements, the result of the network is most satisfying.

After 14844 iterations through the training set, the relation between the desired outputs and network outputs is shown in Table 1. The result of the examination about sensitivity in the correction is shown in Table 2. From Table 2, it is easy to see that back-propagation neural networks can get rid of the sensitivity in the correction. The result of the examination about adjustability in the correction is shown in Table 3. From Table 3, it is obtained that back-propagation neural networks have the ability to adjust or modify the stiffness matrix in the correction by a static test.

Table 1. Network outputs of sample and desired outputs

Samples	1	2	3	4	5	6	7	8
Inputs	0.00520833	0.00260417	0.0104167	0.00260417	0.00520833	0.00520833	0.00520833	0.00520833
	0.100446	0.0502232	0.200892	0.0502232	0.100446	0.100446	0.100446	0.100446
	0.0193452	0.00967261	0.0386904	0.0117187	0.0172991	0.0193452	0.0193452	0.0193452
	0.178571	0.0892857	0.357142	0.128348	0.139509	0.178571	0.178571	0.178571
	0.0401786	0.0200893	0.0803572	0.0275297	0.0342262	0.0386905	0.0401786	0.0401786
	0.234375	0.117188	0.468750	0.184151	0.195313	0.206473	0.234375	0.234375
	0.0654762	0.0327381	0.130952	0.0478050	0.0556176	0.0611979	0.0645461	0.0654762
	0.267857	0.133929	0.535714	0.217633	0.228795	0.239956	0.251116	0.267857
	0.0930060	0.0465030	0.186012	0.0703124	0.0792411	0.0859375	0.0904018	0.0926339
	0.279018	0.139509	0.558036	0.228794	0.239956	0.251116	0.262277	0.273438
Desired Outputs	1	2	0.5	2	1	1	1	1
	1	2	0.5	1	2	1	1	1
	1	2	0.5	1	1	2	1	1
	1	2	0.5	1	1	1	2	1
	1	2	0.5	1	1	1	1	2
Network Outputs	1.00541	1.99872	0.50157	1.99625	0.99684	0.99232	1.00044	0.99563
	1.00240	1.99924	0.50129	1.00625	1.99430	0.99666	0.99548	0.99431
	1.00626	1.99828	0.49967	1.00843	0.99881	2.00108	0.99975	0.98997
	1.00125	2.00333	0.49976	1.01034	1.00642	1.00334	2.00168	1.00097
	0.99797	2.00128	0.49960	1.00650	1.00122	1.00661	1.00221	2.00497

Table 2. Examination about sensitivity in the correction of stiffness matrix

Number	1	2	3	4	5	6
Inputs	0.0052083	0.005208	0.00521	0.0052	0.00521	0.0052
	0.10045	0.1004	0.100	0.10	0.100	0.10
	0.019345	0.01935	0.0193	0.019	0.0193	0.019
	0.17857	0.1786	0.179	0.18	0.179	0.18
	0.040179	0.04018	0.0402	0.040	0.0387	0.039
	0.23438	0.2344	0.234	0.23	0.206	0.21
	0.065476	0.06548	0.0655	0.065	0.0612	0.061
	0.26786	0.2679	0.268	0.27	0.240	0.24
	0.093006	0.09301	0.0930	0.093	0.0859	0.086
	0.27902	0.2790	0.279	0.28	0.251	0.25
Desired Outputs	1	1	1	1	1	1
	1	1	1	1	1	1
	1	1	1	1	2	2
	1	1	1	1	1	1
	1	1	1	1	1	1
Network Outputs	1.00537	1.00587	1.00790	1.00990	1.00663	1.01323
	1.00246	1.00143	1.00399	0.97890	1.00864	1.06468
	1.00615	1.00633	1.02097	1.10307	2.07947	2.12596
	1.00131	1.00071	0.98600	0.90647	0.97486	0.93681
	0.99806	1.00346	1.01258	1.11383	1.01404	1.09771

Table 3. Examination about adjustability in the correction of stiffness matrix

Number	1	2	3	4	5	6
Inputs	0.00520833	0.005208	0.00521	0.00130208	0.001302	0.00130
	0.100446	0.1004	0.100	0.251116	0.02511	0.0251
	0.0179812	0.01798	0.0180	0.00790551	0.007906	0.00791
	0.152530	0.1525	0.153	0.103237	0.1032	0.103
	0.0352183	0.03522	0.0352	0.0212054	0.02121	0.0212
	0.189732	0.1897	0.190	0.159041	0.1590	0.159
	0.0560516	0.05605	0.0561	0.0394346	0.03943	0.0394
	0.223215	0.2232	0.223	0.200894	0.2009	0.201
	0.0791172	0.07912	0.0791	0.0602681	0.06027	0.0603
	0.234376	0.2344	0.234	0.212056	0.2121	0.212
Desired Outputs	1	1	1	4	4	4
	1.5	1.5	1.5	1	1	1
	1.5	1.5	1.5	1	1	1
	1	1	1	0.8	0.8	0.8
	1	1	1	1	1	1
Network Outputs	1.00760	1.00805	1.01254	4.02156	4.02188	4.02350
	1.50358	1.50310	1.49710	1.00132	1.00285	1.00529
	1.50941	1.50950	1.49296	1.00626	1.01254	1.00272
	1.00122	1.00071	1.00629	0.80096	0.80006	0.79821
	0.99646	0.99450	0.98720	0.99788	0.98677	0.99703

5 Conclusions

The general conclusion that can be derived from these preliminary results is that, for this particular pattern recognition problem, training with some noise increases the ability of the back-propagation network to withstand faults within the inputs. Increasing the noise beyond a certain level degrades the fault tolerance of the network.

In finite element method, the approach to establish model by the combination of analysis and test is always an interesting subject among scientists and engineers. There is a sensitive problem in the correction of stiffness matrix by a static test, while this sensitive problem goes out when using back-propagation neural network trained on noisy inputs. This is only preliminary research.

Ongoing work is examining the sensitivity of the fault tolerance increase effect to the amount of noise, so that the optimal noise level can be determined for this problem. Additional work is required to test the noisy training hypothesis on different problems, using networks of different sizes learning different training sets.

Acknowledgments. Project supported by the "333 Project" of Jiangsu Province, China (Grant No. BRA2010004). Project supported by the "Six Talent Peak" Project of Jiangsu Province, China (Grant No. 2008-JX-14). Project supported by the Doctor Foundation of Jinling Institute of Technology, China (Grant No. Jit-DF-2008-01).

References

1. Minnix, J.I.: Fault Tolerance of the Backpropagation Neural Network Trained on Noisy Inputs. In: International Joint Conference on Neural Networks, vol. 1, pp. 847–852 (1992)
2. Phatak, D.S., Koren, I.: Fault Tolerance of Feedforward Neural Nets for Classification Tasks. In: International Joint Conference on Neural Networks, vol. 2, pp. 386–391 (1992)
3. Yang, Z., Si, H.: Integration of Neural Network and Symbolic Inference and Its Application to Detection of Foundation Piles. In: IEEE International Conference on Electrical Engineering and Automatic Control, vol. 2, pp. 390–393 (2010)

Passive Ranging Based on Single-Band IR Sensor without Extinction Coefficient

Xiao-ning Fu and Di Wang

School of Electromechanical Engineering, Xidian University, No. 2, South Taibai Road, Yanta District, Xi'an 710071, China
xning_fu@163.com

Abstract. We proposed a single-band infrared sensor based passive ranging method without using extinction coefficient, in which radiation intensity ratio between adjacent sampling instances is used to eliminate the effects of the atmospheric extinction coefficient. An equation of using the ratio of radiation intensity between two adjacent sampling instances to estimate the distance ratio between them is developed, in which the irrational assumption of constant target radiation intensity over an extended period is removed. Additionally the proposed method is verified by simulation which demonstrates satisfactory distance estimation.

Keywords: Atmospheric propagation, Infrared, Passive remote sensing, Numerical approximation and analysis.

1 Introduction

Passive location is usually adopted in absence of active radiation [1, 2], such as radar signal. Due to the fact that the wavelength of infrared radiation is much shorter than that of electromagnetic shortwave, infrared radiation could provide a higher degree of precision locating. Therefore, passive ranging from infrared radiation becomes a focus in the study of passive ranging [3].

There are dual-band and single-band infrared sensors. Generally, dual-band sensor provides richer information [4] than the single-band one. However, it is usually the case in application that only a certain wave band has a good infrared detection effect. For example, in low-level of sea and land, detection at 8~14μm band is more effective than at 3~5μm band; while at high altitudes with cloudy, detection at wave band of 3~5μm is more effective. In this case, single-band sensor might work as effective the dual-band sensor. Additionally its cost is lower.

Bearing-only passive ranging with single-band sensor was studied in [5-8]. Despite that the mathematical model, observability conditions and the maneuver ways of observer were proposed, there is still quite a distanced away from practical use [1]. We moved forward from the study in [9] and proposed a passive ranging based on single-band infrared sensor without using extinction coefficient. In this paper, we not only introduce the principle but also demonstrate it with simulation and experiment.

G. Lee (Ed.): Advances in Automation and Robotics, Vol. 1, LNEE 122, pp. 17–23.
springerlink.com

2 The Principle of Passive Ranging

Fig.1 shows the movement model of target against observation platform [9], in which $S_n(x_n, y_n, z_n)$, $S_{n+1}(x_{n+1}, y_{n+1}, z_{n+1})$, $S_{n+2}(x_{n+2}, y_{n+2}, z_{n+2})$ is the geography coordinates of observation platform at any three adjacent sampling moments n, n+1, n+2 and T_n, T_{n+1}, T_{n+2} are the trace of target. r_n, α_n, β_n is the distance, azimuth and pitch of the target in the spherical polar coordinates of observation platform at sampling moment n.

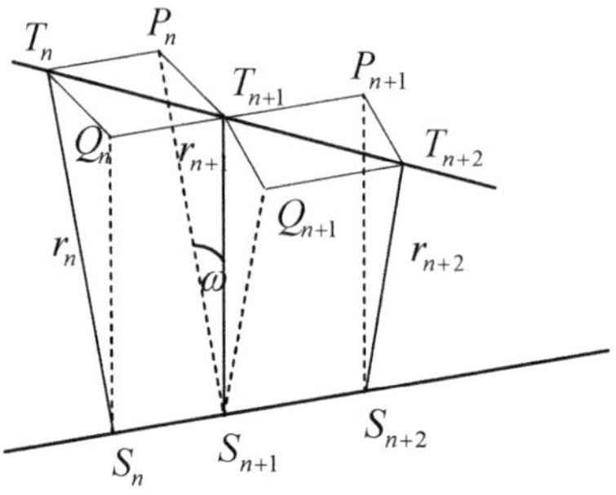

Fig. 1. The 3-D movement model of target against observer

In adjacent three sampling points with small interval, moving target and observer in high speed can be assumed as rectilinear motion. Put it in another way, S_n, S_{n+1} and S_{n+2} are on a straight line and T_n, T_{n+1} and T_{n+2} are also on a straight line [9]. This assumption agrees with the piecewise linearization method in [10].

In Fig.1, $T_nQ_nT_{n+1}P_n$, $T_{n+1}Q_{n+1}T_{n+2}P_{n+1}$ and $Q_nT_{n+1}S_{n+1}S_n$ are parallelograms. In observation coordinates system, let $(l_i, m_i, n_i)^T = (\cos\alpha_i\cos\beta_i \ \sin\alpha_i\cos\beta_i \ \sin\beta_i)^T$, here $i = n$, n+1 and n+2, and $\angle P_nS_{n+1}T_{n+1}=\omega$, then

$$\cos\omega = l_n l_{n+1} + m_n m_{n+1} + n_n n_{n+1} \tag{1}$$

To simplify the analysis but without loss of generality, we assume that the observer moves only on x-axis. When observer moving in the direction to the negative x axis, *i.e.* $x_{n+1} < x_n$, there exists the relationship below [9]:

$$r_{n+1} = r_n \frac{l_n}{l_{n+1}} - \frac{x_{n+1} - x_n}{l_{n+1}}. \tag{2}$$

With $r_n = \rho_{n(n+1)} r_{n+1}$, equation (2) could be re-written as equation (3):

$$r_{n+1} = \left| \frac{x_{n+1} - x_n}{\rho_{n(n+1)} l_n - l_{n+1}} \right|. \tag{3}$$

As a result, the question becomes how to define the distance ratio of target correctly, namely $\rho_{n(n+1)}$.

3 Method Improvement

The studies in [11] find the constraint relationship of distance between the adjacent sampling instant:

$$\rho_{n(n+1)} = \frac{4-\omega^2}{8\cos\omega+2\omega\pm\sqrt{(8\cos\omega+2\omega)^2-4(4-\omega^2)^2}}. \tag{4}$$

Substitute (4) into (2) the target distance could be obtained. However, in practice, this method may introduce large error when ω (*i.e.* the LOS rotation angle between observation vision and the trace of target) is small.

As it is known, the radiation of small target is stable in adjacent observation times if the sampling interval is small. Hence the small target can be view as a pointolite with constant radiation intensity. Then the relationship between irradiance E detected by infrared detector and target distance r can be written as below [12, 13]:

$$E = \tau^r \frac{J}{r^2}. \tag{5}$$

Where J is the radiation intensity of target received by detector which is an unknown quantity and τ is the atmospheric transmittance per unit length. Suppose the distances from target to detector are r_n and r_{n+1} at sampling moment n and n+1. Then the corresponding irradiances of target are as below:

$$E_n = \tau^{r_n} \frac{J}{r_n^2}. \tag{6}$$

$$E_{n+1} = \tau^{r_{n+1}} \frac{J}{r_{n+1}^2}. \tag{7}$$

Square root of (7) divided by square root of (6) we get the equation as below:

$$\sqrt{\frac{E_{n+1}}{E_n}} = \tau^{\frac{r_{n+1}-r_n}{2}} \frac{r_n}{r_{n+1}}. \tag{8}$$

According to [14], we can also get equation (9) from equation (7) and equation (6):

$$\begin{aligned} \frac{E_{n+1}-E_n}{E_n} &= -\left(\frac{2}{r_{n+1}}+\ln\tau\right)\cdot(r_{n+1}-r_n) \\ &= -2(1-\frac{r_n}{r_{n+1}})-\ln\tau\cdot(r_{n+1}-r_n). \end{aligned} \tag{9}$$

Substitute equation (8) into equation (9) and let $\rho_{n(n+1)} = r_n/r_{n+1}$, we can get equation (10):

$$\tau^{-\frac{\rho}{\ln\tau}} \rho_{n(n+1)} = \tau^{-\frac{E_{n+1}/E_n+1}{2\ln\tau}} \sqrt{E_{n+1}/E_n}. \tag{10}$$

Equation (10) can be rewritten as equation (11) by using $\tau^{\frac{1}{\ln \tau}} = a$:

$$a^{\rho} \rho_{n(n+1)} = a^{\frac{E_{n+1}/E_n + 1}{2}} \sqrt{E_{n+1} / E_n}. \tag{11}$$

Furthermore, equation (11) can be rewrite as:

$$\rho_{n(n+1)} = \frac{1}{\ln a} ProductLog\left[(\ln a) \cdot a^{\frac{E_{n+1}/E_n + 1}{2}} \sqrt{E_{n+1} / E_n} \right]. \tag{12}$$

Here, E_n and E_{n+1} can be obtained from image processing [15]. The ProductLog function is an infinite series [16]. If the extinction coefficient relative term ln*a* is known, $\rho_{n(n+1)}$ is uniquely determined. Therefore, if we substitute equation (12) into equation (2), the distance is found.

When the sampling interval is small, distance change between target and the observer is limited. For example, for the aircraft target and aircraft observer platform, $|r_{n+1}- r_n|<60$ m is a reasonable assumption at a sampling frequency of 25 Hz. Under this condition we studied the ranging error for extinction coefficient between 0.1 to 1.1 which covers the weather condition from clear, haze, light rain, light snow, to moderate rain etc. Chebyshev approximation is used to remove extinction coefficient. This method does not specify the infrared band.

As shown in Fig. 2, both 7-order approximation and 5-order approximation have the same maximum fitting error. They both don't exceed 1.5%, but the 7-order approximation error distribution is more uniformity than 5-order approximation.

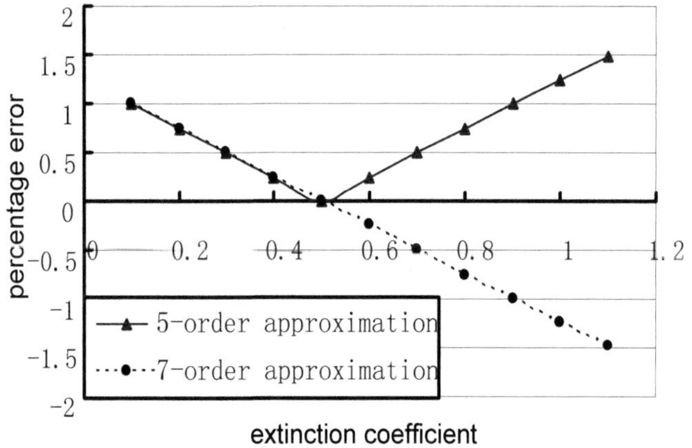

Fig. 2. Chebyshev approximation via extinction coefficient from 0.1 to 1.1

With 5-order approximation, equation (12) could be simplified as equation (13) below:

$$\rho_{n(n+1)} = C_0 + C_1 q + C_2 q^2 + C_3 q^3 + C_4 q^4 + C_5 q^5, q = \sqrt{\frac{E_{n+1}}{E_n}}. \tag{13}$$

Here, C_0=5.052281e-5, C_1=1.012319, C_2=5.312449e-4, C_3=-5.447379e-4, C4=2.792822e-4, C_5=-5.727322e-5.

Table 1 lists the extinction coefficient at 8~12um waveband under typical weather conditions. Since equation (13) is applicable for extinction coefficient from 0.1 to 1.1, the proposed method potentially has great practice value.

Table 1. Extinction Coefficient at 8~12 μm Waveband [17]

Meteorological condition	Clear	Mist	Light Rain	Light Snow	Moderate Rain
Extinction coefficient	0.08	0.105	0.36	0.51	0.69

4 Simulation Results and Conclusion

Simulations are carried out to confirm the validity of proposed method. In the simulation experiment, observers along the negative x-axis for uniform linear motion, the target in a plane parallel to the x-axis to make a speed of 242.6 m/s, acceleration of 2g sinusoidal movement. Radiation intensity is measured by an error of 1%~3%. The description of the target movement is x_t=500, y_{tn}=242.6*i*T, and z_t=2.94275e3*sin(0.08244*i*T); The description of the observer movement is x_n=40000-240*i*T, y_n=z_n=0. While i is the sampling instant number, T =1/25 is the sampling interval.

Fig.3 shows the real distance, the distance estimated by former method and the distance estimated by our proposed method. As it could be seen from Fig. 3, the error of estimated distance is very small when the distance between target and observer is between 4 to 40 km. As a comparison, the error of estimated distance by the method in [9] is significantly higher in this case than that of using our proposed method.

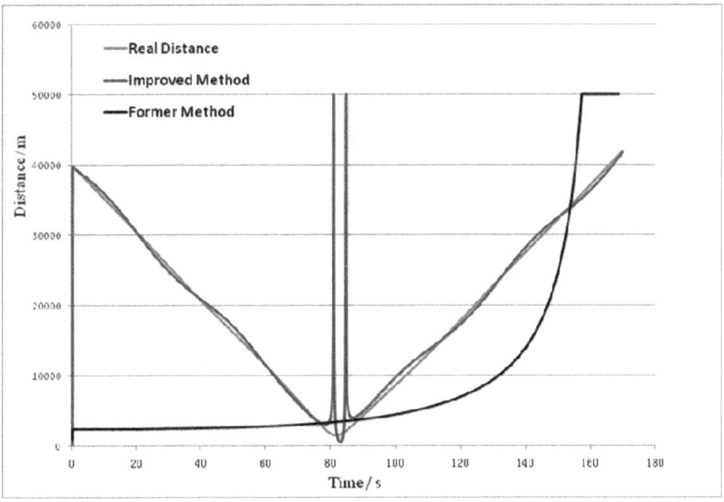

Fig. 3. Simulation results

In brief, the proposed method works well in distance estimation without using extinction coefficient and potentially is of great value in real application.

Acknowledgments. This work was supported both by the National Natural Science Foundation of China under Grant No. 60872136 and by Natural Science Basic Research Plan in Shaanxi Province of China (Program No. 2011JM8002). The authors would like to thank the anonymous reviewers for their valuable advice toward the improvement of this article.

References

1. Fu, X.-N.: Research on Infrared Passive Location Technology from Mono-station. PhD. Thesis, Xidian University (2005)
2. Wang, D.-B.: Research on Infrared Weak Small Targets Detection and Tracking Technology under Complex Backgrounds. PhD. Thesis, Xidian University (2010)
3. Fu, X.-N., Li, X.: Single Observer Passive Location and Achievement with Photoelectric Methods. Journal of Test and Measurement Technology 1, 45–47 (2002)
4. Meitzler, T., Sohn, E., Gerhart, G., Collins, P.: Comparison of the Performance of 3-5 and 8-12 Micron Infrared Cameras. In: Infrared Image Systems: Design, Analysis, Modeling, and Testing, pp. 22–29. SPIE Press (2004)
5. Liu, X.-D., Cheng, X., Zhang, H.: An Algorithm of Tracking a Maneuvering Target with Bearing-only Measurements. Acta Armamentarii 6, 1035–1036 (2006)
6. Xu, Z.-G., Dong, Z.-R.: Bearings-only Target Locating and Tracking Based on a Multi-hypothesis. Journal of HuaiHai Institute of Technology 3, 18–21 (2001)
7. Dong, Z.-R.: Solvability, Observability and Others. Intelligence Command Control and Simulation Techniques 1, 26–28 (2011)
8. Yang, G.-S., Dou, L.-H., Hou, C.-Z.: Observability Criteria Based on Bearing-only Measurements for Three-dimensional Moving Targets. Acta Armamentarii 2, 182–185 (2004)
9. Fu, X.-N., Wu, D.-H.: A-3D Passive Infrared Ranging Algorithm Based on Bearing-only with Monocular. Acta Armamentarii 10, 1189–1193 (2008)
10. Li, H.-R.: Integrated Parameter Identification Method for Targets Maneuvering Linearly at Piecewise Constant Speed under Condition of Passive Direction Finding. Information and Control 5, 621–625 (2005)
11. Chen, Z.-W., Liu, G.-H., Huang, Y., Zhang, S.-J.: Ranging Estimation Model of Infrared Surveillance System in Low Level. Infrared and Laser Engineering 5, 449–452 (2004)
12. Holst, G.C.: Electro-Optical Imaging System Performance, 4th edn., ch. 15. JCD Publishing, Florida (2008)
13. Lu, Y., Ling, Y.-S., Shi, J.-M.: Measurement of Aerial Point Target Distance Using Dual-band Infrared Imaging System. Optics and Precision Engineering 2, 161–164 (2004)

14. Zhu, J.-B., Zhou, H.-C.: Algorithm for Motion Parameters Estimation in ISRT Systems. Journal of Infrared and Millimeter Waves 2, 101–104 (2003)
15. Yang, D.-G., Xiao, S.-P.: Single-band IR Passive Based on IR Radiation Characteristics. Infrared and Laser Engineering 6, 946–950, 1013 (2009)
16. The Mathematical Book Online, `http://reference.wolfram.com/mathematica/ref/ProductLog.html`
17. Accetta, J.S., Shumarker, D.L.: The Infrared and Electro-optical System Handbook: Atmospheric Propagation of Radiation, 2nd edn., vol. 2. SPIE-International Society for Optical Engine (1993)

A New Variable Step Size NLMS Algorithm Based on Decorrelation for Second-Order Volterra Filter

Zhijin Zhao[1,2], Pingping Yan[1], and Lei Shen[1]

[1] Telecommunication School, Hangzhou Dianzi University, Zhejian 310018, China
[2] State Key Lab of Information Control and Security Technology in Communication of No. 36 Research Institute, China Electronic Technology Corporation, Jiaxing 314001, China
yanpp114@163.com

Abstract. When the input signals are strongly correlated, the adaptive algorithm performance of Volterra filter deteriorates. Meanwhile, the correlativity of linear input signals of Volterra filter is different from that of nonlinear input signals of Volterra filter. So, a new variable step size NLMS algorithm based on decorrelation for second-order Volterra filter is proposed. In the new algorithm, the decorrelation is used in linear and nonlinear input signals of Volterra filter, respectively, and the different variable step factors are adopted. Simulation results of noise cancellation show that the correlativity of input noises has no fundamental impact on the proposed algorithm, whether it is strong or weak. And when the input noises are strongly correlated, the new algorithm still has a better noise cancellation performance.

Keywords: second-order Volterra filter, decorrelation, normalized LMS, noise cancellation.

1 Introduction

Since the correlativity of input signals influences the performance of adaptive filtering algorithm, many scholars have done a lot of researches. The decorrelation is used to remove the correlativity of input signals [1]. In order to speed up the convergence of adaptive algorithm for FIR filter, various methods of variable step size are adopted [2-6] and a series of variable step size algorithms based on decorrelation under the background of Gaussian noise are proposed. In reference [7], a variable step size LMS algorithm based on decorrelation for Volterra filter is proposed. And the variable step size method in reference [8] is applied in the algorithm. The algorithm takes the linear and nonlinear input signals of Volterra filter as a whole, and removes its correlativity by decorrelation. However, the correlativity of linear input signals of Volterra filter is different from that of nonlinear input signals of Volterra filter, so the efficacy of decorrelation is limited. In the paper, the decorrelation methods are used in linear and nonlinear input signals of Volterra filter, respectively, and a new variable step size NLMS algorithm based on decorrelation for second-order Volterra filter is proposed. Simulation results of noise cancellation show that the new algorithm has better noise cancellation performance when the input noises are strongly correlated.

G. Lee (Ed.): Advances in Automation and Robotics, Vol. 1, LNEE 122, pp. 25–32.
springerlink.com

2 Proposed Algorithm

Given a second-order Volterra filter, of which the memory size is N, define the linear and nonlinear input signal vectors as

$$\boldsymbol{X}_1(n) = [x(n), x(n-1), \ldots, x(n-N+1)]^T \tag{1}$$

$$\boldsymbol{X}_2(n) = [x^2(n), x(n)x(n-1), \ldots, x(n)x(n-N+1), x^2(n-1), \ldots, x^2(n-N+1)]^T \tag{2}$$

The corresponding weight coefficient vectors are

$$\boldsymbol{W}_1(n) = [w_0(n), w_1(n), \ldots, w_{N-1}(n)]^T \tag{3}$$

$$\boldsymbol{W}_2(n) = [w_{0,0}(n), w_{0,1}(n), \ldots, w_{0,N-1}(n), w_{1,1}(n), \ldots, w_{N-1,N-1}(n)]^T \tag{4}$$

So, the output signal is written as

$$y(n) = \boldsymbol{X}_1^T(n)\boldsymbol{W}_1(n) + \boldsymbol{X}_2^T(n)\boldsymbol{W}_2(n) \tag{5}$$

The output error is given by

$$e(n) = d(n) - y(n) \tag{6}$$

Therefore, the update equations of weight coefficient vectors for Volterra VLMS algorithm[9] are as follows

$$\boldsymbol{W}_i(n+1) = \boldsymbol{W}_i(n) + 2\mu_i e(n)\boldsymbol{X}_i(n), \quad \text{for} \quad i = 1, 2 \tag{7}$$

And the update equations of weight coefficient vectors for Volterra normalized LMS algorithm (VNLMS) are given by

$$\boldsymbol{W}_i(n+1) = \boldsymbol{W}_i(n) + \frac{2\mu_i e(n)\boldsymbol{X}_i(n)}{\zeta + \sum_{j=1}^{2} \boldsymbol{X}_j^T(n)\boldsymbol{X}_j(n)}, \quad \text{for} \quad i = 1, 2 \tag{8}$$

where μ_i denotes the different step factor and the small number ζ satisfies $\zeta > 0$.

As we know the correlativity of input signals influences the performance of algorithm. So we assume that the input signals are correlated and generated by AR(1) model, as follows

$$x(n) = ax(n-1) + v(n) \tag{9}$$

where $v(n)$ obeys a zero-mean and unit variance Gaussian distribution, a satisfies $0 < a < 1$.

Take $N = 2$ for example, the related coefficients of linear input signals are

$$\rho_{1,i}(n) = \frac{E[x(n-i)x(n-i-1)]}{E[x^2(n-i-1)]} = \frac{E[ax^2(n-i-1) + v(n-i)x(n-i-1)]}{E[x^2(n-i-1)]}$$

$$= a + \frac{E[v(n-i)x(n-i-1)]}{E[x^2(n-i-1)]}, \quad \text{for} \quad i = 0,1 \tag{10}$$

The related coefficients of nonlinear input signals with square terms are

$$\rho_{2,i}(n) = \frac{E[x^2(n-i)x^2(n-i-1)]}{E[x^4(n-i-1)]}$$

$$= \frac{E[a^2x^4(n-i-1) + 2av(n-i)x^3(n-i-1) + v^2(n-i)x^2(n-i-1)]}{E[x^4(n-i-1)]}$$

$$= a^2 + \frac{E[2av(n-i)x^3(n-i-1) + v^2(n-i)x^2(n-i-1)]}{E[x^4(n-i-1)]}, \quad \text{for } i = 0,1 \tag{11}$$

And the related coefficient of nonlinear input signals with crossed product terms is

$$\rho_3(n) = \frac{E[x(n)x(n-1)x(n-1)x(n-2)]}{E[x^2(n-1)x^2(n-2)]}$$

$$= \frac{E\{a^2x^2(n-1)x^2(n-2) + [av(n)x(n-2) + av(n-1)x(n-1) + v(n)v(n-1)]x(n-1)x(n-2)\}}{E[x^2(n-1)x^2(n-2)]}$$

$$= a^2 + \frac{E\{[av(n)x(n-2) + av(n-1)x(n-1) + v(n)v(n-1)]x(n-1)x(n-2)\}}{E[x^2(n-1)x^2(n-2)]} \tag{12}$$

From equation (10) to equation (12), we can see that, for second-order Volterra filter the related coefficients of linear input signals are connected with a and the related coefficients of nonlinear input signals are connected with a^2. Thereby, in order to reduce the correlativity of input signals, the decorrelation is used in linear and nonlinear input signals of Volterra filter, respectively, given in equation (13) and (14).

$$c_i(n) = \frac{\boldsymbol{X}_i^T(n)\boldsymbol{X}_i(n-1)}{\boldsymbol{X}_i^T(n-1)\boldsymbol{X}_i(n-1)} \tag{13}$$

$$\boldsymbol{Z}_i(n) = \boldsymbol{X}_i(n) - c_i(n)\boldsymbol{X}_i(n-1), \quad \text{for} \quad i = 1,2 \tag{14}$$

The different step factors used in each order of Volterra filter can improve the performance of algorithm[9], so both the methods of different step factors[9] and variable step size[8] are adopted, as shown in equation (15)~(18).

If $n = kL$,

$$p_i(n) = \beta p_i(n-1) + (1-\beta)[e(n)e(n-1) + \sum_{i=0}^{L-1} \exp(-i)e^2(n-i)] \tag{15}$$

$$A_i(n) = \alpha\mu_i(n) + \gamma p_i^2(n) \tag{16}$$

$$\mu_i(n+1) = \begin{cases} \mu_{i,\max}, & A_i(n) > \mu_{i,\max} \\ \mu_{i,\min}, & A_i(n) < \mu_{i,\min} \\ A_i(n), & \mu_{i,\min} < A_i(n) < \mu_{i,\max} \end{cases} \tag{17}$$

Else,

$$\mu_i(n+1) = \mu_i(n) \tag{18}$$

for $i = 1, 2$, where L denotes the time interval, α and β are very closed to 1, γ is a small positive number, $\mu_{i,\max}$ and $\mu_{i,\min}$ are the upper and lower limits of variable step factors of second-order Volterra filter, respectively. At the initial moment, $\mu_i(0) = \mu_{i,\max}$, $p_i(0) = \mu_{i,\max}$.

Therefore, the new update equations of weight coefficient vectors for second-order Volterra filtering algorithm are obtained as follows

$$\mathbf{W}_i(n+1) = \mathbf{W}_i(n) + \frac{2\mu_i(n+1)e(n)\mathbf{Z}_i(n)}{\zeta + \sum_{j=1}^{2} \mathbf{Z}_j^T(n)\mathbf{Z}_j(n)]}, \quad \text{for} \quad i = 1,2 \tag{19}$$

Equation (1) ~ (6) and (13) ~ (19) form the new variable step size NLMS algorithm based on decorrelation for second-order Volterra filter.

3 Simulation Results

Volterra series models can describe the transfer characteristics of many nonlinear systems, so they have been widely applied to noise cancellation, system identification, adaptive channel equalization and so on[10]. The second-order Volterra adaptive noise cancellation system, which is composed of primary and reference channels, is shown in Fig.1, where s is the ideal signal, x is the input noise independent from s, and r denotes the noise to be removed and is correlated with x. The primary channel receives wanted signal s and noise r. While, the reference channel receives noise x only, which is input to Volterra adaptive filter to obtain the output signal y. Then y is subtracted from the output signal d of primary channel to obtain output error e. By adaptive adjusting the weight coefficient vectors of Volterra filter, the output signal y is very closed to noise r, so the noise in primary channel is removed.

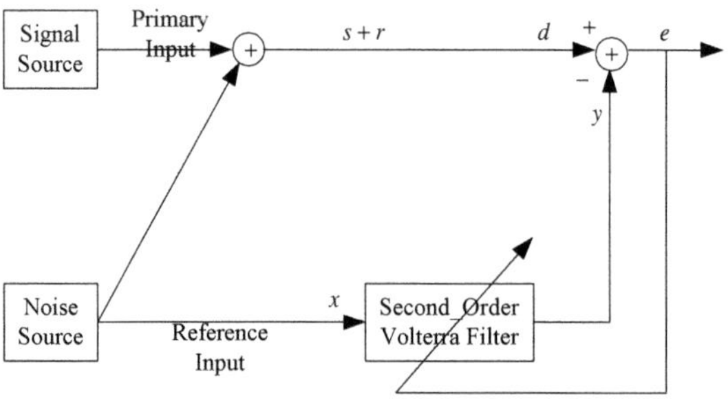

Fig. 1. Second-Order Volterra adaptive Noise Cancellation System

To compare the performances, the VLMS algorithm[9], VNLMS algorithm, VSDVLMS algorithm[7] and proposed algorithm are applied to adaptive noise cancellation. Where, the input signal $s(n)$ is a sine signal with unit amplitude, noise

$r(n) = 0.1x(n) + 0.2x(n-1) + 0.3x(n-2) + 0.4x^2(n) + 0.5x^2(n-1) + 0.6x^2(n-2) + 0.7$ $x(n)x(n-1) + 0.8x(n-1)x(n-2) + 0.9x(n)x(n-2)$, input noise $x(n)$ is generated by $x(n) = ax(n-1) + v(n)$, and $v(n)$ obeys a zero-mean and unit variance Gaussian distribution. In simulations, $L = 10$, $\alpha = 0.95$, $\beta = 0.9$, $\gamma = 10^{-6}$, $\zeta = 10^{-6}$, and each curve is the average of 30 independent experiments. The step factor values of four algorithms for different a, which are given in Table 1, are chosen when the performance of each algorithm gets to the best.

Table 1. The step factor values of four algorithms for different a

		$a = 0.3$	$a = 0.6$	$a = 0.9$
VLMS	μ_1	0.005	0.002	0.00003
	μ_2	0.004	0.015	0.00005
VNLMS	μ_1	0.1	0.03	0.1
	μ_2	0.2	0.09	0.15
VSDVLMS	μ_{max}	0.85	0.75	0.75
	μ_{min}	0.0001	0.0001	0.0001
Proposed	$\mu_{1,max}$	0.35	0.3	0.35
	$\mu_{1,min}$	0.0001	0.0001	0.0001
	$\mu_{2,max}$	0.65	0.55	0.6
Algorithm	$\mu_{2,min}$	0.0001	0.0001	0.0001

When the value of a is 0.3, 0.6 and 0.9, the noise cancellation results of four algorithms are shown in Fig.2~Fig.4, where original sine signal, mixing signal and signals extracted by cancellation systems based on four different algorithms are given from up to down. From the figures, it can be seen that with the increase of a, the correlativity of input noises gradually strengthens. However, the performance of VLMS algorithm continuously deteriorates and the signal extracted by cancellation system based on VLMS algorithm is not the sine signal for $a = 0.9$. Compared to other three algorithms, the signal extracted by cancellation system based on proposed algorithm is more closed to original sine signal and the noise cancellation performance is the best.

In order to better illustrate the noise cancellation performances of different algorithms, when the correlativity of input noise is different, the output errors of noise cancellation systems based on different algorithms are shown in Fig.5~Fig.7. These figures further illustrate that the noise cancellation system based on the proposed algorithm has a much better cancellation performance.

From Fig.2 to Fig.7, we can see that the correlativity of input noises has no fundamental impact on the proposed algorithm, whether it is strong or weak. Thereby, it is shown that the efficacy of decorrelation used in linear and nonlinear input signals respectively are much better.

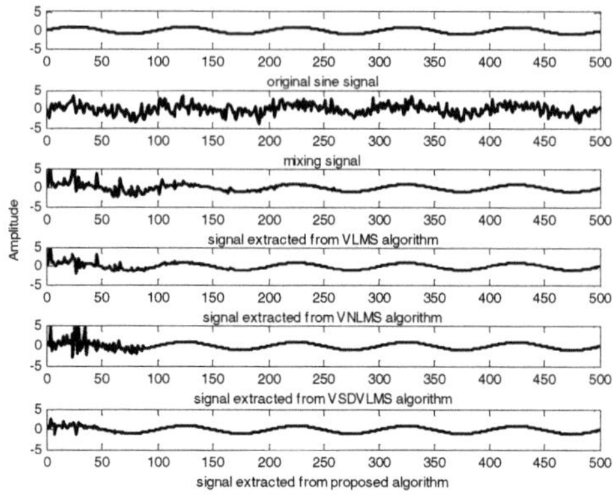

Fig. 2. For $a = 0.3$, the noise cancellation results of different algorithms

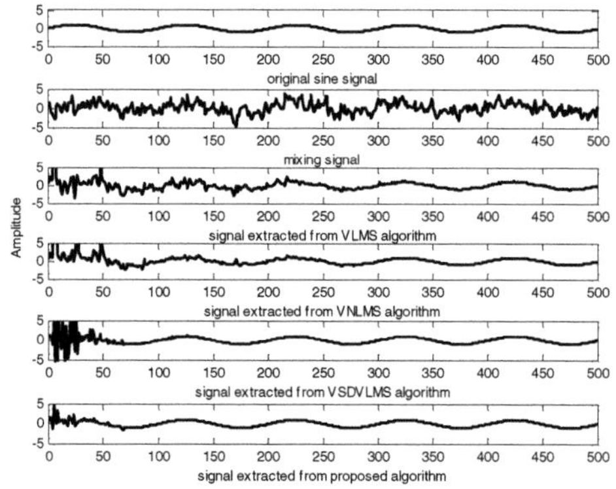

Fig. 3. For $a = 0.6$, the noise cancellation results of different algorithms

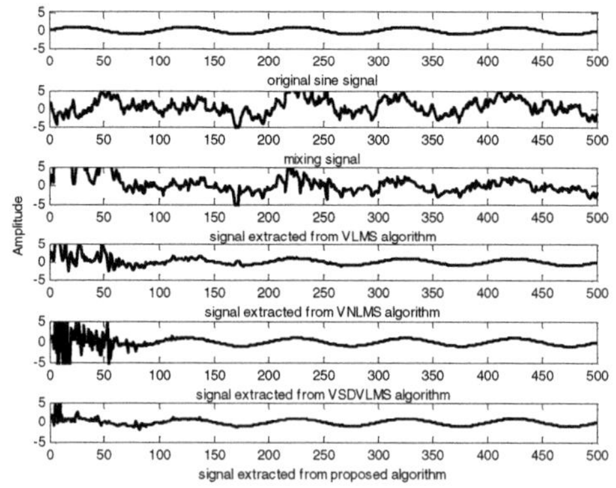

Fig. 4. For $a = 0.9$, the noise cancellation results of different algorithms

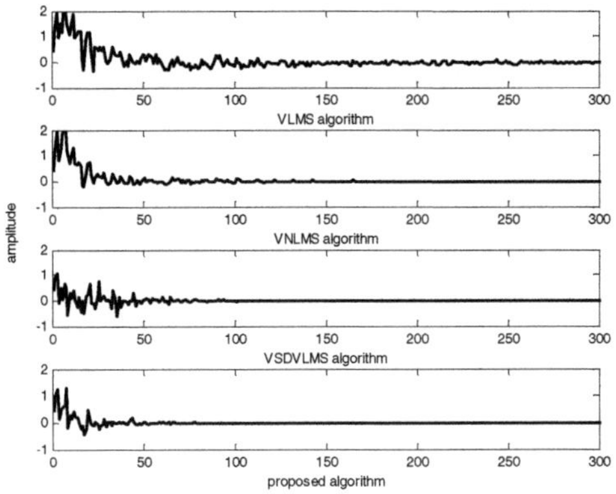

Fig. 5. For $a = 0.3$, the output errors of noise cancellation systems based on different algorithms

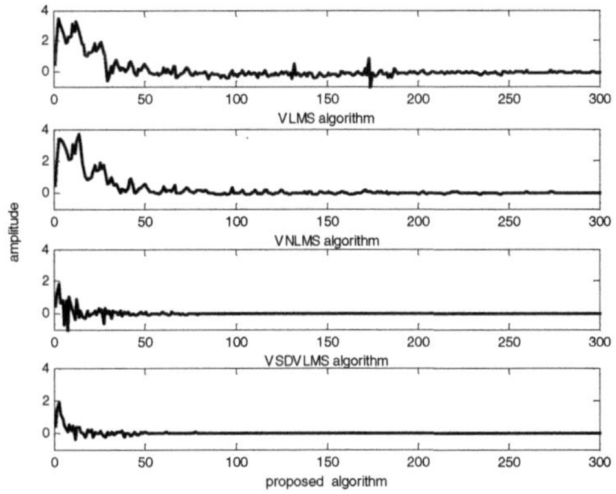

Fig. 6. For $a = 0.6$, the output errors of noise cancellation systems based on different algorithms

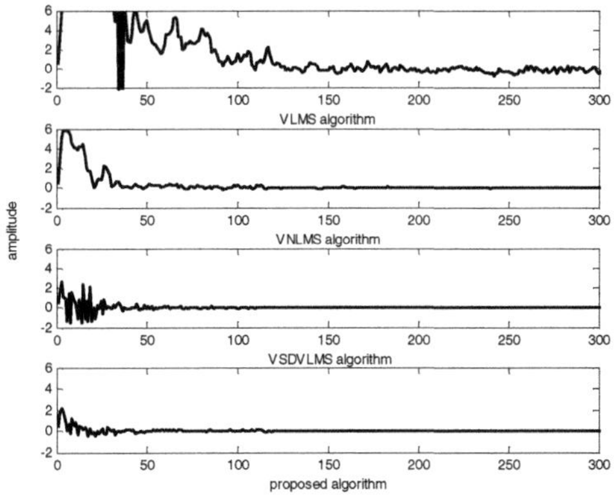

Fig. 7. For $a = 0.9$, the output errors of noise cancellation systems based on different algorithms

4 Conclusion

Volterra filters have been widely applied to noise cancellation, system identification, adaptive channel equalization, and so on. In Volterra adaptive noise cancellation system, since the correlativity of input noise leads to the falling of noise cancellation performance, a variable step size NLMS algorithm based on decorrelation for second-order Volterra filter is proposed. Simulation results show that the proposed algorithm has a better noise cancellation performance when the input noises are strongly correlated.

Acknowledgement. We would like to thank the national science foundation of P.R. China for fund support. The grant of the foundation is 60872092. Meanwhile, we wish to express our gratitude to the reviews for their careful reading and help in improving the paper's quality.

References

1. Doherty, J.F., Porayath, R.: A robust echo canceler for acoustic environments. IEEE Transactions on Circuits and Systems II: Analog and Digital Signal Processing 44(5), 369–389 (1997)
2. Peng, J.D., Duan, Z.H., Wang, Z.Z.: A Variable Step-size Uncorrelated LMS Algorithm. Computer Engineering and Applications 40(30), 136–144 (2004)
3. Long, J.Y., Wang, Z.Z., Xia, S.H., Duan, Z.H.: An Uncorrelated Variable Step-Size Normalized LMS Adaptive Algorithm. Computer Engineering and Science 28(4), 60–62 (2006)
4. Long, Y., Wang, Z., Xue, X.N.: New variable step-size LMS-Type algorithm and simulation based on Decorrelation method. Electronic Measurement Technology 30(1), 52–55 (2007)
5. Han, L., Qiu, F., Sun, A.Q.: An Improved Variable Step-Size LMS Algorithm Based on Decorrelation. Fire Control Radar Technology 38(3), 47–50 (2009)
6. Li, J.H., Li, J.: A novel variable step size LMS algorithm based on decorrelation. In: 3rd International Congress on Image and Signal Processing (CISP 2010), vol. 7, pp. 3291–3294 (2010)
7. Zhang, X.M., Zhao, Z.J., Shang, J.N.: Variable step size decorrelation Volterra LMS algorithm. Computer Engineering and Applications 46(28), 157–159 (2010)
8. Hu, H., Liu, Z.L., Zhang, Y.Q., Zhou, X., Ying, H.: A New Variable Step-Size LMS Algorithm and Its Simulations. Electronic Science and Technology 11, 48–50 (2006)
9. Kon, T., Powers, E.J.: Second-order Volterra filtering and its application to nonlinear system identification. IEEE Transactions on Acoustic, Speech and Signal Processing 33(6), 1445–1455 (1985)
10. Alexandre, G., Gerard, F., Regine, L.B.J.: Nonlinear acoustic echo cancellation based on volterra filters. IEEE Transactions on Speech and Audio Processing 11(6), 672–683 (2003)

Research and Design of the Integrated Administration Management System for Enterprises

Junhong Zheng

Department of Physical Engineering and Information Technology
Wuhan Institute of Physical Education Wuhan, China 430079
Zhengjunhong4329@126.com

Abstract. Administration management is an important part of enterprise management. Based on the China's mainland small and medium private enterprises and combined with their management model and mode, the design problems of the integrated administration management software system for enterprises are described. The design model of the object-oriented integrated administration management system for enterprises is established, and an overall design of software system is proposed, which laid the method basis for the research and development work of the integrated administration management system for enterprises. It also has a great guiding role in the subsequent software development and a certain referential meaning for the integrated management system in other fields.

Keywords: administration management, enterprise management, information management system.

1 Introduction

Administration and personnel management is an important part of enterprise management. In the Mainland of China, because of their special nature, many small and medium private enterprises basically incorporate the personnel and administration management into one department for unified management. With the popularization of computer applications and higher levels of social information, these companies have increasingly high demand for internal management in needs of their own development and social competition, and expect through a special software management system to improve management efficiency, standardize management practices, and conserve resources in order to mobilize the initiative of internal departments and create a good economic benefits.

Database systems are used as the supporting platform in this paper, and with the combination of management model and mode of small and medium private enterprises in the Mainland of China, the design of the integrated administration and personnel management software system for enterprises is studied. The appropriate software model is also established, and the technical approach to the design of the administration and personnel management software system for enterprises is proposed.

G. Lee (Ed.): Advances in Automation and Robotics, Vol. 1, LNEE 122, pp. 33–40.
springerlink.com

2 The Analysis and Decomposition of Administrative Activities in Enterprises

The core target of administration is to meet the development needs of companies, which relies on administrative organization and manage enterprises according to the administrative channels. Enterprise administration mainly includes administrative management, office management, and human resources management, which specifically cover the establishment and the implementation and promotion of relevant systems, routine office management, office materials management, clerical data management, conference management, foreign affairs management, welfare, traffic, safety and health.

Routine office management includes daily scheduling, organization and implementation, communication, coordination and control, inspection summary , punishment and reward and other aspects of management; office materials management includes distribution, use, storage, and purchasing of office materials, as well as the corresponding establishment of the system ; clerical data management includes seals management, document management, records management, and publications management; conference management includes preparation before the conference, service in the conference and work after the conference.

3 Analysis on the Designed Model of Administration and Personnel Management System for Enterprises

3.1 Object Model of Administration and Personnel Management System for Enterprises

In order to accurately represent and store the knowledge of enterprise administration, the design of software system adopts object-oriented software engineering (OOSE), using the "object" as a unit to organize and manage the enterprise administration knowledge.

In this paper, OMT method is used to establish the design objects and the object model (shown in Figure 2) of administration management software system for enterprises (hereinafter referred to as ETOM, as shown in Figure 1). ETOM consists of three parts, in which attribute field contains ETOM attributes, namely: basic data value of ETOM, etc.; methods filed includes the basic operation of ETOM, namely: ETOM method, the corresponding operation, etc.; contact field contains ETOM relationships, namely: inheritance, derivation, and other relationships between ETOM.

The design object of administration management software system for enterprises ETOM
Property: basic property (Basic data value of ETOM) (other properties of ETOM)
method : corresponding operation 1 (the basic operation behavior) (other operation behavior)
relation: corresponding correlation 1 (inheritance relationship between ETOM) (other relationship between ETOM)

Fig. 1. The design object of administration management software system for enterprises

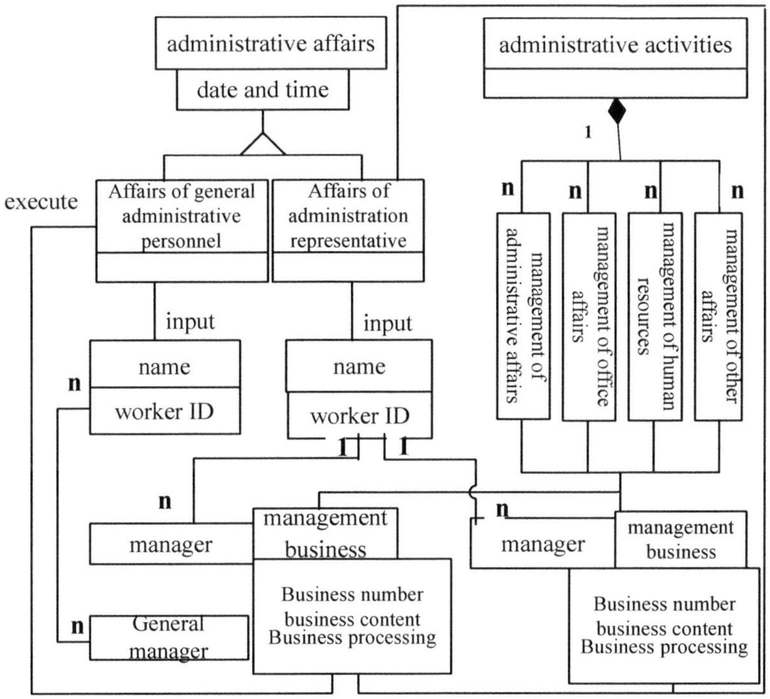

Fig. 2. The object model of administration management system

3.2 Dynamic Model of the Administration Management System for Enterprises

According to the main task and management control process of administration management for enterprises, software system is to retrieve the role and authority of the corresponding management personnel from the database, thus conducting specific management on the administrative operations of the enterprises, coupled with the timely archiving for the data level document. The dynamic model of software system behavior is as shown in Figure 3.

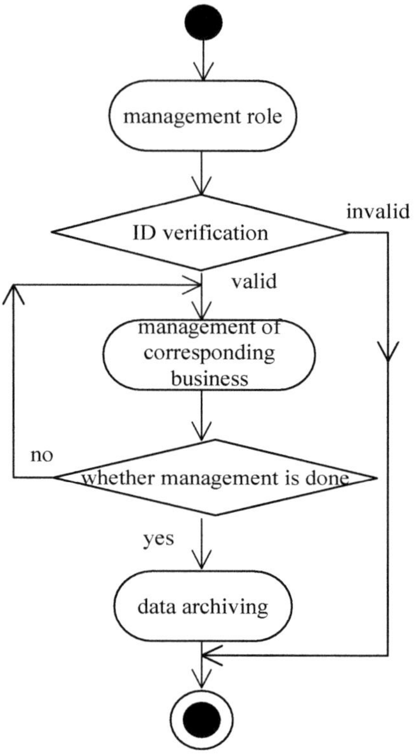

Fig. 3. The dynamic model of the behavior of the administration management software system for enterprises

3.3 Analysis on the Functional Model of Administrative Management System for Enterprises

According to the specific management activities and the management scope, the administrative management system can be divided into a series of structural units which are composed by several functional modules. The functional hierarchy of the software is as shown in Figure 4.

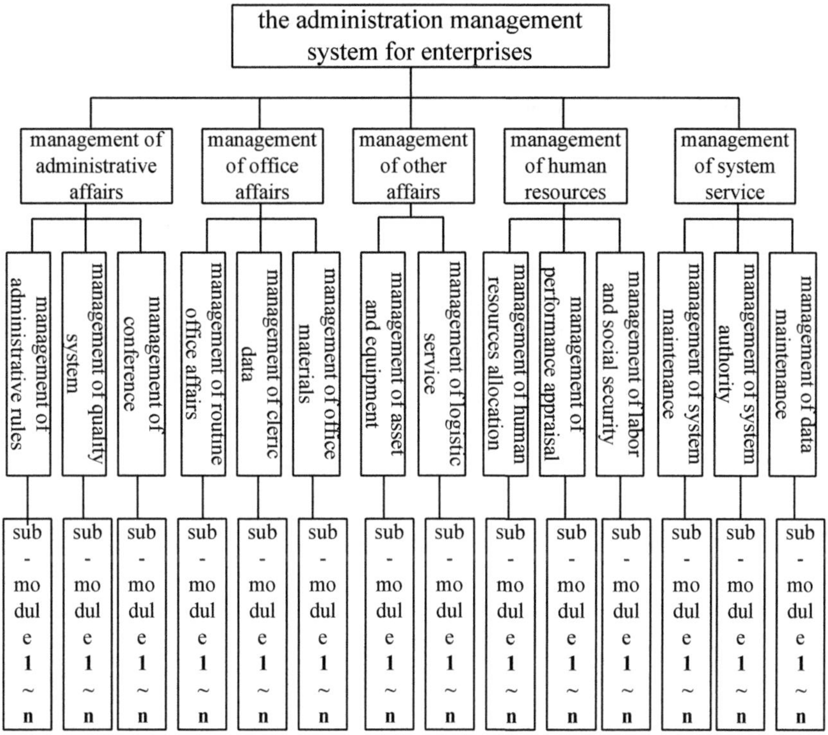

Fig. 4. Function model of administration management system for enterprises

The administration system of the enterprises includes the subsystem of administrative transaction management, subsystem of office transaction management, subsystem of other transaction management subsystem, subsystem of human resources management and systemic service management. The subsystem of administrative transaction management contains 3 parts: administrative system management, quality system management and conference management; subsystem of office transaction management contains routine office management, clerical data management and office materials management; subsystem of other transaction management includes 2 modules, assets equipment management and logistics management services; subsystem of human resources management includes 3 modules, allocation of resources management, performance appraisal management, and labor and social security management. Every module mentioned above contains the corresponding sub-modules, which are relatively independent and interfering with each other, containing the appropriate properties, methods, rules and content. These sub-modules are conductive for the administration management of enterprises. The subsystem of systematic service management consists of 3 service behaviors and the corresponding functional modules, which can provide auxiliary service for the maintenance, operation and management of the entire software system.

3.4 Data Storage and Processing of Administration Management System of Enterprises

Storage mode of data information is very important for the software system, for it may directly determine the speed of data retrieval and extraction. In view of the practical application, the best way is to achieve the data's synchronous updates and seamless transfer. Therefore, the software system selects the Microsoft's default desktop database management system Microsoft Access for data storage and processing. Supporting distributed computing environment, Microsoft Access is a database management system of high performance, multiuser and relation-base, which can achieve the interconnection and interoperability of data tables under the network environment. Moreover, it also owns the functions such as data fault tolerance, integrity inspection and security encryption, etc, which is conductive to simplifying the data processing system and improving data processing efficiency.

4 Structural Analysis of the Administration and Personnel Management System for Enterprises

Administration of enterprises involves several departments, therefore, it is requested that the administration management software system must be able to run multi-department and in the local area network system of enterprises.

4.1 The Hardware Architecture of Administration Management System for Enterprises

It is necessary to build an appropriate network hardware environment in order to run the administration management software system for enterprises normally. Hardware is usually an office-based LAN composed by the server (server), printer, a number of terminal sets (client), and switches. The terminals within network are linked by star topology, and the terminal sets are linked to the server through the switch. The server is directly linked to the Internet through the external network interface of enterprises. The topology of hardware system is shown in Figure 5.

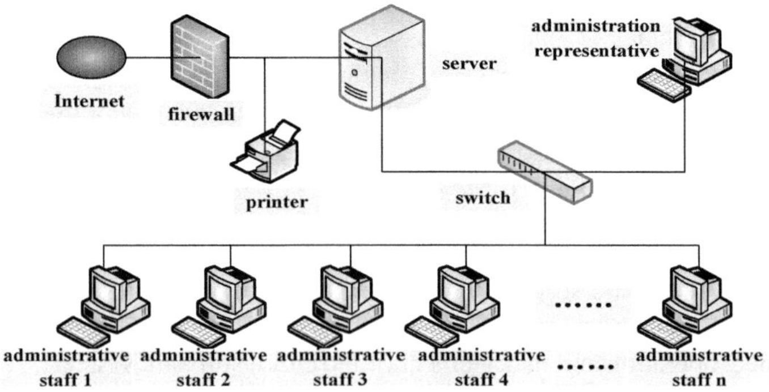

Fig. 5. Topology of hardware system

4.2 Software Architecture of Administration Management System for Enterprises

In order to ensure the security of the software system, C / S (Client / Server) architecture should be used, and the software system includes client applications, server applications, and databases. The client application mainly provides users operations of interface, establishing a connection with the server and submitting a request to the server, the data input and output, and verification. Server application is used to manage users, deal with client requests, maintain system security, and manage data, etc. The architecture of system is shown in Figure 6.

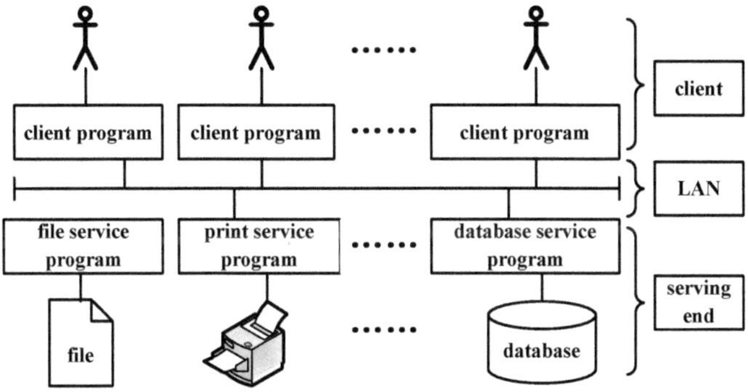

Fig. 6. Architecture diagram of software

5 The Implementation of Administrative and Personnel Management System for Enterprises

To ensure the scalability and maintainability of administrative management software system for enterprises, this system should be an open system. Therefore, the object-oriented software development method should be applied in the design, while the parallel retrieval and dynamic access should be applied when making data processing; moreover, to guarantee the smooth data processing, the ADO.NET data access technology should be utilized to conduct retrieval and access for data information. In addition, when designing the software system, the following key issues should be paid attention to.

5.1 Database Security

As the administrative system involves some commercial secrets and management mode, so we must ensure the security and stability of the database. In order to reduce losses and ensure the security of the system, the encryption processing should be carried out for the database, that is, utilize the key function to dynamically generate the key for databases to ensure the security of data; meanwhile, apply the technology of data table locking to "lock" the data table, which cannot be properly open and utilized until it is fully unlocked.

5.2 Database Backup and Recovery

To ensure the security and integrity of data, data service program should have the function of database auto backup, that is, backup the database files to a specified path. If suffering the system failure or database file's damage, personnel of the system can timely recover the data information through the data service program at any time.

6 Conclusions

The design problems of administration management software system for enterprises are described from the perspective of design models of object-oriented software, and the design model based on object-oriented software system is established. This research has laid a foundation for the development of administrative software system, coupled with the guiding role in the follow-up study of this software system as well as the development of integrated management system in other fields.

References

1. Pan, J., et al.: Software Development Technology. Xidian University Press, Xi'an (2006)
2. Deng, L., et al.: Software Engineering. Xidian University Press, Xi'an (2006)
3. Zhang, H.: Introduction to Software Engineering. Tsinghua University Press, Beijing (2008)
4. Yang, Z.: Analysis and Design of Information Systems. Mechanical Industry Press, Beijing (2007)
5. Ma, X., et al.: Management information system and its development technology. Tsinghua University Press, Beijing (2006)
6. Du, Y., et al.: Design and implementation of VB-based Personnel Management System of Enterprises. China Science and Technology Information (8) (2008)
7. Cai, Q., et al.: Design of Labor and Personnel Management Information System in Enterprises. Ergonomics (2) (1996)

An Improved Iterated Local Search Algorithm for the Permutation Flowshop Problem with Total Flowtime

Xingye Dong[1], Ping Chen[2], and Houkuan Huang[1]

[1] School of Computer and IT, Beijing Jiaotong University, 100044 Beijing, China
[2] TEDA College, Nankai University, 300457 Tianjin, China
xydong@bjtu.edu.cn, chenpingteda@nankai.edu.cn,
hkhuang@bjtu.edu.cn

Abstract. Iterated local search (ILS) algorithm is a powerful metaheuristic for the permutation flowshop problem with total flowtime objective. ILS is based on a local search procedure and the procedure needs to be restarted from another solution when the procedure is trapped into a local optima. In the literature, the solution is often generated by slightly perturbing the best solution found so far. By doing so, the search space is relatively narrow and it is easy to lead the search stagnant. In order to improve this situation, a strategy is proposed in this paper to allow the restart solution be generated from a group of solutions, which are drawn from the local optima solutions found in the search process. An ILS algorithm named MRSILS is proposed. Its performance is evaluated on a set of benchmarks. Comparisons show that the MRSILS is significantly better than or comparable to several state-of-the-art metaheuristics.

Keywords: Scheduling, Permutation flowshop, Total flowtime, Metaheuristic, Iterated local search.

1 Introduction

The flowshop scheduling is a well-known combinatorial optimization problem in the operations research and has attracted many researchers' attention since Johnson's pioneering work [1]. Among various flowshop problems, the permutation flowshop problem (PFSP) has been extensively studied. The PFSP is a simplified version of the flowshop problem and Garey et al [2] proved that the PFSP with total flowtime criterion is strongly NP-complete even with two machines. In the literature, most of the research is to minimize the makespan. Recently, however, more and more researchers begin to concentrate their efforts on the total flowtime minimization. In this paper, the problem discussed is the PFSP with the total flowtime objective.

In order to find a much better solution in a short acceptable time, many metaheuristics were proposed in the literature. These metaheuristics include several ant colony algorithms by Rajendran and Ziegler [3, 4], particle swarm optimization by Tasgetiren et al [5], genetic algorithms by Zhang et al [6] and Tseng and Lin [7, 8], estimation of distribution algorithm by Jarboui et al [9], differential evolution algorithm by Pan et al [10]. In the above algorithms, different local search algorithms were incorporated. We also proposed an iterated local search algorithm and showed

G. Lee (Ed.): Advances in Automation and Robotics, Vol. 1, LNEE 122, pp. 41–48.
springerlink.com

its good optimization ability on the PFSP with flowtime criterion [11]. More recently, Tasgetiren et al [12] illustrated an artificial bee colony algorithm and a discrete differential evolution algorithm. In both the algorithms, a well designed local search procedure was used. And in the local search procedure, the advantages of both the iterated local search [11] and the iterated greedy algorithm [13] were combined. Obviously, local search procedures act as an important role in many metaheuristics.

ILS is a metaheuristic providing a simple but powerful frame for improving the performance of local search. It has attracted much attention of researchers for its simplicity, effectiveness and efficiency. Recently, Ruiz and Stützle proposed a very effective iterated greedy algorithm for the PFSP with makespan criterion [13], which is quite similar to ILS. We applied the ILS to the PFSP with total flowtime objective, and showed its performance is better than several other metaheuristics [11], including PACO and M-MMAS by Rajendran and Zeigler [3], ACO2 by Rajendran and Zeigler [4] and the PSO_{vns} by Tasgetiren et al [5]. Additionally, we also applied a variant of this algorithm to capacitated vehicle routing problem and showed its good performance [14].

In ILS algorithm, it needs to choose a restart point when the local search is trapped into a local optima and this restart point has significant effect on the performance of the ILS. Currently, the method to generate the restart point is quite simple, e.g., it is generated by perturbing the best solution found so far in our previous work [11]. So it is necessary to study various perturbing methods and observe whether they can improve the ILS algorithm. In this work, we design a scheme that the restart solution is randomly generated from several selective solutions, then propose an improved ILS algorithm which is comparable to or even better than several state-of-the-art metaheuristics. The algorithm is named multi-restart ILS (MRSILS for abbreviation).

The remainder of this paper is organized as follows. In Section 2, the formulation of the PFSP with total flowtime criterion is presented. In Section 3, the proposed MRSILS algorithm is discussed in detail. Then the computational results are presented in Section 4 and the paper is concluded in Section 5.

2 Problem Formulation

The PFSP with total flowtime objective can be formally defined as follows. A set of jobs $J = \{1, 2, ..., n\}$ available at time zero needs to be processed on m machines, where $n \geq 1$ and $m \geq 1$. Each job has m operations, each of which needs an uninterrupted processing time. The processing time of the ith operation of job j is denoted by p_{ij}, where $p_{ij} \geq 0$. The process sequence of these m operations is that the first operation needs to be processed on the first machine, the second operation needs to be processed on the second machine, and so on. An operation of a job starts to be processed only if the previous operation of the job is completed and the requested machine is available. Each machine processes these jobs in the same order and only one operation can be processed at a time. This problem is usually denoted by $F_m|prmu|\Sigma C_j$ [15], where C_j denotes the completion time of job j. Let π denote a permutation, which represents a job processing order, on the set J. Let $\pi(k)$, $k = 1, ..., n$, denote the kth job in π, then the completion time of job $\pi(k)$ on each machine i can be computed through a set of recursive equations:

$$C_{i,\pi(1)} = \Sigma^{i}_{r=1}\, p_{r,\pi(1)} \qquad i = 1, ..., m \tag{1}$$

$$C_{1,\pi(k)} = \Sigma^{k}_{r=1}\, p_{1,\pi(r)} \qquad k = 1, ..., n \tag{2}$$

$$C_{i,\pi(k)} = max\{C_{i-1,\pi(k)}, C_{i,\pi(k-1)}\} + p_{i,\pi(k)} \qquad i = 2, ..., m;\ k = 2, ..., n \tag{3}$$

Then $C_{\pi(k)} = C_{m,\pi(k)}$, $k = 1, ..., n$. The total flowtime is $\Sigma C_{\pi(k)}$, i.e. the sum of the completion time on machine m of all the jobs. PFSP with total flowtime criterion is to minimize $\Sigma C_{\pi(k)}$.

3 The Proposed MRSILS

Iterated local search is a simple but powerful metaheuristic. Unlike some other sophisticated metaheuristics, its frame is very simple. The pseudo code is illustrated in Fig. 1, where s^* denotes the best solution found in the search process.

1. Generate an initial solution s; let $s' = s$; let $s^* = s$;
2. Generate $s'' = LocalSearch(s')$; update s^* if better solution is found in the process;
3. Let $s = Accept(s, s'')$ (this function can be a simulated-annealing one or some other types);
4. If the termination criterion is not satisfied, then generate $s' = Perturb(s)$, go to step 2; otherwise, output s^*.

Fig. 1. Pseudo code of an ILS frame

In the ILS frame, several points need to be specified, i.e., the method to generate the initial solution in step 1, the local search procedure in step 2, the acceptance criterion in step 3, and the termination criterion and the method to perturb solution s in step 4.

The proposed MRSILS is firstly illustrated in Fig. 2, and then it is discussed in detail as follows.

Firstly, parameters of the MRSILS are set. The counter parameters *cnt* and i are initialized. The *flag* indicates whether the best solution π^* is improved and this parameter is initially set to *false*. The parameters *pool* and *pool_size* are used for generating a restart solution for the local search, which are initialized empty and a proper number, respectively.

As for the method to generate the initial solution, we discussed six methods in our previous work [11] and chose the H(2) by Liu and Reeves [16] as the initial method for it can construct a solution in a negligible time and the corresponding ILS performs quite well. In this work, it is still chosen as the initial method.

In this work, the local search procedure and the acceptance criterion are the same as those in our previous work [11]. The local search procedure corresponds to the steps 5 and 6, and this is imbedded in the loop in step 4. The basic idea of this procedure is trying to find a better solution by inserting a job to another position in the sequence. The acceptance criterion is that only better solution can be accepted. This corresponds to step 8 in Fig. 2.

In the literature, the termination criterion can be set as the maximum iteration for the local search procedure or the maximum CPU time which can be consumed. In order to make fair comparisons with other metaheuristics, we will use both criterion in this work, respectively. This criterion is implemented in step 3.

```
1. Set cnt = 0, i = 1, flag = false, pool = Φ, pool_size;
2. Generate an initial permutation π, set π* = π;
3. while(termination criterion is not satisfied)
   {
4.    for(i = 1; i <= n; i++)
      {
5.       Find k, which satisfies π(k) = π*(i);
6.       Insert job π(k) into other n 1 positions in π, respectively, and let π' be the best one
           among the n−1 generated permutations;
7.       if(π' is better than π)
              Set π = π', cnt = 0;
           else
              Set cnt = cnt +1;
8.       if(π is better than π*) Set π* = π, flag = true;
9.       if(cnt == n)
         {
10.         if(flag == true) Set pool = Φ, flag = false;
11.         if(π is not in pool) Add π into pool;
12.         if(|pool| > pool_size) Delete the worst schedule from pool;
13.         if(|pool| < pool_size)
                 Perturb π* to generate a new π;
              else
                 Select one schedule randomly from pool and perturb it to generate a new π;
14.         if(π is better than π*) Set π* = π;
15.         Set cnt = 0;
         }
      }
   }
   Output π* and stop.
```

Fig. 2. Pseudo code of the proposed MRSILS

With respect to the method to perturb the solution s, it is necessary to ensure that the solution s should not be disturbed much [11, 12]. In our previous work [11], the s is always set to the best solution found so far and the perturbation method is swaping six random pairs of adjacent jobs. The reason for doing so is that the generated solution by this method could be considered near to a better solution, and then this restart solution would help the local search procedure find it. However, this method would cause a relatively small search space and difficult to escape from local optima, and so it would not good for searching better solutions in a long run. So, we consider to extend the search space by allowing the restart solution be generated from a set of good solutions. In order to achieve this, a number of good solutions need to be pooled. This set of good solutions is denoted by *pool* in Fig. 2 and its max size is limited to *pool_size*. When the search is trapped, i.e. the counter *cnt* equals to *n*, the local optima is added to *pool* if it is not in the *pool* yet. In order to reserve better solutions, it is necessary to delete the worst one in the *pool* if the size of the *pool* exceeds the size limit *pool_size*. When the *pool* is not full, we prefer to generate the restart solution from the best solution π^* found so far; otherwise, a solution is randomly selected from the *pool* and perturbed to generate the restart solution.

Further, the perturbation method also impacts the performance. In our previous work [11], the method is swapping randomly selected adjacent jobs for certain times. However, Tasgetiren et al [12] found that the performance is quite good by inserting one or two randomly chosen jobs to other randomly chosen positions. In our experiments of this work, we found this perturbation method is almost the same when running 1000 iterations, but slightly better than our previous method for long time search, e.g., running for 5000 iterations. In the proposed MRSILS, the perturbation method is inserting one randomly chosen job to another randomly chosen position.

As for the time complexity of the MRSILS, it is $O(mn^3)$ for it is determined by the local search procedure used in the algorithm. The analysis can be made reference to our previous work [11].

4 Computational Results

In this section, the size of the *pool* is determined by experiments, and then the proposed MRSILS is compared with several state-of-the-art metaheuristics. The benchmarks are taken from Taillard [17], which consist of 120 quite hard instances with 12 different sizes. For each size, there are 10 instances provided. The scale of these problems varies from 20 jobs and 5 machines to 500 jobs and 20 machines. In the literature, many metaheuristics were only tested on the first 90 instances [3-12], so is the proposed MRSILS.

The MRSILS is implemented in C++, running on a PC with an Intel Core2 Duo processor (2.99 GHz) with 2G main memory. Though the computer has two processors, there is only one used in the experiments for there are no any parallel techniques used.

The MRSILS has only one parameter *pool_size*. Obviously, if it is set to 1, the MRSILS will almost be our previous ILS algorithm other than the perturbation method. In order to choose a better parameter value, the *pool_size* is changed from 1 to 20, respectively. For each case, five independent runs are performed for each benchmark instance and the best solution among these five runs is recorded. Then the relative percentage deviation (RPD) is calculated by Eq. (4).

$$RPD = (F - F_{best})/F_{best} \times 100 \tag{4}$$

where F is the result found by the proposed MRSILS and F_{best} is the best result provided by BEST(LR) [16], M-MMAS and PACO [3] for Taillard's benchmarks

[17]. Here, BEST(LR) denotes the best performing heuristic among several heuristics provided by Liu and Reeves [16]. Based on this experiment, we find the *pool_size* should be limited between 2 and 12. In the following experiments, it is set arbitrarily to 5.

In order to make a comparison with our previous ILS algorithm [11], we run both the ILS and the MRSILS 1000 iterations, 3000 iterations and 5000 iterations for five times, respectively, and then record the best solutions among the five runs. The comparisons in RPD, which is computed by Eq. (4), are listed in Table 1. Additionally, the CPU time in seconds for both algorithms by running once in this experiment is almost the same for all items, so only the CPU time consumed by the MRSILS is listed in Table 1 in columns *t*.

Table 1. Comparisons between the ILS and the proposed MRSILS

n|*m*	1000			3000			5000		
	ILS	MRSILS	*t*	ILS	MRSILS	*t*	ILS	MRSILS	*t*
20|5	-0.168	-0.168	0.13	-0.175	-0.175	0.40	-0.168	-0.168	0.65
20|10	-0.035	-0.039	0.29	-0.031	-0.039	0.92	-0.039	-0.039	1.46
20|20	-0.058	-0.068	0.58	-0.068	-0.068	1.74	-0.068	-0.068	2.90
50|5	-0.451	-0.593	1.74	-0.592	-0.721	5.21	-0.691	-0.775	8.67
50|10	-0.955	-1.026	4.15	-1.014	-1.169	12.57	-1.180	-1.224	20.80
50|20	-0.780	-0.863	8.73	-0.932	-0.999	26.33	-1.024	-1.104	43.67
100|5	-0.458	-0.430	12.68	-0.549	-0.609	38.18	-0.609	-0.707	63.39
100|10	-0.814	-0.837	31.13	-1.035	-1.141	93.44	-1.077	-1.250	155.79
100|20	-1.063	-1.072	67.15	-1.241	-1.313	201.22	-1.353	-1.483	335.70
avg.	-0.531	-0.566	14.06	-0.626	-0.693	42.22	-0.690	-0.757	70.33

From Table 1, it can be seen that the average RPDs for the MRSILS on all instances are always smaller than that of the ILS, e.g., the RPDs are -0.531 and -0.566 for the ILS and the MRSILS with 1000 iterations, respectively. This indicates that the MRSILS performs better than the ILS. Additionally, for each run is done independently and there are some randomness in the algorithms, the performance for the same group of instances will change slightly, i.e., it is not necessarily better with more iterations on the same group of instances. For example, for the 20 jobs and 5 machines instances, the RPD for the ILS with 3000 iterations is -0.175, while it is -0.168 with 5000 iterations. However, the average performance on all the instances has little fluctuation. In our several runs, the average performance for both algorithms with 3000 iterations is always better than the counterparts with 1000 iterations.

In order to make more rigorous comparisons with respect to the performance, a group of one-sided paired-samples *t*-tests are carried out between the ILS and the proposed MRSILS with the 1000, 3000 and 5000 iterations, respectively. The tests are done for all the instances and so the freedom degree is 89. The results show the MRSILS is significantly better than the ILS.

Very recently, Tasgetiren et al [12] proposed an artificial bee colony algorithm (denoted by DABC) and applied the artificial bee strategy into the discrete differential evolution algorithm by Pan et al [10] to form a new algorithm called hDDE. Both algorithms were shown better than the estimation of distribution algorithm by Jarboui et al [9], the genetic local search algorithms by Tseng and Lin [7, 8] and an implementation of the traditional iterated greedy algorithm [13]. Both the DABC and the hDDE were implemented in Visual C++ and run independently ten times on an Intel Pentium IV 3.0 GHz PC with 512MB memory. The best results were reported for both algorithms with a short-term search, for which the CPU time was limited to $0.4 \times n \times m$ seconds for each instance per run, and a long-term search, for which the CPU time was limited to $3 \times n \times m$ seconds for each instance per run. The long-term search is too time consuming, which will take more than 20 days for one algorithm. So, only the short-term search is carried out in this work. The comparison results with the hDDE and DABC are arranged in Table 2, where the RPDs are computed by Eq. (4). The RPDs indicate that all these three algorithms perform well on 20 jobs instances. As for the 50 jobs and 100 jobs instances, the MRSILS performs generally better on larger instances and worse on smaller instances. Paired-Samples *t*-tests show that the MRSILS is significantly better than the DABC, but not the case for the hDDE. The MRSILS is comparable to the hDDE and the average RPDs on all instances are -0.858, while it is -0.844 for the hDDE. Moreover, the MRSILS is more attractive for it is rather simple compared to the hDDE and easy to be implemented.

Table 2. Comparisons results for the MRSILS with the hDDE and the DABC

n|*m*	hDDE	DABC	MRSILS
20|5	-0.175	-0.175	-0.175
20|10	-0.039	-0.039	-0.039
20|20	-0.068	-0.068	-0.068
50|5	-0.984	-0.961	-0.956
50|10	-1.433	-1.426	-1.436
50|20	-1.191	-1.252	-1.194
100|5	-0.918	-0.704	-0.874
100|10	-1.313	-1.132	-1.367
100|20	-1.477	-1.508	-1.616
avg.	-0.844	-0.807	-0.858

5 Conclusions

ILS is a powerful metaheuristic. It bases on a local search procedure and the local search procedure needs to be restarted from another solution when the search is trapped into a local optima. In this work, a strategy is proposed to allow the restart solution be generated from a group of good solutions which are drawn from the local optima solutions found in the search process. The performance of the ILS is improved by doing so. Experiments on benchmarks show that the proposed MRSILS is significantly better than or comparable to several state-of-the-art metaheuristics.

Acknowledgments. This work is supported by The Fundamental Research Funds for the Central Universities (Project Ref. 2009JBM018, Beijing Jiaotong University).

References

1. Johnson, S.M.: Optimal two and three-stage production schedule with setup times included. Naval Research Logistics Quarterly 1, 61–68 (1954)
2. Garey, M.R., Johnson, D.S., Sethi, R.: The complexity of flowshop and jobshop scheduling. Mathematics of Operations Research 1, 117–129 (1976)
3. Rajendran, C., Ziegler, H.: Ant-colony algorithms for permutation flowshop scheduling to minimize makespan/total flowtime of jobs. European Journal of Operational Research 155, 426–438 (2004)
4. Rajendran, C., Ziegler, H.: Two ant-colony algorithms for minimizing total flowtime in permutation flowshops. Computers and Industrial Engineering 48, 789–797 (2005)
5. Tasgetiren, M.F., Liang, Y.-C., Sevkli, M., Gencyilmaz, G.: A particle swarm optimization algorithm for makespan and total flowtime minimization in the permutation flowshop sequencing problem. European Journal of Operational Research 177, 1930–1947 (2007)
6. Zhang, Y., Li, X.P., Wang, Q.: Hybrid genetic algorithm for permutation flowshop scheduling problems with total flowtime minimization. European Journal of Operational Research 196, 869–876 (2009)
7. Tseng, L.-Y., Lin, Y.-T.: A hybrid genetic local search algorithm for the permutation flowshop scheduling problem. European Journal of Operational Research 198, 84–92 (2009)
8. Tseng, L.-Y., Lin, Y.-T.: A genetic local search algorithm for minimizing total flowtime in the permutation flowshop scheduling problem. International Journal of Production Economics 127, 121–128 (2010)
9. Jarboui, B., Eddaly, M., Siarry, P.: An estimation of distributation algorithm for minimizing the total flowtime in permutation flowshop scheduling problems. Computers and Operations Research 36, 2638–2646 (2009)
10. Pan, Q.-K., Tasgetiren, M.F., Liang, Y.-C.: A discrete differential evolution algorithm for the permutation flowshop scheduling problem. Computers and Industrial Engineering 55, 795–816 (2008)
11. Dong, X.Y., Huang, H.K., Chen, P.: An iterated local search algorithm for the permutation flowshop problem with total flowtime criterion. Computers and Operations Research 36, 1664–1669 (2009)
12. Tasgetiren, M.F., Pan, Q.-K., Suganthan, P.N., Chen, A.H.-L.: A discrete artificial bee colony algorithm for the total flowtime minimization in permutation flow shops. Information Sciences 181, 3459–3475 (2011)
13. Ruiz, R., Stützle, T.: A simple and effective iterated greedy algorithm for the permutation flowshop scheduling problem. European Journal of Operational Research 177, 2033–2049 (2007)
14. Chen, P., Huang, H., Dong, X.: Iterated variable neighborhood descent algorithm for the capacitated vehicle routing problem. Expert Systems with Applications 37, 1620–1627 (2010)
15. Pinedo, M.: Scheduling: theory, algorithms, and systems, 2nd edn. Prentice-Hall (2001)
16. Liu, J.Y., Reeves, C.R.: Constructive and composite heuristic solutions to the $P//\Sigma C_i$ scheduling problem. European Journal of Operational Research 132, 439–452 (2001)
17. Taillard, E.: Benchmarks for basic scheduling problems. European Journal of Operational Research 64, 278–285 (1993)

A Framework on Halal Product Recognition System through Smartphone Authentication

Cik Ku Haroswati Che Ku Yahaya[1], Murizah Kassim[1], Muhammad Hafiz bin Mazlan[1], and Zamri Abu Bakar[2]

[1] Faculty of Electrical Engineering
[2] Faculty of Computer and Mathematical Sciences
Universiti Teknologi MARA, 40450 Shah Alam, Selangor, Malaysia
haroswati@salam.uitm.edu.my, murizah@salam.uitm.edu.my,
apismazlan@yahoo.com, zamri@salam.uitm.edu.my

Abstract. Muslim consumer faces a problem in identifying the halal products in the market especially in Malaysia. A fake halal logo and unknown detail ingredients is the major problem for Muslim consumer. This framework study will be presented and with the implemented proposed system would help consumer to verify through Smartphone barcode capture straightaway on time with the applied halal product alert database system. All the information regarding halal logo, company, product name, barcode number, ingredients, license number and expiry date will be found in the database. This study focuses on 1-D European Article Numbering (EAN)-13 barcode since it is used internationally for tagging retail goods. However other types of 1-D and 2-D barcodes can still be readable and application will be developed on Android 2.1 (Eclair_MR1) Operating System. Android 2.1 is chosen since most of Android Smartphone is operating at minimum on Android 2.1.

Keywords: Halal Product, MySQL Database, Smartphone Authentication, Barcode Scan, Halal Recognition System.

1 Introduction

Back in 2007 at the world halal forum, Malaysia's former prime minister Tun Abdullah Haji Ahmad Badawi said in his opening speech about government's aim to establish Malaysia as a global halal hub. Halal industry has been the fastest growing global business and this occurrence is an opportunity for Malaysia to set up the world's first development corporation in making halal its major focus. This industry also attracted non-muslims due to its hygienic and contamination-free principles is food production.

However in today's Malaysia market industries, Muslim consumer still faces problem in identifying and verifying the halal products not only food but other usable products in the industries. Although the Malaysian Muslim authority on halal product like Jabatan Kemajuan Islam Malaysia, JAKIM has implemented and produced the halal logo certificate as the halal verifications to the request market traders agencies but still cases of the misused of the halal logo by the irresponsible traders has been

G. Lee (Ed.): Advances in Automation and Robotics, Vol. 1, LNEE 122, pp. 49–56.
springerlink.com

brought up or complained by the consumers. The other constrains is, the method of checking the originality of halal products takes longer time either through JAKIM's website or telephone call to JAKIM office.

One of the approaches to verify halal product in a faster way is by using mobile or Smartphone gadget. This new technology gadget provides camera and barcode readers that could be used to identify and verified the information of the products or the halal status faster and reliable to the consumer on time while they are shopping. Not many system were develop to make used of this existed technology where the product information can be pass through the server mobility and verified for the halal status. One of the reason is due to the gadget technology is new and still it is on the testing process.

In order to address this difficulty, this research will design a halal product database on an open source platform. All data on halal product will be collected from JAKIM and detailed ingredients will be kept as inventories. Database structured will be design according to its product name, halal and non halal ingredients criteria, barcode number, colour coding and halal products alerts. .

2 Previous Researches

Previous research has presented the analysis of Halal products and services due to its challenging verification by consumer for multinational companies in Malaysia [1]. The research presented that Malaysian producers believe that the Halal industry is one of a high potentiality to succeed for its economic growth. It also presented the satisfaction rate of Halal products and service comparison between two sample groups in Kuala Lumpur that is Muslim and non-Muslim.

2.1 RFID Technology

A study focusing on halal status also has been done where RFID technology is used [2]. The process and prototype system on RFID is developed to authenticate the halal logo produce by JAKIM. User evaluation also has been conducted to measure satisfaction on the implemented system. Similar research in detecting illegal and unapproved halal logo also has been done in [3]. Instead of RFID, the researcher used image processing and recognition module from a digital camera. The result came with 91.67% of detection accuracy.

2.2 2-D Barcode

Another halal logo detector is proposed in [4]. The approach is to detect halal product by 2-D barcode and detect fake halal logo by using UV hologram. The scanned image is decoded and used to match with database. The halal verification will be shown on the screen. Prototype hardware was also designed for the purpose. A research in detecting alcohol has also been conducted in [5]. The study used infrared radiation in detecting alcohol in consumer's beverages. The hardware involved is infrared LED, lens, and photodiode and was developed by using Microsoft Excel and Microsoft

Visual Basic. The system is proven to be able to detect alcohol at 1680 nm wavelength.

2.3 Stochastic Petri Net (SPN)

Formal producing and quality certification system can improve the producing efficiency of Halal animal products [6]. Halal animal products have significant meaning for agriculture in Northwest China and other Muslim regions. One of the tools used is called Stochastic Petri Net (SPN) which is an effective tool for performance analysis, but it do not adapt for Halal animal products. This research has proposed the definition and analysis methodologies of the Halal Elements Extended SPN Model (HESPNM). Secondly, it gave the Computer Independence meta-architecture and Platform-Independence architecture of performance calculating software based on HESPNM through methods of Oriented-Object System Analysis and design.

2.4 MyMobiHalal 2.0

Another research also has been done where a mobile-based support application for Muslims to identify the Halal status [8]. The research used mobile device but verifications of the Halal process is only opposed to text entry in SMS. User will receive MMS as a result for their query. This paper also discussed the barcode concept and its applications in consumer product industry. It also discussed the framework and design of the system called MyMobiHalal 2.0.

All the research presented shows that the important of halal product research, verifications and determinations especially in Malaysia. System in [4],[8] looks similar to this research but this research would focus detailed on a new database structure on detailed ingredients, halal verifications alert and new applications technique using Smartphone without accessing network which will be the novel contribution in this research.

3 Implementation of the Framework

Application development in Android platform is growing and has large community of developers mainly because of its features as mentioned in [9]. Four incomparable advantages of Android include openness, equal for all applications, quick and easy development, and application freedom (no boundaries). The required tools and software were JAVA Development Kit (JDK), Android Software Development Kit (Android SDK), Android Virtual Device (AVD), Eclipse Integrated Development Environment (Eclipse IDE), and Android Development Tools Plugin for Eclipse (ADT Plugin). ZXing has been known for its barcode processing library for Smartphones [10]. It is an open-source barcode image processing library implemented in Java. It uses the camera on mobile phones to scan and decode barcodes on the device, without communicating with a server. SQLite is adopted in the development of this project. It is a lightweight embedded database provided in Android system [11]. SQLite database can be viewed and created using SQLite Database Browser for easy user interface.

3.1 System Flow

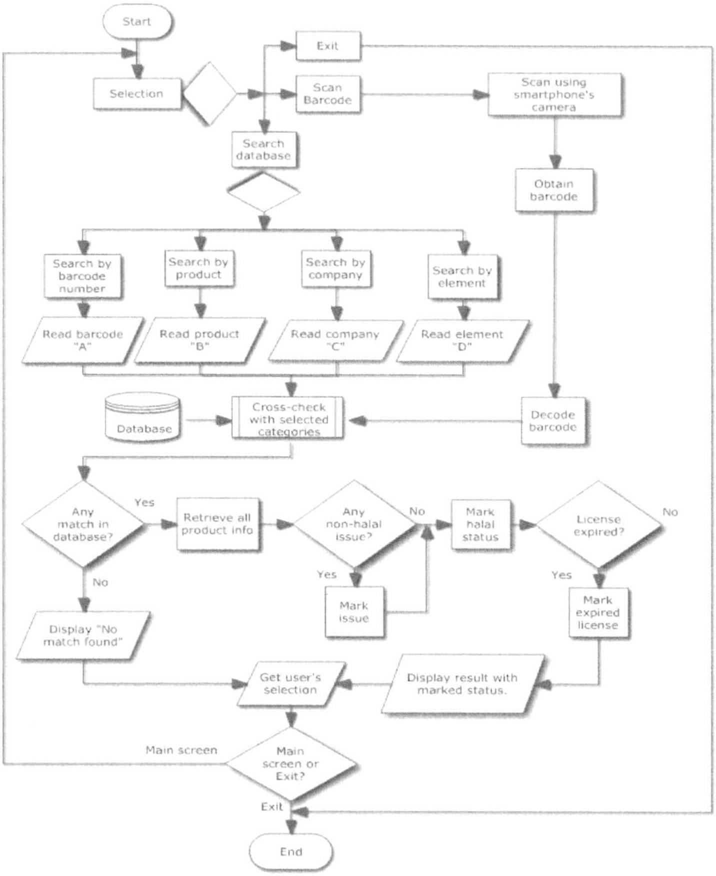

Fig. 1. System Flow

3.2 Hardware

Smartphone that will be used for this project is Android Smartphone that can operate on Android 2.1 (Eclair_MR1) Operating System at minimum. Application will be developed on Android 2.1 since most of new Android Smartphone is operating at minimum on Android 2.1. Android applications are generally forward-compatible with new versions of the Android platform and higher Application Programming Interface (API) levels [12]. The application should be able to run on all later versions of the Android platform. Generally, ZXing Barcode Scanner application can run on all Android devices with:

i. Android 1.5 ("Cupcake") or higher
ii. A camera

There were some issues on the following devices which avert them from running Barcode Scanner properly [13]:

i. HTC Desire
ii. HTC Evo
iii. T-Mobile Pulse
iv. GeeksPhone

Examples of Android Smartphone that run on Eclair OS, to name just a few available models in Malaysia, are:

i. HTC: Aria / Desire / Wildfire – HTC Desire as mentioned above have issues with ZXing Barcode Scanner
ii. Motorola: Milestone / Defy
iii. Sony Ericsson: Xperia X8 / X10 / X10 mini / X10 mini pro

3.3 Tool and Software

Java Development Kit (JDK)
JDK provides Java developer to develop and set up Java applications on desktops and servers [14]. It helps developer to meet demands for embedded environment nowadays. Java offers the rich user interface, performance, flexibility, portability, and security that applications require.

Android Software Development Kit (Android SDK) and Android Virtual Device (AVD) Manager
Android SDK and AVD Manager enable developers to build applications for Android platform [15]. It provides source code, development tools, emulator, and required libraries. 2.1 OS.

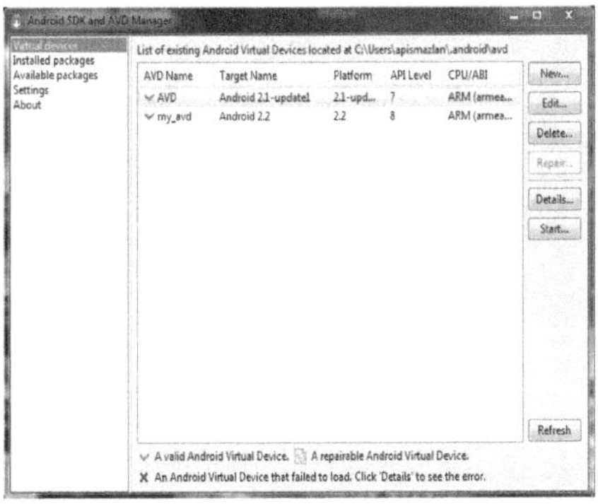

Fig. 2. Android SDK and AVD manager

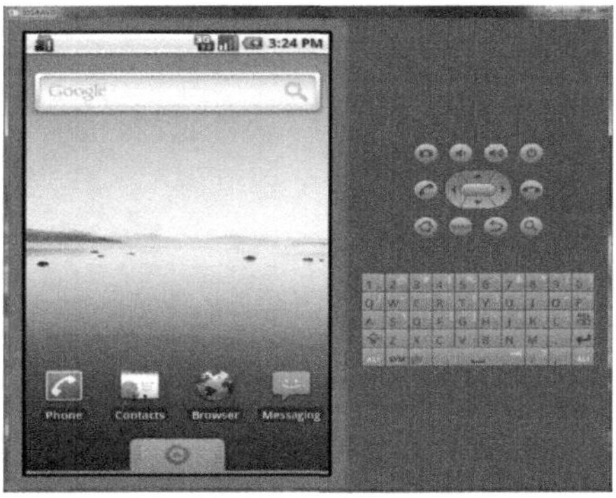

Fig. 3. Android Virtual Device on Android 2.1

Eclipse Integrated Development Environment (Eclipse IDE)
The Eclipse IDE contains what Java developers need to build Java applications. It is considered to be the preferred Java development tool available. Eclipse IDE provides

better Java editing with validation, incremental compilation, cross-referencing, and code assist [16]. Figure 4 shows Eclipse IDE program.

Fig. 4. Eclipse IDE

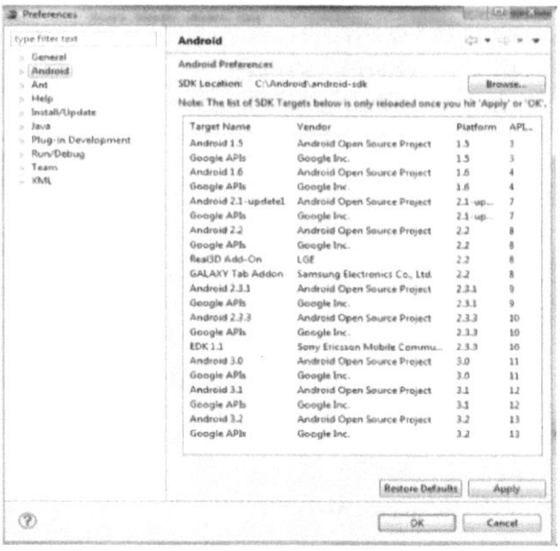

Fig. 5. Installing ADT Plugin

Android Development Tools Plugin for Eclipse (ADT Plugin)

ADT is a plugin for the Eclipse IDE that is designed to give Java developers a great, integrated environment to build Android applications. ADT extends the capabilities of Eclipse to swiftly set up new Android projects, create an user interface application, insert components based on the Android Framework API, debug applications using the Android SDK tools, and even export signed (or unsigned) .apk files to distribute created application [17]. Figure 5 shows Developers Tools which is the ADT plugin to be installed on Eclipse. Figure 6 shows Android tab which is accessible after installing ADT plugin.

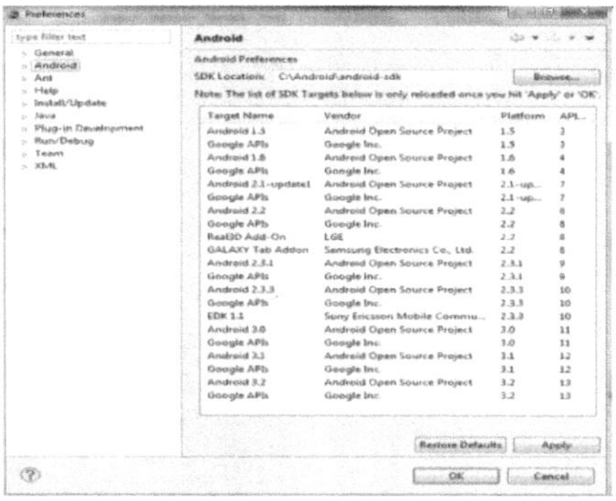

Fig. 6. Android tab on Eclipse

SQLite Database Browser

SQLite Database Browser is a freeware open source visual tool used to create, plan and edit database files compatible with SQLite [18]. It is used by users and developers to create, edit, and search database data using a familiar spreadsheet-like interface, without the need to learn complicated SQL commands. Figure 7 shows SQLite Database Browser when executed.

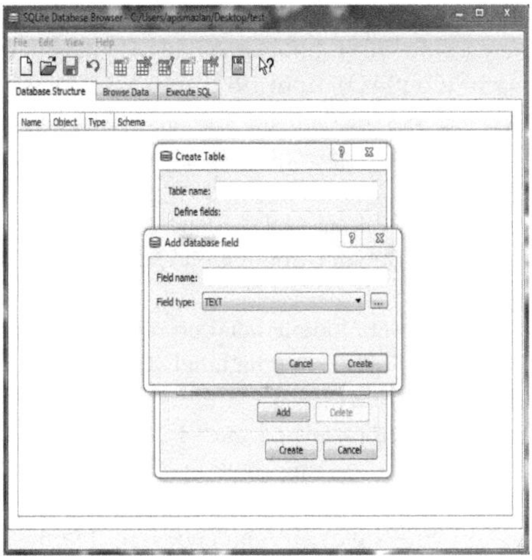

Fig. 7. SQLite Database Browser

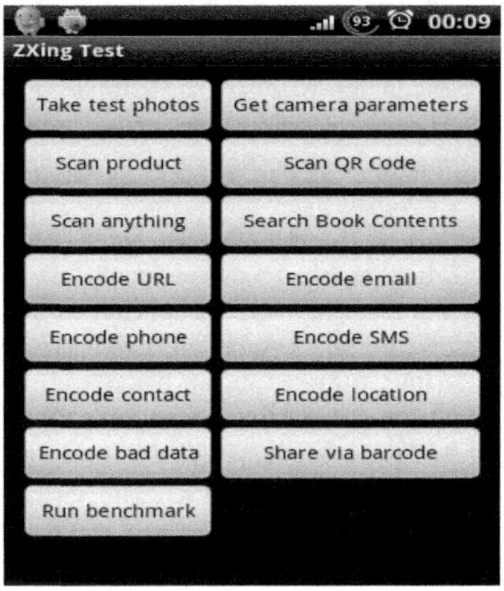

Fig. 8. ZXing test on Android smartphone

Zebra Crossing (ZXing)

ZXing is used as an open-source barcode image processing library implemented in Java. ZXing uses the built-in camera on mobile phones to snap and decode barcodes on the device, without communicating with a server [19]. Figure 8 shows ZXing Test application on Android Smartphone.

4 Conclusion

This system will be the best solution for the muslim consumer especially in Malaysia and has a huge potential to widely used. All the information related to halal products just in their hand. This system will offer an options to facilitate consumer to verify either halal logo, company, product name, barcode number, ingredients, license number and expiry date. Muslim consumer just need to have a smartphone with minimum Android 2.1 as an operating system.

References

1. Danesh, M.M.S., Chavosh, A., Nahavandi, M.: Comparative Analysis of the Muslims And Non-Muslims Satisfaction with Halal Products. In: 2010 IEEE International Conference on Advanced Management Science, ICAMS (2010)
2. Norman, A.A.: Consumer acceptance of RFID-enabled services in validating Halal status. In: 9th International Symposium on Communications and Information Technology, ISCIT 2009 (2009)
3. bin Hj. Mohd, M.N., bin Abd Wahab, M.H., bin Yaakob, A.: 'Halal' Logo Detection and Recognition System. In: 4th International Conference on Information Technology and Multimedia at UNITEN (ICIMU 2008), November 17-19 (2008)

4. Mohd Albakir, S.N.W.S., Mohd-Mokhtar, R.: A conceptual design of genuine Halal logo detector. In: 2011 IEEE International Conference on Imaging Systems and Techniques (IST), May 17-18, pp. 296–301 (2011)
5. Maharom, S.M.M., Hussin, N., Idrus, S.M., Safar, A., Zulkifli, N.: Photonics alcohol detector system for Syariah compliance food and beverages. In: 2011 4th International Conference on Modeling, Simulation and Applied Optimization (ICMSAO), April 19-21, pp. 1–4 (2011)
6. Han, Q., Ding, J., Bao, W.: A performance analysis method for formal producing and quality certification system of halal animal products based on HESPNM. In: World Automation Congress, WAC (2010)
7. Lili, N.A.: Accuracy of Determination for Virtual Animal Halal Slaughtering: A Case Study in Mathematical/Analytical Modelling and Computer Simulation (AMS). In: 2010 Fourth Asia International Conference (2010)
8. Junaini, S.N., Abdullah, J.: MyMobiHalal 2.0: Malaysian mobile halal product verification using camera phone barcode scanning and MMS. In: International Conference on Computer and Communication Engineering, ICCCE 2008 (2008)
9. Li, H., Ding, G., Zhu, T.: The design and the development of the PKM based on the Android platform. In: 2010 International Conference on Information Networking and Automation (ICINA), October 18-19, vol. 2, pp. V2-124–V2-127 (2010)
10. Budde, A., Michahelles, F.: Product Empire — Serious play with barcodes. In: Internet of Things (IOT), November 29-December 1, pp. 1–7 (2010)
11. Jiang, F., Ku, S.: How to Display the Data from Database by ListView on Android. In: 2010 2nd International Workshop on Intelligent Systems and Applications (ISA), May 22-23, pp. 1–4 (2010)
12. Android, API Levels Android Developers, `http://developer.android.com/guide/appendix/api-levels.html`
13. GetTheReader zxing, How to get a compiled reader for your phone Multi-format 1D/2D barcode image processing library with clients for Android, Java. Google Project Hosting (February 2011), `http://code.google.com/p/zxing/wiki/GetTheReader`
14. Java SE Overview at a Glance, `http://www.oracle.com/technetwork/java/javase/overview/index.html`
15. What is Android SDK? - A Word Definition From the Webopedia Computer Dictionary, `http://www.webopedia.com/TERM/A/Android_SDK.html`
16. Eclipse IDE for Java Developers, `http://www.eclipse.org/downloads/moreinfo/java.php`
17. ADT, ADT Plugin for Eclipse Android Developers, `http://developer.android.com/sdk/eclipse-adt.html`
18. SQLite Database Browser (n.d), `http://sqlitebrowser.sourceforge.net/`
19. Zxing, Multi-format 1D/2D barcode image processing library with clients for Android, Java Google Project Hosting, `http://code.google.com/p/zxing`
20. Sulaiman, N.A., Kassim, M.: Developing a customized software engineering testing for Shared Banking Services (SBS) System. In: 2011 IEEE International Conference on Proceeding in System Engineering and Technology (ICSET), June 27-28 (2011)
21. Sulaiman, N.A., Kassim, M.: An approach using RUP test discipline process for shared banking services (SBS) system. Paper presented at the 2nd International Conference on Computer Research and Development, ICCRD 2010, pp. 298–302 (2010)
22. Sulaiman, N.A., Kassim, M., Saaidin, S.: Systematic test and evaluation process (STEP) approach on shared banking services (SBS) system identification. Paper presented at the 2010 2nd International Conference on Education Technology and Computer, ICETC, vol. 5, pp. V5219–V5223 (2010)

Decision Support System for Workload of Design and Research Institutes

Zhizhen Dong and Min Hu

China Shipbuilding NDRI Engineering Co., Ltd, Shanghai 200063

Abstract. Design is a complicated labor-intensive process, not only does it involve mental work, but also manual labor. To a design and research institute, it is a complex issue to properly evaluate the heterogeneous workload for a deign-related job. Therefore, it is crucial to precisely calculate cross-enterprise, heterogeneous workloads to support business goals and service levels. This thesis focuses on the discussion of the relevant calculation methods as well as the decision support system for workload under a cooperative work platform.

Keywords: workload, Production project management system, cooperative platform, decision support system.

1 Foreword

Design is a complicated labor-intensive process. From architecture's point of view, it is an innovation of arts; from building structure's point of view, it is a foundation of safety; from public management's point of view, it is a linkage in a city. Not only does design involve mental work, but also manual labor. It is one of a few jobs that well integrate the two types of labor. For a design institute with its future goal as a contractor, it is crucial to precisely calculate cross-enterprise, heterogeneous workloads to support business goals and service levels. Accurately secluded workload provides the first hand information for the management to ensure a correct cost evaluation process, a sound human resources allocation as well as an ultimate reward program.

2 Production Project Management System

Production project management system is applied to proactively monitor and record all the information in relation to production projects, including operation, contracts, planning, charging, billing, cost and purchasing. By this means, the integrated management of the entire process in production projects has been realized to ensure the application of the chain process from purchasing to transportation. Please refer to Figure 1 for the detailed functions.

G. Lee (Ed.): Advances in Automation and Robotics, Vol. 1, LNEE 122, pp. 57–64.
springerlink.com

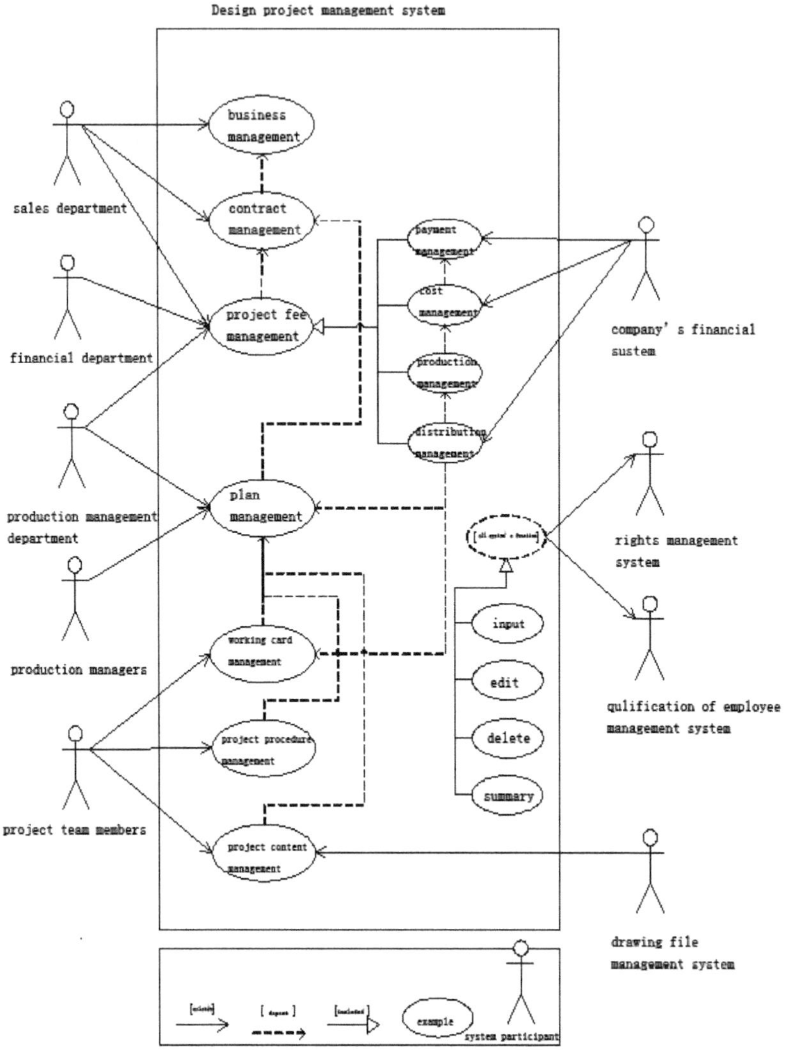

Fig. 1. Functions of Design Projects Management System

The system has well analyzed the specific characteristics of a wild range of projects and the enormous professions involved in design. The analysis mainly focuses on information exchange within different departments and professions, it also involves in inner-departmental and inner-professional cooperation. The system promptly supplies with the dynamic information for the management team, which then makes decisions in keeping with the particular stage during the dynamic progress of a project, such as contracts condition, plan arrangements and adjustments, design progress, work completion status as well as cost allocation. The electric card function allows staff

calculates their work content based on their work hours, The system also offers various functions from raw data analysis to project categorizing, this is to help the management monitor the entire project so as to make final decisions.

3 Collaborative Platform

In the past, the only instrument utilized in the design job is a drawing board. As computer was invented and well developed, Computer-aided design (CAD) has made design a much simpler job, and of which the modification is more continent. However, more and more people start to be aware that the design techniques applied by modern CAD is brittle and inflexible. As a result, the concept of cooperative design has been released.

CSCD refers to Computer-Supported Collaborative Design, which is the process of designing a product through collaboration among multidisciplinary product developers associated with the entire product lifecycle. By synthesizing a collaborative environment for designing a product using the latest network technologies, teams reach a common goal of optimal allocation of resources and distant resource-sharing. In order to reach a common objective, more than two teams (including two teams) work together with the support from the computer network. These multidisciplinary product development teams work on a design task, which was then split into sub-tasks for each team. Through exchanging information and cooperating within the teams, the collaborators interact with each other and with computational resources to reach the common objective. CSCD is collaboration of computer related hardware, software and being supported by different areas, such as social science.

Collaborative design is the solution for ensuring the uniqueness, safety, reliability of design, and cost reduction as well as resource conservation.

A large number of collaborative design platforms have been developed and reported, including Autodesk Buzzsaw, Bentley ProjectWise and some other Chinese platforms.

The former two systems had been developed by foreign companies, and provided us with sufficient tools for the second level development. We have done the second level development on the basis of Bentley ProjectWise, and integrated it into the Auto CAD system, so that automatic platform directory, automatic execution and adjustment of directory right, one key operation for information withdraw and filing can be easily done.

The company has introduced a broad range of professional software to standardize layer arrangements and drawing criteria. This helps collaborate design by AutoCAD consistent with standardized layer arrangements and drawing criteria.

The advantages associated with the function of external reference by Auto Cad are shown as follows:

1) Ensuring the consistency of collaborative design by each profession.

External reference ensures professional design and modification to be conducted simultaneously. For instance, if the architect industry makes a modification in any construction condition, other industries will be able to notice the modification by only re-opening the drawing or re-loading the current drawing. They then will be able to carry out their work in accordance with the latest condition, thereby avoiding the problem caused by mismatching the modified drawings.

2) Reducing the capacity of documents

Documents that contain external reference only record the path of referred documents. Therefore, the capacity of document has not been increased to a large extent. By applying the external reference function, some referred documents can be attached to a relatively small graphic document, so that a complicated graphic document is generated. By doing so, the speed of generating graphics has progressed dramatically. If the external reference function is used in a designing process, a humongous graphic document concerning many professions can be easily handled.

3) Increased drawing speed

Due to the efficiency of the external reference function, drawings from every related profession can be updated at all times, so that other collaborators can act accordingly during the workflow to support the collaborative platform. This helps the team work without the need to print out notices repetitively, that my consequently cause a delay in delivery. In fact, they can increase the product efficiency with minimum time and energy consumption.

4) Optimization of the quantity of designing documents

An externally referred document may be applied by several documents, each of which can apply the referred document repetitively. This process minimizes the quantity of graphic documents, thereby maximizing the efficiency of document management and staff's workload.

5) Easy three-dimensional design and collaborative design.

Three-dimensional design is the direction of the future development. With the current approach of single device for design, three-dimensional design will progressively increase the workload. Only with the external reference function, three-dimensional design can be done without an increase in workload.

4 Automatic Calculation of Workload

Bentley ProjectWise provides us with a powerful and practical platform for cooperative filing management with its excellent filing function, tracking and recovery abilities. The system framework has its advantages, e.g. it classifies functions into modules, which promotes its future development. By connecting it with advanced process management software, it provides ideal functions for workflow management. It also has a outlet for the second level development.

Bentley ProjectWise has a practical tracking function, and provides the customers with an intricate open data system. Over the pat two years of using the system, all the projects on the platform are well operated. By collaborating it with Auto Cad, we have ensured our customers with the function of automatic saving after the save button is clipped each time. This helps us collect useful information with respect to customer operation to files without concern on its safety.

On top of our research on interface and database for the second level development, we have developed a project information calculating system. This system is used to calculate quantitative information in accordance with drawings, updated files and information exchange on the basis of predefined information and workflow management then categorize them.

Table 1. Simplified table of calculated information for a project

Project information	Start date	Due date	Profession	Anticipated date
101-11021lProject 1	2011-04-08	2011-07-18	Construction	2011-05-20
101-11021Project 1	2011-04-08	2011-07-18	Structure	2011-06-02
101-11021lProject 1	2011-04-08	2011-07-18	Elec.	
101-11021lProject1	2011-04-08	2011-07-18	HAVAC	2011-07-07
101-11021lProject1	2011-04-08	2011-07-18	Water supply and drainage	2011-07-10
101-11021lProject 1	2011-04-08	2011-07-18	Automation	

Project information	No. of Drawings	No. of Paper files	Times of updates	Filing information
101-11021lProject 1	57	35	6232	4
101-11021Project 1	53	6	5757	3
101-11021lProject 1	76	4	7750	0
101-11021lProject1	18	9	2371	3
101-11021lProject1	37	5	4507	2
101-11021lProject 1	13	2	2074	0

Project information also includes information such as project profile and project size, etc.

We collect information accordingly, and categorize the projects according to its characteristics and nature. Then, this information will be added to the database, which will weigh the data according to various weighted value. Current weighted value for drawing file is 5, for file numbers is 3, for file updates is 1, and for information withdraws is 1. Based on these weighted values, workload can be calculated.

At present, the calculus method has being used for testing purpose. As the information collecting system runs smoothly for a period of time, and number of projects is accumulated to a certain level, enough data will be generated to support the system operation.

5 Decision Platform

Decision platform is a data analysis system designed for decision markers. By analyzing the collected data and categorized data on the EXCEL platform for the second development, the management can eventually anticipate the current condition of project status, workload, work hours as well as cost, etc. This work is done using the illustrated bar charts and pie charts, etc. E.g. Figure 2, Figure 3, Figure 4.

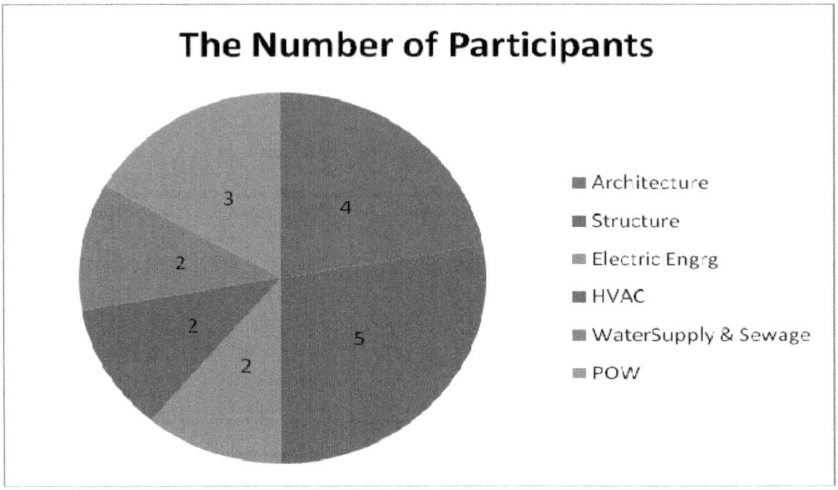

Fig. 2. Pie chart of Project Personnel

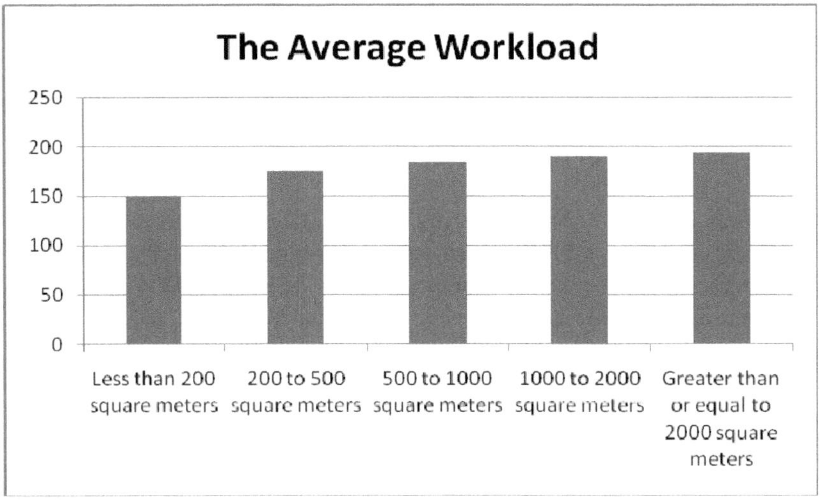

Fig. 3. Average workload of Projects

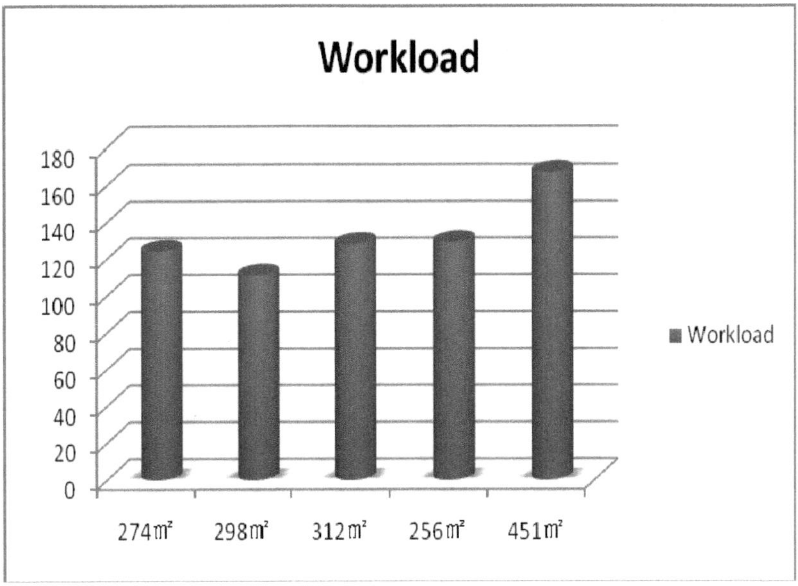

Fig. 4. Bar chart 4 Comparison between workload in different project of similar size

6 Application

Decision platform has provided useful quantitative analysis for workload. By incorporating it along with the electric work cards, rated working hours and our attendance management system, a rational analysis and decision-making process on workload for design related jobs in our company has been generated.

Currently our company has been doing well in many project tenders because of the excellent cost management benefited from the first hand information provided by this platform. Moreover, a high-quality human resources allocation and control in project progress has also been done, thereby reaching an ideal balance of control in project progress, cost and staff allocation.

References

1. He, J., Zhou, J.: About External Reference of Auto CAD. Theory and Practice of Higher Education in Building Material Science 20(2), 104–105 (2001)
2. Zhao, H.: AutoCAD external graphics reference implementation method. Gansu Science and Technology 19(3), 24–25 (2003)
3. Wu, W., Liang, X., Wang, C.: Cooperative Design System for Architecture: CodeFAR3.0. In: The Eighth National Construction Information Technology and Multimedia Aided Engineering Academic Conference Collection

4. Rosenman, M., Wang, F.: A component agent based open CAD system for collaborative design. Automation in Construction 10(4), 383–397 (2001)
5. Ren, X., Wang, G., Yu, J.: A Survey of Work Load of Nursing Staffs in Large General Hospital and Its Authorized Staff Estimation. Chinese Nursing Research 17 (April 2003)
6. Morton, S., Michael, S.: Management decision systems: Computer-based support for decision making (1971)
7. Daellenbach, H.G.: Systems and decision making: A management science approach (1994)

Research and Development of a Ping-Pong Robot Arm

Xiaowen Yu, Yu Zhao, Chenglong Fu, and Ken Chen

Institute of Manufacturing Engineering,
Department of Precision Instrument and Mechanology,
Beijing 100084, China
{yxw09,yu-zhao09}@mails.tsinghua.edu.cn

Abstract. In order to design a ping-pong robot arm system of low cost and high performance, this paper firstly designed the arm's mechanical structure according to the requirement of the robot arm, and then analysised the kinamatics and dynamatics of the system. In addition, dynamic simulation as well as control scheme are described in detail. Results of analysis and experiment illustrate that the design can meet the functional requirement. The ping-pong robot arm can realize high-speed ball hitting, ball balancing techniques at a very low cost.

Keywords: robot, kinematic analysis, dynamics, prototype development.

1 Introduction

Ping pong robot is a platform which focuses on high-speed visual tracking, trajectory prediction and control technique. In table tennis game, ping-pong ball is flying at a very high speed up to over 10m/s. Playing ping-pong by robots involves problems of fast ball tracking, trajectory prediction and accuracy of ball hitting. High-speed visual tracking and trajectory prediction has a very broad application prospect and of great significance.

Ping-pong robot has been developed for years in many countries Supported by 863 program, a national hi-tech program, several universities and institutions in China have started the research and development on ping-pong robot recently. Robotics and Automation Laboratory (RAL) of Tsinghua University successfully resolved several key problems of ping-pong robot and developed a ping-pong robot arm. Compared to some domestic or foreign ping-pong robots, the ping-pong robot developed by RAL is of light weight, more flexibility, low cost and other advantages. Also, it has a multiple sensors based intelligent control system which is capable of a series of actions, such as ball bearing, ball hitting and ball balancing.

2 Mechanical Structure Design of Ping-Pong Robot Arm

2.1 Overall Structure

According to the characteristics desired for ping-pong robot, we design a ping-pong robot arm as shown in Fig.1.

G. Lee (Ed.): Advances in Automation and Robotics, Vol. 1, LNEE 122, pp. 65–71.
springerlink.com

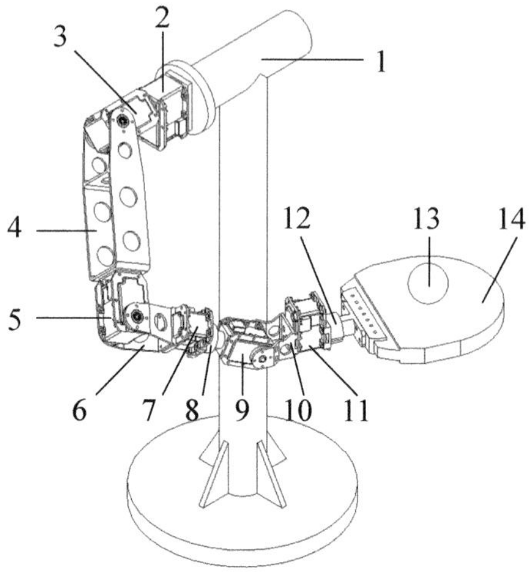

Fig. 1. Structure of ping-pong robot arm, *2,3,5,7,9,11*: actuators; *1*: base; *14*: racket; *4,6,8,10,12*: links; *13*: ping-pong ball

The ping-pong robot arm consists of 6 rotational joints, the first three joints(joints *2,3,5*) determine the position of the end effecter. The purpose of such design is mainly to maximize the workspace. Three rotational joints (joints *7, 9, 11*) are located at the end of the forearm, simulating a human arm and determine the orientation of end effecter. In the kinematic structure, this kind of robot is mostly like human arm, thus has the most compact structure and is of maximum flexibility and workspace, which even can bypass obstacles to reach the target point, therefore is the most promising one among other robots. At the end of the rotational joint(joint 11)is mounted a racket to hit the ping-pong ball. All 6 joints are driven by 6 Dynamixel actuators produced by Robotis Inc. The Dynamixel series robot actuator is a smart, modular actuator that integrates a gear reducer, a precision DC motor and a control circuitry with networking functionality in a single package. The first three joints (shoulder and elbow) are driven by Dynamixel RX-64 with a maximum torque of 64 kgf.cm, and joints of spherical wrist are driven by Dynamixel RX-28 with a maximum torque of 28 kgf.cm.

2.2 Verification of the Dynamic Model

In order to verify the dynamics of the arm, dynamic model of the arm are constructed by inputting structure and dynamic parameters into MATLAB and MATLAB robot toolbox. After establishing the model, dynamic simulation is calculated by programming. Because there is no inverse kinematic module in robot toolbox, the inverse kinematic solution has to be programmed for simulation. Simulation results of a given trajectory, which are angles, acceleration, torques of all joints and velocity of end effecter are shown in Fig.2.

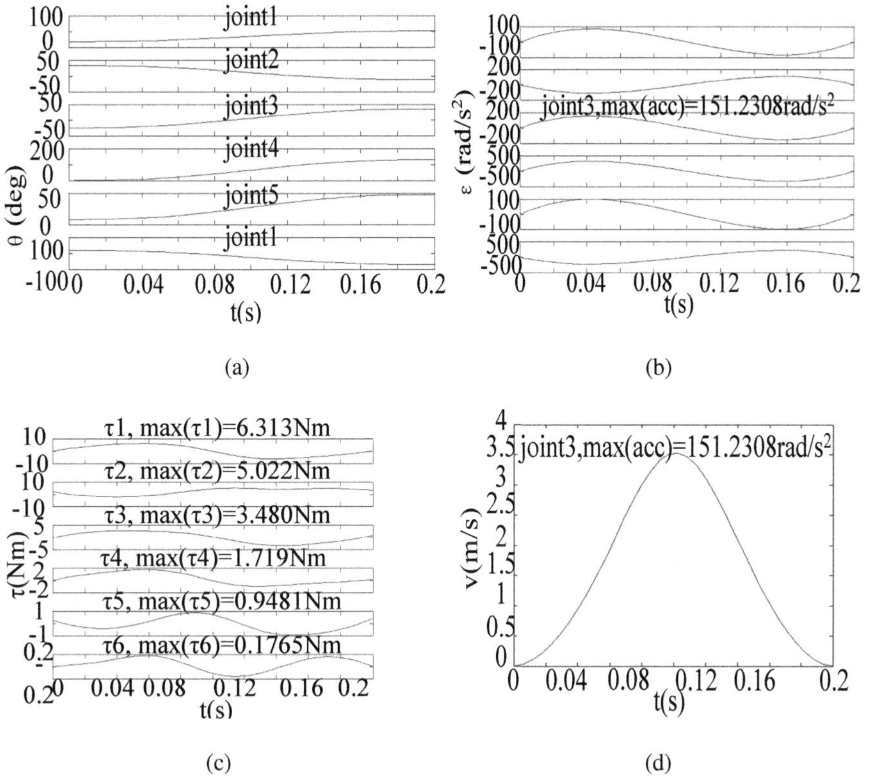

Fig. 2. (a) Positions of joints; (b) angular accelerations of joints; (c) torque of joints; (d) velocity of racket

According to the results, angular acceleration of elbow joint achieved 151.2rad/s^2, velocity of end effecter achieved 3.531m/s. At the same time, torques of all joints is under limitation, i.e., torque of joint *1,2* and *3* is less than 6.400Nm, and torque of joint *4,5* and *6* is less than 2.800Nm. The dynamic simulation in MATLAB shows that both kinematics and dynamics of ping-pong robot arm satisfy the requirement of design.

3 Control System of Ping-Pong Robot Arm

3.1 Hardware System Design

ATmega128 MCU is adopted as the main controller. The main reason of such choice is that the full-featured MCU development board in market only at a price of 200 RMB. ATmega128 is of the highest configuration in 8-bit series of ATMEL Company, which is of a maximum frequency 16MHz that is able to fulfill the requirement in the experiment. There are RS232 serial port, RS485 serial port and key board in MCU, which are required in the experiment.

MCU communicates with touch screen and ping-pong robot arm by RS232 port and RS485 port respectively. Detailed hardware configuration is shown in Fig.3. Touch screen is not required in ball hitting experiment, so we just connect the MCU with Dynamixel actuators by RS485 serial port.

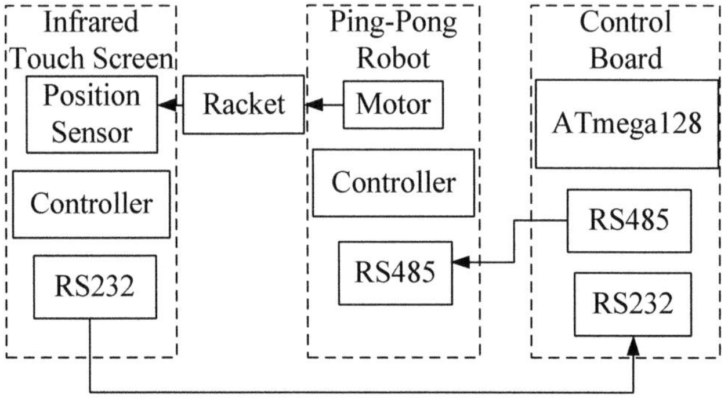

Fig. 3. Hardware of control system

3.2 Control Program Design

In the process of ball balancing, signal of touch screen has to be obtained for real-time control. During control process, signals from touch screen are firstly processed to get the control quantity of actuators' position, and then according to which to drive actuators. A fuzzy controller is used when process signals from touch screen, and Fig.4 shows the control scheme.

In the ball hitting experiment, only angle and speed sequence are required to input into MCU.

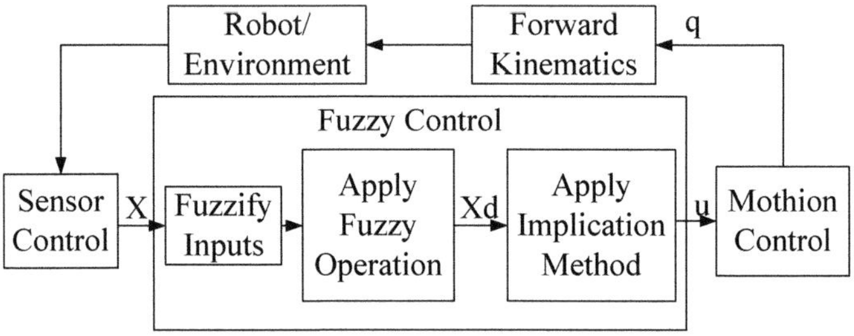

Fig. 4. Control scheme of ball balancing

Software processing of MCU works at an interrupt mode: waiting for occurrence and processing of interruption after initialization. Relevant control is processed through interruption. There are three interrupt subroutines, which are two keys on the MCU and touch screen interruption, and control is achieved in subroutines. terrupt.

4 Experiment

Ball hitting, ball balancing and bearing capacity experiments are done after development of ping-pong robot arm as shown in Fig.5. According to the ball hitting experiment shows that at a condition that a server serves the ball at a speed of 5 m/s, the success rate of hitting ball to a fixed region, whose radius is less than or equal to 150 mm, is greater than 70%. The result show that the ping-pong robot arm has reached the design target and is of reliable and stable performance.

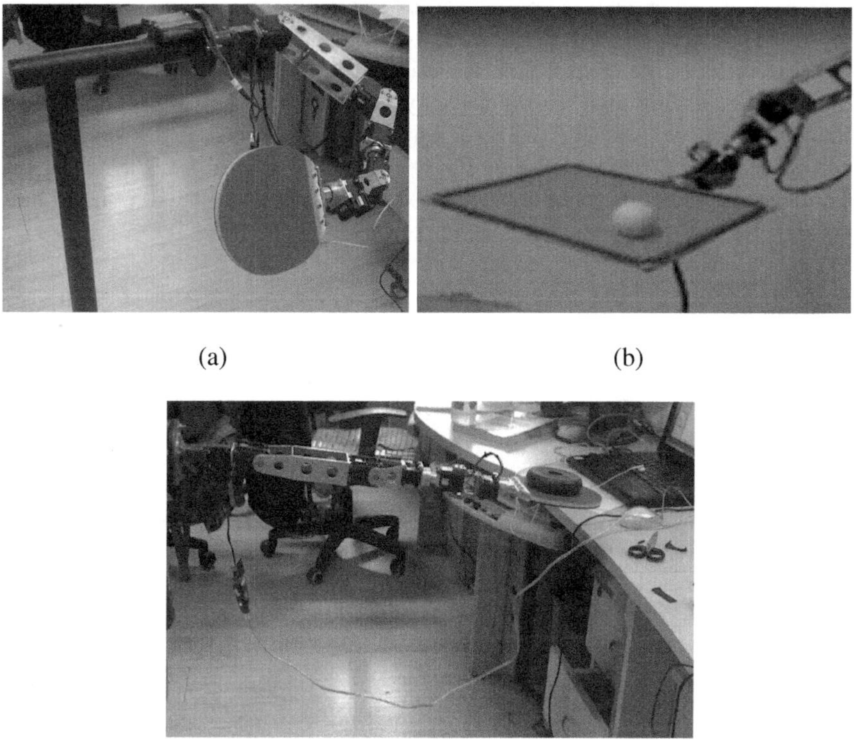

(a) (b)

(c)

Fig. 5. (a) Ball hitting experiment; (b) ball balancing experiment; (c) raising up experiment

In the ball balancing experiment, ping-pong robot arm can stabilize ping-pong ball in a relatively short period of time(15s), which shows its ability of ball balancing. The result of bearing capacity experiment indicates that ping-pong robot arm is able to hold up 0.5 kg. All experimental results above show that the ping-pong robot arm realizes the requirement for both low cost and high performance, and which serves as a good foundation for the follow-up goal that fighting with a real person.

5 Conclusion

In this paper, kinematic modeling and analysis have been constructed to a ping-pong robot arm to verify that the mechanical structure can meet the design requirement. Also, dynamic simulation of a given trajectory problem and static force analysis are done by MATLAB. Simulation results show that all joints are within torque limits when speed of end effecter is up to 3.531m/s and angular acceleration of elbow joint is up to 151.2rad/s^2 as well as the static bearing capacity is larger than 10N. This paper also includes the control scheme design and three experiments, which are ball hitting, ball balancing and bearing capacity respectively. The results show that the ping-pong robot arm has achieved its design target, and specifically, it can stabilize a ping-ping ball in a relatively short time (15s). Both results from simulation and experiment show that the ping-pong robot arm can realize the motion of ball hitting and balancing with its low cost and high performance.

References

1. Hongrui, W., et al.: Tracking Control of Ball and Plate System with a Double Feedback Loop Structure. In: International Conference on Mechatronics and Automation, ICMA 2007, pp. 1114–1119 (2007)
2. Kostamo, J., et al.: Ball balancing system: an educational device for demonstrating optimal control. In: Proceedings of 2005 IEEE International Symposium on Computational Intelligence in Robotics and Automation, CIRA 2005, pp. 379–384 (2005)
3. Ng, K.C., Trivedi, M.M.: Fuzzy logic controller and real-time implementation of a ball balancing beam, Orlando, FL, USA, pp. 261–272 (1995)
4. Kwang-Kyu, L., et al.: Basketball robot: Ball-On-Plate with pure haptic information. In: IEEE International Conference on Robotics and Automation, ICRA 2008, pp. 2410–2415 (2008)
5. Cordón, O., et al.: Applicability of the fuzzy operators in the design of fuzzy logic controllers. Fuzzy Sets and Systems 86, 15–41 (1997)
6. Awtar, S., et al.: Mechatronic design of a ball-on-plate balancing system. Mechatronics 12, 217–228 (2002)
7. Park, J.H., Jong Lee, Y.: Robust visual servoing for motion control of the ball on a plate. Mechatronics 13, 723–738 (2003)
8. Wang, H., et al.: Nonlinear control for output regulation of ball and plate system. In: 27th Chinese Control Conference, CCC 2008, pp. 382–387 (2008)
9. Lea, R.N., et al.: Fuzzy logic based robotic arm control. In: Second IEEE International Conference on Fuzzy Systems, San Francisco, CA, USA, March 28-April 1, pp. 117–133 (1993)
10. Su, J., et al.: Research on trajectory tracking of ball-and-plate system based on supervisory fuzzy control. In: 25th Chinese Control Conference, CCC 2006, Harbin, China, August 7-11, pp. 1528–1532 (2006)

11. Yuan, D.-H.: Pneumatic servo ball plate system based on touch screen and oscillating cylinder. In: 2009 International Workshop on Intelligent Systems and Applications, ISA 2009, Wuhan, China, May 23-24, Hubei University of Technology; Huazhong University of Science and Technology; Huazhong Normal University; IEEE Harbin Section (2009)
12. Wang, H., et al.: Tracking control of ball and plate system with a double feedback loop structure. In: 2007 IEEE International Conference on Mechatronics and Automation, ICMA 2007, Harbin, China, August 5-8, pp. 1114–1119 (2007)
13. Mukherjee, R., et al.: Motion planning for a spherical mobile robot: Revisiting the classical ball-plate problem. Journal of Dynamic Systems, Measurement and Control, Transactions of the ASME 124, 502–511 (2002)
14. Svinin, M., Hosoe, S.: Simple motion planning algorithms for ball-plate systems with limited contact area. In: 2006 IEEE International Conference on Robotics and Automation, ICRA 2006, Orlando, FL, United States, May 15-19, pp. 1755–1761 (2006)
15. Knuplez, A., et al.: Modeling and control design for the ball and plate system. In: 2003 IEEE International Conference on Industrial Technology, ICIT Proceedings, Maribor, Slovenia, December 10-12, pp. 1064–1067 (2003)
16. Moarref, M., et al.: Mechatronic design and position control of a novel ball and plate system. In: 2008 Mediterranean Conference on Control and Automation, MED 2008, Ajaccio-Corsica, France, June 25-27, pp. 1071–1076 (2008)

A New Programming Language for PLC Using Bond Graph

Ahmad Naebi[1], Masoud Hassanpour Aghdam[2], Ali Ghalehban Zanjanab[3], Esmaeil Nourani[4], and Saeed Hassanpour Aghdam[5]

[1] Qazvin Islamic Azad University, Qazvin, Iran
[2] Saveh Islamic Azad University, Saveh, Iran
[3] Assumption University of Thailand
[4] Department of Computer Engineering, Kaleybar Branch, Islamic Azad
[5] Tabriz Islamic Azad University, Tabriz, Iran
Ahmad.Naebi@gmail.com, Mhassanpour110@gmail.com, Esmaeil.Nourani@gmail.com, aligalehban@gmail.com, Saeed.powerel67@gmail.com

Abstract. In this paper a new programming language for PLC with using of bond graph is introduced. Bond graph is a system for modeling in Mechatronics systems. The advantage of using bond graph for PLC programming, Is the use of a common language for hybrid mechanical, electrical systems. For a better understanding we use Ladder which is a Circuit definition language. First we've used the bond graph symbols in the Ladder program. Ladder program has been implemented in MATLAB Simulink software as an application program.

Keywords: Programming language, bond graph, PLC.

1 Introduction

A programmable logic controller (PLC) or programmable controller is a digital computer used for automation of electromechanical processes, such as control of machinery on factory. It's a software controller that receive binary information in system entrance and process them according program in its memory, and sends output of operation as commands to receivers and Performers in output. In other words, a logic controller which we can define to it, logic of control by program and in required situation we can change it easily. previously, relay circuits were in responsibility of PLC functions that now days are not used in New industrial environments. The first problem that appears in this circuit is that, with increasing of relays number, circuit size and weight will grew more, and also its price will increase too. To resolve this problem, electronic circuits were built, but despite this, a small change in trend or the machine's performance cause to change much hardware's in control system in other word connections and the circuit elements must change (Fig. 1.A) [1, 2, 3].

With using of PLCModification in a production processor the operation of a machine is easy to do, because we don't need to change whole system control hardware and wiring, just we must wrote a few line code and send the program to PLC then our expected control will done. PLC allows to controlling system designers

G. Lee (Ed.): Advances in Automation and Robotics, Vol. 1, LNEE 122, pp. 73–80.
springerlink.com

that realize its idea with spending of less time and cost, in fact practice status is the fastest way to specify that an idea can be implemented or not, which eventually result will leads to system upgrade [1, 2, 3].

PLC manufacturers are a programming language innovator companies for own PLC systems, PLC systems manufacturers uses unique programming language for programming of own PLC systems, but generally all of them follow a series of logical rules and alone difference is in the coding and symbols that used in.

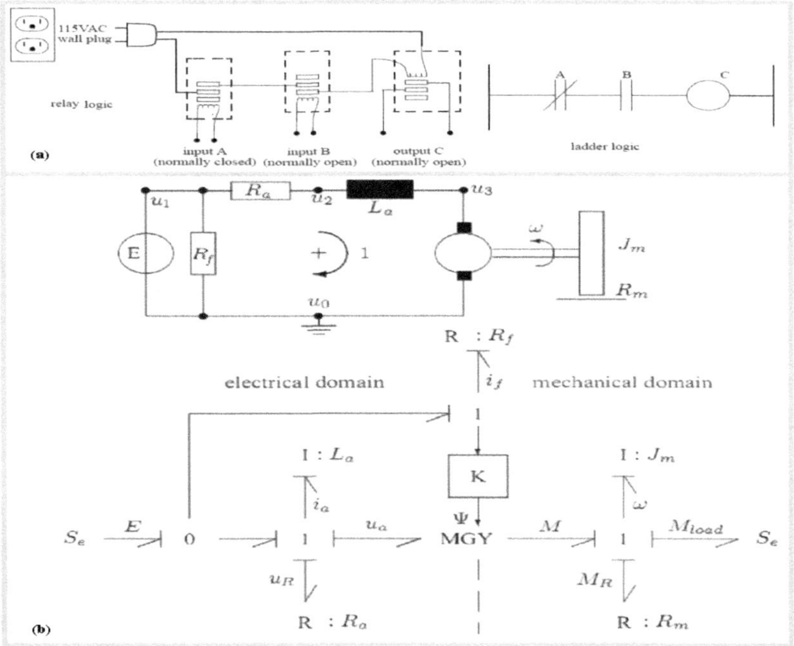

Fig. 1. A) A relay controller and PLC controller. B) Circuit schematics of a motor and bond graph modeling.

Bond graph defined as a modeling tool for unified method for modeling and analyzing dynamic systems especially for Multi-area systems in end of 1950. Bond graph is a diagram base method for hybrid Multi-area systems containing of mechanical, electrical, pneumatic, chemical and etc parts (Fig. 1.B). In fact, it is a better method in infinitely designed studies that explicitly shows the model topology that is includes a set of specific elements for modeling systems that e system is will modeled by connecting the elements together. Bond Graph also used as a Modeling and Simulation tool [4, 5, 6, 7].

Mechatronics systems modeling has been done (mechanics, electronics, etc.) with bond graph previously and because we want to doing Mechatronics systems modeling and programming is with a single language we decide to introduce a new method of mechatronics systems programming flow chart Using Bond Graph in this case we will be able to do Mechatronics systems modeling and programming with unique language (Fig. 2) [8].

Previously programming language with using of bond graph technique is similar to other programming language and The only difference is in the coding and symbols that are used in. This isn't only a limited programming language for PLC programming rather It is a language for Mechatronics Systems, That people can do programming for PLC and modeling for Mechatronics Systems and The PLC software programming as a component will be added to the Mechatronics compiler, major goal of article is to create a mechatronics compiler that will done in the near future.

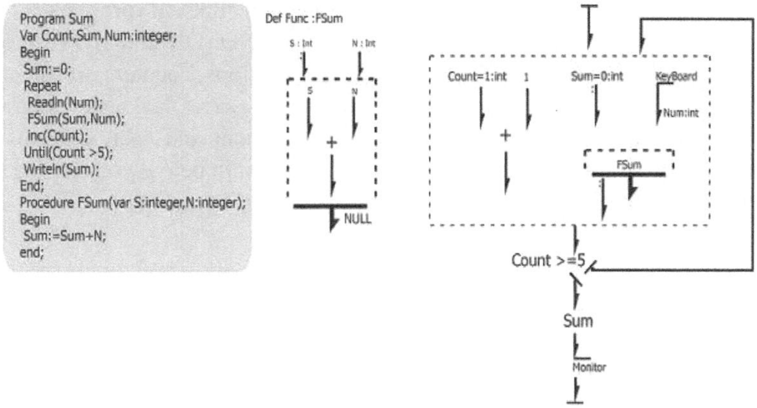

Fig. 2. An example using bond graph programming [8]

Now we will present summary of PLC and the Bond Graph previous works and then will provide proposed method, and then will present a simple controller with Ladder, flow charts and bond graphs language. Finally, we will conclude by comparing methods.

2 Related Works

So far, no PLC programming is done with the bond graph. So for make easy understanding of next sections idea we will explain briefly current used methods.

2.1 PLC

A controlling program is a set of instructions that can issue commands to PLC for control of processes, so it should be written with specific language that its rules and instructions are understandable for the PLC. An example of this language is called SETP5 [1, 2, 3]. These languages are represented in different ways such as ladder, flowchart and script language.

In ladder, any command line of program is shown as a symbol of connection and/or relay coil circuit, as a result, program is structured in this way is almost will be like a relay control circuits. This representation is common in old relay systems. Command circuit maps are often drawn in this way so this method of representation is easy to understanding. A map that be drawn on this way will be like electrical circuit of same system. Some of the symbols used in this program are listed below (Fig. 3.A) [9].

In the flowchart view, program will be shown as a set of symbols (blocks) and rectangular. Some of the symbols used in this program are listed below. In This type logic operation done in each block and inputs and outputs of each block is also shown, this view is in accordance with the logical circuits drawing in electronic form (Fig. 3.B) [9].

In the script method, program is a set of commands that every line of this commands is called Statement and any command line in program is include combination of logical mathematic compounds of AND, OR and NOT. In addition, combinations of other software tool such as flag, flip-flops, counters, timers, etc exist. They are part of the programming language. In the script language programming, program write in several rows which called a Segment and each segment will do certain act. A program is comprised of many Segments. In below you can see a simple example of program which is written in script method (Fig. 3.C) [9].

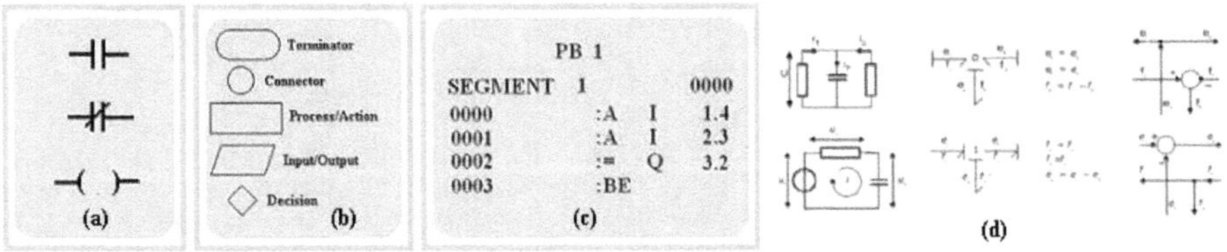

Fig. 3. A) Ladder symbols. B) Flowchart symbols. C) Statement method. D) The sample of bond graph symbols.

2.2 Bond Graph

Given that bond graph describe a wide range of systems physical behaviour in the visual expression. 4 elements of bond graph make it a best option for multi-area modelling and system design at both macro and micro scale which are as bellow [9, 10]: 1) Inventing new systems due to its multi-field and multi-environment behavior. 2) Performance estimation and evaluation of alternative design. 3) The combination features of link and node components for generation of alternative design. 4) Simply of mapping the process of engineering design.

Most physical systems are composed of several parts and when they be divided into the subsections. Each of them will express an aspect of energy and power. Capacitor for energy storage, mechanical damper for energy and battery power is an example of energy component (Fig. 3.D).

3 Proposed Method

Considering to this fact that the world needs a unique programming language in the field of mechatronics and PLC, as language in the field of industrials the same in structure with the various languages, we have uses a mechatronics special language for PLC.

Our Idea is to provide the bond graph language rather than ideas that other languages previously have been express for PLC. Parts of the languages are including as ladder, flowchart and statement. We will define these three language elements with bond graph symbols.

In ladder, any command line of programs shown as a symbol of connection and or relay coil circuit, in result program is structured in this way is almost will be like a relay control circuits that this presentment is common in old relay system, command circuit maps are often drawn in this way and because of This display method understanding of this maps is easy, a map that be drawn on this way will be like electrical circuit of same system. Some of the symbols used in this program are listed below (Fig. 4).

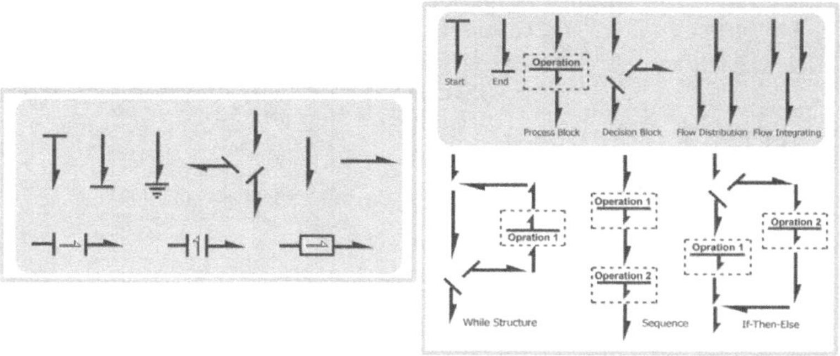

Fig. 4. Left) Symbols of bond graph Ladder. Right) Symbols of bond graph flowchart.

In the flowchart view, program will be shown as a set of symbols and blocks. Some of the symbols used in this program are listed in the figure 4.right. In this type logic operation done in each block and inputs and outputs of each block is also shown, this view is in accordance with the logical circuits drawing in electronic form (Fig. 4) [9].

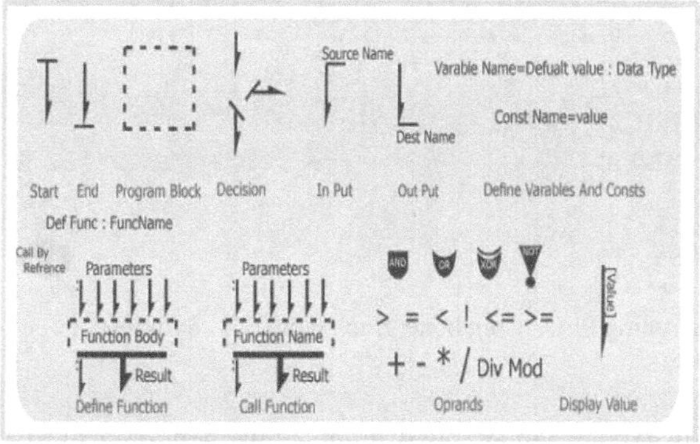

Fig. 5. Symbols of bond graph programming

In the script method, program is a set of commands that every line of its commands is called Statement and any command line in program is include combination of logical and mathematical operations. Some of the symbols are showed in figure 5. [8].

4 Experiments

In this paper 3 methods which are ladder, flowchart and statement method are explained by bond graph theoretically. To express our idea, we only implement ladder elements in MATLAB Simulink software. Which these defined elements in MATLAB, can be used as a pack in PLC in same application. For example we implement a simple program that is mentioned in paper [11] and also in our proposed method. Figure 6 show a normal operation and illustrates the working cycle that a schematic diagram of a wiring system and pneumatic system and PLC programming is required (Fig. 6) [11].

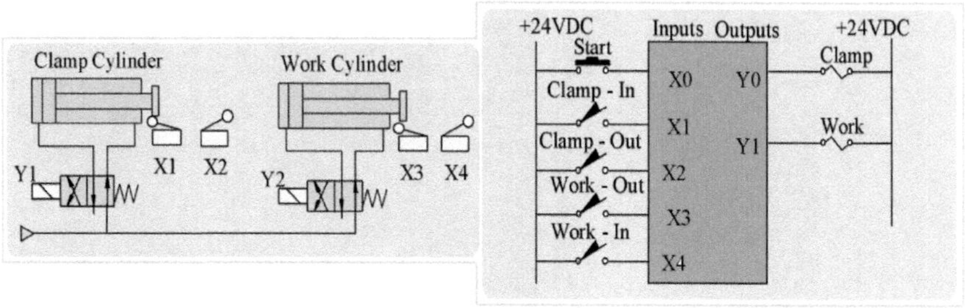

Fig. 6. PLC Program – Analysis

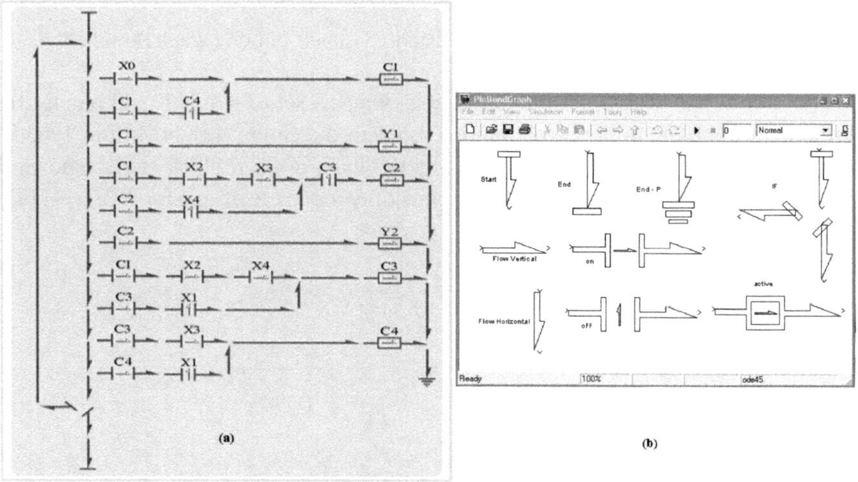

Fig. 7. A) Programming PLC, using bond graph ladder. B) Symbols of bond graph Ladder in Matlab Simulink

First we will explain ladder bond graph implementation in visual status and for better understanding briefly will describe it. Program begin with bond graph start element then it will catch data from input after this steps it will move to next step, processing of received data. After data processing (logical and other operations) to be active or inactive status of PLC ports will be specified. Output specific data will send to PLC output port, then program will face with condition element, if program running time is over, program will finish with bond graph finish element, else program will run while it's running time become over (Fig. 7.A).

Another difference between our method and previous methods except common elements, are elements that are shown which in previous methods they shown in hidden status. In other word start element is starter of all devices and finish element will be finalizing element of all devices.

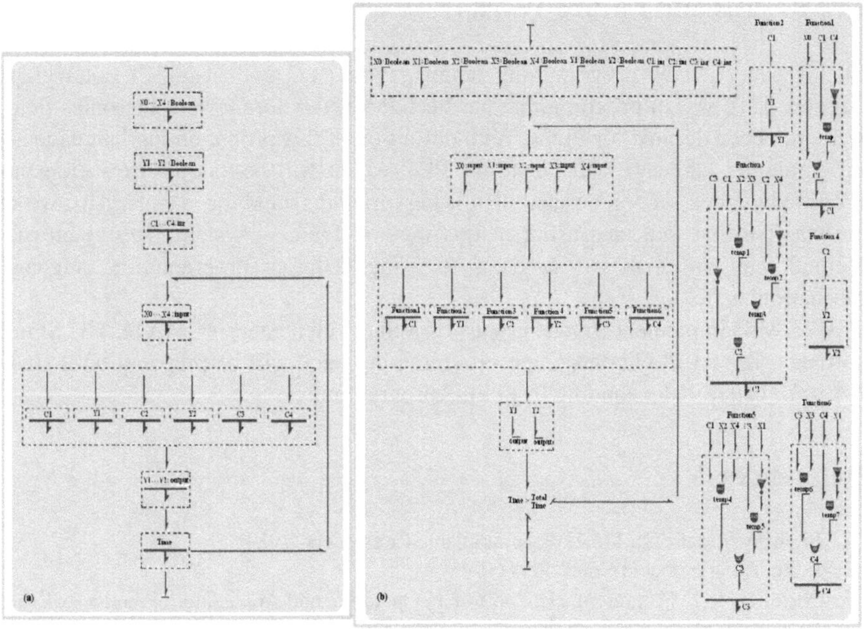

Fig. 8. A) PLC program using bond graph Flowchart. B) PLC program using bond graph Programming.

Figure 7.B shows ladder bond graph elements which are implemented in MATLAB Simulink software. This elements will be used as PLC element in Simulink in simple word this elements create PLC Pack for Matlab Simulink software.

Bond graph flowchart mostly gives an overall explanation from program that need to mix with 3rd programming language method for implement and perform (Fig. 8.A).

Last method "statement method", is a programming language which can be written by many languages such as STEP5, C++ and etc.

In this step we will present system control program code that is written in bond graph programming language which has special elements and after our brief explanation about program you will be able to understand types of bond graph programming language elements. First, program begin with start element and then enter in variables blocks which many variables with different types is defined by "Variable definition element" there, then will move to next block that is called "data receiver block" which in this block data receiver elements catch data from input and store them in relevant variables, after this steps, "data processing block", will send data to functions for data processing then processed values return to main program. Now specific output elements will send result of processes in PLC output ports, the last element is condition element. In that element if program running time is over program will finish with bond graph finish element, else program will run while it's running time become over. (Fig.8. B)

5 Results and Future Works

In this paper a new programming language for PLC with using of bond graph is defined. This type of programming has been defined before for mechatronics field but it has not been defined for PLCs. With definition of this programming language a new programming language will create for PLC which will be unique. New elements in proposed method section defined, that is easy to understand for readers. Also we show an example that indicates using of this method for new system. This is an overall method that will make us needless to using different programming language in mechatronics field.

We will implement bond graph flowchart elements in MATLAB Simulink software. In fact PLC bond graph programming pack will be prepared until students use it for better understanding from PLC programming.

References

1. Siemens-Simatic S5-115 U. Programmable Controllers (1989)
2. Festo. Programmable Controllers (1995)
3. Warnock, I.G.: Programmable Controllers Operation and Application. Prentice-hall (1988)
4. Breedveld, P.: Bond Graph. Introduction for Bond Graph
5. Gawthrop, P.J., Bevan, G.P.: Bond-Graph Modeling. 24 IEEE Control System Magazine (April 2007)
6. Broenink, J.F.: Introduction to Physical Systems Modeling with Bond Graphs (1998)
7. Gawthrop, P.J., Smith, L.P.S.: Metamodeling: Bond Graphs and Dynamic Systems. Prentice Hall, Hemel Hempstead (1996)
8. Naebi, A., Khalegi, F., Hosseinpour, F., Zanjanab, A.G., Khoshravan, H., Kelishomi, A.E., Rahmatdoustbeilankouh, B.: A New Flowchart and Programming Technique Using Bond Graph for Mechatronic Systems. In: 2011 UKSIM 13th International Conference on Modeling and Simulation, Cambridge, United Kingdom, March 30-April 01 (2011) ISBN: 978-0-7695-4376-5
9. Zalata, M.A., Alia, M.: Realizing Sequential Processes Using Programmable Logic Controllers (PLCS). Simulation 5(3-4), ISSN 1473-804x online, 1473-8031 print
10. Fan, Z., Seo, K., Rosenberg, R.C., Hu, J., Goodman, E.D.: Exploring Multiple Design Topologies using Genetic Programming and Bond Graphs. In: Proceedings of the Genetic and Evolutionary Computation Conference, GECCO 2002, New York, pp. 1073–1080 (2002)
11. Parker, J.K.: Introduction to Programmable Logic Controllers in a Mechanical Engineering Instrumentation Course. In: Proceedings of the 2001 American Society for Engineering Education Annual Conference & Exposition Copyright (C) 2001. American Society for Engineering Education (2001)

Physical and Computer Base Human Brain Simulations

Zhi Zun Zhou[1,*], Jing Ding[2], Yan Fan Yang[1], and Bao Cui[1]

[1] Medical Image School, Mudanjiang Medical University, 157011 Heilongjiang, China
[2] Pharmacy School, Mudanjiang Medical University, 157011 Heilongjiang, China
zzho007@hotmail.com, dingjing009@hotmail.com,
yangyanfang001@hotmail.com, cb2010china@hotmail.com

Abstract. Objective The simulated CSF and brain were set up. The aim of this project is to use them to study characteristic of real human brain in laboratory condition. **Method** Edible gelatine was used to simulate the brain. A concentration of 6g/100ml gelatine had physical characteristics similar to real brain. A series of different salt jelly concentrations were tested. **Results** The frequency and temperature characteristic of simulated CSF conductivity were measured at room temperature, body temperature and above the body temperature. **Conclusion** Kinds of direct comparisons of the results of mathematical calculations with experimental results is done using tangential and radial dipoles insert into the simulated SCF and simulated brain, to check on the validity of the parameters of simulated brain.

Keywords: SCF, simulated brain, dipole.

1 Introduction

Electric sources in the cortex can be modeled as equivalent dipoles. Both mathematical modeling and experimental measurements of the fields produced by dipoles in saline are used to show that the relevant patterns would be produced by dipoles located approximately 1 – 2 mm apart in the cortex[3][4]. In order to solve many magnetic and electric potential patterns of human head, the CSF (Cerebrospinal Fluid) and brain were constructed and simulated in this paper[1]. A human head is constructed of different tissues, with different conductivities[5][6]. Conductivity of scalp and brain is same, 0.33S/m[7]. The values of conductivity used in most head models are based on the values published by Geddes and Baker (1967)[8]. Conductivity of the skull is very small, only 0.0042S/m. Skull conductivity is only one eighth that of the brain. Cerebrospinal fluid conductivity is 1.4S/m. the conductivity of cartilage is 0.4S/m[8].

2 Methods

In this project, edible gelatine was used to simulate the human brain. A concentration of 6g/100ml gelatine physical characteristics similar to real brain when it was totally

* Correspondence author.

G. Lee (Ed.): Advances in Automation and Robotics, Vol. 1, LNEE 122, pp. 81–86.
springerlink.com

set. A series of different salt jelly concentrations were tested and it was found that for a conductivity of 0.33S/m, the necessary concentration of NaCl was 0.075g/100ml in 6% gelatine. Physiological saline is 0.9g/100ml and its conductivity was measured at 1.4S/m (same as that of cerebrospinal fluid).

A circuit was build for the conductivity measurements. A chamber was made of plastic and each end consisted of a stainless steel electrode, held in place by a screw-on cap so that no fluid could leak out of the chamber. The internal diameter of the cylinder was 25mm and the length 100mm. 5V, 100 Hz sinusoid power supply was applied across the electrodes. V_R (Voltage across a voltage-limited resister) and V_C (Voltage across the chamber) were measured. Alternating current was used for this measurement because direct current caused polarization of the electrodes, which gave a continuously changing voltage reading. Conductivity σ was calculated using the formula below.

$$\sigma = \frac{LV_R}{100V_C A} = 2.038\frac{V_R}{V_C}. \tag{1}$$

First, body of the brain was constructed by edible Gelatine, starting from 1g/100ml to 10g/100ml gelatine. If the quantity of gelatine in the hot water was too small the gelatine will take long time to set (normally two hours). After series testing we find that the 6g/100ml concentration jelly its physical characteristic was nearly same as real human brain. The conductivity of 6g/100ml pure gelatine was 0.198S/m. So we can use it by adding certain amount of salt to reach the 0.3S/m conductivity value. It cost more time to testing the series different salt jelly concentrations. And finally find that 0.33S/m brain conductivity salt concentration is 0.075g/100ml in 6% jelly. It was very important to follow the procedures which we summarized in this experiment.

The series certain concentration of salt jelly was made and tested. The concentration of NaCl use to simulate brain was 0.075g/100ml in 6g/100ml gelatine and its conductivity was 0.3S/m at room temperature and 100Hz power supply.

Table 1. Final experimental measurement figures of NaCl concentration for the brain, scalp, and conductivity of cerebrospinal fluid

Tissue Simulated (g/100ml)	Conductivity (S/m)	Concentration NaCl
Brain, Scalp	0.33	0.075 in 6% gelatine
Cerebrospinal Fluid	1.4	0.9 in water

3 Result and Conclusion

The frequency and temperature characteristic of simulated CSF conductivity were measured at room temperature (27 ^oC), body temperature (37 ^oC) and above the body temperature (45 ^oC). The frequency range from 5Hz to 500Hz was investigated.

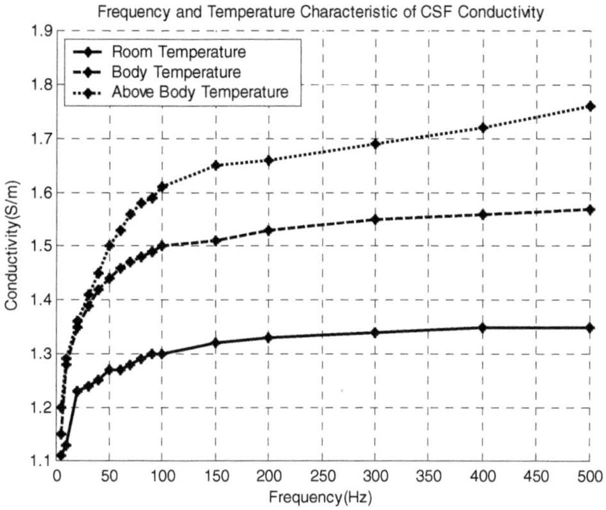

Fig. 1. Frequency and temperature characteristic of simulated CSF conductivity

Figure 1 shows that when temperature increases, the conductivity of the simulated CSF is increased. That means the temperature will play a role in the human head simulation. The ideal temperature for the experimental measurement would be body temperature.

The conductivity varies when the frequency changes from 5Hz to 500Hz. Between 0Hz to 100Hz the conductivity changes sharply (the variation was 0.2S/m). Between 100H to 500Hz, the conductivity tends to saturate even though the conductivity still slightly increases. So the ideal range of frequency is between 100Hz to 500Hz. In this range the conductivity is almost constant during the experiment. The frequency over 500Hz was not considered because the biomedical signal in the human brain is below hundred Hertz. In the low frequency range relevant to human EEG, the conductivity varies with frequency.

Above method again was used for temperature below $50\,^{o}C$. If the temperature was much higher the salt gelatine would not set. We try to make different concentration of jelly and find that for different concentration of jelly the conductivity was different. By a series testing, the concentration of jelly for conductivity 0.33S/m was 10g/100ml water for 100Hz at body temperature. The graph bellow (figure 2) was the temperature characteristic of this jelly conductivity.

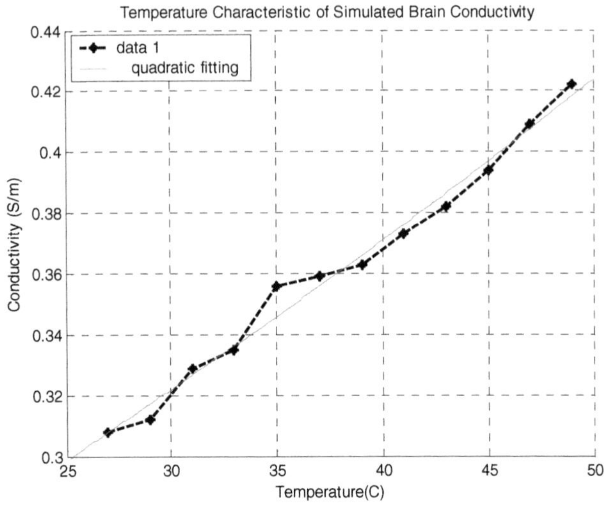

Fig. 2. Temperature characteristic of simulated brain conductivity

Figure 2 shows that when temperature increases, the conductivity was increased. The ideal temperature for the experimental measurement would be body temperature because the human brain was very sensitive to the variation of the temperature. If the temperature in the human brain increases (above the body temperature) the conductivity of the human brain will also increase.

Measurements were made with only one dipole. A series of different measurements was made as the dipole was shifted with respect to the electrode array. The results of various pairs of individual measurements are presented.

Tangential dipole potential distributions were measured with the dipole at a depth of 0.5 cm. In this experiment a 40 Hz, 300mV peak to peak sinusoidal signal was applied to the tangential dipole[2]. A glass container with an array of sintered Ag/AgCl pellet electrodes was used for recording. The reference electrode was at the edge of the container. The distance from centre to centre between the electrodes in the recording array was 10 mm and from edge to edge was 5mm. The distance from the dipole to the array was set at 5mm. For this experiment the concentration of NaCl in the water was 0.9 g/100 ml. Polarity was taken into account by measuring the voltage at a particular time after the start of data acquisition.

One tangential dipole, 0.5cm away from the electrode array, was moved along the array and the results from various dipole positions were compared. The purpose was to see how close two dipoles can be before they cannot be distinguished.

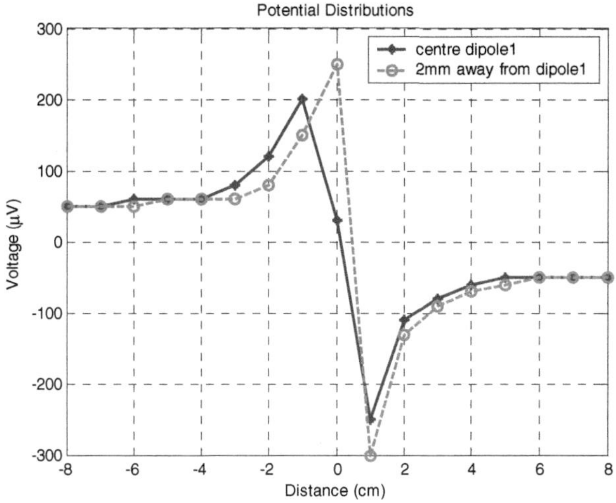

Fig. 3. Tangential dipole resolutions testing in simulated CSF In this situation, when this tangential dipole was moved 8mm, its potential distributions from this testing array can be totally separated

Everything was same as above except using a radial dipole to replace a tangential dipole (as figure 4).

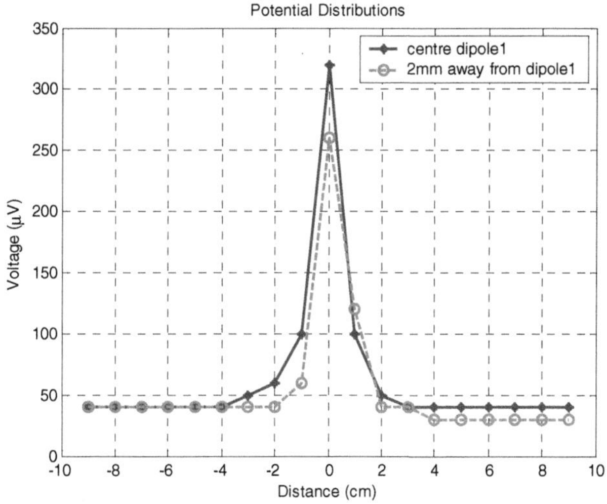

Fig. 4. Radial dipole resolutions testing in simulated CSF

Kinds of direct comparisons of the results of mathematical calculations with experimental results were done using tangential and radial dipoles insert into the simulated SCF and simulated brain. The experimental results show that the parameters of simulated brain were valid and true. The simulated CSF and brain were very useful to study characteristics of real human brain in laboratory condition.

Acknowledgments. This paper was supported by Heilongjiang provincial technical department science and technology special funds for overseas study and Repatriation person, LC07C16 and Mudanjiang medical university student research project.

References

1. Ruiz, Y., Pockett, S., Freeman, W.J., Gonzalez, E., Guang, L.: A Method to Study Global Spatial Patterns Related to Sensory Perception in Scalp EEG. Journal of Neuroscience Methods 191, 110–118 (2010)
2. Pockett, S., Zhou, Z.Z., Brennan, B.J., Bold, E.J.G.: Spatial Resolution and the Neural Correlates of Sensory Experience. J. Brain Topography 21(6), 1–6 (2007)
3. Freeman, W.J., Viana di Prisco, G.: Relation of olfactory EEG to Behavior: Time Series Analysis. J. Behavioral Neuroscience 100, 753–763 (1986)
4. Freeman, W.J., Baird, B.: Relation of Olfactory EEG to Behavior: Spatial Analysis. J. Behavioral Neuroscience 101, 393–408 (1987)
5. McConnell, K.A., Nahas, Z., Shastri, A., Lorberbaum, J.P., Kozel, F.A., Bohning, D., George, M.S.: The Transcranial Magnetic Stimulation Motor Threshold Depends on the Dstance from Coil to Underlying Cortex: A Replication in Healthy Adults Comparing Two Methods of Assessing the Distance to Cortex. Biological Psychiatry 49, 454–459 (2001)
6. Menon, V., Freeman, W.J., Cutillo, B.A., Desmond, J.E., Ward, M.F., Bressler, S.L., Laxer, K.D., Barbaro, N., Gevins, A.S.: Spatio-Temporal Correlations in Human Gamma Band Electrocorticograms. Electroencephalography and Clinical Neurophysiology 98, 89–102 (1996)
7. Camelia, G., Sami, G.: Compilation of the Dielectric Properties of Body Tissues at RF and Microwave Frequencies (1996), `http://www.emfdosimetry.org/dielectric/Report/Report.htm#Introduction`
8. Duck, F.A.: Physical Properties of Tissue: A Comprehensive Reference Book. Academic Press, Harcourt Brace Jovanovich (1990)

Fault Diagnosis of Excavator Hydraulic System Based on Expert System

Jian Yin[1,2] and Liting Mei[2]

[1] Mechnical Engineer Department, Tongling Univercity, China
[2] Shanghai Univercity, China
yinjianshanghai@163.com

Abstract. For complex electromechanical and hydraulic configuration, there are more and more types and forms of faults in Hydraulic Excavator. It is difficulty to exact diagnosis various faults using a single discrimination method, because its mechanical, electrical, hydraulic and control subsystems have strong interact and coupling. The thesis studied the fault diagnosis system based on expert system through engineer practicing. A fault diagnosis research framework is proposed and the whole system scheme is designed.

Keywords: Excavator, Hydraulic System, Fault Diagnosis, Expert System.

1 Introduction

Hydraulic Excavator has the complex structures with mechanical electrical-hydraulic integration. Its functions become more and more strong. While, its faults have the multi-level, fuzzy indefinite factors and it is difficulty for exact diagnosis various faults using a single discrimination method, because its mechanical, electrical, hydraulic and control subsystems have strong interact and coupling. So the fault requirement for the maintenance operators should increase accordingly. Traditional diagnosis methods are difficult to realize accurate and effective judgment.

1.1 Survey of Past Works on Fault Diagnosis

Various methods of fault diagnosis have been proposed in the literatures. Fault tree analysis provides a systematic procedure for identifying failures which could lead to undesirable events such as fire, explosion, etc. A signed directed graph method, whose nodes correspond to process variables, alarm conditions and whose branches represent causal dependencies between the nodes, is assumed that all the system symptoms can be attributed to a single causal source. The neural network approach to fault diagnosis was used extensively now because of its capability to classify nonlinear patterns among feature vectors. Expert systems can adequately capture the knowledge based symbolic reasoning strategies of expert plant operators. Diagnostic expert system used process specific, complicated knowledge based on past experience [1].

G. Lee (Ed.): Advances in Automation and Robotics, Vol. 1, LNEE 122, pp. 87–92.
springerlink.com

2 Hydraulic System Fault Analysis of Excavator

Hydraulic system is an important part of Excavator, which can be divided into boom hydraulic subsystem, Bucket Rod hydraulic subsystem, scoop hydraulic circuit, walk subsystem, etc, according to working part. It is composed of hydraulic pump, control valve, actuator and other auxiliary elements, etc, according to its function. Liquid under certain pressure is produced by hydraulic pump in the hydraulic system. Power is transferred to actuator such as hydraulic motor and hydraulic cylinder, etc, through control valves and oil circuits. Than, excavator can realize various actions such as soil excavating, loading and walking, etc. The hydraulic system is a complex structure composed of mechanical electrical-hydraulic integration and the subsystems are interfering and coupling mutually. So the fault reasons are many aspects. When the system is operating, various components are often occurred wear, fatigue and aging phenomenon for different design, manufacture, work condition and being affected by harsh environment.These lead to performance failure and made the equipment not work normally. The changing of components working environment or working parameters change , improper installation position lead to system deviation from normal station. The typical hydraulic system fault reasons are showed in Table 1.

3 Fault Diagnosis Method Based on Expert System

Based on above reasons, we choose the expert system as our research method in this paper. The fault diagnosis method based on expert system is supported by knowledge base and database and proceeds series reasoning combined of various rules. Then the final or most possible fault was been diagnosed. Using expert system for hydraulic excavators diagnosis, first a relative knowledge base and rule base was been built according to failure mode, fault mechanism and historical data. Then fault characteristic signal was extracted through information processing technology or hydraulic system fault was reasoned combined with rule base according to field fault phenomena. The schematic diagram of fault diagnosi system based on expert system is shown in Figure 1. The system is usually suitable for situations that there are more fault rules and fault reasoning rules are clearly [2].

Expert system is a computer program that imitating expert's behaviours, which can use reasoning mechanism of expert and apply knowledge and reasoning to solve complex fault diagnosis problems. In order to solve these problems, it need fault forms combined with diagnosis experiences and the Artificial Intelligence reasoning process. These can be considered as an effective empirical model. To understand working principle and function of the diagnosis objects is also very important except for experiences in the equipment fault diagnosis system. As an important method in Artificial Intelligence, Expert system is widely applied in the diagnosis on construction machinery, aeronautics and astronautics and medical equipment, etc.

Table 1. Typical hydraulic system fault reasons and countermeasure

fault phenomenon	reasons	countermeasure
oil cannot return to pump	suction pipe or core filter blocking oil temperature is too low or viscosity is too high loss ability oil adsorption for over-wearing of pump ail intake in oil suction pipe	clean core filter and oil lines, replace oil preheating to suitable temperature(30~60℃) and then loading use recommended viscosity oil repair or replace wear parts, change new pump check oil lines joints
pressure cannot increase	relief valve is not working normal oil supply lines leakage severely	the setting pressure is irrelevancy needle valve is failure piston operate adversely Spring is failure pump is damaged
pump produce queer sound	air is in oil tank or air is inhaling from oil line connection booster pump damage	supply new oil check the oil line joints whether there are air leakage repair booster pump
oil temperature increase quickly	oil is insufficient pressure of relief valve is setting too high	supply new oil check pressure gage and adjust the oil height setting value
leakage of pressure oil lines	oil viscosity is too low sealing device is damaged line joints are loosening	cooling the hydraulic oil replace seal tighten joints
pressure is abnormal in lubrication system	internal obstruction of pump obstruction of diverter valve or pipeline	check the obstruction system locate the obstruction consult inspection manual

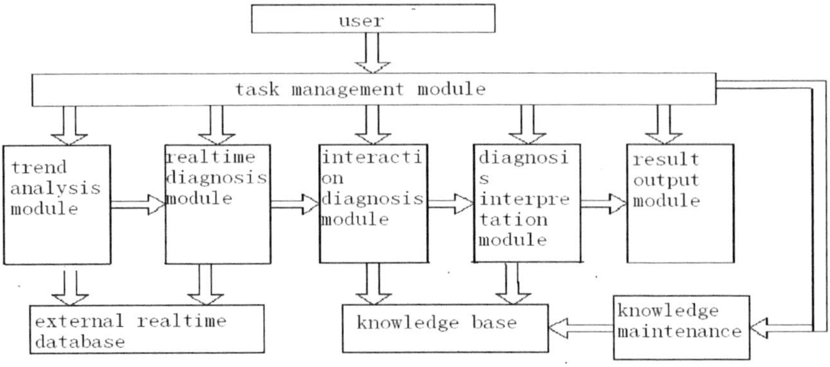

Fig. 1. Schematic diagram of fault diagnosi system based on expert system

4 Knowledge Process and Reasoning Mechanism

Expert system has obtained remarkable effect especially in the fault diagnosis, equipment monitoring, maintenance decision aspects.Hydraulic excavator is a complex system, which have a lot of linkage control, many control signals and huge logical relationship network. The establishment of its knowledge base is a tedious and difficulty task. The perfect and accuracy of system knowledge directly influenced the reliability of the diagnosis result. Therefore, we should investigate previous maintenance experience and be familiar with maintenance mechanism according to existing excavating equipment data. Then we proceeded the analysis and research work of subsystems. Although there are many factors restrict the performance of expert system. The indeed determined action is the knowledge owned by the expert system. The height of the knowledge representation played a decisive role for the expert system. The establishment of knowledge data should considerate following three aspects: (1) Hydraulic excavator is composed of various mechanical electrical-hydraulic equipments and its descriptions of fault diagnosis knowledge are different greatly. Even for same coil fault equipment, the fault expression forms on an excitation coil of motor and on a magnetic valve coil are difference.▯ 2▯ The fault occur of hydraulic excavator has some certain randomness. While the stochastic breakdowns is difficult to completely grasped through theory analysis. (3) With the multi-project extension and application, we should add continuously, improve original knowledge and realize accumulation of expert system experiences [3], [4].

The knowledge is core and key for expert system, which is a stating point to solve problems. To acquire knowledge information is solving the quantity and quality problem obtained by expert system [5]. This directly influenced the fault problems solving level and efficiency of expert system. The acquiring process of expert knowledge is transferring the professional knowledge from knowledge source to knowledge base. First the expert knowledge is been input to expert system knowledge base. Then the system working conditions are been dynamic measuring control through real-time signal detection system, which made the machine learning obtain and accumulate knowledge from processing problems. The flow chart of acquiring knowledge is shown in Figure 2.

Acquiring the maintenance experiences knowledge of maintenance engineer is a key problem, it is also the most difficulty work in this phase. The diagnosis knowledge will test for the system under the large range domain knowledge and many cases analysis. The system judged whether there is independent or mutual contradiction between the rules. It also judged whether the rules are completed or conclusion is correct and whether the fault process is according with practical application. The expert system took necessary modification to these problems and test them again. After the experimental simulation test, the system also need accept field test and make the reliability increase [6], [7].

As a complex system, The fault diagnosis problems of excavator need various theory depth domain knowledge, especially experience knowledge, causal knowledge, static structure knowledge, behavior knowledge and functional knowledge. The system is difficult to use the first law knowledge, because a uniform model based on first law knowledge whether function model, structural model, causal model and behavior model is difficult to obtain. The organization model of knowledge is closely related with reasoning strategy and both are interdependence, mutual restriction. In order to reduce the complexity of fault diagnosis, the selected reasoning strategy should support hierarchical diagnosis compared with hierarchical organization model of fault analysis tree [8]. In addition , Incomplete information of some devices caused uncertainty of some diagnosis rules. So the reasoning strategy should support the uncertainty reasoning.

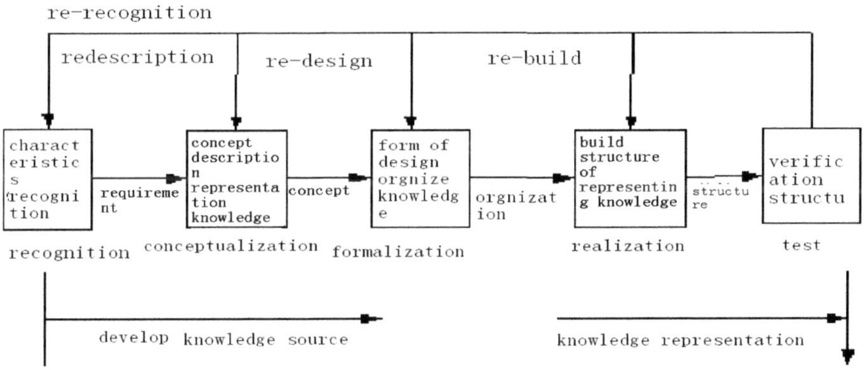

Fig. 2. Flow chart of knowledge acquiring

According to the task of the excavator fault diagnosis, there are two aspects. One is condition monitoring under equipments operation and the other is fault diagnosis when the device state is abnormal. The condition monitoring used the real-time data to complete, which ensure certain real-time performance. Its function is arbitrarily, so it is suitable for forward reasoning. However, the function of equipment fault diagnosis is suggestion and the diagnosis procedure is a plausible reasoning process. So its control strategy is based on hypothesis-test forward and reverse hybrid reasoning.

5 Conclusions

Comprehensively, the excavator fault diagnosis system has adopted an interactive diagnosis model based rule, which has a friendly interactive machine interface and complete knowledge base and the system can use for analysis and judgement on the fault phenomenon, fault cause and fault location.

Based on obtaining relevant information, the fault diagnosis system imitated the expert thinking and got the optimization results through recursive consultation for user. Combined with intelligent diagnosis with expert system, we have designed the general diagnosis scheme through further study on the occurrence mechanism of hydraulic excavator and collect occurred reasons, location and processed experiences from engineering practice. A set of operation convenient and fast and strong practicality software system has been designed scientific and rational combined with construction situation presently. The software system can make the expert knowledge widely service for engineering construction no limits of time and space. We seriously and deeply summarized the related professional knowledge and experiences in the engineering construction and promoted the subject development in this domain.

Acknowledgments. This research was supported by Nature Science Research Project of education department in Anhui province-Research of construction machinery fault diagnosis system. No KJ2011Z376.

References

1. Burrell, P., Inman, D.: An Expert System for the Analysis of Faults in an Electricity Supply Network: problems and achievements. Computers in Industry 37, 113–123 (1998)
2. Harada, M., Igarashi, Z.: Development of Expert System-Supported Construction Planning for a Shield-Tunneling. In: Method Innovative Applications of Artificial Intelligence, pp. 66–79. AAAI Press, Menlo Park, CA (2002)
3. Zhang, E.W.Y., Tor, S.B., Britton, G.A., Deng, Y.-M.: EFDEX: A Knowledge-Based Expert System for Functional Design of Engineering Systems. Int. Journal of Advanced and Manufacturing Technology 17, 339–353 (2001)
4. Sheremetov, L.B., Contreras, M., Valencia, C.: Intelligent Multi-agent Support for the Contingency Management System. Expert Systems with Applications 26(1), 57–71 (2004)
5. Chang, P.H.: A straight-line Motion Tracking Control of Hydraulic Excavator system. Journal of Mechatronics 12, 119–138 (2002)
6. Lee, S.-U.: Control of a Heavy-duty Robotic Excavator using Time Delay Control with Integral Sliding Surface. Journal of Control Engineering Practice 10, 697–711 (2002)
7. Haga, M.: Digging Control System for Hydraulic Excavator. Journal of Mechatronics 11, 665–676 (2001)
8. Hall, A.S.: Robust Bucket Position Tracking for a Large Hydraulic Excavator. Journal of Mechanism and Machine Theory 40, 1–16 (2005)

Geometric Errors Measurement of a Large Scale Five-Axis Gantry Type High Speed Machine Tool

Jian Yin[1,2], Ming Li[1], and Changkai Xu[1]

[1] CIMS & Robert Center, Shanghai Univercity, China
[2] Tongling Univercity, China
yinjianshanghai@163.com

Abstract. With the requirement of higher productivity and increased automation, precision five-axis machine tools are essential element of an industrial society. So enhancing the accuracy is become increasingly important on five-axis machine tool. The position dependent systematic geometric errors are 30 and the position independent systematic errors are 9 in a five-axis machine tool. It is necessary to measure all these errors components precisely in order to utilize software error compensation for accuracy improvement. A laser interferometer system and DBB test system have been built for a five-axis machine tool geometrical error measurement in this paper.

Keywords: Five-axis machine tool, error measurement, laser interferometer, Double-Ball-Bar.

1 Introduction

Machine tools with three translations axes have the ability to manufacture a large variety of products with relativity simple geometry to a satisfactory accuracy. With the requirement of higher productivity and increased automation, precision five axis machines are essential element of an industrial society. Modern industrial is critically dependent on precision machines ranging from integrated circuits to optical components, to automobiles, to aviation. In order to machine workpieces with complex shapes, such as impeller blades, engine blocks, aerofoil, etc, five axis machine tools are preferred due to excellence of simultaneously positioning and orienting the tool respect to the workpieces. Although a three-axis machine tool is usually applicable for machining most sculptured surfaces, however a five-axis machine tool can machine in a single setup with versatile positions of end effector [1, 2, 3].

Most five-axis machine tools have three translational axes and two rotary axis, the two rotary axes are used to adjust the tool in an optimal orientation relative the cutting surface. With the introduction of rotary axes, mutual interaction between the five axes is totally different from traditional machine tools with three orthogonal linear axes. The benefits of five-axis machine tools, such as small amounts of jigs and fixtures, less setup time, etc, which lead to efficient manufacturing for complicated shapes with significant lower costs, less setup and preparatory cost. But on the other hand due to more degree of freedom, the five-axis machine tool has poor stiffness which

G. Lee (Ed.): Advances in Automation and Robotics, Vol. 1, LNEE 122, pp. 93–99.
springerlink.com

affects the accuracy and reliability. The accuracy of a five-axis machine tool is generally not as high as that of tradition three-axis counterpart. So enhancing the accuracy is become increasingly important on five-axis machine tool. The error compensation becomes a necessary and practical solution for improving the accuracy of the machine tool [2, 4, 5, 6].

Errors in position and orientation in a five-axis machine tool are the results of errors in the individual links of the machine and the interactions between them. Errors are caused by numerous factors including geometrical and kinematic errors, thermally induced errors, static and dynamic deflections, etc. All these errors contribute to a combined error at the tool tip through error propagation in a machine tool. This synthesized error will directly affect the accuracy of the machine tool. If all these errors can be determined accurately, this information can send to the machine tool for error correction. It is necessary to measure all these errors components precisely in order to utilize software error compensation for accuracy improvement.

1.1 Survey of Past Works on Error Measurement

Over the last decades a number of engineers have become interested in the accuracy of machine tools and many different tests for assessing machine errors have been developed. The error mapping methods can be categorized into two groups: direct and indirect methods. The direct method also called parametric method is based on measuring each component error independently. Each component error is measured by conventional equipment such as laser interferometer and electronic level. The indirect method is realized by measuring errors with some type of artifacts or kinematic reference standards, such as step gage, a ball bar, etc, and estimating component errors based on the inverse kinematic analysis. There was no suitable measurement device to measure the overall errors of five-axis machine tool until now. It is clear that the key step toward effective accuracy enhancement of five-axis machine tool is the identification of these errors [7].

2 Objectives and Scope of the Research

The main objectives of this paper are to achieve the following two aspects for five-axis machine tool: To propose a method based on laser interferometer measurement to give a direct evidence of accuracy. To develop an inverse kinematic analysis method to quickly and effectively calibration the rotary joints errors of machine tool based on the double ball bar test. In this paper, a large scale five-axis gantry type high speed machining tool which is produced by Hantian precision Machine factory shown in Fig .1 is used to measure. It is a numerically controlled machining center with a movable beam frame and a fixed table, especially designed for high speed milling. In addition to the main three translation axes A and C on the cutter spindle, the working unit has two rotary axes, providing the inclination and orientation of the tools in any direction in the space. The major advantage of this type is that heavy parts can be machined easily because they do not need rotating and titling. Its working parameters are 20 meters stoke in X direction, 3 meters stroke in Y direction and 800 millimeter stroke in Z direction. An axis is rotating from -110°to 110° and C axis can rotate 360°around its axial line. However, the required accuracy of X, Y, Z axis is less than 0.06 micrometer and 0.02 micrometer respectively. Depending on the error avoidance to improve accuracy through careful design and manufacturing is very costly and not feasible. So error compensation is the only suitable method to enhance accuracy. As the machine tool is composed of links and joints, which have unavoidable errors during their position and movements. These errors can be calculated as; position dependent systematic geometric errors are 30, whereas the position independent systematic errors are 9. The ANSI/ASME standard provided procedures for the performance evaluation of CNC machining centers and requires some instruments, such as laser interferometers, electronic levers and capacitance gages, to measure errors.

Fig. 1. Picture of the large scale five-axis gantry type high speed machining tool

3 Mapping Geometrical Errors of the Machine Tool

As a machine tool producing factory, accuracy is one of the most important factors to be considered. Machine tool accuracy is most important for its performance assurance. A detailed analysis of machine tool accuracy helps to identity the impact of new design features on machine tool performance. After manufacturing a machine tool, a thorough verification is carried out for performance inspection. Here, we choose laser interferometer to map the geometrical errors of the five-axis gantry type high speed machining tool. The laser interferometer can be used to map geometric errors, such as positioning errors, straightness errors, angular errors, and squareness errors. The laser interferometer records the relative motion between two optics in the measured error direction. The advantage of using this device is its high accuracy (usually 0.1 μm) and non-contact measurement ability. In this paper, a renishaw XL-80 laser interferometer system was used for measurement errors. The XL-80 offered the ultimate in high accuracy, repeatable and traceable measurement, using externally mounted interferometers [8, 9].

3.1 Errors Measurement of Prismatic Joints

Fig.2 shows the laser interferometer components and the path of laser beam took for the various errors measurement setup used. In our case, X-axis is controlled by two drives and two feedback system. A second laser and optics, coupled with dual axis software, provides the capability to automatically capture data of parallel axes simultaneously. The laser interferometer measured the change in position of the return mirror with respect to the linear interferometer. The return mirror moved with the axis while the linear interferometer was fixed on the machine rigidly. When measuring angular, the laser interferometer actually measure the change in length between the two mirrors in the return mirror block. Straightness measurement was made by comparing the change in length of two different paths between the straightness interferometer and straightness reflector. Since the laser interferometer is not able to measure the roll error of machine slides, an electronic level was used to measure the horizontal roll error component as showed in Fig.3, which utilized the principle of the pendulum to reference measurement to gravity. With regarding to the vertical axis, the electronic level is not available to measure the roll error. Here, a renishaw RX-10 was used to the vertical axis roll error measurement as showed in Fig.4, which can save more than 5 hours over traditional methods such as auto-collimators and optical polygons.

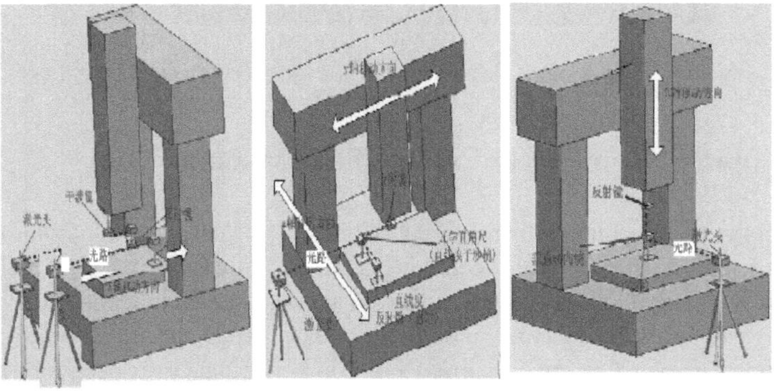

Fig. 2. Laser interferometer setup for position, pitch, yaw and straightness errors

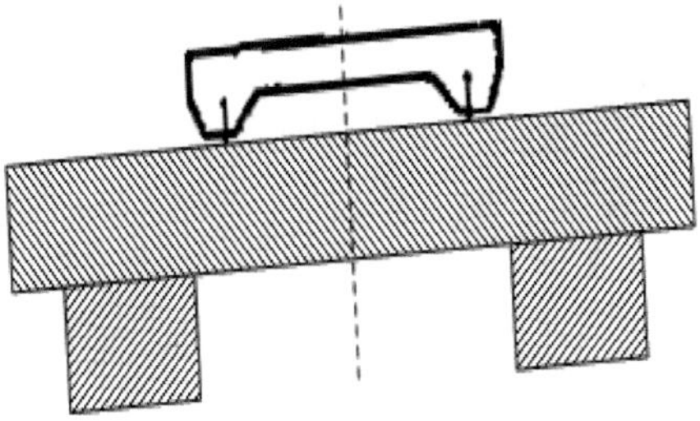

Fig. 3. Electronic level setup for horizontal axis roll errors

Fig. 4. Renishaw RX-10 setup for vertical axis roll errors

3.2 Geometrical Errors Measurement of Rotary Joints

Rotary joints are rotary indexing table, dividing head, spindle etc. These rotary joints contain three linear and three angular errors as theory of rigid body motion assumption. The laser interferometer is difficulty to measure these errors. The double-ball-bar method can measure the dynamic performance of machine tool and its circular test reflected the errors of rotary joints. The operation of DBB test is simple and rapid and its measurement is very high. So in this paper, we choose the DBB method to measure the errors of rotary joints. As shown in Fig.5, a DBB connects the spindle and the work table through two high precision spheres and two magnetic sockets. The test machine is programmed to move along circular trajectories at the center sphere with nominal radius equal to the length of the DBB and the relative motion in radial direction between the tool and the work table is detected by a linear variable difference transducer inside the DBB. Circular tests are performed with more than one axis motion involved, errors measured are the combined effects of the errors components. Therefore, we use inverse kinematic analysis to identify these errors components [9].

Fig. 5. DBB setup for rotary joint errors

Based on the procedure established in this paper, some experimental errors results are showed in Fig.6

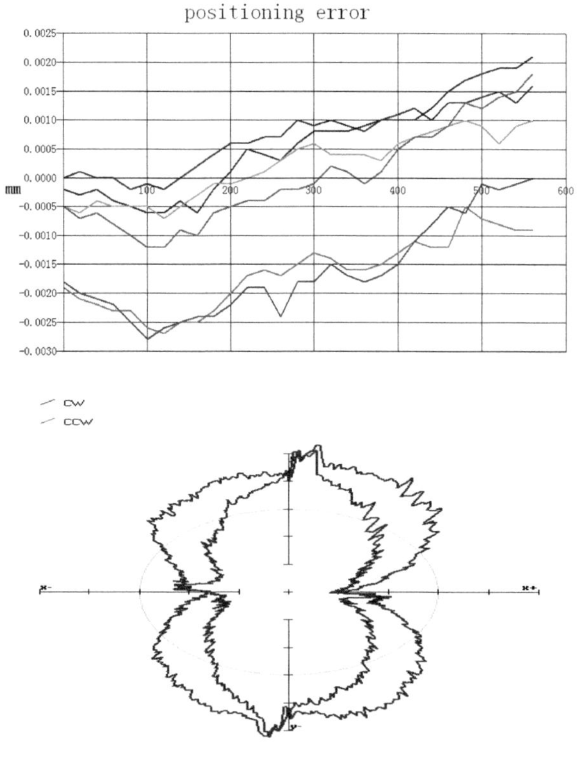

Fig. 6. Experimental results of position errors and rotary errors

4 Conclusions

This paper has focused on the measurement all errors components precisely in order to utilize software error compensation for accuracy improvement. A laser interferometer system and DBB test system have been built for a five-axis machine tool error measurement.

Acknowledgments. This project is supported by High Grade Numerical Control Machine Tool and Basis Manufacturing Equipment Key Special Research of China. Grant No 2009ZX04001-111.

References

1. Han, N.: Machine tool accuracy enhancement by inverse kinematic analysis and real time error compensation. University of Michigan (1995)
2. Barman, S.: Enhancement of Accuracy of Multi-axis Machine Tools through Error Measurement and Compensation of Errors Using Laser Interferometry Technique. Journal of Metrology Society of India 25(2), 79–87 (2010)
3. Tsutsumi, M.: Identification and Compensation of System Deviations Particular to 5-axis Machining Center. Journal of Machine Tools& Manufacture 43, 771–780 (2003)
4. Wang, S.-M.: An Efficial Error Compensation System for CNC Multi-axis Machines. Journal of Machine Tool & Manufacting 42, 1235–1245 (2002)
5. Dassanayakek, M.: Simultaneous Five-axis Motion for Identifying Geometric Deviations through Simulation in Machining Centers with a Double Pivot Head. Jounal of Advanced Mechanical Design, System, and Manufacturing 2, 47–58 (2008)
6. Bohez, E.L.J.: System Geometric Rigid Body ErrorIdentification of 5-axis Milling Machines. Jounal of Computer-Aided Design 39, 229–244 (2007)
7. Remus Tutunea, O.: Configuration Analysis of five-axis Machine Tools Using a Generic Kinematic Model. Jounal of Machine Tools& Manufacture 44, 1235–1243 (2004)
8. Lee, D.-M.: Identification and Measurement of Geometric Errors for a Five-axis Machine Tool with a Tilting Head using a Double Ball-bar. Journal of Precision And Manufscturing 12, 337–343 (2011)
9. Delbressine, F.L.M.: Modelling Thermomechanical Behaviour of Multi-axis Machine Tools. Jounal of Precision Engineering 30, 47–53 (2006)

Identical Predictions for Different Singular Mixed Models

Shiqing Wang

College of Mathematics and Information Sciences, North China University of Water Resources and Electric Power, Zhengzhou, 450011, China
wangshiqing@ncwu.edu.cn

Abstract. In linear mixed models theory one is assumed to know the structure of random effects covariance matrix. The suggestions are sometimes contradictious, especially if the model includes interactions between fixed effects and random effects. Mols [3] presented conditions under which two different random effects' variance matrices will yield equal estimation and prediction results during the paper it is assumed that X is of full column rank. Wang [11] weakened the conditions of his theorem, and obtained the same results as his. Wang [12] extended Mols's [3] results to situation that X is deficient in rank. Wang [13] gave a series of results it is assumed that X is possibly deficient in rank. They contain some necessary and sufficient theorems. We extend Wang's [11-13] results to singular linear mixed models in this paper.

Keywords: linear mixed models, best linear unbiased estimator, best linear unbiased predictor, robustly predictable, mean square error.

1 Introduction

In linear mixed models theory one is assumed to know the structure of random effects' covariance matrix. It is not always an easy task to determine the correct covariance matrix or correct covariance structure. Even in some relatively simple situations different authors can give contradictory suggestions. Consider, as an example, interactions between a random factor and a fixed factor. Some authors, like Searle [8], and SAS software [6] use unconstrained parameter model, suggesting that these interaction terms should be treated as independent random variables. Others, like Scheffe [7] and Neter, Wasserman & Kutner [4] suggest that the interaction terms should be restricted to sum zero over the levels of fixed factor. These two approaches lead to quite different structures for G, if the mixed model considered contains these types of interactions. The problem has been recently addressed by Voss [9]; Wolfinger & Stroup [10]; Hinkelmann [2].When treating complex practical situations it is often even harder to choose a correct covariance structure. Therefore, it may be of interest to know, which different covariance structures can lead to the same answers. This question is considered in great detail for estimating fixed effects. For example Puntanen & Styan [5] give an impressive list of conditions when Ordinary Least Squares Estimator (OLSE) is equal to the best linear unbiased estimator (BLUE). Harville [1] gives a condition when two different covariance (weight) matrices lead to

G. Lee (Ed.): Advances in Automation and Robotics, Vol. 1, LNEE 122, pp. 101–107.
springerlink.com

the same estimators for fixed effects. However, question when two different covariance structures lead to the same prediction results for linear combinations involving both random and fixed effects has remained unanswered. In this paper we intend to shed some light to this particular problem.

Let us consider the following mixed effects model

$$Y = X\beta + Z\gamma + \varepsilon$$

where Y is a vector of n observable random variables, β is a vector of p unknown parameters having fixed values (fixed effects), X and Z are known matrices and γ (random effects) and ε are vectors of length k and n. We suppose that $E(\gamma)=0$, $E(\varepsilon)=0$ and

$$Var\begin{pmatrix}\gamma\\ \varepsilon\end{pmatrix}=\begin{pmatrix}G & 0\\ 0 & R\end{pmatrix}$$

The variance matrix of Y can be expressed as $V = R + ZGZ^T$. The BLUE of fixed effects is denoted as $\hat{\beta}(V) = (X^T V^+ X)^- X^T V^+ Y$ where A^- is the generalized inverse (g-inverse) and A^+ Moore-Penrose inverse of A. The best linear unbiased predictor (BLUP) of random effects is denoted as $\hat{\gamma}(V) = GZ^T V^+ (Y - X\hat{\beta}(V))$. $C(A)$ denotes column space of a matrix A. In the following we consider two mixed models differing from each other only by having different variance matrices $V_1 = R_1 + ZG_1Z^T$ and $V_2 = R_2 + ZG_2Z^T$ .The main purpose of the paper is to understand the conditions under which these two different models lead to identical prediction results. Let V and R be nonsingular matrices, X matrix of full column rank. Mols [3] and Wang [11] gave the following results, respectively.

THEOREM 1. (Mols [3]) If $C(Z(G_1 - G_2)Z^T) \subset C(X)$ and $C(R_1 - R_2) \subset C(X)$ then BLUP predictors $Z\hat{\gamma}(V_1)$ and $Z\hat{\gamma}(V_2)$ are equal.

THEOREM 2. (Wang [11]) If

$$C(Z(G_1 - G_2)Z^T) \subset C(X) \text{ and } C(R_1 - R_2) \subset C(X) \tag{1}$$

then

$$C(V_2^{-1}R_2 - V_1^{-1}R_1) \subset C(V_i^{-1}X) \text{ and } C(V_1^{-1}X) = C(V_2^{-1}X) \tag{2}$$

where i = 1 or 2.

THEOREM 3. (Wang [11]) If $C(V_2^{-1}R_2 - V_1^{-1}R_1) \subset C(V_i^{-1}X)$ where i = 1 or 2, and $C(V_1^{-1}X) = C(V_2^{-1}X)$ then BLUP predictors $Z\hat{\gamma}(V_1)$ and $Z\hat{\gamma}(V_2)$ are equal.

Theorem 2 isn't sufficient and necessary since the equation (1) can't be obtained from (2). Wang [12] extended Mols's [3] results to situation that X is deficient in rank.

Wang [13] gave a series of results. They contain some necessary and sufficient theorems. In the following let V and R be singular, X possibly deficient in rank.

2 Derivation

LEMMA 1. Let A be symmetric nonnegative definite matrix. Then

$$A^{1/2}X(X^TAX)^-X^TAX = A^{1/2}X .$$

Proof

$$A^{1/2}X(X^TAX)^-X^TAX = A^{1/2}X(X^TA^{1/2}A^{1/2}X)^-X^TA^{1/2}A^{1/2}X = A^{1/2}X$$

LEMMA 2. (Harville [1])

$$X\hat{\beta}(V_1) = X\hat{\beta}(V_2) \Leftrightarrow X(X^TV_1^+X)^-X^TV_1^+ = X(X^TV_2^+X)^-X^TV_2^+$$

THEOREM 4. If $C(V_2^+R_2 - V_1^+R_1) \subset C(V_1^+X)$, then

$$Z\hat{\gamma}(V_1) = Z\hat{\gamma}(V_2)$$

$$\Leftrightarrow ZG_2Z^TV_2^+(X\hat{\beta}(V_1) - X\hat{\beta}(V_2)) = (V_1V_1^+ - V_2V_2^+)(Y - X\hat{\beta}(V_1))$$

Proof

$$Z\hat{\gamma}(V_1) = Z\hat{\gamma}(V_2) \Leftrightarrow ZG_1Z^TV_1^+(Y - X\hat{\beta}(V_1)) = ZG_2Z^TV_2^+(Y - X\hat{\beta}(V_2))$$

$$\Leftrightarrow (V_1V_1^+ - R_1V_1^+)(Y - X\hat{\beta}(V_1)) = (V_2V_2^+ - R_2V_2^+)(Y - X\hat{\beta}(V_2))$$

$$\Leftrightarrow V_1V_1^+(Y - X\hat{\beta}(V_1)) - R_1V_1^+(Y - X\hat{\beta}(V_1))$$

$$= V_2V_2^+(Y - X\hat{\beta}(V_2)) - R_2V_2^+(Y - X\hat{\beta}(V_2))$$

$$\Leftrightarrow V_1V_1^+(Y - X\hat{\beta}(V_1)) + R_2V_2^+(X\hat{\beta}(V_1) - X\hat{\beta}(V_2)) = V_2V_2^+(Y - X\hat{\beta}(V_2))$$

since $(R_2V_2^+ - R_1V_1^+)(Y - X\hat{\beta}(V_1)) = 0$ from $C(V_2^+R_2 - V_1^+R_1) \subset C(V_1^+X)$ and Lemma 1 noting V_i^+ is symmetric nonnegative definite.

$$\Leftrightarrow (V_1V_1^+ - V_2V_2^+)(Y - X\hat{\beta}(V_1)) + R_2V_2^+(X\hat{\beta}(V_1) - X\hat{\beta}(V_2))$$

$$= V_2V_2^+(X\hat{\beta}(V_1) - X\hat{\beta}(V_2))$$

$$\Leftrightarrow ZG_2Z^TV_2^+(X\hat{\beta}(V_1) - X\hat{\beta}(V_2)) = (V_1V_1^+ - V_2V_2^+)(Y - X\hat{\beta}(V_1))$$

THEOREM 5. If $C(V_2^+R_2 - V_1^+R_1) \subset C(V_2^+X)$, then

$$Z\hat{\gamma}(V_1) = Z\hat{\gamma}(V_2)$$

$$\Leftrightarrow ZG_1Z^TV_1^+(X\hat{\beta}(V_1)-X\hat{\beta}(V_2))=(V_1V_1^+-V_2V_2^+)(Y-X\hat{\beta}(V_2))$$

The proof is similar to Theorem 4.

COROLLARY 1. If $C(V_2^+R_2-V_1^+R_1)\subset C(V_i^+X)$ where i = 1 or 2, $V_1V_1^+=V_2V_2^+$ and $X\hat{\beta}(V_1)=X\hat{\beta}(V_2)$ then $Z\hat{\gamma}(V_1)=Z\hat{\gamma}(V_2)$

Let V_1,V_2 and ZG_1Z^T or ZG_2Z^T be nonsingular matrices, then one obtains Wang's [13] Theorem 2.1 and Theorem 2.2 from Theorem 5. Further let V_1,V_2 be nonsingular matrices and X of full column rank then one obtains Wang's [11] Theorem 2.1 from Corollary 1.

THEOREM 6. If $C(R_i)\subset C(X)\quad(i=1,2)$, then

$$Z\hat{\gamma}(V_1)=Z\hat{\gamma}(V_2)\Leftrightarrow V_1V_1^+(Y-X\hat{\beta}(V_1))=V_2V_2^+(Y-X\hat{\beta}(V_2))$$

Proof

$$Z\hat{\gamma}(V_1)=Z\hat{\gamma}(V_2)\Leftrightarrow ZG_1Z^TV_1^+(Y-X\hat{\beta}(V_1))=ZG_2Z^TV_2^+(Y-X\hat{\beta}(V_2))$$

$$\Leftrightarrow (V_1V_1^+-R_1V_1^+)(Y-X\hat{\beta}(V_1))=(V_2V_2^+-R_2V_2^+)(Y-X\hat{\beta}(V_2))$$

$$\Leftrightarrow V_1V_1^+(Y-X\hat{\beta}(V_1))=V_2V_2^+(Y-X\hat{\beta}(V_2))$$

since $R_iV_i^+(Y-X\hat{\beta}(V_i))=0\ (i=1,2)$ from $C(R_i)\subset C(X)$ and Lemma 1.

COROLLARY 2. If $C(R_i)\subset C(X)\quad(i=1,2)$, $V_1V_1^+=V_2V_2^+$ then

$$Z\hat{\gamma}(V_1)=Z\hat{\gamma}(V_2)\Leftrightarrow V_1V_1^+X\hat{\beta}(V_1)=V_2V_2^+X\hat{\beta}(V_2)$$

Let V_1,V_2 be nonsingular matrices, then one obtains Wang's [13] Theorem 2.3 from Theorem 6. Further let X be of full column rank then obtains Wang's [13] Corollary 2.2 from Theorem 6.

3 Sampling Variance: Known V

DEFINITION 1. Linear combination of fixed and random effects is robustly predictable if it can be expressed in the form of $l(X\beta+Z\gamma)$ for some vector l.

THEOREM 7. If $R_1=R_2=R$, $C((V_2^+-V_1^+)R)\subset C(V_1^+X)$ then the mean square error (MSE) of any robustly predictable linear combination

$$D(l(X\hat{\beta}(V_1)+Z\hat{\gamma}(V_1))-l(X\beta+Z\gamma))$$

$$=D(l(X\hat{\beta}(V_2)+Z\hat{\gamma}(V_2))-l(X\beta+Z\gamma))$$

$$\Leftrightarrow L_1 + M_1 + M_1^T + RV_1^+W_1^T + W_1V_1^+R + R(V_1^+W_1^T - V_2^+W_2^T)V_2^+R$$
$$= L_2 + M_2 + M_2^T + RV_2^+W_2^T + W_2V_2^+R$$

where

$$L = WV^+W^T - VV^+W^T - WV^+V + VV^+W^TV^+V \tag{3}$$

$$M = V(V^+ - V^+W^TV^+)R \tag{4}$$

$$W = X(X^TV^+X)^-X^T \tag{5}$$

Proof. MSE of the combined parameter vector is

$$D\begin{pmatrix} \hat{\beta}-\beta \\ \hat{\gamma}-\gamma \end{pmatrix} = \begin{pmatrix} D_{11} & D_{12} \\ D_{12}^T & D_{22} \end{pmatrix}$$

where

$$D_{11} = Cov(\hat{\beta}-\beta) = (X^TV^+X)^-X^TV^+X[(X^TV^+X)^-]^T \tag{6}$$

$$D_{12} = Cov(\hat{\beta}-\beta, \hat{\gamma}-\gamma) = -(X^TV^+X)^-X^TV^+ZG \tag{7}$$

from Lemma 1 and noting that γ is random variable and

$$Cov(\hat{\beta},\gamma) = (X^TV^+X)^-X^TV^+Cov(Y,\gamma) = (X^TV^+X)^-X^TV^+ZG.$$

Therefore

$$D_{22} = Cov(\hat{\gamma}-\gamma) = Cov[(UZ-I)\gamma + U\varepsilon]$$

$$= (UZ-I)G(UZ-I)^T + URU^T = UVU^T - UZG - GZ^TU^T + G \tag{8}$$

where

$$U = GZ^TV^+(I - WV^+)$$

It is sufficient to show that quantity $D(X\hat{\beta} + Z\hat{\gamma} - X\beta - Z\gamma)$ will be the same for two covariance matrices satisfying the requirements of the theorem. From (6)-(8) one can prove that

$$D(X\hat{\beta} + Z\hat{\gamma} - X\beta - Z\gamma)$$

$$= D\left[(X:Z)\begin{pmatrix} \hat{\beta}-\beta \\ \hat{\gamma}-\gamma \end{pmatrix}\right] = (X:Z)\begin{pmatrix} D_{11} & D_{12} \\ D_{12}^T & D_{22} \end{pmatrix}\begin{pmatrix} X^T \\ Z^T \end{pmatrix}$$

$$= XD_{11}X^T + ZD_{12}^TX^T + XD_{12}Z^T + ZD_{22}Z^T$$

$$= WV^+W^T - ZGZ^TV^+W^T - WV^+ZGZ^T$$

$$+ZUVU^TZ^T - ZUZGZ^T - ZGZ^TU^TZ^T + ZGZ^T \tag{9}$$

where

$$ZGZ^TV^+W^T = (V - R)V^+W^T = VV^+W^T - RV^+W^T \tag{10}$$

$$ZUVU^TZ^T = ZGZ^TV^+(I - WV^+)V(I - V^+W^T)V^+ZGZ^T$$
$$= ZGZ^TV^+ZGZ^T - ZGZ^TV^+WV^+ZGZ^T \tag{11}$$

since

$$(I - WV^+)V(I - V^+W^T) = V - VV^+W^T - WV^+V + WV^+W^T.$$

And

$$ZUZGZ^T + ZGZ^TU^TZ^T$$
$$= ZGZ^T(V^+ - V^+WV^+)ZGZ^T + ZGZ^T(V^+ - V^+W^TV^+)ZGZ^T$$
$$= 2ZGZ^TV^+ZGZ^T - ZGZ^TV^+WV^+ZGZ^T - ZGZ^TV^+W^TV^+ZGZ^T \tag{12}$$

Then from (9)-(12) one obtains

$$D(X\hat{\beta}(V) + Z\hat{\gamma}(V) - X\beta - Z\gamma)$$
$$= L + M + M^T + RV^+W^T + WV^+R - R - R(I - V^+W^T)V^+R \tag{13}$$

$$D(l(X\hat{\beta}(V_1) + Z\hat{\gamma}(V_1)) - l(X\beta + Z\gamma))$$
$$= D(l(X\hat{\beta}(V_2) + Z\hat{\gamma}(V_2)) - l(X\beta + Z\gamma))$$
$$\Leftrightarrow L_1 + M_1 + M_1^T + RV_1^+W_1^T + W_1V_1^+R - R(I - V_1^+W_1^T)V_1^+R$$
$$= L_2 + M_2 + M_2^T + RV_2^+W_2^T + W_2V_2^+R - R(I - V_2^+W_2^T)V_2^+R \tag{14}$$

from (13)

$$\Leftrightarrow L_1 + M_1 + M_1^T + RV_1^+W_1^T + W_1V_1^+R + R(I - V_1^+W_1^T)(V_2^+ - V_1^+)R$$
$$= L_2 + M_2 + M_2^T + RV_2^+W_2^T + W_2V_2^+R - R(V_1^+W_1^T - V_2^+W_2^T)V_2^+R$$
$$\Leftrightarrow L_1 + M_1 + M_1^T + RV_1^+W_1^T + W_1V_1^+R + R(V_1^+W_1^T - V_2^+W_2^T)V_2^+R$$
$$= L_2 + M_2 + M_2^T + RV_2^+W_2^T + W_2V_2^+R$$

since $R(I - V_1^+W_1^T)(V_2^+ - V_1^+)R = 0$ from $C((V_2^+ - V_1^+)R) \subset C(V_1^+X)$ and Lemma 1.

THEOREM 8. If $R_1 = R_2 = R$, $C((V_2^+ - V_1^+)R) \subset C(V_2^+X)$ then MSE of any robustly predictable linear combination

$$D(l(X\hat{\beta}(V_1) + Z\hat{\gamma}(V_1)) - l(X\beta + Z\gamma))$$
$$= D(l(X\hat{\beta}(V_2) + Z\hat{\gamma}(V_2)) - l(X\beta + Z\gamma))$$

$$\Leftrightarrow L_1 + M_1 + M_1^T + RV_1^+W_1^T + W_1V_1^+R + R(V_1^+W_1^T - V_2^+W_2^T)V_1^+R$$
$$= L_2 + M_2 + M_2^T + RV_2^+W_2^T + W_2V_2^+R$$

Proof. The proof is similar to Theorem 7.

COROLLARY 3. If $R_1 = R_2 = R$, $C((V_2^+ - V_1^+)R) \subset C(V_i^+X)$ where i = 1 or 2, and $X\hat{\beta}(V_1) = X\hat{\beta}(V_2)$ then

$$L_1 + M_1 + M_1^T = L_2 + M_2 + M_2^T \Rightarrow$$
$$D(l(X\hat{\beta}(V_1) + Z\hat{\gamma}(V_1)) - l(X\beta + Z\gamma))$$
$$= D(l(X\hat{\beta}(V_2) + Z\hat{\gamma}(V_2)) - l(X\beta + Z\gamma))$$

Proof. $W_1V_1^+ = W_2V_2^+$ since $X\hat{\beta}(V_1) = X\hat{\beta}(V_2)$ from Lemma 2.

References

1. Harville, D.A.: Matrix Algebra from a Statistician's Perspective. Springer, Heidelberg (1997)
2. Hinkelmann, K.: Resolving The Mixed Models Controversy, Comments. Am. Stat. 54, 228 (2000)
3. Mols, M.: Constraints on Random Effects and Mixed Linear Model Predictions. Acta. Appl. Math. 79, 17–23 (2003)
4. Neter, J., Wasserman, W., Kutner, M.H.: Applied Linear Statistical Models, 3rd edn. CRC Press (1990)
5. Puntanen, S., Styan, P.H.: The Equality of the Ordinary Least Squares Estimator and the Best Linear Unbiased Estimator. Am. Stat. 43, 153–164 (1989)
6. SAS Institute Inc.: SAS Onlinedoc. Version 8. Cary, NC, SAS Institute Inc. (1999)
7. Scheffe, H.: The Analysis of Variance. Wiley (1959)
8. Searle, S.R.: Linear Models. Wiley (1971)
9. Voss, D.T.: Resolving the Mixed Models Controversy. Am. Stat. 53, 352–356 (1999)
10. Wolfinger, R., Stroup, W.W.: Resolving the Mixed Models Controversy, Comments. Am. Stat. 54, 228 (2000)
11. Wang, S.Q.: Constraints on Random Effects and Linear Mixed Model Predictions. In: Proceedings of the 7th Conference on Biological Dynamic System and Stability of Differential Equation, pp. 260–263. World Academic Press, Liverpool (2010)
12. Wang, S.Q.: Constraints on Random Effects and Mixed Model Predictions with the Coefficient Matrix of Deficient in Rank. In: Proceedings of Third International Conference on Modelling and Simulation. Applied Mathematics and Mathematical Modelling, vol. 5, pp. 88–91. World Academic Press, Liverpool (2010)
13. Wang, S.Q.: The Conditions with Different Models Having Identical Predictions. In: Advances on Probability and Statistics, pp. 183–186. World Academic Press, Liverpool (2010)

Probing Concepts of Sustainable Package Design

Min Jie

Nanchang University, 235, Nanjing East Road, Nanchang, Jiangxi, Prov., 330047, PRC
jieqianxiban@sina.com

Abstract. Taking ideas of sustainable development as starting point, and based on the analysis of cutting-edge concepts related to sustainable design, this paper discusses the principles in sustainable package design and predicts the future trend of as well.

Keywords: Package Design, Sustainable Concepts, Sustainable Design.

Population explosion, environment pollution and resource shortage are three major challenges which human beings have to face in 21st century. Due to the devastating traditional mode of production and consumption during the long course of social development, human society is now plagued with serious problems which, in the long run, will threat the existence of human race as a whole. So far, the ever worsening environment pollution and resource shortages have given rise to concerns from all around the world. Strategy of Sustainable Development, which advocates saving natural resources and protecting environment, was first proposed in World Conservation Strategy in 1980 [1]. Sustainable design is a new mode of design which borrows its concepts from general principles of sustainable development. This new mode of design integrates user, machine and environment into one system and creates a harmonious relationship between human and nature. By now, the concept of sustainable design is being adopted in various fields of design. With the introduction of new concepts of living, production, energy, transportation and consumption, the ever-increasing demand for sustainable design will create a new field for designers to try their talents. As an indispensible part of production which plays a vital role in economic activities, packaging design calls for more new and creative ideas to ensure its sustainable nature.

1 Basic Ideas of Sustainable Design

Design is meant to create order out of chaos. Sustainable design, which, as a new idea evolved into being in the long process of social development, is closely related to the thoughts of Victor Papanek in his Theories on Resource Shortage. Taking modern ecological theories as guidelines and aiming at maintaining ecological balance, sustainable design gives full consideration to environmental benefit in every decision-making stage in order to minimize its negative effects on environment. In doing so, sustainable design not only creates social values and maintains ecological

G. Lee (Ed.): Advances in Automation and Robotics, Vol. 1, LNEE 122, pp. 109–114.
springerlink.com

balance which guarantees a harmonious relationship between human and nature, it also reduces the consumption of energy and raw materials, which in return cuts down carbon emission, and realizes easy recovery and recycle of different parts of products [2]. Sustainable design, in short, can be regarded as the consummation and perfection of green design which focuses mainly on the solutions to some particular problems which exert ill effects on environment, problems like the efficiency of energy and resource consumption, the reduction of waste, etc. By contrast, while attaching more importance to the recycling feathers of the whole system of product, sustainable design also focuses on how to extend the life circle of product, the main purpose of sustainable design being encouraging a socially healthy and environment-friendly mode of consumption. One of the key principles of sustainable design is to adopt a holistic approach which, unlike traditional approach to design, doesn't limit itself to the much narrower business-product domain. It is this holistic approach which attaches importance to the basic benefits of human beings and aims at re-establishing a balance between human and nature.

2 The Process of Sustainable Package Design

Since sustainable package is the embodiment of ideas and principles of sustainable design, it is harmless in terms of recycle and exerts little negative effect on environment. Besides minimizing the use of resources and cutting down production cost, sustainable package employs renewable or regenerative resources in order to facilitate the recycle and reuse of waste materials. Based on the features mentioned above which guarantees the sustainability of packaging process, five steps of sustainable package design are fully discussed in the following chapters.

2.1 Step One: Design

In this step, a comprehensive market survey is carried out and relevant information collected. Then, environment-friendly materials are decided and a systematic design completed and adjusted according to necessity. This systematic design includes innovative ideas, relevant parameter (mass and variable data, planned capacity or allowable variation), environmental factors and artistic expression. As an integrated part of sustainable design, environmental factors have to be taken in to account when it comes to the employment of packing materials. It is important to use degradable, harmless and environment-friendly materials to facilitate recycle and reuse of package. To this end high-performance synthetic materials and natural materials are strongly suggested. With respect to artistic expression, sustainable design, like any type of design, has to make every effort to create aesthetic structures, patterns and illustrations with corresponding colors and words. Any kind of visual pollution generated in design shall be avoided.

2.2 Step Two: Production

In this step, full consideration is attached to the improvement of techniques in order to shorten production process while maintain the quality of products. According to recent researches, every measurement unit of consumed materials and time in production process goes in proportion to the environmental costs our society has to pay. In this sense, successful sustainable design shall be able to continuously improve techniques, cut down production costs and guarantee qualities like safety, convenience and eye-pleasing appearance, etc.

2.3 Step Three: Transportation

Since the improvement of package directly determines the efficiency of transportation, it is crucial to tailor package design in accordance with the different natures, characteristics, functions and potential damages of products in order to facilitate transportation. Sustainable package shall be able to minimize the pollution and energy consumption caused in transportation and to lessen transportation-related burdens to environment.

2.4 Step Four: Consumption

The major concern of this step is to save energy, reduce pollution, facilitate the maintenance and extend the life cycle of package products, aiming at encouraging a healthy and sustainable mode of public consumption.

2.5 Step Five: Recover and Recycle

Sustainable package design, on one hand, integrates ideas of how to ensure easy recover so that the parts and materials of packaging products can be easily disassembled and classified. On the other hand, materials used in sustainable package must facilitate the purpose of recycle, for some synthetic materials are difficult or even impossible to decompose.

3 Principles of Sustainable Package Design

3.1 Principle of System

A system is a group of interacting, interrelated, or interdependent elements forming a complex whole. At the heart of a system is a holistic view which emphasizes the whole rather than inter-dependent parts. The principle of system in sustainable package is based on sustainable concepts and attaches importance to the realization of a systematic circle which includes the use of raw materials, production process, consumption and recycle. Only by means of systematic thinking and scientific approach can we solve the problems concerning the production, transportation, consumption and recycle of packing products and minimize their negative effects on human bodies and environment. To this end, designers shall be armed with environmental awareness and consciously develop systematic thinking. It is the systematic thinking which eventually leads to an opener and wiser mind in which creative ideas are generated[3]. In order to establish a holistic design with optimal systematic features, it is necessary to integrate sustainable concepts into the early stage of package design while taking environment protection measures into full consideration. In short, the utmost goal of sustainable package is to strike an optimized balance between technology and arts, function and form, economy and environment, nature and society. Only by means of adopting systematic package design, which takes into consideration production, transportation, consumption and recycle from a holistic view, can non-sustainable package be permanently avoided.

3.2 Principle of Economy

The principle of economy in sustainable package includes two major aspects: one concerns the fact that package, after fulfilling its basic functions (give protection, facilitate transportation, and boost sale, etc.), should be minimized and its redundant parts removed, in order to save energy and natural resources while reduce packing waste as much as possible. Another aspect concerns the use of packing materials, which should be environment-friendly, high-performance and harmless to human bodies, and the consideration of potential environment pollution related to packing products. In order to reduce the consumption of energy and natural resources during the processes of production, transportation, consumption and abandonment of packing materials, it is crucial to minimize the use of energy and natural resources and cut down effluents and carbon emission caused in the designing of package. The principle of economy is one of the major principles of sustainable package which works in accordance with the national strategy of building a conservation-minded society. Under the influences of sustainable concepts, the principle of economy is embodied not only in the reducing of resource consumption and simplified structures of design, but also in the optimized allocation of raw materials and equipments. In addition to this, the economy principle of sustainable package also can be implemented by using fewer materials to make more packing products of good quality.

3.3 Principle of Waste Recovery

The third principle of sustainable package refers to its easy solution to waste recovery. According to this principle, purer materials are more favorable than mixed materials in terms of package products, because mixed materials are not likely to be easily decomposed for recycle. When using composite materials, package shall be designed with features which can be easily disassembled and which facilitate recycle and recovery. Once sustainable package is no longer in use, it can be recovered to make new regenerative products, or burnt to collect heat energy, or used to make compost to serve recycling purpose. The principle of waste recovery puts emphasis to the recycling of package product or, if possible, in order to use resources fully and effectively and reduce as much waste as possible. With respect to package recovery, Finland set up a excellent example with its bottle package recovery. In Finland, where all glass and plastic bottles produced with standardized size, a system of waste recovery for bottled soft drinks and liquors has been set up. As a result, standardized bottles, no matter when and where they are first produced, can be recovered and reused by any other soft drink companies as these companies have refilling equipments which match the standardized sizes of recovered bottles.[4] In doing so, sustainable concepts are integrated in the design of bottle package and due emphasis is attached to recoverability and sustainability. As a result, all used glass and plastic bottles are recovered and reused easily with very little waste caused in the whole process. In this case, sustainable package design makes the formerly short-lived package products extend their life cycles and realize the sustainability of package products.

3.4 Principle of Recycle

Recycle, otherwise termed resurrection, refers to the process in which materials of used products are recovered and reused to make new products[5]. The principle of recycle in sustainable package attaches importance to the maximized exploitation of discarded wastes in order to make good use of materials of discarded products. The principle of recycle can be regarded as a method of procedure which aims at extending the life circle of a package product and its service. In other words, while preventing a product from being discarded as a piece of garbage, the principle of recycle encourages the maximized use of the product to serve as many purposes as possible. As an intermediate link, products connect consumers and suppliers and both should adopt the principle of recycle to serve their different purposes. Suppliers shall seek in-depth cooperation and elaborated division of labor to extend the cycle of product-garbage transformation and achieve the optimized efficiency of resource. For consumers, the principle of recycle means methods of procedure shall be adopted to maximize the use of products and extend their life cycle, which in turn helps consumers to develop a healthy and sustainable mode of consumption. In order to establish a benign closed circle within a living system, sustainable package design must take into consideration the relationship between the whole process of production and the optimized efficiency of material energy. The theory of closed circle in ecological system lays out the logical foundation of the principle of recycle, which turns out to be the groundwork for sustainable design.

4 The End

Being a new approach to thinking which takes Scientific Outlook on Development as guideline, sustainable package design attaches more importance to the people-oriented values and adopts a holistic view which puts more weight in energy conservation and environment protection, making steady contributions to ensure a coordinated and sustainable development of the nation. The introduction of sustainable concepts in the field of package opens up a new approach to green thinking for designers and paves the way for its future development. For package designers, there is still a long way to go when it comes to the refinement of the technology and aesthetic ideology of sustainable package design. It is universally believed that sustainable package design, reusable packing materials, embodiment of people-oriented values in package design and high-tech packing technology will become the main stream in packaging industry in years to come.

References

1. Shi, J.: An Analysis on Green Packaging. Packaging Engineering 26(5), 238 (2005)
2. He, K.: On Sustainable Development and Its Influence on Green Packaging Design. Art Panorama 08, 122 (2009)
3. Liu, J.: A Introduction to Design of Art, pp. 126–127. Tsinghua University Press (2004)
4. Denison, E.: Packaging Prototypes 3: Thinking Green, pp. 8–42. Shanghai People's Fine Arts Publishing House (2004)
5. Huang, H.: Principles of Design, p. 199. Southeast University Press (2005)

Variable Sparse Multiple Kernels Learning for Novelty Detection

Xiankai Chen, YingDong Ma, Liu Chang, and George Chen

Center for Digital Media Computing
Shenzhen Institutes of Advanced Technology Chinese Academy of Sciences
PO Box 158055, Shenzhen, China
xiankaichen@gmail.com

Abstract. Novelty detection from multiple information sources is an important problem and selecting appropriate features is a crucial step for solving this problem. In this paper, we propose a novel data domain description algorithm which is inspired by multiple kernel learning and elastic-net-type constrain on the kernel weight. Most Multiple kernel learning algorithms employ the 1-norm constraints on the kernel combination weights, which enforce a sparsity solution but maybe lose useful information. In contrast, imposing the p-norm(p>1) constraint on the kernel weights will keep all the information in the base kernels, which lead to non-sparse solutions and brings the risk of being sensitive to noise and incorporating redundant information. To address this problem, we introduce an elastic-net-type constrain on the kernel weights. It finds the best trade-off between sparsity and accuracy. Furthermore, our algorithm facilitates the grouping effect. The proposed algorithm can be equivalently formalized as a convex-concave problem that can be effectively resolved with level method. Experimental results show that the proposed algorithm converges rapidly and demonstrate its efficiency comparing to other data description algorithms.

Keywords: Multiple kernel learning, Data description, Level method, Sparsity.

1 Introduction

Data domain description refers to a description of a set of data points, this description covers the class of objects represented by the training set, and rejects all other possible data points in rest space. Data domain description is usually applied to anomaly detection, novelty detection and function estimation. It is also called one-class problem.[1]

Many methods have been proposed for data description, including statistical law for the outlying patterns [2], Parzen density estimation, and a mixture of Gaussians on the normal class[3]. The main disadvantage is that it requires a large dataset and uses high dimensional feature vectors.

Support vector domain descriptions(SVDD) is one of the few methods that can actually handle example outlier objects[4]. It discriminates a set of target class from outlier class. The sphere is put such that the volume of the sphere is minimized, while

G. Lee (Ed.): Advances in Automation and Robotics, Vol. 1, LNEE 122, pp. 115–124.
springerlink.com

still keeping all (or most of the) target class inside. Another data description algorithm, SVND, is a one-class classification, which try to estimate a function so that its value is positive on training set and negative on other sets [5], It has been applied in many fields and applications.

Multiple kernel learning (MKL) is one of the primary methods to learn from multiple sources. MKL is a hot topic in machine learning [1, 6-10]. It has been considered as a promising technique for identifying the combination of data sources or feature subsets and has been applied in a number of applications [11-16].

MKL aims at finding a best combination of base kernel functions or matrices so that the generalized performance can be reached. Typical measures studied for multiple kernel learning include maximum margin classification errors [1, 6, 17], kernel target alignment [18], Fisher discriminative analysis [19], and estimated error bounds [20, 21].

Most researches on MKL focus on four directions. The first is faster the implementation of MKL algorithm [22-24]. The second direction is to improve the accuracy of MKL by exploring the possible combination of base kernels[9, 25]. The third direction is to expand the idea of MKL to other kernel algorithms[26]. The fourth direction is to find more applications for MKL[13, 15].

In this paper we focus on the third direction. We will expand the idea of MKL to support vector novelty detection algorithm and then propose a new support vector novelty detection algorithm with multiple kernels. First of all we construct a new Reproduce Kernel Hilbert Space (RKHS) which is embedded by many other Hilbert spaces. The samples are mapped to these RKHSs. Then an objective function is given, so that we can learn the optimal solution from multiple Hilbert spaces. As we know that traditional MKL, such as l-norm, does not make sense when it is extend to SVDD and SVND. If we select Gaussian kernel function as base kernel, the biggest bound width is always select. One Improvement is to employ the non sparse combination between kernels [26]. It causes MKL lose sparsity and may mix noise kernels to model. Another improvement is to find the best tradeoff between accuracy and sparsity, which combines l1-norm and l2 norm regularization items. Although it can be used to control the sparsity of kernel weight, implementation of this method is more complex. In order to handle these problem, we propose sparse multiple kernel learning for SVND. We construct the constraints of kernel weight by combining l1-norm and l2 norm regularization items. The weight between l1-norm and l2 norm can be controlled by a tunable parameter, then solve the objective function through two step way by employing level method. That is, update the kernel weight by level method and dual variables by solving a SVND with single kernel.

The rest of this paper is organized as follows. Section 2 presents the formulation of our algorithm, which is followed by the optimization method, the level method. Section 3 shows the experimental results and section 4 concludes this paper.

2 Variable Sparsity Multiple Kernel Learning for SVND

Before present the formulation with multiple kernels, we introduce a definition which is similar to the one presented in [1]. Consider M kernel matrixes are combined,

$\mathcal{H}_m$ corresponds to the m_{th} reproducing kernel feature space with the kernel $K = K_m(x, y)$. Define a new Hilbert space $\mathcal{H}$:

$$\mathcal{H} = \bigoplus_{m=1}^{M} \mathcal{H}_m^{'} \tag{1}$$

Where

$$\mathcal{H}_m^{'} = \left\{ f \in \mathcal{H}_m : \frac{\|f\|_{\mathcal{H}_m}}{d_m} < \infty \right\} \tag{2}$$

For convenience, we say $\|f\|_{\mathcal{H}_m} = 0$, if $d_m = 0$ Given $f, g \in \mathcal{H}_m^{'}$, define the inner product for the m_{th} Hilbert space $\mathcal{H}_m^{'}$:

$$\langle f, g \rangle_{\mathcal{H}_m^{'}} = \frac{1}{d_m} \langle f, g \rangle_{\mathcal{H}_m} \tag{3}$$

It can be easy to prove that $\mathcal{H}_m^{'}$ is a reproducing kernel feature space

$$f(x) = \langle f(\cdot), d_m K_m(x, \cdot) \rangle \forall f \in \mathcal{H}_m^{'} \subseteq \mathcal{H}_m \tag{4}$$

After constructing the kernel feature space, we get:

$$K(x_i, x_j) = \sum_{m=1}^{M} d_m K_m(x_i, x_j) \tag{5}$$

$$\|f\|_{\mathcal{H}_m}^2 = \sum_{m=1}^{M} \frac{1}{d_m} \|f_m\|_{\mathcal{H}_m}^2 \tag{6}$$

Consider the following constraint:

$$\mathcal{D} = \left\{ \mathbf{d} \in \mathbb{R}^+ : p\|\mathbf{d}\|_1 + (1-p)\|\mathbf{d}\|_2^2 \le 1 \right\} \tag{7}$$

where the adjustable parameters $p \in [0,1]$ balance between two terms in constrain. It is used to control the sparsity of kernel weight $\mathbf{d}$. Especially, if $p = 1$, we call it l1-norm of kernel weights. If $p = 0$, we call it l2-norm of kernel weight. Therefore, the primal problem of SVND with multiple kernels is shown as following:

$$\begin{aligned} &\min_{\mathbf{d}, R, \{\xi_i\}, c} -\frac{1}{2} \sum_{m=1}^{M} \frac{1}{d_m} \|f_m\|_{H_m}^2 + \frac{1}{vl} \sum_{i=1}^{l} \xi_i - b \\ &s.t. \sum_{m=1}^{M} f_m(x_i) \ge b - \xi_i \ , \forall i \\ &\xi_i \ge 0, \forall i \\ &p\|\mathbf{d}\|_1 + (1-p)\|\mathbf{d}\|_2^2 \le 1 \\ &\mathbf{d} \ge \mathbf{0} \end{aligned} \tag{8}$$

Note that when $p = 1$, (8) is l1-norm constraint on the vector $\mathbf{d}$ which is a sparsely constraint forcing some d_m to be zero and encouraging sparse basis kernel expansions. Unfortunately, it tends to choose one kernel for learning, gauss kernel with biggest width parameter. The same problem occurs in [27]. When $p = 0$, (8) leads to a non-sparsity solution for $\mathbf{d}$. It may integrate all kinds of kernel information. This causes the risk of being sensitive to noise and incorporating redundant information. To tackle this problem, we employ tunable parameter p which has benefit from the sparsity property and also facilitates the grouping effect.

According to (8), the formulation of variable sparsity multiple kernel learning for SVND can be obtained by replacing $J(\mathbf{d})$ as SVND has single kernel:

$$\min_{\mathbf{d}} J(\mathbf{d}) \quad s.t. \quad \mathbf{d} \in \mathcal{D}$$

$$J(\mathbf{d}) = \begin{cases} \max\limits_{f,b,\xi} -\frac{1}{2}\sum_{m=1}^{M}\frac{1}{d_m}\|f_m\|_{\mathcal{H}_m}^2 + \frac{1}{vl}\sum_{i=1}^{l}\xi_i - b \\ s.t. \ \sum_{m=1}^{M} f_m(x_i) \geq b - \xi_i \ , \forall i \\ \xi_i \geq 0 \ , \forall i \end{cases} \tag{9}$$

The Lagrange of problem $J(\mathbf{d})$ is

$$\mathcal{L} = -\frac{1}{2}\sum_{m=1}^{M}\frac{1}{d_m}\|f_m\|_{\mathcal{H}_m}^2 + \frac{1}{vl}\sum_{i=1}^{l}\xi_i - b - \sum_{i=1}^{l}\alpha_i\left(\sum_{m=1}^{M} f_m(x_i) - b + \xi_i\right) - \sum_{i=1}^{l}\beta_i\xi_i \tag{10}$$

where $\alpha_i, \beta_i \geq 0$ are the Lagrange multipliers of the constraints related to the usual support vector domain description problem. When setting the gradient of (10) to zero with respect to the primal variables, we get the follows:

$$\frac{\partial \mathcal{L}}{\partial f_m} : \frac{f_m}{d_m} = \sum_{i=1}^{l}\alpha_i K_m(x_i, \bullet) \tag{11}$$

$$\frac{\partial \mathcal{L}}{\partial \xi_i} : \alpha_i + \beta_i = \frac{1}{l} \tag{12}$$

$$\frac{\partial \mathcal{L}}{\partial b} : \sum_{i=1}^{l}\alpha_i + 1 = 0 \tag{13}$$

From (11), (12) and (13), we can derive the dual form of $J(\mathbf{d})$, then (9) can be rewritten as:

$$\min_{\alpha} I(\mathbf{d}, \boldsymbol{\alpha}^*) \quad s.t. \quad \mathbf{d} \in \mathcal{D}$$

$$I(\mathbf{d}, \boldsymbol{\alpha}^*) = \begin{cases} \max\limits_{\boldsymbol{\alpha}} -\frac{1}{2}\sum_{i=1}^{l}\alpha_i\alpha_j\sum_{m=1}^{M} d_m K_m(x_i, x_j) \\ s.t. \ 0 \leq \alpha_i \leq \frac{1}{vl} \ , \forall i \\ \sum_{i=q}^{l}\alpha_i = 1 \end{cases} \tag{14}$$

where $\boldsymbol{\alpha} = (\alpha_1, \alpha_2, ..., \alpha_l)^{\mathrm{T}}$ is Lagrange multiplier. Then the decision plane can be stated as:

$$g(x) = \operatorname{sgn}\left(\sum_{i=1}^{l}\alpha_i\sum_{m=1}^{M} d_m K_m(x_i, x) + b^*\right) \tag{15}$$

We can solve (14) under a unified framework, as shown below:

$$\min_{\mathbf{d}} I(\mathbf{d}, \boldsymbol{\alpha}^*) \quad s.t. \quad \mathbf{d} \in \mathcal{D} \tag{16}$$

where $I(\mathbf{d}, \boldsymbol{\alpha}^*)$ is SVND , $\mathcal{D}$ is defined as (7).

Note that (16) is a highly unified framework for multiple kernel learning, almost all the wrapped-based multiple kernel learning are a special case of it. For instance, it can be other algorithms with single kernel, such as support vector machine for classification (binary and multi-class) or regression. $\mathcal{D}$ is the constrain for the kernel weights. Various $\mathcal{D}$ will derive deferent multiple kernel learning algorithms. For example, l_1-norm constrain leads to traditional multiple kernel learning, l_p-norm(p>1) constrain leads to non-sparsity multiple kernel learning.

For the problem of (14), the solvability can be guaranteed by the following theorem:

Theorem 1: Consider the problem of (14), let the kernel $\{K_m\}_{m=1}^{M}$ being positive semi-definite. Then there exist optimal solution $\mathbf{d}^*$, such that it satisfies the condition $p\|\mathbf{d}^*\|_1 + (1-p)\|\mathbf{d}^*\|_2^2 = 1$.

Proof of this theory is similar to the theorem in [28]. Two key points are (a). $I(\mathbf{d},\boldsymbol{\alpha}^*)$ in (14) is an monotonically decreasing function, (b).The constrain $p\|\mathbf{d}\|_1 + (1-p)\|\mathbf{d}\|_2^2 \leq 1$ is a element wise and monotonically increasing function. The problem of (14) can be solved by many methods and the detail of proof is not show here due to the space limitation. The optimal solution of (14) can be solved with the unified framework (16). For example it can be solved by gradient-based techniques and quadratic constrained quadratic programming [25]. In this paper we employ the level method to tackle this problem.

3 Optimization with Level Method

One possible approach for solving problem (14) is the wrapped-based optimization algorithm [1]. One of the method is alternatively algorithm [29]. But it lacks convergence guarantees and may lead to numerical problems. Another approach is the quadratic constrained quadratic programming which is supported by many softwares, such as Mosek [30]. But it is suitable for middle scale problems. The wrapping-based methods have been applied to handle the multiple kernel learning models, e.g SimpleMKL [1], SILP[7], level method[23]. Its main idea is to update the kernel weight and dual variables or primal variables alternatively. Some results show that level method is efficient in machine learning. It has been successfully applied to solve multiple kernel learning problems. In this paper, we use the level method to solve this problem.

Level method contains three steps. Firstly, construct the corresponding lower bound and upper bound of the object function. Secondly, construct the level set. Thirdly, update the kernel weight.

After employing the level set conditions, on one hand, it can be solved in a faster way, since the solution space is smaller. On the other hand, it makes the kernel weight more stable. The convergence rate of level method is $O(1/\varepsilon^2)$. But a better rate $O(l\log(1/\varepsilon))$ can be obtained in practice as discussed in [28].

4 Numerical Experiment

In this section we study properties of the proposed sparse multiple kernel learning for SVND. We experiment on synthetic dataset and real world dataset to demonstrate that our algorithm having the character of sparsity, efficiency, and accuracy.

4.1 Experiment Setting

We compare our algorithm on four algorithms: we derive three kinds of MKRs, called L1-MKL when the value of p as 1, L2-MKL when p is 0 and L12-MKL when p is 0.3, respectively. Another algorithm is un-weight MKL(UW-MKL), which combines kernels with the same weight. It can be seen as a single kernel algorithm. The optimization of constructing cutting plane of (24) and updating the kernel weight of (25) are solved by a optimization software, Mosek[30]. The tune parameter μ is set to 0.9 initially and increased to 0.99 when $\Delta' / M' < 0.01$. All MKL models use the same SVND solver with default setting. UW-MKL is solved by standard SVND with single kernel. The algorithm achieves the stop criteria if the max number of iteration ~~is~~ exceeds 100, or if the gap between upper bound and lower bound is lower than 0.001. The information of dataset is shown in Table 1.

Table 1. Information of datasets: c is the number of class, d is the sample dimension, k is the number of used kernels, s is the sample size

Dataset	c	d	k	s
Syn 1	2	20	41	300
Syn 2	2	20	41	300

4.2 Synthetic Experiment

In this section, we discuss the characters of our algorithm on two synthetic datasets, which are also used by Haiqin [28]. We do not show how to generate the synthetic data, more detail information can be found in [28]. But the different is that x is normal distribution in $[0,1]^{N\times 20}$. After that, the positive class contains more samples, than the negative class. We generate 300 samples on two synthetic datasets. We call two synthetic datasets as Synthetic Dataset1 and Synthetic Dataset2, respectively. We randomly extract 100 samples from the positive class and the rest are used as testing samples. All experiments on two datasets run 20 times.

We adopt the following settings which are similar to [1] to construct the base kernel matrices, i.e.,

Gaussian kernels with 41 different widths (0.01 to 10) on all features and on each single feature. Figure 1 shows the value of kernel weights learned by the L12-MKL. The results show that our algorithm can change the sparsity of kernel weight by tuning parameter p. When p increasing, the kernel weight becomes non-spare. On the contrary, the kernel weight becomes spare, e.g. L12-MKL keeps the sparsity and useful kernels as shown in (a) and (e). L1-MKL keeps the sparsity but lose some useful kernels, which are illustrated in (b) and (f). L2-MKL keeps all useful kernels but lose sparsity see (c) and (g). UW-MKL keeps all kernels with the same kernel weight.

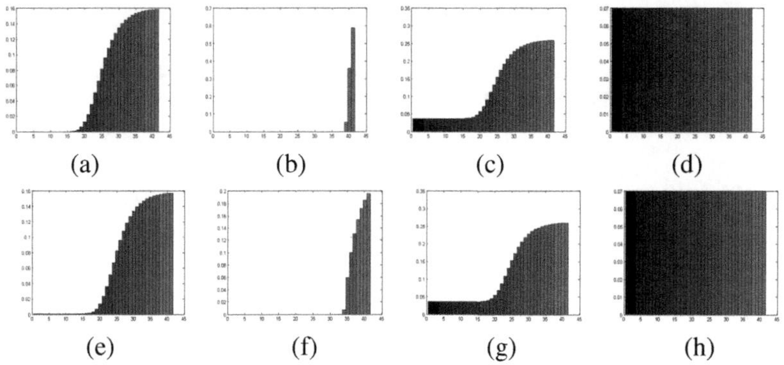

Fig. 1. The value of kernel weights learned by the L12-MKL, the L1-MKL, the L2-MKL, and the UW-MKL on syn 1 and syn 2. (a). L12-MKL on the synthetic data1; (b).L1-MKL on the synthetic data1; (c).L2-MKL on the synthetic data1; (d).UW-MKL on the synthetic data1; (e).L12-MKL on the synthetic data2; (f).L1-MKL on the synthetic data2; (g).L2-MKL on the synthetic data2; (h).UW-MKL on the synthetic data2.

Table 1 shows the testing performance of deferent methods on the syn 1 and syn 2 datasets. Firstly, the results indicate that L1-MKL need more training time than any other algorithms. L2-MKL need the least training time. Note in these experiments the training time will increase if p is larger. Secondly, we found that L12-MKL achieves the best accuracy. L1-MKL achieves less accuracy because some useful information is lost. L2-MKL achieves worse accuracy due to some mixed redundant kernls. UW-MKL achieves the worst accuracy which shows that we need to select appropriate kernels for training model. Thirdly, L1-MKL finds the least kernels. L2-MKL selects all kernels for training. L12-MKL selects more kernel than L1-MKL but less than L2-MKL. According to these observations, L12-MKL finds the best sparsity and accuracy because it achieves sparsity and also keeps the sparsity of kernels weight.

Figure 2 shows the number of learned kernels and the accuracy achieved by the proposed L12-MKL, Figure 2.(a) indicates that the number of selected kernels will decrease if tunable parameter v increasing. Figure 2.(b) indicates that the best accuracy for all dataset is achieved when the value of v is 0.3. There is no significant change when v more then 0.3. The accuracy increases when the value of v is less than 0.3.

Figure 3 shows the convergence rate of our algorithms. Figure 3.(a) indicates that the convergence rate is stable and the upper bound changes faster than the lower bounds. Figure 3.(b) indicates that if the algorithm iterates more times, the max kernel weight becomes larger.

According to the synthetic experiments, we obtain the following conclusions. Firstly, our algorithm can control the sparsity of kernel weight. Secondly, while data contains noise, which comes from normal distribution, L12-MKL will outperform L1-MKL and L2-MKL. L12-MKL can find the best trade-off between accuracy and sparsity of kernel weight. Thirdly, our algorithm is stable.

Table 2. The testing performances of different methods on the synthetic dataset1 and the synthetic dataset2.

Dataset	Algorithm	Times	Accuracy	Kernels
Syn 1	GMKL	1.30 ± 0.10	**78.70 ± 2.50**	24.80 ± 0.40
	L1-MKL	1.80 ± 0.10	76.30 ± 2.80	8.00 ± 0.00
	L2-MKL	0.90 ± 0.10	69.70 ± 3.30	41.00 ± 0.00
	UW-MKL	0.10 ± 0.00	56.30 ± 4.50	41.00 ± 0.00
Syn 2	GMKL	1.40 ± 0.00	**81.40 ± 2.00**	24.90 ± 0.30
	L1-MKL	1.80 ± 0.10	78.20 ± 2.30	8.00 ± 0.00
	L2-MKL	1.00 ± 0.10	72.10 ± 2.50	41.00 ± 0.00
	UW-MKL	0.10 ± 0.00	59.30 ± 3.40	41.00 ± 0.00

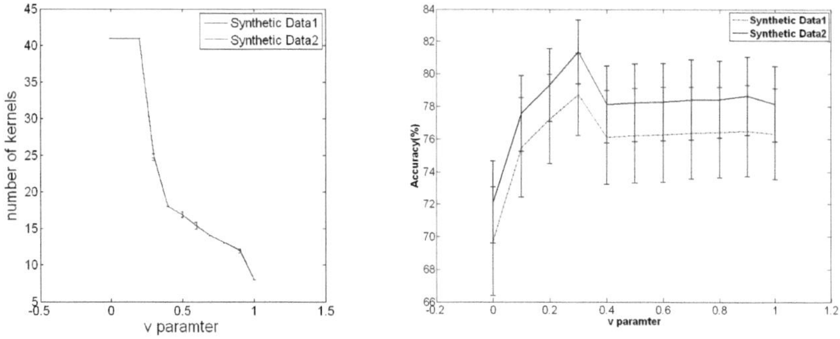

(a) (b)

Fig. 2. The number of learned kernel and the accuracy of our algorithm on the synthetic data1 and the synthetic data2. (a). the number of kernels. (b). the accuracy. The best accuracy for all datasets is achieved while v=0.3. The number of learned kernels will decrease when v increase.

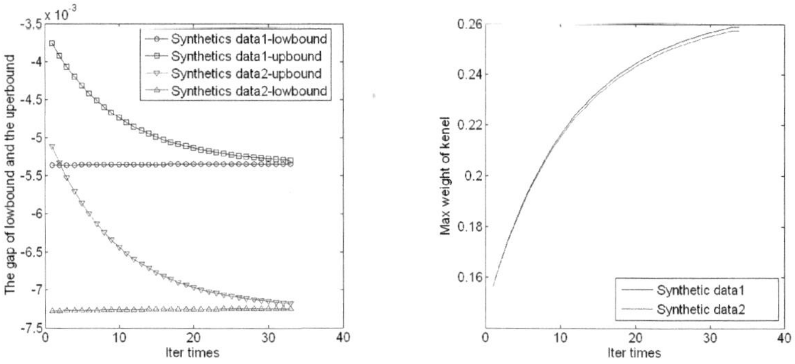

(a) (b)

Fig. 3. Evolution of the object value and the variant of weights. (a) evolution of the gap between upper bound and lower bound; (b) evolution of the variant of max weight of kernels.

5 Conclusion

In this paper, we propose a novel multiple kernel learning for data domain description. The algorithm can control the sparsity of kernel weight and then find the best trade-off between accuracy and sparsity. We introduce an elastic-net-type constrain on the kernel weights, design the constraints of kernel weight by combining the l1-norm and l2-norm regularization items. The experimental results show that our algorithms outperform most up-to-date algorithms.

There are several future works. Firstly, design an appropriate formulation with multiple kernels for SVDD algorithm. Secondly, extend our algorithm to include the UW-MKL as a special case. Thirdly, the selected kernels of our algorithm can be further improved by finding more appropriate formulation for SVND.

References

1. Rakotomamonjy, A., et al.: SimpleMKL. Journal of Machine Learning Research 9, 2491–2521 (2008)
2. Ritter, G., Gallegos, M.T.: Outliers in statistical pattern recognition and an application to automatic chromosome classification. Pattern Recognition Letters 18(6), 525–539 (1997)
3. Tarassenko, L., et al.: Novelty detection for the identification of masses in mammograms. In: Proceedings of the 4th IEE International Conference on Artificial Neural Networks 1995, pp. 442–447 (1995)
4. Tax, D.M.J., Duin, R.P.W.: Support vector domain description. Pattern Recognition Letters 20 (1999)
5. Scholkopf, B., et al.: Support Vector Method for Novelty Detection. In: Advances in Neural Information Processing Systems (December 2000)
6. Lanckriet, G.R.G., et al.: Learning the kernel matrix with semidefinite programming. Journal of Machine Learning Research 5, 27–72 (2004)
7. Sonnenburg, S., et al.: Large scale multiple kernel learning. Journal of Machine Learning Research 7, 1531–1565 (2006)
8. Aflalo, J., et al.: Variable Sparsity Kernel Learning. Journal of Machine Learning Research 12, 565–592 (2011)
9. Kloft, M., et al.: l(p)-Norm Multiple Kernel Learning. Journal of Machine Learning Research 12, 953–997 (2011)
10. Mehmet, G., Ethem, A.: Multiple Kernel Learning Algorithms. Journal of Machine Learning Research 12, 2211–2268 (2011)
11. Lampert, C.H., Blaschko, M.B.: A multiple kernel learning approach to joint multi-class object detection. In: Annual Symposium of the Deutsche-Arbeitsgemeinschaft-fur-Mustererkennung (DAGM), Munich, Germany (2008)
12. Longworth, C., Gales, M.J.F., IEEE: Multiple Kernel Learning for speaker verification. In: 33rd IEEE International Conference on Acoustics, Speech and Signal Processing, Las Vegas, NV (2008)
13. Qiu, S., Lane, T.: Multiple Kernel Support Vector Regression for siRNA Efficacy Prediction. In: Măndoiu, I., Wang, S.-L., Zelikovsky, A. (eds.) ISBRA 2008. LNCS (LNBI), vol. 4983, pp. 367–378. Springer, Heidelberg (2008)
14. Lin, Y.Y., Liu, T.L., Fuh, C.S.: Multiple Kernel Learning for Dimensionality Reduction. IEEE Transactions on Pattern Analysis and Machine Intelligence 33(6), 1147–1160 (2011)

15. Yeh, C.Y., Huang, C.W., Lee, S.J.: A multiple-kernel support vector regression approach for stock market price forecasting. Expert Systems with Applications 38(3), 2177–2186 (2011)
16. Suard, F., et al.: Model selection in pedestrian detection using multiple kernel learning. In: IEEE Intelligent Vehicles Symposium, Istanbul, Turkey (2007)
17. Bach, F.R.: Consistency of the group Lasso and multiple kernel learning. Journal of Machine Learning Research 9, 1179–1225 (2008)
18. Cristianini, N., et al.: On kernel-target alignment. In: Advances in Neural Information Processing Systems 14, vol. 1,2, pp. 367–373 (2002)
19. Ye, J.P., Ji, S.W., Chen, J.H.: Multi-class discriminant kernel learning via convex programming. Journal of Machine Learning Research 9, 719–758 (2008)
20. Chapelle, O., et al.: Choosing multiple parameters for support vector machines. Machine Learning 46(1-3), 131–159 (2002)
21. Srebro, N., Ben-David, S.: Learning Bounds for Support Vector Machines with Learned Kernels. In: Lugosi, G., Simon, H.U. (eds.) COLT 2006. LNCS (LNAI), vol. 4005, pp. 169–183. Springer, Heidelberg (2006)
22. Bach, F.R., Lanckriet, G.R.G., Jordan, M.I.: Multiple kernel learning, conic duality, and the SMO algorithm. In: Proceedings of the Twenty-first International Conference on Machine Learning, pp. 6–13. ACM, Banff (2004)
23. Xu, Z., et al.: An extended level method for efficient multiple kernel learning. In: Advances in Neural Information Processing Systems, vol. 21, pp. 1825–1832 (2009)
24. Orabona, F., Jie, L.: Ultra-Fast Optimization Algorithm for Sparse Multi Kernel Learning. In: Proceedings of the 28th International Conference on Machine Learning, Bellevue, Washington (2011)
25. Kloft, M., Brefeld, U., Laskov, P.: Non-sparse multiple kernel learning. In: NIPS Workshop on Kernel Learning: Automatic Selection of Optimal Kernels (2008)
26. Kloft, M., Nakajima, S., Brefeld, U.: Feature Selection for Density Level-Sets. In: Buntine, W., Grobelnik, M., Mladenić, D., Shawe-Taylor, J. (eds.) ECML PKDD 2009. LNCS, vol. 5781, pp. 692–704. Springer, Heidelberg (2009)
27. Shawe-Taylor, J., Hussain, Z.: Kernel learning for novelty detection. In: Proceedings of the NIPS Workshop on Kernel Learning 2008 (2008)
28. Haiqin, Y., et al.: Efficient Sparse Generalized Multiple Kernel Learning. IEEE Transactions on Neural Networks 22(3), 433–446 (2011)
29. Grandvalet, Y.: Least absolute shrinkage is equivalent to quadratic penalization. In: Niklasson, L., Bod´en, M., Ziemske, T. (eds.) Proceedings of the 8th International Conference on Artificial Neural Networks, pp. 201–206 (1998)
30. Mosek. The MOSEK Optimization Software (2011), `http://www.mosek.com/index.php?id=2` (cited 2011)

An Approach for Determining the Next Best View Based on Voxel Map and Simplex Method

Shihui Zhang[1,2] and Xiaowei Qu[2]

[1] State Key Laboratory of Robotics and System (HIT), Harbin 150080, China
[2] College of Information Science and Engineering, Yanshan University, Qinhuangdao 066004, China
sshhzz@ysu.edu.cn

Abstract. Aiming at the problem of the next best view, an approach for determining the next best view is proposed by using voxel map and simplex method. Firstly, a range image of the scene is obtained, then a voxel map is created according to the range data and related voxel information is obtained. Secondly, an objective function which considers area, quality, overlap, occlusion, navigation and other factors is defined, and the problem of the next best view is abstracted to the optimization solving problem of objective function. Finally, the simplex method is applied to optimize the objective function. The direction and position corresponding to the optimal solution of the objective function is the next best view. Experimental results show that the proposed approach is feasible and effective.

Keywords: range image, voxel map, next best view, objective function, simplex method.

1 Introduction

Next best view (NBV) planning is always an important issue in the fields of three-dimensional reconstruction, machine navigation, object recognition and others. It means to determine the next sensor direction and position which will extract the greatest amount of unknown scene information according to the current viewing information, thus the complete information of the object surface can be obtained at the fewest viewpoints. Various NBV planning approaches have been proposed by researchers. For example, a combination approach for determining the NBV was proposed by Banta [1], but the method considered the occlusion factor only and ignored the influence of other factors. Chen [2] presented an approach for automatic sensor placement for model-based stereo vision system, which set many constrains for images and developed a viewpoint planner to generate the sensor placement plan, but with large calculation amount and long time consuming. Li [3] presented an information entropy-based viewpoint planning approach, in which the next best view was the viewpoint that contained maximal information about the object. But it was not suitable for the models with complex surface, especially the models with holes. Scott [4] presented a multi-phase, model-based and modified measurability matrix view planning approach, which needed to obtain the prior knowledge about scene and did not

G. Lee (Ed.): Advances in Automation and Robotics, Vol. 1, LNEE 122, pp. 125–132.
springerlink.com

consider overlap factor. He [5] presented a new automatic view planning approach, which incorporated the visual region of monocular laser vision system with the limit surface of unknown model, but without considering navigation factor.

Taking those considerations and the actual need of research into account, we propose a next best view determining approach which is based on voxel map and applies the simplex method to optimize the objective function, considering area, quality, overlap, occlusion, navigation and other factors. The approach can obtain the 3-dimensional information of object surface at fewer viewpoints and acceptable quality, and is suitable for the objects with complex shape. Experimental results show that the proposed approach is feasible and effective.

2 Determining Next Best View

2.1 The Approach Overview

The proposed approach is based on the scene information obtained at the current sensor placement and determines the next best view by some standards. The objective is to obtain the greatest amount of the scene information that can not be obtained at the current sensor placement. Common scene representation methods contain mesh, occupancy, voxel map and so on. Because the voxel map can be placed anywhere in space and every voxel can store more related information, the voxel map is applied to represent the scene. The idea of this paper is to obtain the range image of the scene at the current sensor placement; create voxel map according to the range data and obtain the related voxel information; define the objective function according to the voxel information and abstract the NBV problem to objective function optimization solving problem; optimize the objective function with the simplex method. Finally, the sensor placement corresponding to the optimal solution of the objective function just is the next best view. Details will be described below.

2.2 Obtaining Voxel Information

Firstly, a range image of the scene is obtained at the current sensor placement. Then the coordinates x and y of the range image data are taken as the coordinates x and y of the voxel map, and the depths of the range image data are taken as the coordinate z of the voxel map. If z is 0, then the voxel is marked as empty, else marked as occupied. Shoot a ray from each occupied voxel in the direction pointing away from the viewpoint and mark all the voxels encountered on the way as unseen. Meanwhile, mark the voxel adjacent with any of its six faces to an empty voxel as occluded. And count the number of each kind of voxels when marking them. Each voxel stores information by means of three attributes: a label indicating its type, a local surface normal corresponding to the range data point and a quality value. The surface normal is calculated by plane fitting method and the quality value (between 0 and 1) reflects the confidence of voxels. The quality of a voxel is the cosine of the angle formed between the surface normal and the sensor optical axis. Additionally, the surface normal and quality value are only defined for occupied voxels. The amount of voxels, the surface normal and a quality value will be used to define the objective function related with the next best view.

2.3 The Definition of Objective Function

The next best view problem is abstracted to objective function optimization solving problem. The basic idea is depending on the voxel information obtained at the current sensor placement to define the following objective function, which comprehensively considers the area factor, quality factor, occlusion factor [6], navigation factor [7] and the overlap structure factor of the scene voxel map obtained at the next view, in order to evaluate the next view.

$$f = f_{area} * (1 + f_{quality} + f_{occlusion} + f_{navigation} + f_{structure}) \ . \tag{1}$$

f_{area} represents area factor which provides the area information of unseen and overlap region at the next view. And the value of f_{area} is the sum of percentage function of occluded voxels (corresponding to unseen region) and occupied voxels (corresponding to overlap region). Both calculation methods are following.

$$f(x,\alpha) = \begin{cases} f_1(x,\alpha) = -\dfrac{2}{\alpha^3}x^3 + \dfrac{3}{\alpha^2}x^2, x \le \alpha \\ f_2(x,\alpha) = -\dfrac{2}{(\alpha-1)^3}x^3 + \dfrac{3(\alpha+1)}{(\alpha-1)^3}x^2 - \dfrac{6\alpha}{(\alpha-1)^3}x + \dfrac{3\alpha-1}{(\alpha-1)^3}, x > \alpha \end{cases} \ . \tag{2}$$

Where x is the percentage of occluded voxels or occupied voxels. The percentage of a voxel type is calculated as the amount of voxels of this type divided by total number of voxels in the ray tracing, and empty voxels are not considered. α is the optimal percentage for occluded voxels or occupied voxels. Generally, α =0.8 for occluded voxels and α =0.2 for occupied voxels.

$f_{quality}$, whose value is the average of quality values for all occupied voxels, reflects the confidence of occupied voxels. And the value is the closer to 1, the better.

$f_{occlusion}$, the occlusion factor, aims to solve occluded areas. Its value is the ratio of the occluded voxels amount and the range image resolution, and it reflects the visibility of scene.

$f_{navigation}$ aims to evaluate the navigation distance between the current viewpoint and the next view. Suppose x is the orthodromic distance defined as the shortest distance between two points on a sphere ($x \in [0,1]$); ρ is a minimum value of the next view with the largest distance. $\rho \in [0,1)$ and ρ =0.2 generally. Then

$$f_{navigation} = (\rho - 1)x^2 + 1 \ . \tag{3}$$

$f_{structure}$, the structure factor of overlap region, aims to favor the sensing of areas where the surface has non-degenerate shape, thus easing the registration. The non-degenerate region means the region has a nonzero area and can not degenerate to point or line. $f_{structure}$ is calculated by the variance of the surface normals at the new sensed points. It reflects the deviation the surface normals of the new sensed voxels away from the overlapping region surface normal at the current view, namely it reflects the similarity of overlapping region. Suppose $norm_i$ is the surface normal of the new sensed occupied voxel i at the next view; n is the number of the new sensed occupied voxels at the next view; $norm$ is the surface normal of overlap region at the current view, whose value is the average of the occupied voxels surface normals at the current view. Then

$$f_{structure} = \frac{1}{n}\sum_{i=1}^{n}(norm_i - norm)^2 \quad . \tag{4}$$

The objective function is a combination of the previous factors. Because the main objective of the view planner is to obtain the greatest amount of scene information that can not be obtained at the current view, this combination considers the area factor as a basic factor and the others as secondary factors, thus the objective function is defined as formula (1).

2.4 The Optimization of the Objective Function

The information obtained at the current view is applied to initialize the objective function. Depending on the information obtained at the next view, the simplex method is applied to optimize the objective function [8]. A local maximum of the objective function will be obtained by optimization, and the viewpoint corresponding to the value is taken as the next best view. An icosahedron is used to determine the direction of the sensor. The centroids of faces of an icosahedron are taken as the values to test. These directions are evenly distributed around a sphere and always point to the sphere center (approximate to the barycenter of object). Simplex method is used to determine the position of the sensor. A simplex in 3D space is a tetrahedron, in which each vertex corresponds to a sensor position. The vertices of the initial simplex are set as follows: the first vertex is set as the current sensor position, and the other three are set as the points choosing from the centroids of an icosahedron faces randomly. At each vertex, the voxel information of scene is obtained at 20 directions by the approach mentioned in 2.2 section and the values of the objective function at 20 directions are calculated. The maximum of the 20 objective function values will be the value at that position and the direction corresponding to the maximum is the direction of sensor at that position. Based on these, the best vertex, the worst vertex and the two great vertices are determined. Then a simplex evolves in 3D space changing its shape, size and position to obtain large values of the objective function. The optimization approach used in this paper is as follows.

Suppose that the initial simplex BG_1G_2W is shown in Fig.1 (a). Where B is the best vertex, W is the worst vertex, G_1 and G_2 are great vertices, M is the centroid of the best face. The usable operations in the simplex optimization process contain:

(1)Reflection: along the edges in the tetrahedron pointing from W to B, from W to G_1 and from W to G_2, the objective function values increase progressively. So take the triangle that the three vertices B, G_1 and G_2 located as the boundary, and the objective function value of the vertex opposite to W is larger. In order to choose the tested point R, a segment from W to M is drawn, whose length is d. Then the extended line of the segment is drawn from M, and its length is d. Thus the vertex R is obtained, as shown in Fig.1 (b).

(2)Expansion: if the value at R is larger than that at W, the direction of solving is right, and maybe the position of maximum is a little farther than R. So the segment between M and R is extended to E to constitute a new expanded tetrahedron BG_1G_2E. The extended length is d, as shown in Fig.1 (c). If the value at E is larger than that at R, the E is better than R.

(3)1D Contraction: if the values at R and W are equal, another point needs to be tested. Maybe the value at M is larger, but in order to constitute a tetrahedron W can not be replaced by M. Take the midpoints C_1 and C_2 of the two segments WM and WR into account respectively, as shown in Fig.1 (d). The point with larger value is C, and the tetrahedron BG_1G_2C is constituted.

(4)3D Contraction: if the value at C is not larger than that at W, then G_1, G_2 and W will contract towards B, as shown in Fig.1 (e).

The simplex gets smaller progressively as it contracts towards vertices where the objective function value is larger. The simplex optimization is terminated when the range of change of the objective function value between the best and the worst vertices is below a threshold. The viewpoint corresponding to the maximum of the objective function is the next best view.

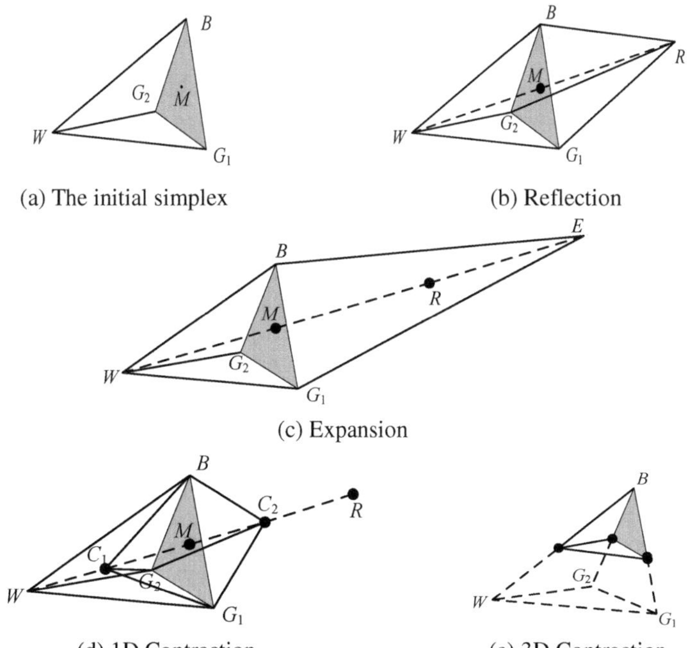

(a) The initial simplex (b) Reflection

(c) Expansion

(d) 1D Contraction (e) 3D Contraction

Fig. 1. Possible operations of simplex method in each iteration

3 Experimental Results

The range images used in this paper are chosen from Stuttgart Range Image Database at http://range.informatik.uni-stuttgar.tde/htdocs/html. The size of voxel map is 400×400×400, and the size of each voxel is 1×1×1. The placement of the sensor is represented as (X, Y, Z, x, y, z), where X, Y, and Z are the coordinates of the sensor position, and (x, y, z) is the unit vector representing the direction of the sensor. The experimental progress is following.

(1)Choose a range image of Bunny from the range image database randomly, as shown in Fig.2. The viewpoint corresponding to the image is (-139.486,-45.742, 100.761,0.783,0.257,-0.566).

(2)Create the voxel map according to the range data, mark the voxels and obtain the related information of each type of voxels. As shown in Fig.3, the amount of occupied voxels (marked as black) is 38362, the amount of occluded voxels (marked as red) is 12244, and the total number of voxels in the ray tracing is 166375.

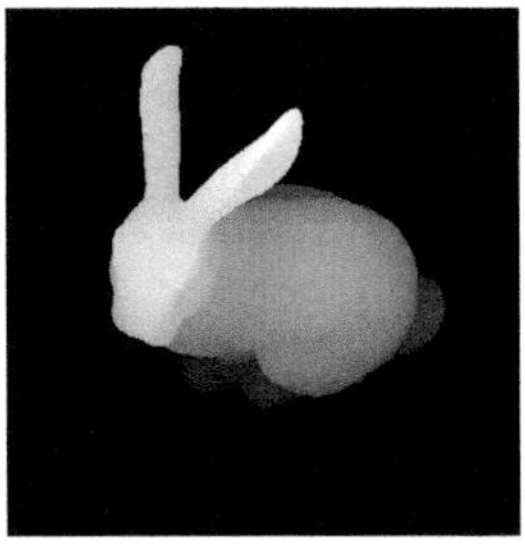

Fig. 2. Bunny's range image at a certain view

Fig. 3. The voxel map

(3)Calculate the value for each factor at the current sensor placement. f_{area} = 0.482, $f_{quality}$ = 0.617, $f_{occlusion}$ = 0.077, $f_{navigation}$ = 1, $f_{structure}$ = 0.052, and the value of objective function is 1.324, which is the initial value for the simplex optimization.

(4)Choose three centroids of faces from an icosahedron randomly as the candidate sensor positions. They are (280.341,59.018,89.027), (28.974,-176.325,240.977) and (79.354,278.381,-78.784), respectively. At each position, the values of the objective function at 20 directions are calculated. The maximum of the 20 objective function values will be the value at that position and the direction corresponding to the maximum is the direction of sensor at that position. The directions are (-0.935,-0.197, -0.297), (-0.097,0.588,-0.803) and (-0.265,-0.928,0.263), respectively.

(5)Evaluate the vertices of the current simplex according to the values calculated above. The best vertex is (79.354,278.381,-78.784,-0.265,-0.928,0.263), the worst vertex is (28.974,-176.325,240.977, -0.097,0.588,-0.803), and the better vertices are the current sensor placement and (280.341,59.018,89.027,-0.935, -0.197,-0.297). Then the simplex changes its shape, size and position to optimize the objective function. Specific changing progress is shown in Table 1.The simplex optimization is terminated when the change value of the objective function between the best and the worst vertices is below the threshold 0.001, namely the simplex contracts to a point. The sensor placement corresponding to the maximum of the objective function (528.341,-91.008,322.385,-0.854,0.166,-0.494) is the next best view.

Table 1. Part of optimization progress for the objective function

Iteration	Objective function maximum	Sensor placement corresponding to maximum	Changing progress of simplex
0	1.324	The current sensor placement	
1	1.390	(79.354,278.381,-78.784, -0.265,-0.928,0.263)	initial simplex
2	1.439	(103.624,368.957,38.445,-0.269,-0.958,-0.099)	expansion
3	1.558	(174.531,326.897,75.183,-0.462,-0.865,-0.199)	expansion
4	1.659	(-53.679,-101.458,-29.147,0.453,0.857,0.246)	reflection
5	1.778	(-89.456,-230.374,-76.125,0.346,0.891,0.294)	expansion
6	2.021	(-124.873,-298.754,-51.079,0.381,0.911,0.156)	expansion
7	2.269	(-63.582,-174.256,46.478,0.333,0.911,-0.243)	reflection
...	...	...	...
61	5.492	(528.356,-90.998.,322.393,-0.845,0.146,-0.515)	reflection
62	5.492	(528.341,-91.008,322.385,-0.854,0.166,-0.494)	3D contraction
63	5.492	(528.341,-91.008,322.385,-0.854,0.166,-0.494)	3D contraction

(6)Move the sensor to the position and direction according to the obtained next best view. The corresponding range image is shown in Fig.4.

Fig. 4. Bunny's range image at the NBV

4 Conclusions

This paper makes full use of the advantages of range image and voxel representation. The next best view problem is abstracted to the objective function optimization solving problem. Additional, an objective function is defined, which considers area, quality, overlap, occlusion, navigation and other factors. The simplex method is applied to optimize the objective function. The next best view is chosen by expansion, reflection and contraction of the simplex. The placement corresponding to the final optimal solution of the objective function is taken as the next best view. Because many factors are considered in the proposed approach, it makes the obtained next best view more reasonable and establishes the foundation for performing the related vision task better.

Acknowledgments. The project is supported by State Key Laboratory of Robotics and System (HIT, Grant No. SKLRS-2010-ZD-08) and Natural Science Foundation of Hebei province(Grant No. F2010001276).

References

1. Banta, C.J., Wong, L.M., Dumont, C., Abidi, M.A.: A Next-best-view System for Autonomous 3-D Object Reconstruction. IEEE Transaction on Systems, Man and Cybernetics, Part A 30(5), 589–598 (2000)
2. Chen, S.Y., Li, Y.F.: Automatic Sensor Placement for Model-based Robot Vision. IEEE Transaction on Systems, Man and Cybernetics, Part B 34(1), 393–408 (2004)
3. Li, Y.F., Liu, Z.G.: Information Entropy—based Viewpoint Planning for 3-D Object Reconstruction. IEEE Transaction on Robotics 21(3), 324–337 (2005)
4. Scott, W.R.: Model-based View Planning. Machine Vision and Applications 20(1), 47–69 (2009)
5. He, B.W., Zhou, X.L.: View Planning Method Based on Visual region of Monocular Vision System. Chinese Journal of Scientific Instrument 31(2), 390–396 (2010)
6. Albalate, M.T.L., Devy, M., Martí, J.M.S.: Perception Planning for an Exploration Task of a 3D Environment. In: Proceedings of International Conference on Pattern Recognition, vol. 16(3), pp. 704–707 (2002)
7. Vasquez-Gomez, J.I., Lopez-Damianand, E., Sucar, L.E.: View Planning for 3D Object Reconstruction. In: International Conference on Intelligent Robot Sand Systems, St. Louis, MO, United States, pp. 4015–4020 (2009)
8. Sanchiz, J.M., Fisher, R.B.: A Next-best-view Algorithm for 3d Scene Recovery with 5 Degrees of Freedom. In: British Machine Vision Conference, Nottingham, UK, pp. 163–172 (1999)

Research of Self-organization Wireless Sensor Network

Yulan He and Jie Yuan*

School of ESE, Nanjing University, China
yuanjie@nju.edu.cn

Abstract. To solve the problem of the limitation of computing, power consumption and network resources that faced by self-organization wireless sensor network, this paper forwards a method of dynamic adjustment of nodes to reach a flexible transformation of the router and endpoint nodes. This algorithm bases on a set of system platform that combined of hardware and software and takes the advantages of self-organization network itself. Experimental results show that the algorithm reduces network node redundancy and node isolation, which reduce the energy loss of network.

Keywords: Wireless sensor networks, Self-organization network, Dynamic nodes adjust.

1 Introduction

Wireless sensor network consists of a large number of micro and inexpensive wireless nodes. Each node is responsible for data acquisition, energy management functions, data transmission, and other functions. With the help of the information including time and location of these nodes, the mutual control of the sensor nodes and data exchange are realized to ensure the accuracy and real-time of the whole detection and control system. Through various integrated micro-sensors, wireless sensor network makes real-time monitoring, collection and perception of information, which then can be wirelessly sent to the user endpoints, to achieve the connectivity of the physical world and human society[1-2].

However, wireless sensor networks faced with the constraints of computing, energy consumption and network resources, and self-organization network is a self-organization and self-healing network without any central entity. It can be used immediately without relying on the existing fixed communication network infrastructures. The network nodes cooperate with each other and via wireless links to communicate and exchange information to achieve the sharing of information and services. When two network nodes cannot directly communicate with each other, they can form a multi-hop mode of communication with the help of other nodes for packet forwarding [3]. The principle of self-organization network that we proposed is to use a new algorithm to solve the problem of network energy loss which is caused by redundancy network nodes and isolated nodes. In addition, the zigbee protocol stack

* Corresponding author.

G. Lee (Ed.): Advances in Automation and Robotics, Vol. 1, LNEE 122, pp. 133–139.
springerlink.com

used in small footprint greatly reduces the cost. Therefore, the research on self-organization wireless network is of great significance.

2 Self-organization Network Theory

In the field of wireless sensor networks, router and endpoint nodes are the most common devices. The key technology is the routing path selection method and the energy loss algorithm which have a decisive impact on the application and development of wireless networks. This paper releases that self-organization wireless network can be formed by the means of dynamic adjustment with the help of software, so that these nodes can achieve type transformation in different conditions, which can reduce the hardware redundancy and network energy consumption, and increase network flexibility [4-7].

The self-organization network theory we proposed mainly depends on the method of dynamic adjustment of nodes, which includes transformation of router and endpoint node.

The transformation procedure of router can be divided into the following steps:

Step 1, create a network through a coordinator, initialize the nodes for routing, and add the described the network that created by the coordinator;

Step 2, determine whether the router can satisfy the state of transformation;

Step 3, if the router meets the condition, there is no new nodes wants to join the router, so the router transforms, otherwise, it is not converted, and exits.

Step 4, initiate the endpoint and request to join the parent node of the network;

The transformation procedure of endpoints includes the following steps:

Step 5, the endpoint nodes determine the status of the parent routing node, and determine whether to transform the node;

Step six, endpoint nodes transform into a router, then add the parent node of the network and create their own endpoint node as described in network.

The condition that a router can transform into an endpoint is:

$$Tw \geq 5\text{s} \quad (1)$$

Where Tw is the time router waits the endpoint nodes to join in.

The condition the endpoint node can satisfy the router is:

$$(N{=}{=}5) \text{ and } (R \geq 1) \quad (2)$$

Where N is the number of child nodes. R is the number of the request to join the network.

3 System Components

3.1 Hardware Platform

Wireless sensor network systems include hardware and software components, in which hardware includes control module, transmitter module, sensor module and the computer.

3.2 Software Platform

The software is formed by basic network protocols and related applications software. In this system adopts the ZigBee protocol stack as the basic network protocol. The physical layer of ZigBee protocol stack and MAC layer use the communications protocol of IEEE 802.15.4, and set a comprehensive specification on network layer and application layer as well as other aspects of information security.

4 System Architecture

4.1 Formation Process of the Network System

The system of the network is of the mesh type and is composed of the coordinator, routers, and endpoint nodes. The coordinator is responsible for creating the network, routers and endpoint join the network and the router recreates its own network to wait for other devices to join in. Figure 1 shows the process to form a specific network. A network has only one coordinator, which is responsible for communication with the computer through the basic network information showed by PC software [8-9].

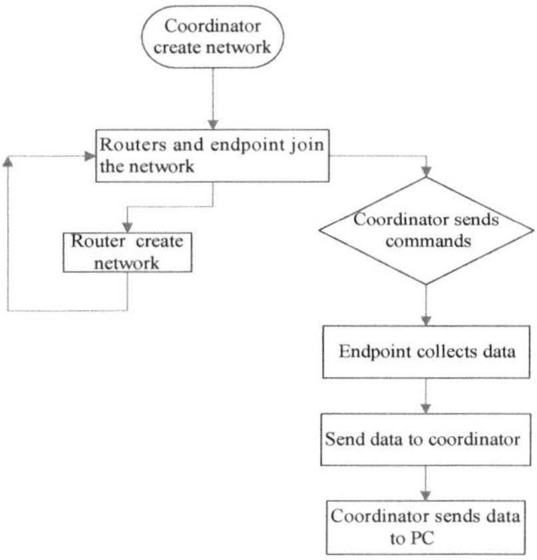

Fig. 1. Flow chart of system network

4.2 Transformation of Router Nodes

The first step is to create an initial network by the coordinator and allow other nodes to join. Endpoint nodes join to form a mesh network through the router. Then set power on the node, initialize the protocol stack, including the initialization of physical layer, MAC layer, network layer and application layer, and the creation of routing tables. Then join the existing network to determine their neighbor table. The router makes the PanID and the short address as the constant identifier of the equipment in the network. The short address comes from the parent node distribution. Then the router creates a network, and set up communication channel of the network, and set PanID and the maximum number of child nodes to five. Then the router waits for other nodes to join in.

The second step, check the number of child nodes in the neighbor table, and within the settled time (such as 5 seconds, can even be set to 0.1 seconds), if the number of child nodes is zero, then prepare to transform the router node into the endpoint node. The node notifies its parent and sibling nodes to modify their nodes content on the neighbor table, including device type, the relation with this device, and the ability of data forwarding and the like.

The third step is to transform the node into the endpoint type, and modify the certain protocol tasks. Mainly modify the network layer and application layer, and ultimately achieve the endpoint function. The conversion process as showed in Figure 2, in this process, it is necessary to delete the routing table, routing discovery table and neighbor table. Parent routing table records the PanID and short address through the long address of its neighbors table to facilitate the use of the following steps. The nodes realize the reset of protocol stack, and make the endpoint node function well through software reset. The use of software reset in this process can better reduce energy consumption.

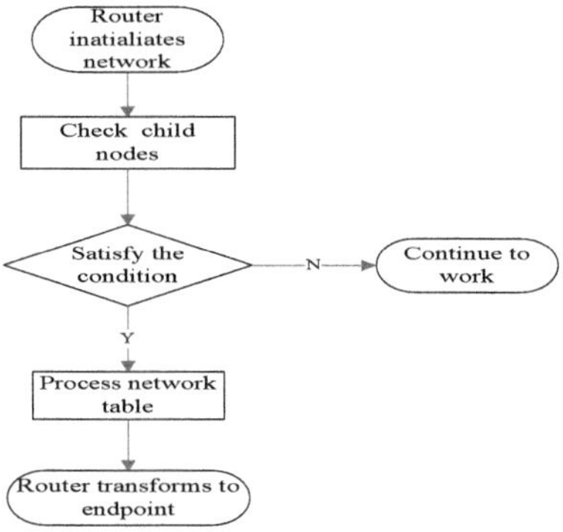

Fig. 2. The procedure of router transformation

4.3 Transformation of Endpoint Nodes

During normal working hours, if the number of child nodes reaches the maximum 4, there are still some nodes request to join into the network, and they will fail to join in. As is showed in Figure 3, the node E5 wants to join in the parent node R3, when the child nodes has reached the maximum, node E5 will fail to join the network. At this time the parent node will notify the designated child node (the example in this paper , select No.1 as the child node E1) to convert into a router. This router must inherit the relationship between the parent router and the parent-child node, and notify the parent and sibling nodes to modify the information about this node in their neighbor table respectively, which mainly including the device type and the relationship with the transformed node.

Reset the stack and initiate the node into a router, then re-create the routing tables. Join to the original routing network, and assign to the previous PanID and short address, initiate the node routing protocol stack to create its own network. Notify the initially failed endpints to join its network. At this point, the endpoint node has achieved a dynamic conversion of a router, as is showed in Figure 4. It can be seen in

Figure 5, the original child node E1 has changed into a router R4. R4 and child node E1 have the same PanID and short addresses, and still keep the same relationship with R3. After R4 created a network, it received a join request from the node E5, and then it sent a confirmation signal to accept the join of the node E5 to its network.

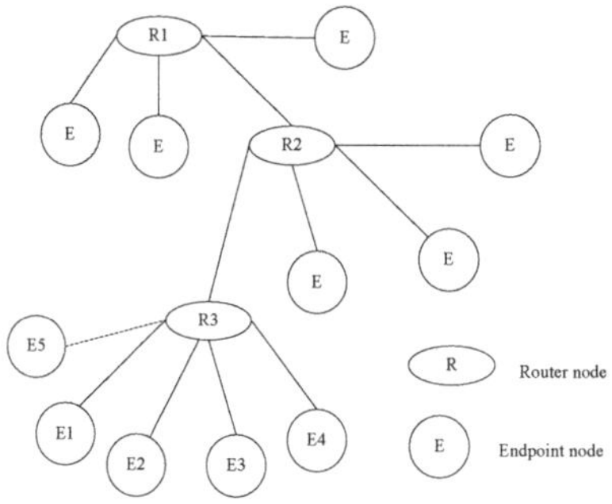

Fig. 3. The request of endpoint to join

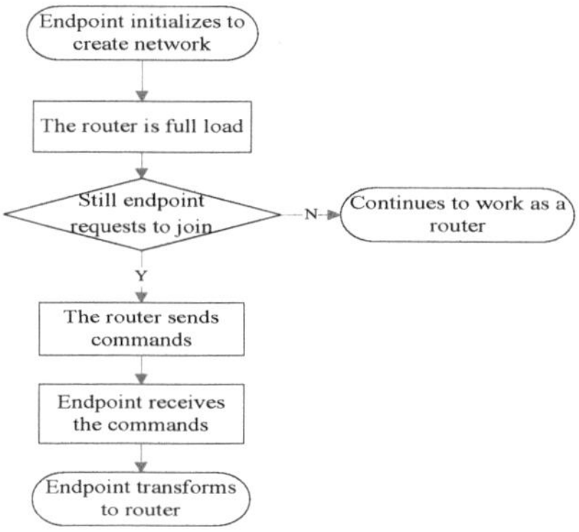

Fig. 4. The procedure of endpoint transformation

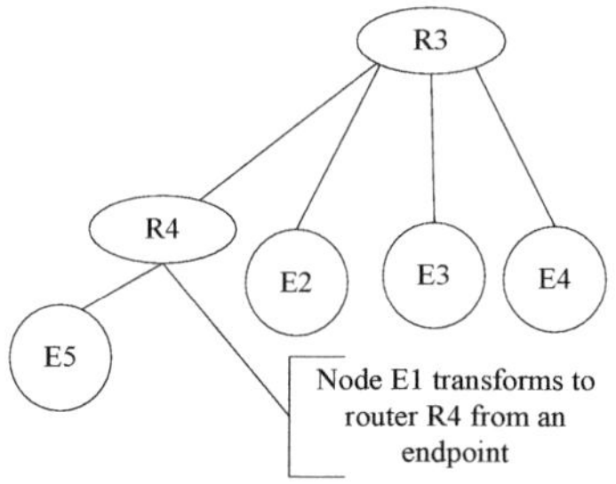

Fig. 5. The Schematic diagram of endpoint into router

4.4 Experimental Results and Analysis

On the base of ZigBee protocol stack, it realized one to one and one to many data sending and receiving. Figure 6 shows the signal that collected by the protocol analyzer. There is a coordinator and endpoint nodes in the network. It can be seen that the two endpoints nodes have sent data to the coordinator.

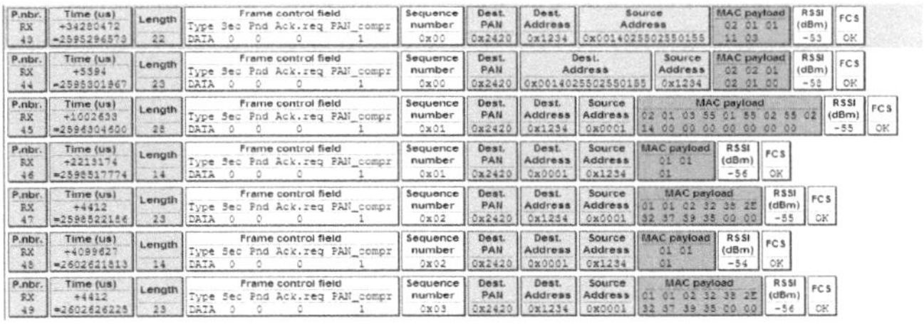

Fig. 6. The result of PC

The above algorithms, based on the hardware platform and the ZigBee protocol stack improves the software, determine and control chip reset operation mainly through the number of child nodes in the neighbour table, and finally achieve the conversion between the routing and terminal nodes. It is achieved through the coordinator to relax conversion command to the terminal end. Before and after nodes conversion, PANID, the addresses of parent and sibling nodes are stored in the EEPROM, which well preserved the information. Experimental results show that, this algorithm can reduce network energy consumption and increase network flexibility.

5 Summary and Outlook

Wireless sensor network has broad application prospects in intelligent medical care, intelligent household and industrial control and other areas and has significant research value, which conform to the tendency of net related, information based, virtually, and intelligent requirement of modern advanced manufacturing technology. The advantages of self-organization network, such as dynamic establishment of topology, needless to pay for infrastructure, the instant configuration time are in line with the real-time and fast requirements of wireless sensor networks which better reduces network energy consumption.

Therefore, the study of self-organization network is necessary. This paper made researches and forwarded the principle of self-organization network, conducted experiments based on the already designed and completed software and hardware platform of ZigBee protocol stack, and found ways to effectively improve the network flexibility and reduce energy loss. However, this self-organization wireless network system in some other aspects still need to be improved, which include: 1)the transmission distance of the designed transmitter module is relatively short, even with the help of power amplifier chip the longest distant is only 20 meters. Our next goal is to extend the working distance to 200 meters; 2) as the number of nodes is terminate, currently it is not yet capable to conduct experiment on power consumption of large-scale networks and multi-hop routing test, which deserves our further research.

References

1. Lu, B.G., Yuan, J.: The Three-dimensional Positioning Algorithm for Wireless Sensor Network. Computer Engineering 37(14), 97–99 (2010)
2. Gao, J., Fang, B., Yin, J.Y.: The Node Design and Network Implementation of ZigBee Wireless Communication Network. Computer Measure & Control 16(12), 1912–1914 (2008)
3. Marsh, D., Tynan, R.: Autonomic Wireless Sensor Networks. Engineering Applications of Artificial Intelligence 17, 741–748 (2004)
4. Jing, B., Zhang, J., Zhang, Z.L.: Research on Applications and Core Technology of Distributed Wireless Sensor Networks. Journal of Test and Measurement Technology 19(2), 181–185 (2005)
5. Akyildiz, I.F., Su, W., San, Y., Cayirci, E.: Wireless Sensor Networks: a survey. Computer Networks 38, 393–422 (2002)
6. Liu, D., Qian, Z.H., Liu, Y.: Characteristics of Node-Based Routing of ZigBee Network Energy Optimization. Northeast Normal University 20(10), 118–121 (2009)
7. Yuan, P.Y., Liu, P., Tang, Y.M.: Wireless Network Strategies in a Controlled Flooding. Development of Computer and Technologies 16(12), 1912–1914 (2010)
8. Pen, M.G.: Next-generation Broadband Mobile Communication System Network. Self-organization Technology Telecommunications Technology 5, 71–73 (2010)
9. Travis, C.C., Taylor, C.: Self-organization in Sensor Networks. J. Parallel Distrib. Comput. 64, 866–873 (2004)

The Design of the Composite Wheeled Mining Robots

Liu Yun

College of Mechanical and Electrical Engineering, Weifang University, Shandong, China
wflyun@126.com

Abstract. In allusion to complex and changing sediment environmental characteristics of cobalt-rich crusts and hydrothermal sulfide survey area in deep sea, designed a new high over-obstacle Composite wheel mining robots. It is made by 4-composite wheel set connected with an articulated frame. The design of composite wheel set makes the mining robots have both active and passive obstacle model under the auxiliary fuel tank. Through analyzing, the results showed that the Composite wheeled mining robots has a strong terrain self-adaption and over-obstacle capacity in the complex and changing deep sea environment.

Keywords: Composite Wheeled mining robots, Composite Wheels, Over-obstacle Capacity.

1 Introduction

The using of the deep-sea cobalt-rich crusts and the hot metal sulfides is becoming a developing industry in the century. In order to occupy a better position in the research of deep-sea, countries all over the world dedicated to the development of the deep-sea technology, the oceans have become other fight hot spots worldwide fallowing the moon, mars.

According to their moves method, the deep-sea mining robots can be divided into three types: tracked, wheeled, legged and wheel-leg. Considering the flexibility of movement, and the complexity of power and control, the most of deep-sea mining robots are wheeled. The ability to adapt to the terrain and the obstacle is a key index to the performance of the deep-sea mining robots. But the deep-sea mining robots widely used now haven't a satisfied performance of the obstacle and adapt to the soft terrain, generally their obstacle ability is lower than the height of wheels and the diameter of the wheels. Aiming at a better adapt ability to the deep seabed and better movement and obstacle performance, this thesis designed a deep-sea composite wheeled mining robot with both active and passive obstacle capacity.

2 The Design of the Deep-Sea Composite Wheeled Mining Robot

2.1 The Overall Design of the Mining Robot

Deep-sea mining composite wheeled robot is composed of four sets composite wheeled mechanism and articulated compression-type seal tank frame. The model shown in Fig 1:

G. Lee (Ed.): Advances in Automation and Robotics, Vol. 1, LNEE 122, pp. 141–146.
springerlink.com

Fig. 1. Structure of composite wheeled submarine walking mechanism

2.2 The Design of the Composite Wheels Group

Compared with the traditional wheels, the composite wheeled mechanism (shown Fig 2) has the characteristic of the traditional wheeled mechanism and highly active control of intelligent high-performance over-obstacle. With the assistance of the cylinder 1 and cylinder 2, the three wheels W1, W2 and W3 could changes the structure, thus the mining robot has both active and positives obstacle model. In the active model, the robot could get across the barrier which is higher than the vertical radius of the tire; in the positives model the robot could climb 30° slopes and adapt to the normal sea-bed environment, ECT.

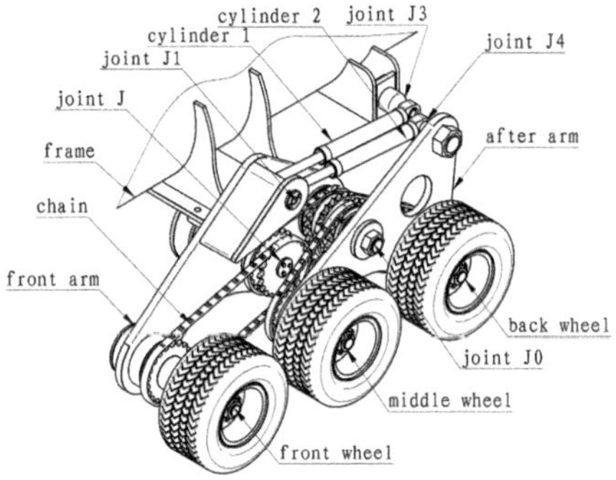

Fig. 2. Structure of composite wheels

2.3 The Design of the Transmission

The deep-sea mining composite wheeled robots mostly work in harsh environment, the parts are often stained with oil and water, considering the structure and application of the robot, the chain is chosen. The structure of the transmission is shown in Fig 3.

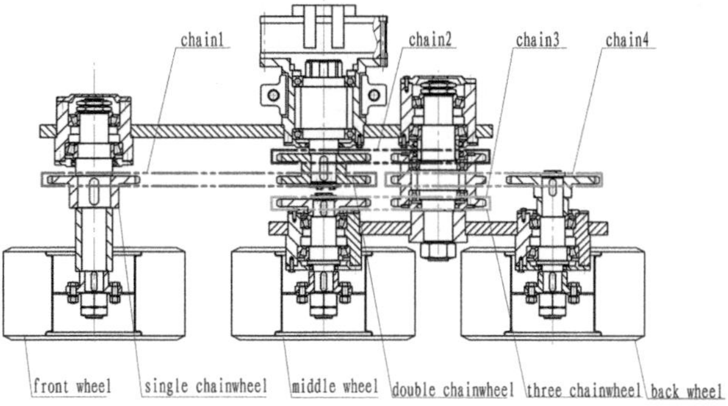

Fig. 3. Structure of the transmission system

The transmission includes three chains, three sprockets, a dual sprocket, a triple sprocket, three tires (front, middle and rear), and shafts and so on. Hydraulic motor translates the rotator movement to the dual sprocket through the input shaft, the dual sprocket delivers the force to the triple sprocket and the single sprocket through the chain 1 and the chain 2, the triple sprocket drives the single sprocket 2 and 3, through the chain 3 and chain 4, the single sprocket 1, 2 and 3 fixed on the front wheel, middle wheel and rear wheel respectively, thus drives the tires, each wheel on the wheels group is driving wheel, the entire mining robot is robot 12x12-wheel drive.

3 The Analysis to the Obstacle Capability of the Deep-Sea Mining Robot

3.1 Climbing the Stairs Which Is Higher Than the Radius of the Wheels

The mining robot will takes active model when facing the obstacle which is higher than the radius of the wheels. The process is shown Fig 4.

When the mining robot driving in the ground, as soon as it detects the level obstacle, the robot draws back cylinder1 to lifter the front wheels each side of the front wheels group, as the front wheels climb on the obstacle, the cylinder1 is stretched, thus the front, middle, rear wheels in the front wheels group will stay in a line, the cylinder3 removes to the original, then the cylinder 3 and cylinder 4 will be self-locking, the robot moves on, the wheels drive the middle wheels and rear wheels of the front wheels group to the obstacle, the cylinder of the fronts wheels group unlocks(release hydraulic), the three wheels in front wheels group are in the adaptive state, all contact on the ground. The rear wheels group repeats the action of the front wheels group, thus all will climb to the obstacle.

The cylinders cooperate with the front, middle, rear wheels of the wheels group to avoid a serious impact. The front wheels of the front wheels group keep locking before reach the stair to maintain rigidity. The mining robot moves on until the front, middle wheels drive through the stair. The cylinder1 elongates, thus the front wheels group turning along the rear axis until the front wheels touch the ground, and then the cylinder maintain rigid (Fig 4-8). The mining robot moves on, the cylinder unlocks when the rear wheel9 of the front wheels group drives through the stair, the front wheels group turns down along the axis of the front wheel, the three wheels are in the self-adaptive state. The rear wheels group repeats the action of the front wheels group.

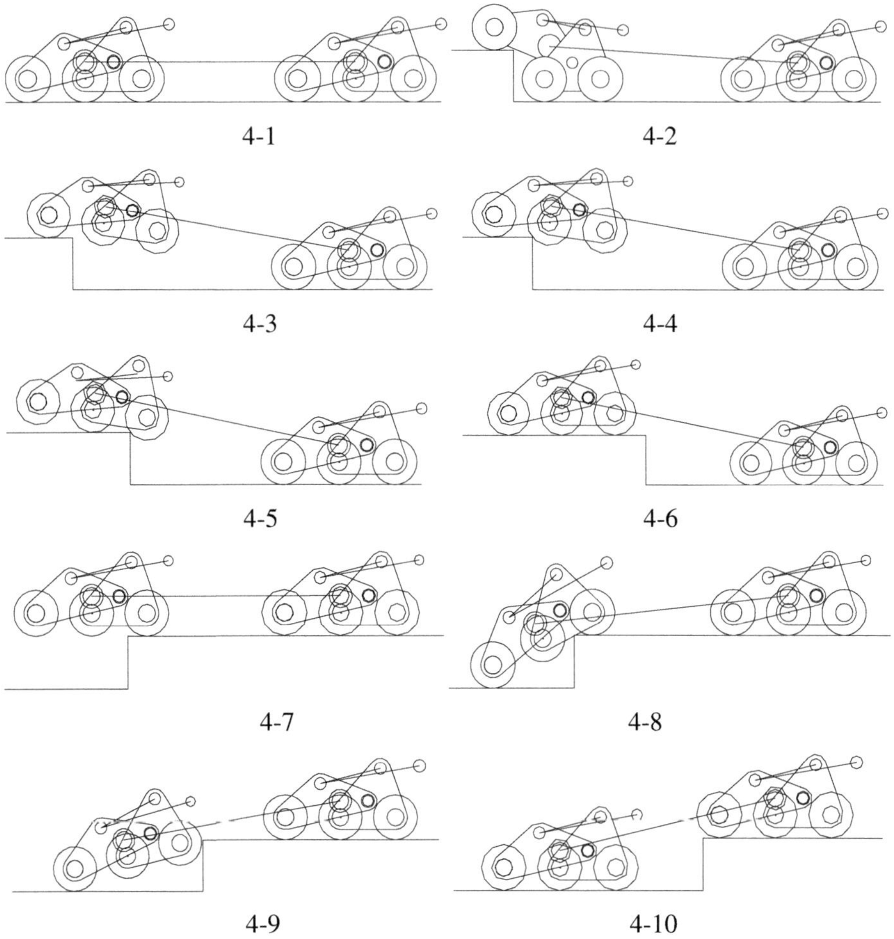

Fig. 4. The obstacle-surmounting course than tire radius sports diagram

3.2 Step across the Ditch

When the mining robot facing ditch, and detects the width is narrower than the length between the axis of the front and rear wheels. The front wheels group keeps self-locking thus the front, middle, rear wheels and the frame keep a rigid state. The mining robot keeps a constant velocity in the process, at least a wheel of each wheels group touch the ground to ensure the robot gets across the ditch. (Fig 5-1 to Fig 5-5)

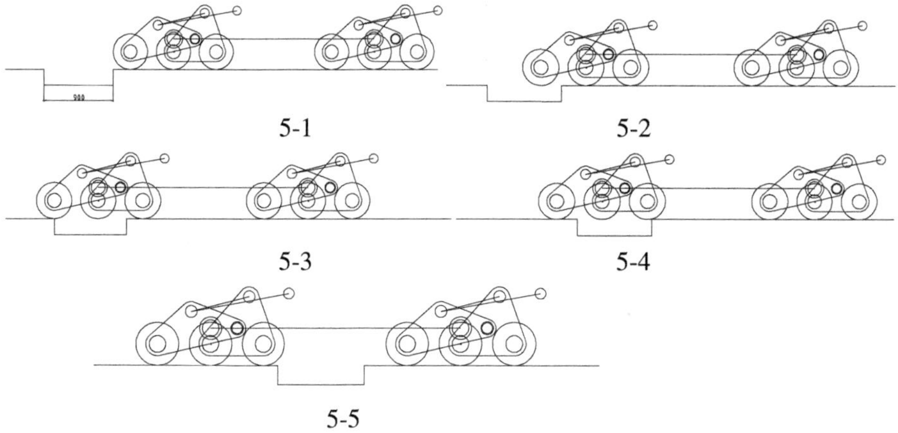

Fig. 5. Ditches the obstacle-surmounting course sports diagram

3.3 Self-adjust Driving in the Normal Ground

When the robot driving in the normal ground (the height of the obstacle is lower than the radius of the wheel), the cylinders of the wheels group unlock (release the pressure), the cylinder is floating each wheels group keep in self-adjust state. The robot adjusts cylinders to fit for the ground. The wheels all contact with ground, thus the gravity of the robot is distributed to the three wheels to make all use of the adhesion. Effectively improve the adhesion and efficiency in the complex terrain in the deep-sea.

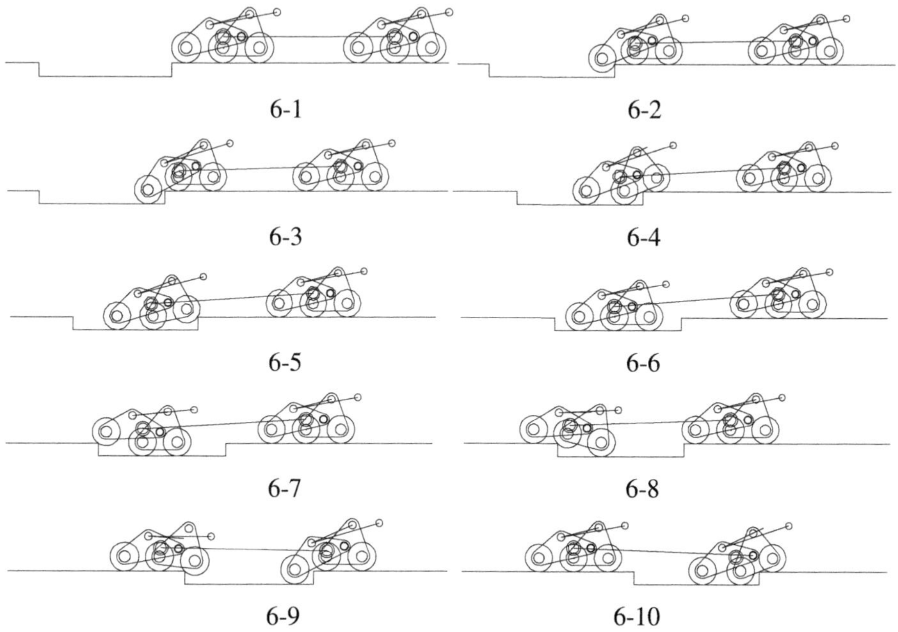

Fig. 6. Adaptive terrain the obstacle-surmounting course sports diagram

4 Conclusions

The study of deep-sea mining robot is very essential for China's deep-sea technology, and has a great importance of the research of deep-sea cobalt-rich crusts and hydrothermal sulfide resource.

This thesis puts forward a new deep-sea composite wheels robot with active and positive obstacle model. This robot not only can self-adjust to the normal deep-sea terrain but also can get across the stairs higher than the radius of the wheels by active model to get across the moat which is higher than the diameter of the wheels, that make up for the poor performance of the traditional robot. The application will not only provide a theoretical reference and technical support in the deep-sea mining vehicle research, but also can play an important role in space expedition, disaster relief for risk, military reconnaissance, act, thus this research is of great meaning in academic and engineering.

References

1. Feng, Y.: Obstacle Performance of a Planetary Wheel Mining Vehicle in the Deep Sea Bed. Journal of University of Science and Technology Beijing (2008)
2. Deng, Z., Li, S., Gao, H.: Research and Development of Passive Rocker Suspension of Planetary Exploration Rover. Journal of Astronautics 29, 1655–1700 (2008)
3. Randel, A., Brain, D., Ary, M., et al.: Mars Exploration Rover Mobility Development: Mechanical Mobility Hardware Design, Development, and Testing. IEEE Robotics and Automation Magazine 13, 19–26 (2006)
4. Exploration for Manganese Crusts in the Southwestern Pacific. Annual Report, pp. 102–112 (2002)
5. Huang, Z., Liu, S., Xie, Y.: Obstacle Performance of Cobalt-enriching Crust Wheeled Mining Vehicle. Computer Simulation 23 (2006)
6. Chen, S.: Development of Deep-sea Mining Technology in China. Mining Research and Development 10, 40–48 (2006)
7. Deng, X.: Kinematic Analysis of an Omni-Directional Mobile Robot. Robot 26, 49–53 (2004)
8. Qiao, F., Yang, R.: Analysis on the Stair-climbing Ability of Six-wheeled Mobile Robot. Robot 26, 301–305 (2004)

Hall Systems Based on Vehicle Integrated Life-Cycle Project Management Model

Yongtao Yu[1,*,**] and Ying Ding[2]

[1] Institute of Stratege and Policy Naval Avademy of Armament,
Box 1303-17, Beijing, China, 100161
y_yt128@yahoo.com.cn
[2] Beijing Sino-Ocean JIYE Property Management CO., LTD, Beijing, China, 100038

Abstract. With the progress of vehicle projects, project objectives, planning, control and management of the main targets have also changed, such as information exchange projects in the participating units close enough to some extent, project management is characterized by a certain degree of continuity and mutual independence of the Non-continuous. In order to maximize project management efficiency, to build a vehicle project on the Hall project life-cycle concept of integrated management model for the formation of a vehicle project integration management structure, the completion of the project process analysis. Hall systems based on the use of vehicle of project life-cycle concept model to achieve integrated management of complex multi-attribute and the role of dynamic vehicle project and effective management of the participating units, thus solving a class of high-investment, high-interest, long-period large-scale project management issues.

Keywords: Life-cycle, integration, management.

1 Introduction

Vehicle project and more for large investment, high interest, long-period large-scale project, its technical and economic uncertainties that many, and the complex internal and external environmental conditions, project management is characterized by a certain non-continuous and mutual independence. Therefore, only set up the system, complete vehicle integration project life-cycle management, engineering practice in order to meet the urgent needs of the project management requirements and higher [1].

Currently, the industry on project management theory and applications of basic and key issues, such as the composition of the whole life cycle, management, and management tools and other objects, carried out extensive research and made a large number of results. At the same time, while scholars often the potential problems of project management gains, vehicle project integration management researchers often

---

* Author: Yongtao Yu, (1974 -), male, research direction for the systems engineering; Ying Ding, (1973-), female, undergraduate, research direction for the business administration.
** Corresponding author.

G. Lee (Ed.): Advances in Automation and Robotics, Vol. 1, LNEE 122, pp. 147–154.
springerlink.com

do not pay sufficient attention. This leads to the Hall system life-cycle integration of project management implications have not been adequately studied.

This article attempts to study hall system in the project management in the use of vehicle, vehicle of a typical project cycle management as the object of study to build a conceptual model of integrated management of vehicle projects, set up a management organization, analysis of management processes, and as a design and implementation vehicle project based on integrated management information system.

This chapter is organized as follows: Section 2 outlines Hall systems theory. Section 3 to build a vehicle system on the Hall project life-cycle concept of integrated management model, given the organizational structure of the integrated management of vehicle projects, completed the project management process integration analysis in Section 4 vehicle project of integrated management information system examples. Section 5 summarizes the full text.

2 Hall Systems Engineering Overview

Hall, Three-dimensional structure, also known as Hall systems engineering, systems engineering experts in the United States Hall (A • D • Hall) in 1969 as a systems engineering methodology, and their descendants compared with the soft system methodology, known as a hard system methodology. It appears to solve large and complex system of planning, organization, management provides a uniform way of thinking, which has been widely applied [2].Hall, Three-dimensional structure of the systems engineering activities throughout the convergence process is divided into before and after the close of the seven phases and seven steps, and also considered for the completion of these stages and steps required for a variety of professional knowledge and skills. Thus, the formation by the time dimension, and knowledge-dimensional logical dimension consisting of three-dimensional spatial structure. Among them, the time dimension that systems engineering activities from the beginning to the end of the chronological order of the whole process is divided into planning, programming, development, production, installation, operation, update the seven time periods. Logical dimension is the time dimension at each stage of work to be undertaken within the content and should follow the thinking process, including identifying problems, setting goals, systems integration, and systems analysis. Optimization, decision-making, implementation of the seven logical steps. Lists need to use the knowledge dimension, including engineering, medicine, architecture, business, law, management, social sciences, arts, and other kinds of knowledge and skills[3].Three-dimensional architecture vividly describes the system engineering framework, of which at any stage and every step can be further expanded to form a hierarchical tree system. By the time dimension, logic and knowledge-dimensional three-dimensional structure composed of dimension [4].

3 Vehicle and Integrated Management of the Project Life-Cycle Concept Model

Vehicle integration project life-cycle management model is defined by the user units (relevant decision-making bodies, etc.) led unit has overall responsibility for professional advice, from all the major units involved in the election of one or two experts were composed of one project life cycle management group, the whole life cycle of the main units involved, the management of content, the project management phases combine to achieve the organization, resources, objectives, responsibilities, benefits and other integration-related information between the participating units to communicate effectively and fully shared to the user units and other stakeholders to maximize the value of the project units to provide the product. Integrated management of the project life-cycle groups in the composition of vehicle project life cycle, are not immutable, should be mainly involved with the phase adjustment unit changes, add new items in units of experts, to facilitate the achievement of overall project effective management[5].

3.1 Conceptual Model of the Integrated Management of Design

Start of the study, the first three-dimensional structure model based on the Hall theory, a vehicle project life-cycle management integration shown in Fig. 1. Vehicle project life-cycle management model integration covers three main areas: integration of the participating units, elements of management integration, integration of the management process [6].To achieve the integration of the participating units, the units will help to break the project time, scope and content of the work on the boundaries, to promote the integration and management elements of the management process integration. Management process to achieve integration, but also requires management to break the phase interface, the implementation of integrated management elements played a role in promoting. The management factor also in turn promotes the implementation of the integration process of integration. On this basis, operational processes, organizational structure and information platform is an integrated vehicle project life-cycle management of the three basic elements.

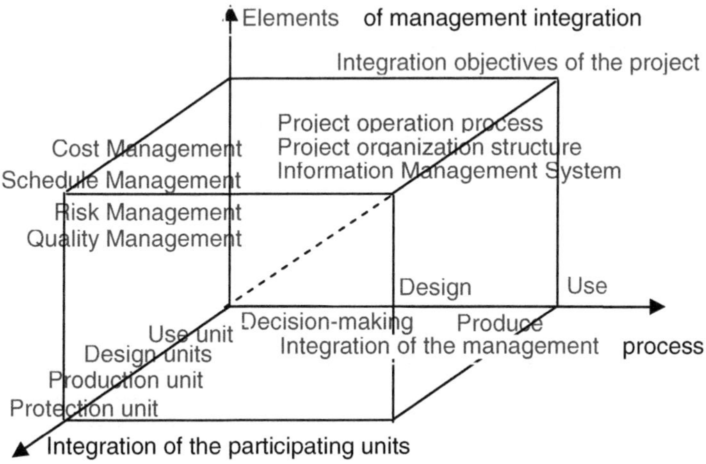

Fig. 1. Schematic diagram of an integrated management model weapons program

3.2 Integrated Management Structure Set

Vehicle involved in the project management organization project management and responsibilities, authority and mutual relations division of a team arrangements and facilities, including user units, consulting units and other units involved in the project management work for the project management organization established[7].Vehicle project life-cycle management can be integrated organizational structure shown in Fig. 2. User units (relevant decision-making bodies, etc.) as the project's top decision-makers to oversee and manage the project life-cycle integrated management group, the project has the ultimate decision-making control over the final decision on project implementation and contract parties, while organization, leadership and supervision of the work to fulfill the user unit review, approval and authorization of the power to make the project goals. Project Life Cycle Management Group is an integrated unit commissioned by the user phase of the project management of the main vehicle, each unit is also responsible for the coordination of the various stages, the project participants through the project life-cycle units in the integrated management of groups involved in the project ahead of schedule, full participation to complete its task.

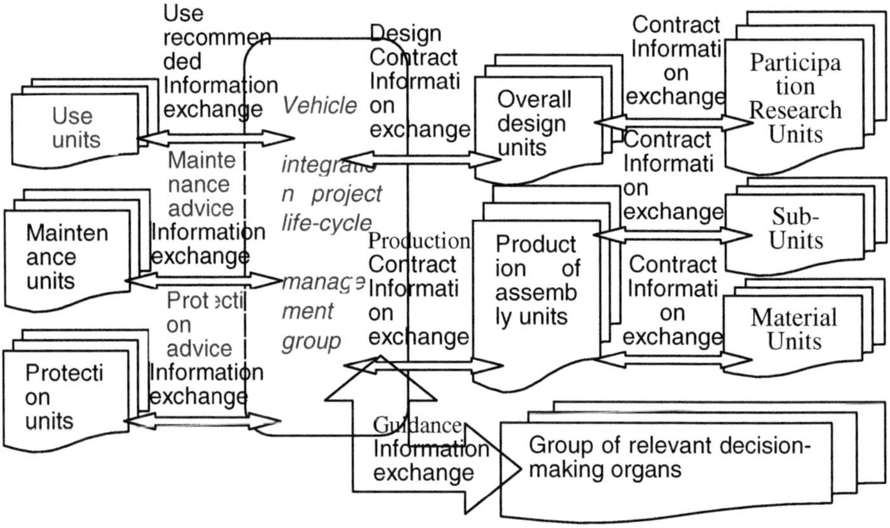

Fig. 2. Projects Organization Chart

Vehicle integration project life-cycle management group is to achieve the integration of vehicle project life-cycle management of the basic means. It ensures that all stages of the participating units to keep abreast of progress of the project, timely and effective participation in project implementation, to make up due process of phasing the project work is not due to the continuity of the process to achieve the integration of vehicle and items.

3.3 Integrated Management Process Analysis

Vehicle and integrated management of project life-cycle mode operation of the project process flow with the traditional operation of the project has some similarities, but more emphasis on the balance of the project in the target unit, effective flow of information and the application of concurrent engineering.

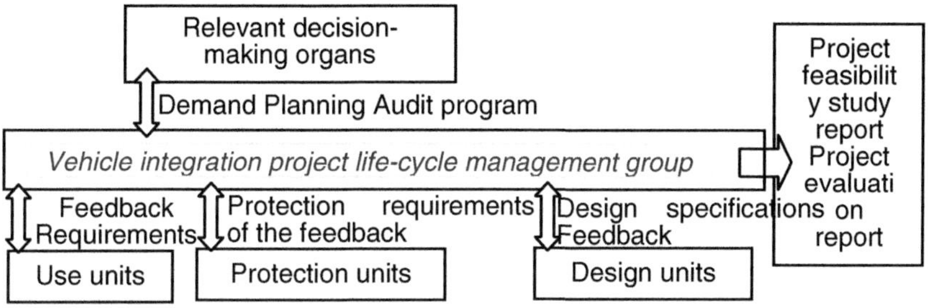

Fig. 3. Project decision-making stage Flow chart

Project decision-making phase of the operational flow of vehicle shown in Fig. 3. Project Life Cycle Management Group as the main responsibility for integration and coordination unit, responsible for collecting information from all participating units, to determine the initial program and feedback to the relevant decision-making bodies. Relevant decision-making body considering the building carries out mission requirements, unit structure and the preparation of financial constraints and other conditions, to determine the optimal program. Project management team to refine the optimal solution demonstration, seek advice and design units, and timely analysis and collation of information of all kinds, and finally made vehicle project feasibility study report and project evaluation reports.

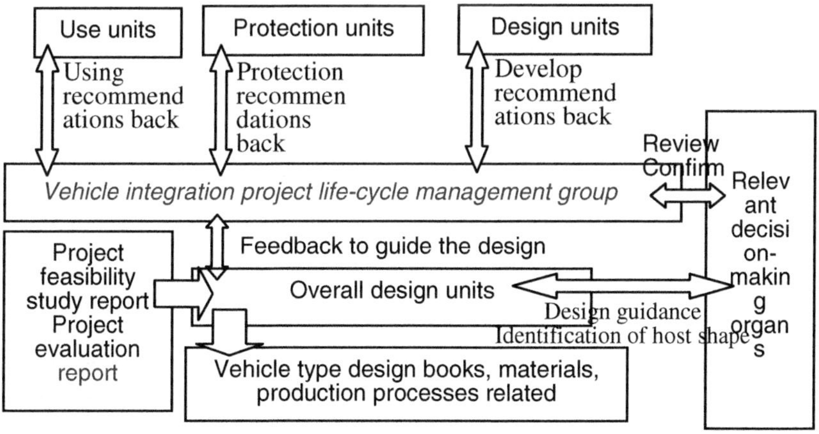

Fig. 4. Project design phase Flow chart

Vehicle design phase of the project operation process shown in Fig. 4, the overall design of the system unit as the main responsible unit, to feasibility studies, conceptual design and planning requirements as the main design basis, project life cycle through integration with other management groups The participating research units in repeated discussions on the design, determine the design that meets your needs, access to relevant decision-making bodies recognition and identification after completion of the design shape. The model design document and related materials, production technology transfer production assembly units.

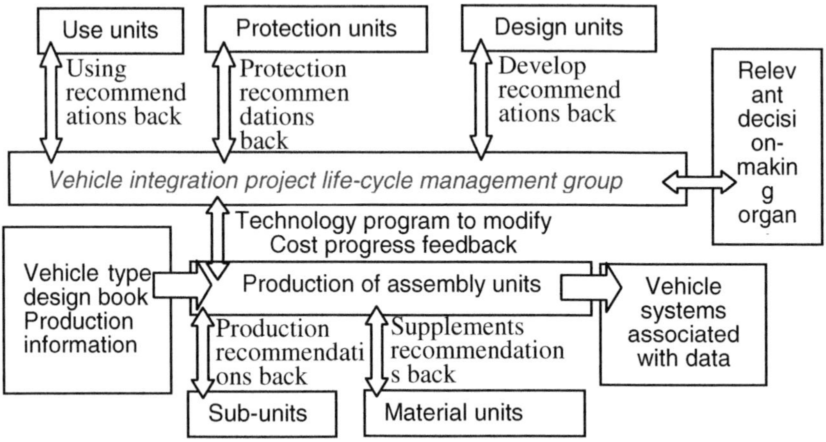

Fig. 5. Project production phase Flow

Stage of the operational flow of vehicle produced by the project as shown in Fig. 5, the main responsibility for production and assembly units for the coordinator to design the book and related types of vehicle manufacturing materials as the main production basis, production process, considering the use and protection units Use protection recommendations, feedback is given after repeated discussions execution. If the production process required to modify the design, via the integrated management group to submit changes to the design units in aggregate demand by the design unit to make design changes after production assembly unit in order to implement specific changes.

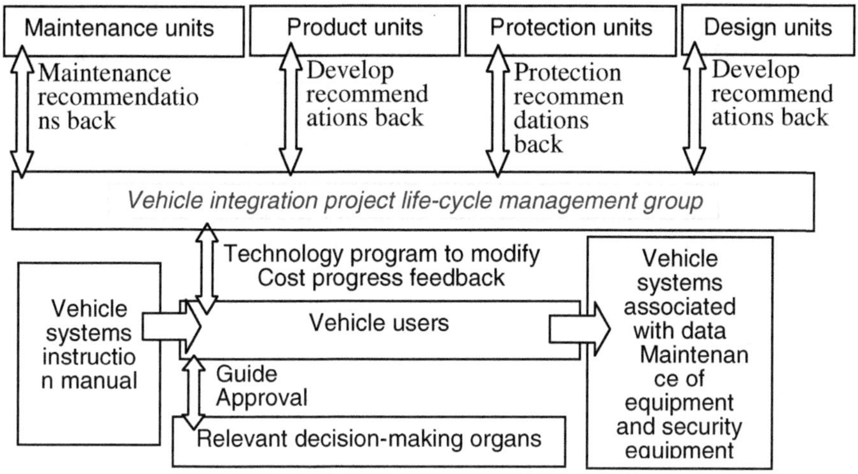

Fig. 6. Stage equipment project Flow chart

Stage of the operational flow of vehicle and equipment items as shown in Fig. 6, the use of units as the main responsible person in the first few stages of the project to collect data based on the use of vehicle systems, combined with equipment used to protect and maintenance of a comprehensive assessment of the vehicle project. And evaluation results back to the design unit, does not meet the requirements of items by the production unit to improve the final delivery of qualified vehicle systems.

4 Integrated Management Information Systems Vehicle Project Examples

Information in the normal flow of the project and the correct treatment is the key to the smooth implementation of the project, so the vehicle project management is based on the information. Vehicle to achieve integrated management of the project life-cycle model units, the various stages of effective communication and sharing of information, we must first achieve the information integration. Vehicle project integrated information management system to ensure accurate and timely complete the project decision-makers to grasp fully reflect the implementation of the project information, and the participants of the project to achieve the flow of information and data sharing, to help project decision makers to make scientific decisions. Integrated information management system project structure design shown in Fig. 7.

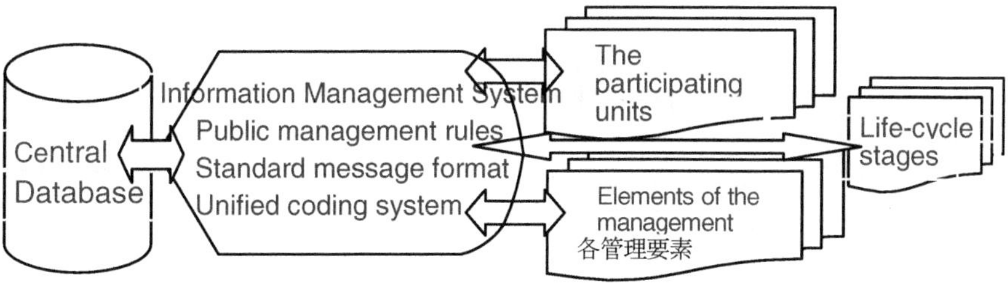

Fig. 7. Integrated information management system project structure

The system can make use of computer and information processing technology, by building a unified central database and integrated information management platform project, the use of common, unified management of language and rules, the various stages of project implementation, including vehicle, different functional blocks of data information integration, and the participants of the project to provide personalized information resources. Its implementation will help project the participating units to share information, exchange and transfer, making the project all participating units to communicate between the co-operations into an orderly, but also to ensure the project's centralized data management, digital, integrity and consistency. For the vehicle of the integrated management of the project life-cycle model of the project breakdown structure, coding system and exchange standards and how to apply modern information technology to achieve integrated information management systems need to be further studied.

5 Conclusion

This reference life-cycle concept first proposed to build vehicle based on the Hall system integration project life-cycle management concept model of the vehicle for an objective analysis of the project management process; then the conclusions based on analysis of project management of vehicle, vehicle projects to form an integrated management structure, completion of the project, including vehicle, the decision-making phase, design phase, production phase and the phase of the integrated management of equipment, process analysis, project running in arms during the whole life continued uninterrupted implementation of effective management; to achieve maximum benefit for the pursuit of vehicle project .

Next, will continue to enrich, update and deepen the analysis of the participating units, vehicle and items to complete project life-cycle integrated vehicle management conceptual model, and the introduction of intelligent database research to improve the integrated management information systems vehicle project efficiency.

References

1. Bai, S.: Modern project management. Beijing Machinery Industry Press (2003)
2. Chen, H.: Project management. China Building Industry Press (June 2001)
3. Lv, Y.: Systems engineering. Beijing Northern Jiaotong University Press (2003)
4. Qiu, W.: Project management. Beijing Science Press (2001)
5. Tan, Y.: The system engineering principles. National Defense University Press (2003)
6. The institute of Internal Auditors, Applying COSO's Enterprise Risk Management—Integrated Framework (September 2004)
7. Committee of Sponsoring Organizations of the Treadway Commission, Enterprise Risk Management —Integrated Framework, Executive Summary (September 2004)

An Intelligent Fault Diagnosis Method for Oil-Immersed Power Transformer Based on Adaptive Genetic Algorithm

Youwen Li, Demei Bao, Cun Luo, Lijun Huang, and Liming Cao

GUODIAN NANJING AUTOMATION CO., LTD Research Institute,
46 New Mofan Road, Nanjing, 210003
leiyuziqq@163.com, {lucybao,luocun,hlj,clm}@sac-china.com

Abstract. For the information and knowledge in oil-immersed power transformer fault diagnosis is randomness and uncertainty, this paper firstly analyzed the relationship between DGA and the fault categories; and then combined it with adaptive genetic algorithm, and proposed a adaptive genetic algorithm based fault rules intelligent learning model, so the model was used for oil-immersed transformer fault diagnosis. It combined adaptive genetic algorithm, credit evaluation system and experts' guidance, because of the global search properties of genetic algorithm and the artificial expert prior knowledge, the system will search out the optimality fault rules finally. At last, the experiments also proved that the proposed fault rules intelligent learning mode used in oil-immersed power transformer diagnosis has a high rate of identification for faults.

Keywords: Power Transformer, Adaptive Genetic Algorithm, Dissolved Gas Analysis, Fault Diagnosis.

1 Introduction

The transformer is a very valuable equipment for power transmission in electric power industry.The method of Dissolved Gas Analysis (DGA) is an effective measure for oil-immersed transformers,In recent years, the artificial neural network algorithm, fuzzy inference, gray theory, expert system and other intelligent algorithms were applied to dissolved gas analysis transformer fault diagnosis, and also achieved good effects. Domestic and foreign researchers established BP network model for power transformer fault detection by improving the standard BP algorithm[1]; through the introduction of self-organization and self-learning, they proposed a preparation transformer fault diagnosis system by using fuzzy algorithms[2,3]; and based on ideas of integration and complementary, they combined gray relational analysis theory with information theory together for transformer fault diagnosis[4,5];the expert system had an advantage when the information is incomplete or uncertain, it still could get a more reasonable conclusion because of the priori knowledge,therefore foreign scholars made a wide range of applications by research transformer fault diagnosis from perspective of expert system[6].

G. Lee (Ed.): Advances in Automation and Robotics, Vol. 1, LNEE 122, pp. 155–162.
springerlink.com

Genetic algorithm (GA) is based on evolutionary theory, which means survival of the fittest, eliminate of the unfittest[7]. GA can be used for search and optimiza tion, and also as a machine learning technique for influencing behavior of the system after learning rules from environment. This paper combined GA and DGA together, and proposed an intelligent fault learning model based on adaptive genetic algorithm, which is useful for helping solve the problem of power transformer fault diagnosis.

2 Fault Categories of Oil-Immersed Transformer

With the pyrolysis temperature increased, the order of pyrolysis gas appeared was alkanes→alkenes→alkynes, and also the quantity would be increased quickly with the time elapsed. It included some lower hydrocarbon gases, such as CH_4, C_2H_2, C_2H_4 ,C_2H_6, and also CO, CO_2, H_2[8,9]. With the faults development, the gases will be diffused and dissolved in the oil and each different types of faults, the density of gases were different, so the amount of each gas could be used as feature vectors for indicating status of the transformer[10].According to the different gas types and content, the fault categories can be roughly divided into the following seven types.

1. Normal
2. General overheating: Produce some C_2H_4 and extraordinary amount of and C_2H_6, CH_4, generally, no C_2H_2, or very little.
3. Severe overheating: When the temperature was above 800℃, it will produce a small amount of C_2H_2, but its content does not exceed the 10% of the maximum of C_2H_4. CO, CO_2 is also higher, and the greater of the ratio CO/CO_2 is, the higher of the overheating temperature is.
4. Partial discharge: Produce more H_2, followed by some CH_4, when the discharge energy is higher, it will produce a small amount of C_2H_2.
5. Spark discharge: Include C_2H_2 mainly, H_2 with lower total hydrocarbon content
6. Arc discharge: Include C_2H_2 mainly, H_2 with higher total hydrocarbon content
7. Damp water: More H_2. Much water come into the oil, and is electrolyzed to H_2, but also because of the chemical reaction ($3H_2O+2Fe \rightarrow Fe_2O_3+3H_2$).

From the above analysis between DGA and the characterizations for different types of faults, rules can be constructed similar to production rules as "if <condition> then <action>", such as: $if\ (T_1)\ and\ (T_2)...\ and\ (T_n)\ then\ class = C_i$.

Hereinto, (T_1) and …… is the part of condition, and class=Ci is the part of result, also treated as an action. T_i is a triple, like (Attribute, Operator, Value), which means Attribute and Value are connected by Operator. Attribute including H_2、CH_4、C_2H_2、C_2H_4、C_2H_6、CO、CO_2 and also the sum of hydrocarbon (sum of CH_4、C_2H_2、C_2H_4、C_2H_6, ST for short). Operator including greater than (>), less than (<) and equal (=). Value is the content of the gas, unit is ul/L. For example (CH_4, <, 100) means the content of CH_4 is less than 100 ul/L.

C_i is the fault categories, including Normal(C_0)、General overheating(C_1)、Severe overheating(C_2)、Partial discharge(C_3)、Spark discharge(C_4)、Arc discharge(C_5) and Damp water(C_6).

Take Rule【1】as an example, it indicated that the fault category was C_2 when the content of H_2、CH_4、C_2H_2、C_2H_4、CO、CO_2、ST exceed 10 ul/L、40 ul/L、5 ul/L、60 ul/L、40 ul/L、20 ul/L、170 ul/L respectively and C_2H_6 less than 10 ul/L, so it's a severe overheating transformer, must be taken attention.

RULE【1】:

$$\textit{if } (H_2 ,> ,10) \textit{ and } (CH_4 ,> ,40) \textit{ and } (C_2H_2 ,> ,5) \textit{ and } (C_2H_4 ,> ,60) \textit{ and } (C_2H_6 ,< ,10)$$
$$\textit{and } (CO ,> ,40) \textit{ and } (CO_2 ,> ,20) \textit{ and } (ST ,> ,170) \textit{ then class} = C_2$$

3 Intelligent Fault Diagnosis Method Based on Adaptive Genetic Algorithm

3.1 Fault Rules Intelligence Learning Model

Generally, a classification system based on genetic algorithm contains three parts: the implementation subsystem, evaluation subsystem and rule discovery subsystem, and the structural framework were shown in Figure 1.

The implementation subsystem interacted with the system environment directly, just like an expert system based on production rules, and each rule worked as a classifier, the condition part and action part in the rules were strings which played a role of passing messages between rules when evaluating rules [11].

The classification system trained itself by obtaining feedback information from the environment, which was realized through the evaluation correctness and efficiency of rules. The work of evaluation was completed by evaluation subsystem, and Bucket Brigade Algorithm (BBA) was used in this paper for estimating the rules.

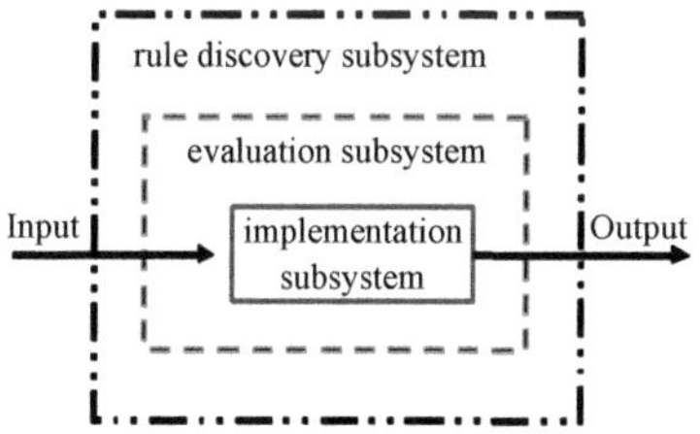

Fig. 1. Framework of Classification System

Rule discovery subsystem is at the highest level of this framework, which generates new rules to replace the inefficient rules in system. New rules in this paper are produced by adaptive genetic algorithm and also with the expert's guidance. Each rule has a strength parameter S (t) that indicates the rule's effectiveness at time t, the greater of the value is, the more effective of the rules is, on the contrary, the smaller effectiveness of the rules is.

According to framework of classification system in Figure 1, this paper proposed a fault rules intelligence learning model as Figure 2.The function and structure in each subsystem were expanded, and also an expert system was added, and the processes of how to learn the fault rules in this model were given below.

Step 1: The detector encoded the output (DGA composition and content) from environment into external messages and put it into message queue. External message was meant to make contact with system directly, also called environmental message.

Step 2: Matching all rules in message queue with fault rules set, and put the matched rules into matching rules set, then checked rules with conflict detector.

Step 3: After checked rules with conflict detector, activated the survival rules in matching rules set at last, and sent the action part of the rule (just the fault category part) as a new message to actuator. The inactivated rules called internal message were put into message queue. Internal message was meant to the action part of the rules which matched with fault rules set at least one time, and also could be treated as the premise of the rules to match rules, they did not contact with system environment.

Step 4: Cleaning up the matching rules set. If the message in Step 3 was an external message, it would act on the environment by actuator; and if it's an internal message, it would be put into message queue, so we could get an internal message loop chain, and the process repeated Step 2 to Step 4 until all messages were external message in Step 3.

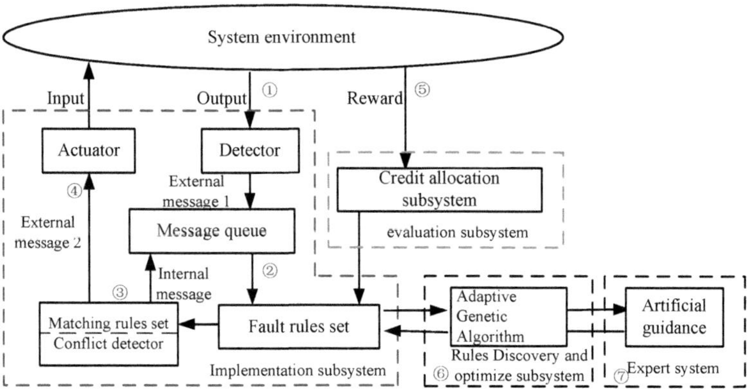

Fig. 2. Fault Rules Intelligence Learning Model Based on Adaptive Genetic Algorithm

Step 5: Using BBA Algorithm for credit allocation. If the activated rule was an external message, it got reward from system environment directly; but if it's an internal message, its reward would be got from the rules matched in Step2, and also added reward at the followed matching circle when it became an external message.

Step 6-7: The adaptive genetic algorithm was restarted to optimize the set of rules each time of T_{ga}. When the rules were produced by using adaptive genetic algorithm, before put into fault rules set they were got guidance from expert system with the prior knowledge, so the incorrect, non-representative rules would be discarded for improving the accuracy of algorithm.

Step 8: Return to Step1.

3.2 Intelligent Fault Diagnosis Based on Adaptive Genetic Algorithm

The "Value" in RULE【1】 was continuous real data, such as (CH_4, <, 100) means the content of CH_4 is less than 100 ul/L, that was to say all value less than 100 could be match it, so it cannot make a distinction between value equal to 1 ul/L and value equal to 99 ul/L, it made the scope expanded. In order to reflect the differences between different levels of gas in DGA, we normalized the collected data to make the range between [0, 1] for reducing the difference between the sample data. After normalization, the value was discretized by N intervals between 0 to 1 equality. For example, the content of a gas was "x" after normalization, it could be discredited to "i" when x subjected to $\frac{i}{N} \le x < \frac{i+1}{N}$ $(0 \le i < N-1)$. After normalization and discretization the triple could be converted to a two-tuples as (Attribute, Value). This paper used Michigan coding for constructing a rule, it's shown in Table 1, and one chromosome meant one rule, also was a classifier.

Table 1. Rules of Coding Structure

Gene(1)		......	Gene(i)		......	Gene(n)	
Attribute(1)	Value	......	Attribute (i)	Value	......	Attribute(n)	Value

The use of genetic operators (such as Selection, Crossover and Mutation) acting on the rules set can result in a new generation of groups[11].

- Selection: Selected the better individuals from previous generation to the next generation according to the individual's fitness S(t) and certain rules.
- Crossover: Made each individual in the group into pairs randomly, and crossover part of the genes between individuals at a ratio Pc.
- Mutation: For each individual, its chromosome may be changed to one or a few value of the other allele.

The strength of classifiers S(t) was used for the fitness function in genetic algorithms, we selected the rules with higher value of S(t) to add into fault rules set. In order to prevent evolutional process getting premature or reducing convergence speedily, it used adaptive genetic algorithm [12], which was proposed by Srinvivas firstly. The great advantage of adaptive genetic algorithm was its crossover rate *Pc* and mutation rate *Pm* is not immutable, but changing with evolution process.

When the largest fitness *Smax(t)* in the group is near to the average fitness *Savg(t)*, the algorithm should increase *Pc* and *Pm* because the algorithm is trend to be convergence; On the contrary, the algorithm should decrease because the individual in group is diversiform, that is to say *Pc* and *Pm* have inversely proportional to *Smax(t)-Savg(t)*; and also, in order to prevent destroying the better gene structure, the individual with higher fitness should have smaller *Pc* and *Pm*; the individual with lower fitness should have larger *Pc* and *Pm*, that is to say *Pc* and *Pm* have directly proportional to *Smax(t)-S (t)*, so the following formulas were used for adjusting parameters *Pc* and *Pm* dynamically. *Smax(t)* is the higher fitness in the two crossover individuals' fitness; *Sm(t)* is the fitness of the individual which will be mutated.

In formula (1), when *Smax(t)>Savg(t)*, we can get *(Smax(t)-Scmax(t))/ (Smax(t)-Savg(t))*<1, so $Pc<k_1$. If *Smax(t)*$\leq$*Savg(t)*, then $Pc<k_2$. For the individual with *Scmax(t)>Savg(t)*, we hope its structure reserved but not destroy, so the ratio of crossover rate *Pc* will be decreased. On the contrary, for the individual with *Scmax(t)* $\leq$*Savg(t)*, the crossover rate will be increased, and we can get $k_1<k_2$, $k_1, k_2 \in (0,1)$.

$$P_c = \begin{cases} k_1(S_{\max}(t) - S_{\mathrm{cmax}}(t)) / (S_{\max}(t) - S_{\mathrm{avg}}(t)),\ S_{\mathrm{cmax}}(t) > S_{\mathrm{avg}}(t) \\ k_2, S_{\mathrm{cmax}}(t) \leq S_{\mathrm{avg}}(t) \end{cases} \tag{1}$$

$$P_m = \begin{cases} k_3(S_{\max}(t) - S_{\mathrm{m}}(t)) / (S_{\max}(t) - S_{\mathrm{avg}}(t)),\ S_{\mathrm{cmax}}(t) > S_{\mathrm{avg}}(t) \\ k_4, S_{\mathrm{m}}(t) \leq S_{\mathrm{avg}}(t) \end{cases} \tag{2}$$

4 Experiments and Analysis

In order to make the experiments have universalities, we need to collect as many as different faults for each types of transformer, and in this paper we had gathered 570 samples. They were divided into training set and test set according to the ratio about 3:1, and the distribution was shown in Table 2.

Table 2. Sample Distribution in Each Fault Category

	Sample distribution							
Fault category	C_0	C_1	C_2	C_3	C_4	C_5	C_6	SUM
Samples	112	83	90	80	69	58	78	570
Training samples	84	62	68	60	52	44	58	428
Test samples	28	21	22	20	17	14	20	142

The results of were shown in Table 3. The algorithms of combined neural network, decision tree and fuzzy neural network which mentioned in reference 13 could be used for vertical comparison. We also did several experiments for horizontal comparison, such as Experiment [A]: Use the original genetic algorithm only but without the intelligent learning model; Experiment [B]: Use the original genetic algorithm and with the intelligent learning model; Experiment [C]: Use the adaptive genetic algorithm and with the intelligent learning model which was proposed here.

As can be seen from Table 3, the average accuracy of combined neural network was only 78.9%, and the reason was related with its own working method, when the levels of neural network reached a certain number, it couldn't improve the recognition rate for similar faults any more. The decision tree algorithm and fuzzy neural network algorithm could overcome the shortcomings but they heavily rely on the training sets, the average accuracy were 81.7%, 83.8% respectively.

Analysis of comparative Experiment [A] can be seen that the average accuracy was just 71.1%; it's mainly because of that the original genetic algorithm may be converged to local optimum at the process of searching optimal solution. In Experiment [B], when combined genetic algorithm with intelligent fault rules learning model together the average accuracy was quickly improved to 90.8%, so it demonstrated that the model of this paper was very useful. Further analysis can be seen in the Experiment [C], its average accuracy was 94.4% because of introducing adaptive genetic algorithm.

Table 3. Results Comparison between This algorithm and Common Fault Diagnosis Algorithm

		Results							
Fault category		C_0	C_1	C_2	C_3	C_4	C_5	C_6	SUM
Training samples		84	62	68	60	52	44	58	428
Test samples		28	21	22	20	17	14	20	142
Combined neural network	Correct	24	17	16	16	12	10	17	112
	Accuracy	85.7%	81.0%	72.7%	80.0%	70.6%	71.4%	85.0%	78.9%
Decision tree algorithm	Correct	28	17	16	17	11	10	17	116
	Accuracy	100.0%	81.0%	72.7%	85.0%	64.7%	71.4%	85.0%	81.7%
Fuzzy neural network	Correct	28	18	18	16	13	11	15	119
	Accuracy	100.0%	85.7%	81.8%	80.0%	76.5%	78.6%	75.0%	83.8%
Experiment [A]	Correct	23	16	14	14	14	9	11	101
	Accuracy	82.1%	76.2%	63.6%	70.0%	82.4%	64.3%	55.0%	71.1%
Experiment [B]	Correct	28	17	19	18	16	14	17	129
	Accuracy	100.0%	81.0%	86.4%	90.0%	94.1%	100.0%	85.0%	90.8%
Experiment [C]	Correct	28	19	20	19	16	14	18	134
	Accuracy	100.0%	90.5%	90.9%	95.0%	94.1%	100.0%	90.0%	94.4%

The proposed method in this paper achieved an average accuracy rate at 94%, it reached an acceptable level of field applications. And in the actual field application, you can always continue to add amended samples to the training sample sets (e.g. addition of new samples, removed redundant samples, etc.), with the time of self-studying increased in this model, it would reach a higher accuracy, and the results of fault diagnostic would also be better.

5 Conclusions

The paper divided the faults into seven categories for oil-immersed transformers according to dissolved gas content. Method proposed for intelligent fault rule learning model was trained by the given training set and tested by the testing set, and the rules were very simple for judging the fault category, so it made the mechanism for fault diagnosis become more reasonable clearly. Through vertical comparison with other methods, we can see that the proposed algorithm can significantly improve the fault recognition rate; and with the horizontal comparison of the original genetic algorithm, we can see that adaptive genetic algorithm based intelligent fault rules learning model have a strong ability to learn, it also greatly improved the accuracy, which proved the validity of the method in this paper.

References

1. Cai, X., Rui, J.: On-line detection of gases dissolved in transformer oil and the faults diagnosis. In: 30th Symposium on Electrical Insulation Materials, pp. 1109–1112 (1998)
2. Surya, K.M., Ravindranath, R.B., Singh., B.P.: Transformer fault diagnosis using fuzzy logic and neural network. In: Annual Report Conference on Electrical Insulation and Dielectric Phenomena, CEIDP 2005, pp. 486–489 (2005)
3. Li, J., Sun, C., Liao, R., Zhou, Q.: Study on Analysis Method About Fault diagnosis of Transformer and Degree of Grey Incidence Based on Fuzzy Clustering. Chinese Journal of Scientific Instrument 25(5), 587–589 (2004)
4. Dong, M., Yan, Z., Taniguchi, Y.: Fault diagnosis of Power transformer based on modeling-diagnosis with grey relation. In: Proceeding of the 7th International Conference on Properties and Applications of Dielectric Materials, Nagoya, vol. 3, pp. 1158–1161 (2003)
5. Lv, G.-Y., Cheng, H.-Z., Zhai, H.-B., Dong, L.-X.: Fauit Diagnosis Of Power Transformer Based On Improved Grey Relation Analysis. Proceedings of The Chinese Society For Electrical Engineering 24(10), 121–126 (2004)
6. Purkait, P., Chakravorti, S.: Time and frequency domain analysis based expert system for impulse fault diagnosis in transformers. IEEE Trans. on Dielectrics and Electrical insulation 9(3), 433–445 (2002)
7. Goldberg, D.E.: Genetic Algorithms in Search, Optimization, and Machine Learning. Addison-Wesley, New York (1989)
8. Fault diagnosis against oil-immersed transformer based on PNN and GM(1,1). In: IEEE International Conference on Integration Technology, Shenzhen, China, pp. 394–397 (2007)
9. Zhao, W.: Transformer Fault Diagnosis and Prediction Research Based on Data Mining. North China Electric Power University, Baoding (2009)
10. IEC Publication 599. Interpretation for the Analysis of Gases in transformers and other Oil-Filled Electrical Equipment in Service (1978)
11. Ji, W.-Y., Zhou, A.-Y., Zhang, L., Jin, W.: A Method to Optimize Classifiers by Using Genetic Algorithms. Journal of Software 13(2), 245–249 (2002)
12. Srinivas, M., Patnaik, L.M.: Adaptive Probabilities of Crossover and Mutation in Genetic Algorithm. IEEE Trans. on Systems, Man and Cybernetics 24(4), 656–667 (1994)
13. Peng, N.: Research on an Intellectual Fault Diagnosis System for Power Transformer Based on Dissolved Gas Analysis Technology. School of electrical engineering Wuhan University, Wuhan (2004)

A Novel Approach to Edge Detection of Color Image Based on Quaternion Fractional Directional Differentiation

Chaobang Gao[1,2,3], Jiliu Zhou[2,3], Fangnian Lang[1,3], Qiang Pu[1,3], and Chang Liu[1,3]

[1] College of Computer Science and Technology, Chengdu University, Chengdu 610106, China
[2] School of Computer Science, Sichuan University, Chengdu 610064, China
[3] Key Laboratory of Pattern Recognition and Intelligent Information Processing, Sichuan Province, Chengdu 610106, China
kobren427@163.com

Abstract. In this paper, we denote a color image by a quaternion function, then find edge points by solving the maximum of quaternion fractional directional differentiation(QFDD)'s norm. This method is called edge detection based on QFDD. Experiments indicate that the method has special advantages. Comparing with Canny, LOG, Sobel, and general fractional differentiation, we discover that QFDD has fewer false negatives in the textured regions and is also better at detecting edges which are partially defined by texture, which means we will obtain better results in the interesting regions by QFDD and these results are more consistent with the characteristics of human visual system.

Keywords: quaternion, fractional differentiation, QFDD, norm, similarity.

1 Introduction

Edge detection is one of basic problems in image processing and computer vision. Edge points are associated with the boundaries of objects in the image. The criteria proposed by Canny is used as the performance evaluation standard of edge detection algorithm [1]. Canny's criteria is an ideal standard when it is used to evaluate the accuracy of edge detection algorithm. For a particular edge detection operator, the three aspects of the criteria are mutual restraint and difficult to take into account, therefore each edge detection algorithm has its own scope and focus. For example, Martin et al at Berkeley University think that a good edge detection result should be consistent with human vision [2,3]. In this paper, we hope to find a novel approach to edge detection of color images whose edges are more consistent with human vision.

In recent years, quaternion has been used to process color image [4], in addition, the fractional calculus has been applied to digital image processing because of its good characteristics of dealing with non-stationary signal in many articles [5-7]. Based on the theories of quaternion and fractional calculus and their specific applications in digital image processing, the authors proposed the theory of quaternion fractional directional

G. Lee (Ed.): Advances in Automation and Robotics, Vol. 1, LNEE 122, pp. 163–170.
springerlink.com

differentiation (QFDD) and applied it to image enhancement in [8]. Now, we apply QFDD to edge detection of color image.

2 Theory of QFDD

In seeking to construct quaternion fractional directional differentiation, in [8], the authors first gave the definition of quaternion fractional directional derivative. For a monadic signal denoted by quaternion function with respect to one variable t : $F(t)=f_r(t)+f_i(t)i+f_j(t)j+f_k(t)k$ ($f_r(t)$, $f_i(t)$, $f_j(t)$, and $f_k(t)$ are real numbers), there exists continuous derivative of order $m+1$. When $v>0$, m is no less than $[v]$, so v order derivative could be expressed as:

$$ {}_a^G D_t^v F(t) = \lim_{h\to 0} \sum_{m=0}^{\left[\frac{t-a}{h}\right]} \frac{(-1)^m \Gamma(v+1)}{h^v m! \Gamma(v-m+1)} F(t+mh), \tag{1} $$

written as $d^v F / dt^v$, hence (1) is called the quaternion fractional derivative of $F(t)$. This is a generalized definition of G-L. Its geometric meaning is the fractional slope and the physical meaning is the generalized amplitude-and-phase modulation of the signal $F(t)$, that is, the fractional amplitude-and-phase modulation [5-9].

Similarly, the authors of [8] gave the definition of quaternion fractional directional derivative for a quaternion function $F(x,y)=f_r(x,y)+f_i(x,y)i+f_j(x,y)j+f_k(x,y)k$, where $f_r(x,y)$, $f_i(x,y)$, $f_j(x,y)$, and $f_k(x,y)$ are real numbers.

$$ \frac{\partial^v F}{\partial n_1^v} = \lim_{h\to 0} \sum_{m=0}^{N} \frac{(-1)^m \Gamma(v+1)}{h^v m! \Gamma(v-m+1)} F(x+mh, y), \tag{2} $$

$$ \frac{\partial^v F}{\partial n_2^v} = \lim_{h\to 0} \sum_{m=0}^{N} \frac{(-1)^m \Gamma(v+1)}{h^v m! \Gamma(v-m+1)} F(x-mh, y), \tag{3} $$

$$ \frac{\partial^v F}{\partial n_3^v} = \lim_{h\to 0} \sum_{m=0}^{N} \frac{(-1)^m \Gamma(v+1)}{h^v m! \Gamma(v-m+1)} F(x, y+mh), \tag{4} $$

$$ \frac{\partial^v F}{\partial n_4^v} = \lim_{h\to 0} \sum_{m=0}^{N} \frac{(-1)^m \Gamma(v+1)}{h^v m! \Gamma(v-m+1)} F(x, y-mh), \tag{5} $$

$$ \frac{\partial^v F}{\partial n_5^v} = \lim_{h\to 0} \sum_{m=0}^{N} \frac{(-1)^m \Gamma(v+1)}{h^v m! \Gamma(v-m+1)} F(x+mh, y+mh), \tag{6} $$

$$ \frac{\partial^v F}{\partial n_6^v} = \lim_{h\to 0} \sum_{m=0}^{N} \frac{(-1)^m \Gamma(v+1)}{h^v m! \Gamma(v-m+1)} F(x+mh, y-mh), \tag{7} $$

$$\frac{\partial^v F}{\partial n_7^{\ v}} = \lim_{h\to 0}\sum_{m=0}^{N}\frac{(-1)^m\Gamma(v+1)}{h^v m!\Gamma(v-m+1)}F(x-mh,y+mh), \tag{8}$$

$$\frac{\partial^v F}{\partial n_8^{\ v}} = \lim_{h\to 0}\sum_{m=0}^{N}\frac{(-1)^m\Gamma(v+1)}{h^v m!\Gamma(v-m+1)}F(x-mh,y-mh), \tag{9}$$

where $N = \left[(t-a)/h\right]$, $n_1, n_2, \cdots, n_8$ denote the positive x-coordinate direction, negative x-coordinate direction, positive y-coordinate direction, negative y-coordinate direction, right downward diagonal direction, left downward diagonal direction, right upward diagonal direction, and left upward diagonal direction.

By (2-9), the quaternion fractional directional differentiation (QFDD) of $F(x, y)$ is defined:

$$d_v F = \sum_{i=1}^{8}\frac{\partial^v F}{\partial n_i^{\ v}}dn_i^{\ v}. \tag{10}$$

It shows the change of quaternion function $F(x, y)$ at (x, y) along the direction $n_i, i = 1, 2, \cdots, 8$. And the norm of $d_v F$ is:

$$N(d_v F) = d_v F \cdot \overline{d_v F} = \sum_{i=1}^{8}\sum_{j=1}^{8} a_{ij} dn_i^v dn_j^v, \tag{11}$$

where $a_{ij}, i = 1, 2, \cdots, 8, j = 1, 2, \cdots, 8$, denotes the coefficient of $dn_i^v dn_j^v$. Although QFDD is a linear operator, its norm is not linear clearly. In addition, QFDD defined in (10) does not possess rotation invariance (isotropy), but its norm (11) possesses the property.

Since edges can be defined as where gradient of image intensity function reaches its local maximum, in other words, edge points are points in the image where pixel brightness changes drastically, we can also find edges by the maximum of QFDD's norm. The authors of [8] gave a way to solve the maximum of $N(d_v F)$ as follows. First view (11) as a quadratic type:

$$N(d_v F) = n' A n, \tag{12}$$

where $A = \left(a_{ij}\right)_{8\times 8}$, $n = \left(dn_1^v, dn_2^v, dn_3^v, dn_4^v, dn_5^v, dn_6^v, dn_7^v, dn_8^v\right)'$. Then give the following theorem and corollary [8,9]:

Theorem: Let A be a symmetric matrix of order m and its eigenvalues arranged in order: $\lambda_1 \ge \lambda_2 \ge \cdots \ge \lambda_m$. For any column vector $\alpha \ne 0$, then

$$\lambda_m \le \frac{\alpha' A \alpha}{|\alpha|^2} \le \lambda_1. \tag{13}$$

It is easy to obtain the following corollary from Theorem:

Corollary: Let A be a symmetric matrix of order m and its eigenvalues arranged in order: $\lambda_1 \ge \lambda_2 \ge \cdots \ge \lambda_m$. For any unit column vector α, then

$$\lambda_m \leq \alpha' A \alpha \leq \lambda_1 . \tag{14}$$

Consequently, to solve the maximum of $N(d_v F)$ in (11), just need to find the largest eigenvalue and the corresponding eigenvectors of A in (12). Since A is an 8×8 matrix, it is easy to find the corresponding eigenvalue and eigenvector. In other words, for a color image, RGB channels are viewed as three components of a quaternion function, next only consider the largest variable of this quaternion function at each point by the norm $N(d_v F)$. In addition, in [8], the authors gave the numerical implement of QFDD by backward difference expression. The corresponding numerical calculation mask is shown in Fig.1.

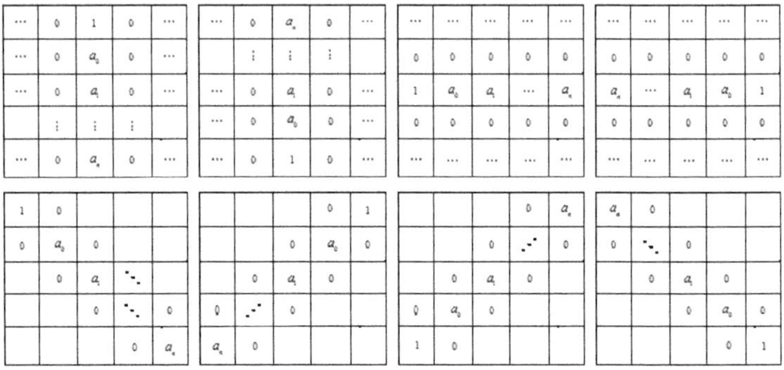

Fig. 1. Quaternion fractional directional derivative masks

In Fig.1, the mask coeffcients $a_0 = -v$, $a_1 = (v - v^2)/2$, $a_2 = (v^3 - 3v^2 + 2v)/6$, $\cdots$, $a_n = \Gamma(n-v-1)/[(n-1)!\Gamma(-v)]$. To calculate the approximate value of quaternion fractional directional derivative of two-dimension signal $F(x, y)$ by Fig.1, we can move the mask point by point on the image plane.

3 Experimental Results

For a color image, we may choose 5×5 masks just as Fig.1 to compute quaternion fractional directional derivative of $F(x, y)$, then find a_{ij} according to (11), next need to solve the maximum in (11) by Theorem. At last, we can discriminate edge points by a threshold. To compare with integer-order difference, we choose Canny, Sobel and LOG operators to do edge detection respectively. The results are shown in Fig.2. In general, when the order $v \in [0,1]$, QFDD can be very accurate to extract texture information of image, and when $v \in [1,2]$, it can be very accurate to extract the edge information which is consistent with human visual system, while $v > 2$, the amount of image information will reduce sharply and also make edge information reduce drastically, which is not useful to edge detection. Therefore, we choose $1 < v < 2$ as edge detection in general.

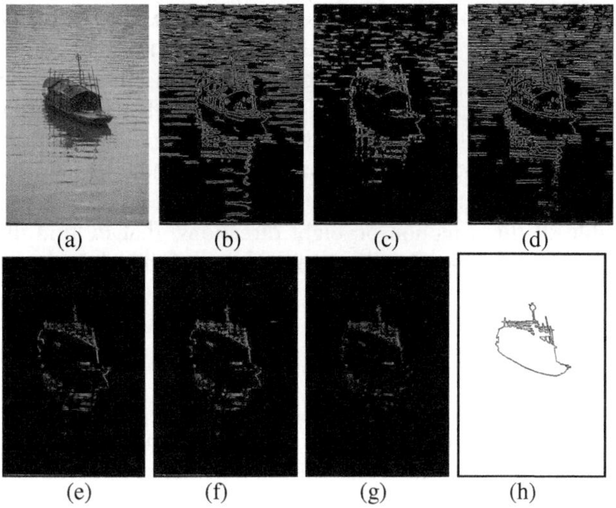

Fig. 2. Information images by QFDD, Canny, Sobel, LOG. (a)original, (b)Canny, (c) Sobel, (d)LOG, (e) v=1.2 QFDD, (f) v=1.5 QFDD , (g) v=1.9 QFDD, (h) Huamn.

From Fig.2, by eye's qualitative analysis, we note that QFDD also enhances the edge information of high-frequency as same as integer-order differentiation. But its enhancement effect is poorer than that of integer order differentiation. Moreover, Fig.2 (b)-(d) contain a lot of false negatives in the textured regions. While QFDD is different, it almost completely maintain contour information of the original image on the subject, at the same time, it also weakens false negatives in the textured regions. With the increasing of order v, the false negatives in the textured regions can even be completely removed, which is more consistent with the characteristics of human visual system.

The contradistinctive results between general fractional differentiation (GFD) and QFDD are shown in Fig.3.

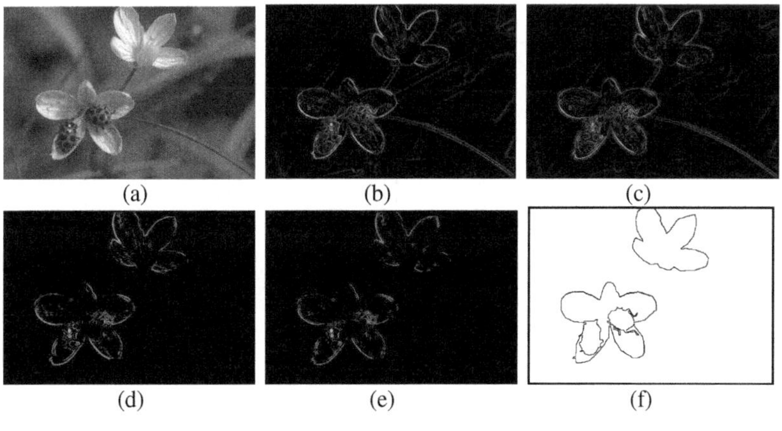

Fig. 3. Information images by GFD and QFDD. (a)original, (b)v=1.3 GFD, (c)v=1.4GFD, (d)v=1.4QFDD, (e)v=1.8 QFDD, (f) Human.

By contrast, there are some differences between them. Let us first perform GFD to every channel of color image separately, then mix three kinds of edges to form edges. By this method, it can do the max change of gray-scale in the gradient direction at every point for RGB channels, however, it is possible that the value mixed by three maximums from R, G and B channels may not be the maximum of the whole image at the point. That is, it may not be an edge point of the whole image. Because RGB are viewed as three components of a quaternion function in this paper, only need to find the max variable of this function on eight directions, that is, find the maximum of QFDD's norm, then the point must be an edge point of the whole image under certain conditions. From Fig.3(d)(e), we see the results are better.

4 Results Analysis

To measure results of different algorithms for image segmentation, we compare these results with human edge images in Berkeley color images database [2,3]. The authors of [10] associate a measure of texture with each point in an image in order to suppress contour processing in textured regions and vice versa. Martin et al. [2,3] think that this segmentation is meaningful and true rather than blindly image processing. And they have put forward two indicators to evaluate the effectiveness of different edge filters---GCE (global consistency error) and LCE (local consistency error) [2]. In [11], the authors have also given two indicators to evaluate the results of detection edge-DER (the detection error rate) and DCR (the detection common rate). Moreover, in [9], the authors have given one new indicator (DCS) to evaluate the results of detection edge.

We first do the image binarization of edge information images by the self-adaptive threshold (OTSU), thus the edges of image are obtained. Some binarization edge images are shown in Fig.4-5.

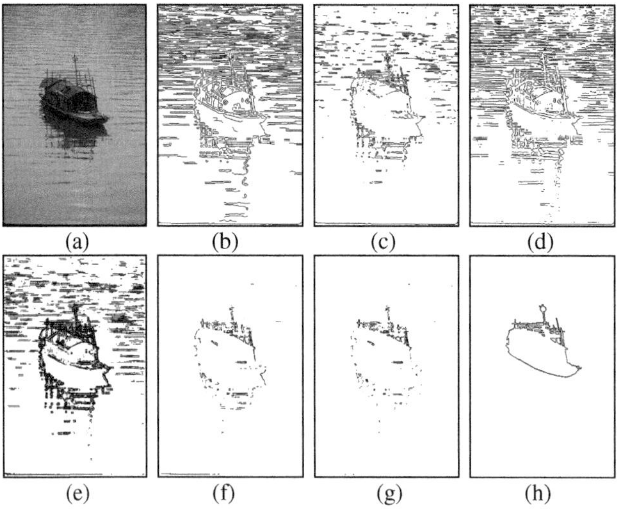

Fig. 4. Edge images. (a)original, (b)Canny, (c)Sobel, (d)LOG, (e) v=1.4GFD, (f)v=1.5 QFDD, (g)v=1.9 QFDD, (h) Huamn.

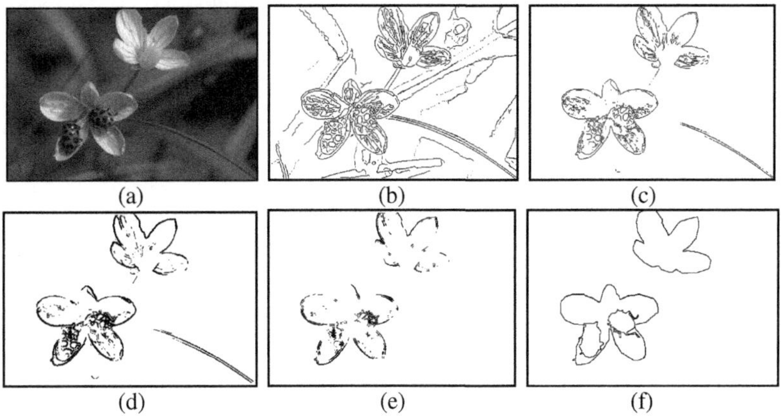

Fig. 5. Edge images. (a)original, (b)Canny, (c)Sobel, (d)LOG, (e) v=1.4GFD, (f)v=1.5 QFDD, (g)v=1.9 QFDD, (h) Huamn.

From Fig.4-5 and by eye's qualitative analysis, we observe that QFDD can suppress nearby weak parallel edges in some degree and it is good at enhancing edges among different textures, that is, it has fewer false negatives in the textured regions and is also better at detecting edges which are partially defined by texture. On the other hand, we do not see thin edges as well as non-maximal suppression applied to Canny, Sobel, LOG. By contrast, Canny, Sobel and LOG operators are slightly better at precision of certain edges. Compared with human image in Berkeley database, the edges obtained by QFDD are more similar than the other methods.

To compute the DCS [9], we view the image in Berkeley segmentation database as the benchmark, i.e. as E_2, the edge images of different algorithms as E_1, we have listed the graphs of DCS with respect to the order v in Fig.6.

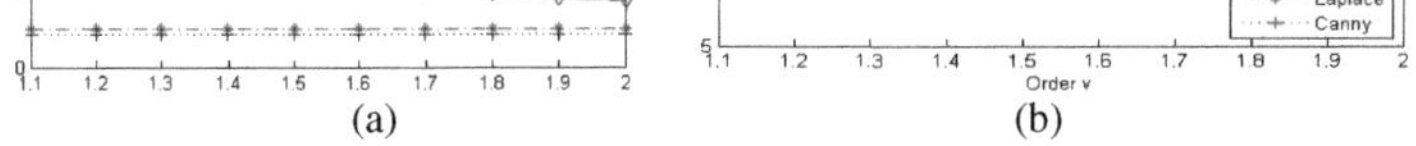

Fig. 6. The graphs of DCS with respect to v. (a) DCS of images in Fig.4, (b) DCS of images in Fig.5.

Since the greater DCS is, the more similar E_1 and E_2 are, which implies the results will be better and also more consistent with principles of human visual system. From Fig.6, we can see intuitively that the curves of edge images obtained by QFDD are clearly above on those of edge images obtained by others methods, which means that the similarity of edges obtained by QFDD is better than by Canny, Sobel, LOG, and GFD, that is, the edge images obtained by QFDD are closer to the Berkeley image database.

5 Conclusions

In order to make edge detection more consistent with the characteristics of human visual system, a novel detector, the QFDD detector, are obtained though the generalization of general fractional differentiation and quaternion. Experiments indicate that this method has special advantages and much better results compared with Canny, Sobel, LOG and general fractional differentiation.

Acknowledgment. This work is supported by the Natural Science Foundation of China (No.60773168), the Project of Science Technology Department of Sichuan Province (No.2011JY0077, No. 2010JQ0032, 2010SZ0004).

References

1. Canny, J.: A computational approach to edge detection. IEEE Transaction on Pattern Analysis and Machine Intelligence 8(6), 679–698 (1986)
2. Martin, D., Fowlkes, C., Tal, D., Malik, J.: A database of human segmented natural images and its application to evaluating segmentation algorithms and measuring ecological statistics. In: Proceedings Eighth International Conference on Computer Vision, vol. 2, pp. 416–423 (2001)
3. Martin, D., Fowlkes, C.C., Malik, J.: Learning to detect natural image boundaries using local brightness, color, and texture cue. IEEE Transaction on Pattern Analysis and Machine Intelligence 26(5), 530–549 (2004)
4. Sangwine, S.J., Ell, T.A.: Colour image filters based on hypercomplex convolution. In: IEE Proceedings-Vision, Image and Signal Processing, vol. 147 (2), pp. 89–93 (2000)
5. Mathieu, B., Melchior, P., Oustaloup, A., Ceyral, C.: Fractional differentiation for edge detection. Signal Processing 83, 2421–2432 (2003)
6. Pu, Y.F., Zhou, J.L., Yuan, X.: Fractional differential mask: a fractional differential based approach for multi-scale texture enhancement. IEEE Transactions on Image Processing 19(2), 491–511 (2010)
7. Gao, C.B., Zhou, J.L., Zheng, X.Q., Lang, F.N.: Image enhancement based on improved fractional differentiation. Journal of Computational Information Systems 7(1), 257–264 (2011)
8. Gao, C.B., Zhou, J.L.: Image enhancement based on quaternion fractional directional differentiation. Acta Automatica Sinica 37(2), 150–159 (2011)
9. Gao, C.B., Zhou, J.L., Hu, J.R., Lang, F.N.: Edge Detection of Color Image Based on Quaternion Fractional Differential. IET Image Processing 5(3), 261–272 (2011)
10. Malik, J., Belongie, S., Leung, T., Shi, J.: Contour and texture analysis for image segmentation. International Journal of Computer Vision 43(1), 7–27 (2001)
11. Hua, J., Wang, J., Yang, J.: A novel approach to edge detection based on PCA. Journal of Image and Graphics 14(5), 912–919 (2009)

Cost Distribution of Multi-path Based on Vehicle Speed on Cross-Section

Wei Cai[1], Anguo Dong[2], Hongzhi Xue[2], Xueli Song[2], and Bo Chen[3]

[1] ChangAn University, Xi'an, China 710064
[2] School of Science, ChangAn University, Xi'an, China 710064
[3] School of Economy and management, ChangAn University, Xi'an, China 710064

Abstract. In order to solve the toll distribution problem of highway, a multi-path cost partition method for the highway is proposed. The probability distribution of the time is analyzed by velocity distribution of the vehicle on the path of each section and the length of each section. By means of mixed probability model, the probability distribution of travel time from the entrance to the exit is decomposed into the linear combination of the probability distribution for travel time of each path, where the combination coefficient is partition one. Finally, the simulation results show that the partition coefficient is consistent with the vehicle segregation coefficient.

Keywords: Multi-path cost distribution, cross-section vehicles speed, the shortest path, mixed probability model.

1 Introduction

The high speed road networks have gradually formed a ring form, which brings the driver the choice of multiple paths. Meanwhile, the highway properties investment subject, and diversification of the management system, make the charge mode in a networked collection of tolls and split the problem complicated. The shortest path methods are based on theory driving the vehicle by the shortest route path selection. However, problems of multi-path selection make this assumption not need practical cases, which causes path selection unreasonable.

The main purpose of path identification is to solve the collection and partition problem of the vehicle toll. According to the road network structure, investment, management and other factors, two solutions are proposed: ①The shortest path charges, the probability of partition; ②Driving path charges, accurate partition. The first program employ a multi-scale model of the path to get the multi-path traffic the probability distribution on sites in the path of multi-path traffic, See literature [1-9]. In detail, these methods are based on statistical investigation or traffic assignment model. However these models have any real-time information, not self-adaptive. In addition, parameters in the model can not fully reflect all the factors affecting the traffic choice, which cause the allocation ratio deviate the actual case. The second program is based on accurate identification of the vehicle path. Hence, the accurate identification is the core of the program. At present, the methods mainly employ electronic tags to identify

G. Lee (Ed.): Advances in Automation and Robotics, Vol. 1, LNEE 122, pp. 171–178.
springerlink.com

RFID-based, vehicle license plate recognition technology, supplemented by mutual confirmed. Although the individual technology in the system has been mature, the overall product technology is not mature, especially compatibility problems of equipments, identification rates. In addition, the hardware construction and equipment management needs a lot of money. For the cost partition issue, so many devices and systems equipment in the highway road is clearly not economical.

Based on the above problems, the paper proposes the multi-path cost partition method based on the cross-section vehicle speed. In detail, without any additional investment in facilities, we make full use of the highway topology information, cross-sectional vehicle speed timing information, time information of the entrance and the exit and the use of hybrid probabilistic model [10-11] to obtain mixing coefficient by least squares approximation. We regard these coefficients as multi-path cost partition coefficients. Theoretical analysis and experimental results demonstrate that the method derived from the program is consistent with the ratio of real path choice. The method enjoys high accuracy, good stability and strong real-time and don't need additional survey data.

The multi-path cost partition method based on the cross section speed is based on the following information: 1. topology structure of the road network; 2. the vehicle speed data derived from highway measuring speed equipment (without vehicle identification); 3. moment of vehicles entering and going out the road network and the type of vehicle.

2 The Probability Distribution of the Vehicle Traveling Time

Suppose that there are n vehicles from A to B every day three available paths, because the travel time T of the n vehicles on the network (from the moment of entrance to moment of the exit) is known and the probability distribution of vehicles in each path travel time $T_i = (i = 1,2,3)$ is determined by road section speed and section length.

As shown in Fig.1, since the length and the travel speed for each path are different, the probability distribution of the travel time for all paths can be calculated by using the speed distribution of each path.

Fig. 1. A path from the nodes i to j

We denoted by $V_{ijmt} (t = 1,2,\cdots,|L_{ijm}|)$ the speed of the vehicle on each section of the m-th path L_{ijm} from i to j, where $|L_{ijm}|$ is the number of sections of L_{ijm},

$d(L_{ijm})$ denotes the distance of the path L_{ijm}, and $d_t(L_{ijm})$ the length of each section of the path L_{ijm}.

Theorem 1: Suppose $V \sim N(u,\sigma^2)$[14–15]. Then $T = \frac{l}{V}$ is obeyed the inverse Gaussian distribution with the density function $f(t) = \frac{1}{\sqrt{2\pi\sigma t^2}} e^{-\frac{(l-ut)^2}{2\sigma^2 t^2}}$

Proof: Set $F(t)$ is the distribution function of T. Then

$$F(t) = p(T \le t) = p(\frac{l}{V} \le t) = p(V \le \frac{l}{t}) = \int_{-\infty}^{l/t} \frac{1}{\sqrt{2\pi\sigma}} e^{-\frac{(x-u)^2}{2\sigma^2}} dx,$$

$$f(t) = \frac{dF(t)}{dt} = \frac{1}{\sqrt{2\pi\sigma t^2}} e^{-\frac{(l-ut)^2}{2\sigma^2 t^2}}.$$

Theorem 2: Let D_1 and D_2 denote two sections (which also denote their lengths), and V_1 and V_2 the random variable of the running speed on them. Then the average speed on the path $L_1 + L_2$ is $\overline{V} = \frac{D_1 + D_2}{D_1V_2 + D_2V_1} V_1V_2$.

Proof: We denote by t_1 and t_2 the running time on D_1 and D_2, respectively. Then $\overline{V} = \frac{D_1 + D_2}{t_1 + t_2} = \frac{D_1 + D_2}{D_1/V_1 + D_2/V_2} = \frac{D_1 + D_2}{D_1V_2 + D_2V_1} V_1V_2$.

Corollary 1: If $V_1 \approx kV_2$ (The speed ratio for each vehicle on a different section is a constant), then $\overline{V} = \frac{D_1 + D_2}{D_1 + kD_2} kV_2$; If $V_i \sim N(u_i,\sigma_i^2)$ $(i = 1,2)$, then $\overline{V} \sim N(\overline{u},\overline{\sigma}^2)$, where $\overline{u} = \alpha u_2, \overline{\sigma} = \alpha\sigma_2$ and $\alpha = \frac{D_1 + D_2}{D_1 + kD_2} k$.

Corollary 2: Suppose the values of V_1 and V_2 are similar (the difference of the speed for each vehicle on the different section is very small). Then $\overline{V} = \frac{D_1}{D_1 + D_2} V_1 + \frac{D_2}{D_1 + D_2} V_2$. If $V_i \sim N(u_i,\sigma_i^2)$ $(i = 1,2)$, we have $\overline{V} \sim N(\overline{u},\overline{\sigma}^2)$, where $\overline{u} = \alpha_1 u_1 + \alpha_2 u_2$, $\overline{\sigma} \approx \alpha_1\sigma_1 + \alpha_2\sigma_2$, $\alpha_1 = \frac{D_1}{D_1 + D_2}$ and $\alpha_2 = \frac{D_2}{D_1 + D_2}$.

By Theorem 1 and Theorem 2 we can obtain the following conclusions.

Theorem 3: Suppose $V_i \sim N(u_i,\sigma_i^2)$ $(i = 1,2)$, and T to be the running time on the path $D_1 + D_2$. If $V_1 \approx kV_2$, T is obeyed the inverse Gaussian distribution, whose density function is $f(t)=\frac{l}{\sqrt{2\pi\bar{\sigma}t^2}}e^{-\frac{(l-\bar{u}t)^2}{2\bar{\sigma}^2t^2}}$, where $\bar{u}=\alpha u_2$, $\bar{\sigma}=\alpha\sigma_2$, $\alpha=\frac{l}{D_1+kD_2}k$, and $l=D_1+D_2$.

Theorem 4: Suppose $V_i \sim N(u_i,\sigma_i^2)$ $(i=1,2)$ and T to be running time on the path D_1+D_2, If the values of V_1 and V_2 are similar, then T is obeyed the inverse Gaussian distribution, and its density function is $f(t)=\frac{l}{\sqrt{2\pi\bar{\sigma}t^2}}e^{-\frac{(l-\bar{u}t)^2}{2\bar{\sigma}^2t^2}}$, where $\bar{u}=\alpha_1u_1+\alpha_2u_2, \bar{\sigma}\approx\alpha_1\sigma_1+\alpha_2\sigma_2$, and $l=D_1+D_2$.

The similar conclusion can be obtained if there are more than 2 sections on the path.

It is easy to prove the above Theorem and Corollary. However, their conditions are approximate, such as $V_1 \approx kV_2$ is defined clearly. Now we only verify the above conclusions by some experiments.

Example 1: There are two sections for some path, where the length of first section is 80, and the length of second section is 30. The above and the below parts for each figure in Fig.2 can be obtained by Theorem 3 and Theorem 4 respectively. The red curve is the use of road section speed (normal distribution random generation[1]) actual travel time distribution, and the blue curve can be obtained by theoretical analysis. (a) shows that the error of Theorem 4 is larger if the difference of the mean is larger while the result for The Theorem 3 is better, (b) shows that the result of Theorem 4 are both better when the difference of the mean is not larger .(c) shows that the error of Theorem 3 is larger and the result of Theorem 4 is better if the difference of the mean is not larger but the rations of the mean and the mean-square deviation is not consistent.

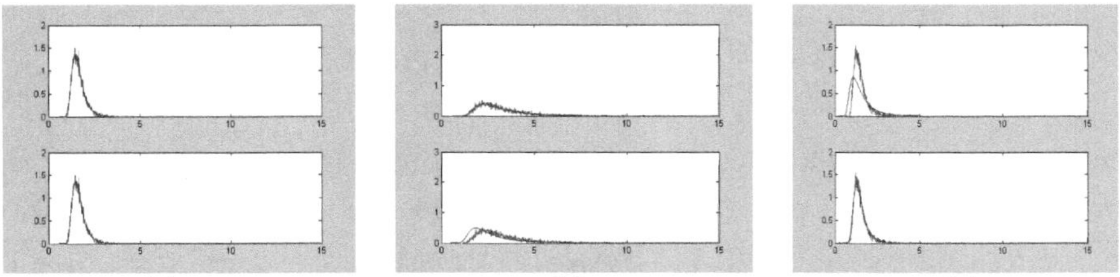

(a) (b) (c)

Fig. 2. Actual path travel time and theoretical travel time of the frequency distribution

Remark: (a) corresponds to the $V_1 \sim N(30,13^2)$, $V_2 \sim N(85,39^2)$

(b) corresponds to the $V_1 \sim N(65,13^2)$, $V_2 \sim N(85,18^2)$

(c) corresponds to the $V_1 \sim N(75,13^2)$, $V_2 \sim N(85,38^2)$

3 The Cost Partition Coefficient Based on Time Distribution of Path

Algorithm calculating the first k-shortest paths of two nodal points A and B is discussed in [13]. Our paper only considers the first 3-shortest paths, that is, we calculate the ratio of vehicles from A to B choosing these 3 paths. By the analysis of the above section, we can get the probability distribution of the travel time $T_i(i=1,2,3)$ of each path. Suppose the corresponding probability density function is $f_i(t)(i=1,2,3)$. Since the moments that the vehicles enter into A and get away form B is known, we can get the running time of all vehicles entering into A and get away from B during a period of time is random variable T. Suppose there exist n vehicle. The corresponding running time forms a sample $t_i(i=1,2,\cdots,n)$. Suppose the ratio of the vehicles passing these three paths is $\alpha_i(i=1,2,3)$, The we can get the following statement.

Theorem 5: (Mixed probability distribution) Let the density function of T be $f(t)$. Then $f(t)=\alpha_1 f_1(t)+\alpha_2 f_2(t)+\alpha_3 f_3(t)$, $\alpha_1+\alpha_2+\alpha_3=1$.

Proof:
$$P(T<t)=P(A_1)P(T<t/A_1)+P(A_2)P(T<t/A_2)+P(A_3)P(T<t/A_3)$$
$$=\alpha_1 P(T_1<t)+\alpha_2 P(T_2<t)+\alpha_3 P(T_3<t).$$

Where $f(t)=\alpha_1 f_1(t)+\alpha_2 f_2(t)+\alpha_3 f_3(t)$, $A_i(i=1,2,3)$ means the event vehicle passes the i-th path, respectively. Because the n samples $t_i(i=1,2,3)$ for t were known, we can denote by the frequency distribution $\hat{f}(t)$ the estimate of $\hat{f}(t)$. Sampling $\hat{f}(t), f_1(t), f_2(t), f_3(t)$ at $t_i(i=1,2,\cdots,m)$, we can get the samples $\hat{f}(t_i), f_1(t_i), f_2(t_i), f_3(t_i), (i=1,2,\cdots,m)$. Suppose the proportion of vehicles for three paths is $(\alpha_1,\alpha_2,\alpha_3)$. Then $\hat{f}(t) = \alpha_1 f_1(t) + \alpha_2 f_2(t) + \alpha_3 f_3(t)$.

By the above samples, we can get fitting residual squared sum is $(\alpha_1,\alpha_2,\alpha_3)^T=(M^T M)^{-1}(M^T b)$, where

$$M=\begin{pmatrix} f_1(t_1) & f_2(t_1) & f_3(t_1) \\ f_1(t_2) & f_2(t_2) & f_3(t_2) \\ \cdots & & \\ f_1(t_m) & f_2(t_m) & f_3(t_m) \end{pmatrix},\ b=(\hat{f}(t_1)\ \ \hat{f}(t_2)\ \ \cdots\ \ \hat{f}(t_m))^T.$$

4 Simulation Analysis

4.1 Experimental Plan Design

This paper mainly calculates the split coefficient based on the speed of the vehicle running on the different section of the path. Hence, we can not obtain all the data of the running speed and we get these data by simulating method. Firstly, the plan is designed as follows:

Suppose there are 3 paths to be chosen for the entrance A and the exist B: and there exist n vehicles from A to B. Given probability values $p_1, p_2, p_3 (p_1 + p_2 + p_3 = 1)$, we can choose the i-th $(i = 1,2,3)$ paths based on the probability p_i for each vehicle, and obtain a running speed on each section of the path. Then the running speed of each vehicle can be derived, that is the n samples $t_i (i = 1,2,\cdots,n)$ of T can be get, and the actual number of the vehicles passing every path can also be obtained. We can get the real proportion of vehicles $(\beta_1, \beta_2, \beta_3)$. Moreover, there are speed samples for each section of a path (here, we dose not use the number of vehicle because the different path may pass the same section). From these samples, we can estimate the probability distribution of the speed for the vehicle on this path. By the distributions and Theorem 3 or Theorem 4 we can estimate the time probability distribution $f_1(t), f_2(t)$ and $f_3(t)$ for each path. Furthermore, using the partition algorithm on Section 3 we can get proportion coefficient $(\alpha_1, \alpha_2, \alpha_3)$. Lastly, compared with the real proportion $(\beta_1, \beta_2, \beta_3)$, the accuracy of the partition algorithm can be verified.

4.2 Simulation

For the road network shown in Fig.3 (the number in edge denotes the index of the edge), the shortest path, the shorter path, and the short path of the nodes 7-9 are (9,11), (15,18), and (15,10,11) respectively. Take $n = 100000$; for the number of vehicles, and the real proportion coefficient of the vehicles on these three paths is $(\beta_1, \beta_2, \beta_3) = (0.4522839, 0.3065552, 0.2411609)$. We can calculate the proportion coefficient $(\alpha_1, \alpha_2, \alpha_3) =$ (0.4522803,0.3065628,0.2411568) . The corresponding time probability distribution is shown in Fig 4.

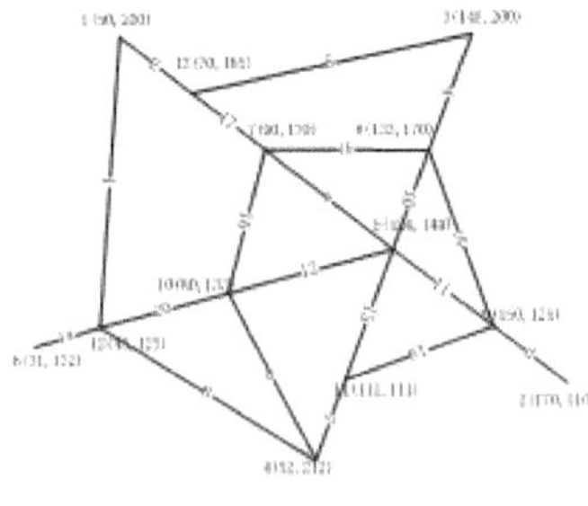

Fig. 3. Highway network

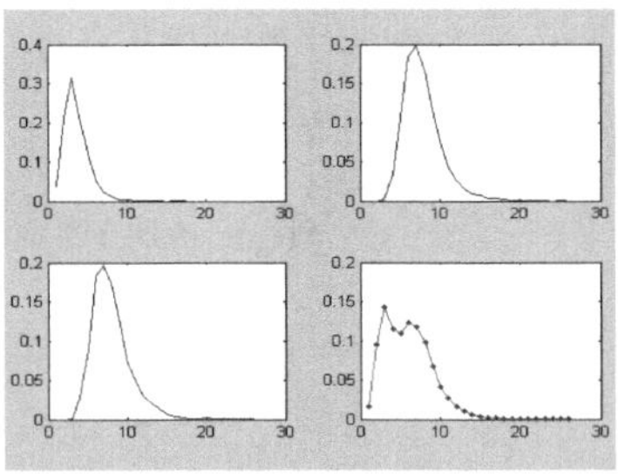

Fig. 4. The density function of travel time

Remark: The first three curves are the density function of travel time for three paths, respecting the blue line the last curve denotes the travel time distribution for the real vehicle and the red curve is the distribution curve by mixed the travel time of vehicles traveling on these three paths based on the proportion of calculated results.

5 Conclusion

Cost partition plan of section vehicle speed based on multi-path is low cost, high precision split. Implementation requires only the use existing speed equipment, records the vehicle speed (do not need to record vehicle identity) and combines with the moment of vehicle into the network and away the network. The partition coefficient of relative path is calculated, and we find the partition coefficient of relative sections. It provides the basis for the cost split of whole network. In the process of speed, if some vehicles are not detected, that will not influence the calculation accuracy. This algorithm is not use the number of vehicles, only use cross sections vehicle speed of probability distribution. Therefore, the proposed partition scheme enjoys high stability. This paper considers only the three shortest paths; In general, if the k-path is considered, similar algorithm in this paper, split coefficient can be obtained, where path determination algorithm in ref [12-13]. Moreover, this calculation does not consider the vehicle to enter service area time to rest, this impacts on determining the split coefficient. By the distribution of time of special entrance and exit (only one path) in road network, probability distribution of section speed separate service area rest time, we will further research which the time is introduced into the partition algorithm.

References

1. Li, S.-G., Ju, Y.-F., Ji, J.-G.: Path Flow Estimation Method for Highway Toll Distribution Application. Journal of Beijing University of Technology 35(8), 1028–1032 (2009)
2. Cao, C.-Z., Zheng, Q.-J.: Expressway path notarizing technique discussion. Computer and Communications 18(5), 4–8 (2000)
3. Du, H., Zhang, Y., Song, J.: Allocating algorithm for expressway network with ring structure. Journal of Highway and Transportation Research and Development 20(1), 110–113 (2003)
4. Bell, M.G.H., Shield, C.M., Busch, F., et al.: A stochastic user equilibrium path flow estimator. Transportation Research C 5(1), 197–210 (1997)
5. Anthony, C., Piya, C.: Development of a path flow estimator for deriving steady state and time dependent origin destination trip tables. Institute of transportation studies, University of Califorina, California (2004)
6. Zhong, L.-D., Li, X.-W., Rong, J., et al.: Confirmation of the capacity of urban expressway basic segment. Journal of Beijing University of Technology 32(7), 605–609 (2006)
7. Nie, Y., Zhang, H.M.: Models and algorithms for the traffic assignment problem with link capacity constraints. Transportation Research Part B 38(2), 285–312 (2004)
8. Larsson, T., Patriksson, M.: An augmented lagrangiandual algorithm link capacity side constrained traffic assignment problems. Transportation Research B 29(2), 433–455 (1995)

9. Li, S.-G., Ju, Y.-F., Ji, J.-G.: Estimation model and algorithm of multi-mode path flow. Journal of Traffic and Transportation Engineering 9(1), 87–91 (2009)
10. Sheng, W., et al.: Gaussian-mixture Probability Hypothesis Density Filter Based Multitarget Tracking Algorithm for Image Plane of Scanning Optical Sensor. Acta Eronauticaet Astronautica Sinica 32(3), 479–506 (2011)
11. Ba-Ngu, V., Wing-Kin, M.: The Gaussian mixture probabili-ty hypothesis density filter. IEEE Transactions on Signal Processing 54(11), 4091–4104 (2006)
12. Chen, Y.-L.: Finding the first shortest paths in a time-window network. Computers & Operations Research 31, 499–513 (2004)
13. Eppstein, D.: Finding the k shortest paths. SIAM Journal of Computing 28, 652–673 (1998)
14. Zhou, H.-M., Ma, Y.-C., et al.: Study on Operating Speed Distribution of Highway Cross-section. Journal of East China Jiao tong University 25(5), 32–35 (2008)
15. Yan, Y., et al.: Research on Section Operating Speed Distribution Characteristics of Expressway. China Safety Science Journal 18(7), 171–176 (2008)

The Study of Applying the Fuzzy Overall Evaluation Method to Solve the Satisfaction of the Objective in the Separating-Planting Mechanism*

Li Dai[1,2], QunWei Hu[3], Xiong Zhao[3], and Yun Zhao[3]

[1] School of Biology System Engineering and Food Science, Zhejiang University, China
[2] School of Economics and Management, Zhejiang Sci-Tech University, China
[3] School of Machine and Automation Controlling, Zhejiang Sci-Tech University, China
dlzist@163.com

Abstract. In this paper, according to the points in ambiguity, nonlinear, multi-objective and the other characteristics of the separating-planting mechanism, it combines fuzzy multi-objective optimization algorithm, applies the fuzzy overall evaluation method of the fuzzy mathematics theory to solve the objective satisfaction of the separating-planting mechanism of the rice transplanter, and accordingly obtains the better objective value and more feasible solution. On the platform of the "Star Judging Method" of the separating-planting mechanism aided analysis software, through the parameters optimization to establish the objective initial judgment matrix, through the expert knowledge to establish evaluation set, through the digital design to establish judgment set, and then evaluate a better feasible solution among all of the solutions. This method provides a new evaluation method for further optimization of the separating-planting mechanism.

Keywords: Separating-planting mechanism, the fuzzy overall evaluation method, satisfaction, optimization.

1 Introduction

At present, the optimization design of the separating-planting mechanism is one of the main research contents of the transplanter researchers. The separating-planting mechanism has strong coupling, fuzziness, non-linear, multi-parameter, multi-objective characteristics. In the existing optimization methods of the separating-planting mechanism, the human-computer interactive optimization method is the most main optimization method, the main functions of this system are parameters optimization, virtual experiment, virtual test, etc.[1-3]. Because of the complexity of the separating-planting mechanism, the research of the optimization method of the mechanism also has been improvement and optimization in constant [4].

* This project is supported by: National Natural Science Foundation of China(No.51005214), Scientific research project of education department of Zhejiang Province(No.Y201009991).

In this paper, the elliptical planetary gears transplanting mechanism is as an example, it relies on the "Star Judging Method", through the optimization to obtain multiple groups of feasible objective values, then analyzes the satisfying kinematics objective values, through the fuzzy mathematics theory, in view of the points in the fuzziness characteristic of the parameter optimization of the separating-planting mechanism, applies the fuzzy overall evaluation method to optimize the separating-planting mechanism, obtain the satisfactions of the objective values of the kinematics and then find out the highest satisfaction evaluation parameter among them. This method provides an assessment gist for parameter optimization, as to further improve the mechanism performance purpose.

2 The Elliptical Planetary Gears Transplanting Mechanism

2.1 Introduction

The elliptical planetary gears transplanting mechanism is one of the high-speed rice transplanter commonly used in rotary transplanting mechanism, it has the characteristic of stable transmission, simple structure, good reliability, The elliptical planetary gears transplanting mechanism is sketched in Fig.1.

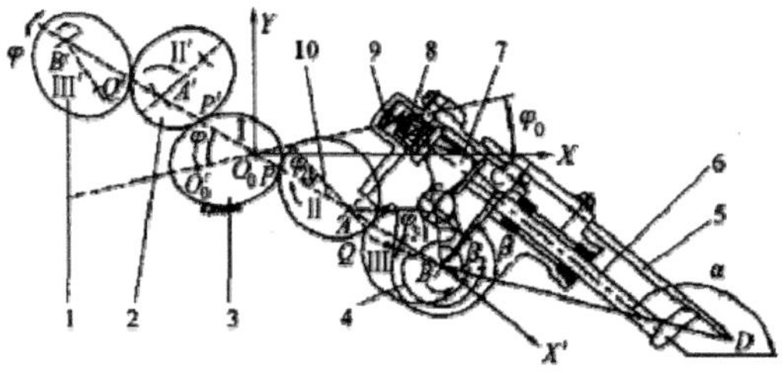

Fig. 1. The sketch of transplanting mechanism with elliptical planetary gears

There are five identical elliptic gears initially installed in the same phase. Two pairs of elliptic gears are located symmetrically at both sides of the sun gear 3. The transplanting-arm is fixed to the planetary gear 1, and the gearbox 10 shares the same shaft with the sun gear. The seeding-pushing device consists of cam 4, seeding-pushing rod 6 and transplanting–arm case 9. The transplanting–arm is composed of seeding-separating needle 5, spring 8, shifter 7 and seedling-pushing device. During operation, the sun gear is fixed with no movement while the gearbox is rotating. Two idle gears 2 rotate around the sun gear, driving the two planetary gears librating during operation. The seedling separating needle on the transplanting–arm performs a complex motion: An implicational clockwise rotary motion of the gearbox plus a relative anticlockwise rotary motion of the planetary gear. The motion results in the special locus and pose of seeding-separating needle[1,5].

2.2 The Trajectory of the Mechanism

According to the description in refs[1], then set up the coordinate system XOY which is shown in Fig. 1. According to the eight known parameters(shown in Table 1) to set up the displacement equations of a few key points that influences the transplantation quality which are shown in refs[1] (the theoretical analysis and derivation are shown in refs[5-6])) , then set up the kinematics displacement equation and angular displacement equation etc., further simulates the trajectory of transplanter on the platform of the software which is shown in Fig. 2.

Table 1. The instructions of the main input parameters

Input parameters	Descriptions
a	the long semi-axis of planetary gear
$\varphi 0$	the initial assemble angle of gearbox
$\delta 0$	the initial angle between the line that the shaft of planetary gear with the shaft of middle gear and the line that the shaft of middle gear with the shaft of sun gear
$\alpha 0$	the angle between initial assemble position of gearbox and the line from seeding-separating needle tip D to the shaft of planetary gear
k	the ratio of the elliptic short axis and long axis
S	the distance between the needle tip and the planet round turn center
H	the distance between seedlings
W	the angular velocity of the center axis

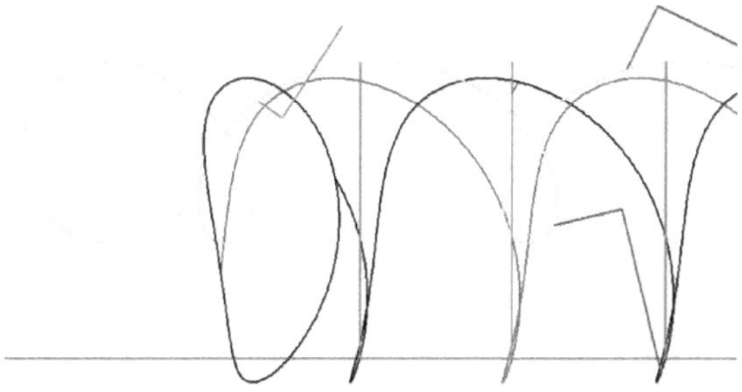

Fig. 2. The initial position trajectory diagram

This text is on the software platform in refs[1], inputs eight initial parameters, gets multiple satisfying objective values, then draws multiple trajectories which like the trajectory showed in Fig.2. But in those feasible goal solutions, *we can't judge which one is the best*, only the experts can identify the best solution.

Based on this, this paper applies the fuzzy overall evaluation method, finds out the satisfaction of all the possible objective values, so as to get the best solution. This method can not only further optimize the structure parameters, but also provide a more universal platform for the general farm machinery workers.

3 Fuzzy Overall Evaluation Method

3.1 Fuzzy Transform

Fuzzy transform is one of the base theories of fuzzy overall evaluation method.

The U and V are two sets, A is a limited fuzzy subset in U, and B is a limited fuzzy subset in V. A is expressed by fuzzy vector as: $A=(a_1,a_2,\cdots,a_n)$, a_i is the corresponding membership $\mu_A(u_i)(i=1,2,\cdots,n)$. Fuzzy set A is be transformed by fuzzy set T, then gets fuzzy set B, written for: $T(A)=B$.

When U and V are limited, the fuzzy transform T is fuzzy mapping T. Any to $R\in F(U\times V)$, as to be the only fuzzy transform that from U to V, written for: $T=T_R:F(U)\to F(V)$, that makes for any $A\in F(U)$, will get $T_R(A)=A\circ R\in F(V)$ [7].

3.2 Fuzzy Overall Evaluation

In production, scientific research and daily life, people often need to evaluate the quality of some thing, evaluate its effect degree, so that to do the corresponding processing. For example, evaluate some new product for its performance, evaluate some parameters for its reasonable, so as to improve product design, improve product quality. Because some thing has multiple attributes or multiple influence factors, so when evaluating the thing, we should give consideration to each aspect, and must consider all the related factors, that is the overall evaluation. If the factors involves fuzzy factors, that is *fuzzy overall evaluation* [8].

The optimization problem of the separating-planting mechanism is a typical problem which multiple factors influence multiple objectives problem, by its fuzzy characteristic, the fuzzy overall evaluation method can just be used to evaluate mechanism parameters, and then obtains the bad or good degree. The next we will analyze the multiple groups' solutions of the separating-planting mechanism through *the first level fuzzy overall evaluation method.*

Establish Factor Set. The factor set is a set that its elements are composed of all the factors which affect the judge object, usually be expressed as U, that is $U=(u_1,u_2,\cdots,u_n)$, each element $u_i(i=1,2,\cdots,n)$ represents each effect factor.

According to the theory analysis of the separating-planting mechanism in the refs[5], the optimization design of the separating-planting mechanism need eight input parameters (known parameters are shown in Table 1) and twelve kinematic objectives (constraint conditions are shown in Table 2). When a set of results meets the requirement of the kinematic trajectory, then it is a feasible solution.

Among the twelve objectives, only the object1 and object2 can be exact judged, the others all have fuzzy characteristic, namely: each objective constraint is digital to the objective value which is not only a value, but a scope, the values in the scope all can satisfy the requirements of the kinematic work. For example, the object5 is: the difference between the seedling-separating angle and the seedling-pushing angle should be 55°~60°, but for 54.5°, we can't judge that the transplanting trajectory can't finish the transplantation process.

Table 2. The meanings of the kinematics objectives

Kinematics objective	The constraint descriptions
object1	two transplanting–arms do not collision together
object2	seedling box does not touch the transplanting–arms
object3	take down the rectangular seedling block
object4	not push the seeding and not connect together of the adjacent seedlings
object5	the difference between the seedling-separating angle and the seedling-pushing angle is 55°~60°
object6	trajectory height is more than 260 mm
object7	seedling-pushing cave is less than 30 mm
object8	seedling-pushing angle is 65°~80°
object9	seedling-separating angle is -10°~20°
object10	the planetary gear center does not push and damage the seedling
object11	the gear modulus is more than 2.5 mm
object12	the distance of the disc from the floor is more than 200 mm

To sum up, in this case, the evaluation factor set is: $U=(u_1,u_2,\cdots,u_9,u_{10})$, each element is defined as follows: $u_1=Object3$, $u_2=Object4$, $u_3=Object5$, $u_4=Object6$, $u_5=Object7$, $u_6=Object8$, $u_7=Object9$, $u_8=Object10$, $u_9=Object11$, $u_{10}=Object12$.

Establish Factor Weight Set. According to the important degree of each factor $u_i(i=1,2,\cdots,n)$, we evaluate a corresponding weight to each factor $a_i(i=1,2,\cdots,n)$, the set being constituted of all the weight $a_i(i=1,2,\cdots,n)$ is called factor weight set, written for: $A=\{a_1,a_2,\cdots,a_n\}$. According to the arithmetic requirement, each $a_i(i=1,2,\cdots,n)$ should satisfy unitary and negative, its value require be subjective to confirm in general, also can be determined on its membership.

In this paper, according to the relationship between the kinematic objectives which obtained on the platform of the separating-planting mechanism aided analysis software and the trajectory, and scoring after calculation of the experts experience, we determine a group of weights:

$$A = \{0.1037,0.1185,0.1333,0.0889,0.1259,0.0741,0.0593,0.0444,0.1481,0.1037\}$$

Owing to the space is limited, the derived process here is omitted.

Establish Evaluation Set. The evaluation set is composed of the total evaluation results which people judge the evaluation objects, usually is shown in capital letter V , $V = \{v_1, v_2, \cdots, v_m\}$, and each element $v_i (i = 1, 2, \cdots, m)$ represents each evaluation result, the fuzzy overall evaluation purpose is based on the overall consideration all factors, draws a better evaluation result from the evaluation set, apparently the relationship of $v_i (i = 1, 2, \cdots, m)$ to V is a common set relationship. To improve the precision of the evaluation, and the scope of kinematic objectives , in this paper, the evaluation set V includes ten elements those mean ten levels, namely: $V = \{1,2, 3,4,5,6,7,8,9,10\}$.

Single Factor Fuzzy Evaluation. From one factor $u_i (i = 1, 2, \cdots, n)$ to judge, then to determine the membership which the judge objective to the evaluation set element $v_i (i = 1, 2, \cdots, m)$, that is called single factor fuzzy evaluation, which can be expressed simply as: $R_i = (r_{i1}, r_{i2}, \cdots, r_{im})$.

The memberships of the single factor judge set generate the matrix which is called single factor judgment matrix:

$$R = \begin{bmatrix} r_{11} & r_{12} & \cdots & r_{1m} \\ r_{21} & r_{22} & \cdots & r_{2m} \\ \vdots & \vdots & & \vdots \\ r_{n1} & r_{n2} & \cdots & r_{nm} \end{bmatrix}_{n \times m}$$

This paper is based on the separating-planting mechanism aided analysis software (shown in refs[1]), according to the "Star Judging Method" to get the data set of the judgment matrix. The "Star Judging Method" devises a three-star-representation with the red star "★"、 "★★"、 "★★★" (shows the degree of meeting the objectives) respectively or the black star "★" (shows the result is not good) which means the current situation of all the objectives [1].

This paper optimizes twenty group feasible solutions through the "Star Judging Method". For each set of solution, the digital value of each objective from no star to three stars is transferred into the number between one and ten. Then according to the ten levels of the evaluation set V , defines each objective membership corresponding to the discrete value of the evaluation set V , and each set of solution can form a matrix with ten lines (ten objectives) and ten columns (membership of each target corresponding to the ten levels). For example, the single element evaluation matrix of the first feasible solution $\underset{\sim}{R}_1$ is followed:

$$\underset{\sim}{R}_1 = \begin{bmatrix} 0 & 0 & 0 & 0 & 0 & 0 & 0 & 0 & 1 & 0 \\ 0 & 0 & 0 & 1 & 0 & 0 & 0 & 0 & 0 & 0 \\ 1 & 0 & 0 & 0 & 0 & 0 & 0 & 0 & 0 & 0 \\ 0 & 0 & 0 & 0 & 0 & 0 & 0 & 0 & 0 & 1 \\ 0 & 0 & 0 & 0 & 0 & 0 & 0 & 0 & 0 & 1 \\ 0 & 0 & 0 & 0 & 0 & 0 & 0 & 1 & 0 & 0 \\ 0 & 0 & 0 & 0 & 0 & 0 & 0 & 0 & 0 & 1 \\ 0 & 0 & 0 & 0 & 1 & 0 & 0 & 0 & 0 & 0 \\ 0 & 0 & 0 & 0 & 0 & 0 & 0 & 0 & 0 & 1 \\ 1 & 0 & 0 & 0 & 0 & 0 & 0 & 0 & 0 & 0 \end{bmatrix}$$

Similarly the twenty groups feasible solutions by this method can be transformed a judgment matrix R with two hundreds lines and ten columns. Due to the "Star Judging Method" is a kind of fuzzy algorithm itself, so R is a fuzzy relation matrix.

Fuzzy Overall Evaluation. Single factor evaluation only reflects the influence of one factor to the evaluation objectives, which obviously is not enough for multi-objective optimization problem, we must consider the synthesis influence of all the factors, the fuzzy overall evaluation method can achieve this effect.

According to the single factor evaluation matrix, we can see the first i line of R reflects that the first i factor affects the degree of the evaluation objective taking each element of evaluation set V , the first j column of R reflects that all factors affects the degree of the evaluation objective taking the first j element of evaluation set V , so we can sum each column elements:

$$R_j = \sum_{i=1}^{n} r_{ij} (j = 1, 2, 3, \cdots, m)$$

Then combine the weight of each element, it can reasonably reflect the overall influence of all the factors, and then adopt the operation of $M(\bullet, +)$ [8], the overall evaluation result can be expressed as:

$$B = A \circ R = (a_1, a_2, \cdots, a_n) \circ \begin{bmatrix} r_{11} & r_{12} & \cdots & r_{1m} \\ r_{21} & r_{22} & \cdots & r_{2m} \\ \vdots & \vdots & & \vdots \\ r_{n1} & r_{n2} & \cdots & r_{nm} \end{bmatrix}$$

$$= (b_1, b_2, \cdots, b_m)$$

Among them: $b_j = \overset{n}{\underset{i=1}{\vee}} (a_i \wedge r_{ij}), (j = 1, 2, 3, \cdots, m)$

Thus the overall evaluation result is: $E = B \circ V^T$.

This example calculates twenty group sample data, then obtains twenty overall evaluation results after being programmed in the Matlab, the results are: 6.6811, 7.0884, 9.1696, 7.7921, 8.5326, 9.1770, 7.8810, 8.0364, 8.8067, 9.1622, 8.9473, 7.9032, 8.6363, 8.8364, 7.6217, 8.0587, 8.2958, 6.7848, 8.1549, 8.1994.

Through the comparison of the values that the highest evaluation is 9.1770, the corresponding initial parameters are: a =22, K=0.941, $\alpha 0$ =-68, $\varphi 0$ =17, $\delta 0$ =19, S=184, H=150, W=200 that is the optimal solution among the twenty sample data.

4 Conclusions

This paper applies the fuzzy overall evaluation principle to establish the mapping relationship between the objective result of the separating-planting mechanism and the kinematic satisfaction based on the fuzzy mathematics theory, and calculates all the satisfactions of all the feasible solutions, then gets the best optimal solution.

This method makes the non-professional transplanter workers to have the ability to judge the degree in multiple feasible solutions on the satisfaction, also provides an excellent tool for developer of the separating-planting mechanism. So this evaluation method has more important reference value for the research and development of the separating-planting mechanism.

References

1. Dai, L., Zhao, X., Zhao, Y.: The software development on kinematics parameters optimization of rice transplanter separating-planting mechanism. In: The International Conference on Consumer Electronics Communications and Networks, CECNet 2011, CECNET5 E48026, Xianning, China, pp. 4568–4571 (2011)
2. Wu, C., Zhao, Y., Chen, J.: Optimization Design of Rice Transplanter Separating-planting Mechanism with Visualization Human-computer Interaction Method. Chinese Journal of Mechanical Engineering, 46–49 (2008) (in Chinese)
3. Yu, G., Sun, L., Zhao, Y.: Parameters Optimization Based on Human-computer Conversation of Transplanting Mechanism with Planetary Spur Gears and Elliptical Gears.Transactions of the Chinese Society for Agricultural Machinery, 47–50 (2008) (in Chinese)
4. Zhao, Y., Zhao, X., Zhang, W., Dai, L.: Modern Design Theory and Method of Rice Transplanter.Transactions of the Chinese Society for Agricultural Machinery, 65–58, 43 (2011) (in Chinese)
5. Chen, J.: Dynamic performance analysis, parameter optimization, and proof for transplanting mechanism with planetary elliptic gears. Zhejiang University (2004) (in Chinese)
6. Zhao, Y.: Analysis and Synthesis of Agricultural Machinery. Publishing House of Machine Industry, Beijing (2009) (in Chinese)
7. Xie, Q., Luo, Y., Li, Q.: Mechanical engineering fuzzy optimization method. Publishing House of Machine Industry, Beijing (2002) (in Chinese)
8. Wang, T., Wang, Y.: The fuzzy mathematics and its application. Publishing House of Northeastern University, Shenyang (2009) (in Chinese)

The Learning of Morphological Principles: A Statistical Learning Study on a System of Artificial Scripts

Qiuyan Wu[1,2], Xiaoping Fang[3], Quanjing Chen[3], Yaoran Li[3], and Yuan Deng[1]

[1] Key Laboratory of Behavioral Science, Institute of Psychology, Chinese Academy of Sciences, Beijing 100101, P. R. China
[2] Graduate University of Chinese Academy of Sciences, Beijing 100049, China
[3] School of Psychology, Beijing Normal University, Beijing, 100875, P. R. China

Abstract. By using a familiarization-preference procedure, the current study investigated the statistical learning of two morphological principles without providing any phonological cues. A series of experiments was conducted by varying learning duration. After a relatively short period of exposure to the learning stimuli, participants firstly learned the co-occurrence regularity; in comparison, the learning of graphic position required more exposure to the learning stimuli. These findings suggest that these two morphological principles could be acquired without phonological cues, but they were not equally easy to learn.

Keywords: statistical learning, non-alphabetic language, co-occurrence frequency, position frequency, morphological principles.

1 Introduction

Human language has much regularity with respect to both its spoken and written aspects. Previous studies have shown that adults as well as infants are capable of segmenting a continuous speech stream into words based on statistical cues derived from the distributional analysis of patterns of words [1-3]. For example, by using an artificial language corpus, Saffran et al [4] constructed four trisyllabic nonsense words, which were concatenated into a continuous speech stream without acoustic cues to indicate word boundaries. During the learning phase, 8-month old infants were exposed to this speech stream for two minutes. In the subsequent preference test, they showed significantly higher preferences for real words (presented in the learning phase) than for the non-words (novel tri-syllabic strings created by randomly re-constructing familiar syllables), or for the part-words (less predictable trisyllabic strings with one syllable spanning a word boundary). In a later study, Aslin et al [5] controlled the frequencies of words and part-words, and found that infants still could reliably discriminate words from part-words, suggesting that infants can definitely extract transitional probability. These studies confirmed that child learners can rely on statistical cues (i.e., non-linguistic information) in a speech stream to achieve word segmentation.

G. Lee (Ed.): Advances in Automation and Robotics, Vol. 1, LNEE 122, pp. 187–196.
springerlink.com

In terms of spelling, several studies of school children have found that children show sensitivity to the principles of the legal combinations of letters after learning how to spell; specifically, they showed sensitivity to spelling (i.e., graphotactic patterns), the appropriate spelling choice between several plausible representations of a given sound (i.e., morphological patterns), and the mappings between letters and sounds [6-8]. These findings suggest that statistical learning of spelling-related principles occurs at several different levels, and could be induced by different linguistic cues. However, little is known about whether there is a similar mechanism(s) for learning a non-alphabetic language. Generally, in a typical non-alphabetic language such as Chinese, a visual character is used as the basic unit of writing to map onto a spoken mono-syllable [9]. That means there are no phonological cues to indicate the selection of written components (morphological patterns). It is therefore unclear how learners of non-alphabetic languages grasp the underlying statistical principles of the language at a morphological level, and which cue(s) are used by learners.

The current study examines the statistical learning of morphological principles at the lexical level in a language with a non-alphabetic writing system. A visual symbol system was created by following two morphological principles that are particularly important for the recognition of non-alphabetic words [10-12]. More specifically, one principle was indicated by the cue of co-occurrence frequency, which reflects how frequently the sub-components of a visual character appear together; another principle was indicated by the cue of position frequency, which reflects the frequency of a certain component occurring in a given position of a word. The created set of visual symbols was considered an imitation of a non-alphabetic language. To investigate the perception of morphological principles, no other linguistic information (e.g., pronunciation or meaning) was provided during the learning task, when participants were asked to learn the symbols. The study therefore attempts to answer the following two questions: 1) whether the two aforementioned morphological principles can be extracted after only a short amount of learning, and, if so, 2) whether they can be extracted in parallel.

2 Experiment 1

2.1 Method

Participants. Twenty-eight native Chinese speakers were paid to participate in Experiment 1. None had been previously exposed to the Tibetan language. Written consent was obtained from all participants.

Materials and Apparatus. Thirty Tibetan letters were chosen to construct the stimuli characters; these letters had no familiarity or preference bias when rated by a group of native Chinese speakers. Ten of these letters were used to generate the characters for the learning part of the experiment, while the other 20 were constructed into 10 buffer characters that were presented either prior to or after the presentation of the learning part characters.

For the learning part, each character consisted of two Tibetan letters and had a left-right structure. Based on the morphological constraints of Tibetan, three types of characters were used as learning stimuli (see Fig. 1): 1) high co-occurrence frequency and high position frequency (HC-HP) characters, in which two components of a specific compound character appeared together more frequently than other combinations, and each subcomponent appeared in the most frequently-presented position across the whole corpus, i.e., the left or right side; 2) low co-occurrence frequency and high position frequency (LC-HP), and 3) low co-occurrence frequency and low position frequency (LC-LP). Each word type had five tokens. Different repetition times for each type produced the differences in the co-occurrence frequency and position frequency; the ratio of high to low co-occurrence frequency was 4:1, and the ratio of high to low position frequency was 5:1 (see Table 1). Take "A1" as an example. As shown in the table, component "A" co-occurred with component "1" 20 times. Furthermore, it can be seen that component "A" appeared in the left position 25 times in total, while component "1" appeared in the right position 25 times in total. Thus, "A1" was regarded as a HC-HP character.

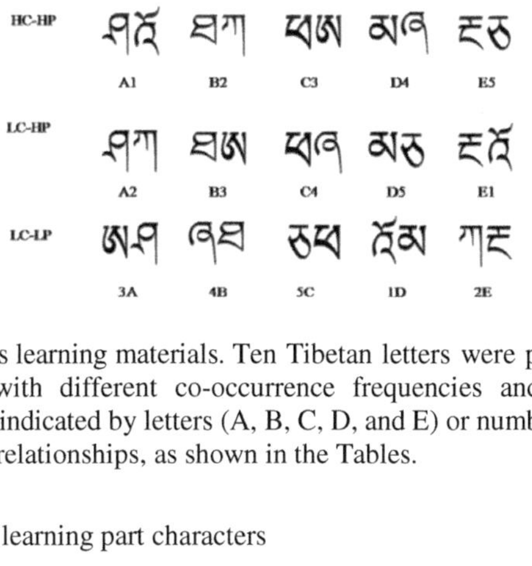

Fig. 1. Characters used as learning materials. Ten Tibetan letters were pair-wise combined into three types of words with different co-occurrence frequencies and position frequencies. Components were either indicated by letters (A, B, C, D, and E) or numbers (1, 2, 3, 4, and 5) to illustrate their statistical relationships, as shown in the Tables.

Table 1. Statistics of the learning part characters

Type	Repetition (times)	Tokens				
HC-HP	20	A1	B2	C3	D4	E5
LC-HP	5	A2	B3	C4	D5	E1
LC-LP	5	3A	4B	5C	1D	2E

To test the learning effect, a set of novel characters was created, with each character belonging to one of the following two categories based on how the components were arranged: a) the positions of a learned character's components were switched, resulting in an unchanged co-occurrence frequency but a reversed position frequency; b) the individual components of a learned character were recombined, resulting in new characters with sub-components that never appeared together during the learning part of the experiment (zero co-occurrence frequency, ZC) (see Fig. 2).

Fig. 2. Novel characters used in the test phase

Procedures. A familiarization-preference procedure was employed [13] in which participants were initially familiarized with the learning stimuli (familiarization) and then tested for specific learning effects (preference test). The experimental procedures used for learning and testing were computerized using E-prime software [14].

During familiarization, a total of 25 characters were presented individually, including 15 training characters repeated multiple times (5 HC-HP characters were repeated 20 times, 5 LC-HP characters were repeated 5 times, and 5 LC-LP characters were repeated 5 times) and 10 buffer characters repeated only once, resulting in a total of 170 trials. The learning stage characters were presented in random order, with the constraint that the same character never occurred twice in succession. The ten buffer characters were presented in a fixed order at the beginning and also at the end of the learning stage to counterbalance any potential effects of proactive inhibition or retroactive inhibition.

After the familiarization phase, a two-alternative forced-choice (2AFC) preference test was conducted. In each trial, a pair of stimuli was sequentially presented for 800 ms each, with a 1000 ms inter-stimulus interval (ISI). Participants pressed one of two keys on the left and right sides of a keyboard to indicate which character seemed more familiar to them. The test trials were randomized for each participant. The percentage of selection of higher co-occurrence frequency characters, high position frequency characters, and old characters as more familiar was used as the measurement of statistical learning.

To investigate the position effect, comparisons between the high and low position frequency characters were made at zero, low and high co-occurrence frequency levels. Similarly, to investigate the co-occurrence effect, comparisons between the high and low co-occurrence frequency characters were made at low and high position frequency levels, with three pairs of comparisons used within each level. In addition, to examine any possible confounding by the "old/new" effect, comparisons between the "old" and "new" characters were also made at the low and high position frequency levels. There were a total of 11 groups of pairs (see Table 2).

To avoid fatigue effects and awareness of the underlying structure of the stimuli in the test phase, test trials were divided into two sub-sets with roughly equal numbers of pairs, and the test conditions were balanced across these two sets. Participants underwent the same familiarization stage and then were randomly assigned to one of the test sets. There were 12 participants for one test set, and 16 for the other.

Table 2. Eleven groups of pairs used in the preference test

Examined Effect	Constant Levels	Pairs	Examples	Notes
Position Effect	Zero co-occurrence	ZC-HP vs. ZC-LP	A4 vs. 5A	New vs. New
	Low co-occurrence	LC-HP vs. LC-LP	A2 vs. 3A	Old vs. Old
	High co-occurrence	HC-HP vs. HC-LP	A1 vs. 1A	Old vs. New
Co-occurrence Effect	High position	HC-HP vs. LC-HP	A1 vs. A3	Old vs. New
		LC-HP vs. ZC-HP	A2 vs. A4	Old vs. New
		HC-HP vs. ZC-HP	A4 vs. A1	New vs. Old
	Low position	HC-LP vs. LC-LP	1A vs. 3A	New vs. Old
		LC-LP vs. ZC-LP	3A vs. 5A	Old vs. New
		HC-LP vs. ZC-LP	1A vs. 5A	New vs. New
Old/New effect	Low co-occurrence high position	LC-HP vs. LC-HP	A3 vs. A2	New vs. Old
	Low co-occurrence low position	LC-LP vs. LC-LP	2A vs. 3A	New vs. Old

2.2 Results

Performance on the preference test is presented in Fig. 3a. For the position effect, participants did not show any preference for characters with high position frequency compared with those with low position frequency, regardless of co-occurrence frequency levels (all *ps* > 0.05). As for the co-occurrence effect, results showed that at high position frequency level, the mean percentage of selecting high co-occurrence frequency characters over zero co-occurrence frequency characters was significantly above chance [Mean= 58.67%, t (14) = 2.982, p = 0.01]. For the familiarity effect, participants showed a significant preference for old characters at high position frequency level [Mean=58.52%, t (26) = 2.791, $p = 0.01$].

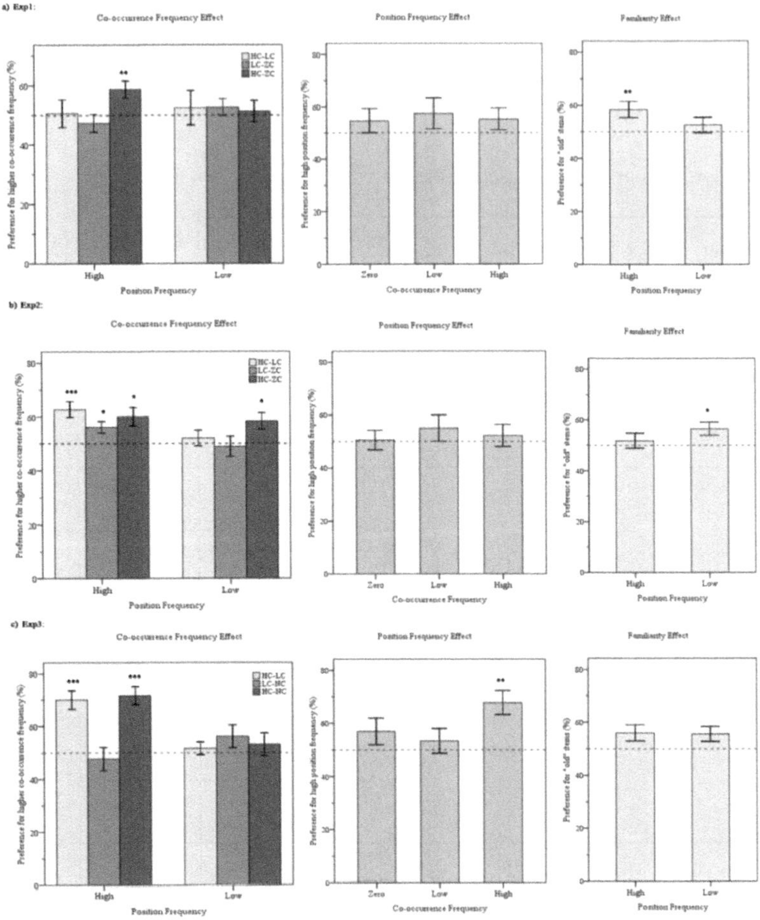

Fig. 3. Preference performance for Experiment 1 (first row), Experiment 2 (second row) and Experiment 3 (third row). For the left column, the bars represent the mean percentages of selecting higher co-occurrence frequency characters at both high and low position frequency levels. For the middle column, the bars represent the mean percentages of selecting high position frequency characters at zero, low and high co-occurrence frequency levels. For the right column, the bars represent the mean percentages of selecting old words compared with new words. The dashed lines indicate chance performance (50%). Standard errors of the mean over participants are also plotted. Single-sample t-tests were conducted. * $p < 0.05$, ** $p < 0.01$, *** $p < 0.001$.

2.3 Discussion

Results indicated that a significant difference for preference was shown for co-occurrence frequency, but not for position frequency. However, the "old/new" effect was also significant, suggesting participants had greater familiarity with trained characters than new characters. Therefore, the preference for characters with higher co-occurrence frequency was likely confounded by the higher familiarity with old characters. The failure to extract these two kinds of statistical cues may have been due to the short exposure time. Indeed, evidence has shown that insufficient familiarization tends to worsen participants' performance in the extraction of statistics [15]. Therefore, in our subsequent Experiment 2, we wanted to see if extended exposure could facilitate the extraction of co-occurrence and position statistics.

3 Experiment 2

3.1 Method

Another group of 28 participants were recruited for this experiment. The procedure was similar to that of Experiment 1, with the exception that all of the learning stimuli were repeated one additional time during familiarization (for a duration of about 8 minutes). The preference test was the same as that used in Experiment 1, and the test trials were similarly divided into two sub-sets. Ten participants completed one test set, while the rest completed the other.

3.2 Results

Performance on the preference test is presented in Fig. 3b. Participants still showed no significant preference for characters with high position frequency compared with those with low position frequency, despite the extended familiarization time (all *ps*> 0.05). However, longer exposure was found to further boost the co-occurrence effect. At the high position frequency level, characters with relatively higher co-occurrence frequency were significantly chosen as more familiar: high vs. low co-occurrence frequency [Mean = 62.78%, $t(17) = 4.254$, $p < 0.01$]; low vs. zero co-occurrence frequency [Mean = 56.11%, $t(17) = 2.829$, $p < 0.05$]; high vs. zero co-occurrence frequency [Mean = 60.00%, $t(17) = 2.838$, $p < .05$]. At the low position frequency level, the co-occurrence effect also reached significance for the comparison between high and zero co-occurrence frequency [Mean = 57.78%, $t(17) = 2.613$, $p < 0.05$]. Finally, an analysis of the "old/new" effect showed that this effect was significant at the low position frequency level [Mean = 54.63%, $t(27) = 2.489$, $p < .05$].

3.3 Discussion

Even after a longer exposure time to the learning stimuli, the position effect was still not observed, whereas the co-occurrence effect became much stronger. Similar to the results of Experiment 1, the familiarity effect was also evident in Experiment 2. However, the observed significant co-occurrence effect cannot be simply explained by this confounding effect, because some comparisons in testing the co-occurrence effect were made between two old characters or two new characters. That is, the characters in these comparisons were equally familiar or equally unfamiliar to participants. As mentioned above, the position frequency involves detailed information about a character's construction, i.e., the regularity of the positions of its components. Thus, it might take a longer time for participants to learn position regularity than to recognize co-occurrence regularity. To induce the extraction of position regularity, we continued to prolong the exposure time for learning in a subsequent experiment.

4 Experiment 3

4.1 Method

A new group of 25 participants were recruited for this experiment. The procedure was similar to that of Experiment 1, with the exception that all of the learning stimuli were repeated four times during familiarization (for a duration of about 16 minutes of exposure). The preference test was the same as that used in Experiment 1, and the test pairs were again divided into two sub-sets. Thirteen participants were randomly assigned to complete one test set, while the rest were assigned to the other.

4.2 Results

Performance on the preference test is presented in Fig. 3c. The position effect became significant at the high co-occurrence frequency level. Characters with their subcomponents displayed at a high frequency position were significantly chosen as more familiar than those with subcomponents placed at a low frequency position [Mean = 67.69%, $t(24) = 3.888$, $p < 0.01$]. As for the co-occurrence effect, results showed that characters with a relatively higher co-occurrence frequency were significantly chosen as more familiar at high position frequency level: high vs. low co-occurrence frequency [Mean = 70%, $t(24) = 5.586$, $p < 0.01$]; high vs. zero co-occurrence frequency [Mean= 71.45%, $t(24) = 6.062$, $p < 0.01$]. Unlike in Experiment 1 and Experiment 2, the familiarity effect was not found to be significant in this experiment ($ps > .06$).

4.3 Discussion

Compared with Experiments 1 and 2, after a longer exposure to the learning stimuli, the co-occurrence effect remained pronounced. At the same time, the position effect also became significant. This pattern suggests that the two distinct morphological principles were not learned in parallel. Specifically, it seems to be more difficult and to require more time for learners to learn position regularity compared with co-occurrence regularity.

5 General Discussion

In the current study, we investigated the statistical learning of two morphological principles in an artificial system without providing any phonological cues. Results showed that co-occurrence statistics (within-item level) are extracted earlier than position statistics (across-item level). Consistent with the results of previous studies, the current findings suggest that different levels of statistics might not always be acquired in parallel [15, 16].

In the current study, as the learning duration gradually increased, the learning effect of co-occurrence regularity and position regularity became more robust, while the familiarity effect became less so. The familiarity effect seems to have largely reflected the result of memorizing a specific stimulus, while the extraction of co-occurrence and position regularities seem to depend on an awareness of the underlying rules. Therefore, it is possible that learners could eventually extract the regularities underlying the learning stimuli, independent of the memory ability for specific learning characters.

There are two possibilities for the observed different rates of extraction of co-occurrence regularity and position regularity: 1) the mechanisms for extracting these two regularities are separate; 2) the two mechanisms of regularity extraction may be the same, but their efficiency is different. Unlike statistical learning studies in the auditory domain, in which the stimuli are usually continuously presented as a speech stream [4, 17-19], the current study used visual characters as the learning stimuli, in which visual characters were discretely presented to participants. By using this method of presentation, the segmentation cues between characters were already provided. The co-occurrence cue is always presented within an individual word, and thus it can be considered to definitely bootstrap statistics extraction. However, the acquisition of position statistics entails not only realizing that two components are in conjunction, but also figuring out their position regularity. Therefore, the recognition of position regularity for a certain component may involve additional cognitive steps, such as spatial analysis or working memory. In addition, the awareness of position regularity is especially indispensable in non-alphabetic languages because there is no grapheme-phoneme (G-P) mapping to indicate the organizational pattern of the visual components within a character [9]. Taken as a whole, the results of this study suggest that the second possibility described above best explains the data, that is, that the two mechanisms of regularity extraction are the same, but their efficiency is different.

Acknowledgements. This research was supported by Chinese Natural Science Foundation (30700232) to Yuan Deng.

References

1. Saffran, J.R., Newport, E.L., Aslin, R.N., Tunick, R.A., Barrueco, S.: Incidental Language Learning: Listening (and Learning) out of the Corner of Your Ear. Psychol. Sci. 8, 101–105 (1997)
2. Christiansen, M.H., Allen, J., Seidenberg, M.S.: Learning to Segment Speech Using Multiple Cues: A Connectionist Model. Lang. Cogn. Process. 13, 221–268 (1998)
3. Brent, M.R., Cartwright, T.A.: Distributional Regularity and Phonotactic Constraints Are Useful for Segmentation. Cognition 61, 93–125 (1996)
4. Saffran, J.R., Aslin, R.N., Newport, E.L.: Statistical Learning by 8-Month-Old Infants. Science 274, 1926–1928 (1996)

5. Aslin, R.N., Saffran, J.R., Newport, E.L.: Computation of Conditional Probability Statistics by 8-Month-Old Infants. Psychol. Sci. 9, 321–324 (1998)
6. Pacton, S., Fayol, M., Perruchet, P.: Children's Implicit Learning of Graphotactic and Morphological Regularities. Child Dev. 76, 324–339 (2005)
7. Pacton, S., Perruchet, P., Fayol, M., Cleeremans, A.: Implicit Learning out of the Lab: The Case of Orthographic Regularities. J. Exp. Psychol.-Gen. 130, 401–426 (2001)
8. Treiman, R., Kessler, B.: Spelling as statistical learning: Using consonantal context to spell vowels. J. Educ. Psychol. 98, 642–652 (2006)
9. Perfetti, C.A., Liu, Y., Fiez, J., Nelson, J., Bolger, D.J., Tan, L.H.: Biling.-Lang. Cogn. 10, 131–146 (2007)
10. Taft, M., Zhu, X.P., Peng, D.L.: Positional Specificity of Radicals in Chinese Character Recognition. J. Mem. Lang. 40, 498–519 (1999)
11. Taft, M., Zhu, X.P.: Submorphemic Processing in Reading Chinese. J. Exp. Psychol. Learn. Mem. Cogn. 23, 761–775 (1997)
12. Ding, G.S., Peng, D.L., Taft, M.: The Nature of the Mental Representation of Radicals in Chinese: A Priming Study. J. Exp. Psychol. Learn. Mem. Cogn. 30, 530–539 (2004)
13. Jusczyk, P.W., Aslin, R.N.: Infants' Detection of the Sound Patterns of Words in Fluent Speech. Cogn. Psychol. 29, 1–23 (1995)
14. Schneider, W., Eschman, A., Zuccolotto, A.: E-Prime: User's Guide. Psychology Software Inc. (2002)
15. Fiser, J., Aslin, R.N.: Statistical Learning of Higher-Order Temporal Structure from Visual Shape Sequences. J. Exp. Psychol. Learn. Mem. Cogn. 28, 458–467 (2002)
16. Fiser, J., Aslin, R.N.: Unsupervised Statistical Learning of Higher-Order Spatial Structures from Visual Scenes. Psychol. Sci. 12, 499–504 (2001)
17. Saffran, J.R., Newport, E.L., Aslin, R.N.: Word Segmentation: The Role of Distributional Cues. J. Mem. Lang. 35, 606–621 (1996)
18. Johnson, E.K., Jusczyk, P.W.: Word Segmentation by 8-Month-Olds: When Speech Cues Count More Than Statistics. J. Mem. Lang. 44, 548–567 (2001)
19. Newport, E.L., Aslin, R.N.: Learning at a Distance I. Statistical Learning of Non-Adjacent Dependencies. Cogn. Psychol. 48, 127–162 (2004)

Research on Multi-Agent Intelligent Decision Support System Based on Blackboard

Yongyong Sun

School of Computer Science and Engineering, Xi'an Technological University,
Xi'an, Shaanxi, China

Abstract. Multi-Agent Intelligent Decision Support System based on blackboard is designed that uses of blackboard system and Agent, it describes the framework and gives the detail about the each Agent, at the same time, the function of various base in resources layer is suggested in the system, and the structure of database is provided. In addition to, the action about Agent is described. The system can help to cooperate in Multi-Agent about complicated works.

Keywords: blackboard system, Agent, Intelligent Decision Support System, action.

Introduction

As a part of artificial intelligence, Agent is the combination of artificial intelligence and object entity[1], it is a program that can replace the user to perform a specific task independently and automatically. Agent has the autonomy, collaboration, learning, induction, reasoning and other features[2], these features can support the decision making and problem solving in various stage, and also can enhance the functions of traditional decision support systems. The architecture based on Agent provides the flexibility, variability and robustness, that suitable to solve some of the dynamic, uncertain and distributed problems. And Agent can solve large-scale complex problems through mutual scheduling and mutual cooperation among Agents. In today's society, the majority of work is completed by collaboration of various components in a particular group environment. Most work was done that associated with the occurrence of decision-making process, whether it is good or not will influence the successful completion of tasks, a networking, intellectualized, integrated and coordinating human-machine intelligent decision support system can be achieved by Agent technology[3,4].

In this paper, make use of the blackboard system and multi-Agent technology, a multi-Agent Intelligent Decision Support System based on blackboard is provided on the basis of traditional IDSS. In this system, blackboard can offer public workspace for distributed system, each Agent can use the blackboard to exchange the information, data and knowledge adequately, access to the blackboard at any time and query the contents of release, then each Agent extracts the required work information in order to complete their task, and through the blackboard to coordinate between the Agent, the

G. Lee (Ed.): Advances in Automation and Robotics, Vol. 1, LNEE 122, pp. 197–206.
springerlink.com

Agents work together to solve a given issue, at the same time, emphasizing the combination of the decision-makers' wisdom and system.

1 Architecture of Multi-Agent Intelligent Decision Support System Based on Blackboard

In the multi-Agent Intelligent Decision Support System based on blackboard, each decision task or function may be completed by an independent, and various Agent analyzes the problem by a corresponding standard from their point of view, each Agent has different capability, will and belief so that they have different knowledge and skills about problem-solving areas. Different Agent inspects decision problem from their own point of view, and completes certain sub-tasks independently, then, through the blackboard collaboration to achieve a common goal. Due to its autonomy and materialization[5], each Agent can join or leave a problem-solving system at any time, making the decision-makers can participate easily in the decision-making process, and hence making the system flexible. As is shown in Figure 1, the three-layer architecture of interactive layer→ intelligence decision-making layer → resources layer is established in this paper. In this system, interactive layer is made up of intelligent interface Agent and decision maker, intelligence decision-making layer is composed of blackboard, function Agent and decision Agent, and resources layer is composed of model base and model base management system, knowledge base and knowledge base management system, method base and method base management system, database and multi-base collaborative control.

2 Interactive Layer of Multi-Agent Intelligent Decision Support System Based on Blackboard

Decision maker and intelligent interface Agent compose the interactive layer of e multi-Agent Intelligent Decision Support System based on blackboard. Intelligent interface Agent is the component that is associated with decision maker in the architecture of multi-Agent Intelligent Decision Support System based on blackboard[6,7]. It can continue to run independently, and systems can communicate with decision-makers through it. In addition, it can help the user fulfill tasks more efficiently by learning the user's goals, preferences, habits, experiences, behaviors and so on. It replaces the traditional man-machine interactive interface, emphasizing autonomy and learning of the Agent, and it can detect environmental changes initiatively, through continuous learning, access to user's certain features of the knowledge during the interaction decision-making process that user and it work together. Accordingly, it can provide a suitable user interface according to perceive user behavior in decision-making process, it can make a strategy conform to user's will independently. The global knowledge of Intelligent interface Agent includes problem domain knowledge, user model or user knowledge, self-knowledge and knowledge of the other Agents' capabilities. It can learn knowledge through simulating and discovering, and also it can learn knowledge through accessing to the user's feedback of forward and reverse, then, it can access to knowledge by the user's instruction, and communicate with other Agent to acquire knowledge.

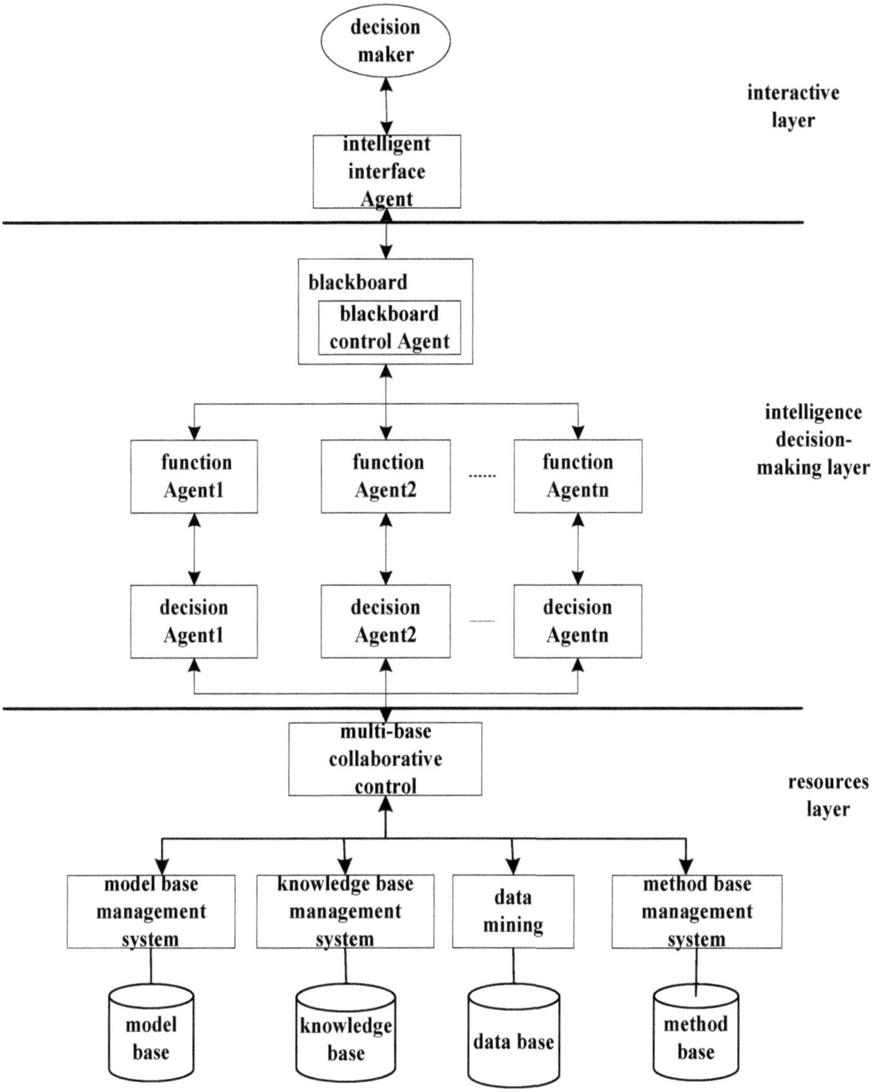

Fig. 1. Architecture of Multi-Agent Intelligent Decision Support System Based on Blackboard

3 Intelligence Decision—Making Layer of Multi-Agent Intelligent Decision Support System Based on Blackboard

3.1 Function Agent

Function Agent monitors the changes of data plane on the blackboard, and it matches the activation conditions of decision Agent with the plane information of blackboard, then, the information on the blackboard will be sent to the relevant decision Agent. Function Agent makes use of system message and the queue mechanism to send information on the blackboard, and is responsible for reporting the operation of the system, the completion of the task was reported to the users, at the same time, it provides explanations and queries. Function Agent has an internal activation condition table that every decision Agent reads the blackboard, and it will be checked that is related to changes on the layer when the information changes on every layer of blackboard, if they are matched, then, further analysis to determine whether activate the corresponding decision Agent directly to read the blackboard to accomplish the task or to do other operations, such as providing reports and explanations. It also provides queries about the implementation of the system. Actually, Function Agent is a reactive Agent that it can perceive the information and changes on the blackboard, whenever the information changes on the blackboard, it was triggered to check the data plane operation of blackboard.

3.2 Decision Agent

Decision Agent has the knowledge and skills of problem solving in a particular areas. Each decision method is designed as a decision Agent, many decision Agent can solve complex decision problems through mutual coordination and cooperation under the control and monitor of function Agent, Such Agent is extensible, with the development and progress of the decision theory and intelligent decision methods, more and more decision Agent can be developed. Decision Agent has not model about the external environment and the other Agent's knowledge, but still has the ability to interact with the other Agent.

3.3 Blackboard and Blackboard Control Agent

In multi-Agent intelligent decision support system based on blackboard, the combination of different function Agents and decision Agents can be regarded as various knowledge sources. And these knowledge sources interact through the blackboard, which acts as an integrated database that stores data, transmits information and processes methods[7]. The blackboard is the global workspace in the system and can be divided into several data planes according to different knowledge areas top-down. Event table is the most important data plane. Event can be the results of operation of the knowledge sources on the original event, it also can be the processing procedure from the outside. According to the event feature that enter the event table, the appropriate operation will be provided under the boot of monitoring. The type of event mainly has the certain thresholds of controlled process, the operator's instructions, the new assumptions of conditions about controlled process, the modification of original assumptions, and the requests that change control mode and so on. Assumption table is another important data plane. Assumption is derived from the facts and evidence to get the middle of the state, which is the understanding and speculation about the running state of controlled process, and it is complement as the current set of facts. Assumption table can be gotten that organized assumption appropriately. The changes of contents on blackboard can trigger the corresponding event, and it also is the generation of new assumption or changes of old content, the results of operation will change the blackboard again.

The blackboard model makes use of blackboard control Agent to improve the traditional blackboard. The blackboard control Agent is the brain of system, that controls the blackboard directly and the other Agents indirectly in the system. It divides the problem into appropriate sub-problems in light of the self-knowledge and the principle of collaboration, which are assigned to the corresponding data plane in blackboard, and the blackboard control Agent administers the every data plane as a whole, then, eliminates conflicts between different decision Agent. The blackboard control Agent should be designed to become an initiative control Agent in implementation of system, its behavior completes under the support of the internal model, ability and belief, initiative Agent has the perception, intentions, beliefs, abilities and learning. The blackboard control Agent determines the type of problems according to users' requirements, and uses the knowledge base and model base to decompose the problem, then, sends data and control information. At the same time, the blackboard control Agent assigns the sub-problems to the appropriate decision Agent. Until the data and results are written in the blackboard that are sent from decision Agent, that is, the contents of every data plane on the blackboard are changed, the blackboard control Agent eliminates conflicts between different decision Agents, coordinates every decision Agent, and feedbacks the synthesis of results to users.

3.4 Description of Agent Action

The main purpose is decision-making that Agent was constructed, which is a process that from perception to action. In this paper, the decision function of Agent was divided into perception function and action function, *see* is perception function and *action* is action function. Agent has an internal state, I is the set of all the internal state on Agent, Per is the set of non-null perceptions, and decision process are based on the information, perception function the mapping was achieved that from a state of external environment to perception by *see*:

$$see : E \to Per$$

Action function (*action*) is defined as the mapping from internal state to action:

$$action : I \to Ac$$

As a additional function, *next* is introduced, which implements mapping from internal state and perception to internal state:

$$next : I \times Per \to I$$

Agent action can be described as following:

Agent begins to observe the state of the environment (e) from an initial internal state (i_0), and produces a perception ($see(e)$), then updates the internal state of the Agent through the *next* function to become $next(i_0,see(e))$. Agent chooses an action by the $action(next(i_0,see(e)))$, which will lead Agent into another cycle, continue to perceive external world through *see*, update the state through *next*, and choose action to perform by *action*. The process is shown in Figure 2.

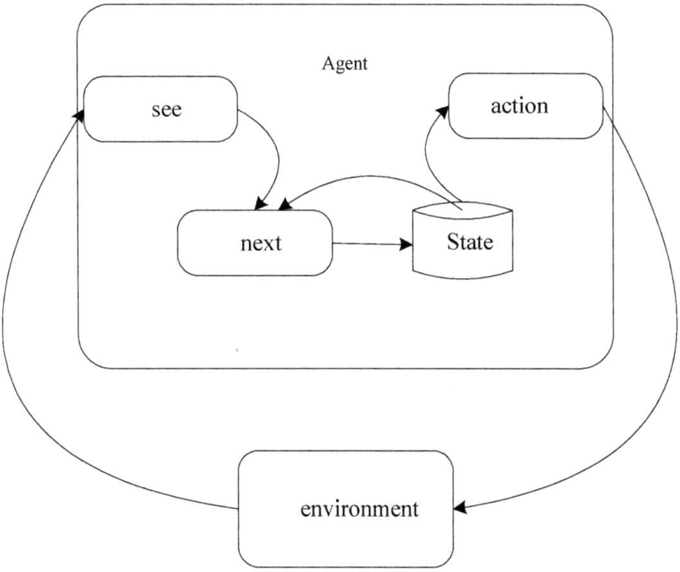

Fig. 2. Process of Agent Action

4 Resources Layer of Multi-Agent Intelligent Decision Support System Based on Blackboard

There are four bases in the resources layer, that is the model base, knowledge base, data base and method base, which are managed through the model base management system, knowledge base management system, method base management system and data mining and are controlled globally through the multi-base collaborative control. Multi-base collaborative control has the total control scheduling, collaborative planning, communication and so on, and can be achieved the management of each base and resources sharing and coordinated operation among the bases.

4.1 Model Base and Model Base Management System

Model base is the central part in decision support system, it is used to store code of model, and it is a set of various models, in which model is an abstract description about the things, phenomena, processes or systems in the world. It is need the diversity of Model to combine complex models to solve complex problems, and also to expand the assisted decision-making capabilities. In the model base, models can be reused to avoid redundancy. Model base is made up of model dictionary base and model file base. Model dictionary is used to store the description of the model information (such as restrictions, constraints, parameters models, etc.) and abstract data of models, that is the instructions of data access on models. Model dictionary can also be stored the information that help the user learn to use the model, such as model structure, model performance, condition of model application, model solution techniques, the meaning of model input-output and the reliability of model. In order to control and manage model base intensively, model base management system manage all of the model, then realize the sharing of model and create a new model. Its main functions are configuration model management, the model access management and model operation management.

4.2 Knowledge Base and Knowledge Base Management System

Knowledge base is the source of intelligent in system, it is mainly used to store rules. The knowledge is classified according to certain principles, and they are stored and organized hierarchically. Each block and each layer can also be subdivided. This hierarchy can easily schedule and search for knowledge, and speed up reasoning. The representation of knowledge uses production rule, production rule has strong structure. For rules:

IF antecedent THEN action CF

In the expression, antecedent is premise of the rule, action is the rule's conclusion under the premise is right, CF is the credibility factor, , CF = 1 when make certain reasoning.

Creation, deletion, reorganization and maintenance on knowledge base, entry, query, update, optimization on knowledge, its integrity, consistency, redundancy checking, security protection and other work, all of them are management on knowledge, which is guarantee that improve whole performance and efficiency about system. Management of the knowledge is accomplished by knowledge base management system.

4.3 Method Base and Method Base Management System

Method base is a collection of method, it storages various methods that models used, and it provides the algorithms basis and method support for the model solving. In the intelligent decision system, method base stores a number of standard functions and processes that are generated under the decision-making, it can also store common analysis methods and algorithms. Methods base collects methods organically in light of certain principles, which are controlled and managed intensively by method base management system, and various methods are queried, retrieved, added and deleted, etc, it can also select the appropriate method according to different decision-makers' needs.

4.4 Data Base

Data base is a organized form of data storage, it gets original data from initial data source, and reorganizes data in accordance with the requirements of decision-making to form an integrated data layer with different granularity. At the same time, while the data base also controls and manages the data in order to support decision-making.

The overall hierarchy of data base should be made up of the basic function layer of data base, management layer of data base and environment support layer of data base so that realize the functions of data base. The basic function layer of data base should contain data extracted from the data source, the data will be filtered and cleaned, which are loaded into the data base, then set up a data mart according to the needs of users, and complete functions that complex query, decision analysis and knowledge mining for data base. Management layer of data base including data management and metadata management, it is mainly responsible for extraction, cleaning, loading, updating and refreshing of the data in data base. Only these operations complete normally, new data sources are provided continuously for data base and data base is used accurately to decision analysis and knowledge mining. Environment support layer of data base mainly has data transmission r and data base basis. The structure of data base is shown in Figure 3. Build up data base gradually through independently developed data mart. Data mart is a data base that has specific application, which supports users to manage decision-making that use existing data according to specific demand. They are designed to be the components of main data base so that ensure data organization, format and structure are consistent in the whole data base, and the phenomenon of data reconstruction is avoided in data mart. When using the data base, query and analysis of various decisions will be done for data mart or data base through OLAP. And model extraction and knowledge discovery of data are done through data mining in data base, which provides the better decision support for data base. Data mining based on data base is better that meets the requirements of high-level strategic decision-making. The basic process of data mining is shown in Figure 4.

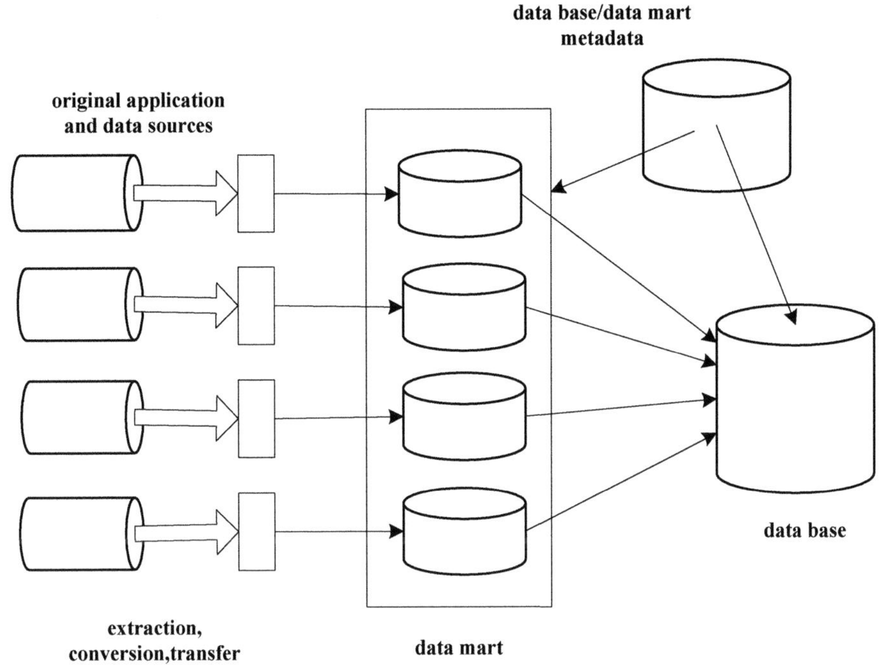

Fig. 3. Structure of Data Base

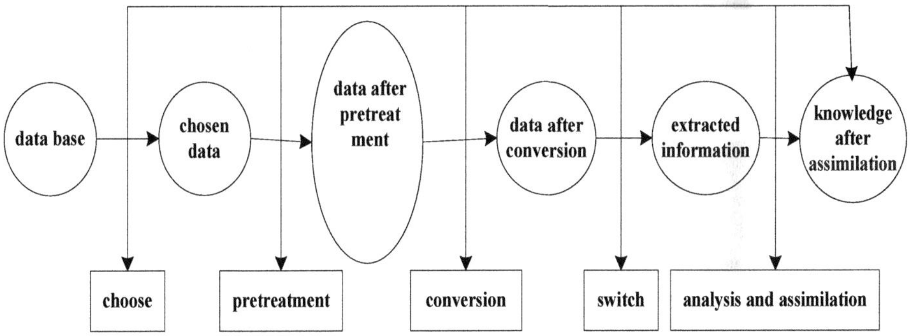

Fig. 4. The Basal Process of Data Mining

Conclusion

In this paper, each Agent has its own autonomy in Multi-Agent Intelligent Decision Support System based on blackboard, Agents share information through the blackboard and make use of the information in the blackboard to determine their own behavior, collaborate to solve complex problem. As the Agent's own uncertainty, the effective coordination need to be solved between Agent for multi-Agent system. Therefore, establish a unified coordination mechanism so that the Agent can coordinate work effectively, and improve overall system performance in practical applications that need to be further improved.

References

1. Mao, H., Tang, H.: Development of the Research on Intelligent Decision Support Systems. MINI—MICRO SYSTEMS 24(5), 874–879 (2003)
2. Gottinger, H.W., Weimann, P.: Intelligent Decision Support System. Decision Support System 8, 317–332 (1992)
3. Fetzer, J.H.: Pepople are not Computers(most) thought Processes are not Computational Procedures. Journal of Experimental & Theoretical Artifical Intelligence 10, 371–391 (1998)
4. Xiang, Y., Shen, H.: Intelligent Decision Support System Based on Agent. Computer Engineering 32(16), 167–169 (2006)
5. Chang, Y., Chen, M., Yuan, J.: Study on the Architecture of MAS-Based Distributed Group Decision Support System. Computer Engineering and Applications 18, 100–102 (2003)
6. Wooldridge, M.: An Introduction to Multi-Agent Systems (2002)
7. Toledo, C.B.E., Bourne, R.A., Jennings, N.R.: Reasoning About Commitments and Penalties for Coordination between Autonomous Agents (2001-2006)

Consensus Control for a Class of High Order System via Sliding Mode Control

Changbin Li, Yi He, and Aiguo Wu

School of Electrical and Automation Engineering, Tianjin University, Tianjin, China, 300072

Abstract. In this paper, consensus problem for a class of high order multi-agent systems is studied by using the so called sliding mode control technique. For directed connected multi-agent systems, both the fixed topologies and switching topologies are considered in this paper. By using the sliding mode control technique, a sliding mode surface is designed, each agent are coupled according to the prescribed sliding surface so that they can reach the same sliding surface in finite time and therefore converge to their consensus value. This consensus control technique can be very easily used in the consensus protocol design.

Keywords: consensus control, high order, multi-agent systems, time-varying topologies.

1 Introduction

In recent years, many researchers have drawn great attention to the consensus control of multi-agent systems. The consensus problem has many applications in our society such as team formation, flocking, rendezvous and so on[1-9]. There are plenty results of the consensus problems especially for the first-order and second order multi-agent systems.

Unfortunately, due to de difficult exist in the unknown connection topologies and high order even nonlinear dynamics bring by the multi-agent systems, there are still many challenge works to do for the high order multi-agent systems. In [5, 6], consensus problems are considered for high order multi-agent systems with directed and fixed topologies. If the connection between multi-agent systems is undirected or directed but balanced, [7] gave a consensus control approach for passive systems. This approach is also effective if the balanced topologies are time varying. Recently, by using the compare theorem, Zhihua qu[9] gave cooperative control strategy for high order multi-agent systems. Cooperatively stability and Lyapunov stability can be guaranteed if the communication topology is uniformly sequentially complete over time.

On the other hand, sliding mode control is a simple and effective technique for the control of nonlinear systems. Our aim of this paper is to design a consensus protocol for multi-agent systems. By using the sliding mode control technique, we studied consensus problem for a class of high order multi-agent systems. The communication topology considered in this paper is directed, and it may not be balanced. Both the

G. Lee (Ed.): Advances in Automation and Robotics, Vol. 1, LNEE 122, pp. 207–213.
springerlink.com

fixed topologies and switching topologies are considered. A sliding mode surface is designed, each agent are coupled according to the prescribed sliding surface. We proved that if the multi-agent is connected, then all the agents will reach to the same sliding surface and therefore converge to their consensus value.

2 Preliminary

In this section, we present some notations and some preliminary results of algebraic graph theory and multi-agent system.

A. Graph Theory

Let $\mathcal{G}(t) = (\mathcal{V}, \mathcal{E}(t), \mathcal{A}(t))$ be a weighted directed graph of order N, where $\mathcal{V} = \{v_1, \cdots, v_N\}$ is the sets of nodes, $\mathcal{E}(t) \subseteq \{(v_i, v_j) | v_i, v_j \in \mathcal{V}\}$ is the sets of edges, and $\mathcal{A}(t) = [a_{ij}(t)]$ is a weighted adjacency matrix, defined as $a_{ii}(t) = 0$ and $a_{ij}(t) \geq 0$, where $a_{ij}(t) > 0$ if and only if $(v_i, v_j) \in \mathcal{E}(t)$. We assume that $\mathcal{A}(t)$ is piecewise continuous and bounded. The weighted directed graph $\mathcal{G}(t)$ is connected at time t if there exists a node v_i such that all the other nodes of the graph are connected to v_i via a path that follows the direction of the edges of the directed graph. A node v_i is said to be connected to node v_j in the internal $I = [t_a, t_b]$ if there is a path from v_i to v_j which respects the orientation of the edges for the directed graph ($v_i, \bigcup_{t \in I} \mathcal{E}(t), \int_I \mathcal{A}(\tau) d\tau$). A weighted directed graph $\mathcal{G}(t)$ is said to be uniformly connected if there exists a time horizon $T > 0$ and an index i such that for all the nodes $v_j (j \neq i)$ are connected to the node v_i across $[t, t+T]$. If a directed graph $\mathcal{G}(t)$'s topology is time invariant, then it is said to be connected if and only if it has a spanning tree. A spanning tree of a directed graph is a connected directed tree with the same vertex set but with an edge set which a subset of the edge set of the directed graph.

B. Multi-agent Systems

Consider N agents, interacting over a network whose topology is given by a graph $\mathcal{G}(t) = (\mathcal{V}, \mathcal{E}(t), \mathcal{A}(t))$.

The dynamics of agents i are given by

$$\dot{y}^i = \sum_{j=1}^{N} A_{ji}(t) F(y^j - y^i) \tag{1}$$

Definition 1: A C^1 function $k:\mathbb{R}\rightarrow\mathbb{R}$ is locally passive, if there exist $\sigma^- > 0$ and $\sigma^+ > 0$ such that $yF(y) > 0$ for all $y\in\left[\sigma^-,\sigma^+\right]\subset\mathbb{R}$ apart from $y = 0$, where $F(0) = 0$.

Definition 2: Consider the multi-agent system in (1). If for any initial state y_0, the states of agents satisfy

$$\lim_{t\rightarrow\infty}\left\|y^j(t)-y^i(t)\right\|=0$$

for all $i, j\in \underline{N}$, then we say the multi-agent system solves a consensus problem asymptotically. Furthermore, if there exists $y^*\in\mathbb{R}^n$ such that

$$\lim_{t\rightarrow\infty}\left\|y^i(t)-y^*\right\|=0$$

for all $i\in \underline{N}$, then we call y^* as the consensus state of the multi-agent system.

Now, we study the consensus problem of multi-agent system (1). First, we consider a time invariant topology case, i.e. the multi-agent system (1) has a fixed topology. Then, we have the following theorem.

Proposition 1 (Fixed Topologies)**:** Consider a directed network of agents under dynamics in equation (1) with fixed topologies and $F(\bullet)$ is local passive. Then, the system globally converges to the equilibrium point (its consensus value) if, and only if the weighted directed graph is connected (has a spanning tree). Moreover, convergence occurs in finite-time if there exists a constant $\varepsilon > 0$ such that $yF(y)\geq\varepsilon,\ y\neq 0$.

Proof: Define a Lyapunov function candidate as

$$V(y)=\max_i(y^i)-\min_k(y^k)=y_{\max}-y_{\min} \tag{2}$$

Note that the Lyapunov function is not smooth and differentiable, but it still can be used to conclude the attractivity properties of equilibria. The right hand side Dini derivative of $V(y)$ of the system can be defined as

$$\dot{V}(y)=\limsup_{r\rightarrow 0^+}\frac{1}{r}\left[V(y(t+r))-V(y(t))\right]$$

Suppose v_i is one of the agent with the maximum value at time t, then the Dini derivative of the candidate Lyapunov function along (1) is

$$\begin{aligned}\dot{V}(y)&=\dot{y}_{\max}-\dot{y}_{\min}\\&=\sum_{j=1}^{N}A_{ji}(t)F(y^j-y_{\max})-\sum_{j=1}^{N}A_{jk}(t)F(y^j-y_{\min})\end{aligned}$$

Since $A_{ji}(t)>0$ and $A_{jk}(t)>0$, we have $\dot{V}(y)=\dot{y}_{\max}-\dot{y}_{\min}\leq 0$ immediately.

Note that $\dot{V}(y)=\dot{y}_{\max}-\dot{y}_{\min}=0$ holds if and only if all the parent nodes of v_i and v_k hold both the maximum value and the minimum value, also so do their respective parents.

Now we prove the finite time stability of system (1).

Let $\min\{A\}$ denote the minimum positive element of A_{ji}, then we have $\dot{V}(y)=\dot{y}_{\max}-\dot{y}_{\min}<\min\{A\}\times\varepsilon$ holds all the time except all parent nodes and their respective parents hold both the maximum value and the minimum value, form the finite-time stability presented in [2], we can know the consensus reached in finite time.

Remark 1: The finite time consensus protocol presented in Proposition 1 is a discontinuous consensus strategy. Actually, continuous but non-smooth consensus strategy such as $F(\bullet)=\operatorname{sgn}(y^j-y^i)\left|y^j-y^i\right|^{\alpha}$, $0<\alpha<1$ is also valid to guarantee finite time consensus.

Next, we consider a switching topologies case for multi-agent system (1).

Proposition 2 (Switching Topologies)**:** Consider a network with a time varying topology $\mathcal{G}(t)=(\mathcal{V},\mathcal{E}(t),\mathcal{A}(t))$. If the switching topology is uniformly connected and $F(\bullet)$ is local passive, then the system globally converges to the equilibrium point (its consensus value) if, and only if the weighted directed graph has a spanning tree. Moreover, convergence occurs in finite-time if there exists a $\varepsilon>0$ such that $yF(y)\geq\varepsilon,\ y\neq 0$.

Proof: Consider Lyapunov function in Eq. (2) for the switching topology case. Although the set of neighbors and the directed weight matrix change with time, according to change in the topology, in every time interval $[t,t+T]$ or in its sub-domain, the network is connected. Therefore, there is at least one $y_{\max}$ or $y_{\min}$ changed according to its neighbors' status in a sub-domain of $[t,t+T]$ (not necessarily in the whole time interval).

As a result, $\dot{V}(y)=\dot{y}_{\max}-\dot{y}_{\min}$ is decreasing along the sub-domain (or the whole time interval $[t,t+T]$) and consequently $\dot{V}(y)=\dot{y}_{\max}-\dot{y}_{\min}\leq 0$, which means the Lyapunov function in Eq. (2) vanishes and the system achieves convergence.

Similarly, if we have $\dot{V}(y)=\dot{y}_{\max}-\dot{y}_{\min}<\min\{A\}\times\varepsilon$ in every time interval $[t,t+T]$, then consensus reached in finite time. This completes the proof.

3 Main Theoretical Results

In this section, we give the consensus control strategy of a class of high order multi-agent systems.

Consider a class of N identical linear dynamic systems, which can be of any order n, are given as

$$\begin{aligned} \dot{x}_1^i &= x_2^i \\ \dot{x}_2^i &= x_3^i \\ &\vdots \\ \dot{x}_n^i &= f(x^i,t)+u^i+d^i(x^i,t) \end{aligned} \tag{3}$$

where $x^i=(x_1^i,x_2^i,\cdots,x_n^i)\in R^n$ represents the state vector of the i-th node; $u^i\in R^1$ is the control input.

We assume the consensus state vector to be $\overline{x}^*$. Define the consensus error state of agent v_i as $e^i=x^i-\overline{x}^*$. The consensus control objective is to achieve $e^i=0,\ i=1,\cdots,N.$

The first step is to design a sliding surface function. Let the sliding- surface function be selected as

$$S^i=C^Te^i \tag{4}$$

where $C=[c_1,c_2,\cdots,c_n]^T$. If an appropriate consensus control law is available, then the sliding mode will be obtained in finite time. In the sliding mode, the error dynamics of node v_i will be

$$c_ne_n^i+c_{n-1}e_{n-1}^i+\cdots+c_1e_1^i=0\,.$$

Those $c_1,c_2,\cdots,c_n$ are positive constants such that the polynomial $\phi(\lambda)=c_n\lambda^{n-1}+c_{n-1}\lambda^{n-2}+\cdots+c_1$ is Hurwitz. The choice of C decides the decaying rate of the consensus error. The expected consensus dynamic should be a stable dynamic.

The second step is to design a consensus control law such that all agents can reach and sustained to the same sliding mode in a finite time.

From (3) and (4), we have

$$\begin{aligned} \dot{S}^i &= C^T\dot{e}^i \\ &= c_n\dot{e}_n^i+c_{n-1}\dot{e}_{n-1}^i+\cdots+c_1\dot{e}_1^i \\ &= c_n(f(x^i,t)+u^i+d^i(x^i,t))+c_{n-1}x_n^i+\cdots+c_1x_2^i \end{aligned}$$

Therefore, the controller can be designed as follows:

$$u^i = u_0^i + u_s^i \tag{5}$$

where $u_0^i = -f(x^i, t) - c_n^{-1}(c_{n-1}x_n^i + \cdots + c_1 x_2^i)$

and $u_s^i = \sum_{j=1}^{N} A_{ji} F(S^j - S^i)$

where $F(\cdot)$ is a local passive function and $\exists \varepsilon > 0,\ yF(y) \geq \varepsilon,\ y \neq 0$.

Then, we have the following theorem.

Theorem 1: Consider multi-agent system (3) with directed and fixed topologies. Given the sliding function (4) and control law (5), the sliding mode is guaranteed to be reached in finite time if and only if the graph is connected. Therefore, the multi-agent system will reach consensus by consensus protocol (5).

Proof: From (3), (4) and (5) , we have

$$\begin{aligned}\dot{S}^i &= \sum_{j=1}^{N} A_{ji}(t) F(S^j - S^i) \\ &= \sum_{j=1}^{N} A_{ji} F(S^j - S^i)\end{aligned}$$

Note that $F(\cdot)$ is local passive and $\exists \varepsilon > 0,\ yF(y) \geq \varepsilon,\ y \neq 0$. From proposition 1, it easily to obtain that all S^i for all $i = 1, \cdots, N$ will converge to their consensus value S^* in finite time. After that, all agents will reach to the same sliding mode and converge to their consensus state. This completes the proof.

Remark 2: It is worth noting that the passive function $F(\cdot)$ has plenty of choices. For example, $F(\cdot) = k_1(S^j - S^i) + k_2 \operatorname{sgn}(S^j - S^i)$ can be chose as a candidate function.

Remark 3: The final consensus value of system (3) over protocol (5) is unknown previously and can be determined by many factors. The initial value of all agents, the sliding surface selected, the communication topologies will all affect the final consensus value. The final consensus value of the multi-agent systems will be $\overline{x}^* = (\overline{x}_1^*, 0, \cdots, 0) \in \mathbb{R}^n$.

In the following, we consider the directed topology is changing over time. Similarly to Proposition 2, we have the following theorem for directed multi-agent systems with switching topologies.

Theorem 2: Consider multi-agent system (3) with directed and switching topologies. Given the sliding function (4) and control law (5), the sliding mode is guaranteed to be reached in finite time if and only if the graph is connected. Therefore, the multi-agent system will reach consensus by consensus protocol (5).

Proof: The proof is similar to Proposition 2 and is omitted here.

4 Conclusions

This paper is concerned with the consensus control of a class of high order nonlinear system. Both the fixed directed topologies and switching directed topologies are considered. By using the so called sliding mode control technique, we proposed a new consensus control protocol which pledged all agents can reach to the same consensus value under fixed and switching directed topologies. The final consensus value is determined by both the initial state of the multi-agent system and the pre-designed sliding surface. Our future objective is to design a consensus dynamic for high order multi-agent systems over applicable partial communication topologies.

References

1. Papachristodoulou, A., Jadbabaie, A., Münz, U.: Effects of delay in multi-agent consensus and oscillator synchronization. IEEE Trans. Autom. Control 55(6), 1471–1477 (2010)
2. Hui, Q., Haddad, W.M., Bhat, S.P.: Finite-time semistability, Filippov systems, and consensus protocols for nonlinear dynamical networks with switching topologies. Nonlinear Analysis: Hybrid Systems 4(3), 557–573 (2010)
3. Lin, P., Jia, Y.: Consensus of a Class of Second-Order Multi-Agent Systems With Time-Delay and Jointly-Connected Topologies. IEEE Trans. Autom. Control 55(3), 778–784 (2010)
4. Scardovi, L., Sepulchre, R.: Synchronization in Networks of Identical Linear Systems. Automatic 45, 2557–2562 (2009)
5. Li, Z., Duan, Z., Chen, G., Huang, L.: Consensus of Multiagent Systems and Synchronization of Complex Networks: A Unified Viewpoint. IEEE Trans. Circ. Syst. I 57(1), 213–224 (2010)
6. Seo, J.H., Shim, H., Back, J.: Consensus of high-order linear systems using dynamic output feedback compensator: Low gain approach. Automatica 45, 2659–2664 (2009)
7. Chopra, N., Spong, M.W., Lozano, R.: Synchronization of bilateral teleoperators with time delay. Automatica 44(8), 2142–2148 (2008)
8. Jiang, F., Wang, L.: Consensus seeking of high-order dynamic multi-agent systems with fixed and switching topologies. International Journal of Control - Int. J. Contr. 83(2), 404–420 (2010)
9. Qu, Z.: Cooperative Control of Dynamical Systems. Springer (2009)
10. Utkin, V.I.: Sliding modes in control and optimization. In: CCES. Springer, New York (1992)
11. Huang, J.Y., Gao, W., Huang, J.C.: Variable structure control: A survey. IEEE Trans. Ind. Elect. 40, 2–22 (1993)
12. Huang, Y.-J., Kuo, T.-C., Chang, S.-H.: Adaptive Sliding-Mode Control for Non-linear Systems With Uncertain Parameters. IEEE Transactions on Systems 38(2), 534–539 (2008)

Design and the Position Analysis of a Novel Translational Orthogonal Parallel Micromanipulator

Yuzhu Wang[1] and Zhenlin Jin[1,2]

[1] College of Mechanical Engineering, Yanshan University, Qinhuangdao, 066004, China
yzwang86@163.com
[2] State Key Laboratory of Mechanical System and Vibration, Shanghai, 200240, China
zljin@ysu.edu.cn

Abstract. A novel translational orthogonal parallel micromanipulator of unique design is proposed, it is the variation of DELTA robot. The structure and layout features of the parallel micromanipulator are described. Based on its architecture feature, direct and inverse kinematics equations are developed. The micro-displacement of moving platform is simulated by loading the displacement of a one-way branch, after establishing the finite element analysis model of the micromanipulator. Comparing the simulation result with the theoretical value, the results show that it has many advantages such as better isotropic properties, perfect movement decoupling and high precision. Therefore, the micromanipulator can be widely applied to the fields of precision machinery, genetic engineering and microelectronic packaging, etc.

Keywords: parallel micromanipulator, orthogonal, position analysis, movement decoupling, finite element.

1 Introduction

Micromanipulator which is an important branch in the field of MEMS has been closely concerned by scholars at home and abroad, it has wide application foreground on the areas of precision machinery, connecting fiber, genetic engineering and microelectronic packaging [1, 2]. Parallel mechanism which has compact structure, good symmetry and little error accumulation is suitable for the configuration of micromanipulator [3, 4]. For example, Culpepper (a professor at MIT) designed a Heflex planar six-axis compliant parallel micromanipulator [5], professors of Tsinghua University developed a micromanipulator based on the six-DOF Stewart platform [6], a novel three-DOF (two translations and one rotation) parallel micromanipulator was put forward by Ping'an Liu [7], a tree-link six-DOF parallel micromanipulator was designed by Lining Sun of Harbin institute of technology [8], etc. A new kind three-DOF translational orthogonal micromanipulator which has reasonable layout is presented by the author[9], it is the variation of DELTA robot, direct and inverse kinematics equations are developed, its micro-displacement error is analysed by using finite element simulation, the paper can provide theoretical basis for designing and controlling the micromanipulator.

G. Lee (Ed.): Advances in Automation and Robotics, Vol. 1, LNEE 122, pp. 215–222.
springerlink.com

2 The Structures and Characters

The unitary structure diagram of the micromanipulator is shown in figure 1(a), it is consist of the moving platform 1, three kinematic chains 2 of the same structure and the base 3. The kinematic chain is shown in figure 1(b), which is consist of the parallel plank translational joint 4, piezoelectric actuator 5, handspike 11 with two flexible revolute pairs 6 and 7 at both ends, the T swing rod 12 with flexible revolute 13 at one end, a pair parallel pillar 9 with flexible spherical joints 8 and 10 at both ends. The main body of the micromanipulator isn't assemblies, which is once machining. The centres of 3 pairs spherical joints, which are connected by the moving platform and the 3 kinematic chains, separately distributed on three mutually perpendicular surfaces of the moving platform. The lines of its centres are mutually perpendicular in space. The 3-DOF translation of the moving platform is implemented by driving corresponding the parallel plank translational joint with PZT drivers.

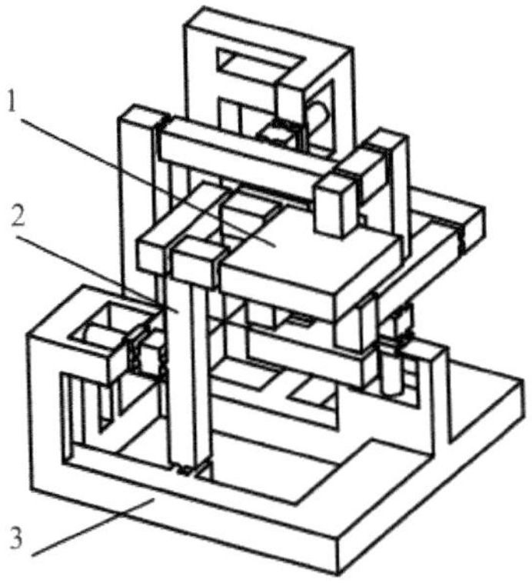

Fig. 1. (a) The structure schematic

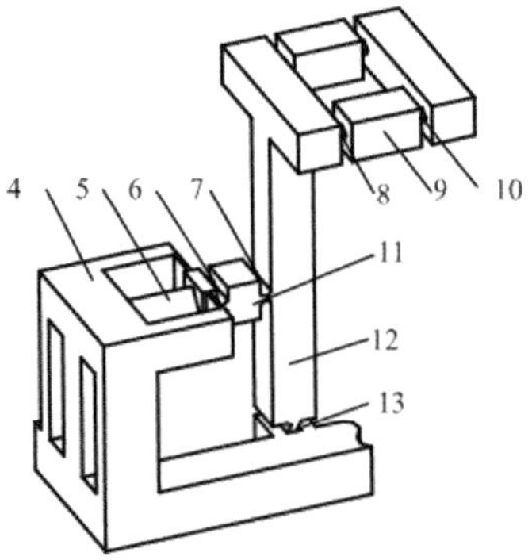

Fig. 1. (b) The kinematic chain schematic

3 The Position Analysis

As fig.2 shows, it is the parallel mechanism schematic, the static coordinate $\{V\}$: O-xyz is fixed within the base of micromanipulator, and the moving coordinate $\{U\}$: O_1-$x_1y_1z_1$ whose origin is at the center of the moving platform is fixed within the moving platform and its three-dimensional coordinate axis separately parallel to the three mutually perpendicular lines of the moving platform. When the micromanipulator is in its initial posture, the moving coordinate and the static coordinate is coincident. Assumed that : $2a$ is the width and length of the moving platform, $2h$ is the hight of the moving platform, b is the length of the pillar, $2s$ is the distance of the mutually pillars, c is the length of the T-swing rod, d is the distance of the flexible revolute pairs R_{i1} and R_{i2} $(i=1,2,3)$.

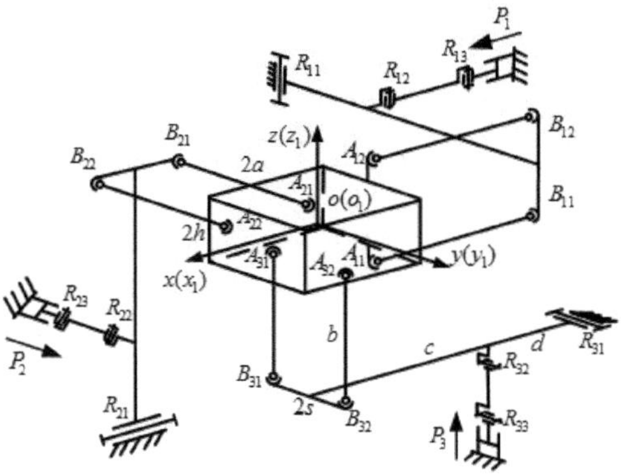

Fig. 2. The coordinate schematic

Assumed that:

$$\Delta \boldsymbol{m} = (\Delta x, \Delta y, \Delta z)^{\mathrm{T}} \tag{1}$$

$$\Delta \boldsymbol{l} = (\Delta l_1, \Delta l_2, \Delta l_3)^{\mathrm{T}} \tag{2}$$

Where

$\Delta \boldsymbol{m}$ =the position vector of the moving platform's reference point O_1,

$\Delta \boldsymbol{l}$ =the input micro displacement of the parallel plank translational joint.

In the moving coordinate $\{U\}$: O_1-$x_1y_1z_1$, the space position vector $\boldsymbol{A}_{\mathrm{U},i1}$, $\boldsymbol{A}_{\mathrm{U},i2}$(i=1, 2, 3) of the flexible spherical joints A_{i1} , A_{i2} are showed as follows:

$$\begin{cases} \boldsymbol{A}_{\mathrm{U},11} = (-a \quad 0 \quad -s)^{\mathrm{T}} \\ \boldsymbol{A}_{\mathrm{U},12} = (-a \quad 0 \quad s)^{\mathrm{T}} \\ \boldsymbol{A}_{\mathrm{U},21} = (-s \quad -a \quad 0)^{\mathrm{T}} \\ \boldsymbol{A}_{\mathrm{U},22} = (s \quad -a \quad 0)^{\mathrm{T}} \\ \boldsymbol{A}_{\mathrm{U},31} = (0 \quad -s \quad -h)^{\mathrm{T}} \\ \boldsymbol{A}_{\mathrm{U},32} = (0 \quad s \quad -h)^{\mathrm{T}} \end{cases} \tag{3}$$

In the static coordinate $\{V\}$: O-xyz , the space position vector $\boldsymbol{A}_{\mathrm{V},i1}$, $\boldsymbol{A}_{\mathrm{V},i2}$(i=1, 2, 3) of the flexible spherical joints A_{i1} , A_{i2} are showed as follows:

$$\begin{cases} \boldsymbol{A}_{\mathrm{V},i1} = \boldsymbol{A}_{\mathrm{U},i1} + \Delta \boldsymbol{m} \\ \boldsymbol{A}_{\mathrm{V},i2} = \boldsymbol{A}_{\mathrm{U},i2} + \Delta \boldsymbol{m} \end{cases} \tag{4}$$

When the space position vector of the moving platform is $\Delta \boldsymbol{m} = (\Delta x, \Delta y, \Delta z)^{\mathrm{T}}$, assume that θ_i(i=1, 2, 3) is the angular displacement of the flexible revolute pairs R_{i1} in the chain i. In the static coordinate $\{V\}$: O-xyz the space position vector $\boldsymbol{B}_{\mathrm{V},i1}$, $\boldsymbol{B}_{\mathrm{V},i2}$(i=1, 2, 3) of the flexible spherical joints B_{i1} , B_{i2} are showed as follows:

$$\begin{cases} \boldsymbol{B}_{\mathrm{V},11} = \left(-(b+a)+c\sin\theta_1 \quad -c+c\cos\theta_1 \quad -s\right)^{\mathrm{T}} \\ \boldsymbol{B}_{\mathrm{V},12} = \left(-(b+a)+c\sin\theta_1 \quad -c+c\cos\theta_1 \quad s\right)^{\mathrm{T}} \\ \boldsymbol{B}_{\mathrm{V},21} = \left(-s \quad -(b+a)+c\sin\theta_2 \quad -c+c\cos\theta_2\right)^{\mathrm{T}} \\ \boldsymbol{B}_{\mathrm{V},22} = \left(s \quad -(b+a)+c\sin\theta_2 \quad -c+c\cos\theta_2\right)^{\mathrm{T}} \\ \boldsymbol{B}_{\mathrm{V},31} = \left(-c+c\cos\theta_3 \quad -s \quad -(b+h)+c\sin\theta_3\right)^{\mathrm{T}} \\ \boldsymbol{B}_{\mathrm{V},32} = \left(-c+c\cos\theta_3 \quad s \quad -(b+h)+c\sin\theta_3\right)^{\mathrm{T}} \end{cases} \tag{5}$$

According to geometry relation:

$$\begin{cases} \left|\boldsymbol{A}_{\mathrm{V},i1} - \boldsymbol{B}_{\mathrm{V},i1}\right| = b \\ \left|\boldsymbol{A}_{\mathrm{V},i2} - \boldsymbol{B}_{\mathrm{V},i2}\right| = b \end{cases} \tag{6}$$

According to its characteristic of the structural layout, two pairs of equation (6) are equivalent. Therefore, select one of equation (6) and take equations (4) and (5) into it.

$$\begin{cases} (\Delta x+b-c\sin\theta_1)^2 + (\Delta y+c-c\cos\theta_1)^2 + \Delta z^2 = b^2 \\ \Delta x^2 + (\Delta y+b-c\sin\theta_2)^2 + (\Delta z+c-c\cos\theta_2)^2 = b^2 \\ (\Delta x+c-c\cos\theta_3)^2 + \Delta y^2 + (\Delta z+b-c\sin\theta_3)^2 = b^2 \end{cases} \tag{7}$$

According to the simplification:

$$A_i\sin\theta_i + B_i\cos\theta_i + C_i = 0 \quad (i = 1, 2, 3) \tag{8}$$

Where

$$\begin{cases} A_1 = -2\Delta xc - 2cb \\ B_1 = -2\Delta yc - 2c^2 \\ C_1 = \Delta x^2 + \Delta y^2 + \Delta z^2 + 2\Delta xb + 2\Delta yc + 2c^2 \\ A_2 = -2\Delta yc - 2cb \\ B_2 = -2\Delta zc - 2c^2 \\ C_2 = \Delta x^2 + \Delta y^2 + \Delta z^2 + 2\Delta yb + 2\Delta zc + 2c^2 \\ A_3 = -2\Delta zc - 2cb \\ B_3 = -2\Delta xc - 2c^2 \\ C_3 = \Delta x^2 + \Delta y^2 + \Delta z^2 + 2\Delta zb + 2\Delta xc + 2c^2 \end{cases} \tag{9}$$

The geometry relation of the angular displacement θ_i $(i = 1, 2, 3)$ of the flexible revolute pairs R_{i1} in the chain i and the input micro displacement Δl_i is showed as follows:

$$\Delta l_i = d\sin\theta_i \tag{10}$$

For the micro-nanometer working capability of micromanipulator, the rotating angles of the above formulas are very small, so there are approximate relations as follows:

$$\sin\theta_i \to \theta_i, \cos\theta_i \to 1 \ (i = 1,2,3) \tag{11}$$

The equations (8) – (10) are solved under neglecting the high-order infinitesimal, its positional reverse solving equation is presented as follows:

$$\begin{cases} \Delta l_1 = \dfrac{d}{c}\Delta x \\ \Delta l_2 = \dfrac{d}{c}\Delta y \\ \Delta l_3 = \dfrac{d}{c}\Delta z \end{cases} \tag{12}$$

So the micromanipulator positional direct solving equation is obtained as follows:

$$\begin{cases} \Delta x = \dfrac{c}{d}\Delta l_1 \\ \Delta y = \dfrac{c}{d}\Delta l_2 \\ \Delta z = \dfrac{c}{d}\Delta l_3 \end{cases} \tag{13}$$

According to the equation (12), the output micro displacement along x, y, z direction of the micromanipulator only relation to input micro displacement of the corresponding parallel plank translational joint, so the micromanipulator is motion decoupling. As result of the same construction dimensions of three kinematic chains, the micro displacements of the micromanipulator transfer fully isotropy.

4 The FEA Analysis of Micromanipulator

4.1 The Finite Element Model of Micromanipulator

The elastomer structure of micromanipulator is monolithic, its material is 65Mn, and the material's relevant parameters are following: Elastic modulus $E = 206\text{GPa}$, Density $\rho = 7800\text{kg/m}^3$, Poisson's ration $\gamma = 0.3$. The structure geometric parameters of micromanipulator are following: $2a$=50mm, $2h$=12mm, $2s$=38mm, b=25mm, c=98mm, d=42mm. After constructing the 3D model in Pro/E and through applying the ANSYS software, the finite element model is constructed as fig.3.

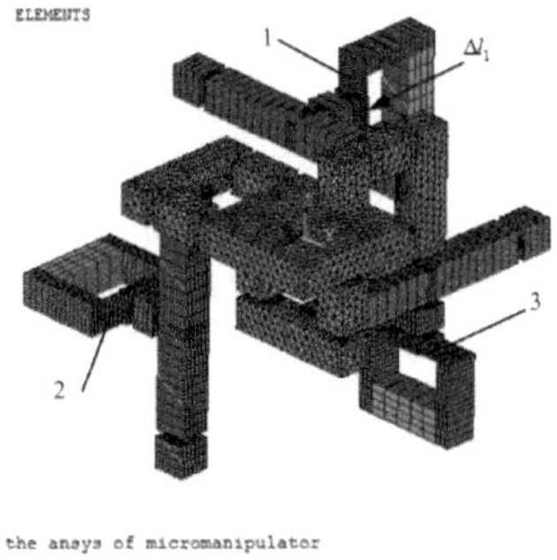

Fig. 3. The finite element model of micromanipulator, 1- the parallel plank translational joint along *x* direction, 2- the parallel plank translational joint along *y* direction, 3- the parallel plank translational joint along *z* direction

4.2 The Micro Displacement Finite Element Simulation Analysis

In order to obtain the micro displacement characteristics of micromanipulator, the three kinematic chains should be analyzed lonely, but the structure of manipulator has the generalized symmetries, the micro displacement characteristic along *x* direction is analyzed only.

As fig.3 shows, when the parallel plank translational joint along x direction is loaded Δl_1 (0~30μm) and the parallel plank translational joints along y, z direction aren't loaded ($\Delta l_2 = \Delta l_3 = 0$), the micro displacement of the moving platform's reference point O_1 is obtained while the parameter Δl_1 takes different values. Comparing the theoretical value with the finite element analyzing result, the line displacement along *x* direction of the moving platform is shown in fig4(a), the line displacement along *y*, *z* direction of the moving platform is shown in fig4(b), the angle displacement along *x* axis of the moving platform is shown in fig4(c) , and the angle displacement along *y, z* axes of the moving platform is shown in fig4(d).

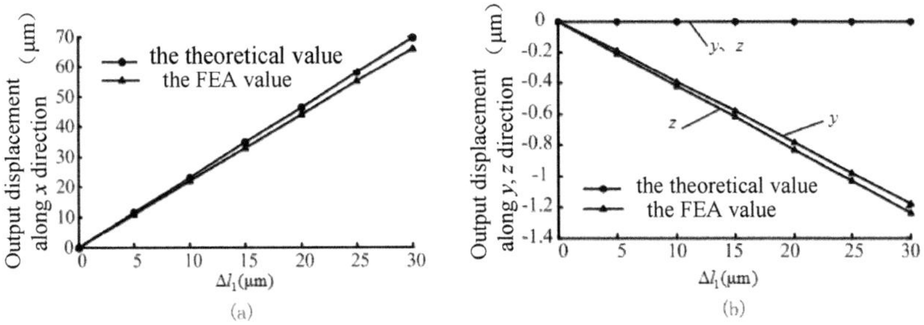

Fig. 4. The micro displacement of moving platform when the parallel plank translational joint along x direction is loaded Δl_1

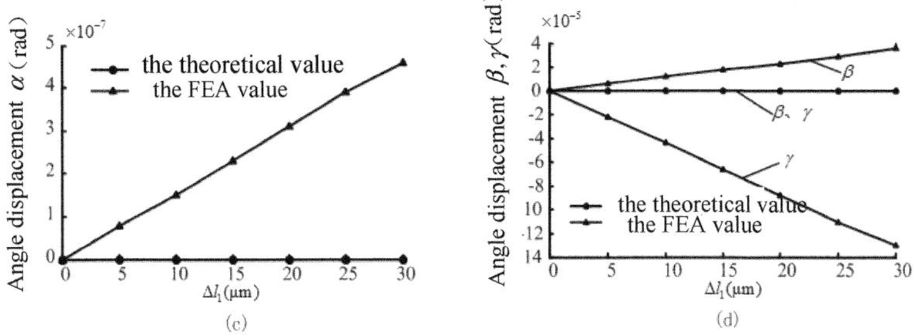

Fig. 4. (*continued*)

According to the fig.4, the micromanipulator exists error of the second kind which includes the micro displacement component of the moving platform along y, z direction and the angle displacement (the angles along x, y, z axes are α, β, γ) of the moving platform's reference point O_1 in static coordinate, this will unavoidably affect the micro displacement of robot's end effector. If an end effector is fixed on the moving platform, its end's position vector in moving coordinate is ${}^{U}\boldsymbol{P} = (0 \quad 0 \quad 100)^{T}$ (unit: mm), the micro displacement error of the end effector is showed in fig.5.

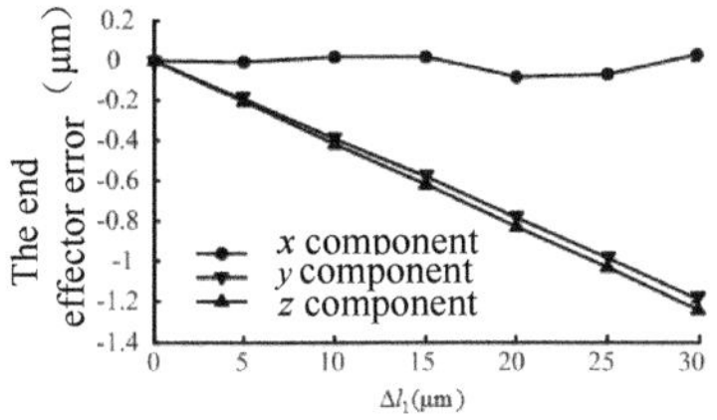

Fig. 5. The micro displacement error of the end effector when the parallel plank translational joint along x direction is loaded Δl_1

The full-range error is defined as percentage of the maximum absolute error and the full-range value in the micromanipulator workspace. So it can be used for the evaluation method of the micromanipulator's positioning accuracy. From the fig.4 and the fig.5, the full-range error of the end effector along x direction is 0.1%. In the error of the second kind, the full-range error of the end effector along y direction is 1.7% and the full-range error of the end effector along z direction is 1.8%. The theoretical analysis for the transfer ration of micro displacement basically coincides with the finite element simulation results. The same conclusion is obtained when the parallel plank translational joints along y, z direction is loaded separately.

From the above discussion, the characteristics of the micromanipulator concludes small micro displacement error, better decoupling and the approximating same transfer ration of micro displacement.

5 Conclusions

Based on the DELTA robot, a novel 3-DOF translational orthogonal parallel micromanipulator is designed in the paper. The kinematic direct equation and inverse position equation of the micromanipulator are given explicitly in analytical form, and each equation has a unique solution. By using finite element simulation analysis, the result shows that: the location error of the micromanipulator is less than 2%F.S., the theoretical analysis basically coincides with the finite element simulation results. The results show: it has many advantages such as better isotropic properties, perfect movement decoupling and high precision etc. The micromanipulator technology will be widely applied on the areas of biomedicine and precision machine etc.

References

1. Yun, Y., Xu, Q., Li, Y.: Survey on Parallel Manipulators with Micro/Nano Manipulation Technology and Applications. Chinese Journal of Mechanical Engineering 44(12), 12–23 (2008)
2. Wang, H., Sun, L., Rong, W., et al.: Research on Key Technologies of Micromanipulation for Life Sciences. Chinese Journal of Scientific Instrument 27(1), 106–110 (2006)
3. Ellis, G.W.: Piezoelectric micromanipulators. Science Instruments and Techniques 138, 84–91 (1962)
4. Jin, Z., Gao, F.: A Novel Parallel Micro-motion Manipulator based on 6-PSS Three-dimmensional Platform Mechanism. Chinese Journal of Scientific Instrument 22(6), 566–569 (2001)
5. Culpepper, M.L., Anderson, G.: Design of a Low-cost Nano-manipulator Which Utilizes a Mnonolithic, Spatial Compliant Mechanism. Precision Enigineering 28(4), 469–482 (2004)
6. Li, J., Chene, K., Dong, Y., et al.: Study on the Static Stiffness of the Parallel Flexure Joint Robot. Journal of Tsinghua University and Technology 39(8), 16–20 (1999)
7. Liu, L., Fang, Y., Wu, S.: A Simple Modeling Method for 3 DOF micro-motion Parallel Manipulator. China Mechanical Engineering 17(24), 2551–2553 (2006)
8. Rong, W., Wang, L., Sun, L.: Static Stiffness Analysis of 3-PPSR Parallel Micromanipulator. Chinese Journal of Mechanical Engineering 44(1), 13–18 (2008)
9. Jin, Z., Yue, Y.: A Full-isotropic Spherical 3-DOF Parallel micromanipulator. China, 200710185234. 8 (April 16, 2008)

Conformal Minimal Convex Analysis of Geodesic Active Contours

Guohua Chen

School of Med-Info Engineering
Guangdong Pharmaceutical University, Guangzhou 510006, China
ghchen2007@yahoo.com.cn

Abstract. Geodesic Active Contour Model (GACM) is a great improvement of the traditional Snake Model. It had converted the problem of finding extremity of an energy functional to the problem of finding a geodesic curve in an appropriate Riemannian space. But (geodesic) active contour models are nonconvex in general, so the solutions of the Euler-Lagrange equation of its energy functional are only critical points, but were often taken as minima of the energy functional as granted. This paper focused on the convex analysis of Geodesic Active Contour Models and had proposed a condition to ensure the convexity of the energy functional of GACM. We proved that if the Gaussian curvature K of the image surface I is less than 0 everywhere along a normal geodesic curve $C(t)$, then its energy functional $E(C)$ will be convex near C. Further more, it can be induced from our theorem that if $\Delta \ln h > 0$ along $C(t)$, then the energy functional $E(C)$ will also be convex near C if the image space has its metric tensor taking the form $g_{ij} = h(|\nabla I(x)|)^2 \delta_{ij}$.

Keywords: Image processing, Convex analysis, Geodesic Active Contour Model, Image as Surface., Second Variational formula.

1 Introduction

An active contour [16] is a curve $C(t) = (x(t), y(t))$, $t \in [0,1]$, which moves in an image to make the energy $E(C)$ minimized:

$$E(C) = K_0 \int_0^1 \|C'(t)\|^2 dt + K_1 \int_0^1 \|C''(t)\|^2 dt + \int_0^1 P(x(t), y(t))dt \tag{1}$$

where $P(x,y)$ is the potential function, K_0 and K_1 are tension and rigidity parameter of the curve respectively. The variation theory[10] tell us that the curve C that minimize $E(C)$ must satisfy the Euler equation[12,13]:

$$-\nabla P(C(t)) + 2K_0 C''(t) - 2K_1 C''''(t) = 0 \tag{2}$$

here we suppose that for t=0 and t=1, $C(\cdot)$ and its derivatives are known.

Active Contour Models are well known in the image edge detection area. A main potential weakness of these methods is that their convergence properties are poorly understood. The two central questions are: will the active contour algorithm converge

G. Lee (Ed.): Advances in Automation and Robotics, Vol. 1, LNEE 122, pp. 223–231.
springerlink.com

to a unique solution, and if so, will this solution be near the truth? Some grasp of the answers to these questions is essential if active contours are to be used for either a scientific or clinical objective. Several authors have begun to study convergence ([3,4],[7]-[9]). It has been determined that active contour models are nonconvex in general and that a solution may be a local minima but no sure. This paper will focus on the convex analysis of geodesic active contour models. We had find a condition to ensure the energy functional of an active contour to be convex and so its solution will be a true minimizer.

From theoretical viewpoint, we can use the second variation of the functional $E(C)$ to analyze its convexity[2,10,17]. Some authors acclaimed that this method is unpractical[7]. So they developed a discrete version of active contour model and analyzed its convexity. Davatzikos et. al. ([7]-[9]) started by modeling an active contour as a collection of points $\vec{d}=(x_1, y_1, \cdots, x_{N-1}, y_{N-1})$ and obtained a discrete approximation $E(\vec{d})$ of $E(C)$. The necessary condition for $\vec{d}$ to minimize $E(\vec{d})$ is that $\nabla E(\vec{d})$=0. If $E(\vec{d})$ is strictly convex, then this condition is also sufficient and $\vec{d}$ is the unique minimal of $E(\vec{d})$ [7]. It was noticed that $E(\vec{d})$ is strictly convex if the Hessian matrix of $E(\vec{d})$ is positive definite. So we need only to focus our efforts to the analysis of the Hessian matrix of $E(\vec{d})$. But this kind of discrete calculation dose not easy to be done in general. We have found very limited uses of this method. It had only be used to some very special cases of active contours, such as ACAR(Active Contour Algorithm for Ribbons) ([7]-[9]). Here we have provided a substitute method for convex analysis: we treat an image I as a surface M ([19]-[22]), and then calculate the Hesse of the functional $E(C)$ directly. i.e. we calculate the second variation of the energy functional to determine the convexity of an active contour. We had obtained the following theorem:

Theorem: If $C(t)$ is a normal geodesic curve, and the Gaussian curvature K of M is less than 0 everywhere along $C(t)$, i.e. K(t) <0 on $C(t)$, then the functional $E(C)$ is convex near $C(t)$. So $C(t)$ is a true minimizer of $E(C)$.

To prove the theorem, we need to introduce some conceptions.

2 Image as Surface and Its Conformal Coordinates

We view a grayscale image $I(x, y)$ as a surface M in R^3 defined on the parametric plane $U(x^1, x^2)$ [1,11,19]. Let x^1=x, x^2=y. The parametric plane $U(x^1, x^2)$ is mapped via X into a surface M in R^3:

$$X : U \rightarrow M \subset R^3$$
$$(x^1, x^2) \mapsto (x^1, x^2, I(x^1, x^2)) \in R^3$$

U and M can be considered as Riemannian manifolds with suitable Riemannian metrics. U is the domain space, M is the image space, (x^1, x^2) is also the coordinate chart on M . The metric on U is the ordinary metric $ds^2 = (dx^1)^2 + (dx^2)^2$. But there are many different metrics on M , the conical one is induced from R^3 as follows:

$$ds^2 = (dx^1)^2 + (dx^2)^2 + dI^2$$
$$= (1 + I_1^2)(dx^1)^2 + 2I_1 \cdot I_2 dx^1 dx^2 + (1 + I_2^2)(dx^2)^2$$

where $I_i = \frac{\partial I}{\partial x^i}$. Let $\frac{\partial}{\partial x^1} = (1, 0, \frac{\partial I}{\partial x^1})$, $\frac{\partial}{\partial x^2} = (0, 1, \frac{\partial I}{\partial x^2})$, then $\frac{\partial}{\partial x^1}$, $\frac{\partial}{\partial x^2}$ form a basis on the tangent plane of M . The induced metric on M can also be denoted as $ds^2 = \sqrt{g_{ij} dx^i dx^j}$, here we had used the Einstein summation convention. The covariant variables g_{ij} on manifolds are defined as follows:

$$g_{11} =< \frac{\partial}{\partial x^1}, \frac{\partial}{\partial x^1} >_{ds^2} = 1 + I_1^2 \qquad g_{22} =< \frac{\partial}{\partial x^2}, \frac{\partial}{\partial x^2} >_{ds^2} = 1 + I_2^2$$

$$g_{12} = g_{21} =< \frac{\partial}{\partial x^1}, \frac{\partial}{\partial x^2} >_{ds^2} = I_1 \cdot I_2 \qquad g = (g_{ij}) = \begin{pmatrix} g_{11} & g_{12} \\ g_{21} & g_{22} \end{pmatrix}$$

g^{ij} are the components of the contravariant metric tensor (the inverse of the covariant metric tensor), G=det(g_{ij}) = $g_{11}\ g_{22}$ - $g_{12}\ g_{21}$. Any curve $C(t) \in X(U)$ is mapped from a curve $\tilde{C}(\tilde{t}) \in U$, i.e. each point p$\in$ $C(t)$ is a mapping of a corresponding point $\tilde{p} \in \tilde{C}(\tilde{t})$ given by p= $X(\tilde{p})$. If the parameter t is proportional to the arc length, then it is denoted by s and is called a normal parameter. See Fig.1[19].

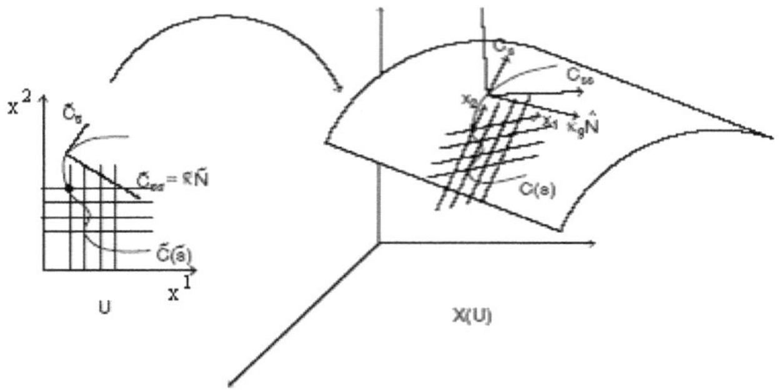

Fig. 1. The curve $C(t)$ on the surface M and its origin $\tilde{C}(\tilde{t})$ on the parameterization plane U [19]

According to above definitions, the derivative of $C(t)$ on M with respective to t is C_t, which is tangent to the curve $C(t)$, and can be represented by:

$$C_t = X_*(\dot{x}^1(t), \dot{x}^2(t)) = \dot{x}^1(t)\frac{\partial}{\partial x^1} + \dot{x}^2(t)\frac{\partial}{\partial x^2}$$

Now we will introduce the conception of Conformal Mapping[2,15]. Let S_1 and S_2 be two regular surfaces, parameterized with (x^1, x^2). Suppose that the mapping $\varphi: S_1 \to S_2$ can be represented as $\varphi(x^1, x^2) = (\varphi^1(x^1, x^2), \varphi^2(x^1, x^2))$. The first fundamental form on S_1 and S_2 can be written respectively as:

$$ds_1^{\,2} = \sum_{i,j} g_{ij} dx^i dx^j, \qquad ds_2^{\,2} = \sum_{i,j} \bar{g}_{ij} dx^i dx^j$$

the pull back metric on S1 induced by φ are:

$$\varphi^* ds_2^{\,2} = \sum_{m,n}\sum_{i,j} \bar{g}_{ij}(\varphi(x^1, x^2)) \frac{\partial \varphi^i}{\partial x^m}\frac{\partial \varphi^j}{\partial x^n} dx^m dx^n$$

Definition 1 (Conformal Mapping) [15]**:** φ is called a conformal mapping between S_1 and S_2, if there exists a positive function $\lambda(x^1, x^2)$ such that

$$ds_1^{\,2} = \lambda(x^1, x^2)\,\varphi^* ds_2^{\,2}$$

Especially, if there is conformal mapping from S_1 to the coordinate plane $U(x^1, x^2)$, then (x^1, x^2) will be said to be a conformal coordinates on S_1, or isothermal coordinates.

Using the conformal coordinates, the metric on S_1 can be formulated as:

$$ds^2 = \lambda(x^1, x^2) * ((dx^1)^2 + (dx^2)^2)$$

Caselles et. al[6] had considered a special "Snake" model, and had introduced a so called general energy functional:

$$E(C) = K_0 \int_0^1 \|C'(t)\|^2 dt + \int_0^1 h(|\nabla I(C(t))|)^2 dt \tag{3}$$

where h:$[0,+\infty] \to R^+$ is a strictly decreasing function, such that $h(\mathrm{r}) \to 0$ when $\mathrm{r} \to \infty$. It had been proved[6,14] that the minima of the general energy functional (3) are geodesic curves in a Riemannian surface. The metric on this Riemannian surface is induced from the image I as $g_{ij} = h(|\nabla I(x)|)^2 \delta_{ij}$. Therefore to seek for the minima of $E(C)$ in equation (3) is equivalent to seek for the minima of the following formula:

$$\int_0^1 \sqrt{g_{ij}\dot{x}^i \dot{x}^j}\, dt \quad \text{(i,j=1,2)} \tag{4}$$

where $C_t = (\dot{x}^1(t), \dot{x}^2(t))$. This had actually introduced a conformal metric into the Image space which makes problem of seeking for the minima of $E(C)$ being converted to an intrinsic minimal problem:

$$Min \int_0^1 h(|\nabla I(C(t))|)|C'(t)|dt \tag{5}$$

This means that, in a new framework, the edge-detection problem is equivalent to the problem of seeking a curve with minimal weighted length in a Riemannian space with conformal metric. The weight to the length is determined by the conformal mapping between the image space M and the domain space U, and can be formulated as follows:

$$g=(g_{ij})=h^2\begin{pmatrix}1 & 0\\ 0 & 1\end{pmatrix},\quad g^{-1}=(g^{ij})=\frac{1}{h^2}\begin{pmatrix}1 & 0\\ 0 & 1\end{pmatrix}$$

The function h need only to satisfy the above condition, but is usually used as the following:

$$h=\frac{1}{1+|\nabla \hat{I}|^p}$$

where p=1 or 2, $\hat{I}$ is the smoothed version of I which can be obtained from Gaussian filter. Under the above conformal metric, the Christoffel symbols and the Riemannian curvature tensor R will be [2,5]:

$$\Gamma^1_{11}=\Gamma^2_{12}=\Gamma^2_{21}=-\Gamma^1_{22}=\frac{\partial \ln h}{\partial x^1},\quad \Gamma^2_{22}=\Gamma^1_{12}=\Gamma^1_{21}=-\Gamma^2_{11}=\frac{\partial \ln h}{\partial x^2}$$

$$R_{1212}=R_{2121}=-R_{1221}=-R_{2112}=G*K$$

here K is the Gaussian curvature of image space M and G=det(g_{ij}).

3 The Second Variation of the Geodesic Active Contours

To escape from notational confusion, we denote the single parameterized closed curve $C(t)$ as $\bar{C}(t)$. Suppose that W is a vector field along $\bar{C}(t)$ on M.

Definition 2[2,5,17]**:** A smooth (first) variation of the closed curve $\bar{C}(t)$ with respect to W is a smooth mapping as follows:

$$C:(-\varepsilon,+\varepsilon)\times[0,1]\to M$$
$$(u,t)\mapsto C(u,t)\in M$$

such that $C(0,t)=\bar{C}(t)$, $\frac{\partial C}{\partial u}(0,t)$=W(t). Here we had supposed that $\bar{C}(t)$ is closed, so we have $\bar{C}(0)=\bar{C}(1)$.

For convenience, let $C_t = \frac{\partial C}{\partial t}(u,t) = \dot{x}^i \frac{\partial}{\partial x^i}$, $C_u = \frac{\partial C}{\partial u}(u,t) = \xi^i \frac{\partial}{\partial x^i}$, $T = C_t\,(0,t)$.

then length function is

$$L(u) = L(C(u,t)) = \int_0^1 \|C_t\| dt = \int_0^1 \langle C_t, C_t \rangle_g^{1/2} dt$$

If we take the parameter t of $\bar{C}(t)$ to be proportional to the arc length, i.e. $\|T\|$=const, then $\bar{C}(t)$ is called normalized, and we have the First Variational Formula[2,10,17]

$$\frac{d}{du} L(u)\,|_{u=0} = -\int_0^1 \langle W, \nabla_T T \rangle dt$$

If $\bar{C}(t)$ is a geodesic curve, then it is a critical point of the functional $L(C(u,t))$. So we have:

$$\frac{d}{du} L(u)\,|_{u=0} = 0$$

The edge-detection problem becomes a problem of finding the critical points of the functional $L(C(u,t))$. This conversion has not simplified the problem of finding geodesics. But it makes it possible for us to use the second variation of the functional $L(C(u,t))$ to judge whether a critical point is a real minima or not. Now we will introduce the second variation of $\bar{C}(t)$. Suppose that W_1, W_2 are two vector fields along $\bar{C}(t)$ on M.

Definition 3[2,5,17]**:** A smooth second variation (or two-variables variation) of the closed curve $\bar{C}(t)$ with respect to W_1, W_2 is a smooth mapping as follows:

$$C : (-\varepsilon, +\varepsilon) \times (-\delta, +\delta) \times [0,1] \rightarrow M$$
$$(u_1, u_2, t) \mapsto C(u_1, u_2, t) \in M$$

such that $C(0,0,t) = \bar{C}(t)$, $\frac{\partial C}{\partial u_1}(0,0,t) = W_1(t)$, $\frac{\partial C}{\partial u_2}(0,0,t) = W_2(t)$. The vector fields W_1, W_2 are generally called the variational vector fields along $\bar{C}(t)$.

If $\bar{C}(t)$ is a normal geodesic, and let $L(u)$ is like above, then the Second Variational Formula[2,10,17], i.e. the Hesse H(L) of the functional $L(u)$ will be

$$\begin{aligned} H(L)(W_1, W_2) &= \frac{dL(u_1,u_2)}{du_1 du_2}\Big|_{(0,0)} \\ &= -\int_0^1 \{\langle \nabla_T \nabla_T W_1^{\perp}, W_2^{\perp} \rangle + \langle R(W_1^{\perp}, T)T, W_2^{\perp} \rangle\} dt \\ &= \int_0^1 \{\langle \nabla_T W_1^{\perp}, \nabla_T W_2^{\perp} \rangle - \langle R(W_1^{\perp}, T)T, W_2^{\perp} \rangle\} dt \end{aligned}$$

where $W_1^{\perp} = W_1 - \langle W_1, T\rangle T$ is the orthogonal components of W_1 on tangent plane of $\bar{C}(t)$, and $W_2^{\perp}$ is similar. R is the Riemannian curvature tensor.

For a normal geodesic $\bar{C}(t)$, if the Hesse H(L)(W, W) of the functional $L(C(u,t))$ is positive definite for any variational vector field W along $\bar{C}(t)$ (we will say that the functional $L(C(u,t))$ is convex), then it can be guaranteed that $\bar{C}(t)$ is a true minima of $L(C(u,t))$. Therefore, to ensure a critical point to be a true minima, we need only to judge whether the energy (arc length) functional $L(C(u,t))$ of $\bar{C}(t)$ is convex or not.

Now we can prove our theorem. It is obvious that if $\bar{C}(t)$ is a normal geodesic and $L(u)$ is like above, then for any variational vector field W along $\bar{C}(t)$, we have

$$\mathrm{H(L)}(W,W) = \frac{d^2L(u)}{du^2}\Big|_{u=0} = \int_0^1 \left\{ \left\| \nabla_T W^{\perp} \right\|^2 - R(T,W^{\perp},T,W^{\perp}) \right\} dt \tag{6}$$

Here R is the curvature tensor of M. For a geodesic curve $\bar{C}(t)$, its principal normal must coincide with the normal N of the surface M, and its binormal normal (see Fig.2) can be written as: $\hat{N} = (-\dot{x}^2(t), \dot{x}^1(t))$. Via direct calculation we know that:

$$R(T,W^{\perp},T,W^{\perp}) = R(T,W,T,W) = K^* \langle W, \hat{N} \rangle_g^2 \tag{7}$$

If the variational vector field W≠T, then $\langle W, \hat{N} \rangle_g^2 \neq 0$. In general, $\frac{\partial C}{\partial u}(0,t) = W$. It was proved[23] that via reparameterization, we have $W \perp T$. So we can alwayse suppose that $\langle W, \hat{N} \rangle_g^2 \neq 0$. From equation (6) it is clear that $\int_0^1 \left\{ \left\| \nabla_T W^{\perp} \right\|^2 \right\} dt \geq 0$. If its second item $-\int_0^1 R(T,W^{\perp},T,W^{\perp}) dt > 0$, then H(L)($W,W$) is positive definite. According to equation(7), if K(t)<0 holds everywhere on $\bar{C}(t)$, then H(L)(W,W) is positive definite. This proves the theorem.

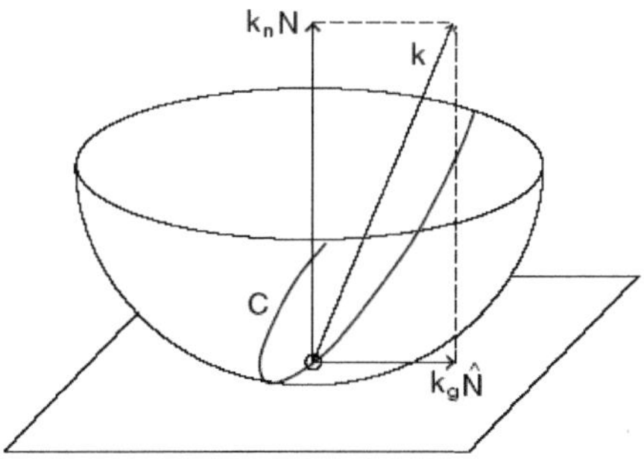

Fig. 2. The curvature has two orthogonal components: the normal and the geodesic curvature[19]

Corollary: If $C(t)$ is a normal geodesic curve as above, the conformal metric on the image M takes the form $g_{ij} = h(|\nabla I(x)|)^2 \delta_{ij}$, and if $\Delta \ln h > 0$ holds everywhere on the curve $C(t)$, then the functional $E(C)$ is convex near $C(t)$.

Proof: For the conformal metric $g_{ij} = h(|\nabla I(x)|)^2 \delta_{ij}$ on M, we have[5,10]:

$$K = -\frac{1}{h^2}\Delta \ln h = -\frac{1}{h^2}\left[\frac{\partial^2 \ln h}{(\partial x^1)^2} + \frac{\partial^2 \ln h}{(\partial x^2)^2}\right]$$

So the corollary is correct.

4 Conclusions

This paper had discussed about the geodesic active contour model and proposed a condition to ensure the convexity of the functional of that model. We proved that if the Gaussian curvature K of M is less than 0 everywhere along a normal geodesic $C(t)$, then its energy functional $E(C)$ will be convex. Further more, from our theorem it can be induced that if $\Delta \ln h > 0$ along $C(t)$, then the energy functional $E(C)$ will also be convex if the image space has its metric tensor take the form $g_{ij} = h(|\nabla I(x)|)^2 \delta_{ij}$.

References

1. Aubert, G., Kornprobst, P.: Mathematical Problems in Image Processing; Partial Differential Equations and The Calculus of Variations; Applied Mathematical Sciences, 2 edn., vol. 147. Springer (2006)
2. Bai, Z., Shen, Y., Shui, N., Guo, X.: An Introduction to Riemann Geometry, April 2002. Higher Edu. Press, Beijing (2002) (in Chinese)
3. Boykov, Y., Kolmogorov, V.: Computing Geodesics and Minimal Surfaces via Graph Cut. In: Proc. IEEE Int. Conf. on Computer Vision (2003)
4. Kolmogorov, V., Boykov, Y.: What Metrics Can Be Approximated by Geo-Cuts, or Global Optimization of Length/Area and Flux. In: Proc. of Int. Conf. on Computer Vision, Beijing, China (October 2005)
5. Carmo, M.D.: Differential Geometry of Curves and Surfaces. Prentice-Hall, New Jersy (1976)
6. Caselles, V., Kimmel, R., Sapiro, G.: Geodesic Active Contour. International Journal of Computer Vision 22(1), 61–79 (1997)
7. Davatzikos, C.A., Prince, J.L.: Convexity Analysis of Active Contour Problems. In: Proc. of CVPR, St. Francisco (June 1996)
8. Davatzikos, C.A., Bryan, R.N.: Using a deformable surface model to obtain a shape representation of the cortex. IEEE Trans. Med. Imag. 15, 785–795 (1996)
9. Davatzikos, C.A., Prince, J.: An active contour model for mapping the cortex. IEEE Trans. Med. 14, 65–80 (1995)

10. Dubrovin, B.A., Fomenko, A.T., Novikov, S.P.: Modern Geometry-Methods and Application, Part I, YPCC, (2001) (Russian)
11. El-Fallah, A.I., Ford, G.E.: On Mean Curvature Diffusion in Nonlinear Image Filtering. Pattern Recognition Letters 19, 433–437 (1998)
12. Evans, L.C.: Partial Differential Equations. In: Proceedings of the American Mathematical Society, Graduate Studies in Mathematics, vol. 19 (1998)
13. Gilbarg, D., Trudinger, N.S.: Elliptic Partial Differential Equations of Second Order, 2nd edn. Springer (1983)
14. Goldenberg, R., Kimmel, R., Rivlin, E., Rudzsky, M.: Fast geodesic active contours. IEEE Trans. Image Processing 10, 1467–1475 (2001)
15. Gu, X., Wang, Y., Yau, S.-T.: Computing Conformal Invariants: Period Matrics. Communication in Information and System 3(3), 153–170 (2004)
16. Kass, M., Witkin, A., Terzopoulos, D.: Snakes:Active contour models. In: Proc. IEEE Int. Conf. on Computer Vision, pp. 259–268 (1987)
17. Milnor, J.: Morse Theory. Princeton Univ. Press, Princeton (1973)
18. Rucklidge, W.: Efficient Visual Recognition Using the Hausdorff Distance. LNCS, vol. 1173. Springer, Heidelberg (1996)
19. Spira, A., Kimmel, R.: Geometric curve flows on parametric manifolds. Journal of Computational Physics 223, 235–249 (2007)
20. Spira, A., Kimmel, R.: Geodesic curvature flow on parametric surfaces. In: Curve and Surface Design: Saint-Malo, Saint-Malo, France, pp. 365–373 (2002)
21. Spira, A., Kimmel, R.: Segmentation of images painted on parametric manifolds. In: Eusipco 2005, Antalya, Turkey (2005)
22. Sochen, N., Kimmel, R., Malladi, R.: From High Energy Physics to Low Level Vision. Report LBNL 39243, LBNL,UC Berkeley, CA 94720 (August. 1996)
23. Yezzi, A., Mennucci, A.: Metrics in the space of curves, arXiv:math/0412454v2 (2005)

Medical Image Denoising Based on Improved Anisotropic Diffusion

Ying Meng

Beijing Institute of Technology Zhu Hai 519085, China

Abstract. This thesis is directed to the most valuable work is the improved algorithms about classical Perona filter, which preserve the image edges more effectively. It is also used to AOS(addictive operator splitting) scheme in anisotropic diffusion equation to improve the competence of edges preservation and achieve region smooth effect. The results of experiments show all improved algorithms have more better performance than unimproved and they fit the needs of medical image denoising. In total, anisotropic diffusion filtering have the best performance. All the methods related in this paper enhance application of digital filter technology and have theoretic and practical meaning.

Keywords: Medical images, Image denoising, Anisotropic, Diffusion Equation.

Anisotropic diffusion equation is based on the theory of partial differential equations, its application on the digital filtering, became a non-linear filtering techniques. This technology makes up for the shortage of traditional digital filtering techniques in a certain extent, it obtained good results in processing of medical image denoising.

1 Principle of Anisotropic Diffusion Filter Denoising

1990, Perona and Malik propose nonlinear anisotropic diffusion filtering equation [1] :

$$I_t = div(c(t, x, y)\ \nabla I) \tag{1}$$

$c(t,x,y)$ is the scaling functions. In the traditional heat conduction equation $c(t,x,y)=1$, at this moment, filter for the image unable to distinguish the boundary region apparently, the result will be loss of image detail. Perona has improved it and given a scaling function $g(x)$ classic choice:

$$g(x) = \frac{1}{1+(x/K)^2} \ (\text{or } g(x) = \exp(-(x/K)^2)) \tag{2}$$

And let $E = \nabla I$, get a Perona diffusion equation as below:

$$\begin{cases} I_t = div(g(|\nabla I|)\nabla I) \\ I_{t=0} = I_0 (I_0 is\ the\ image\ of\ t_0) \end{cases} \tag{3}$$

G. Lee (Ed.): Advances in Automation and Robotics, Vol. 1, LNEE 122, pp. 233–239.
springerlink.com

Use a two-dimensional case as an example to explain the characteristics of Perona Filter, Expand equation (1):

$$I_t = K^2\left(\frac{K^2 - |\nabla I|^2}{\left(K^2 + |\nabla I|^2\right)^2}\right) I_{\eta\eta} + \left(\frac{K^2}{K^2 + |\nabla I|^2}\right) I_{\varepsilon\varepsilon} \tag{4}$$

$I_{\varepsilon\varepsilon}$ and $I_{\eta\eta}$ are second-order partial derivatives of edges and contours.

Perona diffusion equation with a spatial selectivity when filtering on digital images, it able to keep the edge also filters out image noise, with edge-preserving properties. However, Perona diffusion equation at the boundary to be retained while the noise on the boundary be retained also, even the boundary has not been a good regional smoothing, from the perspective of the results, certain areas of the image border is still uneven and disorderly.

1994, Weickert did more further research in this situation, to get anisotropic diffusion equation[2,3,4] base on the Perona's.

$$I_t = div(\mathrm{D} \cdot \nabla I) \tag{5}$$

In this former, local image structure by the structure tensor D to measure. Structure tensor is defined as:

$$J_\rho(\nabla u_\sigma) = G_\rho * (\nabla u_\sigma \nabla u_\sigma^T) \tag{6}$$

$u_\sigma = G_\sigma * u$, u_σ is convolution function of u and standard deviation σ .

J_ρ's eigenvector gives a good direction of the local structure, the corresponding eigenvalue indicates the size of the spread of the direction. A simple calculation can get the eigenvalue of $J_\rho = \begin{pmatrix} J_{11} & J_{12} \\ J_{12} & J_{22} \end{pmatrix}$:

$$\begin{aligned} \mu_1 &= \frac{1}{2}\left(J_{11} + J_{22} + \sqrt{(J_{11} - J_{22})^2 + 4J_{12}^2}\right) \\ \mu_2 &= \frac{1}{2}\left(J_{11} + J_{22} - \sqrt{(J_{11} - J_{22})^2 + 4J_{12}^2}\right) \end{aligned} \tag{7}$$

Feature vector set v_1, v_2, v_1 is parallel to ∇u_σ, v_2 is perpendicular to ∇u_σ, v_1 satisfy the following relationship:

$$v_1 \parallel \begin{pmatrix} 2J_{12} \\ J_{22} - J_{11} + \sqrt{(J_{11} - J_{22})^2 + 4J_{12}^2} \end{pmatrix} \tag{8}$$

Diffusion tensor D has the same eigenvectors of structure tensor J_ρ, Eigenvalues given by the following formula:

$$\lambda_1 : = c_1$$
$$\lambda_2 : = \begin{cases} c_1 & \mu_1 = \mu_2 \\ c_1 + (1 - c_1)\exp\left(\dfrac{-c_2}{(\mu_1 - \mu_2)^2}\right) & \text{others} \end{cases} \tag{9}$$

$c_1 \in (0,1),\ c_2 > 0$

J.weicket model is good at processing the noise at the edge, however, due to J. Weickert model mainly proposed for those who are more texture information image of image enhancement model. For a noisy image, in the smooth region is in accordance with the size of the local eigenvalue spread, this effect will inevitably occur stripes.

2 Two Improved Anisotropic Diffusion Filtering Algorithm

As in the traditional Perona diffusion equation have some deficiencies that exist in filter denoising, this paper will bring our a new algorithm to improve it. For the anisotropic diffusion equation, to use sum decomposition theory and with semi-explicit method, at discrete aspects of the algorithm to make improvements, so that it has a better filtering effect, it also shows advantages and disadvantages of anisotropic diffusion equation's semi-explicit method and explicit method.

2.1 Improved Non-linear Filtering Algorithm of Perona

Perona non-linear filter uses the size of gradient as a standard for judging the image border, but this criterion is not necessarily good. During the research, we noted some idea of the traditional linear filtering in image edge detection algorithm, such as improved adaptive weighted median filter and the Lee filter are stressed the image of the statistical characteristics of the local window, and often use the variance D and mean m of local window, constitute a criterion to determine the image boundary. Lee filter had to use variance of local window and the mean square as a basis for determining the boundary and regional. Here, we will combine Perona non-linear diffusion equation and Lee filter these two ways and take the advantage of them.

This algorithm is variance and mean of local window square ratio D/m^2 as a criterion for distinguishing the boundary image instead of Perona non-linear diffusion equation with the gradient model as a basis for discrimination. It needs to change scaling function gradient mode to $q = \sqrt{\dfrac{D}{m^2}}$ in equation (3). Then get relationship between q and gradient mold as below:

$$q(t, x, y) = \sqrt{\frac{(1/2)(|\nabla I|/I)^2 - (1/16)(\nabla^2 I/I)^2}{[1 + (1/4)(\nabla^2 I/I)]^2}} \tag{10}$$

2.2 The Anisotropic Diffusion Algorithm Based on Additive Decomposition

In order to eliminate J. Weickert model fringe effects also appear to retain the details better, this paper been given an idea based on additive decomposition of the anisotropic diffusion algorithm. It is, for equation (5), the right side of the equation divide into several directions on the isotropic diffusion, the image filtering results can be considered as a proliferation of these directions cumulative average of the results of isotropic diffusion. To put equation (5) into:

$$I_t = div(\mathrm{D}\cdot\nabla I) = \sum_{k=0}^{3} \partial_{e_k}\left(\alpha_k \partial_{e_k} I\right) \tag{11}$$

∂_{e_k} is the partial derivative that I differentiates in the direction $e_k = [\cos\beta_k, \sin\beta_k]$, α_k is a scalar, it has similar meaning with $g(x)$ in Perona non-line filter equation. According to differential two-dimensional function rules can get:

$$\partial_{e_K} I = I_x \cos\beta_k + I_y \sin\beta_k \tag{12}$$

$$\partial_{e_k}\left(\alpha_k \partial_{e_k} I\right) = \alpha_k\left(I_{xx}\cos^2\beta_k + 2I_{xy}\sin\beta_k\cos\beta_k + I_{yy}\sin^2\beta_k\right) \tag{13}$$

Equation (11) expand on left:

$$div(\mathrm{D}\cdot\nabla I) = \begin{bmatrix}\partial_x & \partial_y\end{bmatrix}\cdot\begin{bmatrix} aI_x + bI_y \\ bI_x + cI_y \end{bmatrix} = aI_{xx} + 2bI_{xy} + cI_{yy} \tag{14}$$

Usually take four directions, $\beta_k = -\frac{\pi}{4}, 0, \frac{\pi}{4}, \frac{\pi}{2} (k = 0,1,2,3)$. Expanded equation (11) right as equation (13), input β_k of four directions and compared with equation (14), get a formula as below:

$$\begin{cases} \frac{1}{2}\alpha_0 + \alpha_1 + \frac{1}{2}\alpha_2 = a \\ -\alpha_0 + \alpha_2 = 2b \\ \frac{1}{2}\alpha_0 + \frac{1}{2}\alpha_2 + \alpha_3 = c \end{cases} \tag{15}$$

According to Weicker's proposal $p = \frac{|b| + \min(a,c)}{2}$, so that for each pixel can be under equation (11) get $\alpha_k (k = 0,1,2,3)$.

3 Medical Image Denoising Experiment

This paper applys two kinds of improved anisotropic diffusion algorithm to the practical medical image denosing. From the aspect of image edge preserving and regional smoothing to analysis the results of different algorithms.

Figure1(a) is contain a speckle noise in the type-B ultrasonic image, Figure1(b) contains a speckle noise of the brain nuclear magnetic resonance imaging. Figure 2 is the denoising result by improved Perona's algorithm. Iteration step size $\Delta t = 0.05$, iterations is 80. Figure 3 is the denoising result by anisotropic diffusion algorithm, Gradient threshold $K = 50$, iterations is 15.

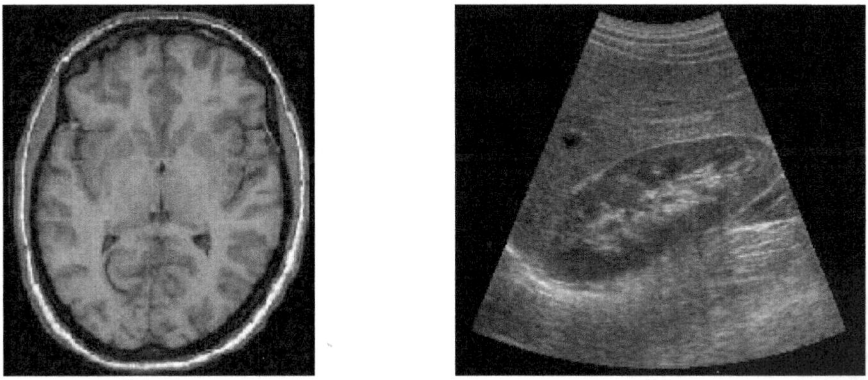

(a) (b)

Fig. 1. Two medical noisy image

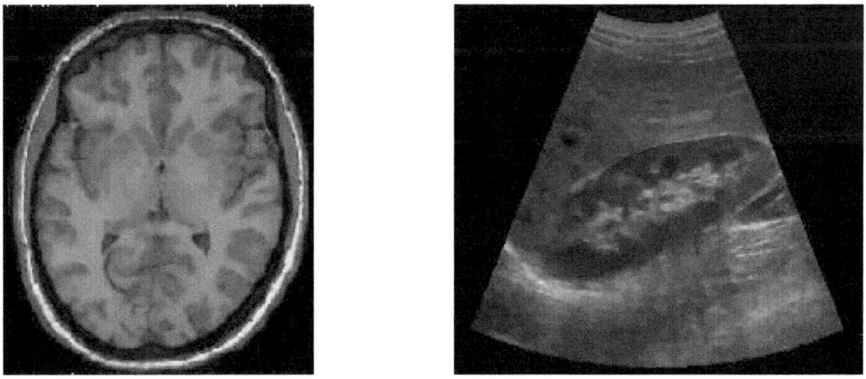

Fig. 2. Improved Perona non-line filter

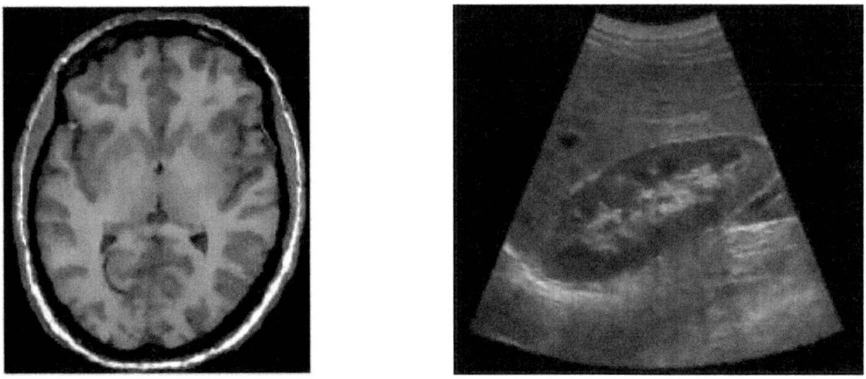

Fig. 3. Anisotropic diffusion algorithm

First, study the performance of edge-preserving, using Canny to extract image edge of some areas in two denoised images, analysis and get that the improved Perona non-linear filtering keeping a ordinary edge, the shape of edge has a certain degree deformation, but the edge remained continuous. Anisotropic diffusion algorithm keeps a well edge, can maintain the basic shape and continuous of the edge. Then use the local edge gradient algorithm to measure the improved algorithm about the freshness of the image edge preserving, after denoising, the edge gradient modulus closer to the original image shows that freshness of the edge preserving is good. The edge gradients of improved non-linear filtering of Perona is 23.3, the edge gradients of Anisotropic diffusion algorithm is 21.2,and the edge gradients of original image is 20.6, so the edge freshness ability of anisotropic diffusion algorithm is better. Finally, by calculating the variance of the image region evaluates the area smoothing degree, the smaller variance, the better region smoothing. The variance of improved Perona non-linear filtering and anisotropic diffusion algorithm is 4.58 and 3.70. Anisotropic diffusion algorithm gets smoother area. The signal-to-noise ratio of Perona non-linear filtering is $SNR = 2.05$, and correlation coefficient is $CORR = 0.72$, the signal-to-noise ratio of anisotropic diffusion algorithm is $SNR = 2.78$, and correlation coefficient is $CORR = 0.78$. From the above data shows that, the results of anisotropic diffusion algorithm proximity the original image as a whole, the correlation coefficient achieves 0.78.

4 Conclusion

This paper discusses the non-linear digital filtering techniques based on the thought of anisotropic diffusion, reviewing classical Perona non-linear diffusion equation, proposed two improved strategies, the improved Perona's non-linear diffusion equation has a better edge protection after denoising and reached the freshness in edge preserving. For the anisotropic diffusion equation, this paper uses the thought of additive decomposition and Semi-explicit form of differential equations, to divide the anisotropic diffusion of image into several single direction, and take average finally. Proved by experiments, the ability of Perona non-linear filter based on partial differential equation and anisotropic diffusion filter is better in filting the speckle noising of medical image. This denoising effect of the two technologies in a draw, the former has strong edge preservation, but the region smoothing is less capacity than latter.

References

1. Perona, P., Malik, J.: Scale-Space and Edge Detection Using Anisotropic Diffudion. IEEE Trans. PAMI 12(7), 629–639 (1990)
2. Weickert, J.: Theoretical foundations of anisotropic diffusion in image processing. Computing Suppl. 11, 221–236 (1996)
3. Weickert, J.: Anisotropic Diffusion in Image Processing. European Consortium for Mathematics in Industry, B.G.Teubner Stuttgart (1998)
4. Weickert, J.: Coherence-enhancing diffusion filtering. International Journal of Computer Vision 31(2-3), 111–127 (1999)

Attitude Measurement of Dynamic Object by Laser Tracking System and Optoelectronic Methodology

Minghua Pan, Chuncao Zhang, and Guoli Zhu

State Key Laboratory of Digital Manufacturing Equipment and Technology, Huazhong University of Science & Technology,Wuhan, 430074, China
pmh24@163.com

Abstract. This paper presents a low-cost laser tracking system to measure attitudes of dynamic object and position of the moving target which only need one feature point fixed on it. An attitude measurement modeling is built up to show the coupling of the three attitudes named yaw, pitch and roll. Installation error and lens distortion error are referred to improve the precision of this system. The improvement of the measurement modeling is transferred to be a nonlinear least-squares problem. A Gauss-Newton iterative method is proposed to solve the nonlinear modeling and optimize error parameters. The experiments successfully examined the effectiveness of the proposed modeling and algorithm in terms of: (1) The attitude measurement modeling perfectly describes the relationship of attitudes and measured parameters by system. (2) The corrected modeling takes installation error and lens distortion error into consideration.(3)The error modeling using G-N iteration optimizes error parameters and availably reduces the error of the system.

Keywords: attitude measurement, laser tracking system, nonlinear least-squares problem, G-N method.

1 Introduction

Position and attitude measurement of dynamic object in real time is required in industrial fields such as construction machinery, unmanned vehicle, walking robot and so on. To a certain extent, the attitude of the objects observed is more important than its position. For example, the azimuth of excavation of tunneling machine has more significant impact in the process of tunnel digging than the coordinates of the tunneling machine head. In order to measure the attitude of the dynamic target, it should be tracked in real time. To meet these demands, techniques about object tracking and attitude measurement have been developed in the recent years. Vision navigation system[1,2] uses video sensor to compute the relative attitude, in which the image of object observed should be recognized clearly. Another system[3] uses three non-collinear feature points fixed on the dynamic object to measure the attitudes of the object. For the reason of the algorithm, the measurement precision are not satisfactory.

This paper presents an effective laser tracking system to measure attitudes and position of the target box which fixed on dynamic object. The target is composed by a

G. Lee (Ed.): Advances in Automation and Robotics, Vol. 1, LNEE 122, pp. 241–248.
springerlink.com

reflector, single inclinometer and CCD camera. The target's three rotation angles of yaw, pitch and roll are closely related and then they should be put in the same system to reflect the coupling of the system. An attitude measurement modeling is built to express the relationship of three rotation angles of dynamic object. Unlike the previous calculation method, the modeling helps to optimize parameters of the system and improves the accuracy of attitude measurement.

For the first time we systematically put forward how to measure the attitude of the moving target used only one feature point. The proposed system comprises one total station and one optoelectronic target. Section 2 is about the principle and modeling to measure the attitude of the target. Section 3 shows the source of error and modeling optimization, meanwhile, Gauss-Newton algorithm, a well-known iterative technique used regularly for solving nonlinear least squares problems, is given in this section. Section 4 is experiments and data processing. Finally we draw some conclusions in section 5.

2 Principle and Modeling

To any rigid body system named Q which is moving in the world coordinate system (WCS), attitude and position can be described by g=(${}^{W}T$, ${}^{W}R$), in which ${}^{W}T$=(Tx, Ty, Tz) is the translation vector, ${}^{W}R$ is the rotation matrix. The world coordinates p of the points $q \in Q$ is calculated by

$$p = {}^{W}T \times {}^{W}R \times q \tag{1}$$

where q is the relative coordinate of the point in the rigid body system. The rotation matrix ${}^{W}R$ and translation vector ${}^{W}T$ are defined by a sequence of rotations and translations about the x, y, or z axes in WCS. In this paper, the rotation angles around the axes are named pitch angle β , roll angle α , and azimuth angle β in sequence of x, y and z as shown in Fig.1.

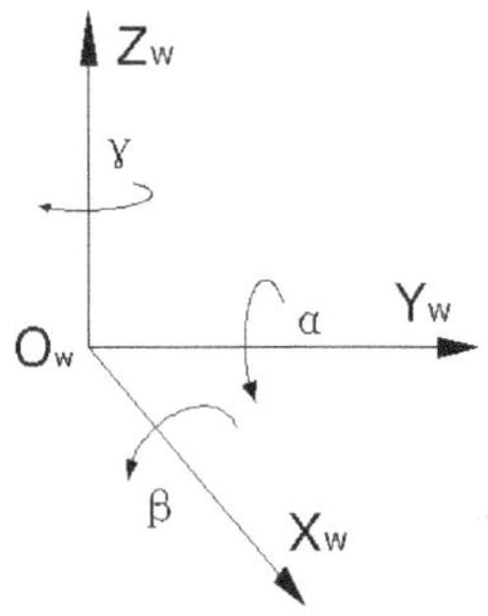

Fig. 1. Schematic diagram of the rotation angles

The transformation corresponding to rotations about x, y and z axes are shown in equation (2), where S and C, respectively, denote sine and cosine. The transformation corresponding to a translation by ${}^{W}T$ is equation (3).

$$R(x,\beta)=\begin{bmatrix}1&0&0&0\\0&C\beta&S\beta&0\\0&-S\beta&C\beta&0\\0&0&0&1\end{bmatrix}\quad R(y,\alpha)=\begin{bmatrix}C\alpha&0&S\alpha&0\\0&1&0&0\\-S\alpha&0&C\alpha&0\\0&0&0&1\end{bmatrix}\quad R(z,\gamma)=\begin{bmatrix}C\gamma&-S\gamma&0&0\\S\gamma&C\gamma&0&0\\0&0&1&0\\0&0&0&1\end{bmatrix} \tag{2}$$

$${}^{W}T=\begin{bmatrix}1&0&0&T_x\\0&1&0&T_y\\0&0&1&T_z\\0&0&0&1\end{bmatrix} \tag{3}$$

As known to all, the order in which the rotations are made is important in the matrix multiplication. As defined in the above, the rotation matrix should be

$${}^{W}R = R(y,\alpha)R(x,\beta)R(z,\gamma) \tag{4}$$

The system we proposed has an optoelectronic target to measure the roll angle α , pitch angle β and azimuth angle γ of itself. The coordinate of the feature point on the target can be defined as the translation vector ${}^{W}T$. When the target is fixed on a dynamic object, feature point will be tracked and the attitude angle measured in real time.

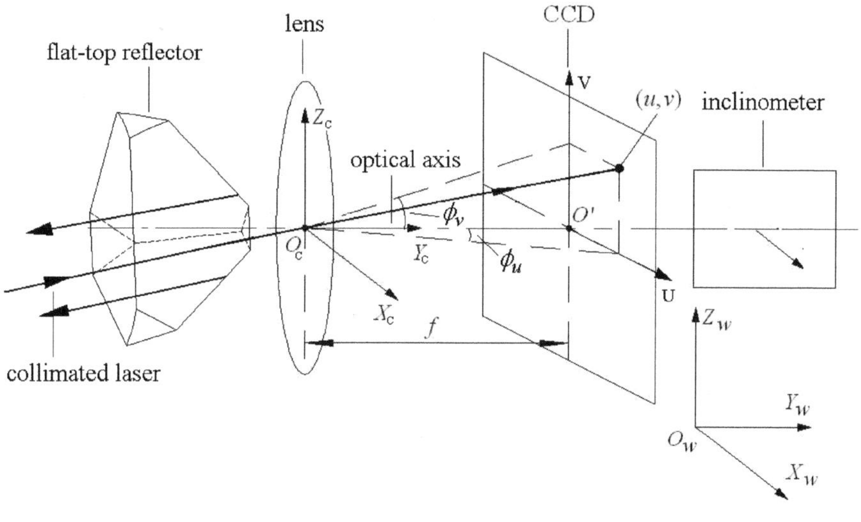

Fig. 2. Schematic diagram of the optoelectronic target system

As shown in Fig.2, the optoelectronic target comprises a flat-top reflector which the plane on the pyramid top parallels to the incident plane of the laser, an industrial camera to capture the laser spot and a single-axis inclinometer for measuring rolling angle α. The origin of the camera coordinate system (CCS) is assumed at the center of the camera lens with the axes OcXc and OcZc respectively parallel to the axes U and V of the CCD plane system, and the axis OcYc perpendicular to the CCD plane. CCD plane is placed at the focal plane of the camera lens and the origin of the CCD plane system is at (0,f,0). The single-axis inclinometer is fixed to the rear of the camera and parallel to axis OcXc of CCS.

When the total station is emitting laser to the reflector, most of the laser returns back to the total station after reflection so as to track the reflector and measure the coordinate of the reflector top, but a little laser pass through the flat top and converged on the CCD plane of the camera. Based on the pinhole principle, the angle between the laser and the axis OcYc can be acquired.

$$\phi = \arctan(\sqrt{u^2 + v^2} / f) \tag{5}$$

where (u,v) denote the 2-D coordinate of the laser spot in the CCD plane and f is the local length.

For the roll angle α around OcYc won't change pitch angle and azimuth angle, it can be assumed to rotate to zero, likes shown in Fig.2. The spot coordinate (u',v') after rotation can be calculated by

$$\begin{bmatrix} u' & f & v' \end{bmatrix} = [u \quad f \quad v] \times R(y,\alpha) \tag{6}$$

Then the incident angle of the laser to the OcYc axis can be calculated by equations (5). Adding to the vertical angle and azimuth angle of the laser which can be acquired from total station, pitch angle and azimuth angle can be acquired.

The attitude angles of roll, pitch and azimuth are feasible to be computed by the measurement value α' of the inclinometer, the coordinate (u',v') of the laser spot, the vertical angle β_T and the azimuth angle γ_T of the laser. So the attitude measurement modeling is given by the form

$$G(\alpha,\beta,\gamma) = T(u,v,\alpha',\beta_T,\gamma_T) \tag{7}$$

3 Error Analysis

3.1 Installation Error between CCD and the Single-Axis Inclinometer

As discussed in the above, a single-axis inclinometer is installed in the target to measure its rotation angle of roll. The ideal installation requires the single-axis of inclinometer paralleling the corresponding axis OcYc of CCD for real-time measurement of roll. Actually, for the axis of CCD plane and inclinometer are virtual axes, can't be measured directly, the actual installation can't meet the requirement.

Fig.3. shows the deviation angles which are errors of installation of the inclinometer. The experiment verifies that horizontal angle $\Delta\varphi$ affects little error and can be neglected, however, the vertical angle $\Delta\theta$ has impact on attitude measurement. For the purpose of reducing installation error, it's important to optimally select the value of $\Delta\theta$. Than the roll angle is corrected by

$$\alpha = \alpha' - \Delta\theta \tag{8}$$

and then the attitudes modeling will be corrected as following.

$$G(\alpha,\beta,\gamma) = T(u,v,\alpha',\beta_T,\gamma_T) = S(\Delta\theta) \tag{9}$$

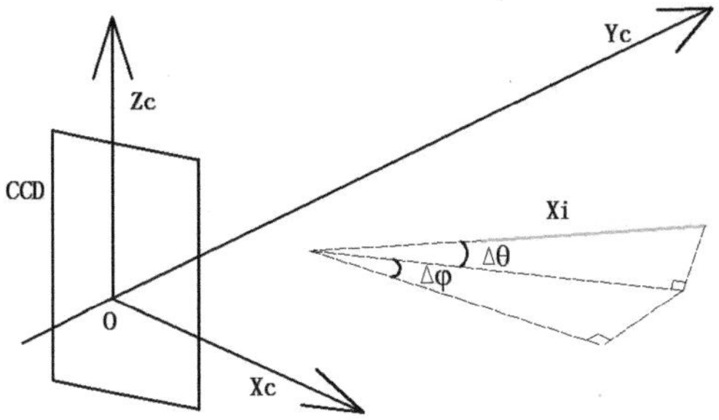

Fig. 3. The deviation angles caused by installation error

3.2 The Error Analysis of Lens Distortion and Focal Length

Due to the manufacturing process problems by the lens, the measurement errors are caused by three main factors: the error of the focal length, spot deviation due to lens distortion and the error of spot center determination. Distortion of the lens makes the spot coordinates not be consistent with pinhole principle. As discussed in Section 2, lens distortion has direct impact on the accuracy of azimuth angle and pitch angle, meanwhile, focal length error also affects them.

It is easy to know that lens distortion and focal length error are systemic errors but the spot center determination error is accidental error. In recent years, many open literatures discussed the technique of spot center determination. As known from them, location precision of laser spot reaches sub-pixel after image processing by special algorithm. In this paper, emphasis is to modify the lens distortion and focal length.

For the purpose of analyzing the spot error resulted from the lens distortion, distortion model should be determined. In general, radial distortion constitutes the main part of the lens distortion. In the Brown model, lens distortion can usually be expressed as follows.

$$\begin{aligned} u_d &= u / (1+k_1 r^2 + k_2 r^4) \\ v_d &= v / (1+k_1 r^2 + k_2 r^4) \end{aligned} \tag{10}$$

In the equation, u and v are the theoretical coordinates of image; r implies the distance from (u, v) to the center of the image plane; k1 and k2 are the coefficients of radial distortions.

For the purpose of reducing error of the angle, the parameters k1, k2 and f should be carefully selected. The attitudes modeling can be expressed as following.

$$G(\alpha,\beta,\gamma) = T(u,v,\alpha',\beta_T,\gamma_T) = S(k_1,k_2,f,\Delta\theta) \tag{11}$$

Combined the installation error and lens distortion error, we can reduce the difference of the computed attitudes and reference attitudes. It is obvious that the difference leads to nonlinear error. The problem of error modeling can be considered as a nonlinear least squares problem. Then we use the Gauss-Newton method[4] to solve the nonlinear least squares problem.

3.3 Error Modeling Using Gauss-Newton Method

The error modeling solution steps:

1) The form of matrix function needed fitting is given by $y = \underset{3\times1}{F}(x_1, x_2, x_3, x_4)$, where $\Delta\theta$ is expressed by variant x_1, k_1 is expressed by variant x_2, k_2 is expressed by variant x_3, focal length f is expressed by variant x_4.

2) Nonlinear equations problem is converted to be a nonlinear least squares problem by $S(x_1, x_2, x_3, x_4) = \Sigma(\underset{3\times1}{F}(x_1, x_2, x_3, x_4) - y)^2$

We will find that solution of problems is the minimum point for the above nonlinear system of equations.

3) Solving the minimum point of multi-function.

We obtain the iterative procedure(Gauss-Newton method)

Step0: Choose an initial x_0

Step1:Repeat until convergence:

Step1.1 Solve $J(x_k)^T J(x_k)\delta_k = -J^T(x_k)G(x_k)$

Step1.2 Set $x_{k+1} = x_k + \delta_k$ where $J(x)$ is the Jacobian matrix of the function $G(x)$

The iterative procedure will stop until the norm of δ_k less than given value.

4 Validation Experiment

For concept demonstrating the validation of the attitude measurement system, a hardware-software integrated prototype system was tested in the extensive lab and field. The attitude measurement system of dynamic object by laser tracking consists of laser device, single-inclinometer and CCD camera. This prototype system device can measure attitudes of dynamic object in real time. In addition, a software program called real-time tracking and optimization was coded using MATLAB and C++, which provide performance for: (1) gathering real-time measurement data; (2) optimization of parameters of $k_1, k_2, f, \Delta\theta$;(3)computing attitudes.

Fig.4. shows the experimental device of the attitude measurement. Reflector is installed in front of the target. Both of them are placed on three-axis turntable which is driven by three high precision servo motors with absolute encoders. Three axis of turntable are perpendicular to each other and accurate calibrate rolling, pitching and yawing attitude changes of the target in the world coordinate system. When the turntable turns to a certain direction, the coordinates of the trajectory of the focal spot (u,v) in the imagine plane are collected to calculate attitudes .The angles of laser β_T, γ_T, and the value α' are respectively measured by laser device and single-inclinometer. At the same time, the angles of the three axes turntable are read to calibrate the attitudes.

The experiment collected groups of data and each group had around 28 points. Compared the computed attitudes from target before correction and the calibration value of attitudes from turntable, error curves can be shown in Fig.5(a). In the group of data, the standard deviations of pitch angle and azimuth angle are separately 0.5818mrad and 0.5793mrad.

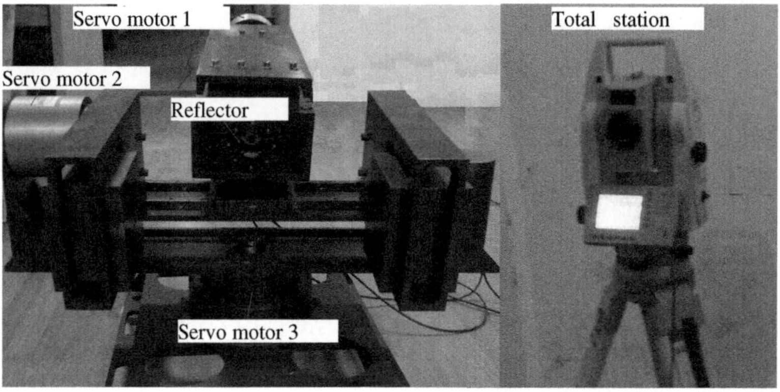

Fig. 4. The experimental device (turntable and total station)

Taking consideration of the installation error and lens distortion error, the coefficients of radial distortions k1and k2, the installation error parameter $\Delta\theta$ and focal length f are optimized to reduce the difference of the computed attitudes and reference attitudes. For the purpose to simple the computation, Gauss-Newton iteration is applied to improve the attitude modeling. After processing by optimization algorithm, the optimal values are as follows: k1=-0.1092, k2=-2.8140, $\Delta\theta$ = -0.17°, f=35.335mm. Fig.5(b) shows the error curves of the attitudes after correction by the optimal parameters above. The standard deviations of pitch angle and azimuth angle are reduced effectively to 0.0927mrad and 0.0389mrad separately.

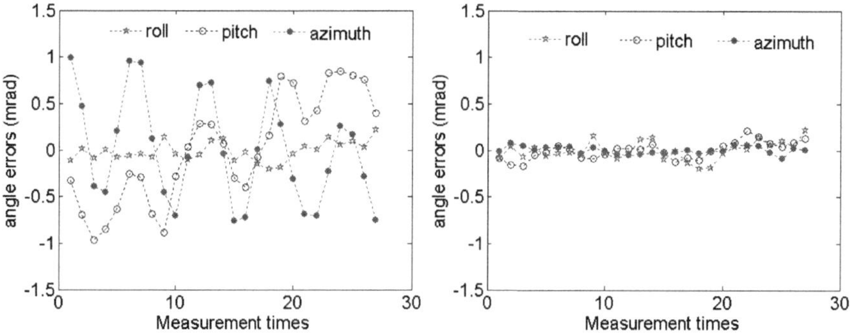

(a) error curves before correction (b)error curves after correction

Fig. 5. The error curves of attitude angles

In order to prove the effect of the optimization, another group of measurement points are analysed. Fig.6(a) and Fig.6(b) are respective error curves before correction and after correction by the same optimal value of k1 , k2 , $\Delta\theta$ and f. The standard deviations of pitch angle and azimuth angle are obviously reduced from 0.6185mrad and 0.5892mrad to 0.1218mrad and 0.0404mrad. The measurement data proves the availability of the optimal value.

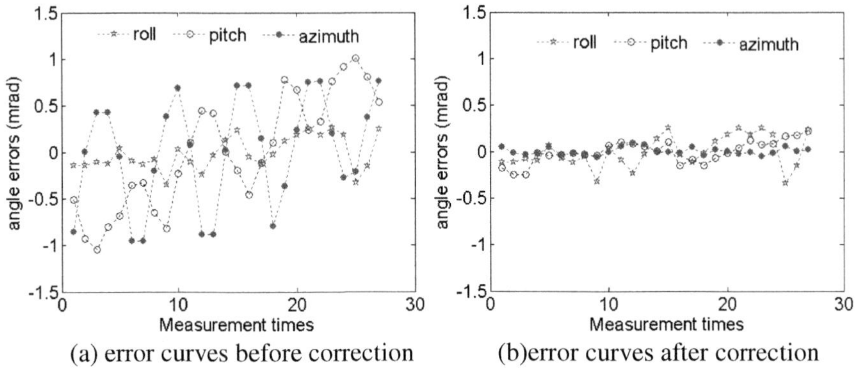

Fig. 6. The error curves of attitude angles in repeated experiment

5 Conclusion

In this paper, we proposed a new optoelectronic methodology to track the dynamic object and measure its attitudes in real time. The model was bulit and the errors were analysed. We used the high precision three axes turntable to calibrate the optoelectronic target and than used Gauss-Newton method to optimize the model. The answer of the optimization shows that the precision of the pitch angles and azimuth angles can reach to 0.2mrad and are well repeatable.

Acknowledgments. This work is supported by the National Basic Research Program of China (No.2007CB714000).

References

1. Zhang, Z., Zhu, D., Peng, Z., Zhang, J.: Improved iterative pose estimation algorithm using three-dimensional feature points. J. Opt. Eng. 46, 127–202 (2007)
2. Ababsa, F., Mallem, M.: Robust camera pose estimation using 2d fiducials tracking for real-time augmented reality systems. In: Proc. ACM SIGGRAPH Int. Conf. on Virtual Reality Continuum and its Applications in Industry, VRCAI 2004, pp. 431–435 (2004)
3. Wang, J., Liu, Y., Hu, Z., Liang, J.: Modeling and algorithm of attitude measurement of moving target by laser tracking systems. J. Opt. Eng. 42, 373 (2003)
4. Volckaert, M., Van Mulders, A., Schoukens, J., Diehl, M., Swevers, J.: Model Based Nonlinear Iterative Learning Control: A Constrained Gauss-Newton Approach. In: Proc. 2009 17th Mediterranean Conference on Control & Automation, MED 2009, vol. 1-3, pp. 718–723 (2009)

Research on Pyrolysis Behavior of Coal Tar Using TG-MS

Lian-yong Wang, Lei Dai, and Jiu-ju Cai

SEP Key Laboratory on Eco-industry, Northeastern University, Shenyang, Liaoning 110819, China

Abstract. Simultaneous thermogravimetry-mass spectrometry (TG-MS) was used to investigate the pyrolysis behaviour of coal tar and the instantaneous evolution characteristic of the gaseous products. The result shows that the gaseous releasing product is mostly H_2O besides a small quantity of C_2H_4 under 400℃. It is mostly CH_4 and C_2H_4 in addition to a small quantity of C_2H_2 between 400-600℃. It is mostly H_2、CO、CO_2 and a small quantity of C between 600-1000℃. It is in evidence that C3、CO、CO2 release between 1000-1300℃.

Keywords: TG-MS, coal tar, pyrolysis.

1 Introduction

Thermogravimetry [1] (TG) is a kind of thermal analysis technology using thermobalance under temperature control of program to measure the relationship between mass and temperature, and it is widely used in coal chemistry field because of the advantages of simple instrument operation, high accuracy, rapid sensitivity, small amount of sample, etc. At present, many researchers [2-9] have used TG to investigate the pyrolysis process of coal tar and the effect of temperature and heating rate on the pyrolysis behaviour of coal tar. It is difficult to make further research on pyrolysis reaction of coal tar and to discuss the reaction mechanism of it, because the volatile component which is released in pyrolysis process can not be measureed with TG. With the help of Thermogravimetry Coupled with Mass Spectrometry [10-14] (TG-MS), physical and chemical change of sample in heating process can be measured directly and accurately, and qualitative and quantitative analysis to reaction product can proceed, and then reaction mechanism can be elaborated better, so TG-MS has become an important measuring technology to investigate the heat stability and thermal decomposition of inorganic, organic, and polymer material. In this paper, TG-MS is used to investigate the pyrolysis behaviour of coal tar, and weight loss characteristic of coal tar and instantaneous evolution characteristic of the gaseous products are investigated, and the pyrolysis process of coal tar is discussed.

G. Lee (Ed.): Advances in Automation and Robotics, Vol. 1, LNEE 122, pp. 249–255.
springerlink.com

2 Experiment

2.1 Experimental Material

Experimental material is solid sample of coal tar in Datong coal gasification after nature airing and particle size of sample is less than 100 mesh. The characteristic of material can be seen in Table 1.

Table 1. Proximate and ultimate analysis of tar

proximate analysis / %				elementary analysis / %					low heating value
water	ash	volatile	fixed carbon	C_{ad}	H_{ad}	O_{ad}	N_{ad}	S_{ad}	/ kJ•kg^{-1}
15.25	1.25	76.13	7.37	82.00	8.93	8.16	1.25	0.56	39854

2.2 Experimental Apparatus and Method

TG-MS adopted in this experiment is made up of thermogravimetric (TG) SETSYS Evolution18 produced by France SETARAM company and quadrupole mass spectrometer (MS) HRP20 QIC produced by British Hiden company. MS is connected to TG with a quartz glass micro-tube which has been heated to 200℃ in order to avoid gaseous product condense in the entrance of MS. Sample is taken from the bottom of coal tar through quartz glass micro-tube, and a little gas is absorbed into ion chamber of MS and then under the influence of high field, gas will generate positive ion which can be accelerated in high voltage electric field and deflected in magnetic field, and when ion finally reaches collector, positive ion can generate signal whoes intensity is proportional to the number of ions arriving, and recorded signal composes mass spectrum. The amount of gas content can be determined according to the partial pressure value of molecular ion of gas. The cyclic measurement of molecule weight is adopted when using mass spectrum in this experiment, and in the range of molecule weight marked from 1 to 50, the abundance of molecular ion of coal tar bombarded by election current is measured every six seconds, and this cycle stops when temperature reaches 1300℃. The advantage of this method is that for each time period and each temperature the molecular weight of whole gaseous substance generated in the pyrolysis process of coal tar can be tracked, and then the molecular formula can be confirmed by a calculation method such as determination based on mass spectrometry or isotope abundance method.

Experimental container: alumina crucible;

Experimental atmosphere: argon, and flow rate is 20ml/min;

Heating rate: 10℃/min, and temperature will be maintained at 1300℃ for 5 minutes.

α-Al_2O_3 is used as calorimetric reference substance in this experiment, considering that heat-conducting property and heat capacity of reference material should be close to that of sampe, so as to reduce TG baseline drift. After each experiment, a blank experiment in same condition is needed to eliminate system error.

3 Experimental Results and Analysis

3.1 The Analysis of TG

Fig.1 shows the curve of TG and the curve of DTG, and the former curve represents that the quality of coal tar varies with temperature variation and the latter curve, which is transient weight loss rate calculated according to the curve of TG, represents the intensity of weight loss at one time. It can be seen from Fig.2 that the weight loss of coal tar appears slowly at about 100℃, and the first weight loss peak appears at 126℃ ,and severe weight loss appears from 200℃ and the second peak appears at 323℃ when weight loss rate reaches maximum of 16.38%/℃, and after that, weight loss rate decreases gradually, and rate decreases obviously after temperature rises to 550℃, and rate increases slowly again when temperature rises to 800℃ and the third peak appears at 877℃ when rate decreases relatively, and then two peaks of weight loss appear at 1084℃ and 1235℃, and thermal decomposition almost completes at 1300℃ and solid residue makes up about 49.8%.

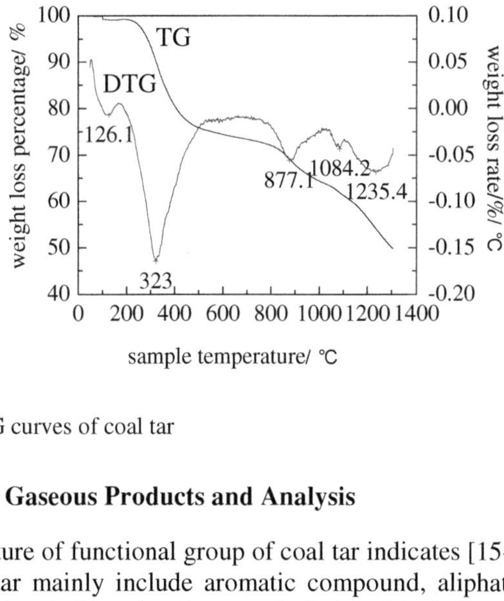

Fig. 1. The TG and DTG curves of coal tar

3.2 The Release of Gaseous Products and Analysis

The research on structure of functional group of coal tar indicates [15-16] that organic components of coal tar mainly include aromatic compound, aliphatic hydrocarbon, plenty of oxygenated chemicals - ether, ketone, alcohol, residual phenol and so on - and heterocyclic compound, nitriles, etc. The amount of functional group is related to the gasification condition. In pyrolytic reaction, coal tar decomposes into solid carbon, gas and free radical which can be cracked further into gaseous product and carbon deposition with the rising of temperature, and carbon deposition can occurs reaction of gasification of steam at higher temperature. The TG-MS spectrogram of gaseous products (H_2O, H_2, CH_4, C_2H_4, C_2H_2, C_3, CO and CO_2) in coal tar pyrolysis can be seen in Fig.2.

It can be seen from Fig.2 (a) that the generation of H_2O in pyrolysis process of coal tar can be divided into three phases, the first releasing peak appears at about 128.5℃, which is close to the temperature of the first weight loss peak in curve of DTG, which is caused by desorption of absorbed water; the second releasing peak is about 316.7℃, which is close to the temperature of the second temperature weight loss peak in curve of DTG, and the most of water generated in this stage is pyrolysis water, that comes from the polycondensation reaction among hydroxy when macromolecule structure of coal tar breaks; the third peak is about 1031.7℃, which is close to the temperature of the fourth peak in curve of DTG, so water in this stage should come from stable oxygen-containing functional group in coal tar.

It can be found from Fig.2 (b) that the releasing of H_2 in pyrolysis process can be divided into two processes, and in the first process between 400℃ and 1050℃ a lot of H_2 is released, and this is because the macromolecule network structure of coal tar begins to break and intense pyrolytic reaction occurs in this stage, and H_2 should come from polycondensation among radical, which is generated in pyrolysis of coal tar. The second stage, in which the generation of H_2 becomes slow, begins from 1050℃, and in this stage, H_2 comes from the water gas reaction ($C+H_2O \rightarrow CO+H_2$) and the shift water gas reaction ($CO+H_2O \leftrightarrow CO_2+H_2$), and two reactions both change to the direction benefiting the generation of H_2, and besides that, it can be known from Fig.2 (a) and (f) that in this temperature range, coal tar is cracked to generate H_2O and carbon, which confirms the source of H_2 in this stage further.

The releasing of CH_4•C_2H_4 and C_2H_2 obeys to the similar regularity, and it can be seen from Fig.2 (c), (d) and (e) that there is a wide range peak which is like gaussian distribution in the releasing of the three. Releasing peak of CH_4 starts from 380℃ and ends at 780℃, and maximum appears at 568.5℃; Releasing peak of C_2H_4 starts from 240℃ and ends at 1000℃, and maximum appears at 475.9℃; Releasing peak of C_2H_2 starts from 350℃ and ends at 600℃, and maximum appears at 488.7℃; the releasing of the three is the result of pyrolysis of naphthenic hydrocarbon in coal tar, bridge bond fracture in coal tar macromolecule (decomposes into small mixture of coal tar molecule) and the fracture of polar functional group in polar aromatic compounds (decreases polarity and molecular weight). It is showed in Fig.2 that the intensity of releasing peak of CH_4 is strongest and that of C_2H_2 is weakest in the three and thus CH_4 and C_2H_4 make up the most in the micromolecule hydrocarbon which is generated in pyrolysis of coal tar.

Fig.2 (f) shows the regularity of separating out C_3 and it can be found that there are three releasing peaks of C_3 between 380℃ and 1300℃ and three peaks appear at about 480.8℃, 713.9℃ and 1183.5℃, and C_3 released in low temperature is fixed carbon generated in the pyrolysis of coal tar, while C_3 released in high temperature is depositional carbon which is generated in further pyrolysis of reaction radical (generated in pyrolysis of coal tar), and the intensity of releasing peak of C_3 in high temperature is high, which indicates that releasing of C_3 is obvious at high temperature.

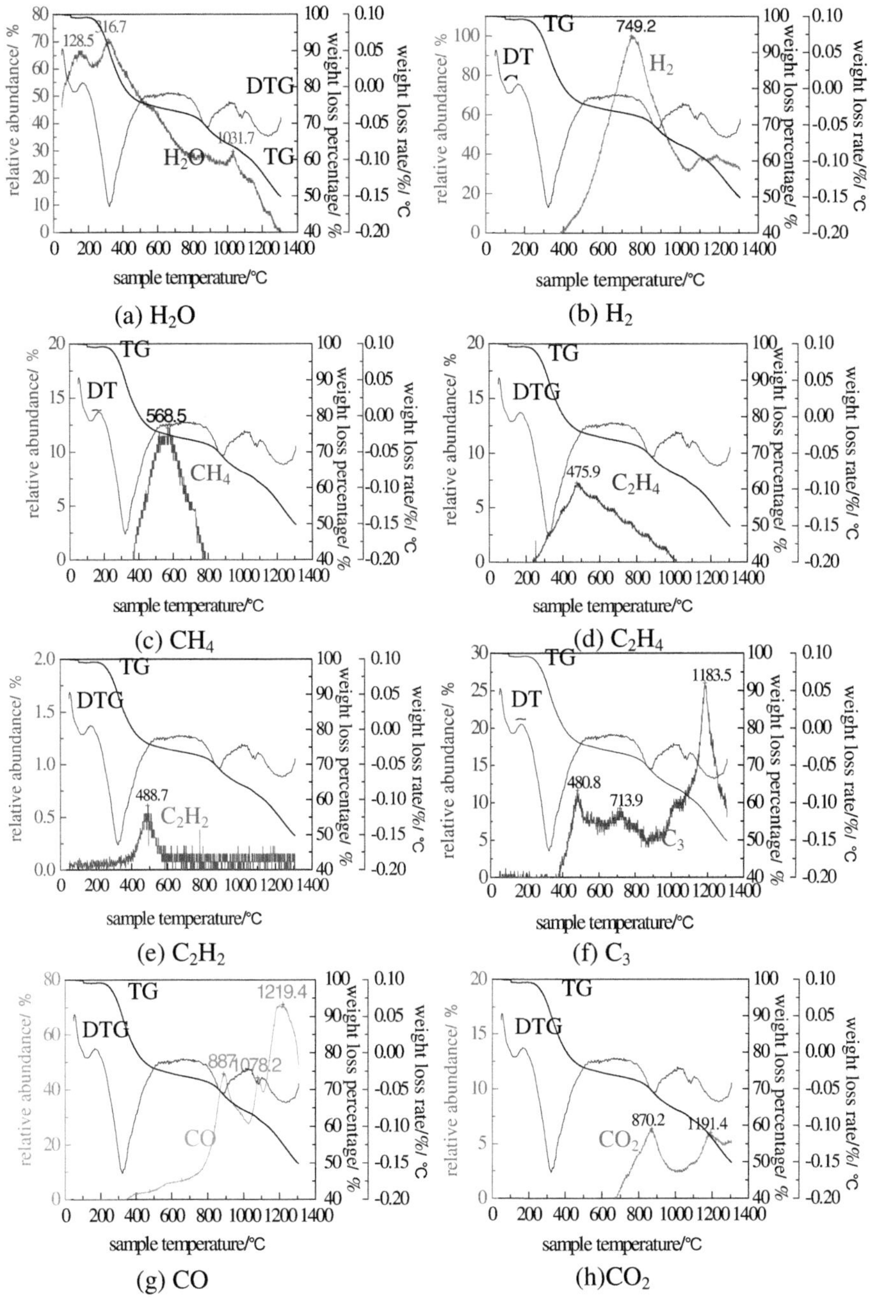

Fig. 2. The TG-MS curves of coal tar during the pyrolysis

Fig.2 (g) shows the releasing regularity of CO and it can be seen that releasing of CO begins at 350℃ with small amount, and when temperature rises above 850℃, a lot of CO separates out and three obvious releasing peaks appear at 887℃,1078.2℃ and 1219.4℃, which is the same with the location of the third, the fourth and the fifth weight loss peak in DTG curve. This is because CO is generated in the pyrolysis process of oxygen-containing functional group and aliphatic hydrocarbon, and in high temperature condition, oxygen-containing functional group and aliphatic hydrocarbon can get enough energy to break chemical bond to generate CO. At the same time, a lot of carbon and a portion of water vapor are generated in pyrolysis of coal tar at high temperature, and water gas reaction of carbon and water vapor happens with the generation of CO and H_2. The effect of water vapor in this condition is of great importance, because water vapor can react with some ingredient in coal tar, which not only generates combustible gas such as CO and H_2, but also reduces the generation of carbon black. Besides, Boudouard reaction of C and CO_2 happens with the generation of CO at high temperature so the releasing of CO is obvious during high temperature phase.

It can be seen from Fig.2 (h) that CO_2 doesn't separate out until 680℃ and its releasing can be divided into two phases: the first releasing peak appears at about 870℃, which is the same with the location of the third weight loss peak in DTG curve, and CO_2 released in this phase should come from unstable oxygen-containing functional group such as hydroxy; the second releasing peak appears at about 1191℃, which is the same with the temperature location of the fifth weight loss peak in DTG curve, and CO_2 released in this phase mainly comes from the gasification reaction ($C+H_2O \rightarrow CO_2+H_2$) of C and the shift reaction of water gas, and two reactions both proceed to the direction of positive reaction with the rising of temperature.

4 Conclusions

TG-MS was used to investigate the pyrolysis behaviour of coal tar and the instantaneous evolution characteristic of the gaseous products, and the mechanism of pyrolysis of coal tar is elaborated. The conclusions include:

(1) The weight loss of coal tar appears slowly at about 100℃ ,and weight loss becomes severe from 200℃ and weight loss rate reaches maximum of 16.38%/℃ at about 323℃, and after that weight loss rate decreases gradually, and the rate increases slowly again when temperature rises to 800℃, and three peaks appear at 877℃, 1084℃ and 1235℃, and the thermal decomposition almost completes at 1300℃, and solid residue makes up about 49.8%.

(2) The releasing of H_2O in pyrolysis process of coal tar can be divided into three phases, in which water belongs to desorption of absorbed water, pyrolysis water that comes from the polycondensation reaction among hydroxy and bond breaking of stable oxygen-containing functional group differently, and the most obvious releasing happens at 316℃. A lot of H_2 separates out between 400℃ and 1050℃, and the generation of H_2 comes into slow phase from 1050℃. A peak like gaussian distribution exists in the releasing of CH_4, C_2H_4 and C_2H_2, and the difference in intensity of releasing peak

indicates that CH_4 and C_2H_4 make up the most in the micromolecule hydrocarbon which is generated in pyrolysis of coal tar. It is in evidence that C_3、CO、CO_2 release at high temperature above 800℃.

Acknowledgments. Authors would like to acknowledge financial support of the Fundamental Research Funds for the Central Universities, China (N090402010) and the Doctor Start-up Fund of Liaoning Province, China(20091024).

References

1. Chen, J.-H., Li, C.-R.: Thermal analysis and its applications, pp. 1–2. Science Press, Beijing (1985)
2. Zhang, Q.-G.: Experimental study of combustion kinetics and fuel characteristics of biomass tar. Dalian University of Technology, Dalian (2006)
3. Li, J.-H.: Study on thermal kinetics characteristics of biomass tar and its distillations. Henan Agricultural University, Zhengzhou (2005)
4. Luo, Z.-Y., Zhang, X.-D., Zhou, J.-S., et al.: Experimental study on catalytic and thermal cracking of tar from biomass pyrolysis. Journal of Chemical Engineering 18, 162–167 (2004)
5. Wu, C.-Y., Cui, Y.-J., Zhang, C.-R., et al.: The relationship between pyrolysis temperature and yield, composition, property of Shenfu coal tar. Shanxi Environment 7, 15–17 (2000)
6. Seshadri, K.S., Shamsia: Effects of temperature, pressure and carrier gas on the cracking of coal tar over a char-dolomite mixture and calcined dolomite a fixed-bed reactor. Industrial and Engineering Chemistry Research 37, 3830–3837 (1998)
7. Hesp, W.R., Waters, R.L.: Thermal Cracking of Tars and Volatile Matter from Coal Carbonization. Ind. Eng. Chem. Prod. Res. 9, 194–202 (1970)
8. Doolan, K.R., Mackie, J.C., Tyler, R.J.: Coal Flash Pyrolysis Secondary Cracking of Tar Vapours in the Range 870K-2000K. Fuel 66, 572–578 (1987)
9. Stiles, H.N., Kandiyoti, R.: Secondary Reactions of Flash Pyrolysis Tars Measured in a Fluidiid Bed Pyrolysis Reactor with Some Novel Design Feature. Fuel 68, 275–282 (1989)
10. Yan, J.-D., Cui, H., Yang, J.-L., et al.: Application advance of Thermogravimetry coupled with Mass Spectrometry (TG-MS). Journal of Instrumental Analysis 22, 105–107 (2003)
11. Chen, J.-J., Yang, H.-Z.: Applications of Thermogravimetry-Mass Spectrometry in environmental protection. Environmental Science and Management 32, 131–134 (2007)
12. Lu, C.-W., Xi, T.-G.: The developing evolution and current situation and prospect for the Thermal Analysis-Mass Spectrometry. Shanghai Measurement and Testing 29, 8–11 (2002)
13. Yan, J.-D., Zhao, H.-J., Yang, J.-H., et al.: Establishment and application of a TG/MS system. Analytical Laboratory 21, 20–22 (2002)
14. Yan, J.-D., Cui, H., Yang, J.-L., et al.: Research on pyrolysis behavior of Yanzhou coal using TG/MS. Journal of China University of Mining and Technology 32, 311–315 (2003)
15. Xiao, R.-H.: Coal Tar Chemical Engineering. Metallurgical Industry Press, Beijing (2002)
16. Wang, S.-D., Guo, S.-C.: Study of structure and components of tar from Shenfu coal. Journal of Fuel Chemistry and Technology 23, 198–204 (1995)

Dynamic Image Segmentation Using Incremental Nyström Method

Hongyu Li and Shicheng Wen

School of Software Engineering
Tongji University, Shanghai, China
hyli@tongji.edu.cn, shicheng.wen@gmail.com

Abstract. Eigen-decomposition plays a critical role in spectral segmentation. However it is often of low efficiency due to the bottleneck of computation. To solve such bottleneck problem, this paper proposes a dynamic image segmentation method with the incremental Nyström approximation. This method consists of two steps: firstly, we segment the low-resolution images with the original Nyström method; secondly, the high-resolution image newly arriving is dynamically dealt with on the basis of the low-resolution segmentation result. In the second step, an incremental Nyström method is designed to efficiently approximate the eigenvectors of spectral kernel matrices. Experimental results have shown that the proposed method can perform well in image segmentation.

Keywords: kernel method, Nyström approximation, incremental learning, matrix decomposition.

1 Introduction

At present, popular image segmentation methods have fallen into two categories, region-based methods and kernel-based methods. In the region-based methods, classical representatives are region merging, region growing, watersheds etc. [1-4]. Since such methods are based on cues in the local neighborhood, the segmentation results are hard to be satisfactory. In the kernel-based methods, both local and global cues are generally utilized to group pixels, therefore resulting in better segmentation performance [5-8]. In this study, we will focus on the latter.

Weiss [5] constructed the affinity matrix with cues from visual inspection. In addition, the author pointed out that a representation that leads trivially to a discrete segmentation can be extracted by using the top eigenvectors of such a matrix simultaneously. In [6], Ng et al. proposed a spectral clustering approach and gave conditions under which eigenvectors could be used to compute a reasonable cluster. In [7], Yu and Shi have applied novel discretization method to find group indicators from obtained eigenvectors. In [8], Cristianini et al provided spectral methods from affinity matrices to kernel matrices and proposed two cost functions to measure the quality of the bipartitioning of a dataset. Although the kernel-based method has shown its strong power in image segmentation, the computation of this method is a challenge.

G. Lee (Ed.): Advances in Automation and Robotics, Vol. 1, LNEE 122, pp. 257–264.
springerlink.com

Low-rank approximation methods, e.g. sampling-based methods, are a good way to solve the computation problem. Such methods can be achieved through spectral decomposition on a small subset of data points. For example, the Nyström method was proposed to accelerate eigen-decomposition of large matrix through downsampling and reducing the size of datasets. In [9], Fowlkes et al. applied Nyström method to image segmentation and video segmentation. In [10], Hongyu et al. constructed an image pyramid of a hyperspectral image by downsampling, and applied Nyström method to the top scale of the image pyramid.

It is worth noting that Nyström methods were designed only for handling batch data. When applications, such as electronic map images, require dynamic update of data and thus eigenvectors, original Nyström methods cannot work in a dynamic and incremental manner. That is, for newly arrived data, the original Nyström method has to behave in a batch manner and recompute the eigen-decomposition with all data, by discarding the previous result of decomposition. So the batch Nyström method needs to be further modified in the dynamic situation. To do this, we propose an increment version of the Nyström method to deal with the dynamically increasing data. In this approach, the original Nyström method is first used to approximate the eigen-decomposition of the kernel matrix constructed with low-resolution images. For dynamic images with high resolution, we approximately update the decomposition based on previous results, which avoids the waste of repetitive computation.

2 Nyström-Based Segmentation for Low-Resolution Images

The basic idea of the Nyström method is to approximate the eigenvectors of a full kernel matrix, such as the weighted affinity matrix *W* of an image, by performing the spectral decomposition on a few samples. The following procedure explains this method in detail. First choose *n* pixels as samples from an image containing *N* pixels, and *m* pixels remain. The numbers *m*, *n*, *N* satisfy that $\mathrm{m} + \mathrm{n} = \mathrm{N}$. Then reconstruct *W* as follows,

$$\mathrm{W} = \begin{pmatrix} \mathrm{A} & \mathrm{B} \\ \mathrm{B}^{\mathrm{T}} & \mathrm{C} \end{pmatrix}, \tag{1}$$

where $\mathrm{A} \in R^{n\times n}$, $\mathrm{B} \in R^{n\times m}$ and $\mathrm{C} \in R^{m\times m}$. Since *n* is usually quite small, the eigen-decomposition of A is efficient and fast,

$$\mathrm{A} = \mathrm{U}\Lambda U^{T}. \tag{2}$$

So the matrix *W* can be approximately decomposed as follows,

$$\mathrm{W} \cong \widetilde{U}\widetilde{\Lambda}\widetilde{U}^{T}, \tag{3}$$

where $\widetilde{U}$ and $\widetilde{\Lambda}$ satisfy the following equations,

$$\widetilde{U} = \begin{pmatrix} U \\ B^T U \Lambda^{-1} \end{pmatrix}, \quad \widetilde{\Lambda} = \Lambda. \tag{4}$$

Obviously, the approximation $\widetilde{W}$ of matrix W can be written as follows,

$$\widetilde{W} = \widetilde{U}\widetilde{\Lambda}\widetilde{U}^T = \begin{pmatrix} A & B \\ B^T & B^T A^{-1} B \end{pmatrix}. \tag{5}$$

It is easy to find from (1) and (5) that $B^T A^{-1} B$ is approximately equal to C. So, the quality of the approximation of the full weight matrix W can be quantified as the norm of the Schur complement,

$$e = \|C - B^T A^{-1} B\|. \tag{6}$$

Since $\widetilde{U}$ in expression (4) may not be orthonormal, we should find an orthogonalized solution. One way to solve this problem [9] is as follows: $Z = \widetilde{U}\Lambda^{\frac{1}{2}}$, $F\Sigma F^T$ denotes the diagonalization of $Z^T Z$ and $V = ZF\Sigma^{-\frac{1}{2}}$. Consequently, the matrix V contains the leading orthogonal eigenvectors of $\widetilde{W}$ and satisfies the condition $V^T V = I$. This orthogonalization procedure is with the computation cost $O(c \times N \times n^2 + d \times n^3)$, where c and d are constants and $n \ll N$ always holds for the low-rank approximation. Therefore, the orthogonalization method is efficient due to the small n.

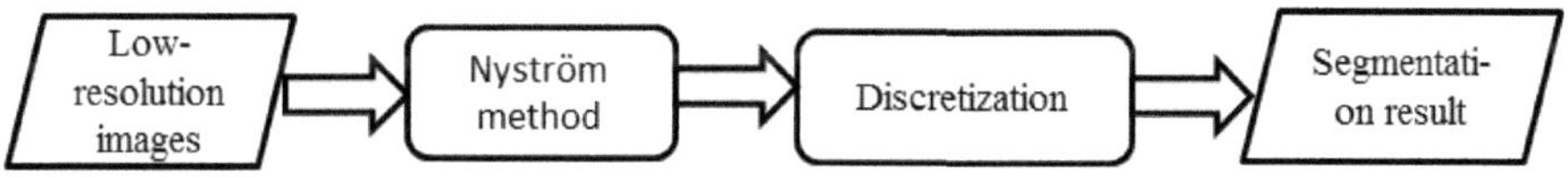

Fig. 1. The two-stage flowchart of low-resolution segmentation

As shown in Fig. 1, we first apply the Nyström method to low-resolution images to approximate eigenvectors, where pixels are usually quite fewer. Then the segmentation results could be obtained by discretizing the eigenvectors. The discretization method adopted here was originally proposed by Yu and Shi in [3], which uses an alternating optimization procedure.

3 Incremental Nyström Method for Dynamic Segmentation

Although the Nyström method has shown its power in image segmentation [9], it cannot work well for the dynamically increasing data, such as electronic map images.

A low-resolution image always arrives first and a high-resolution image comes after a while. As for this situation, the original Nyström method has to compute the two images respectively. Obviously, it is not a wise way. In this section, we extend the original Nyström method and propose an incremental Nyström method to handle the dynamically increasing data.

Assuming that the eigenvectors U_S and eigenvalues Λ_S for a low-resolution image has been obtained with the method stated in section 2, we next apply U_S and Λ_S to the high-resolution image. Let

$$\widetilde{U}_S = \begin{pmatrix} U_S \\ B^T U_S \Lambda_S^{-1} \end{pmatrix}, \tag{7}$$

and define $Z = \widetilde{U}_S \Lambda_S^{-\frac{1}{2}}$ so that $\widetilde{W} = ZZ^T$. The matrix $V = ZF\Sigma^{\frac{1}{2}}$, where F and Σ come from SVD of $Z^T Z$, contains the leading eigenvectors of $\widetilde{W}$. Actually, $\widetilde{W}$ satisfies

$$\begin{aligned} \widetilde{W} &= Z^T Z \\ &= \begin{pmatrix} U_S \Lambda_S^{\frac{1}{2}} \\ B^T U_S U_S^{-\frac{1}{2}} \end{pmatrix} \begin{pmatrix} \Lambda_S^{\frac{1}{2}} U_S^T & \Lambda_S^{-\frac{1}{2}} U_S^T B \end{pmatrix} \\ &= \begin{pmatrix} U_S \Lambda_S U_S^T & U_S U_S^T B \\ B^T U_S U_S^T & B^T U_S \Lambda_S^{-1} U_S^T B \end{pmatrix} \end{aligned} \tag{8}$$

It is easy to find from (1) and (8) that $A \approx U_S \Lambda_S U_S^T$, $B \approx U_S U_S^T B$ and $C \approx B^T U_S \Lambda_S^{-1} U_S^T B$. So, the error of the approximation of the full weight matrix W can be quantified as the norm,

$$e = \|W - \widetilde{W}\|. \tag{9}$$

Figure 2 illustrates the procedure of the proposed method for the dynamic segmentation. The incremental Nyström method is first applied based on the results of the low-resolution image. Then the eigenvectors are discretized to attain the segmentation results.

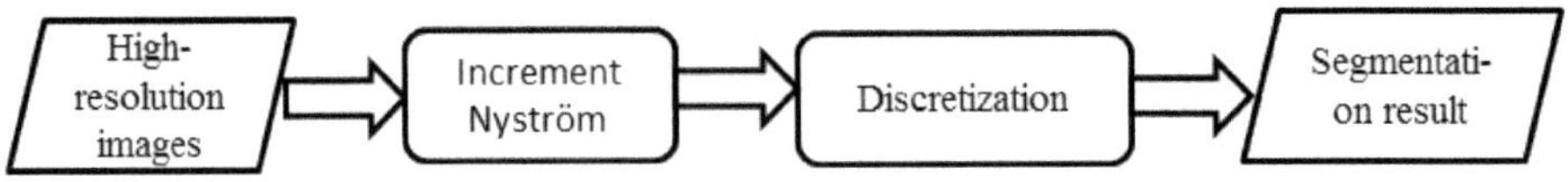

Fig. 2. The flowchart of dynamic segmentation

4 Experimental Results

In this section, we demonstrate the application of the incremental Nyström method to the dynamic segmentation on RGB images. To prevent the boundaries from deteriorating, the size of low-resolution image is required to be not less than 60×60.

Fig. 3. Low-resolution segmentation. (a) the original "sunflower" image of 60×60 size; (b) the segmentation result plotted with pseudocolors; (c)-(f) the first four eigenvectors of matrix W.

The HSV color space is fundamentally different from the widely known RGB color space since it separates out the intensity (luminance) from color information (chromaticity). As for the two chromaticity axes, a difference in the hue of a pixel is found to be visually more prominent compared to that in the saturation. Therefore, we use HSV [11], instead of RGB, to compute the weight matrix of an image. The entry of the weighted affinity matrix is computed in the following way:

Fig. 4. Dynamic segmentation. (a) the input image of 120×120 size; (b) the segmentation result; (c)-(f) the first four eigenvectors.

$$w_{ij} = e^{\frac{-\|F(i)-F(j)\|_2^2}{\sigma_I}} e^{\frac{-\|X(i)-X(j)\|_2^2}{\sigma_X}}, \tag{10}$$

where F(·) represents the HSV values of pixels and $\|\cdot\|_2$ means the Euclidean distance. In our experiments, σ_I and σ_I are set manually. Again, in this paper, we adopt the average association [12] for clustering.

An example on a low-resolution RGB image of a sunflower is shown in Figure 3. The image size is originally 60×60. We set the number of expected segments to be four and the first four leading eigenvectors are shown respectively in Figure 3 (c~f).

Figure 4 shows the example of the dynamic segmentation on a high-resolution image of 120×120. The incremental Nyström method is applied based on the eigenvectors computed in the low-resolution image (in Figure 3). The four eigenvectors are approximately computed, respectively corresponding to Figure 4 (c~f), and the image is finally split into four parts (Figure 4 (b)).

For comparison, we studied the original and incremental Nyström method. In our experiments, both methods are respectively employed to segment a dynamically increasing image. The original method has to recompute the eigenvectors with all pixels each time when the image resolution gets higher, which took long time. However, the incremental Nyström obviously improves computation performance through utilizing the segmentation results of low-resolution images. The comparison of the computation time in segmenting the high-resolution image is illustrated in Figure 5.

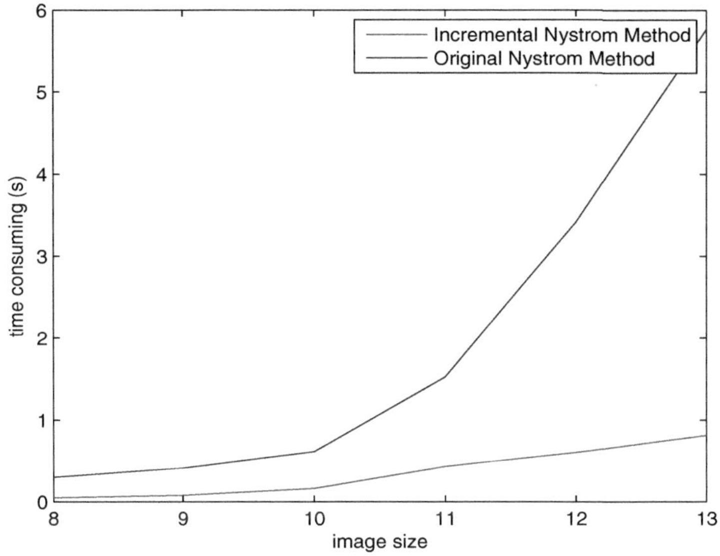

Fig. 5. The comparison of the computation time. The horizontal axis is the image size, for example, '12' means that the image resolution is 120×120. The vertical axis represents the computation time in seconds.

In our experiments, however, it was found that results of the dynamic image segmentation are not ideal enough either. There could be other ways to improve the segmentation results. For example, postprocessing after segmentation or choosing other image features as input. That could be the future topic of our work.

5 Conclusions

In this work, we propose a dynamic image segmentation method using the incremental Nyström method. Experimental results demonstrate that the incremental Nyström method efficiently improves the computational performance, compared with directly applying the original Nyström method to high-resolution images.

Acknowledgment. This research was partially supported by National Natural Science Foundation of China under grant 60903120 and Shanghai Natural Science Foundation under grant 09ZR1434400.

References

1. Peng, B., Zhang, L., Zhang, D.: Automatic Image Segmentation by Dynamic Region Merging. To Appear in IEEE Trans. Image Processing (2011)
2. Peng, B., Zhang, L., Zhang, D., Yang, J.: Image Segmentation by Iterated Region Merging with Localized Graph Cuts. To Appear in Pattern Recognition
3. Zhang, K., Zhang, L., Song, H., Zhou, W.: Active contours with selective local or global segmentation: a new formulation and level set method. Image and Vision Computing 28(4), 668–676 (2010)
4. Zhang, K., Song, H., Zhang, L.: Active Contours Driven by Local Image Fitting Energy. Pattern recognition 43(4), 1199–1206 (2010)
5. Weiss, Y.: Segmentation using eigenvectors: a unifying view. In: IEEE International Conference on Computer Vision, pp. 975–982 (1999)
6. Ng, A.Y., Jordan, M.I., Weiss, Y.: On spectral clustering: analysis and an algorithm. In: Proceedingsof the Neural Information Processing Systems, pp. 849–856 (2001)
7. Yu, S.X., Shi, J.: Multiclass spectral clustering. In: IEEE International Conference on Computer Vision, pp. 313–319 (2003)
8. Crisianini, N., Shawe-Taylor, J., Kandola, J.S.: Spectral kernel methods for clustering. In: Proceedings of the Neural Information Processing Systems, pp. 649–655 (2001)
9. Fowlkes, C., Belongie, S., Chung, F., Malik, J.: Spectral grouping using theNyström method. IEEE Transactions on Pattern Analysis and Machine Intelligence 26, 214–225 (2004)
10. Li, H., Bochko, V., Jaaskelaimen, T., Parkkinen, J., Shen, I.-F.: Kernel-based spectral color image segmentation. Journal Optical Society of America 25 (2008)
11. Shi, J., Malik, J.: Normalized cuts and image segmentation. IEEE Transactions on Pattern Analysis andMachine Intelligence 22, 888–905 (2000)
12. Sarkar, S., Boyer, K.L.: Quantitative measures of change basedon feature organization: Eigenvalues and eigenvectors. In: IEEE Conf. Computer Vision and Pattern Recognition (1996)

Research of the Spatial Position and Attitude Detection Technology about New Deep Hole Measurement Device

Bao Jiandong and Wang Changming

Department of Precision Instrument, Nanjing University of Science & Technology, Nanjing, Jiangsu, China, 210094
baojd025@163.com

Abstract. The paper regarded deep hole with groove ----rifled gun as example, in connection with the characteristics of movement theory and deep hole detection, introduced a new measurement system of deep hole, including elements of measurement system and driving form of analogous planetary gear, and, carrying out kinematics analysis. The system is accurate, reliable and steady, which is useful for deep hole detection and very important practical value, according to validating with lots of experiments.

Keywords: Deep hole kinematics, mode of motion, Automatic detection.

1 Introduction

Deep hole measurement is widely used at present, such as many kinds of large, medium, small tubes, holes, shafts and rail products in the industry of petroleum, chemical, iron and steel. For special deep-gun barrel, after it used as high temperature and gravity of the barrel itself, it will bend down. So it must be re-tested and proofed on time, or it will launch a dramatic effect on the accuracy and security in the real firing of gun. Hence, fast and accurate measurements of the gun barrel bore parameters have a key role in improving shooting accuracy and predicting the remaining life of gun.

It can be seen that the barrel bore gun parameter measurement test is the key technologies in production, while the gun barrel of quality and service life are essential. It has continued without interruption for the study of gun barrel bore measurements detection of parameters. However, due to the constraints and limitations factors of the complexity of the measurement environment, the limitations of application space, and the reliability of measuring system and so on, it makes the measurement system actual used in production process fail the requirements of fast and accurate measurements. So it is very important to study the deep hole measurement device and the corresponding detection technology.

G. Lee (Ed.): Advances in Automation and Robotics, Vol. 1, LNEE 122, pp. 265–271.
springerlink.com

2 Principles of Measurement Systems and Machine

2.1 Composition of Measurement Systems

According to the function and meet specifications of measurement system to be achieved, the carrier of the entire system by the measurement system —— crawling machines, measuring detection devices, industrial machines and data processing software, as shown in Figure 1.

The system works: portable measuring device in the crawling machine barrel (rifled or smooth-bore) within the screw forward, driven by the traction unit and power unit. It can crawl freely in the pipe body movement, real-time tracking crawling machine in the position of the barrel to measure the location of the specified parameters of the barrel bore by the laser range finder. The measuring system has the following advantages: (1) body tube system miniaturization; (2) system installation simple and easy to operate; (3) system smooth motion; (4) easy to adjust the sensor measurements.

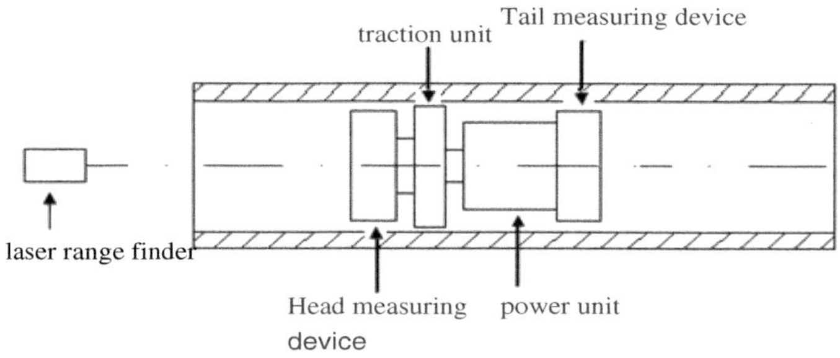

Fig. 1. System hardware components

2.2 Machine Theory

There are many movement patterns of deep hole measurement machine, It can be broadly divided into two categories: linear and spiral movement. The first movement pattern can also be divided into two kinds: direct and indirect linear motion. The movement pattern of the direct linear motion is simple, and the drive pattern can be internal and external (depending on bore size). The movement pattern of the indirect linear motion is complex. Its crawling body is segmented connection, and the drive mechanism and the testing organizations separate from each other. The system uses a more specific way of movement (taking into account noise and barrel wear) the use of planet-wheel drive mode, as shown in Figure 2. According to several rounds of experimental, the system uses polyurethane materials, as shown in Table 1. The Polyurethane wheel is installed between the direction parallel to each other and the wheel axis with the barrel axis is angle 20 ° (according to the speed of crawling machine wheel wear and polyurethane materials to determine this angle). It is also important to choose the materials of Planet-round. Through a variety of material samples (copper, nylon, rubber, polyurethane, etc.) experiments, mainly relatively long hours of work under the conditions of wear, and other factors into account, the final choice is polyurethane.

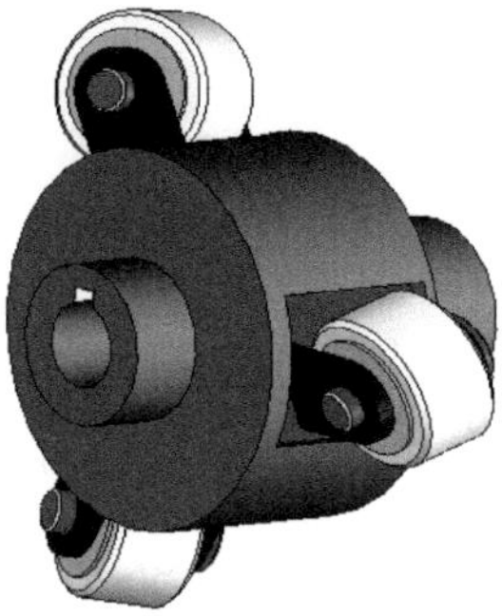

Fig. 2. Planet-wheel drive mechanism

Table 1. The conditions of wear of the planet-wheel

Serial number	the conditions of wear work in the rifled gun		
	Diameter before work /(mm)	Diameter after work (4 hours)/(mm)	Can continue to work
1	30	28.66	YES
2	30	29.13	YES
3	30	27.94	NO
	the conditions of wear work in the smoothbore gun		
	Diameter before work/(mm)	Diameter after work (4 hours)/(mm)	Can continue to work
1	30	29.88	YES
2	30	29.78	YES
3	30	29.90	YES

It is also important to select the appropriate motor speed and polyurethane wheel rotation parameters for measuring bore gun. The factors need being considered have a lot: 1) If the motor speed is too high, the robot will cause severe vibration, and will have a lot of noise, measurement error also increases. 2) If the motor speed is too low, it will affect the measurement of efficiency. Taking 6m of the barrel for example, Figure 3 shows the motor speed and the running time curve. When the motor speed is 180rpm, the barrel needs 20s running off. When the motor speed is 60rpm, the barrel needs 66s running off. All aspects taken into account, the motor speed is 100rpm.

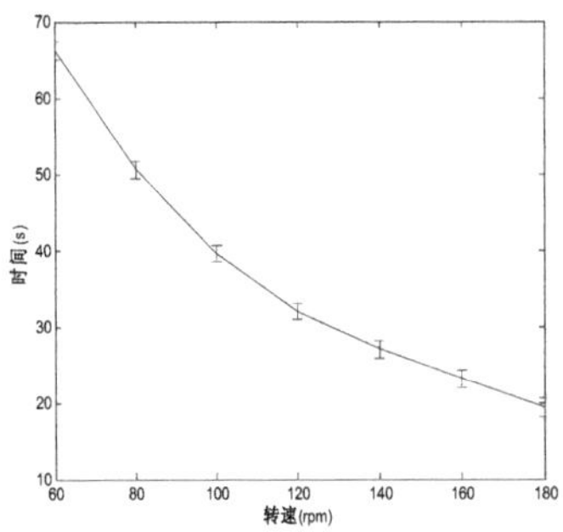

Fig. 3. The motor speed and the running time curve

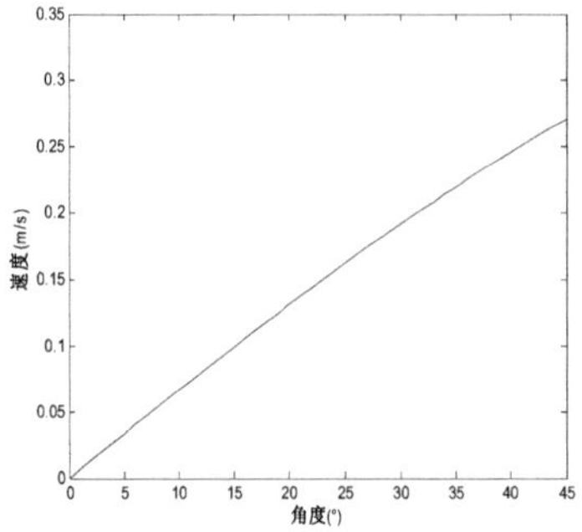

Fig. 4. Planet-rotation angle and the speed curve

Planet-rotation angle range is 0 ~ 45°. The robots with angle range of 45 ~ 90° have the features contrast with the former. Hence there is only the analysis of rotation angle 0 ~ 45 ° with the robot's speed curve, shown in Figure 4. The larger angle is, the faster speed will be, but the greater wheel wear also is.

2.3 Kinematic Analysis

The motion state of crawl machine in the gun deep tube shown in Figure 5: Planet-wheel is counterclockwise rotation, and the crawling bodies are spiral back. As figure 5 (a) shows, we set the motor speed (drive axle) as ω_L, wheel speed as ω_r, wheel radius as r, barrel radius as R, rifling and barrel angle θ, angle of drive shaft with the axis of the barrel as ψ.

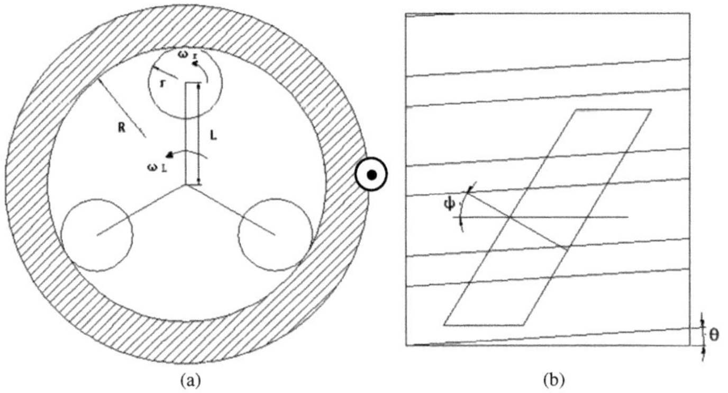

Fig. 5. The motion state of the driving device

Assuming the driving wheel and the barrel wall slip phenomenon does not exist. It gets wheel speed:

$$\omega_r = \frac{\omega_L R}{r}$$

As the driving wheel and the barrel axis form at an angle, as shown in Figure 5 (b) below, the movement of wheel can be divided rotation movement along the circumference of the barrel and forward movement along the barrel axis. Assume that the speed of system measurement is V, there is: $V = \omega_r r \sin\Psi = \omega_L R \sin\Psi$.

We could know that forward speed of the climb machine has nothing to do with the size of the driving wheel, only with the installation angle of the wheel. If deep hole with groove is taken into account, the main consideration is the design of slot wheel width. The size of driving wheel too small will make the Crawling machine vibrate, affecting the measurement accuracy. The size of the driving wheel gets the best value after many experiments.

3 Detection Methods

Take the rifled cannon bore measurement for example. Crawling machine is show in Figure 6. It consists of: 1, Support copper rings, playing a supportive role throughout the body crawling; 2, sensor adjustment nut, for adjusting the volume of stretch sensor; 3, the inductive displacement sensor, for measuring the diameter of artillery; 4, plungers, ensuring that crawling bodies to movement along the rifling; 5, power plant, the motor and reducer; 6, driving shaft; 7, driving wheel, consisting of planet- wheel formed by polyurethane. Detection field test is shown in Figure 7, and the proceeds of the test data is shown in Table 2.

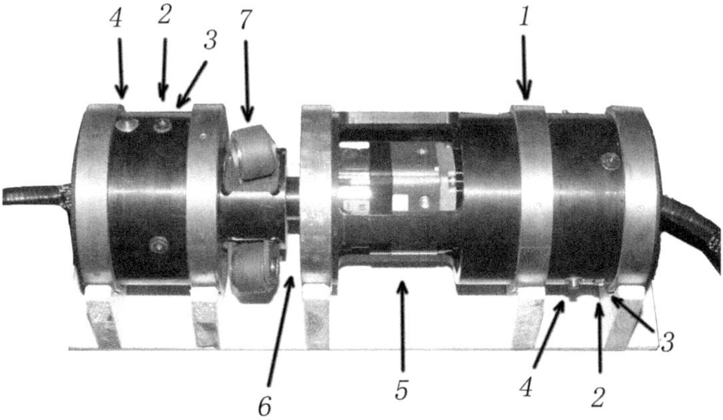

1—Support copper rings; 2—sensor adjustment nut;
3—the inductive displacement sensor; 4—plungers; 5—power plant;
6—driving shaft; 7—driving wheel.

Fig. 6. The device of measuring crawling machine and pipe

Fig. 7. Detection field test

Table 2. The result of test

position (mm)	Land diameter (mm)	Groove diameter (mm)	position (mm)	Land diameter (mm)	Land diameter (mm)	position (mm)	Land diameter (mm)	Land diameter (mm)
25	105.086	107.491	2000	105.059	107.425	4500	105.048	107.437
150	105.139	107.468	2500	105.109	107.388	4608	104.998	107.439
300	105.128	107.445	3000	105.128	107.393			
500	105.073	107.441	3500	105.001	107.379			
1000	105.171	107.463	4000	105.01	107.383			
1500	105.14	107.446	4350	105.094	107.441			

Through on-site inspection, the accuracy of system measurement could be up to 3 μm, the repeatability of measurement could be up to 5 μm, and it is stable, reliable and easy to operate, etc. The detection system has been applied in the enterprise. According to the survey, besides this products, factories in the country has reached the application of similar products is still very rare.

4 Conclusions

This article discusses the movement pattern and detecting techniques of a new deep hole measuring devices. The measuring system has a high degree of automation, high accuracy measurement, and high efficiency, easy to use and reliable. By actual measurement of detecting system, it achieves the design features to meet the required performance. According to the special circumstances of the barrel, it designs adaptable drive devices. And it adopts the appropriate sensor by the different requirements of measurement parameters. Thus, besides measuring gun barrel bore parameter measurement test, the detection system can be extrapolated to a variety of large, medium, small tubes, holes, shafts of automatic real-time measurement of detection by a little improvement, and it has broad prospects for development and application.

References

1. Liu, X.: The Design of the Multi-Parameter Intelligent Detecting and Measuring Platform for Artillery s Chamber. Changchun University of Science and Technology Master's thesis (December 2005)
2. Wang, J., He, X.: Artillery Summary. weapon industry press, Beijing (1992)
3. Wei, Z., Du, F.: Pipe Robot for Measurement of Artillery Barrel. Journal of ballistics 20(2), 88–90 (2008)
4. Bao, J.-D., Wang, C.-M., He, Y.-F., Li, J.-S.: Multifunctional automatic detection robot for artillery bore. Mechanical Design 25(12) (2008)

Data Collection and Transmission System with a Vibrating Stress Meter Based on GPRS

Xin He and Kun Liu

Department of Information Management
Beijing Institute of Petrochemical Technology, China
hexin@bipt.edu.cn, mcqueenlk@sina.cn

Abstract. Based on the rapid development of The Modern Mobile Network and full Coverage, This paper designs a Data collection and transmission system with a stress meter based on GPRS. Use the Internet access capabilities of GPRS, it can collect and transmit data real-timely at anywhere. This paper describes the structure of the system and working principle in detail. The system has more advantages, such as real-time communication, reliability and low cost.

Keywords: Vibrating wire stress meter, GPRS, Data Collection and transmission, Network Center.

1 Introduction

Over the years, the security risks of projects, such as subside of highway, bridges broken, collapse of earth dam and other geological disasters dominated by landslides, have made serious effects on nations' economy and people's living. Thus, people stress more and more on the prevention and monitoring to projects as well as geological disasters. With the huge progress of modern mobile network technology, the transmission technique based on GPRS begins to gradually enter various fields, and it is also a trend to apply it to the stress-monitoring realm [1-2]. It has been showed by practices that the communication method that regarding monitoring host to remote terminal equipment as the communication channel for the GPRS network has the advantages of long communication distance with high reliability and low cost of overall investment, so the data collection and transmission system with a stress meter based on public network owns broad while in-depth application outlook [3-4].

2 The Overall Designing of System Structure

The whole system is made up of the following three parts: Field data collection and transmission unit (FDCTU), mobile operators and network center servers. What we need to do research in is the development of FDCTU and network center servers, and

G. Lee (Ed.): Advances in Automation and Robotics, Vol. 1, LNEE 122, pp. 273–281.
springerlink.com

in addition the fast transmission of data between them. The system structure is illustrated by Fig.1. System Structure:

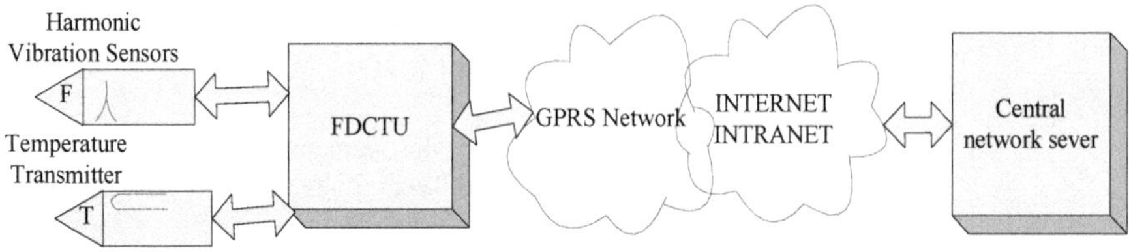

Fig. 1. System Structure

The working process of system: FDCTU completes data collection and storage to vibrating wire stress meters and temperature transmitters, and then transmit data according to TCP/IP protocol. Then network center servers finish receiving, saving and analyzing the data from FDCTU, which allows users to analyze the on-site running condition timely [5].

3 The Designing of Field Data Collection and Transmission Unit

The hardware structure of FDCTU uses modular designing, whose system structure is illustrated by Fig.2. the hardware structure of FDCTU The transmission module of GPRS adopts HUAWEI GTM-900C, the core processor makes use of ATMEL MEGA128, and the power supply utilizes one LM2596 and one LM1117 to respectively provide 4.2V and 5V power, so as to protect the supplement to system's power.

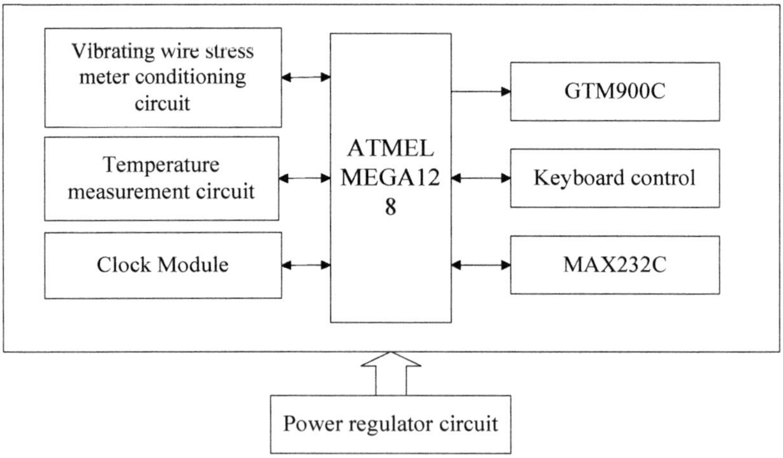

Fig. 2. The hardware structure of FDCTU

For a long period of time, vibrating wire stress meter has its own unique advantages at measuring the sensitivity, stability and anti-interference ability of stress. The working principle of vibrating wire stress meter is showed by Fig.3. Sensitive component vibrating string-2 is a tensed metal wire that is placed in electromagnet-3. Its one end is fixed to the support-1, and the other one is connected with the movable part-4.The static tension T acting on the movable part-4 to make the string tensed. Object-3 is an electromagnet, and when there adds pulse current at the two ends of the electromagnet's coil, due to the electromagnetic force the iron will drive the metal wire to vibration. Along with the uneven forces that are exerted at the two ends of its inside vibrating strings, the resonant frequency of these vibrating strings will accordingly change. And by a series of pulse with different vibration frequency, string will be stimulated to vibrate, causing vibrating wire's resonance, so by measuring the frequency of such resonance, it is available to calculate the sensor's forcing condition as well as the engineering value [6].

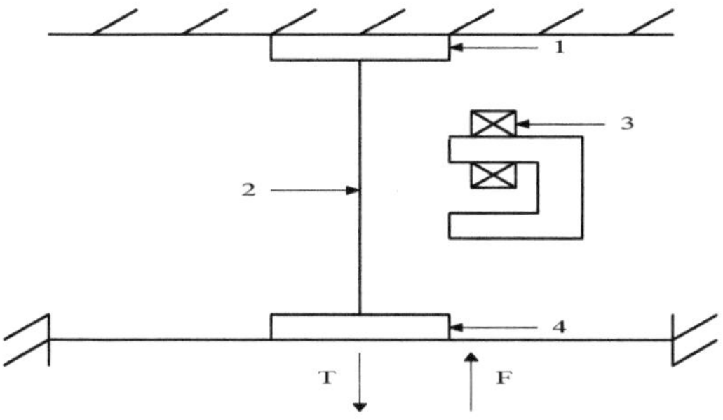

Fig. 3. The working principle of vibrating wire stress meter

Vibrating wire stress meter is a sensor whose sensitive components are steel strings, and who regards force as input quantity and frequency signal as output quantity. Its strings' natural frequency of vibration is related to tension - when the vibrating string's length is given, the altered amount of string's vibration frequency can express the quantity of tension. After calculation based on string vibration equation's derivation, the equation (1) below will be drawn:

$$f = \frac{1}{2l}\sqrt{\frac{T}{\rho}} \tag{1}$$

Where: *T stands for string's tension;*
ρ stands for string's liner density;
l stands for string's length.

As $T = S\sigma$, $\sigma = \varepsilon E$, then

$$f = \frac{1}{2l}\sqrt{\frac{S\sigma}{\rho}} = \frac{1}{2l}\sqrt{\frac{S\varepsilon E}{\rho}} = \frac{1}{2}\sqrt{\frac{E}{\rho_\gamma}\varepsilon} \tag{2}$$

Where: S stands for string's cross-sectional area;
σ stands for the stress suffered by the string;
ε stands for string's strain;
ρ_γ stands for string's density.

The string's strain can be expressed by

$$\varepsilon = K f^2 \tag{3}$$

Where $K = \dfrac{4 l^2 \rho_\gamma}{E}$

Thus, the change of strain is:

$$\Delta \varepsilon = K(f^2 - f_0^2) \tag{4}$$

Where: f_0 stands for the initial frequency.

At present, all vibrating wire stress meters takes this equation as the basis of theoretical calculating model, which makes wide-ranging sense. While in practice, there is another fitting formula (5) that is typically used when calibrating the senor's performance:

$$\Delta \varepsilon = A(f^2 - f_0^2) + B(f^2 - f_0^2) \tag{5}$$

Generally, the density of string ρy, elastic modulus ε along with the string's length can all be considered constant. Therefore, as long as the frequency of string's resonance can be measured, the tension suffered by vibrating string T, can be derived indirectly. The calculation formula of stress σ and string's strain is showed by formula (6):

$$\sigma = \varepsilon E \tag{6}$$

Vibrating wire stress meter's own mechanical structure features lead to its good long term stability - the general standard of zero frequency stability is about 0.025% each year. Vibrating wire stress meter's output is frequency signal, which can be transmitted in cable farther and more stably. This characteristic allows it to be superior to other electronic sensors. In addition, compared to 4-20mA output sensors, to a large extent, the electric part of vibrating wire stress meter can be more insensitive to the interference of lightning, which is very important to sensors that are working in the field environment for long-term.

Vibrating wire stress meter's conditioning circuit mainly includes two parts: excitation circuit and frequency detection circuit. After vibrating wire stress meter is affected by excitation circuit, it will output electric signals whose frequency is the same as oscillation frequency of vibrating string, and which is transformed into standard square wave signal by frequency signal detection circuit's filtering, amplification and shaping. Then by the processing of anti-jamming circuit, such standard signal will be sent to and counted by SCM, completing the output frequency measurements of Vibrating wire stress meter. The specific designing principle is demonstrated by Fig.4. Excitation circuit's principle picture:

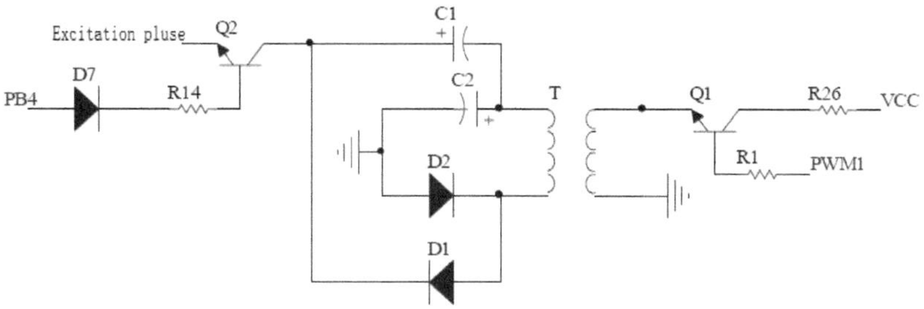

Fig. 4. Excitation circuit's principle picture

By the effect of excitation circuit, vibrating wire stress meter can create reliable start-up and oscillation signal, and after the picking up of pick-up coil inside the sensor, such signal will be sent into detection circuit and be processed through filtering, amplifying and shaping. Transmitting the standard square wave that are created after this process, into the SCM's counter to do counting, finally the measured output's frequency value of vibrating wire stress meter can be got [7].

AT command is a standard command that is applied to the communication between MT and TE program, namely the connection and communications between terminal device and PC applications. And it has a requirement to the transmitted data packet size: that is, for sending AT commands, in addition to the two characters- AT, the length that can be received at most is 1056 characters. It includes some standards such as GSM Rec.07.07 standard AT command, ITU-T Rec. V25ter AT command and standard GPRS AT command etc.

HUAWEI GTM-900C is a an integrated TCP / IP protocol and special modules, which can control modules with standard AT commands, so as to complete the access to IP network, truly attaining the direct connection between embedded systems and INTERNET IP network. Module has its own integrated SIM card interface pins, so users can directly wire in the PCB. Users just need to do some related operations on module with switching commands and serial AT instructions, then by connecting mobile network through CMNET with fixed IP address and port number of central server, the function of transmitting data will then be achieved [8-9]. The AT commands access to network and server-side handshaking process mainly used in this design is shown in Fig.5. the basic flow chart of GPRS transparent transferring command:

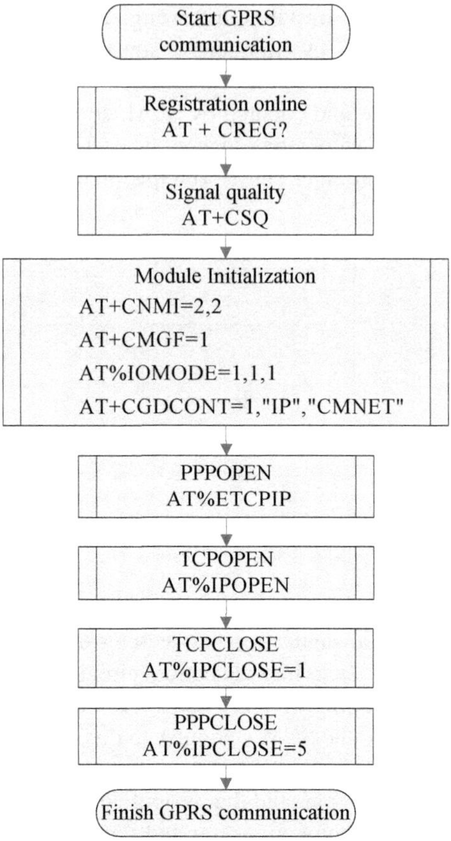

Fig. 5. The basic flow chart of GPRS transparent transferring command

4 The Design of Network Center Sever

Network center is the external "window" of system, which is responsible to present data that is collected and processed by system to monitor staff in a variety of intuitive ways, and meanwhile receive instructions from monitor staff in a "friendly" way, allowing monitor staff to control the running of the entire system. Thereby, position machine monitoring software plays a crucial role in the design of whole system, and its design has direct impact on users' general impression on the entire monitoring system. Network center sever mainly fulfills the tasks of receiving distant data, saving, analyzing and managing users, and it is able to conduct on-site stress analysis based on users' demand.

The design uses Microsoft's Visual C++ compiler environment, with Microsoft's Web-based winsock class-library to complete the function of data reception. In the MFC class library, a socket class CAsyncSocket is provided, and it is very convenient to implement such function [10]. In windows system, with Socket there are two socket approaches to process network communication: datagram socket and stream socket. Datagram socket is based on UDP protocol, which can transmit data wirelessly, and its strength is high speed of transmitting. Yet it cannot ensure the reliability, orderliness and non-repeatability of data transmission. While stream socket applies TCP protocol to provide bit stream transmission that is orderly, reliable, bidirectional, connected and not repeated. This design uses stream socket, and its working procedure is illustrated by Fig.6:

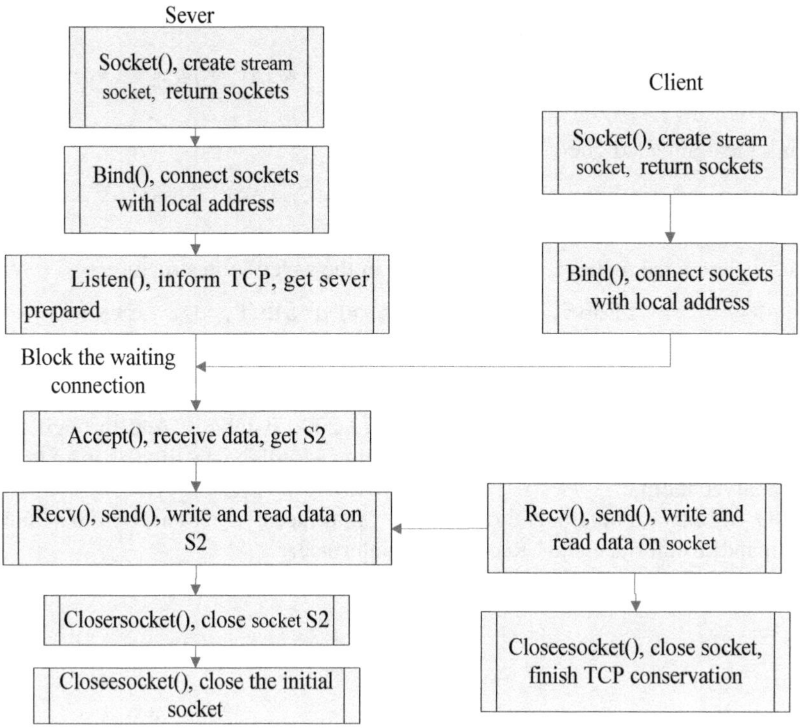

Fig. 6. The working procedure of stream socket

In order to fulfill that many FDCTU can be totally received by severs, the reception end makes use of Overlapped I/O technique, which is also called Asynchronous I/O model. Asynchronous I/O is different from synchronous I/O: to synchronous I/O, the program will be suspended and not get controlled, until the I/O is processed completely. Whereas to asynchronous I/O, a function will be called to inform OS to do I/O operation, then get returned immediately without waiting for the completion of I/O and continue the program's execution. When OS finishes I/O, it will send message to users. Overlapped I/O is merely a model, and it can be attained by kernel object, event kernel object, the asynchronous procedure calls and completed ports (I/O completion) [7].

Its main codes are listed as below:
Create the I/O completion ports:

```
CreateIoCompletionPort (INVALID_HANDLE_VALUE, NULL, 0,
0);
```

Create some threads:

```
CreateThread (NULL, 0, &ThreadFunction, NULL, 0, NULL);
```

Receive clients' request in threads:

```
sClient = WSAAccept(sListen,(sockaddr *)&m_Client_addr,
&len, NULL, 0);
```

Link the connected client and completion port:

```
CreateIoCompletionPort((HANDLE)sClient,CompletionPort,(
DWORD)PerHandleData,0);
```

Receive data from client, which is stored in the saved loop structure:

```
WSARecv (sClient, &PerIoData->DataBuf, 1, &dwRecv,
&Flags, &PerIoData->Overlapped, NULL);
```

In the program, by the means of ADO, the access of data will be achieved with reading and writing operation on SQL Server 2000 database. And through creating different forms for different sites respectively, it will be facilitated to extract data when analyze them.

ADO library contains three basic interfaces: _ConnectionPtr interface, _CommandPtr interface and _RecordsetPtr interface.

```
class ADOConn
{
public:
    ConnectionPtr   m_pConnection;
RecordsetPtr    m_pRecordset;
public:
ADOConn();
           virtual ~ADOConn();
           void OnInitADOConn();
    RecordsetPtr&GetRecordSet(_bstr_t bstrSQL);
    BOOL ExecuteSQL(_bstr_t bstrSQL);
void ExitConnect();
};
```

5 Conclusion

With the accomplishment of data collection and transmission with a vibrating stress meter based on GPRS, it is available to effectively resolve the current troubles that exist in measuring stress, such as wide distribution, too many test points, poor test environment and higher costs etc., and in addition provide users for abundant on-site and real time data. Relying on present and historical data, users can do on-site and real-time monitoring to the key parts of the current stress conditions, which facilitates users to analyze the running condition of this facility.

References

1. Han, Y., Feng, J.: Application of wireless data transmission technology in geological disaster monitoring application. Science and Technology of Survey (May 2007) (in Chinese)
2. Li, Q., Wen, Y., Yuan, J.: Design of Remote Monitoring Node of Landslides Disaster Based on GPRS. Science Educator, 252–253 (2006) (In Chinese)
3. Huang, L., Chen, G.: Research on GPRS-based Remote Safety Monitoring System. Master's Thesis of China University of Petroleum in East China (2007) (in Chinese)
4. Zhu, X., Zhu, X.: Research in the Application of Remote Monitoring System Based on GPRS. Dissertation of Southeast University (2005) (in Chinese)
5. Lin Chin, E., Li, C.-C.: A real-time GPRS surveillance system using the embedded system. American Institute of Aeronautics and Astronautics Inc. (2004)
6. Zhang, H.: Encyclopedia of Sensor Technology. Beijing University of Aeronautics and Astronautics Press (2007) (in Chinese)
7. Xu, J., Liu, H.: Signal Collection and Data Process of Vibrating Wire Sensors. Industrial Safety and Dust-Proof, 16–19 (January 2000) (in Chinese)
8. Yu, H.M., Pang, Z.F., Ruan, D.R.: The Design of the Embedded Wireless Vehicles Monitoring Management System Based on GPRS. Evidence from China, IEEE (2008)
9. Lim, K.P., Wu, D., Wu, S., Susanto, R., Lin, X., Jiang, L., Yu, R., Pan, F., Li, Z., Yao, S., Feng, G.: Video Streaming On Embedded Devices Through Gprs Network, July 6-9. IEEE Computer Society (2003)
10. Wang, G., Lin, L.: Programming Based on Windows TCP/IP.Tsinghua University Press (2002) (in Chinese)
11. Su, X., Gao, P.: Research in Remote Monitoring Method Based on TCP/IP. Qingdao University Journal (2004) (in Chinese)
12. Shen, C.: Research and Design of GPRS Wireless Network Communication. Master Thesis of Anshan University of Science and Technology (2005) (in Chinese)

High-Precision Measurement Method of Time-Interval Based on Pseudo-Random Sampling

Minshuang Huang, Junfen Huang, and Yinqi Feng

Opto-Mechatronic Equipment Technology Beijing Area Major Laboratory,
Beijing Institute of Petrochemical Technology, Beijing, China
huangjunfen@bipt.edu.cn

Abstract. Based on pseudo-random sampling a high-precision measurement method of time-interval for pulsed laser ranging is presented for the first time. In the method, time-intervals of laser pulses are controlled by a chirp signal generated using a Voltage-Controlled Crystal Oscillator. The time-intervals are unequal pseudo-random sequences. Pulse laser signals reflected from the object become sampling signals after they pass through a time discriminating circuit. Characteristics of measured signals are represented by the mapping of sampling time points on a referenced sinusoid signal. Then high-precision measurement of time-intervals can be implemented using a phase estimation algorithm. This method is simple in theory and easy to realize. In experiments using 15MHz reference clock, it is achieved that the measurement precision of time is 16ps.

Keywords: Pseudo-random sampling, Time-interval measurement, Phase estimation.

1 Introduction

In a high-precision flight time laser pulse ranging system, the combination of Coarse Counting and Fine Measurement [1] is a solution at present. In the technique, direct counting is used in Coarse Counting. Its principle is that trigger points are first recorded, and measurement results are then obtained by using high-speed pulse counting method in which the reference clock frequency is usually about 15MHz. Although wide dynamic measurement range could be achieved in Coarse Counting, its time resolution is only on a nanosecond level. In Fine Measurement, a time interpolation technique is used. The basic idea is that a reference clock cycle in Coarse Counting are divided into M equal parts using an appropriate method, which is equivalent to increasing the clock frequency M times.

Time Interpolation techniques mainly are the analog interpolation[2], the delay-line interpolation[3-6], the cursor interpolation[7-8] and the pulse compression interpolation[9]. A high precision time interpolation technique using high-speed FPGA gate delay features is reported by a few references[10-12]. However, it is difficult to obtain a precision which is above picosecond due to the dispersion of FPGA gate delay and other issues. In reference[13], a measurement method which uses phase estimation to achieve picosecond-level high-precision time interval is

G. Lee (Ed.): Advances in Automation and Robotics, Vol. 1, LNEE 122, pp. 283–291.
springerlink.com

presented. Since the phase estimation accuracy is related with the sampling frequency and the sampling frequency is required to be greater than the Nyquist frequency, A/D must be provided with high converting speed which will cause the sampling circuit too complicated to realize. Therefore, a novel high-precision time interval measurement method based on pseudo-random sampling is proposed.

2 The Principle of Fine Time Measurement

The principle of high-precision fine time measurement based on pseudo-random sampling in pulsed time-of-flight laser ranging is shown in Figure 1. The light source is a pulsed laser diode, and its pulse width is 30ns. Its peak power is 40w and the pulse repetition frequency is approximate 8.55 kHz. The time-interval of emitted laser pulses is controlled by the linear frequency modulation signal generated by a Voltage-Controlled Crystal Oscillator (VCXO). The return laser pulse signals are detected by an APD and discriminated by a time discriminator. Then the exact time is obtained for sampling. The reference signals are sampled at those exact times and the sampling values are sent to a phase estimator.

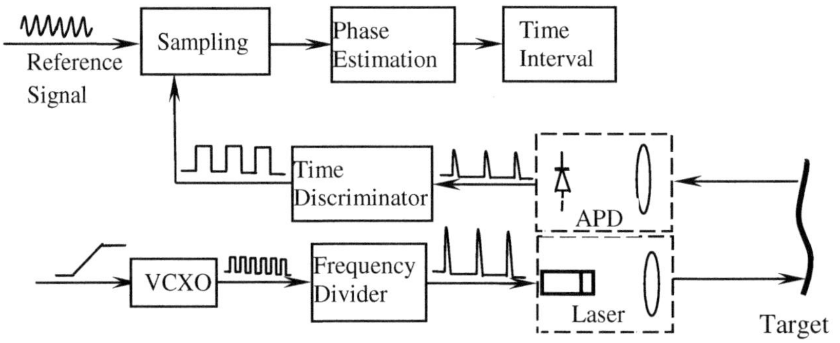

Fig. 1. The principle of high-precision fine time measurement based on pseudo-random sampling

Assume that a reference signal $s(t)$ is a sinusoidal wave with a single frequency. When $s(t)$ is regarded as a time counting base, the phase shift φ of the reference signal $s(t)$ from the pulse transmission time t_{start} to the reception time t_{stop} can be expressed as

$$\varphi = 2\pi k + \varphi_0 = 2\pi(k + \Delta k) \tag{1}$$

where k can be zero or a positive integer and obtained by a direct counter. $\Delta k = \varphi_0 / 2\pi$ is a decimal, thereinto $0 \le \varphi_0 \le 2\pi$. φ_0 is the initial phase of the reference signal sampled at t_{stop} under the condition that the phase of the reference signal at t_{start} is equal to zero. Since the relationship between a sinusoidal value and its phase is nonlinear, the method of figuring out an initial phase from a sinusoidal value results in a low resolution and great errors. The reason is that a large phase change can be caused by a tiny amplitude change when the signal amplitude is near to the peak value. In addition, a large phase error can be generated by a fluctuation of the sinusoidal signal amplitude. To improve the resolution of an initial phase measurement, a basic thought is to seek the initial phase φ_0 by measuring a series values of $s(t)$. Therefore, the problem measuring initial phase φ_0 is transformed into the problem estimating the initial phase φ_0.

The output frequency $f_{vcxo}(n)$ of VCXO is modulated linearly by the input voltage, and it can be expressed as:

$$f_{vcxo}(n) = f_{s0} + a \cdot (n-1). \quad n = 1,2,3.,\cdots,N \tag{2}$$

Where f_{s0} denotes a initial frequency of VCXO; a denotes a scanning slope; N denotes the total number of samples. When a reference signal is sampled, it is assumed that the sampling time interval is m times of the output clock cycle of VCXO. It means that a laser pulse is sent after every m VCXO clock cycles. The phase difference between the (n-1)th sample point and the nth sample point on $s(t)$ is

$$\Delta\varphi(n) = \frac{2\pi m f_0}{f_{vcxo}(n)} - 2\pi \cdot INT\left(\frac{m f_0}{f_{vcxo}(n)}\right). \tag{3}$$

Where f_0 is the frequency of the reference signal $s(t)$. INT() is an integral function. Considering that $s(t)$ is a periodic function, $0 \le \Delta\varphi(n) \le 2\pi$ can be obtained by a rounding operation. The phase of the nth sample point of $s(t)$ can be confirmed by Eq.(3) as

$$\varphi(n) = \begin{cases} \varphi(n-1) + \Delta\varphi(n) - 2\pi \cdot INT\left(\dfrac{\varphi(n-1)+\Delta\varphi(n)}{2\pi}\right) & n = 2,3,4,\cdots,N \\ \varphi_0 & n = 1 \end{cases} \tag{4}$$

The rounding operation is carried in Eq.(4) in order to meet $0 \le \varphi(n) \le 2\pi$. Set $t(n) = \varphi(n)/2\pi$ and $\tau(n) = \Delta\varphi(n)/2\pi$, then after the reference signal is normalized, $t(n)$ and $\tau(n)$ denote the sampling time and the sampling time interval between two adjacent points, respectively. The parameters used in the experiment system are as following: $f_0 = 14.985480MHz$, $f_{s0} = 15.13397MHz$, $m = 1770$. According to Eq.(2), Eq.(3) and Eq.(4), the curves of $\tau(n)$ and $t(n)$ can be calculated and shown in Fig.2 and Fig.3. Because a current sampling time is determinated by a previous sampling time, which can be expressed as $t(n) = t(n-1) + \tau(n)$, where $\tau(n) = k_n \cdot T_0 + \Delta\tau(n)$, k_n is an integer, $T_0 = 1/f_0$ is the period of reference signal, and $\Delta\tau(n)$ is a decimal, a periodic non-uniform sampling is actually used in this kind of sampling method, and the sampling time interval $\tau(n)$ is about k_n times longer than a period of the reference signal, which means that the sampling frequency is k_n times lower than the signal frequency. As the value of f_0 is close to that of f_{s0}, $k_n \approx 1770$ in our experimental system. However, $\tau(n) = \Delta\tau(n)$ can be obtained by transforming using Eq.(3). The sampling time intervals equivalent and sampling times equivalent vs. sampling points are represented in Figure 2 and Figure 3, respectively.

In Figure 2, it can be seen that $\tau(n)$ is periodic. The values of $\tau(n)$ in the zone of [0, 1] are approximately obeyed to a uniform distribution, which means that a probability of a sampling point located in [0, 1] zone is almost equal although every sampling point is selected on purpose. That is, the sampling time is a pseudo-random number. Therefore, the phenomenon of frequency aliasing couldn't occur under the condition that sampling frequency is far lower than the signal frequency.

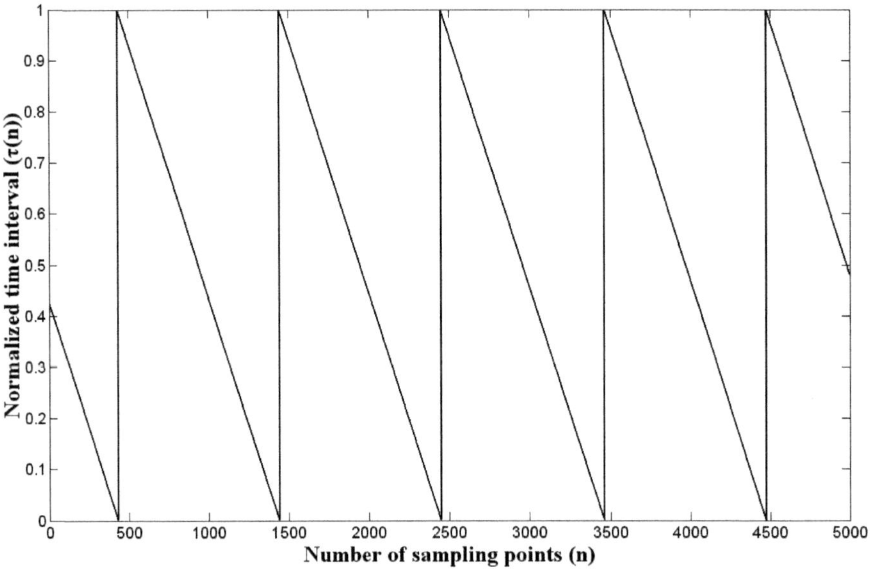

Fig. 2. Changes of sampling time intervals equivalent vs. sampling points

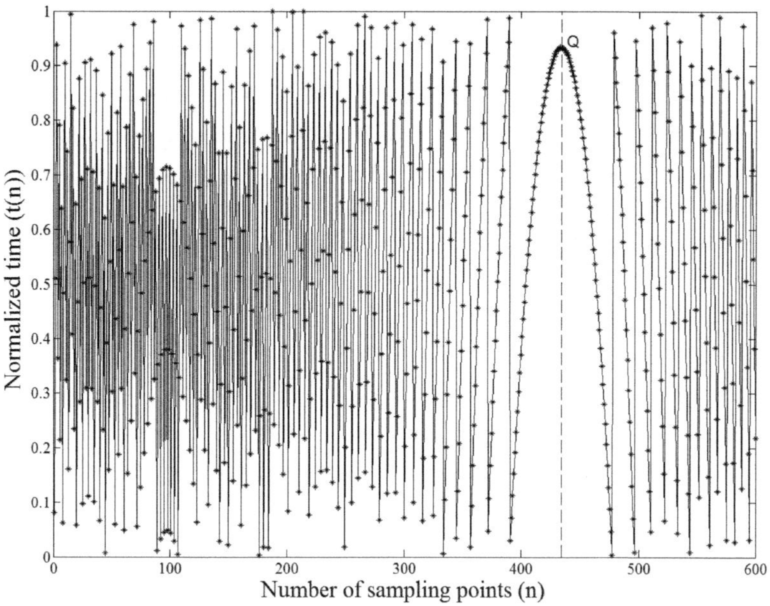

Fig. 3. Changes of sampling times equivalent vs. sampling points

In Fig.3, sampling points near the feature point Q are distributed densely. A number of sampling points lie in the zone where the values of $t(n)$ increase from 0 to 1 or decrease from 1 to 0. So some data can be sampled in a signal period. It means that the sampling frequency come to be several times that of reference signal tested through above transformation. As it meets the condition of Shannon sampling theorem, the initial phase φ_0 could be estimated accurately by sampling *N* sinusoidal values of the reference signal. It is obvious that the requirement to A / D sampling rate are reduced greatly by using this kind of pseudo-random sampling technique. The system structure is simple and easy to implement. And its cost is low.

3 Estimation of Initial Phase

The frequency f_0 of a reference signal could be treated as a constant, as it is produced by a high-precision temperature-compensation crystal oscillator (TCXO). Its frequency stability is better than $\pm 1\times 10^{-8}$ in a short period. So the effect on the precision about estimating of initial phase can be ignored. On the side, the central frequency will be changed after longer time as TCXO vs. Aging is $\pm 1\times 10^{-8}$. When a change of the central frequency goes beyond a permission range, the central frequency could be proofread and estimated accurately using sampled data obtained by the system. Therefore, a ranging process can be simplified as an estimating of an initial phase for a known sinusoidal wave signal, and the ranging speed can be improved through the simplification. The specific steps are shown as following:

1) To calculate the initial frequency f_{s0} of VCXO. Since f_0 ,m and a are constants and the output frequency f_{s0} of VCXO is uncertain, using sampling data the feature point Q could be found and the abscissa "n" at the point Q could be solved. Then f_{s0} is calculated by setting $\Delta\varphi(n)=0$ in Eq.(3).

2) To calculate the time interval $\Delta\varphi(n)$ between two adjacent sampling points. When parameters f_0, f_{s0}, m and a are confirmed, $\Delta\varphi(n)$ could be calculated using Eq.(3) (thereinto $n=1,2,3,\cdots,N$).

3) To determine the range of initial phase φ_0 . Every sinusoidal value is corresponded to two phase values φ_{01} and φ_{02} in the zone of $0-2\pi$. Assume that the limit error of φ_0 measured by an experiment is $\pm\delta$, then the range of initial phase φ_0 could be determined as $\varphi_{01}-\delta\le\varphi_0\le\varphi_{01}+\delta$ or $\varphi_{02}-\delta\le\varphi_0\le\varphi_{02}+\delta$.

4) To estimate the initial phase $\hat{\varphi}_0$. A step size is set to be $2\pi/4096$ (on the assumption of 4096 fold-frequency subdivision) in the range of φ_0 determined by step 3). Given a φ_0 , the theoretical values of phase for each point $\varphi(n)$ could be calculated by formula (4). Assume the sampling sinusoidal values of the reference signal are $x(n)$, thereinto $n=1,2,3,\cdots,N$. ε could be calculated using following equation:

$$\varepsilon=\sum_{n=1}^{N}\left(\sin(\varphi(n))-x(n)\right)^{2} . \tag{5}$$

All of the possible values of φ_0 are taken over and one of values of φ_0 corresponding to the minimum of ε is the desired estimation value of initial phase $\hat{\varphi}_0$.

4 Experiment Results

In the experiment system, a Fine Measurement circuit is mainly constituted by a VCXO, a temperature-compensation crystal oscillator, an A/D converter and a FPGA. The sampling rate of the A/D is 5MHz with a 12-bit resolution. In the beginning of a measurement, the VCXO is driven by a linear scan voltage. Set the number of sample points N=600, then the total scan time should be greater than $m\cdot N/f_{s0}\approx 71ms$ in order to ensure that measured data are in the linear section of the pullability curve of VCXO. The sampling result of the sinusoidal values of the reference signal is shown in In Fig.4, it can be seen that the measured waveform in the experiment is consistent with the theoretical result, which indicates that the proposed high-precision measurement method of time interval based on pseudo-random sampling is feasible.

A curve of the phase error for the reference signal is shown in Fig.5. Its abscissas are the sampling numbers, and its ordinates are the errors between the experimental phase values and the theoretical phase values for a reference sinusoidal signal. Gross errors appear occasionally in measurements, which are caused by laser echo pulses whose amplitudes are out of the measuring range. In Fig. 5, the gross errors have been removed and the maximum error is in the midst of ±5%. It can be seen that phase errors are caused by the fluctuation of the amplitude or noise of the reference sinusoidal signal, especially when the signal amplitude approximates its peak. In addition, phase errors caused by changes of echo laser pulse amplitudes are positive or negative. They obey the statistical rule that its expectation value is zero, which means that these errors can be treated as random errors. Therefore, high precision estimation of initial phases can be obtained by random sampling where phase errors counteract each other.

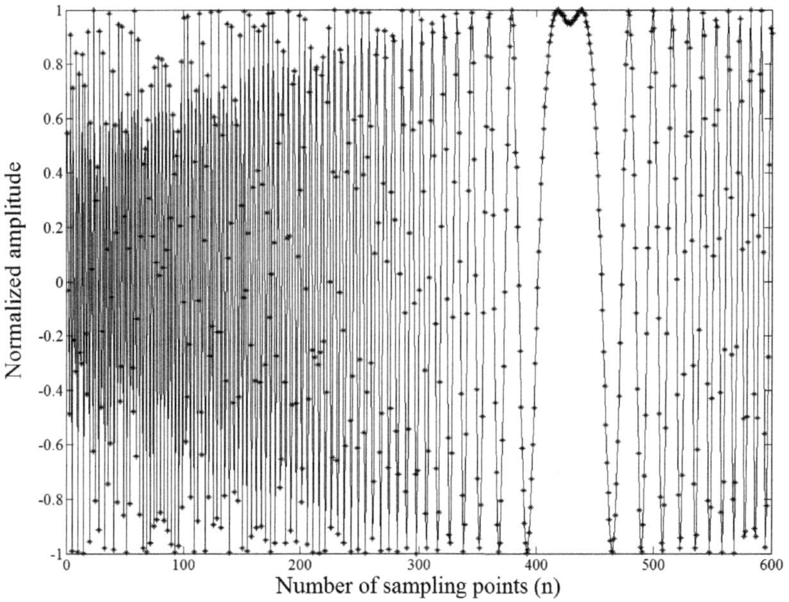

Fig. 4. Experiment result for changes of the reference signal amplitudes vs. sampling points

In the experiment system, a clock with a frequency of about 15MHz is used. That is, 66.7ns is a counting clock period and the counting precision of 66.7ns can be obtained.As the resolution of A/D is 12-bit, counting precision is reduced to 1/4096 of an original frequency after non-linear errors of VCXO pullability curve are compensated accurately. That is, a counting precision of 16ps is obtained corresponding to a ranging precision 2.4mm. The above high-precision measurement technology of time intervals based on pseudo-random sampling has been applied to a high-precision pulsed laser rangefinder. Through measurement, the ranging accuracy of the rangefinder is 3mm +2 ppm, which indicates the high-precision measurment method of time-interval proposed by the paper is feasible.

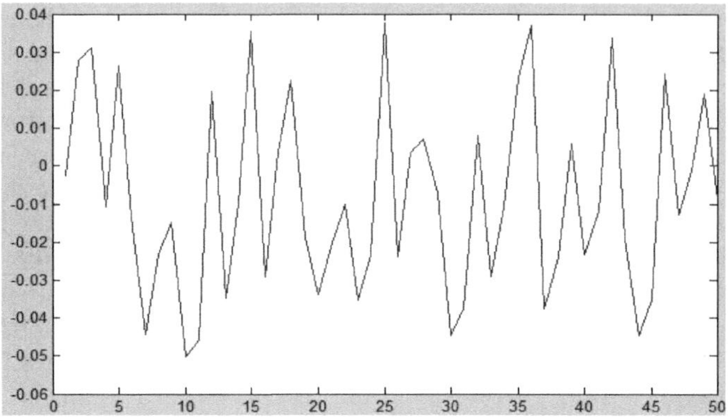

Fig. 5. Errors between experimental measuring phase values and theoretical calculating phase values of a reference sinusoidal signal

5 Conclusion

The high-precision measurement method of time-interval based on pseudo-random sampling is proposed in this paper. It is simple in theory and easy to implement. At present, an attainable measurement precision of time-interval is 16ps. A measurement precision of time-interval higher than 1ps may be obtained by improving sampling rates, enhancing the resolution of A/D or increasing the number of measurement points. In addition, the measurement precision of time-interval could also be further enhanced by ameliorating the algorithm of phase estimating.

References

1. Kalisz, J.: Review of methods for time intevral measurements with Pieoseeond resolution. J. Metrologia 41(1), 17–32 (2004)
2. Koskinen, M., Kostamovaara, J.: An Averaging Mode Time-to-Amplitude Converter with Picosecond Resolution. IEEE Transactions on Instrumentation and Measurement 42(4), 866–870 (1993)
3. Thompson, M.A., Werner, M.W.: Free Running Time to Digital Converter with 1 Nanosecond Resolution. IEEE Transactions on Nuclear Science 35(1), 184–186 (1988)
4. Santos, D.M., Flasck, J.M.: A CMOS Delay Locked Loop and Sub-Nanosecond Time-To-Digital Converter Chip. IEEE Transactions on Nuclear Science 43(3), 1717–1719 (1996)
5. Baronti, F., Fanucci, L.: On the Differential Nonlinearity of Time-to-Digital Converters Based on Delay-Locked-Loop Delay Lines. IEEE Transactions on Nuclear Science 48(6), 2424–2432 (2001)
6. Boujrada, A., Bloyet, D., Tripon, M.: A Digital TDC with a Reduced Number of Delay Line Cells. Nuclear Instruments and Methods in Physics Research, 803–812 (2002)

7. Moyer, G.C., Clements, M., Liu, W.: Precise Delay Generation Using the Vernier Technique. Electronics Letters 32(18), 1658–1659 (1996)
8. Dudek, P., Hatfield, J.V.: A High-Resolution CMOS Time-to-Digital Converter Utilizing a Vernier Delay Line. IEEE Transactions on Solid-State Circuits 35(2), 240–247 (2000)
9. Rahkonen, T.E., Kostamovaara, J.T.: The Use of Stabilized CMOS Delay Lines for the Digitization of Short Time Intervals. IEEE Journal of Solid State Circuits 28(8), 887–894 (1993)
10. Kalisz, J., Szplet, R.: Field-Programmable-Gate-Array-Based Time-to-Digital Converter with 200-ps Resolution. IEEE Transactions on Instrumentation and Measurement 46(1), 51–55 (1997)
11. Szplet, R., Kalisz, J., Szymanowski, R.: Interpolating Time Counter with 100 ps Resolution on a Single FPGA Device. IEEE Transactions on Instrumentation and Measurement 49(4), 879–883 (2000)
12. Zielinski, M., Chaberski, D., Kowalski, M., Frankowski, R., Grzelka, S.: High-resolution time-interval measuring system implemented in single FPGA device. Measurement 35(3), 311–317 (2004)
13. Zhu, X., Sun, G., Yong, S., Zhuang, Z.: A High-Precision Time Interval Measurement Method Using Phase-Estimation Algorithm. IEEE Transactions on Instrumentation and Measurement 57(11), 2670–2676 (2008)

Simulation of Logistics System with Aspect of Pallet Requirements Optimization Based on Digital Factory

Zailin Guan[1,2], Luguang Cao[3], ChuangJian Wang[1,2], Yanyan Cui[3], and Xinyu Shao[1,3]

[1] State Key Lab of Digital Manufacturing Equipment and Technology, Hubei Wuhan 430074
[2] HUST-SANY Joint Laboratory of Advanced Manufacturing Technology Key, Hubei Wuhan, 430074
[3] Huazhong University of Science and Technology, Hunan Changsha, 410079

Abstract. Digital Factory Technology helps not only to reduce the uncertainty from the planning to workshop execution, but also to increase reliability and capability of the facilities to be installed or re-built. Moreover, it keeps the production facility investment and logistics costs under control. This thesis first gives a process-analysis based assessment of the dynamic demands of pallets for component dispatching to an assembly line. Then, by using simulation with Delmia/Quest for Digital Factory, the demand forecast of pallets turns to be much more accurate. As a result, the total investment on the dispatching pallets in the new facility to be installed effectively goes down.

Keywords: Digital Factory, Shop-floor Logistics, Simulation Analysis, Material Handling, Dispatching Pallet.

1 Introduction

Nowadays, with the increasingly intensive competition in global market, how to effectively reduce the cost of the whole logistic system becomes critical element to keep core competition advantage in the target market. As one of the most commonly used tools for handling assembling units in modern production logistics operation, pallet can merge (or split) storage (or delivery) items into standard size and unified weight and volume. This makes it possible for easier transportation and better storage with some palletizing solutions, and helps to dramatically reduce the operational cost[1].

Pallet requirement planning becomes therefore an important task in a practical logistics system, yet the relevant research report on this is up to now rare to see.

Normally, pallets can be clarified into general purpose and special purpose ones. As can be see from the names, general pallets are universal tools, and special pallets are designed for materials with particular size, weight or volume[2].

Generally speaking, pallets are wildly used in storage, picking, and distribution of production and supplying items. Therefore, the discussion in this paper will focus

G. Lee (Ed.): Advances in Automation and Robotics, Vol. 1, LNEE 122, pp. 293–302.
springerlink.com

mainly on the requirement of pallets for use in the picking area, the distribution process, as well as for the storage on site along the production lines. The total pallet demands can be assessed by analyzing its requirements in every single stage. For general purpose pallets, only the demands of each separate tray can be obtained through the process analysis, and the consideration on situations of shared pallet usage are missed in this. Meanwhile, the feasibility of the results from such a process-based analysis is highly dependent on the reliable execution of the delivery process. To make the analysis more exact and comprehensive, the use of digital factory simulation techniques is proposed in this paper[3].

The rest of this paper is organized as follows. In Section 2, the method of simulation based analysis is presented for the pallet requirement planning in a logistics system, while in Section 3 the application of this proposed method in an assembly workshop is described. Finally, some conclusions are drawn in Section 4.

2 Pallet Requirement Planning Based on Digital Factory Simulation

Logistics systems normally involve complex discrete events, which make the traditional analysis methods hard to achieve an optimal solution in the design and control process. Simulation technology gives an effective way to solve this kind of problems by establishing a mathematical model for the logistics system and simulate its execution process in computer. Currently, it has been recognized as one of the most powerful tools for the design and operation of a logistics system[4].

Simulation is necessary in at least the following two stages in the planning and operation of a logistics system.

(1) Design and planning stage

Simulation technology can be used as a useful tool to the planning and design of logistics systems. It can help provide validated model for industrial system design and information system development. By comparing alternative solutions for possible improvements, it gives possibilities of making a strategic plan into feasible operation. Also, it's helpful for system analysis and training, allows for easy understanding and identification of system bottlenecks[5] even before we can run the real operation[5].

(2) Operational stage

Simulation can be run as a production planning and control auxiliary tool. It's wildly applied in virtually every operation process including system design, production planning and scheduling, material flow path optimization, as well as bottlenecks control and problem solving, etc. The establishment of connection between real-time manufacturing operation and simulated system, greatly enhances the management capacity of balancing the production rhythm, making the production management scientific, accurate and reliable, as well as improving the production efficiency, and strengthening the production and logistics management.

The digital factory technology comes with the development of simulation technology. With it, detailed simulation data can easily be implemented in dynamic real-time execution, which providing excellent means to analyze a variety of work shop related activities effects on production operations. The simulation technology based on digital factory is a powerful tool for development of enterprise logistics system, providing much more detailed and accuracy information to support simulation of design and control process.

The software Delmia/Quest which is based on digital factory as a main simulation tool, owns the capacity of schedule-driven, the main characteristics are as following: a strong 3D visual simulation capability, intelligent transportation system template, reusability, openness and rapid interactivity. The simulation strategy of the complex logistic system is not only about discrete time but also about discrete events[6].

With the help of Delmia/Quest software, the production line logistics process can be modeled in 3-D environment, on which the layout, process and logistics solutions are analyzed, validated and optimized. The specific implementation process is shown in Figure 1.

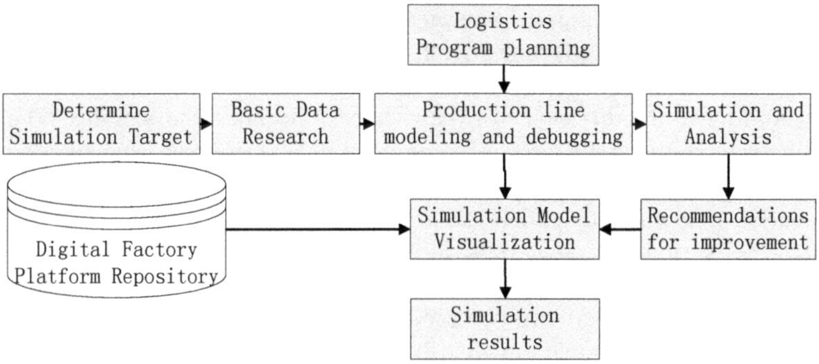

Fig. 1. Logistics Simulation Implementation Flow

(1) Through communicating with business decision-making department, we can decide logistics simulation target. We can analysis the simulation system and decide the content and limit of simulation based on simulation target[7].

(2) Research on simulation system. Obtain the production line technology, working hours and equipment operating parameters etc., and develop a preliminary program of logistics simulation.

(3) Production line modeling with Delmia/Quest. Establish physical models and the connection between each other and apply SCL language to control the element flow strategy among models.

(4) Simulation test. Delmia/Quest adopts an events calling simulation strategy based on time step. The simulation process is as follows:

1) Initialization
   a) Set simulation start time t_0, end time t_f, and simulation step h.
   b) Set the initial status of each entity.
   c) Set initial events and their occurrence time
2) Simulation clock time=t_0+h.
3) If time $\leq t_f$, determine the type of events in simulation time, and sort according to results, or end the simulation.

4) Execute events in order and generate the type and time of subsequent events.
5) Time=time+h, jump 3).
6) Simulation end.

Change the logistics transport equipments, logistics transport path, logistics vehicle and other parameters, analyze the working statuses of equipments and logistics path as well as the logistic cost and thus evaluate the logistics planning schemes.

(5) Propose the optimization scheme. Analyze and validate the programs according to the index and constraints to find the most optimal program and offer the corresponding feasible suggestions.

3 Application in a Digital Assembly Workshop

Take an assembly workshop from the engineering machinery industry for example. The workshop has a normal and an automated warehouse system (AS/RS) as supplier of materials for two assembly lines. Both assembly lines have ten work-stations and the takt-times are both sixty minutes. Materials are stored separately in two source points according to the difference of their shapes, volumes and weights. The normal flat warehouse mainly stores large size components and irregular shaped components, using special pallets; while the AS/RS stores small size and standard type components, using standard pallets. Distribution inside the workshop adopts the so-called table set system, i.e., materials in pallets of a single distribution are assembled into products exactly of the same certain number of fixed takt-time(s) in the assembly production. In our case, there are two kinds of distribution modes including one table set distribution and two table set distribution, according to the number of sets of components to be packaged in pallets of a single distribution.

Table 1. Number of pallets distributed in a single cycle

Types and sources of pallets	One table set	Two table sets
General pallet (AS/AR warehouse)	5	8
Special pallet (flat warehouse)	4	10

The number of pallets stored in the buffer areas of the warehouse mainly depends on the way the pallets are packaged, while the storage modes in the buffer area depend on the off-shelf modes; four table sets are taken off the shelves each time from the AS/RS warehouse.

In order to ensure that the assembly line has enough inventory to carry on its production, the workshop adopts a non-JIT distribution mode, in which two neighboring placing points on-site of the assemble line are demanded for each kind of pallets (or the so-called 2N mode).

3.1 The Estimation of Number of Pallets Based on Process Analysis

The process analysis method can help figure out the number of pallets in the distribution process. Because of the dynamic distribution process on which the method was based, it was called process analysis method. It involves several modes including the off-shelf mode, buffer area storage mode, distribution mode and pallet line storage mode. By analyzing the turnover number of pallets in a distribution cycle, the number of pallets will be figured out. Process analysis of the number of pallets in a workshop is based on the following premises:

(1) Carry out four table sets off-shelf mode in the AS/RS warehouse and flat warehouse, and finish the picking out from all the pallets exactly in four takt times. At the beginning of picking out, in respect of the one table set pallets, make sure there are four empty pallets in readiness in pallet allocation areas. Similarly, in respect of the two table set pallets, make sure there are two empty pallets in readiness in the pallet allocation areas.

(2) Production line has just exhausted the materials in the pallets at the same time of the end of the takt time;

(3) The picking out and pallet allocation in warehouse starts at time 0, and at time 0 of takt time, there must be a loaded pallet in line storehouse for both modes of one table set and two table sets. For the materials of two table sets, there must be a loaded pallet for distribution in the allocation area of warehouse, and for the materials of one table set, there must be four loaded pallets for distribution.

The analysis of the number of pallets for two table sets and one table set are shown in Figure 2 and Figure 3.

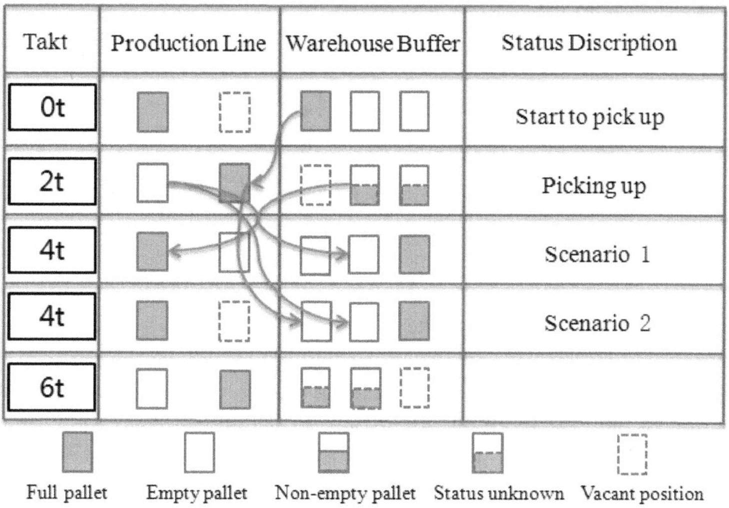

Fig. 2. Analysis demand of two sets pallets

Four table sets are taken off from the shelves every single time in the AS/RS warehouse, and two table sets pallets for one work-station require three storage places in the buffer area of the AS/RS warehouse. At the time 0 of production, there is one loaded pallet ready for using and distributing in the production line and the buffer area of the AS/RS warehouse respectively which can take two takt times, and it begins off-shelving and sorting in the AS/RS warehouse at the second takt time (2t), the loaded pallets in line storehouse have used up. At the same time a loaded pallet from the buffer is delivered to the place, and the empty one is then recycled. At 4t, there are two possibilities: Case 1: At the beginning of 4t, empty pallets in line storehouse are not recycled and taken to the buffer area of the AS/RS warehouse, where picking out works are waiting for those pallets. Therefore at this moment 5 containers are demanded to meet the requirements; Case 2: At the end of 4t, empty pallets are recycled and taken to the buffer area. At this moment only 4 of every kind of special containers are demanded, provided that: recycling time of empty pallets + allocation time + AGV distribution time + all waiting time<2t.

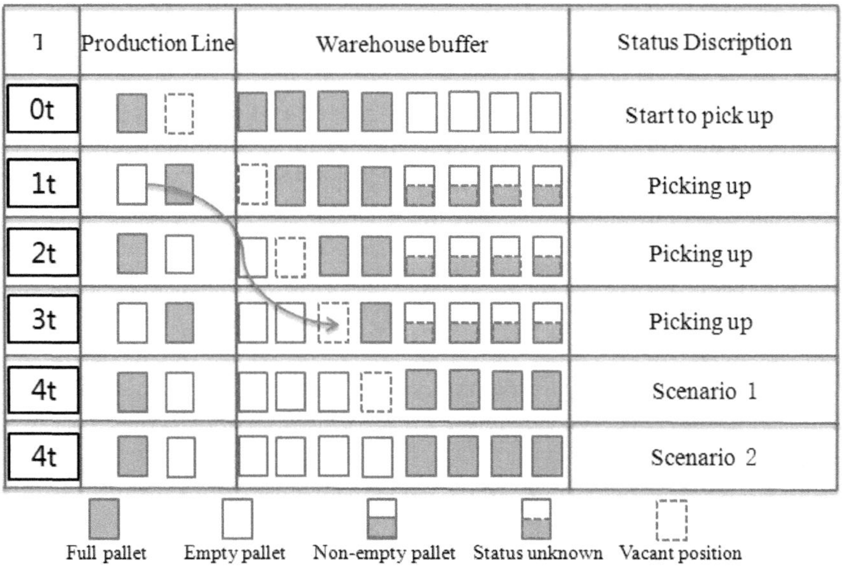

Fig. 3. Analysis demand of single set pallets

Analysis of the number of one table set pallet shown in Figure 3, and the consumption in the first three takt-times are shown above. Two situations occur at the time of 4t: Case 1: At 4t, the last exhausted pallet has been taken to the buffer area, and starts allocation. At this time, just 8 special containers for one table set are needed, provided that: recycling time of empty pallets+ pallet allocation time + AGV distribution time + all waiting time <t; Case 2: At 4 t, the last exhausted pallet has not yet been recycled and taken to the buffer area. Meanwhile, the buffer area starts the picking works for the next distribution take-time. In this case one more pallet is demanded in addition to that of case 1, i.e., 9 pallets are needed in total.

3.2 Estimation of Number of Pallets Based on Delmia / Quest Simulation Analysis

Workshop logistics and distribution is a complex discrete event process. The optimal result of the number of pallets is hard to get by using process analysis method. At the same time, there must be some waste of pallet situations happened in the preliminary analysis to determine the number of pallets, because it's a preliminary estimation limited to certain number of pallets of certain production lines, some standard pallets which can be shared in some circumstances are unable to be taken into consideration.

In consideration of the weaknesses of the process analysis method, we introduce digital factory simulation technology, to simulate the logistics process of factories, workshops and production lines in a virtual environment by using Delmia / Quest, so as to analyze accurately the number of pallets required in distribution. In order to adopt Delmia / Quest in simulating warehouse, distribution, and production line, we need to build geometric models based on the layout and process provided by the platform of digital factory. We then determine the connection between these models and its corresponding simulation parameters and establish their inherent logics of the simulation models by using SCL language.

3.2.1 Delmia / Quest Simulation Modeling

In order to be visualized, in the Delmia / Quest modeling and simulation environment, the corresponding models in the process of warehousing, distribution and production are built to observe the logistics properties and the states of distribution straight forwardly and simultaneously. According to the process and allocation of resources of the existing warehouse and production line, a simulation model of material distribution was established and shown in Figure 4.

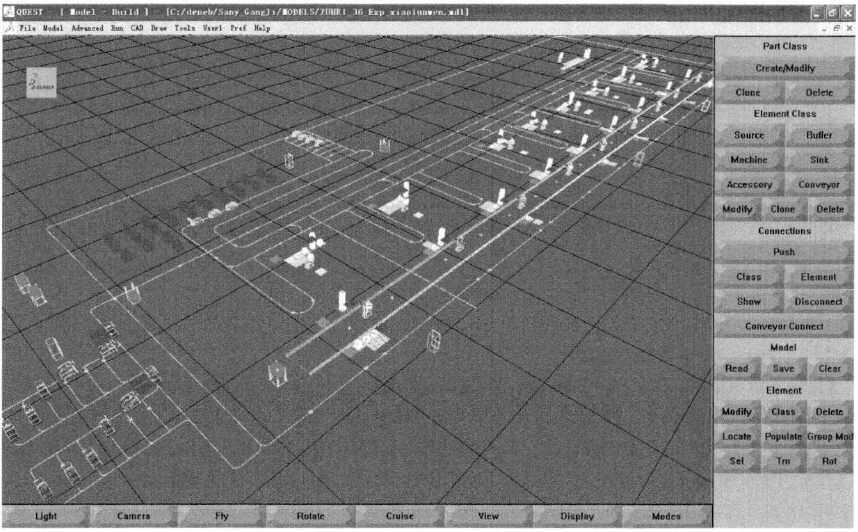

Fig. 4. Distribution simulation model of production lines

Simulation model includes warehouse delivery module, distribution module and production line module. When the basic simulation model has been built in accordance with layouts provided by the digital factory platform, we can associate the elements between the geometric models with Delmia / Quest (connection).

In order to map the operation of production system and distribution operations accurately to the models, and to ensure the trueness and accuracy of the simulation results, some production line and distribution related information and parameters are used in those geometric models. Delmia / Quest provides interactive interfaces including parameter variables of production system, through which simulation parameters can be set on the related interfaces by inputting information.

Some parameters of the simulation include: the simulation start time, simulation time step, the simulation end time, production take-time, speed and acceleration of distribution tools, and loading time. Basic parameters used in the model are as following:

(1) Production line starts at time 0, ends 240 minutes later as a distribution cycle, simulation time step is 1 second;

(2) Four table sets was used as the off-shelf mode;

(3) For the storage of buffer area of warehouse: two places are set aside for four table sets pallets; three places are set aside for two table sets pallets; eight places are set aside for one table set.

(4) T = 4 * t: T stands for the distribution takt time (240 minutes), t represents the production takt time; 4 production takt time equals a distribution takt time(240 minutes);

(5) At the initial time the state of pallets in line storage and buffer area of warehouse and the initial number of pallets in the model are shown as in theTable 2;

(6) The speed of tool distribution inside the workshop is 1.5m/s, loading and unloading time are both 1 minute.

Table 2. The initial number of pallets in the simulation models

Type of pallet	One table set	Two table sets
General pallet	8 or 9	4 or 5
Special pallet	9	5

Coefficient of special pallet: because in no circumstances special pallets are used in common, the coefficient equals the number.

Coefficient of general pallet: because general pallets are kind of pallets used in common, its coefficient equals to the number of general pallets divided by the number of types of pallets.

3.2.2 The Logic Modeling

In Delmia/Quest, decision-making about the production resources in the real production system is called logic, which controls and completes the selection and scheduling of models. In most cases, Delmia / Quest's standard logic or combination of these logics are already satisfied with the requirements. Although in certain cases, the standard logic can not define the functions we need, we can use Delmia / Quest which provides a language similar to Structured Control Language (SCL) to compile.

3.2.3 Simulation Results and Analysis

After established geometric models and logic models of the production line based on Delmia / Quest, we are able to simulate the operations of elements in real system and demonstrate it visually by computer, and get the rules of elements and index the changing with time. This helps to identify an optimal solution in terms of the number of pallets among different solutions.

The simulation runs by initializing a number of pallets in accordance with the above-mentioned data, and dynamically activating a number of distribution tools, to distribute materials according to the needs of the assembly line. In the simulation process, constantly change the number of forklifts and pallets on the premise that the distribution are timely, then results from a variety of solutions are approached as shown in Table 3.

Table 3. Comparison of simulation results of several solutions

Number of distribution tools (rate of utilization)	Coefficient of One table set general pallet	Coefficient of One table set special pallet	Coefficient of two table sets general pallet	Coefficient of two table sets special pallet
5(65%)	8.7	9	4.6	5
6(55%)	8.3	8	4.4	4
7(40%)	8	8	4.2	4

The simulation results of several solutions are shown above. Because of no sharing circumstances for special pallets, there are two fixed values existing. For general pallets, the number changes between two fixed values and decreases as the number of distribution tools increased. As the number of distribution tools increase, the utilization rates of distribution tools go down, that's because that the vacant time of pallets is fixed in line storage. So when the distribution tools reach to a certain number, the empty pallets can be recycled on time. It would not be significant to increase the number of pallets by increasing distribution tools.

For the above three solutions, it's unable to determine directly which program is the best, but we can evaluate them in terms of logistics costs. The number of distribution tools are set as N; values of distribution tools, special pallets for one table set, general pallets for one table set, special pallets and general pallets for two table are set respectively as M、M_1、M_2、M_3 ; utilization rate of distribution tools, coefficient of one table set special pallets, coefficient of one table set general pallets, coefficient of two table sets special pallets and two table sets general pallets are set respectively as φ、φ_1、φ_2、φ_3、φ_4; the type of one table set special pallets and general pallets, the type of two table set special pallets and general pallets are set respectively as L_1 、 L_2 、 L_3 、 L_4 , the logistics costs of those solutions are $M = N \times M / \varphi + \sum_{I=1}^{4} L_I \times \varphi_I \times M_I$. Input the reference values given by the company into the above formula, and we can get a result that the minimal logistics cost only occurred when the number of distribution tools is six, and that solution is the best.

4 Conclusion

Pallet as the main tool in enterprise logistics and distribution, accounts for a growing proportion in logistics costs. The estimation of the number of pallets before planning a production line can be taken as an important reference for the company managers. It is impossible to estimate accurately the number of pallets using traditional process analysis. But with the help of digital factory platform, using Delmia / Quest for the production distribution simulation, we can visually and graphically simulate the operation of production line and distribution, and estimate the number of pallets in workshops more precisely. The simulation provides a convenient, fast and effective way to analyze and reinvent the production distribution system, and is practical to be adopted.

Acknowledgments. This thesis is co-funded by the State Hi-Tech R&D Program of China (grant No. 2007AA04Z110)□ the National Nature Science Foundation of Outstanding Young Fund (No.51035001) and Key Project (No.50825503).

References

1. Xiang, Z., Zhao, Q., Yao, S.: The design of reverse logistics network in pallets recycling on the circumstance of random numbers. Logistics Technology (4) (April 2010)
2. Reckiek, B., Pierre, D.L.: Designing mixed-product assembly lines. IEEE Transactions and Automation 16(3), 268–280 (2000)
3. Fleischmann, B., Gnutzmann, S.: Dynamic vehicle routing based on online traffic information. Transportation Science 38(4), 135–148 (2004)
4. Zhu, W.-F., Fei, Q.: Modeling and Simulation of a Complex Logistics. In: Lan, H. (ed.) Proceedings of 2002 International Conference on Management Science & Engineering, Moscow, Russia., October 22-24, vol. I, pp. 270–275. Harin Institute of technology Press, Harin (2002)
5. Ji, J., Lu, J., Song, Z., Zhang, H.: The application of digital factory technology in layout planning of automotive manufacturing. The Application of Digital Factory Technology in Layout Planning of Automotive Manufacturing 31(11) (2009)
6. Jiang, L.: The research on the network planning of shipway assembly line and its simulation. Master's Thesis, Shanghai Jiao Tong University (2008)
7. Wang, W.: The research on rapid modeling and simulation technology of production lines. NanJing University of Aeronautics and Astronautics, NanJing (2007)

On the Convergence of Iterative Identification of Hammerstein Systems

Guangshe Zhao[1], Guoqi Li[2,3], Changyun Wen[2], and Feng Yang[2]

[1] School of Aerospace, Xi'an Jiaotong University, Shaanxi, Xi'an, 710049, P.R. China
zhaogs@mail.xjtu.edu.cn

[2] School of Electrical and Electronic Engineering, Nanyang Technological University, 639798, Singapore

[3] Data Storage Institute, A*STAR, 117608, Singapore
li_guoqi@dsi.a-star.edu.sg

Abstract. An iterative identification algorithm of Hammerstein systems needs a proper initial condition to guarantee its convergence. In this paper, we propose a new algorithm by fixing the norm of the parameter estimates. The normalized algorithm ensures the convergence property under arbitrary nonzero initial conditions. The proofs of the property give a geometrical explanation on why the normalization guarantees the convergence. An additional contribution is that the static function in the Hammerstein system is extended to non-odd functions.

Keywords: Hammerstein systems, Normalized Iterative identification, Parameter estimation, Convergence.

1 Introduction

One class of block-oriented nonlinear systems with a static nonlinear block followed by a linear dynamic system is called Hammerstein systems. The identification of Hammerstein systems has been extensively studied, see for examples [1]-[10].

One kind of iterative identification approach divides the unknown parameters into two sets, the linear part and the nonlinear part. At each iteration, estimates of parameters for one set are computed while the other is fixed. Then the two sets alternate and their final parameters estimates are obtained iteratively. Such an iterative algorithm was first proposed to estimate Hammerstein systems in [5]. However, its convergence needs proper initialization as pointed out in [1] [2] and [4]. Liu and Bai [4] established the convergence under a given initial condition when the static function is represented by odd basis functions. But the convergence for the case of non odd functions and arbitrary nonzero initial conditions is still an open issue as pointed out in [4]. In this paper, we propose a normalized iterative algorithm to address this issue. It will be shown that the unknown true parameters of the Hammerstein system correspond to the unique partial optimum point of a cost function under the constraint that the norm of the estimates is fixed. With this result, the proposed normalized algorithm ensures that the estimates converge to the true parameters.

G. Lee (Ed.): Advances in Automation and Robotics, Vol. 1, LNEE 122, pp. 303–312.
springerlink.com

2 Normalized Iterative Algorithm

Consider the Hammerstein system below:

$$\begin{aligned} y_t &= d_1 y_{t-1} + ... + d_n y_{t-n} + b_1 x_{t-1} + ... + b_m x_{t-m} + v_t \\ x_t &= f(u_t) = a_0 k_0(u_t) + a_1 k_1(u_t) + ... + a_l k_l(u_t) \end{aligned} \tag{1}$$

where u_t is the input signal, $f(.)$ is a nonlinear function represented by the combination of known basis functions and unknown coefficients $a_0, ..., a_l$, x_t and y_t are the input and output of the linear sub-system with known structure but unknown parameters $d_1, ..., d_n$ and $b_1, ..., b_n$, v_t denotes the noise.

Assumption 1. *Basis functions $k_i(u), i = 0, ..., l$ are orthonormal basis functions on a given interval $[-U_0, U_0]$ and $k_0(u) = 1$. In addition, the parameters of $b_1, ..., b_m$ satisfy that $\sum_{i=1}^{m} b_i \neq 0$.*

Based on (1), we get

$$\begin{aligned} y_t = d_0 + d_1 y_{t-1} + ... + d_n y_{t-n} + b_1(a_1 k_1(u_{t-1}) + ... + a_l k_l(u_t)) \\ + ... + b_m(a_1 k_1(u_{t-m}) + ... + a_l k_l(u_{t-m})) + v_t \end{aligned} \tag{2}$$

where $d_0 = a_0 \sum_{i=1}^{m} b_i$. The identification purpose is to estimate the unknown parameters $d = [d_0 \ ... \ d_n]'$, $b = [b_1 \ ... \ b_m]'$ and $a = [a_0 \ ... \ a_l]'$ in model (1) and (2) based on the observed input and output data $\{u_t, y_t\}$, $t = -r, ...0, 1, ..., N$ where $r = max(m, n)$ for sufficiently large N.

Remark 1. In this paper, we do not need the assumption that $k_i(u)$, $i = 0, ..., l$ are odd symmetrical functions assumed in [4]. Note that every set of basis functions in a given interval can always be orthonormalized. For example, polynomial basis $1, x, x^2, x^3 ...$ including the odd basis $x, x^3, ...$ as a subset basis is easy to be orthonormalized. The assumption on $\sum_{i=1}^{m} b_i \neq 0$ is to guarantee the identifiability of parameter a_0 as $d_0 = a_0 \sum_{i=1}^{m} b_i$.

Assumption 2. *Input $u_t \in [-U_0, U_0]$ and noise v_t are i.i.d random variables. In addition, $E(v_t) = 0$ and $E(v_t^2) = D(v_t) = \sigma_v^2 < \infty$, where $E(.)$ and $D(.)$ denote the expectation and variance of a random variable.*

Denote that $Y = [y_1 \ ... \ y_N]'$ and $v = [v_1 \ ... \ v_N]'$. The Hammerstein system can be rewritten as the matrix form:

$$Y = \mathcal{G}d + b_1 K_1 a + ... + b_m K_m a + v = \mathcal{G}d + (\mathcal{K} \otimes a)b + v = \mathcal{G}d + (b \cdot \mathcal{K})a + v \tag{3}$$

where $\mathcal{K} = [K_1 \ ... \ K_m]$, $b \cdot \mathcal{K} \triangleq b_1 K_1 + ... + b_m K_m$, $\mathcal{K} \otimes a \triangleq [K_1 a \ \ ... \ \ K_m a]$ and $(b \cdot \mathcal{K})a = (\mathcal{K} \otimes a)b \triangleq b_1 K_1 a + ... + b_m K_m a$, and

$$\mathcal{G} = \begin{bmatrix} 1 & y_0 & ... & y_{1-n} \\ 1 & \vdots & \vdots & \vdots \\ 1 & y_{N-1} & ... & y_{N-n} \end{bmatrix}, K_i = \begin{bmatrix} k_1(u_{1-i}) & ... & k_l(u_{1-i}) \\ \vdots & \vdots & \vdots \\ k_1(u_{N-i}) & ... & k_l(u_{N-i}) \end{bmatrix}, i = 1, ..., m \tag{4}$$

Lemma 1. *Under Assumptions 1-2, for any $\mathcal{K} \in R^{N\times ml}, ml < N$, we have $\lim_{N\to\infty} \frac{\mathcal{K}'\mathcal{K}}{N} = I$ almost surely where I is an identity matrix with dimension $ml \times ml$.*

Proof. As the basis functions $k_0(.), k_1(.), ..., k_l(.)$ are orthonormal with $k_0(.) = 1$, then $\int_{-U_0}^{U_0} k_i(u_t)k_j(u_t)du_t = \delta(i-j)$ where $\delta(.)$ 1 if and only if $i = j$, otherwise, it is zero. When $i = 0$ and $j \neq 0$, $\int_{-U_0}^{U_0} k_j(u_t)du_t = 0$, and then $k_i(u_t)$ is a zero mean variable under Assumptions 1-2. It is known that if u_t and $u_{\tilde{t}}$ $(-r \leq t \neq \tilde{t} \leq N)$ are i.i.d, then $k_j(u_t)$ and $k_j(u_{\tilde{t}})$ for $1 \leq j \leq l$ are all zero mean i.i.d variables. Thus all elements in $\mathcal{K}$ are random variables with zero mean and variance 1. So we have $\lim_{N\to\infty} \frac{\mathcal{K}'\mathcal{K}}{N} = I$ almost surely.

Remark 2. Note that almost surely means that an event occurs with probability 1. In Lemma 1, it is possible that $\mathcal{K}$ is a singular matrix in one realization for a particular sequence $\{u_t\}_{i=-r}^{N} \in [-U_0\ U_0] \otimes [-U_0\ U_0] \otimes ... \otimes [-U_0\ U_0] \subset R^{N+r+1}$ but the measure of such sequences is 0. So such a event occurs with probability 0.

Assumption 3. *$[\mathcal{G}\ \ \mathcal{K}]$ is full column rank.*

Assumption 4. *Either $\|b\|_2$ or $\|a\|_2$ is known and the first nonzero entry of b or a is positive.*

Remark 3. Assumption 3 actually refers to the requirement of the input signals. From Lemma 1, it is noted that when the input signals are i.i.d, it is not hard to guarantee linear independence of $\mathcal{G}$ and $\mathcal{K}$ as they are constructed from input and output signals, respectively. Assumption 3 also implies that, for any $b \neq 0$, $[\mathcal{G}\ \ b\cdot\mathcal{K}]$ is full column rank, and for any $a \neq 0$, $[\mathcal{G}\ \ \mathcal{K}\otimes a]$ is full column rank. Assumption 4 is to guarantee a unique expression of the Hammerstein system model, as any pair λa and b/λ for some non-zero λ provides the same input-output data. More details of illustrating such an assumption can be found in [2]. Both Assumptions 3 and 4 are related to the identifiability of the nonlinear system.

Define a cost function $J_N(.)$ as

$$J_N(\bar{a}, \bar{b}, \bar{d}) = \tfrac{1}{N}(\bar{Y} - Y)'(\bar{Y} - Y) = \tfrac{1}{N}(\bar{Y} - Y^* - v)'(\bar{Y} - Y^* - v) \tag{5}$$

where $\bar{Y} = \mathcal{G}\bar{d} + \bar{b}_1 K_1\bar{a} + ... + \bar{b}_m K_m\bar{a}$ and $Y^* = \mathcal{G}d + b_1K_1a + ... + b_mK_ma$. The estimates $\hat{a}, \hat{d}, \hat{b}$ of a, b and d are determined by minimizing the cost function (5). Denote that $J(\bar{a}, \bar{b}, \bar{d}) = \lim_{N\to\infty} J_N(\bar{a}, \bar{b}, \bar{d})$. Since the noise is independent of the system output and $(\bar{b}\cdot\mathcal{K})\bar{a} = \bar{b}_1K_1\bar{a} + ... + \bar{b}_mK_m\bar{a}$, then

$$\begin{aligned} J(\bar{a}, \bar{b}, \bar{d}) &= \lim_{N\to\infty} \tfrac{1}{N}\|\bar{Y} - Y^*\|_2 + \sigma_v^2 \\ &= \lim_{N\to\infty} \tfrac{1}{N}\|\mathcal{G}(\bar{d} - d) + (\bar{b}\cdot\mathcal{K})\bar{a} - (b\cdot\mathcal{K})a\|_2 + \sigma_v^2 \end{aligned} \tag{6}$$

We now obtain $\hat{d}$ without knowing a and b.

$$\begin{aligned} Y &= \mathcal{G}d + b_1K_1a + ... + b_mK_ma + v \\ &= \mathcal{G}d + [K_1\ ...\ K_m]\begin{bmatrix} b_1a \\ \vdots \\ b_ma \end{bmatrix} + v = \mathcal{G}d + \mathcal{K}\gamma + v \end{aligned} \tag{7}$$

where $\gamma' = [\,(b_1 a)' \ ... \ (b_m a)'\,]$. Let $P_{\mathcal{K}} = \mathcal{K}\mathcal{K}^+$ and $P_{\mathcal{G}} = \mathcal{G}\mathcal{G}^+$ denote projection operators onto span$\{\mathcal{K}\}$ and span$\{\mathcal{G}\}$, respectively, where span$\{.\}$ is the space spanned by the column vectors of a matrix and $\mathcal{K}^+ = (\mathcal{K}'\mathcal{K})^{-1}\mathcal{K}'$ is the generalized matrix inverse. From (7), we obtain $P_{\mathcal{G}}\mathcal{G}d = P_{\mathcal{G}}(Y - \mathcal{K}\gamma) - P_{\mathcal{G}}v$ and $P_{\mathcal{K}}\mathcal{K}\gamma = P_{\mathcal{K}}(Y - \mathcal{G}d) - P_{\mathcal{K}}v$. Note that the noise space is independent of space $span\{\mathcal{K}\} \cup span\{\mathcal{G}\}$ with $E(v) = \{0\}$ based on Assumption 2. Thus the noise space is orthogonal to $span\{\mathcal{K}\} \cup span\{\mathcal{G}\}$. And as $P_{\mathcal{G}}v$ and $P_{\mathcal{K}}v$ are operators projecting the noise to the space $span\{\mathcal{G}\}$ and $span\{\mathcal{K}\}$, we have $P_{\mathcal{G}}v = 0$ and $P_{\mathcal{K}}v = 0$. As $P_{\mathcal{G}}\mathcal{G} = \mathcal{G}\mathcal{G}^+\mathcal{G} = \mathcal{G}$ and $P_{\mathcal{K}}\mathcal{K} = \mathcal{K}\mathcal{K}^+\mathcal{K} = \mathcal{K}$. Thus, it can be obtained that

$$\mathcal{G}d = P_{\mathcal{G}}(Y - \mathcal{K}\gamma) \tag{8}$$

$$\mathcal{K}\gamma = P_{\mathcal{K}}(Y - \mathcal{G}d) \tag{9}$$

From (8) and (9), it is easy to establish that $\mathcal{G}d = P_{\mathcal{G}}(Y - P_{\mathcal{K}}(Y - \mathcal{G}d))$, which gives $(I - P_{\mathcal{G}}P_{\mathcal{K}})\mathcal{G}d = P_{\mathcal{G}}(I - P_{\mathcal{K}})Y$. Let $\hat{d}$ be the estimate of d. Then

$$\hat{d} = H_{\mathcal{G}}^{+} P_{\mathcal{G}}(I - P_{\mathcal{K}})Y \tag{10}$$

where $H_{\mathcal{G}} = (I - P_{\mathcal{G}}P_{\mathcal{K}})\mathcal{G}$. After $\hat{d}$ is obtained, the cost function $J_N(\bar{a}, \bar{b}, \bar{d})$ in (5) becomes $J_N(\bar{a}, \bar{b}, \hat{d})$.

Definition 1. *A point $(a^\star, b^\star)$ is a partial optimum of $J_N(\bar{a}, \bar{b}, d)$ if $J_N(a^\star, b^\star, d) \le J_N(a, b^\star, d)$ when $b^\star$ is fixed and $J_N(a^\star, b^\star, d) \le J_N(a^\star, b, d)$ when $a^\star$ is fixed.*

Now we are at the position to present our normalized iterative algorithm as follows with k denoting the $k-$th iteration. Based on Assumption 2, we fix the norm $\|\hat{b}\|_2 = \|b\|_2$ in the iteration.

Step 1: Obtain estimates $\hat{d}$ by using (10) and choose an arbitrary nonzero initial value $\hat{b}(0)$.

Step 2: Solve the problem $\hat{a}(k) = argmin_{\bar{a}} J_N(\bar{a}, \hat{b}(k-1), \hat{d})$.

Step 3: Find $\hat{b}_{op}(k) = argmin_{\bar{b}} J_N(\hat{a}(k), \bar{b}, \hat{d})$ and let $\hat{b}(k) = \|b\|_2 \hat{b}_{op}(k)/\|\hat{b}_{op}(k)\|_2$. Clearly, $\|\hat{b}(k)\|_2 = \|b\|_2$.

Step 4: If a stopping criterion is satisfied, end. Otherwise, replace k by $k+1$ and go back to Step 2.

Remark 4. 1). The order of Step 2 and Step 3 can be permuted. With permutation, we start the estimation with nonzero initial value $\hat{a}(0)$. 2). There are several ways to define the stopping criterion in Step 4 of the algorithm. For example, one can consider the difference of cost function values $J_N(k)$ and $J_N(k-1)$ since $J_N(k)$ is a decreasing sequence. or the absolute value of the difference between $(\hat{a}(k), \hat{b}(k))$ and $(\hat{a}(k-1), \hat{b}(k-1))$.

3 Convergence Analysis of the Iterative Algorithm

We first prove that $\lim_{N\to\infty}\hat{d}=d$. Then, it is shown that the true parameters correspond to the unique partial optimum point of the cost function and we obtain the convergence of the normalized iterative algorithm.

Lemma 2. *Under Assumption 3, matrix $(I-P_{\mathcal{G}}P_{\mathcal{K}})$ is full rank.*

Proof. From Assumption 3, $\mathcal{K}$ and $\mathcal{G}$ do not have joint column space, i.e, $\text{span}\{\mathcal{K}\}\cap\text{span}\{\mathcal{G}\}=\{0\}$. Assume that $(I-P_{\mathcal{G}}P_{\mathcal{K}})$ is not full rank, then $\exists x\neq 0$ such that $(I-P_{\mathcal{G}}P_{\mathcal{K}})x=0$, i.e, $P_{\mathcal{G}}P_{\mathcal{K}}x=x$. As $P_{\mathcal{K}}$ and $P_{\mathcal{G}}$ are projection operators onto $\text{span}\{\mathcal{K}\}$ and $\text{span}\{\mathcal{G}\}$ respectively, we have $x\in\text{span}\{\mathcal{K}\}\cap\text{span}\{\mathcal{G}\}=\{0\}$, and this contradicts that $x\neq 0$. So this lemma holds.

Theorem 1. *Under Assumptions 1-3, the estimate $\hat{d}$ given in (10) satisfies that $\lim_{N\to\infty}\hat{d}=d$ almost surely.*

Proof. Under Assumptions 1-3, matrix $\mathcal{G}$ is full column rank, then

$$\begin{aligned}\hat{d}&=[(I-P_{\mathcal{G}}P_{\mathcal{K}})\mathcal{G}]^{+}P_{\mathcal{G}}(I-P_{\mathcal{K}})Y\\&=[(I-P_{\mathcal{G}}P_{\mathcal{K}})\mathcal{G}]^{+}P_{\mathcal{G}}[I-\mathcal{K}(\mathcal{K}^{'}\mathcal{K})^{-1}\mathcal{K}^{'}](\mathcal{G}d+\mathcal{K}\gamma+v)\\&=[(I-P_{\mathcal{G}}P_{\mathcal{K}})\mathcal{G}]^{+}P_{\mathcal{G}}(I-P_{\mathcal{K}})[\mathcal{G}d+(\mathcal{K}-\mathcal{K}(\mathcal{K}^{'}\mathcal{K})^{-1}\mathcal{K}^{'}\mathcal{K})\gamma+v]\\&=[(I-P_{\mathcal{G}}P_{\mathcal{K}})\mathcal{G}]^{+}P_{\mathcal{G}}(I-P_{\mathcal{K}})(\mathcal{G}d+v)=A\mathcal{G}d+Av\end{aligned}$$

where $A=[(I-P_{\mathcal{G}}P_{\mathcal{K}})\mathcal{G}]^{+}P_{\mathcal{G}}(I-P_{\mathcal{K}})$. From Lemma 2, $(I-P_{\mathcal{G}}P_{\mathcal{K}})$ is full column rank, and then

$$\begin{aligned}A\mathcal{G}&=[(I-P_{\mathcal{G}}P_{\mathcal{K}})\mathcal{G}]^{+}[\mathcal{G}(\mathcal{G}^{'}\mathcal{G})^{-1}\mathcal{G}^{'}(I-\mathcal{K}(\mathcal{K}^{'}\mathcal{K})^{-1}\mathcal{K}^{'})\mathcal{G}]\\&=[(I-P_{\mathcal{G}}P_{\mathcal{K}})\mathcal{G}]^{+}[\mathcal{G}-\mathcal{G}(\mathcal{G}^{'}\mathcal{G})^{-1}\mathcal{G}^{'}\mathcal{K}(\mathcal{K}^{'}\mathcal{K})^{-1}\mathcal{K}^{'}\mathcal{G}]\\&=[(I-P_{\mathcal{G}}P_{\mathcal{K}})\mathcal{G}]^{+}[(I-P_{\mathcal{G}}P_{\mathcal{K}})\mathcal{G}]=I\end{aligned}$$

So we have $A=\mathcal{G}^{+}$. Thus $\hat{d}=A\mathcal{G}d+Av=d+\mathcal{G}^{+}v=d+(\mathcal{G}^{'}\mathcal{G})^{-1}\mathcal{G}^{'}v$, $E((\hat{d}-d)'(\hat{d}-d))=(\mathcal{G}'\mathcal{G})^{-1}\sigma_v^2$, which gives $\sum_i E((\hat{d}_i-d_i)'(\hat{d}_i-d_i))=trace((\mathcal{G}'\mathcal{G})^{-1})\sigma_v^2$ where $trace(.)$ is the trace of a matrix. Now we show that $\lim_{N\to\infty}\hat{d}=d$ almost surely. Let $\mathcal{G}_N$ denote a matrix $\mathcal{G}$ with dimension $N\times(n+1)$. Note that $\mathcal{G}_N^{'}\mathcal{G}_N$ is a symmetrical positive definite matrix as $\mathcal{G}_N$ is full column rank. Let g_N' be the $N-$th row vector of $\mathcal{K}_N$, then

$$\mathcal{G}_{N+1}'\mathcal{G}_{N+1}=[1\ \ \mathcal{G}_N'\ \ g_{N+1}]\begin{bmatrix}\mathcal{G}_N\\ g_{N+1}^{'}\end{bmatrix}=\mathcal{G}_N'\mathcal{G}_N+g_{N+1}g_{N+1}'. \tag{11}$$

Let $\lambda_i(N)$, $i=0,...,n$, denote the eigenvalues of $\mathcal{G}_N^{'}\mathcal{G}_N$ and assume that $\lambda_0(N)>...>\lambda_n(N)>0$. There exists a matrix P_N such that

$$P_N'\mathcal{G}_N^{'}\mathcal{G}_NP_N=diag[\lambda_0(N)\ ...\ \lambda_n(N)]. \tag{12}$$

Let $\alpha=[1\ 0\ ...\ 0]'$, then $\alpha'P_N'\mathcal{G}_N'\mathcal{G}_NP_N\alpha'=(P_N\alpha)'\mathcal{G}_N'\mathcal{G}_N(P_N\alpha)=\lambda_0(N)$ and

$$\lambda_0(N+1)=\alpha'P_N'\mathcal{G}_{N+1}'\mathcal{G}_{N+1}P_N\alpha=\lambda_0(N)+((P_N\alpha)'g_{N+1})^2 \tag{13}$$

Let $\lambda_0^*(N) = \lambda_0(N+1) - \lambda_0(N) = ((P_N\alpha)' g_{N+1})^2$. Note that $\lambda_0^*(N) = ((P_l\alpha)' g_{N+1})^2$ is positive almost surely. There exists a constant c such that the probability $p(\lambda_0^*(N) > c)$ is nonzero for all N. Thus we obtain $\lim_{N\to\infty} \lambda_1(N) \to \infty$ almost surely. Similarly, $\lim_{N\to\infty} \lambda_i(N) \to \infty$, for $i = 1, ..., n$ almost surely. Therefore,

$$\lim_{N\to\infty} tr((\mathcal{G}_N'\mathcal{G}_N)^{-1}) = \sum_{i=1}^{M} \frac{1}{\lambda_i(N)} = 0 \qquad (14)$$

Then $\lim_{N\to\infty} \sum_i E((\hat{d}_i - d_i)'(\hat{d}_i - d_i)) = \lim_{N\to\infty} \sigma_v^2 trace((\mathcal{G}_N'\mathcal{G}_N)^{-1}) = 0$ almost surely. Thus, $\lim_{N\to\infty} \hat{d} = d$ almost surely and this lemma holds.

As $Y = \mathcal{G}d + (b\cdot\mathcal{K})a + v$, we have $lim_{N\to\infty}\frac{1}{N}a'(b\cdot\mathcal{K})'(b\cdot\mathcal{K})a = \|b\|_2^2\|a\|_2^2 \leq lim_{N\to\infty}\frac{1}{N}\|Y - \mathcal{G}d\|_2^2 + \sigma_v^2 = C$ where C is a constant. So if $\|\bar{b}\|_2$ is fixed in $J(\bar{a}, \bar{b}, d)$, then $\|\bar{a}\|_2$ is bounded. Define the domain $D = \{(\bar{a}, \bar{b})|\ \|\bar{b}\|_2 = \|b\|_2, \|\bar{a}\|_2 \leq M\}$ where M is a constant denoting the bound of $\|\bar{a}\|_2$.

Lemma 3. *Under Assumptions 1-4, $J(\bar{a}, \bar{b}, d)$ has a unique partial minimum point (a, b, d) in the domain $D = \{(\bar{a}, \bar{b})|\ \|\bar{b}\|_2 = \|b\|_2, \|\bar{a}\|_2 \leq M\}$.*

Proof. We first prove that (a, b, d) is a partial optimum point of $J(\bar{a}, \bar{b}, d)$. From (6), we get $J(a + \Delta a, b + \Delta b, d) = \sigma_v^2 + \frac{1}{N}\|\Delta Y\|_2 \geq \lim_{N\to\infty} J_N(a, b, d) = \sigma_v^2$ for any Δa and Δb. This shows that (a, b, d) is a global minimum point and of course a partial optimum point.

Now we prove the uniqueness of the partial optimum point (a, b, d) by contraction. Assume that $(\tilde{a}, \tilde{b}, d)$ is a partial optimum point in D with $\tilde{a} \neq a$ or $\tilde{b} \neq b$. Let $N_{\tilde{a}}$ and $N_{\tilde{b}}$ denote the neighborhoods of $(\tilde{a}, \tilde{b})$ when either $\tilde{b}$ or $\tilde{a}$ is fixed, respectively, and ρ_1, ρ_2 be the respective radii of $N_{\tilde{a}}$ and $N_{\tilde{b}}$. We have $N_{\tilde{a}} = (\tilde{a} + \Delta a, \tilde{b})$ where $\Delta a = \bar{\bar{a}} - a = \rho_1\frac{\bar{\bar{a}} - \tilde{a}}{\|\bar{\bar{a}} - \tilde{a}\|_2}$. Figure 1 gives a geometrical illustration of the neighborhood $N_{\tilde{a}}$ when $\tilde{b}$ is fixed. Similarly, $N_{\tilde{b}} = (\tilde{a}, \tilde{b} + \Delta b)$ where $\Delta b = \bar{\bar{b}} - b = \rho_2\frac{\bar{\bar{b}} - \tilde{b}}{\|\bar{\bar{b}} - \tilde{b}\|_2}$ when $\tilde{a}$ is fixed. Assume that θ_1, θ_2 are the angles between Δa and $a - \tilde{a}$, Δa and $\tilde{a}$; θ_3, θ_4 are the angles between Δb and $b - \tilde{b}$, Δb and $\tilde{b}$, respectively. Assume the clockwise as the positive direction of the angles. From Figure 1, it is easy to establish that

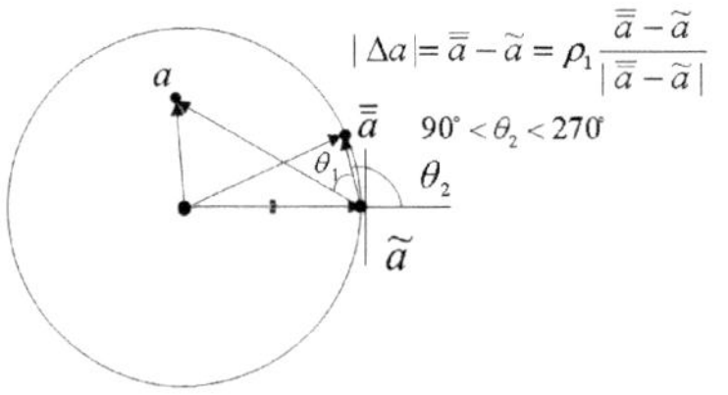

Fig. 1. The geometrical illustration of the neighborhood of $\tilde{a}$ when $\tilde{b}$ is fixed

$$\Delta a\cdot(a - \tilde{a}) = \rho_1(a - \tilde{a})'(a - \tilde{a})cos\theta_1 = \rho_1\|a - \tilde{a}\|_2^2 cos\theta_1,$$
$$\Delta a\cdot\tilde{a} = \tilde{a}'(\bar{\bar{a}} - \tilde{a}) = \rho_1\|\tilde{a}\|_2 cos\theta_2$$

Similarly when $\tilde{a}$ is fixed, we have

$$\begin{aligned} \Delta b \cdot (b - \tilde{b}) &= \rho_2 (b - \tilde{b})'(b - \tilde{b}) cos\theta_3 = \rho_2 \|b - \tilde{b}\|_2^2 cos\theta_3, \\ \Delta b \cdot \tilde{b} &= \tilde{b}'(\bar{\bar{b}} - \tilde{b}) = \rho_2 \|\tilde{b}\|_2 cos\theta_4 \end{aligned}$$

Note that

$$\begin{aligned} & J(\tilde{a} + \Delta a, \tilde{b} + \Delta b, d) \\ &= \lim_{N\to\infty} \tfrac{1}{N} \|\tilde{Y} + \Delta Y - Y^*\|_2^2 + \sigma_v^2 \\ &= \lim_{N\to\infty} (\tilde{J}_N(\tilde{a}, \tilde{b}, d) + \tfrac{1}{N} \|\Delta Y\|_2^2 - \tfrac{2}{N} (Y^* - \tilde{Y})' \Delta Y) + \sigma_v^2 \end{aligned} \tag{15}$$

where $\tilde{Y} = \mathcal{G}d + (\tilde{b} \cdot \mathcal{K})\tilde{a}$, $Y^* = \mathcal{G}d + (b \cdot \mathcal{K})a$ and $\Delta Y = (\tilde{b} + \Delta b \cdot \mathcal{K})(\tilde{a} + \Delta a) - (\tilde{b} \cdot \mathcal{K})\tilde{a} = (\tilde{b} \cdot \mathcal{K})\Delta a + (\Delta b \cdot \mathcal{K})\tilde{a} + (\Delta b \cdot \mathcal{K})\Delta a$ and $Y^* - \tilde{Y} = (b \cdot \mathcal{K})a - (\tilde{b} \cdot \mathcal{K})\tilde{a} = ((b - \tilde{b}) \cdot \mathcal{K})(a - \tilde{a}) + (\tilde{b} \cdot \mathcal{K})(a - \tilde{a}) + ((b - \tilde{b}) \cdot \mathcal{K})\tilde{a}$ From Lemma 1, we have

$$\lim_{N\to\infty} x' \frac{(b \cdot \mathcal{K})'(\tilde{b} \cdot \mathcal{K})}{N} y = b'\tilde{b}x'y \tag{16}$$

Based on (16), we get $\lim_{N\to\infty} \frac{1}{N} \|\Delta Y\|_2^2 = \lim_{N\to\infty} \frac{1}{N} (\Delta Y)'(\Delta Y) = s_1 + s_2 + s_3 + s_4$ where

$$\begin{aligned} s_1 &= \rho_1^2 \tilde{b}'\tilde{b} + \rho_2^2 \tilde{a}'\tilde{a} + 2\rho_1\rho_2 \|\tilde{a}\|_2 \|\tilde{b}\|_2 cos\theta_2 cos\theta_4 \\ s_2 &= \lim_{N\to\infty} \tfrac{1}{N} ((\Delta b \cdot \mathcal{K})\Delta a)'((\Delta b \cdot \mathcal{K})\Delta a) = \rho_1^2 \rho_2^2 \\ s_3 &= \lim_{N\to\infty} \tfrac{1}{N} ((\tilde{b} \cdot \mathcal{K})\Delta a)' ((\Delta b \cdot \mathcal{K})\Delta a) = \rho_1^2 \rho_2 cos\theta_4 \|\tilde{b}\|_2 \\ s_4 &= \lim_{N\to\infty} \tfrac{1}{N} ((\Delta b \cdot \mathcal{K})\tilde{a})' ((\Delta b \cdot \mathcal{K})\Delta a) = \rho_1 \rho_2^2 cos\theta_2 \|\tilde{a}\|_2 \end{aligned}$$

and $\lim_{N\to\infty} \frac{2}{N} (Y^* - \tilde{Y})' \Delta Y = s_5 + s_6 + s_7 + s_8$ where

$$\begin{aligned} s_5 &= 2\rho_1\rho_2 cos\theta_1 cos\theta_3 \|a - \tilde{a}\|_2^2 \|b - \tilde{b}\|_2^2 \\ s_6 &= 2(\rho_1 cos\theta_1 \|a - \tilde{a}\|_2^2 \|\tilde{b}\|_2^2 + \rho_2 cos\theta_3 \|\tilde{a}\|_2^2 \|b - \tilde{b}\|_2^2 \\ &\quad + \rho_1 cos\theta_2 \tilde{b}'(b - \tilde{b}) \|\tilde{a}\|_2 + \rho_2 cos\theta_4 \|\tilde{b}\|_2 \tilde{a}'(a - \tilde{a})) \\ s_7 &= 2\rho_1\rho_2 cos\theta_1 cos\theta_4 \|\tilde{b}\|_2 \|a - \tilde{a}\|_2^2 \\ s_8 &= 2\rho_1\rho_2 cos\theta_2 cos\theta_3 \|b - \tilde{b}\|_2^2 \|\tilde{a}\|_2 \end{aligned}$$

Since only s_6 includes the first order terms of ρ_1 and ρ_2, when $\rho_1 \to 0, \rho_2 \to 0$, we have $[(s_1 + s_2 + s_3 + s_4) - (s_5 + s_6 + s_7 + s_8)] \to -s_6$ and thus (15) becomes

$$\begin{aligned} & J(\tilde{a} + \Delta a, \tilde{b} + \Delta b, d) \\ &= \tilde{J}(\tilde{a}, \tilde{b}, d) + (s_1 + s_2 + s_3 + s_4) - (s_5 + s_6 + s_7 + s_8) \to \tilde{J}(\tilde{a}, \tilde{b}, d) - s_6 \end{aligned} \tag{17}$$

When $\tilde{b}$ is fixed, $\rho_2 = 0$ and $s_6 = 2\rho_1(cos\theta_1 |a - \tilde{a}|^2 |\tilde{b}|^2 + cos\theta_2 \tilde{b}'(b - \tilde{b})|\tilde{a}|)$. Note that $\tilde{b}'(b - \tilde{b}) = \tilde{b}'b - \tilde{b}'\tilde{b} \leq 0$ under $\|b\|_2 = \|\tilde{b}\|_2$. Seen from Figure 1, there always exists an $\bar{\bar{a}}$ such that $-90° < \theta_1 < 90°$ $(cos\theta_1 > 0)$ and $90° < \theta_2 < 270°$ $(cos\theta_2 < 0)$ in $N_{\tilde{a}}$. Thus we have $s_6 > 0$ and $J(\tilde{a} + \Delta a, \tilde{b}, d) < \tilde{J}(\tilde{a}, \tilde{b}, d)$ which means $(\tilde{a}, \tilde{b}, d)$ cannot be a partial optimum point in this case. Similar conclusion can be obtained when $\tilde{a}$ is fixed. So as long as the point $(\tilde{a}, \tilde{b}, d)$ is different from (a, b, d), we can always find certain points in its corresponding neighborhood $N_{\tilde{a}}$ or $N_{\tilde{b}}$ with smaller cost function values. This contradicts with the assumption that $(\tilde{a}, \tilde{b}, d)$ is a partial optimum point. Therefore the conclusion of this lemma holds.

Remark 5. As mentioned in Remark 2, estimating a and b can be permuted. Define $\tilde{D} = \{(\bar{a}, \bar{b}) | \ \|\bar{a}\|_2 = \|a\|_2, \|\bar{b}\|_2 \leq \tilde{M}\}$ where $\tilde{M}$ is a constant denoting the bound of $\|\bar{b}\|_2$. Then under Assumptions 1-4, the cost function $J(\bar{a}, \bar{b}, d)$ has a unique partial minimum point (a, b, d) in the domain $\tilde{D}$.

Theorem 2. *Under Assumptions 1-4, we have* $\lim_{N\to\infty, k\to\infty} \hat{a}(k) = a$ *and* $\lim_{N\to\infty, k\to\infty} \hat{b}(k) = b$ *almost surely.*

Proof. Let $\hat{z}(k) = (\hat{a}(k),\ \hat{b}(k),\ \hat{d}(k))$. Employing the proposed normalized iterative algorithm, we obtain $\lim_{N\to\infty} \hat{d} = d$ almost surely in Theorem 1 in one iteration step. Since sequence $\{J(k)\}$ is positive and decreasing, it is convergent. Then the generated sequence $\{\hat{z}(k)\}$ has at least one accumulation point denoted as $z^* = (a^*,\ b^*, d)$. By Lemma 3, z^* will be the unique partial optimum point if either $\|\hat{b}(k)\|_2$ or $\|\hat{a}(k)\|_2$ is fixed. Thus, $\lim_{N\to\infty, k\to\infty} \hat{a}(k) = a^* = a$ and $\lim_{N\to\infty, k\to\infty} \hat{b}(k) = b^* = b$ almost surely for arbitrary nonzero initial conditions.

Remark 6. In employing the iterative algorithm, if the norm $\|\hat{a}\|_2$ or $\|\hat{b}\|_2$ is not fixed to be $\|a\|_2$ or $\|b\|_2$, the iteration sequence may diverge as explained below. Let $\tilde{a}$ and $\tilde{b}$ denote the current estimates of a and b at the k-th iteration. Assume that $\tilde{b}'(b - \tilde{b}) > 0$, which is possible when $\|\tilde{b}\|_2 < \|b\|_2$. Also seen in Figure 1, we could choose an $\bar{\tilde{a}}$ locating outside the sphere such that $90° < \theta_1 < 270°$ ($cos\theta_1 < 0$) and $-90° < \theta_2 < 90°$ ($cos\theta_2 > 0$). Then $cos\theta_2\tilde{b}'(b - \tilde{b})\|\tilde{a}\|_2 > -cos\theta_1\|a - \tilde{a}\|_2^2\|\tilde{b}\|_2^2$, i.e, $s_6 > 0$. The case that $cos\theta_1 > 0$ means the iteration point moves toward a, while $cos\theta_1 < 0$ corresponds to that the iteration point moves away from a. This implies that there also exist certain directions along which the iteration point moves away from true parameter a while the cost function decreases. Thus, if the norm of $\hat{a}$ is not fixed, the sequence $\{(\hat{a}, \hat{b}, d)\}$ will move to $(\infty, 0, d)$. Similarly, $(0, \infty, d)$ could also be a accumulation point. This explains why a counterexample could be provided by Stoica in [5] and also explains why we need to fix the norm of $\hat{a}$ or $\hat{b}$.

4 Illustrative Examples

Example 1. In this example, we consider the following Hammerstein system given in [4] for the purpose of comparison with existing techniques in the area.

$$\begin{aligned} y_t &= 0.3y_{t-1} + 0.2y_{t-2} + 0.1y_{t-3} + 3x_t - 2x_{t-1} + v_t \\ x_t &= 0.9454u_t + 0.2983u_t^3 \end{aligned}$$

By orthonormalizing the odd function basis u and u^3 with coefficients 0.9545 and 0.2983 to orthonormal polynomials (Legendre polynomials) given as 1, u, $\frac{1}{2}(3u^2 - 1))$ and $\frac{1}{2}(5u^3 - 3u))$ with coefficients 0, 1.1335, 0 and 0.1193, the true parameters of the Hammerstein system in the form of (2) are $a = [a_0\ a_1\ a_2\ a_3] = [0\ 1.1335\ 0\ 0.1193]'$, $b = [b_1\ b_2] = [\ 3\ -2]'$ and $d = [d_0\ d_1\ d_2\ d_3] = [0\ 0.3\ 0.2\ 0.1]'$.

The input u_t is i.i.d in the interval $[-1, 1]$ and the noise v_t is zero mean with standard derivation 0.1. Let $\|b\|_2 = \sqrt{13}$ and consider the l_2 norm of the estimation error $\|e\|_2 = \sqrt{e'e}$ where $e = (\hat{a}, \hat{b}, \hat{d}) - (a, b, d)$.

The comparison results are shown in Figure 2 in which $\|e\|_2$ with respect to the number of data points N is plotted. It can be observed that both methods perform very well. Note that the initial value $\hat{b}(0)$ is randomly given in our method and it is $\hat{b}(0) = [1\ 0\ 0\ ...]'$ in [4].

Example 2. In our proposed algorithm, the nonlinear function is not necessary to be odd even when the linear system is IIR . To illustrate this, we consider the Hammerstein system:

$$\begin{aligned} y_t &= 0.4y_{t-1} + 0.1y_{t-2} + 0.5x_t + 0.3x_{t-1} + 0.2x_{t-2} + v_t \\ x_t &= 0.1 + 0.6u_t + 0.3u_t^2 \end{aligned}$$

where $u_t \in [-1, 1]$ and v_t is white noise with zero mean and standard derivation 0.3. By orthonormalizing the function basis 1, u and u^3 to Legendre polynomials given as $k_1(u) = 1$, $k_2(u) = u$ and $k_3(u) = \frac{1}{2}(3u^2 - 1)$ on the interval $[-1, 1]$, we get $x_t = 0.2k_1(u_t) + 0.6k_2(u_t) + 0.2k_3(u_t)$. Then this Hammerstein system can be rewritten as

$$\begin{aligned} y_t &= 0.4y_{t-1} + 0.1y_{t-2} + 0.5x_t + 0.3x_{t-1} + 0.2x_{t-2} + v_t \\ x_t &= 0.2k_1(u_t) + 0.6k_2(u_t) + 0.2k_3(u_t) \end{aligned}$$

and true parameters to be estimated are $a = [0.2\ 0.6\ 0.2]$, $d = [0\ 0.4\ 0.1]$ and $b = [0.5\ 0.3\ 0.2]$.

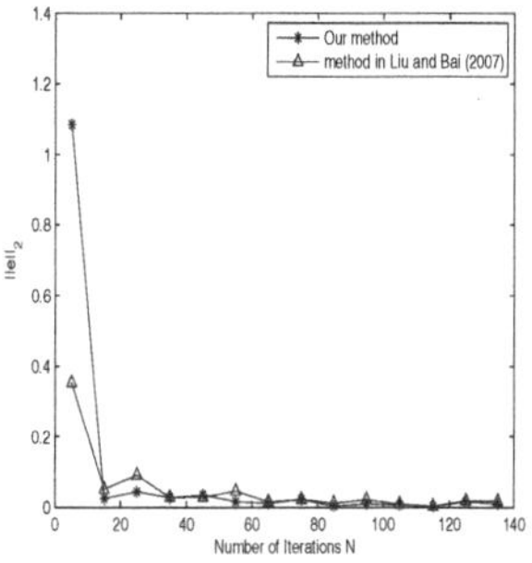

Fig. 2. Estimation error with respect to the number of data points N

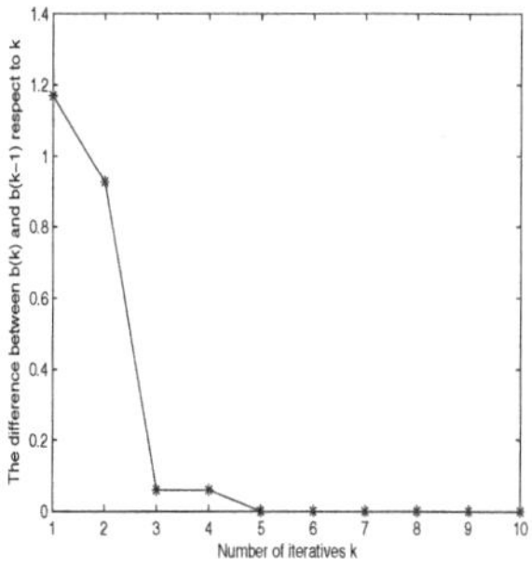

Fig. 3. The illustration that fixed point iteration algorithm converges in a few iterations

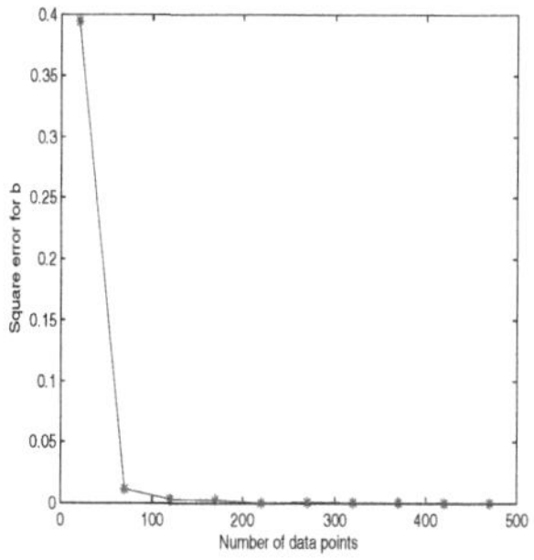

Fig. 4. Estimation error with respect to number of data points

We choose $N = 500$ and fix $\|a\|_1 = |a_1| + |a_2| + |a_3| = 1$ with $a_1 > 0$. Note that both $\|.\|_1$ and $\|.\|_2$ can be used to fix the norm of the parameters vector. The ratio of the standard derivation of the noise v_t to that of the output y_t is $\frac{\sigma_v}{\sigma_y} =$ 66%, which shows a high noise level. By implementing the proposed iterative algorithm, we obtain $\hat{a} = [0.1990\ 0.5980\ 0.2030]$, $\hat{d} = [0.0001\ 0.4110\ 0.0922]$ and $\hat{b} = [0.5015\ 0.2988\ 0.1997]$. Obviously, the estimates are very close to the true values. To show how the estimates converges to the fixed point with respect to the number of iterations, we calculate the difference $\sum_{i=1}^{m} |\hat{b}_i(k+1) - \hat{b}_i(k)|$ at each iteration and use it as a stop criterion. Figure 3 shows that the algorithm converges in only a few iterations. To show how the estimates behave with the number of data points N, we plot the square error of $\|\hat{b} - b\|_2^2$ with respect to N in Figure 4. This figure shows that the square error can be made small enough by choosing N appropriately.

5 Conclusion

In this paper, we revisit the iterative identification method and propose a normalized algorithm for Hammerstein systems. Convergence property is established under arbitrary nonzero initial conditions. As pointed out in [2], if the iterative algorithm converges, it converges fast and is very efficient. The static function is extended to non-odd functions. Examples are also used to illustrate the effectiveness of the proposed scheme.

References

1. Stoica, P.: On the convergence of an iterative algorithm used for Hammerstein system identification. IEEE Transactions on Automatic Control 26, 967–969 (1981)
2. Bai, E.W., Li, D.: Convergence of the iterative Hammerstein system identification algorithm. IEEE Transactions on Automatic Control 49, 1929–1940 (2004)
3. Chen, H.F.: Pathwise convergence of recursive identification algorithms for Hammerstein systems. IEEE Transactions on Automatic Control 49, 1641–1649 (2004)
4. Liu, Y., Bai, E.W.: Iterative identification of Hammerstein systems. Automatica 43, 346–354 (2007)
5. Narendra, K.S., Gallman, P.G.: Continuous time Hammerstein system identification. IEEE Transactions on Automatic Control 11, 546–550 (1966)
6. Vörös, J.: Recursive Identification of Hammerstein Systems with Discontinuous Nonlinearites Containing Dead-zones. IEEE Transactions on Automatic Control 48, 2203–2206 (2003)
7. Zhao, W.X., Chen, H.F.: Adaptive tracking and recursive identification for Hammerstein systems. Automatica 45, 2773–2783 (2009)
8. Zhao, Y.L., Zhang, J.F., Wang, L.Y., Yin, G.G.: Identification of Hammerstein systems with quantized observations. SIAM Journal on Control and Optimization 48, 4352–4376 (2010)
9. Li, G., Wen, C.: Convergence of normalized iterative identification of Hammerstein systems. Systems and Control Letters (2011), doi:10.1016/j.sysconle.2011.07.010
10. Li, G., Wen, C., Zheng, W.X., Chen, Y.: Identification of a Class of Nonlinear Autoregressive with Exogenous Inputs Models Based on Kernel Machines. IEEE Transactions on Signal Processing 59, 2146–2158 (2011)

Learning Biped Locomotion Based on Q-Learning and Neural Networks

Peng Ziqiang, Pan Gang, and Yu Ling

State Key Laboratory of Industrial Control, Zhejiang University,
Hangzhou 310027, China
zqPeng@zju.edu.cn, gpan@iipc.zju.edu.cn, lyu@iipc.zju.edu.cn

Abstract. Robot postures are transformed continuously until an impact occurs. In order to solve the continuous state problem, a Q-Learning controller based on Back-Propagation (BP) Neural Networks is designed. Instead of Q table, a Multi-input Multi-output BP Neural Network is employed to compute Q value for continuous state. Eligibility trace is used to solve time reliability problem in Q-Learning, and we integrate the eligibility trace algorithm to the gradient descent method for continuous state. To avoid dimension explosion, an inverted pendulum pose-energy model is built to reduce the dimension of the input state space. For the sake of balance between "explore" and "exploit" of Q-Learning, we use a new ε-greedy method with a variable stochastic probability, which decreases with the increasing of the step number. Simulation results indicate that the proposed method is effective.

Keywords: Biped locomotion, Q-Learning, eligibility trace, BP Neural Networks.

1 Introduction

In recent years, a number of stable controllers for biped walkers are designed based on ZMP (Zero Moment Point) criterion [1]. ZMP-based method does not take advantage of the inherent dynamics of biped robot, and the high gain actuators are employed to track the desired trajectories, so the energy efficient is much lower than human. Actually, tracking specified trajectories accurately is not a necessary condition for a biped robot to walk stably [2]. The limit cycle biped gait can be realized by using the robot dynamics directly. Nowadays, more and more attentions are paid to the methods of controlling a biped robot with its inherent dynamics [3, 4].

A learning controller is model free, and can be performed online. Q-Learning algorithm [5] can learn a stable biped gait trail by trial without monitor signals, so more and more controllers based on Q-Learning algorithm are designed for biped robot [6-10].

Schuitema et al. [7] proposed a control method based on Q-Learning algorithm, and obtained different optimal gaits on a simulation model with different optimization objectives. Wang Shouyi et al. [8] adopted Q-Learning algorithm to choose the PD controller parameters, and realized biped walking successfully on uneven floor.

G. Lee (Ed.): Advances in Automation and Robotics, Vol. 1, LNEE 122, pp. 313–321.
springerlink.com

Mao Yong et al. [9] designed a quasi-PDW robot, using Q-Learning algorithm based method to get the natural biped gait. But a discrete Q table is used in these Q-Learning algorithms, so the continuous state should be discretized. At the same time, a biped robot has many degrees of freedom, thus the dimension of the Q table is huge and Q-Learning is difficult to converge.

Morimoto et al. [10] demonstrated limit cycle locomotion on real robot based on poincaré-map and Q-Learning. The dimension of input state was reduced by inverted pendulum angle-velocity model. Mao Yong et al. [9] employed phase-energy state vector.

In this paper, a biped locomotion controller is based on Q-Learning and BP Neural Networks [11]. A Multi-input Multi-output BP Neural Network is used to compute Q-value of continuous state. To improve computing efficient, inverted pendulum pose-energy model is applied to reduce dimension of state space. A newε-greedy method with a variable stochastic probability is used to balance "explore" and "exploit". The simulator is a simple 2D biped robot with MACCEPA (Mechanically Adjustable Compliance and Controllable Equilibrium Position Actuator) [12]. The simulation result shows a stable, natural and periodic biped gait.

This article is organized as follows. Section 2 introduces a simple 2D biped robot, and inverted pendulum pose-energy model, which can reduce dimension of input state space. In section 3, the eligibility trace and gradient-descent algorithm for continuous state is expounded concretely. Section 4 is focused on the biped robot control system based on Q-Learning and BP Neural Networks. In section 5, a stable, natural and periodic biped locomotion is shown.

2 2D Biped Robot and Inverted Pendulum Model

2.1 Biped Robot

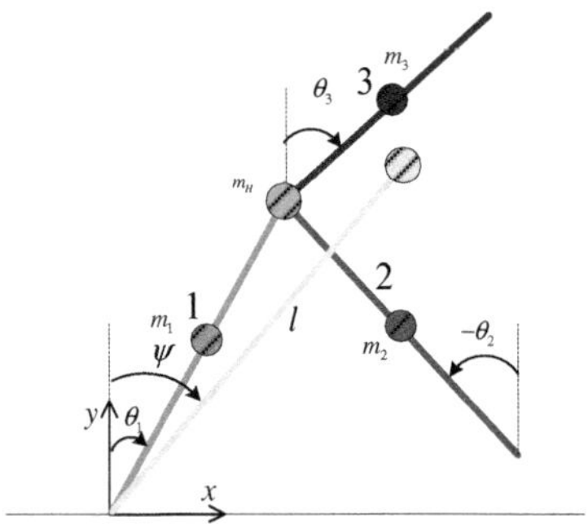

Fig. 1. Two dimensional 3-link model. (x , y) is the local coordinate and clockwise is the positive orientation. Inverted pendulum's pose is defined by (ψ , l). The mass of link is denoted by m , and angle of biped robot is denoted by θ .

This section introduces a simple 2D biped robot, which only consists of hip, torso and two legs (**Fig. 1**). The walking cycle could be supposedly divided into a Single Support Phase (SSP) and a Double Support Phase (DSP). It is assumed that the contact is inflexible, and DSP is transient. So the 3-link model is a hybrid system [13], which consists of two parts: the continuous swing process and the discrete impulse process.

Two torques, τ_1 and τ_2, are applied between the two legs and the torso. MACCEPA takes advantage of robotic nature dynamic feature [12], and biped robot can walk more efficiently. The *i*-th joint's torque function can be defined by

$$\tau_i = -k_i(\phi_i - \varphi_i) - b_i\dot{\phi}_i \tag{1}$$

where k_i is the coefficient of stiffness, b_i is the damping coefficient, φ_i is the equilibrium position, $\phi_i, \dot{\phi}_i$ are the angle and angle velocity between torso and the *i*-th leg.

Damping coefficient b_i is usually constant, and thus the torque of MACCEPA is decided by coefficient of stiffness k_i, equilibrium position φ_i and robot's state vector $(\theta_1, \theta_2, \theta_3, \omega_1, \omega_2, \omega_3)'$. So we take coefficient of stiffness and equilibrium position $(k_1, \varphi_1, k_2, \varphi_2)$ as the action vector of reinforcement learning.

2.2 Inverted Pendulum Model

If Reinforcement Learning (RL) employs robot's state as input state vector, it is difficult to converge considering the large state space dimensions. Actually, biped robot has more degrees of freedom than 3-link simulator, and the state space is more complex. To reduce the dimension of input state vector, biped robot can be equivalent to inverted pendulum. The state of inverted pendulum and biped robot are defined in **Fig. 1**.

The complete inverted pendulum model contains information of pose and velocity $(\psi, l, \dot{\psi}, \dot{l})'$. In order to get a smaller dimension of input state, angle velocity $\dot{\psi}$ and length velocity $\dot{l}$ are characterized by kinetic energy E. In this paper, sate vector for Q-Learning is equivalent to inverted pendulum pose-energy model defined by $(\psi, l, E)'$.

The pose $(\psi, l)'$ of inverted pendulum can be computed by its barycenter's position $G = (G_x, G_y)$. The kinetic energy of inverted pendulum E is equal to the sum of biped robot's kinetic energy. The input state vector $(\psi, l, E)'$ is defined by

$$\begin{cases} \psi = \dfrac{\pi}{2} - \arccos\dfrac{G_x}{\sqrt{G_x^2 + G_y^2}} \\ l = \sqrt{G_x^2 + G_y^2} \\ E = E_1 + E_2 + E_3 + E_H \end{cases} \tag{2}$$

3 Q-Learning Based on BP Neural Networks

3.1 Q-Learning Principle

Q-Learning is model free, and selects the next action based on Q table for state-action pairs. MACCEPAs export torques, and then biped robot reaches a new state. Controller evaluates the last state-action pair according to the new state. Through a series of search in state-action space, controller gets an optimal policy Π^*, which maximizes the action-value function $Q^*(s,a)$ defined by:

$$Q^*(s,a) = \max Q^{\Pi^*}(s,a), \forall s \in S, a \in A \tag{3}$$

where S is an input state space, A is an action set.

Q-learning is a kind of Temporal Difference (TD) control method. At each episode, the update rule of action-value function $Q(s,a)$ is:

$$Q_{t+1}(s,a) = \begin{cases} (1-\beta)Q_t(s_t,a_t) + \beta\left[r + \gamma \max_{a \in A} Q_t(s_{t+1},a)\right], & (s,a) = (s_t,a_t) \\ Q_t(s_t,a_t) & , \forall (s,a) \neq (s_t,a_t) \end{cases} \tag{4}$$

where β is the learning rate, r is the reward, γ is the discount rate.

3.2 BP Neural Networks Approximation

Robot's state is continuous, and discrete state space impossibly contains all actual states. Q-Learning for continuous state space can be realized by BP Neural Networks. The Multi-input Multi-output BP Neural Network (**Fig. 2**) has an output node for each element in action set $A = \{a_1, a_2, \cdots, a_M\}$. The reward r_t evaluates the last state-action pair (s_t, a_t), so only the weights and thresholds corresponding to last action a_t can be updated every episode, instead of all weights and thresholds.

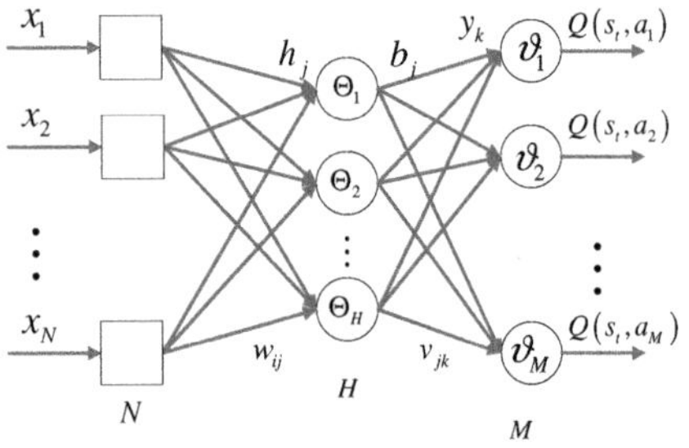

Fig. 2. The Multi-input Multi-output BP Neural Network. w_{ij} is the weight between input node and hidden node, Θ_j is a threshold of hidden node, net_j is the input of hidden node, b_j is the output of hidden node; v_{jk} is the weight between hidden node and output node, ϑ_k is an output layer's threshold, y_k is the input of output node.

TD error is the gap between real output and expected output of the BP Neural Network. The weights and thresholds are updated by TD error, and then the learning experience is saved. For continuous state space learning, TD error is computed by:

$$e=\begin{cases} r_t+\gamma\max_{a\in A} Q(s_{t+1},a,W_t)-Q(s_t,a_t,W_t), a=a_t \\ 0 \qquad ,a\neq a_t \end{cases} \tag{5}$$

where $W_t=\left(w_{ij},\Theta_j;v_{jk},\vartheta_k\right)_t$ is weight and threshold vector.

During continuous walk, current gait is the result of all actions since the beginning of this locomotion. So while updating learning experience, it is needed to take into account the last actions. This problem is named time reliability problem in RL. Q-Learning uses eligibility trace to solve the problem [5]. The increment of weight is the product of learning rate β and the derivatives of TD error mean-square deviation. In this paper, the last increment is added to current increment base on eligibility trace. To adapt continuous state learning, eligibility trace and gradient descent algorithm is mixed. The eligibility trace $\phi_t(a)$ updating rule is:

$$\Delta\phi_t(a)=\begin{cases} \lambda\Delta\phi_{t-1}(a)+\beta e\frac{\partial Q(s_t,a)}{\partial W_t}, a=a_t \\ \lambda\Delta\phi_{t-1}(a) \qquad ,a\neq a_t \end{cases} \tag{6}$$

where λ is the eligibility discount rate, $\Delta\phi_t(a)$ is the eligibility increment, $\frac{\partial Q(s_t,a)}{\partial W_t}$ is the gradient of BP Neural Networks; there is an eligibility for each element of W_t.

Transfer function of hidden layer and output layer, $f(x)$ and $g(x)$ are tan sigmoid function. The gradient between hidden layer and output layer $\frac{\partial Q(s_t,a)}{\partial v_{jk}}$ is computed by:

$$\frac{\partial Q(s_t,a)}{\partial v_{jk}}=f'(y)b_j=\left[1-Q(s_t,a,W_t)\right]\left[1+Q(s_t,a,W_t)\right]b_j \tag{7}$$

Weight increment Δv_{jk} is computed by:

$$\begin{cases} \delta v_{jk}(t+1)=\begin{cases} \beta e_k\left[1-Q_k(s_t,a,W_t)\right]\left[1+Q_k(s_t,a,W_t)\right], a=a_t \\ 0 \qquad ,a\neq a_t \end{cases} \\ \delta' v_{jk}(t+1)=\delta v_{jk}(t+1)b_j(t+1)+\lambda\delta' v_{jk}(t) \\ \Delta v_{jk}(t+1)=\begin{cases} \delta' v_{jk}(t+1), a=a_t \\ 0 \qquad ,a\neq a_t \end{cases} \end{cases} \tag{8}$$

Output layer's threshold ϑ_k is updated by:

$$\begin{cases} \delta\vartheta_k(t+1)=\begin{cases} \beta e_k\left[1-Q_k(s_t,a,W_t)\right]\left[1+Q_k(s_t,a,W_t)\right], a=a_t \\ 0 \qquad ,a\neq a_t \end{cases} \\ \delta'\vartheta_k(t+1)=\delta\vartheta_k(t+1)+\lambda\delta'\vartheta_k(t) \\ \Delta\vartheta_k(t+1)=\begin{cases} \delta'\vartheta_k(t+1), a=a_t \\ 0 \qquad ,a\neq a_t \end{cases} \end{cases} \tag{9}$$

The gradient between hidden layer and output layer $\frac{\partial Q(s_t,a)}{\partial w_{ij}}$ is computed by:

$$\frac{\partial Q(s_t,a)}{\partial w_{ij}}=g'(net_j)x_i\sum_{k=1}^{M}f'(y_k)v_{jk} \tag{10}$$

Weight increment Δw_{ij} updated rule is:

$$\begin{cases} \delta w_{ij}(t+1) = [1-b_j(t+1)][1+b_j(t+1)]\sum_{k=1}^{M}[\delta v_{jk}(t+1)v_{jk}(t+1)] \\ \delta' w_{ij}(t+1) = \delta w_{ij}(t+1)x_i(t+1) + \lambda\delta' w_{ij}(t) \\ \Delta w_{ij}(t+1) = \delta' w_{ij}(t+1) \end{cases} \tag{11}$$

Updating rule of hidden layer's threshold Θ_j is:

$$\begin{cases} \delta\Theta_j(t+1) = [1-b_j(t+1)][1+b_j(t+1)]\sum_{k=1}^{M}[\delta v_{jk}(t+1)v_{jk}(t+1)] \\ \delta'\Theta_j(t+1) = \delta\Theta_j(t+1) + \lambda\delta'\Theta_j(t) \\ \Delta\Theta_j(t+1) = \delta'\Theta_j(t+1) \end{cases} \tag{12}$$

4 Q-Learning Control System

Biped robot RL control system architecture is shown by **Fig. 3**. The policy function is approximated by Multi-input Multi-output BP Neural Networks and gradient descent algorithm in section 3. The input state is transferred to inverted pendulum pose-energy model state $s_{t+1} = (\psi, l, E)'_{t+1}$ by (2). The critic evaluates the last state-action pair (s_t, a_t), and then the actor selects the next action based on a new ε-greedy algorithm.

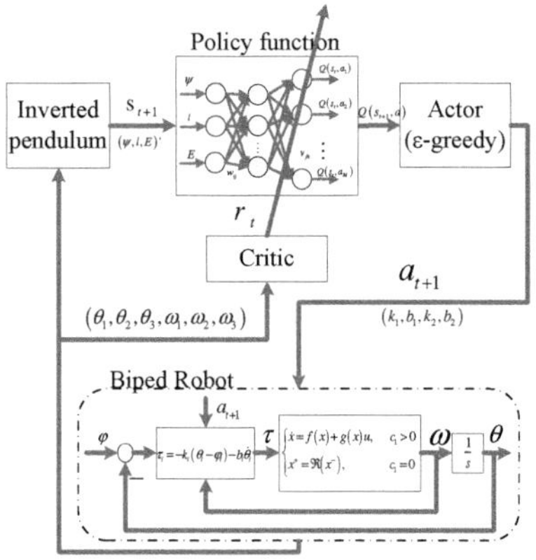

Fig. 3. Biped robot RL control system

At the time of impulse, biped robot's state is detected, and then the dimension of input state is reduced to 3. Q-value for current state s_{t+1} and all actions is computed by BP Neural Network. After the critic outputting reward r_t for the last state-action pair (s_t, a_t), the weights and thresholds are updated. The next action a_{t+1} is selected, and then controller changes MACCEPAs' parameters at the time of impulse.

The critic evaluates the last state-action pair (s_t, a_t) with 3-link state $(\theta_1, \theta_2, \theta_3, \omega_1, \omega_2, \omega_3)'$. The reinforcement reward is defined by:

$$r_t = \begin{cases} -1, \text{fall} \\ +1, \text{walk a step} \end{cases} \tag{13}$$

The next action is selected by actor according to $Q(s_{t+1}, a)$ $a \in A$. For the sake of balance between "explore" and "exploit" [5], a newε-greedy algorithm is used in this paper. The stochastic probability ε of ε-greedy is reduced with the increasing of continuous step number. The stochastic probability is computed by:

$$\varepsilon = \varepsilon_0 \cdot e^{(\mu - step)/5} \tag{14}$$

where ε_0 is the initial stochastic probability of episode, μ is a constant, *step* is the number of continuous steps.

5 Simulation

The mechanical parameters of the biped robot are listed in **Table 1**. MACCEPA will compute torques with action vector $a = (k_1, \varphi_1, k_2, \varphi_2)$ of Q-Learning and 3-link state $(\theta_1, \theta_2, \theta_3, \omega_1, \omega_2, \omega_3)'$. The two torques, τ_1 and τ_2, are computed by:

$$\begin{cases} \tau_1 = k_1(\pi + \theta_1 - \theta_3 - \varphi_1) - b_1(w_1 - w_3) \\ \tau_2 = -k_2(\pi - \theta_2 + \theta_3 - \varphi_2) - b_2(-w_2 + w_3) \end{cases} \tag{15}$$

Where k_1, b_1, k_2 and b_2 are positive; φ_1, φ_2 are a little smaller than the angle between torso and two legs at the next impulse.

Table 1. The mechanical parameters of biped robot

	Stance leg	Swing leg	Torso	Hip
Mass(kg)	5	5	10	15
Length(m)	1	1	1	0

Table 2. The action set

	Max	Min	Number of segmentation
Stance leg's coefficient of stiffness	90	20	71
Swing leg's coefficient of stiffness	90	40	51
Equilibrium position of stance leg	4	3.5	2
Equilibrium position of swing leg	4	4	1

The number of elements of action set (**Table 2**) is 71×51×2×1=7242, and BP Neural Networks has the same number of output nodes. There are 3 nodes in input layer, and 1000 nodes in hidden layer. Considering generalization energy and approximate accuracy, the value field of threshold is [-0.5, 0.5], and the range of weight [14] is computed by:

$$\begin{cases} \widehat{w} = \dfrac{\left|f^{-1}\left(1-\varepsilon^{'}\right)-f^{-1}\left(\varepsilon^{'}\right)\right|}{\sqrt{2}k^{'}n} \\ -\gamma^{'} \le a_i^{'} \le \gamma^{'} \\ w_i = \widehat{w}\sqrt{a_i^{'}+1} = \dfrac{\left|f^{-1}\left(1-\varepsilon^{'}\right)-f^{-1}\left(\varepsilon^{'}\right)\right|}{\sqrt{2}k^{'}n}\sqrt{a_i^{'}+1} \end{cases} \tag{16}$$

where f is an active function, $\varepsilon^{'}$=0.1, $\gamma^{'}$=0.3~0.9, $a_i^{'}$ is a rand, $k^{'}$=2~8, n is the input number of node.

The damping coefficient b_1=b_2=0.1; The discount rate of RL γ=0.95, the learning rate β=0.01, the discount rate of eligibility trace λ=0.1; the initial stochastic probability of ε-greedy algorithm ε_0=0.4, the constant μ=2.5.

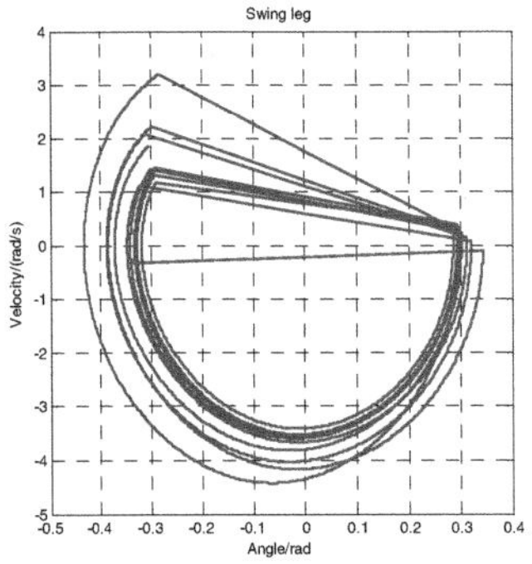

Fig. 4. The limit cycle of swing leg

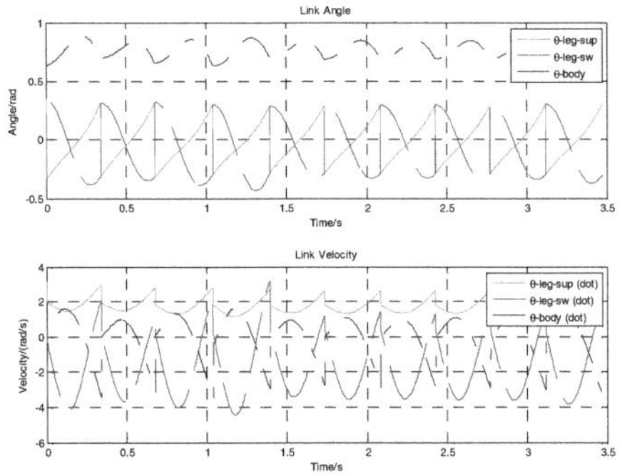

Fig. 5. Angle and angle velocity of 3-link model.

After average 200,000 trials, biped robot can walk by more than 10 steps. **Fig. 4** shows a stable limit cycle of swing leg. The angle and angle velocity curves of biped robot are shown in **Fig. 5**.

Biped robot controller, based on Q-Learning and BP Neural Networks, has learnt a stable, natural and periodic biped locomotion. And the simulation result shows the effectiveness of our method.

6 Conclusions

In this paper, a learning controller has been designed for continuous state based on Q-Learning and BP Neural Networks. A Multi-input Multi-output BP Neural Network was used for computing Q value of current state. In order to solve the time reliability problem in Q-Learning, eligibility trace was adapted to continuous state and mixed with gradient descent method. A newε-greedy algorithm, which contains a stochastic probability decreasing with the increasing of continuous step number, was used by actor. With method in this paper, biped robot controller has learnt successfully a stable, natural and periodic biped locomotion.

The work in this paper is concentrated in simulation of simple model. A 7-link biped simulator will be built, and the simulation experience data and method will be finally applied on a physical robot.

References

1. Vukobratovic, M., Borovac, Branislav: Zero-Moment Point—Thirty Five Years of Its Life. Humanoid Robotics 1(1), 157–173 (2004)
2. Pratt, J.E.: Exploiting inherent robustness and natural dynamics in the control of bipedal walking robots. Massachusetts Inst. Technol., Cambridge (2000)
3. Steve Collins, A.R.R.T.: Efficient Bipedal Robots Based on Passive-Dynamic Walkers. Science 307, 1082–1085 (2005)
4. McGeer: Passive dynamic walking. Robotics Research 9(2), 62-82 (1990)
5. Sutton, R.S., Barto, A.G.: Reinforcement Learning: An Introduction. The MIT Press, Cambridge (1998)
6. Tedrake, R., Zhang, T.W., Seung, H.S.: Stochastic policy gradient reinforcement learning on a simple 3D biped. In: Proceedings of 2004 IEEE/RSJ International Conference on Intelligent Robots and Systems, Sendai, Japan (2004)
7. Schuitema, E., Hobbelen, D.G.E., Jonker, P.P., Wisse, M.A.K.J.: Using a controller based on reinforcement learning for a passive dynamic walking robot. In: Proceedings of 2005 5th IEEE/RAS International Conference on Humanoid Robots, Sendai, Japan (2005)
8. Wang, S., Braaksma, J., Babuska, R., Hobbelen, D.: Reinforcement Learning Control for Biped Robot Walking on Uneven Surfaces. In: 2006 International Joint Conference on Neural Networks, Sheraton Vancouver Wall Centre Hotel, Vancouver, BC, Canada (2006)
9. Mao, Y., Wang, J., Jia, P., Li, S., Qiu, Le Zhang, Z., Han, Z.: A Reinforcement Learning Based Dynamic Walking Control. In: 2007 IEEE International Conference on Robotics and Automation, Roma, Italy (2007)
10. Morimoto, J., Atkeson, C.G.: Learning Biped Locomotion. IEEE Robotics Automation Magazine 14(3), 41–51 (2007)
11. Anderson, C.W.: Learning to control an inverted pendulum using neural networks. IEEE Control Systems Magazine 9(3), 31–37 (1989)
12. Van Ham, R., Vanderborght, B., Van Damme, M., Verrelst, B., Lefeber, D.B.I.: MACCEPA: the mechanically adjustable compliance and controllable equilibrium position actuator for 'Controlled Passive Walking'. In: IEEE International Conference On Robotics And Automation, 345 E 47TH ST, NEW YORK, NY 10017 USA (2006)
13. Grizzle, J.W., Abba, G., Plestan, F.: Asymptotically stable walking for biped robots: analysis via systems with impulse effects. IEEE Transactions on Automatic Control 46(1), 51–64 (2001)
14. Fernández-Redondo, M., Hernandez-Espinosa, C.: Weight initialization methods for multilayer feedforward. In: European Symposium on Artificial Neural Networks, pp. 25-27 (2001)

Using GA and Computation Parallelism for the Optimal Pilot Bus Selection

Michal Kopček and Dušan Mudrončík

Slovak University of Technology,
Faculty of Materials Science and Technology in Trnava,
Institute of Information Technology, Automation and Mathematics, Hajdoczyho 1,
91700 Trnava, Slovakia
{Michal.Kopcek,Dusan.Mudroncik}@stuba.sk

Abstract. The power system is generally characterized in particular by its size, complexity and time varying, beforehand unknown electrical load. Because large electric power is transferred, even slight changes in the settings can produce significant economic savings. Based on the characteristics of the power system and the nature of the secondary voltage control, it is clear that the optimal pilot bus selection plays a key role in the quality of control and related economic impacts. This paper deals with the system for optimal pilot bus selection of supervisory node voltage control in a large – scale power system, so that the key problem could be solved using the principle of parallelism and grid computing structure to make the computer time consumption as low as possible.

Keywords: Optimal pilot bus selection, parallelism, power system.

1 Introduction

This paper describes the system architecture design and proof of concept implementation of the system for optimal pilot bus selection for the need of power system supervisory control in the real time. Since there is a requirement to work in the real time and individual selection methods [3], [5], [6] require the calculation of the objective functions, which is based on simulating the operation of the power system, parallel processing principles (grid computing) were used in the designed architecture. Since optimal pilot bus selection is a problem with extensive state space, as a base of the architecture were chosen the genetic algorithms, which can also very well apply the principles of parallelism.

2 The Design and Implementation of the System Architecture

The architecture [4] is based on the principle “divide et impera”, which integrates the knowledge from before made analysis, which dealt with grid computing [2] and parallel genetic algorithms [1]. The basic idea behind this principle is to divide a complex task into a several independently solvable operations. These operations are

G. Lee (Ed.): Advances in Automation and Robotics, Vol. 1, LNEE 122, pp. 323–328.
springerlink.com

spread over multiple computing resources connected to each other in the computer communication network.

We can assume that the result of this decomposition will be a virtual system with computational power much higher than any solo performance from connected parties. This means that the system is in the architecture divided into three main sections as follows:

- GA Master – separate application used for task definition and genetic algorithm functions maintenance,
- GA Slave – separate application operating with functions associated with the fitness function computation of a given chromosome,
- Database – data storage and processing of some dedicated operations.

Structure of the whole system for optimal pilot bus selection for the needs of power system real time supervisory control captures the block diagram of the architecture shown in Fig. 1.

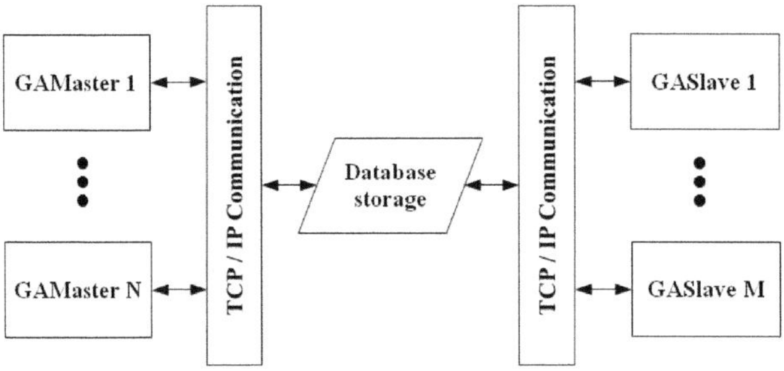

Fig. 1. Block diagram of designed computation system architecture

Blocks GA Master and GA Slave are indirectly linked to each other through database data storage, which besides storing of the chromosomes and mediation of the communication of other parts of the system also performs certain tasks contributing to the increase in computing speed of the whole system. The applications GA Master and GA Slave could run in more than one instance at the same time. Thanks to this, the computation parallelism is achieved and parallel structure of genetic algorithms could be configured. Detailed description of the system follows in the next chapters.

2.1 Chromosome Encoding

Chromosome is a string composed as a sequence of numerical or symbolic values and represents chosen properties or parameters of an individual. It could be represented using binary, integer, real number, symbolic values or as a combination of mentioned types. The choice of representation depends on the nature of the solved problem. [7]

The chromosome is encoded by integer values within the designed architecture, while the range of values is limited by the number of rows of the candidate set file and individual genes represent specific line of this file.

The candidate set file is created on the basis of the sensitivity coefficient matrix. Each row of the file contains the name of the generator node, the name of the load node and the value of their mutual sensitivity coefficient. The candidate set file is stored as a *.csv, which means that the elements are separated by a semicolon.

2.2 GA Master

This separate application integrates functions necessary for the definition of solved task and functions of genetic algorithms. The application window is shown in Fig. 2.

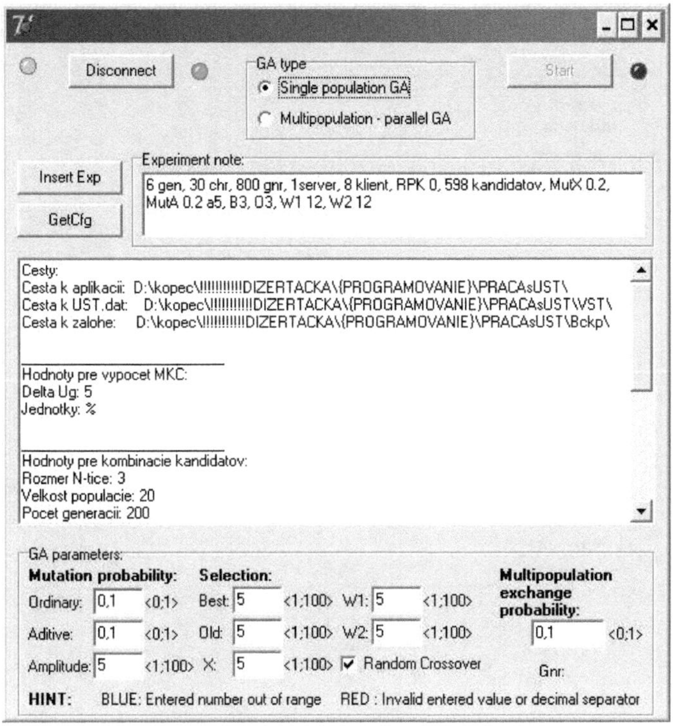

Fig. 2. Application window - GA Master

The definition of the solved task is composed of two parts:

- general configuration parameters settings of the task in the Config.ini file,
- genetic algorithm parameters settings for a particular subpopulation.

Subsequently is the defined task entered into the database and the record also contains embedded files needed for processing the task, namely:

- Config.ini – general configuration parameters of the task,
- Candidates.csv – candidate set file,
- Ust.dat – input file for the power system simulation model.

After the task creation, the system can run and start solving of the task. The application GA Master instance controls the genetic algorithm over the one subpopulation. In case of using the parallel (multipopulation) genetic algorithm structure, the exchange of the best chromosome between two instances is done according to a preset probability.

2.3 GA Slave

This separate application maintains the fitness function calculation of a particular chromosome, which is loaded from the database through stored procedure.

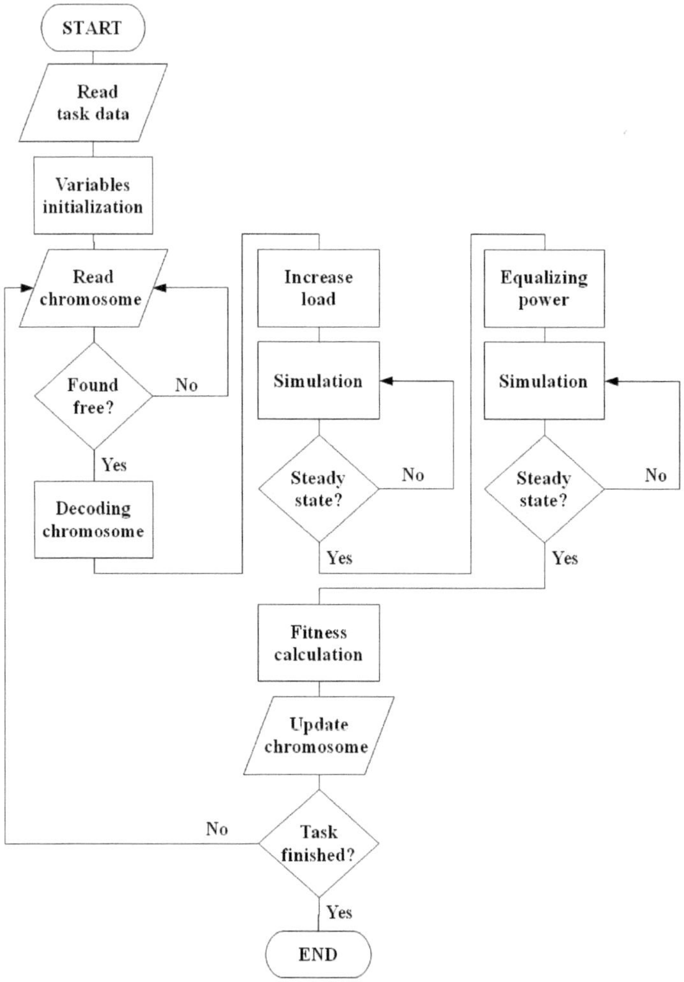

Fig. 3. Flowchart of the GA Slave main algorithm

The algorithm of fitness function evaluation is shown in flowchart in Fig. 3. After the start, the application reads data from the database of the ongoing task, which was previously defined by the GA Master. The parameters of the task and the work files needed to calculate the fitness function are loaded.

The application through stored procedure cyclically selects particular chromosomes from the database, for which should be the fitness function calculated. After processing of the given chromosome its fitness function value is updated in the database and subsequently a stored procedure for population evaluation is called. The fitness function value calculation consists of a several steps.

The selection of the particular chromosome, which should be processed, runs cyclically. Thanks to this, the independent run of the application is achieved and there is no need to intervene in its activities during the whole processing of the defined computing task.

The application GA Slave could run in more than one instance at the same time. Thanks to this, as many chromosomes are evaluated at the same time, as many instances are running. This way is the computation parallelism achieved and the computation time is markedly reduced. The application window is shown Fig. 4.

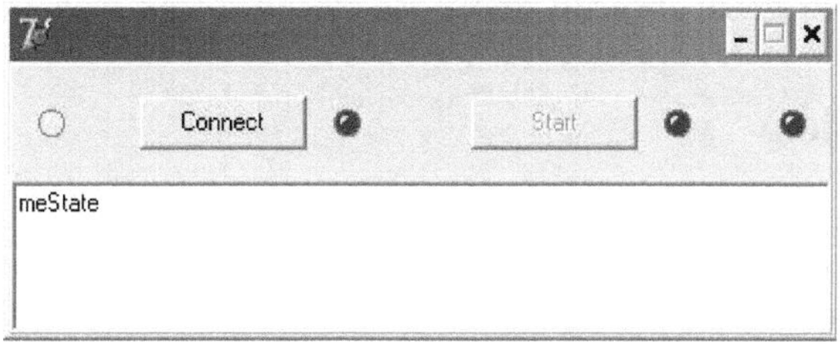

Fig. 4. Application window - GA Slave

2.4 Database

The centerpiece of the designed architecture is the database, which mediate indirect communication between separate running instances of the all other parts of the system. The database is implemented using MySQL 5.1.30, which offers enough resources for implementation of the designed architecture. The physical structure of the database is shown in Fig. 5.

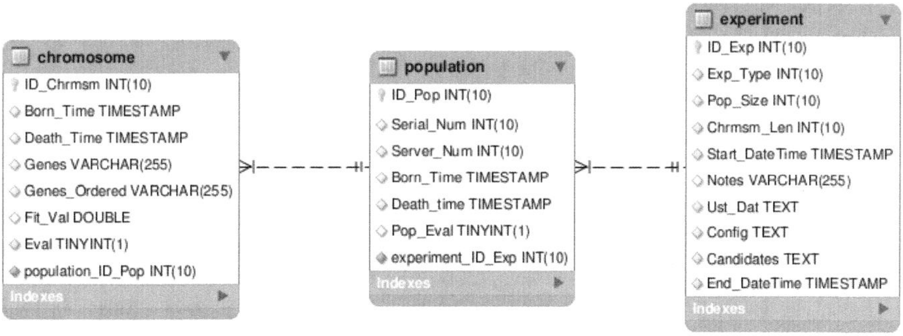

Fig. 5. Physical model of the implemented database

There are several stored procedures running on the database level as mentioned before. Thanks to this the communication burden between partner applications and the database is markedly restricted. Let us consider a situation without the use of the stored procedures. The application GA Slave has to select and read out from database all chromosomes, which represent the whole population, to be able to select exactly one chromosome, for which should be the fitness function evaluated. If the population size is e.g. 40, it means to transfer 40 records through the communication channel. This communication burden will significantly rise the overall computation time.

But when we use the stored procedure to find the free chromosome, then it will be transferred just one specific record in any case.

Another stored procedure checks newly inserted chromosomes. If the chromosome was evaluated somewhere in the past generations, it is not evaluated again and the value of its fitness function is simply updated according to the past evaluation. Thanks to this, the chromosome is evaluated exactly once during the whole computation task, which markedly lowers the overall computation time.

3 Conclusion

The presented proof of concept implementation of the designed architecture represents a significant contribution in the field of optimal pilot bus selection for the need of power system supervisory control in the real time. The very first experimental results proved a significant reduction of the overall computation time and therefore the system is able to find optimal solution in a few minutes. Thanks to this, the architecture could be used to build a complex system capable to operate in the real time.

References

1. Cantú-Paz, E.: A Survey of parallel genetic algorithm, `http://neo.lcc.uma.es/cEA-web/documents/cant98.pdf`
2. Foster, I., Kesselman, C.: The Grid: Blueprint for a New Computing Infrastructure II. Morgan Kaufmann (2004)
3. Kopček, M.: Optimalizácia procesov elektrizačnej sústavy na primárnej a sekundárnej úrovni riadenia. Dissertation thesis, 116 p. KAIA MtF STU, Trnava (2011); Supervisor: Dušan Mudrončík
4. Kopček, M., Mudrončík, D.: Optimal pilot bus selection using GA and computation parallelism. In: Process Control: 9th International Conference. Kouty nad Desnou, 7.-10. 6, University of Pardubice, Pardubice (2010), ISBN 978-80-7399-951-3. - C060b-1-4
5. Kuník, S.: Optimálny výber pilotných uzlov v sekundárnej regulácii zložitej ES. Dissertation thesis, 116 p, KAIA MtF STU, Trnava (2008); Supervisor: Dušan Mudrončík
6. Rakovský, M.: Redukcia výberu pilotných uzlov v sekundárnej regulácii. Dissertation thesis, 101 p, KAIA MtF STU, Trnava, (2006); Supervisor: Dušan Mudrončík
7. Sekaj, I.: Evolučné výpočty – a ich využitie v praxi. IRIS, Bratislava (2005), ISBN 80-89018-87-4

Research on Ontology-Based Relational Data Integration

Zhuang Miao[1], Jingwu Wang[2], Jinpeng Wang[1], Yang Li[1], and Weiguang Xu[2]

[1] Institute of Command Automation, PLA Univ. of Sci. & Tech., Nanjing 210007, China
[2] Chinese Institute of Electronic System Engineering, Beijing 100141, China
Emiao_beyond@163.com

Abstract. To integrate relational data semantically, schema description and query accessing issues are discussed. Ontology is chosen as the global schema to describing semantics. To building domain ontology rapidly, an approach based on subgraph extraction is presented. Then the semantics of SAPRQL queries is defined and algorithm for query rewriting is put forward. Minimal connectable unit is defined to represent a fragment of semantic mapping from the query to the rewriting. Finding and joining minimal connectable units produce semantically correct query plans. Soundness and completeness of query rewriting algorithm is proved. Computational complexity of the algorithm is discussed. Under the worst case, the query rewriting algorithm requires $O(n^m)$ time. Experiment results show that the query processing algorithm can provide nice performance.

Keywords: data integration, ontology, SPARQL, minimal connectable units, query rewriting.

1 Introduction

The explosive growth of data resources is a disturbing trend with the extensive applications of information technology [1]. It is difficult to utilize vast amounts of data stored in heterogeneous relational databases and to interoperate among them [2]. Integrating heterogeneous relational data is a hot research topic in database [3]. Data integration is to provide users a uniform query interface of heterogeneous data sources [4]. Traditional data integration approaches lack semantic description mechanisms to achieve the semantic interoperation.

Ontology is shared conceptualization of some domain and can model domain knowledge. Semantic Web provided several approaches to integrating data using ontology [5]. Ontology-based integration of heterogeneous relational database is an important application, which has received special attention in the database community [6-7]. One of significant obstacles is approach of converting a global query into correct SQL query for relational databases. How to build appropriate domain ontology for data integration is also taken into account.

G. Lee (Ed.): Advances in Automation and Robotics, Vol. 1, LNEE 122, pp. 329–336.
springerlink.com

Systems like Virtuoso [8] and D2RQ [9] rewrite SPARQL [10] queries to SQL. The MySQL SPASQL [11] module compiles SPARQL queries directly into an evaluation structure to be executed by the relational engine. These research works don't use SPARQL for data integration. Early works of relational database integration constructed ontology from relational databases schema using ontology [12, 13]. They cannot fully capture the semantics behind the relational schema, which makes it hard to integrate heterogeneous relational databases.

2 Building Domain Ontology as Global Schema

Domain ontology describes domain knowledge of integration and is the key part of integration system based on semantics. Relational database can be abstracted as a graph model. Hence describing the global schema of integration system using graph model is suitable. RDF(Resource Description Framework) describes knowledge by generating statements about an entity's properties. In this paper, we describe domain ontology using RDF and build RDF domain ontology as the global schema. Building ontology manually takes much time and costs and reusing domain knowledge is essential. Since ontology describes domain knowledge, reusing established ontologies is an immediate way of building domain ontology. Extracting a smaller sub-ontology from large-scale ontology is helpful to meet the demand.

RDF Graph is consisting of a set of RDF triples including subject, predicate and object. RDF triples are also called statements. The values of a statement are URI references. RDF model abstracts knowledge as resource and all the RDF triples form a network of resources.

Definition 1 (Graph). A graph is a pair $G = (N, E)$, where N is a set whose elements are called nodes, and E is a set of edges.

Two edges e_1, $e_2 \in E$ are said to be incident if they share a node ($e_1 \cap e_2 \neq \varphi$).The degree of a node $n \in N$, denoted degree (n), is the number of edges incident to it.

A graph $G_1 = (N_1, E_1)$ is a subgraph of $G= (N, E)$, denoted $G_1 \subseteq G$, if $N_1 \subseteq N$ and $E_1 \subseteq E$.

Definition 2 (Directed graph). A directed graph is a graph where the edges have a direction, i.e., we distinguish between edges connecting the same nodes, but being oriented differently. We choose not to define E as a set of pairs, as above, but rather as a set of edge elements with two maps

from: $E \rightarrow N$; to : $E \rightarrow N$

which yield the source and the target of each edge. Edges e_1, e_2 of a directed graph are incident if to(e_1) = from(e_2).

Definition 3 (RDF ontology). RDF ontology is a directed graph $G = (N, E)$. Where N is a set of nodes representing subjects and objects in the statements; E is a set of edges representing predicates in the statements.

RDFS (RDF schema) has reasoning mechanism in the form of a set of inference rules [14]. Applying the inference rules, new RDF triples can be derived iteratively. We call the set of all RDF triples composed of the original triples and the derived ones the closure of RDF *ontology*.

Definition 4 (RDFS closure). RDFS closure $c(G)$ is a set of RDF triples. Where G is the original RDF triples set. It is obvious that $c(G)$ is unique for any graph G and $G \subseteq c(G)$.

Since the closure $c(G)$ of graph G, we extract sub-ontology from a larger RDF *ontology* G based on the closure $c(G)$. First, we collect proper terms as main concepts of demanded ontology. Secondly, the terms are regarded as the nodes of RDF graph model. Hence, a subgraph consists of the term nodes can be extracted from the closure of source graph and sub-ontology extraction problem is translated a subgraph mapping problem. Finally, editing the extracted sub-ontology and building required domain ontology.

3 Semantics of SPARQL Queries

In the following, we discuss query processing over RDF Graph model and we choose SAPRQL as query language. The SPARQL query language is based on matching graph patterns. Pérez introduced the formal semantics of SPARQL in Ref. [15].

Definition 5 (SPARQL query). A SPARQL query Q = SELECT *vars* WHERE *gp*, where $vars = \{v_1, v_2, \ldots, v_n\}$, lists all the return variables in Q, gp is a set of RDF triples that Q should satisfy. Let var(gp) be all variables in gp, obviously $vars \subseteq \text{var}(gp)$.

Definition 6 (SPARQL query result). The query result of an SPARQL query Q = SELECT *vars* WHERE *gp* above an RDF graph G is a finite set of mappings $\Omega_Q = (\mu_1, \mu_2, \ldots, \mu_n)$, where μ_i maps v_i to an RDF term in G.

Definition 7 (Query containment). Given two SPARQL queries Q_1 and Q_2, $\forall \mu_i \in \Omega_{Q_2}$, $\exists \mu_j \in \Omega_{Q_1}$, where $\mu_i(vars(Q_2)) \subseteq \mu_j(vars(Q_1))$ and $\mu_j(gp_{Q1}) \subseteq \mu_i(gp_{Q2})$, then Q_1 contains Q_2, denoted by $Q_1 \subseteq Q_2$ or $Q_2 \supseteq Q_1$.

Definition 8 (Query equivalent). Given two SPARQL queries Q_1 and Q_2, if $Q_1 \subseteq Q_2$ and $Q_2 \subseteq Q_1$. then Q_1 is equivalent to Q_2, denoted by $Q_1 = Q_2$.

Definition 9 (SPARQL containment mapping). Given two SPARQL queries Q_1 and Q_2, mapping τ from $vars(Q_1)$ to $vars(Q_2)$ is a containment mapping, if

1) $\tau(x) \in vars(Q_2)$, then $x \in vars(Q_1)$, namely, $vars(Q_2) \subseteq vars(Q_1)$;

2) $\tau(gp_{Q1}) \subseteq gp_{Q2}$.

Lemma 1. Given two SPARQL queries Q_1 and Q_2, $Q_1 \supseteq Q_2 \Leftrightarrow$ there is a containment mapping from Q_1 to Q_2.

4 Query Processing

In an ontology-based relational data integration system, users are faced with a global schema. Integration system decomposes global queries into sub-queries according to their semantics, rewrites the sub-queries to consist with the semantic of underlying databases, and transforms the rewriting result to SQL queries. Query processing includes query decomposing, rewriting and transforming. We focus on the query rewriting and propose rewriting algorithm.

4.1 Query Decomposing

According to the semantics of SPARQL queries, complicated graph pattern is composed of base graph patterns joined with operators and variables. Base graph pattern is the minimum unit of query graph pattern. Integration system decomposes complicated graph patterns into base ones. Main idea of Query decomposing is that if the query graph pattern is basic graph pattern then return the query, else if input group graph pattern, alternative graph pattern or optional graph pattern, then construct sub-queries by combining queries with AND, UNION, OPTION and recursive call the query decomposing course until the query cannot be decomposed.

4.2 Query Rewriting

A semantically correct query plan amounts to finding proper relational tables whose sub-graph covers the SPARQL query. We present the query rewriting algorithm to generate semantically correct and executable query plans under the RDF Graph model. Our algorithm proceeds in two stages. In the first stage, we find all relative tables and decompose them to Minimal Connectable Units (MCU) according to source descriptions. In the second stage, we join MCUs to produce the query plans.

An MCU is a subset of RDF triples in source description of a relational table that can be executed on heterogeneous databases. Intuitively, an MCU represents a fragment of semantic mapping from the query to the rewriting of the query.

Definition 10 (Minimal connectable units) Given a SPARQL query Q, a relational table P and its Table Sub-graph G_P, an MCU m for P is a tuple of the form (Y, μ') where Y is a subset of triple patterns in G_P and μ' is a partial mapping from variables appearing in G_P to corresponding columns in P. The algorithm of finding the MCUS is as follows.

```
findMCUs(Q, G_P)
Input: SPARQL query Q;
Output: G_P, table sub-graph of relational table P.
Initialize M = Φ;
for each triple pattern t in Q, do
  for each triple pattern t' in G_P, do
    if exist a mapping τ that map t to t', then
      find the minimal subset(denoted by Y) of triple
patterns of G_P that is connectable.
      find the subset(denoted by μ') of μ that
relative to Y.
      M = M U<Y,μ'>;
  end for
end for
return M.
```

We say a set of triple patterns Y connectable if the following conditions hold: (1) Each return variable of Q is also return variable in Y. (2) If x is not a return variable in Q, then for every triple pattern t that includes x, $t \in Y$.

Then we consider combinations of MCUs, and for each valid combination we create a conjunctive rewriting of the query. The final result is a union of conjunctive queries. Given a set of MCUs M, the actual rewriting is constructed as follows.

```
joinMCUs(Q, M, A)
Input: a SPARQL query Q, a set of MCUs M, M = {m1, …, mn}, where mi = (Yi, μi');
Output: a set of rewritings A.
Initialize A = Φ;
for each minimal subset {m1,…, mk} of M such that
  Y1∪Y2∪…∪Yk cover all triple patterns of Q.
  Create the conjunctive rewriting Q' containing all relative tables to {m1,…, mk}
  Add Q' to A
end for
return A.
```

4.3 Query Transforming

After query rewriting, sub-queries are consistent with underlying relational databases' semantics. To execute on relational databases, the rewriting result should be transformed into SQL queries. The transforming is generally the form of "*SQL* = "SELECT DISTINCT" + *varlist* +"FROM" + graphPatternTranslation(*gp*)". where "graphPatternTranslation(*gp*)" is the graph pattern translating algorithm, which translates given graph pattern *gp* into SQL query and the detail is omitted.

5 Rewriting Algorithm Discussion

5.1 Soundness and Completeness for Query Rewriting

For a SPARQL query Q = SELECT *vars* WHERE *gp*, where *vars* = $\{v_1, v_2, \dots, v_n\}$, lists all the return variables in Q, *gp* is a set of RDF triples that Q should satisfy.

Lemma 2 (Soundness). Each Q_1 the query rewriting algorithm generated satisfy $Q_1 \subseteq Q$.

Proof:

Let $Q_1(\overline{X}) = V_1(\overline{X_1}), V_2(\overline{X_2}), \dots, V_k(\overline{X_k})$ is a rewriting generated by a query rewriting algorithm. Since return variables in Q_1 are consistent with return variables in Q, obviously the first condition in definition 9 is satisfied.

By the definition of MCU combination algorithm,

$$gpq = gpm_1 \cup gpm_2 \cup \dots \cup gpm_k \quad (1)$$

$$gpq' = \psi_1(gpv_1) \cup \psi_2(gpv_2) \cup \dots \cup \psi_k(gpv_k) \quad (2)$$

By definition 10, $\varphi_i(gpm_i) \subseteq gpv_i$, then,

$\varphi_1(gpm_1) \cup \varphi_2(gpm_2) \cup \ldots \cup \varphi_k(gpm_k) \subseteq gpv_1 \cup gpv_2 \cup \ldots \cup gpv_k$, since ψ is the inverse of φ, then,

$$gpm_1 \cup gpm_2 \cup \ldots \cup gpm_k \subseteq \psi_1(gpv_1) \cup \psi_2(gpv_2) \cup \ldots \cup \psi_k(gpv_k) \quad (3)$$

By equation (1), (2) and (3), we get $gpq \subseteq gpq'$, the second condition in definition 9 is satisfied. According to Lemma 1, $Q_1 \subseteq Q$. □

Definition 11 (Redundant Rewriting). Given a rewriting Q_1, let $exp(Q_1)$ is the unfolding of Q_1 following views' definitions. According to Lemma 1, there is a containment mapping τ from Q_1 to $exp(Q_1)$. If $\exists tp \in exp(Q_1)$ for $\forall\ tp_0 \in gpq$, $\tau(tp_0) \neq tp$, then Q_1 is a redundant rewriting of Q, otherwise Q_1 is a non-redundant rewriting of Q.

Lemma 3. Let Q_1 is a rewriting of Q, then there must be a non-redundant rewriting of Q, called Q_2, satisfying $Q_1 \subseteq Q_2$.

Proof:

If Q_1 itself is a non-redundant rewriting, then let $Q_2 = Q_1$, we get $Q_1 \subseteq Q_2$.

If Q_1 is redundant, by its definition, $\exists tp \in exp(Q_1)$ for $\forall\ tp_0 \in gpq$, $\tau(tp_0) \neq tp$, remove tp from the $exp(Q_1)$ we get Q_1', obviously Q_1' is non-redundant. Let $Q_2 = Q_1'$, we get $Q_1 \subseteq Q_2$. □

Lemma 4. Let Q_1 is a non-redundant rewriting of Q, then it can be find by the query rewriting algorithm.

Proof:

Let Q_1 is a non-redundant rewriting of Q, and $exp(Q_1)$ is the unfolding of Q_1 following views' definitions. Since $Q_1 \subseteq Q$, according to Lemma 1, there is a containment mapping τ from Q to $exp(Q_1)$. Let V_i = SELECT $vars_i$ WHERE $gpv_{i,}$, then $gpv_1 \cup gpv_2 \cup \ldots \cup gpv_k = gpq$, $\forall\ i \neq j$, $gpv_i \cup gpv_j = \varnothing$.

Let c_i = <gpv_i, τ_i>, where τ_i is subset of τ that covering $vars_i$, since τ is a containment mapping from Q to $exp(Q_1)$, a Q to exp (Q_1) that contains maps, and τ_i is a subset of τ, $\forall x \in gpv_i$, if $x \in vars(Q_1)$, $\tau_i(x) \in vars(V_i)$, if $\tau_i(x) \notin vars(V_i)$, for $\forall tp$ that contain x, $\tau_i(tp) \in gpv_i$. So c_i satisfy Definition 10.

Construct a query rewriting $Q_1'(\overline{X}) = V_1(\overline{X_1}), V_2(\overline{X_2}), \ldots, V_r(\overline{X_r})$, $Q_1 = Q_1'$. Since Q_1' can be generated by the query rewriting algorithm, and then Lemma 4 is proved. □

Lemma 5 (Completeness).

Proof:

Let $Q_1 \subseteq Q$, if Q_1 is redundant, according to Lemma 3, there is a non-redundant rewriting Q_1' satisfying $Q_1 \subseteq Q_1'$; if Q_1 itself is a non-redundant, let $Q_1' = Q_1$. According to Lemma 4, Q_1' can be found by the query rewriting algorithm. Lemma 5 is proved. □

5.2 Computational Complexity

Given a set of data source $V=\{v_1, v_2, \ldots, v_n\}$ and a SPARQL query Q=SELECT *vars* WHERE *gp*, $|gp|=m$, query rewriting algorithm needs to find all possible subset of V that can form an MCU, which requires $O(n^m)$ time. However, as the first step of query rewriting already rule out the nonrelevant data source, significantly cutting the unnecessary search branches, it can be expected that the actual operating efficiency of the query rewriting algorithm will be more satisfactory.

5.3 Experimental Results

The rewriting algorithm has been implemented on a Pentium Dual 1.79GHz computer with the Java 1.6. We construct 10 different relational databases, containing 100 heterogeneous relational tables and 100 different global queries with length from 1 to 10 to simulate various user queries.

Fig.1(a) shows that the performance of query processing scalable in terms of length of query: the running time is almost exponential to the length of query. An interesting finding time is not sensitive to the length of query when number of data sources is less than 50. Fig.1(b) shows that the performance of query processing scalable in terms of number of data sources: the running time is almost exponential to the number of data sources. But the running time is not sensitive to the length of query when length of query is less than 8.

The experimental results are consistent with the analysis of time complexity. When the length of query is less than 8, but lot of data sources (more than 80), the query processing algorithm can provide nice performance. This is consistent with the practical circumstances, where the query length is usually short.

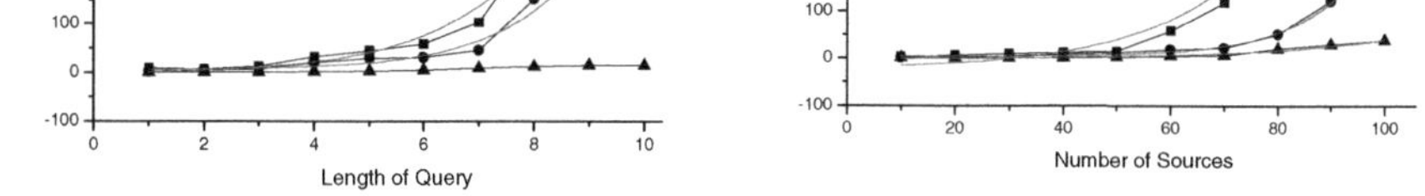

(a) (b)

Fig. 1. Performance test of query processing

6 Conclusions

In this paper we discuss ontology-based relational data integration and focus on domain ontology building and query processing problems. We use ontology as the global schema and building domain ontology by extracting sub-ontology based on closure. The semantic of query rewriting, represented by SPARQL queries, is defined. Algorithm for query rewriting is presented, so that SPARQL queries over global schema can be rewritten into local SQL queries executing on relational databases. The provided approach can be extended for other type of data sources.

References

1. Gantz, J.: The diverse and exploding digital universe. International Data Corporation, Framingham (2008)
2. Bell, G., Hey, T., Szalay, A.: Beyond the data deluge. Science 323(5919), 1297–1298 (2009)
3. Halevy, Y.A., Rajaraman, A., Ordille, J.J.: Data integration: the teenage years. In: Proc. of the 32nd International Conference on Very Large Data Bases, Seoul, Korea, pp. 9–16 (2006)
4. Bernstein, A.P., Hass, M.L.: Information integration in the enterprise. Communications of the ACM 51(9), 72–79 (2008)
5. Lee, B.T.: Relational Databases on the Semantic Web (1998), `http://www.w3.org/DesignIssues/RDB-RDF.html`
6. Dou, D., LePendu, P.: Ontology-based integration for relational databases. In: Proc. of ACM Symposium on Applied Computing, Dijon, France, pp. 461–466 (2006)
7. Zhu, H., Madnick, S.: A lightweight ontology approach to scalable interoperability. In: Very Large Data Bases Workshop on Ontology-Based Techniques or DataBases and Information Systems, Seoul, Korea, pp. 45–54 (2006)
8. Erling, O., Mikhailov, I.: RDF support in the virtuoso DBMS. In: Proc. of the 1st Conference on Social Semantic Web, Leipzig, Germany, pp. 59–68 (2007)
9. Bizer, C., Seaborne, A.: D2RQ – Treating Non-RDF Databases as Virtual RDF Graphs (Poster). In: 3rd International Semantic Web Conference, Hiroshima, Japan (2004)
10. Hommeaux, E., Seaborne, A.: SPARQL query language for RDF (2008), `http://www.w3.org/TR/rdf-sparql-query/`
11. Prudhommeaux E.: SPASQL: SPARQL support in MySQL (2006), `http://xtech06.usefulinc.com/schedule/paper/156`
12. Laborda, d.P.C., Conrad, S.: Bringing Relational Data into the Semantic Web using SPARQL and Relational OWL. In: Proc of the 22nd International Conference on Data Engineering Workshops, Atlanta, GA, USA (2006)
13. Bayardo, J.R., Bohrer, W., Brice, R., et al.: Infosleuth: Agent-Based Semantic Integration of Information in Open and Dynamic Environments. SIGMOD Record 26, 195–206 (1997)
14. Hayes, J.: A graph model for RDF, PhD dissertation, Darmstadt University of Technology, Germany (2004)
15. Pérez, J., Arenas, M., Gutierrez, C.: Semantics and Complexity of SPARQL. In: Cruz, I., Decker, S., Allemang, D., Preist, C., Schwabe, D., Mika, P., Uschold, M., Aroyo, L.M. (eds.) ISWC 2006. LNCS, vol. 4273, pp. 30–43. Springer, Heidelberg (2006)

Research of Neural Network PID Control of Aero-engine

Haiquan Wang, Dongyun Wang, and Guotao Zhang

Department of Electronic Information Engineering,
Zhongyuan University of Technology, Zhengzhou, China
whq036076260@mail.nwpu.edu.cn

Abstract. In order to complete the full flight envelope control of aeroengine, neural network PID control method with excellent self-learning and adaptive capability is introduced in this paper. Through the online learning algorithm with the principle of gradient descent method, the PID controller parameters, as the outputs of neural network, can be adjusted to overcome the influence of engine condition changing. The simulation results show the well tracking performance and robust performance of the designed control system, it is suitable for aero-engine control.

Keywords: Aero-engine, Neural network, PID, Robustness.

1 Introduction

As the heart of aircraft, aeroengine is considered to be the source of power to ensure the aircraft safety. Meanwhile, as a complex aerodynamic thermodynamic process, it also has the characteristics such as time delay, strongly nonlinear, serious coupling and so on[1]. Traditional methods such as PID control and robust control depends on the mathematical models of the object and the parameters of the controller can't self-adjust, thus there exists great limitations to control the object with strong changes in state, especially aeroengine. In recent years, as the development of computer technology and intelligent control algorithms, neural network approach which does not depend on a precise mathematical model of controlled object and possess excellent learning and adaptive capabilities have gradually developed[2]. It has been widely used in optimized calculation, fault diagnosis, especially in complex systems identification and control, etc[3,4,5]. Considering the self-adjust characteristics of neural network, combined with the simple and easy designed PID algorithm, PID controller based on neural network is used to flight envelop aeroengine control.

2 Structure and Principles of BP Network PID Controller

2.1 BP Neural Network

BP neural network is one of the most common neural network which is simple structure, easy to implement. It is the feed-forward network which is composed of

G. Lee (Ed.): Advances in Automation and Robotics, Vol. 1, LNEE 122, pp. 337–343.
springerlink.com

nonlinear units, there are three layers in the network: the input layer, hidden layer and output layer. The input signal propagates forward to the hidden layer, and spread to the output layer through the function. If the network is failured to get the desired output, then the signal will be transferred to back–propagation, it will cause the error signal return along the original connection path and modify the weights of each layer neurons in order to minimize the error signal. Obviously, this network has good self-learning ability, it can fulfill effective control of object with strong state changes.

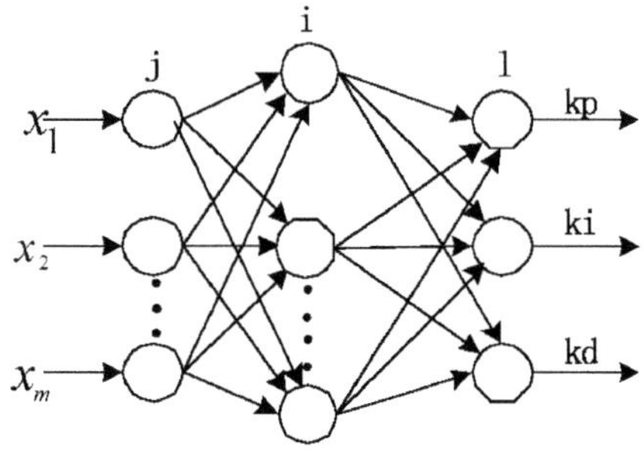

Fig. 1. BP neural network structure

2.2 The Neural Network PID Controller

The classic PID controller can directly applied for closed-loop control, and the structure and design procedure are simple. The deficiency of the method is that it depends on the system' accurate model, and the parameters can't achieve online adjusting to meet the requirements of real-time control. Considering the performance of the engine during the operation process changes strongly, BP neural network is used for online learning of the parameters of PID controller (the outputs of the network correspond to the three parameters of PID controller kp\ki\kd), the structure of the control system as shown in figure 2.

The algorithm of the controller is shown in the following two steps

2.2.1 The Forward Algorithm

(1) The inputs of the BP neural network are $x(j)$, (j = 1 ,2 ,..., M), where M is the number of the inputs which is chosen based on the complexity of the controlled system.

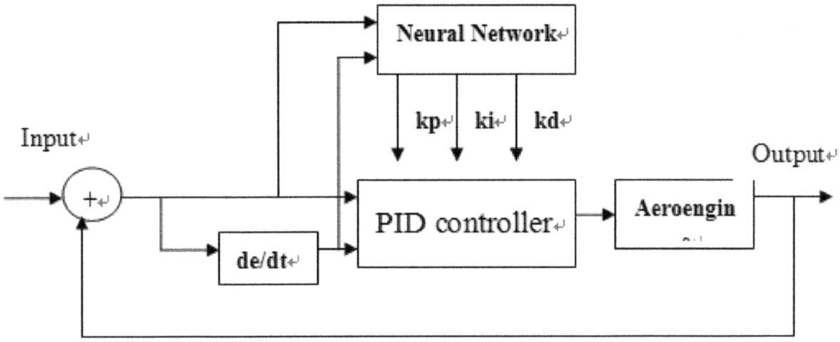

Fig. 2. The structure of BP Neural network

(2) The input and output of the Network hidden layer are.

$$net_i(k) = \sum_{j=0}^{M} w_{ij} x(j) \quad (1)$$

$$O_i(k) = f(net_i(k)) \qquad (i = 1, 2, \ldots, Q) \quad (2)$$

Where Q is the number of the hidden layer, w_{ij} is the hidden layer weights that connects the j-th input layer node with the i-th hidden layer node, and the activation function of the hidden layer neurons is the negative and positive symmetrical sigmoid function $f(x) = \frac{e^x - e^{-x}}{e^x + e^{-x}}$.

(3)The input and output of the three node in the output layer are shown as follows.

$$net_l^{(1)}(k) = \sum_{i=0}^{Q} w_{li}^{(1)} O_i(k)$$
$$O_l^{(1)}(k) = g\left(\sum_{i=0}^{Q} w_{li}^{(1)} O_i(k)\right) \qquad (l=1,2,3) \quad (3)$$

Where $w_{li}^{(1)}$ is the output layer weights that connects the i-th hidden node with the l-th node in output layer, and the activation function of the output layer neurons is the nonnegative sigmoid function $g(x) = \frac{e^x}{e^x + e^{-x}}$. And the outputs of the network are equal to the parameters of the PID controller kp\ki\kd respectively.

2.2.2 Back–Propagation Algorithm

The back-propagation algorithm is based on the error function

$$E(k) = \frac{1}{2}(r(k) - y(k))^2 \tag{4}$$

The gradient descent method is used to adjust the network weights, and the weights of the network output layer could be obtained as

$$\Delta w_{li}^{(1)}(k) = \alpha \Delta w_{li}^{(1)}(k-1) + \eta \delta_l^{(1)} O_i(k) \tag{5}$$

Where $\delta_l^{(1)} = error(k)\,\mathrm{sgn}(\frac{\partial y(k)}{\partial \Delta u(k)})\frac{\partial \Delta u(k)}{\partial O_l^{(1)}(k)} g'(net_l^{(1)}(k))$, and $g'(.) = g(x)(1-g(x))$, α and η represent the inertia coefficient and learning rate of the neural network respectively.

The equation of hidden layer weights learning is shown as follows.

$$\Delta w_{ij}(k) = \alpha \Delta w_{ij}(k-1) + \eta \delta_i x(j)(k) \tag{6}$$

Where $\delta_i = f'(net_i(k)) \sum_{l=1}^{3} \delta_l^{(1)} w_{li}^{(1)}(k) \quad (i = 1, 2, \ldots, Q)$

$f'(.) = (1 - f^2(x))/2$

Through the above process, based on the forward propagates and back propagates, the BP neural network could tune the PID control parameters on-line in order to deal with the control problem when object state changed.

3 Neural Network PID Controller Design of Aeroengine

3.1 Object of Aeroengine

The object mentioned in this paper is a twin spool, mixed flow, after burning gas turbofan engine. The linear model of engine with the fuel flow as its input and low-pressure rotor speed as output has been obtained with fitting method which makes the the linear dynamic response equation fit the nonlinear dynamic response data.The follow equation is the transfer function at H=0 and Ma=0

$$G(s) = \frac{y(s)}{u(s)} = \frac{0.5993s + 0.5026}{s^2 + 3.33s + 2.086} \tag{7}$$

3.2 Algorithm Achieve for Neural Network PID Controller

According to the properties of aero-engine and neural network PID controller, following steps are given for programming realize neural network PID controller.

(1) Determining the initial values of the PID controller parameters :

kp=0.8; ki=0.04; kd=0.1;

Determine the basic structure of BP neural network: three-layer network. The node number of input layer is 4, they are the desired input r (k), the object output y (k), error e(k) and the first derivative of error; The node number of hidden layer is five and the node number of output layer is 3; The initial value of network weights are selected randomly, the learning rate η and inertia coefficients α are selected as 0.2, 0.05.

(2) Sampling system output y (k) and reference input r (k), and calculating the error e (k);

(3) Calculating the inputs and outputs for each neurons of each layer of BP network; and the network output corresponds to the three parameters of PID controller ;

(4) Calculate PID controller output u (k);

(5) Neural network learning, and adjusting weights online to accommodate the object's state change,;

(6) Return to step (1) and continue the next control cycle.

3.3 Control System Performance Verification

For the designed control system, simulation has been done to investigate the performance.

(1) System tracking performance

The uint step response of aero-engine are shown as Fig.3, the results show that as the system output, low-pressure rotor speed can well follow the reference input under the action of the controller, the regulating time is about 0.6 second and there are't any overshoot.

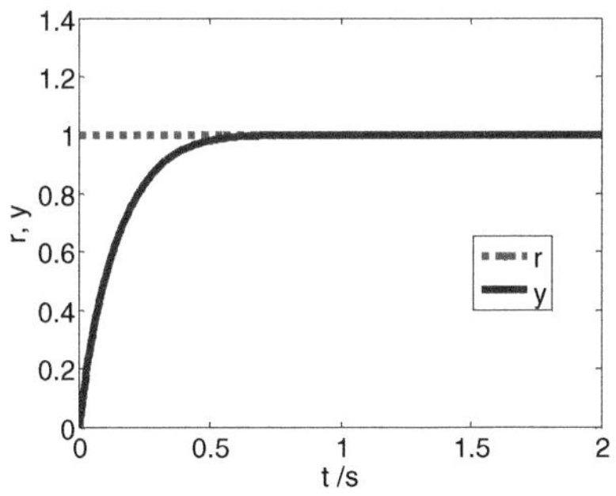

Fig. 3(a). Low speed response curve

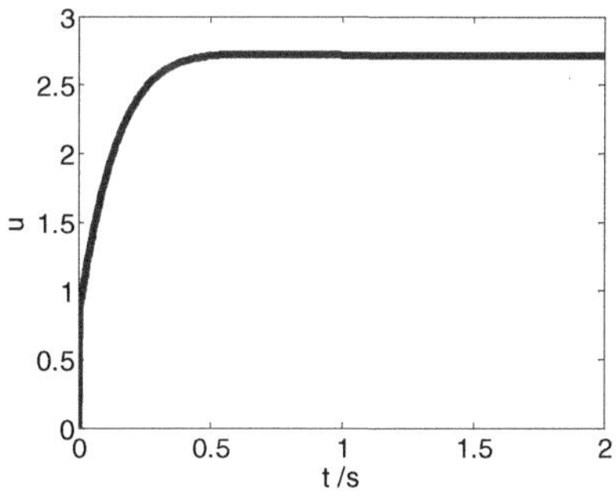

Fig. 3(b). Fuel flow change curves

(2) Robust performance

In order to test the robustness of the controller, simulations have been done at other state points of flight envelope (taking H=4km、 Ma=1.5 as an example), Fig. 4 shows the results that the whole system is still stable and well tracking when the engine state changed strongly.

(3) Ability of anti-disturbance

For investigating the anti-disturbance capability of the control system, a disturbance with the amplitude of 0.1 is added on the low pressure rotor speed during the stable state of the unit step response. The simulation results are shown in Fig.5, obviously the control system has a strong anti-disturbance ability, the adjust time from disturbance back to balance is about 0.8 s.

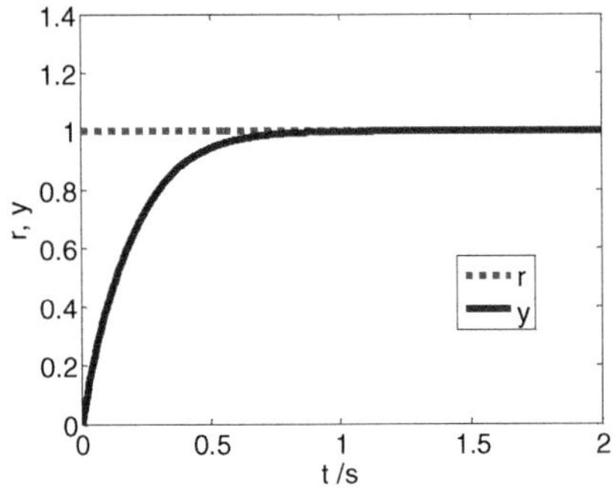

Fig. 4(a). Low voltage speed response curve

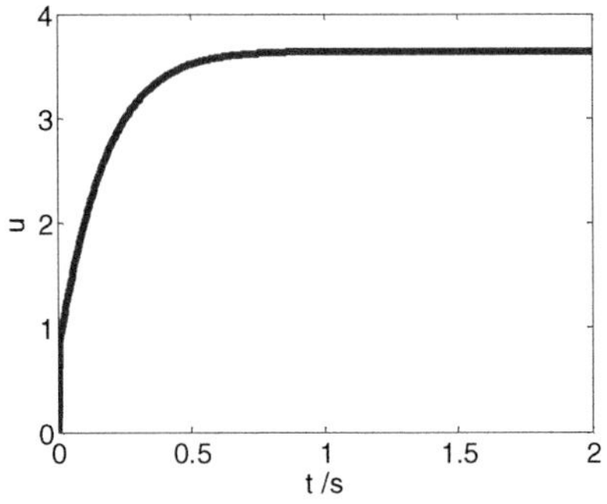

Fig. 4(b). Fuel flow response curve

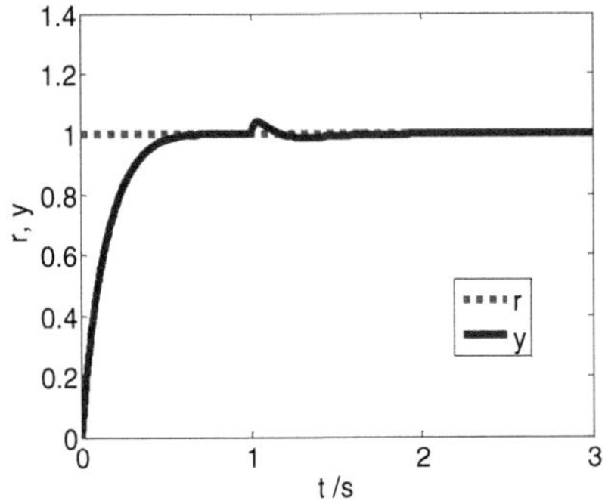

Fig. 5(a). Low voltage speed response curve

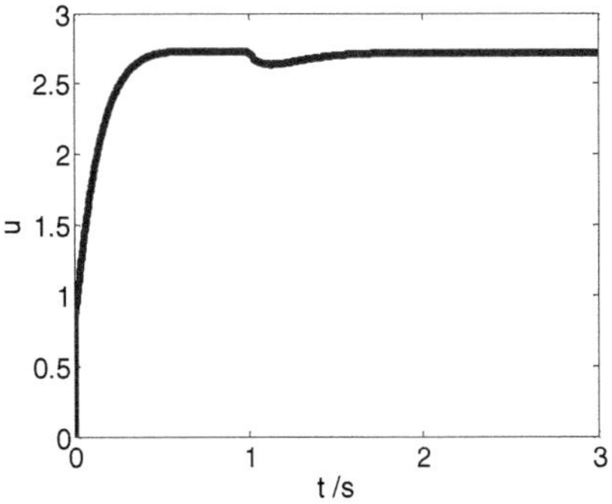

Fig. 5(b). Fuel flow response curve

Through a series of simulation results above, the PID controller based on neural network for aero-engine has a excellent control effect, it has a good ability in robustness, command tracking, anti-disturbance.

4 Conclusion

This paper combines neural network with the PID control algorithm and applies it to aeroengine control in order to make use of the advantages of BP neural network such as simple, online learning and high identification precisio. The simulation results verify that the designed controller could response the command quickly, meanwhile possesses strong robustness and the ability of anti-disturbance.

References

1. Sun, J., Vasilyev, V.: Advanced multivariable control system of aero-engines. In:Beijing University of Aeronautics and Astronautics(2005)
2. Narendra, K.S.: Neural Networks for Control Theory and Practice. Proceedings of IEEE 84(10), 1385–1406 (1996)
3. Zhang, M., Li, W., Liu, M.: Adaptive PID Control Strategy Based on RBF Neural Network Identification. In: The 17th IEEE International Conference on Tools with Artificial Intelligence, vol. 3, pp. 1854–1857 (2005)
4. Zhu, Q.M.: A back propagation algorithm to estimate the parameters of non-linear dynamic rational models. Applied Mathematical Modelling 27, 169–187 (2003)
5. Chen, T., Sun, J., Yang, W.: Self-Organizing Neural Network Based Fault Diagnosis for Aeroengine Gas Path. Acta Aeronautica et Astronautica Sinica 24(1), 46–48 (2004)
6. Liu, Y., Guo, Y.: Aircraft Engine Double Variable Controller Based on a New Neural Network. Aeroengine 31(1), 34–36 (2005)
7. Xu, H., Maj, M.: Robust Nneural Adaptive Control of a Hypersonic Aircraft. In: AIAA 2003, 5641 (2003)
8. Li, Q., Xu, G., Sun, J.: Aero-engine wavelet neural network PID control. Journal of Aerospace Power 24(4), 875–879 (2009)

The Design and Implementation of Soccer Robot for RoboCup Middle Sized League

Wei Li[1,2], Huifeng Wang[1], and Lars Asplund[2]

[1] Meilong Road NO.130 East China University of Science and Technology, 200237, Shanghai, China
liwei10-1@163.com, whuifeng@ecust.edu.cn
[2] Vasagatan NO.44 Mälardalen University, 72215, Västerås, Sweden
lars.asplund@mdh.se

Abstract. This paper mainly illustrates the background and development of soccer robot for RoboCup Middle Sized League, and puts more emphasis on both hardware and software design for the soccer robot based on the microcontroller AT90CAN128. In the hardware design, the sensor module is implemented by high precision laser range finder and electric compass for robot self-localization; USB-CAN converter are designed in the communication module; based on the introduction of main chips in use and workflow of the system, the two most important actuator in the soccer robot are designed separately, including motor module and kicking module. According to the hardware above, the software design shows the working flow of each module, including the laser ranger finder and electric compass in the sensor module, the transition mechanism between USB protocol and CAN protocol and easy-to-use bootloader for AT90CAN128 debugging online. The test shows both hardware and software designs for the soccer robot work well.

Keywords: RoboCup, Middle Size League (MSL) soccer robot, hardware design, software design.

1 Introduction

RoboCup is a standard soccer robot competition mainly focused on scientific research and development in fields like mobile autonomous robotics, multi-agent cooperation and control, computer vision, electronics, mechanics, image processing, artificial intelligence, control theory and many other areas. Besides the football competition in RoboCup consists of Simulation League (SL), Small Sized League (SSL), Middle Sized League (MSL) and Humanoids League (HL) [1]. This paper aims to build a private platform from scratch including both hardware and software for MSL soccer robot.

Generally there are many different ways to construct a soccer robot system both in hardware and software. Therefore, the way of selection of hardware and software could have a great influence on the performance of the corresponding soccer robot. In the past few years, J.H.Kim [3] uses the Intel's 16 bit processor 80196KC which has specialized communication modes for multi CPUs expansion. Liu Yinping [2] uses the TI's DSP

G. Lee (Ed.): Advances in Automation and Robotics, Vol. 1, LNEE 122, pp. 345–354.
springerlink.com

processor TMS320LF2407 as the main board because of the short sampling time, high running speed and lower power consumption. Gordon Wyeth [4] chooses 208MHZ Strong ARM microcontroller together with TMS320F243 microcontroller [5] which is supposed to act as an adapter and filter for the robots internal CAN network. However, they either do not make full use of CAN bus as internal network, or use an additional board instead to implement the CAN network inside the soccer robot. Considering the features of CAN bus, including lower price, high performance and high speed, AT90CAN128 is selected as a main controller in this paper because it has integrated CAN communication which needs lower cost and less effort to realize. Furthermore, as the mechanical characteristic and control characteristic are developed and improved better and better, three wheels drive and electromagnetic drive of kicking device are getting more and more popular among MSL due to more stable control and higher efficiency [8-10].

In this paper, the main contribution is to successfully build a fundamental MSL soccer robot system. Considering the rules for RoboCup is usually improved year by year, the soccer robot should have high compatibility by using CAN bus as an internal communication. Therefore, since the PC on the robot cannot directly connected to CAN bus, a USB-CAN converter is introduced here to send and receive CAN message and USB data in both directions. Finally, this paper proposes a high efficient electromagnetic drive of kicking device. More importantly, to reduce the price cost as well as labor cost, AT90CAN128 is selected as the main microcontroller for each module within all this paper.

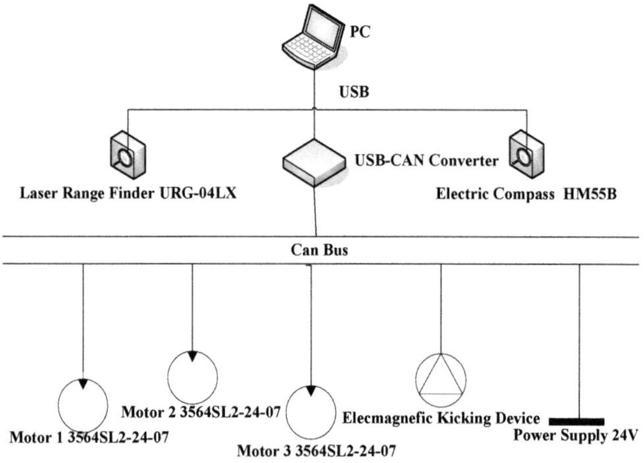

Fig. 1. The overall architecture of soccer robot with CAN internal communication has high compatibility in future

This paper is organized as follows. Section 2 briefly presents the hardware design about the soccer robot, including the sensor, USB-CAN converter, motor driver, kicking device respectively. Afterwards section 3 accordingly shows the software design about the soccer robot, including the sensor, the USB-CAN converter and boot loader. Based on hardware and software design, section 4 gives the result of testing. Finally the conclusion is given in section 5.

2 Hardware Design

2.1 Sensor

In general, two sensors are implemented in this paper. The laser ranger finder can be used for detecting both the direction and distance of obstacles on the way of robot while the electric compass can be used as detecting the position of other soccer robots through magnetic field.

The specifications of URG-04LX is :(1) Power source, 5V DC; (2) Detection range, 20~5600mm; (3) Accuracy, 60~1000mm; (4) Angular resolution, 0.36°; (5) Light source, semiconductor laser diode($\lambda = 785nm$); (6) Scan time, 100msec/scan; (7) Interface,USB2.0,RS-232C; (8) Weight, 160g[11]. From the specification, it is obvious to find the fact that URG-04LX has many benefits for being applied in the soccer robot system. First of all, the precision of the result is good enough for RoboCup which needs to predict the obstacles in a dynamical environment through the competition time. Secondly, it can be easier connected to microcontroller by USB which guarantees the correctness and communication speed between them.

The main features of HM55B are as follows: (1) Sensitive to variations in magnetic field strength; (2) Simplifiers direction by resolving magnetic field measurements into two components axes; (3) Only 30 to 40 ms between start measurement and data-ready [12]. As is seen above, HM55B is quite helpful to get the accurate information of robot's position, especially for the case that it is necessary to know the head direction of robot.

2.2 USB-CAN Converter

Since the PC cannot be connected to CAN bus directly, the USB-CAN converter is proposed to provide the interface transition between the USB bus and CAN bus. Basically it is made up of interface transition chip FT232BL, microcontroller AT90CAN128 and CAN transceiver TJA1040. Therefore PC can send and receive USB data to microcontroller AT90CAN128 via FT232BL [13] and microcontroller AT90CAN128 can send and receive CAN messages to CAN bus nodes via CAN transceiver TJA1040 [14]. The architecture of USB-CAN Converter is given in Fig 2.

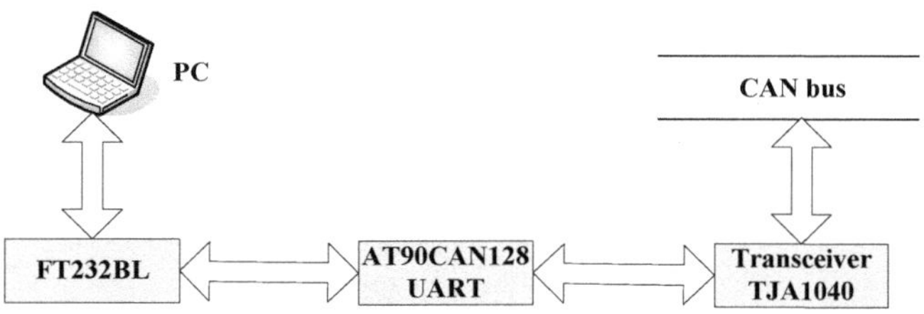

Fig. 2. The architecture of USB-CAN Converter shows the transition mechanism between USB bus and CAN bus

2.3 Motor Driver

The microcontroller AT90CAN128 receives CAN message via transceiver. After analyzing the message, it sends corresponding commands to motor controller MC33033 [15]. The Smart power module FSB0550 will be triggered by MC33033 and driver the connected Brushless DC motor (BLDC). At the same time, MC33039 is monitoring the working status of motor and give feedback signal back to MC33033. As a whole it is a closed-loop motor speed control system which can minimize the difference between the desired motor speed and the actual motor speed and improve the precision of motor speed in control. The architecture of motor driver is shown in Fig 3.

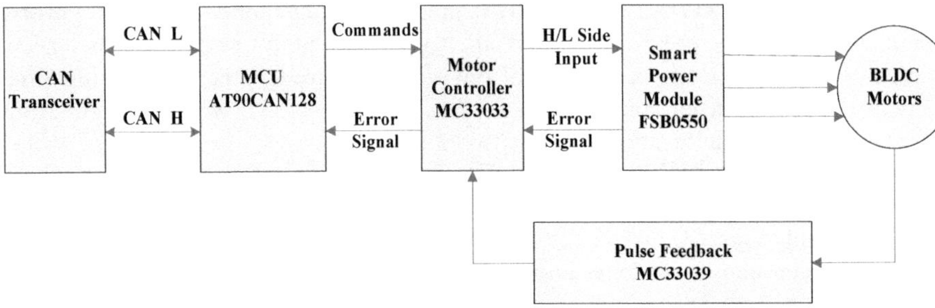

Fig. 3. The architecture of motor driver contains a feedback signal which describes actual motor speed

2.4 Kicking Device

In this work, the microcontroller AT90CAN128 receives commands from the PC and send commands accordingly to control the whole kicking device. The trigger circuit can trigger the electric switch on or off based on the control commands from microcontroller. Since the high voltage and current exists in the electric switch, the DC-DC converter which mainly contains a super big capacitor is introduced to drive the switch. Therefore when it capacitor is discharging, the electric switch turned on and the electromagnetic kicking device will kick the ball immediately. The architecture of kicking device is depicted in Fig 4.

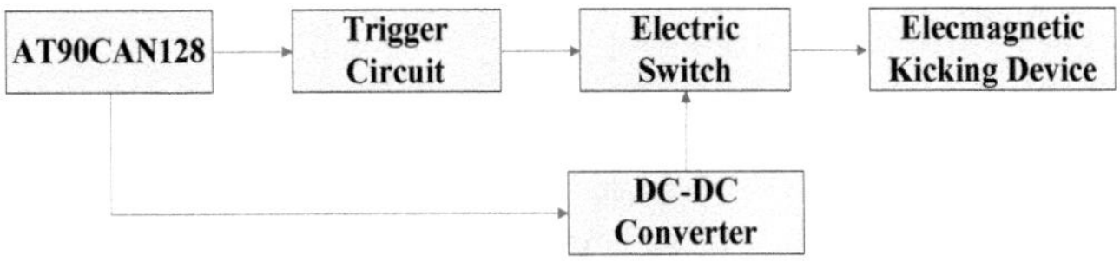

Fig. 4. The new architecture of kicking device is introduced here to increase the efficiency of kicking action

3 Software Design

3.1 Sensor

For the laser range finder, the software design has following several steps :

(1) Initialization. Since the laser range finder is directly connected to the PC via USB, so some USB parameters should be set up from the start, including the baud rate, data bit, stop bit and parity bit.

(2) Search. The laser range finder has the functionality that search within the nearby environment.

(3) Read. PC gets all the data that stored in the buffer of laser range finder via USB.

(4) Analysis. Some unimportant data should be filtered because these data just represent there is no obstacle that can be found currently.

(5) Draw. According to the distance and direction between the laser range finder and obstacles, the environment picture can be drawn.

(6) Repeat the step (2), start the next search period.

For the electric compass, the software design contains following several steps:

(1) Initialization. Since the electric compass is directly connected to the PC via USB, so some USB parameters should be set up from the start, including the baud rate, data bit, stop bit and parity bit.

(2) Read. PC gets four bytes from HM55B module once via USB. The first two bytes describe the value of x coordinate and the last two bytes describe the value of y coordinate.

(3) Analysis. Calculating the angler by x and y value through arctan function and this angle is based on the North direction.

(4) Repeat step (2), start the next search period.

3.2 USB-CAN Converter Mechanism

USB-CAN Converter is connected to the CAN bus through one AT90CAN128 microcontroller and to the USB bus through one FTDI 232BL chipset. Apart from it, the USB-CAN mechanism contains several parts, including serial frame header, handshake and configuration, open and closes the channel, read and write CAN messages.

(1) Serial frame header

The serial frame header contains a number of flags which describe the type of message the frame contains. The number of data bytes a message contains depends on the type of message being sent.

Table 1. The format of frame header

Bit 7	Bit 6	Bit 5	Bit 4	Bit 3	Bit 2	Bit 1	Bit 0
HNDS	-	ACK	CONF	OPEN	CHNL	WRTE	READ

As is seen from table 1, the bit7 HNDS represents the mask will be used to manipulate the correct message. The bit6 is aborted. The bit5 ACK should be set together with another command flag (such as HNDS or OPEN) to acknowledge that the command has been received. The bit4 CONF will result in the message being interpreted as a configuration message. The bit3 OPEN is evaluated in conjunction with the bit2 CHNL flag. When both these two flags are set the USB-CAN converter will enter a connected state. The bit1 WRTE means send a serialized CAN message to CAN bus. The bit0 Read triggers a read request in the USB-CAN converter.

(2) Handshake

To prevent connecting to random devices which contain a FTDI USB controller, there is a handshake procedure that filters unwanted and unavailable devices from the list of possible connections. The header is always prepended by two synchronization bytes to signal the start of a new frame. Once the synchronization bytes have been read and the header has been received, the recipient will determine the number of bytes the actual message contains.

(3) Configuration

It is possible to configure the following aspects of the USB-CAN device: baud rate of CAN bus; baud rate of USB bus; synchronization data; ID mask (CAN rev 2.0A/B); ID tag (CAN rev 2.0A/B).

(4) Opening and closing the channel

It is possible to enable and disable the CAN bus by setting the OPEN flag as described in Table 1. In order to close the channel the same message is sent without OPEN set.

(5) Read CAN messages

When the PC side wants to read a message from the CAN bus it will first try to read from the serial bus, if a message can be read from here it checks the header for the READ and ACK flags, if these are set it is a message that the USB-CAN converter has passed on from the CAN bus.

(6) Write CAN messages

Messages can be sent to the CAN bus by setting the WRTE flag in the serial header and filling the rest of the frame with the appropriate data.

3.3 Boot Loader

During debugging time it takes a lot of inconveniences to plug in Jtag to start downloading the code every time. Therefore AT90CAN128 provides a special bootloader program which can download the code by itself. The main function includes Read/Write Flash and Read/Write EEPROM. By setting the according flags in SPMCSR, Store program Memory Control and Status Register, the content of each page can be completely removed first and written with new content.

4 Results

4.1 Sensor

From the software design illustrated from last section, the result of running a testing code is shown below, which shows the exactly enough environment information close to the robot.

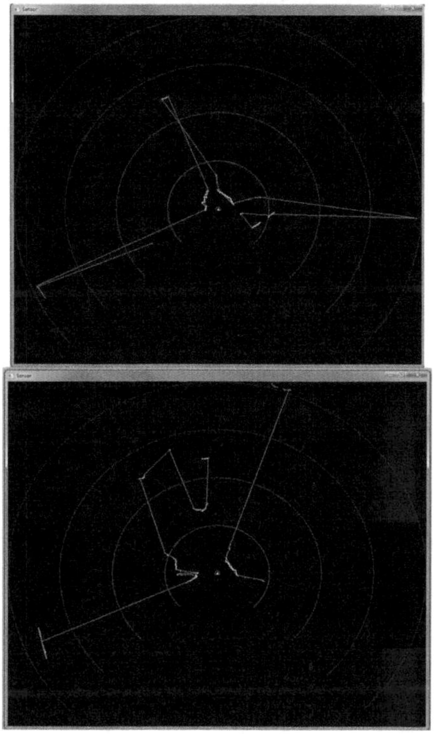

Fig. 5. The picture drawing by experimental data from laser range finder URG-04LX.As is seen, the radius of each circle is 1m within each red line and small white line represents the closest obstacle. The green line means the connection among the obstacles.

4.2 USB-CAN Mechanism

To test the new USB-CAN mechanism, Kvaser USBcan Rugged HS/HS is used here as a main USB-CAN converter. In the testing, on one side it is connected to PC while it is connected to CAN bus on the other side. The testing result shows the USB-CAN converter can successfully send and receive both CAN message and USN data in both ways.

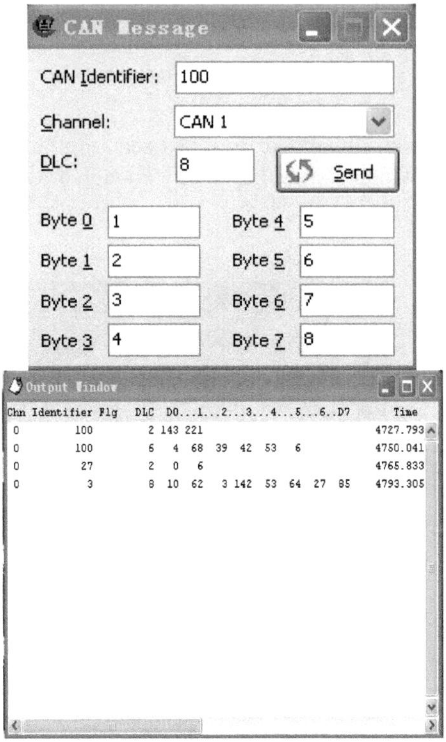

Fig. 6. CANking provides a sending data window for sending message to CAN bus through laptop. On the other side, there is an output window which shows the output of CAN bus.

4.3 Motor Driver

For evaluation, actual speed which is derived from feedback pulse is compared with set speed which is programmed as input value of PWM. The calculation of actual speed is

$$Actual\_speed = \frac{D \times \pi}{T} \tag{1}$$

Where D means a diameter of robot's wheel, T is an average time interval between CAN messages.

In contrast, the robot's speed can be set according to the proportional Pulse Width Modulation (PWM) input value. It can be calculated by the equation (2).

$$Set\_speed = \frac{N \times V_{\max}}{PWM_{\max} \times GR \times s} \tag{2}$$

Where N means input value of PWM, V_{max} is the maximum speed of motor PWM_{max} is the maximum value of PWM, GR is gear ratio of attached gearbox, s is second.

Table 2 presents the data for both actual speed and set speed. It has a slightly difference at the beginning. However, as the increase of set speed, the error increases a lot. Especially when the input value of PWM is set at 180, the maximum error goes up to 0.2485m/s.

Table 2. The data of actual speed and set speed in the test

Actual speed(m/s)	0.3181	0.6155	0.8508	1.1712	1.3849	1.6947
Set speed(m/s)	0.3239	0.6477	0.9716	1.2955	1.6193	1.9432

4.4 Kicking Device

Due to the lack of mechanical kicking device currently, the test just shows the electromagnetic kicking device can be fully charged to 300V within less than 7 seconds. However, it is more than enough to drive a normal mechanical kicking device with this approach.

5 Conclusion

This paper mainly designs and implements both hardware and software of soccer robot system. Most of the work is done during the study of Saga robotic project with classmates from Sweden, China, Thailand, Romania, India, Iran and Pakistan. Therefore Mälardalen University is aiming to participate in the upcoming 2012 Mexico RoboCup Middle Sized League, probably together with East China University of Science and technology. The main work is as follows:

(1) Two sensors are employed in this robot system, laser range finder and electric compass respectively. Within their help, the robot can get all necessary accurate environment information.

(2) From the perspective of cost and compatibility, this paper designs all the modules based on the same microcontroller AT90CAN128. Besides, USB-CAN converter mechanism is introduced to make the internal communication module easy and reliable.

(3) The motor driver module uses AT90CAN128 as the main controller to control motors. Moreover, there is a feedback signal meaning the actual speed for robot. From this design architecture, the control can be more stable, flexible and precise.

(4) Compared to the traditional pneumatic drive of kicking device, electromagnetic drive of kicking device is more efficient, getting a big capacitor fully charged to 300V within 7 seconds.

In addition, there is still much future work to be done for making an advanced soccer robot, such as employing more sensors, implementing PID algorithm on the motor driver module and so on.

References

1. RoboCup Official Website, http://www.robocup.org/
2. Liu, Y.-P.: Design and Implementation of Soccer Robot Control System, Science Technology and Engineering. Modular Machine Tool & Automatic Manufacturing Technique (2004)
3. Kim, J.-H., Shim, H.-S., Kim, H.-S., Jung, M.-J., Choi, I.-H., Kim, J.-O.: A Cooperative Multi-Agent System and Its Real Time Application to Robot Soccer. In: Proceedings of the IEEE International Conference on Robotics and Automation, vol. 1, pp. 638–643 (1997)
4. Wyeth, G., Kee, D., Wagstaff, M., Brewer, N., Stirzaker, J., Cartwright, T., Bebel, B.: Design of an Autonomous Humanoid Robot. In: Proceedings of Australian Conference on Robotics & Automation, pp. 39–44 (2001)
5. Leng, C.: Jiaolong Team Description (2010), http://robocup.sjtu.edu.cn/English/down
6. Min, H., Hui, S., Wang, X.: Water Team Description Paper (2010), http://jdgcxy.bistu.edu.cn/robocup/index_robocup.asp
7. Neves, A.J.R., Azevedo, J.L., Cunha, M.B.: CAMBADA Team Description paper, http://www.ieeta.pt/atri/cambada
8. de Best, J.J.T.H., Bruijnen, D.J.H., Hoogendijk, R.: Tech United Eindhoven Team Description (2010), http://www.techunited.nl/media/files
9. Kappeler, U.P., Zweigle, O., Rajaie, H.: RFC Stuttgurt Team Description (2010), http://robocup.informatik.uni-stuttgart.de/rfc/www/
10. Su, S., Gu, Z., Chen, X.: ENDEAVOR Team Description Paper (2010), http://elc.wzvtc.net.cn/robot/index.html
11. URG-04LX-UG01Datasheet, http://www.active-robots.com/products/sensors/hokuyo
12. HM55B Datasheet, http://www.hobbyengineering.com
13. FT232BL Datasheet, http://www.usconverters.com
14. AT90CAN128 Datasheet, http://www.atmel.com
15. MC33033 Datasheet, http://www.alldatasheet.com

Study of Cloud-Type Recognition Based on Multi-class Features

Ling Yang[1], Zhong-ke Wang[2], Jun Wang[1], and Wen-ting Cui[1]

[1] College of Electronic Engineering, Chengdu University of Information Technology, Chengdu 610225, China
[2] College of Network Engineering, Chengdu University of Information Technology, Chengdu 610225, China

Abstract. According to changeability of cloud, cloud-type recognition was primarily based on single-class feature in previous papers which was restricted to a certain degree. A set of features describing the color, texture as well as the shape features were extracted, then the shape and texture features combination methods were discussed. Here Gray-level co-occurrence matrix(*GLCM*) and Gabor wavelet transform based texture features and Zernike moment based shape features were combined, then support vector machine (*SVM*) was employed to recognize cloud-type. Experimental results showed that the correct recognition rates of altocumulus, cirrus, clear, cumulus and stratus were improved significantly, with the average recognition rate of 88.6%, and clear sky and stratus's recognition rate of 100%.

Keywords: Feature Extraction, Feature Combinations, Cloud-Type Recognition, Support Vector Machine (*SVM*).

1 Introduction

Cloud-type recognition is useful for predicting the tendency of weather change. Total-sky image cloud-type recognition research is a relative new research field. What's more, only a few of the relevant achievement were got by domestic and foreign scholars. Buch et al(1995) [1] achieved the recognition of cloud-type observed in Whole Sky Imager(*WSI*) which could distinguished five different sky types(altocumulus, cirrus, cumulus, stratus and clear sky) in 61% accuracy. Peura et al(1999) [2] used k-means clustering algorithm which could correctly recognize cloud images(Taken from the whole sky imager designed by VAISALA company) in 65% accuracy. Calbo et al(2008) [3] realized the recognition of clear, billow cloud, cirrus, stratiform and cumuliform cloud in 76% accuracy. In China, Sun Xuejin et al(2009, PLA university of science and technology) [4] developed whole sky infrared cloud measuring system, stratus, cumulus, undulatus, cirrus and clear sky which were analyzed by LBP operator and the average recognition rate was 87.2%. However, whole sky infrared cloud measuring system needs to calibrate the instruments beforehand, and the measurement results are impacted by the environment.

G. Lee (Ed.): Advances in Automation and Robotics, Vol. 1, LNEE 122, pp. 355–361.
springerlink.com

In this paper, on the basis of others studies, clouds were separated from its background by color features of clouds. The texture and the shape features of clouds were extracted from cloud areas, respectively. A combination of the texture and the shape features were used as the input of support vector machine (*SVM)* to recognize cloud-type in this paper, while single texture features were used to identify cloud-type in previous work.

2 Feature Extraction

2.1 Clouds Extraction

In this experiment, the visible Total-sky images were selected from Solar Radiation Research Laboratory of American National Renewable Energy Laboratory (2005～2010) [5]. Optical image for 352 x 288/24bit *JPEG* can objectively reflect the total sky information within the 160° field of view. Images taken by instrument were preprocessed before cloud-type recognition. In order to fully extract cloud area, the geophysics information which image edge contains should be removed. The cloud area of the whole images is the target in the research.

In this paper, the sky index [6] method was used to extract cloud area by color features of clouds. The sky index was defined as follow:

$$\text{sky index} = (B-R)/\,(B+R). \tag{1}$$

where B, R respectively presents the values of the blue and red channel in *RGB* color space. In equation (1), the sky index range is (0~1). The sky color is white if the sky index is close to 0. On the contrary, the sky color is blue if the sky index is close to 1. In this paper, the threshold of sky index which was used to estimate the cloud area was 0.3. This meant that if the sky index of one area was less than 0.3, this area was the cloud area. Figure 1 showed the template which was used to remove geophysics information which image edge contained. Figure (a), (b), (c), (d) respectively showed stratus (St), cirrus (Ci), cumulus (Cu), altocumulus (Ac) and clear sky which were taken by the camera. Figure (e), (f), (g), (h) showed the extractions of cloud area when the sky index was 0.3 corresponding to figure (a), (b), (c) ,(d). Because of no cloud in clear sky, the entire background was used as the research object in this paper.

Fig. 1. Template

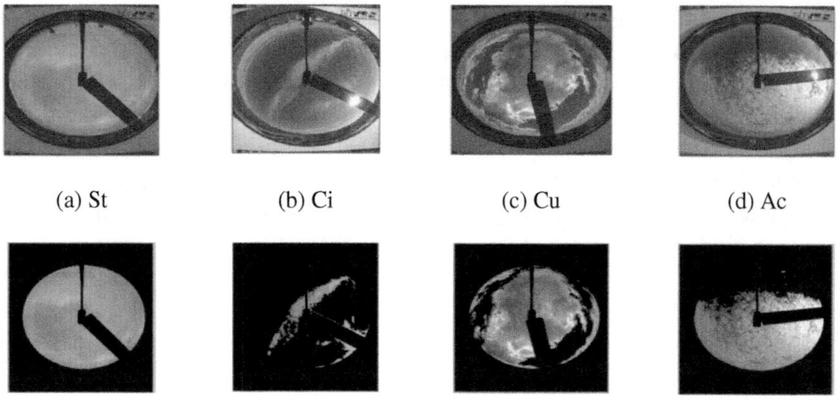

(a) St (b) Ci (c) Cu (d) Ac

(e) cloud area of St (f) cloud area of Ci (g) cloud area of Cu (h) cloud area of Ac

Fig. 2. The four types of clouds and its cloud area

2.2 Feature Analysis

From figure 2, different types of cloud images had different texture features and shape features. In this paper, the texture and shape features were analyzed primarily.

The texture feature of image, which represents the measurement of the pixel-to-pixel in the image, is used to describe the spatial information in the image, such as position, orientation, size, shape and so on. The texture feature is rotational invariance, and anti-noise. The texture features were extracted by employing gray-level co-occurrence matrix [7] and Gabor wavelet transform [8] in this paper.

Shape, which is another important feature to describe objects, is the feature of outline of targets or corresponding regions based on the binary image. The shape feature of clouds is mainly expressed by two aspects: outer contours of clouds, the interior area surrounded by the contours of cloud. Only using single edge [9] couldn't distinguish the clouds because the cloud's shape is complex. Region-based algorithm is more suitable for cloud-type identification as it could extract the edge features and the internal features of the clouds better than other edge detection methods do. All the pixels in the target area are expressed by Zernike moment, which based on the shape of the regional feature extraction algorithm. The Zernike moment is invariant of scale, translation, and rotation and scale. There was a high recognition rate of Zernike moment for identifying severe malformation image [10]. The Total-sky images used in this paper were based on globular projection, so the cloud image which obtained was malformation. Therefore, Zernike moment [11] was selected to extract shape features of cloud in this paper.

3 Feature Combinations

The texture and shape features reflected the features of images from different perspectives, so only one of them which were used to recognize the image's features was not enough [12]. Gray-level co-occurrence matrix mainly was used to count the gray-value data on a small scale in the spatial domain. The gray-level co-occurrence matrix was good at the image with simple texture features but bad at complex features. Gabor wavelet had two properties: multi-scale and multi-direction, so Gabor wavelet can extract more information, especially details. It can be seen that gray-level co-occurrence matrix and Gabor wavelet transform reflected texture features of image from different perspectives. A combination of these two kinds of texture features not only could classify the single image well, but also could extract more detail information. Region-based Zernike moment extracted not only the edge features but also the internal features of the clouds. What's more, it was also helpful for the malformation of cloud image in Total-sky images.

Therefore, this paper presented a algorithm, which was a combination of gray-level co-occurrence matrix, and Gabor wavelet transform based texture features and Zernike moment based shape features to form integrated features of cloud, in order to describe the cloud from different angles.

4 Results

In this experiment, 2500 visible total-sky images [5] were selected to be divided into 5 classes, and in each class, the first 400 were used for training and the last 100 were used for the cloud-type pattern identification. Note that cloud images contained only single cloud type were selected in this paper. Support vector machine shows many special advantages in resolving the small sample, nonlinear and high dimensional pattern recognition and it can also be applied to function-fitting and other machine learning problem. The characteristics of *SVM* and its commonly used kernel function can refer to literature [13]. There are many software packages to implement *SVM* algorithm now. Professor Chih-Jen Lin in Taiwan University has written *LIBSVM* software package using C++ to implement *SVM* algorithm. Its advantages are easy to use, and choose a few parameters. So in this paper, it was selected to complete cloud-type pattern recognition experiment.

In the experiment, as described previously, many features as clarity, energy, correlation, entropy and stability of sample images in 0°,45°,90°,135° four directions were calculated, respectively. Then, mean and variance of each feature in four directions which add up to 8 features were calculated as texture features. The Gabor filters with four scales and eight directions were used to extract two texture features by 64-dimensional mean and variance; Zernike moments were used to extract shape features with 12 characteristic parameters in 5 orders. These features were taken as SVM's inputs to identify cloud-type. A combination of gray-level co-occurrence matrix(GLCM), and Gabor wavelet transform based texture features and Zernike moment based shape features which count 22 features was used as the input features of *SVM* for cloud-type recognition. Results of cloud-type recognition were showed in Table 1.

Table 1. The recognition accuracy of cloud type

Accuracy	Ac	Ci	clear	Cc	St	Accuracy
GLCM	71%	86%	99%	76%	100%	86.4%
Gabor wavelet transform	66%	67%	78%	40%	97%	69.6%
Zernike moment	52%	72%	100%	76%	99%	74.8%
After combined	72%	91%	100%	79%	100%	88.6%

As shown in table1, the five kinds of sky type were able to completely identify by four kinds of feature extraction methods. The methods which based on gray-level co-occurrence matrix and Gabor transform were the easiest method to recognize clear sky and stratus; cirrus the next; cumulus and altocumulus the worst. The region recognition based on Zernike moments had the best recognition accuracy for clear sky and stratus; cumulus and altocumulus the next. The average recognition rate was improved significantly by combining texture and shape feature. Recognition rate of clear sky and stratus were highest which all were 100%; cumulus and altocumulus the follow. The reasons for low recognition rate were showed in Table 2.

Table 2. Recognition results of test images after combined

	Ac	Ci	Clear	Cu	St	Accuracy
Ac	72	5	0	★23	0	72%
Ci	7	91	0	2	0	91%
Clear	0	0	100	0	0	100%
Cu	7	7	0	79	7	79%
St	0	0	0	0	100	100%

In table 2, the first column was artificially recognition, and the first row was *SVM* recognition. Gray shades demonstrated the clouds which recognized by *SVM* and artificially recognition respectively were the same type. Data marked with ★ demonstrated one type which had the largest number of single-class misclassification.

As shown in table 2, the accuracy of cirrus, stratus and clear sky all surpassed 90%; the accuracy of cumulus and altocumulus were 72% and 79%, respectively; clear sky and stratus which were both 100%; the average accuracy reached 88.6%. There occurred a phenomenon of false recognition of cumulus and altocumulus: 23 altocumulus cloud images were mistaken for cumulus; 7 cumulus cloud images were mistaken for altocumulus.

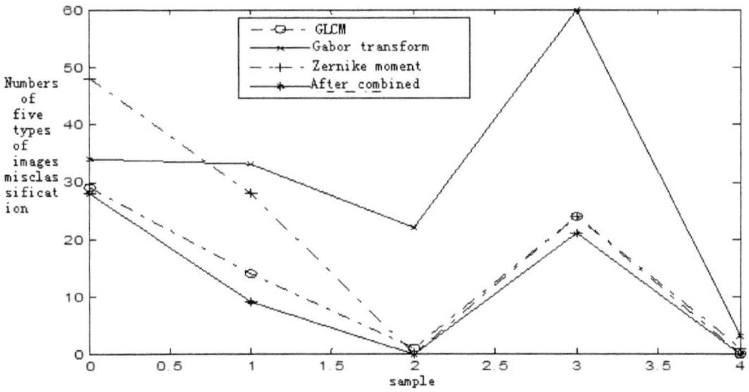

Fig. 3. Numbers of five types of images misclassification

From figure 3, the number of errors reduced obviously after combined. However, the samples from cumulus and altocumulus had large numbers of misclassification. Reasons were as follows: as shown in table 2, cumulus and altocumulus were easy to be misjudged; cumulus and altocumulus had obvious block effects, clear contour line, high contrast value relative to sky, texture heavy and fold obviously showed in images. As a result, cumulus and altocumulus were difficult to identify through machine recognition.

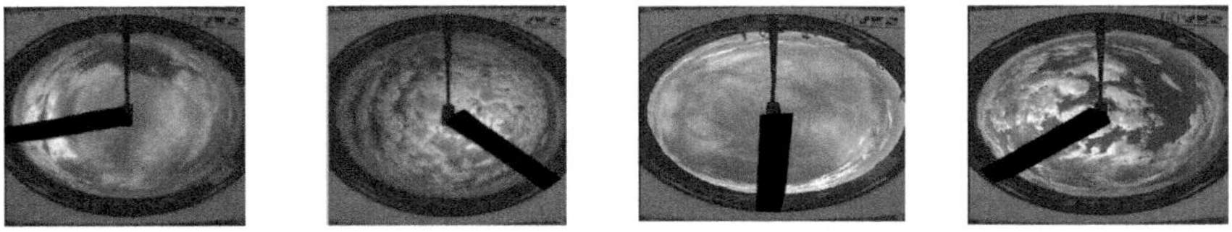

(a) altocumulus (b) altocumulus (c) cumulus (d) cumulus

Fig. 4. Altocumulus images are mistaken for cumulus

Such as altocumulus in cloud-type recognition, as Figure 4(a), (b) shown, the experiments which used the methods based on Gray-level co-occurrence matrix, Gabor wavelet transform and Zernike moment respectively, were misjudged to cumulus(Figure 4(c), (d)). Then the experiment which employed a combination of Gray-level co-occurrence matrix, and Gabor wavelet transform based texture features and Zernike moment based shape features, successfully recognized the altocumulus showed in Figure 4(a), (b). This showed feature combined can improve the cloud-type recognition rate.

5 Conclusions

In this paper, a set of features describing the color, texture as well as shape features were extracted. Clouds were separated from the background using color features of clouds in visible Total-sky image samples. Then texture and shape features which were extracted from cloud area were combined to realize the recognition of altocumulus, cirrus, clear, cumulus and stratus by *SVM*. The total recognition rate was 88.6%. Compared with the results of classification by using different features, it showed that altocumulus and cumulus were difficult to identify because of the similar in shape and texture features. In the single-class identification process, some cloud-type would be misclassification. These phenomena were able to be improved by the combination of gray-level co-occurrence matrix and Gabor wavelet transform based texture features and Zernike moment based shape features in this paper. At the same time, the total recognition rate improved dramatically, and the recognition rate of clear sky and stratus were both 100%, especially.

References

1. Buch Jr., K.A., Sun, C.H., Throne, L.R.: Cloud Classification Using Whole-Sky Imager Data (1995)
2. Tag, P.M., Bankert, R.L., Robin Brody, L.: An AVHRR Multiple Cloud-Type Classification Package (1999)
3. Calbó, J., Sabburg, J.: Feature Extraction from Whole-Sky Ground-Based Images for Cloud-Type Recognition. J. Atmos. Oceanic Technol. 25, 3–14 (2008)
4. Sun, X.-J., Liu, L., Gao, T.-C., Zhao, S.-J.: Classification of Whole Sky Infrared Cloud Image Based on the LBP Operator. Transactions of Atmospheric Sciences 32(4), 490–497 (2009)
5. All-sky imager, `http://www.knmi.nl/samenw/cloudnet/realtime/rt_img_tsi.html`
6. Yamashita, M., Yoshimura, M., Nakashizuka, T.: Cloud Cover Estimation Using Multi Temporal Hemisphere Imageries. In: ISPRS Congress, Istanbul (2004)
7. Li, Y., Ou, Z.: Face Recognition Based on Gabor Wavelet Transform and Support Vector Machines. Computer Engineering 32(19), 181–182 (2006)
8. He, X.-J., Wang, F.-M.: Study of Texture Analysis Method Based on GLCM. Shanxi Electronic Technology 4, 89–93 (2010)
9. Yang, H.-Y., Du, W.-B., Zhang, L.: The Application of Image Recognition Method Based on Hu-moment Invariants in Auto-pouring System. Transactions of Shenyang Ligong University 28(5), 34–45 (2009)
10. Ying, Y.-B., Gui, J.-S., Rao, X.-Q.: Fruit shape classification based on Zernike moments. Journal of Jiangsu University (Natural Science Edition) 28(1), 1–3 (2007)
11. Liu, M., He, Y., Hu, H.: Image Shape Feature Description and Evaluation Measure Based on Zernike Moment in Image Mining. Computer Engineering 32(22), 178–180 (2006)
12. Welch, R.M., Sengupta, S.K., Goroch, A.K., Rabindra, P., Rangaraj, N., Navar, M.S.: Polar Cloud and Surface Classification Using AVHRR Imagery. An Intercomparison of Methods. J. App. Meteor. 31, 405–420 (1992)
13. Jiang, Y., Zhao, Q.: The Machine Learning Approach, vol. 08. Publishing House of Electronics Industry (2009)

A Fast Sparsity Adaptive Matching Pursuit Algorithm for Compressed Sensing

Chun-hui Ma[1], Chun-yun Xu[1], Lei Shen[1], and Shi-lian Zheng[2]

[1] Telecommunication School, Hangzhou Dianzi University, Zhejian 310018, China
[2] State Key Lab of Information Control Technology in Communication System of No.36 Research Institute, China Electronic Technology Corporation, Jiaxing 314001, China
lancer711@163.com

Abstract. Sparsity adaptive matching pursuit algorithm (SAMP) is a compressed sensing signal reconstruction algorithm with good performance. However, as the support set expands one time, the backward pursuit should be processed for many times, in which redundant update calculations are included. To solve the problem, this paper presents a fast sparsity adaptive matching pursuit algorithm (FSAMP). During the operation of the FSAMP, the backward pursuit is performed only once when the support set expands. After the support set is fixed, the backward pursuit shall be executed for multi times to guarantee the reconstruction precision. Simulation results for signal reconstruction show that the FSAMP has the slight higher reconstruction precision and the much faster reconstruction speed compared with the SAMP.

Keywords: compressed sensing, subspace pursuit, sparsity adaptive matching pursuit, image processing.

1 Introduction

Compressed sensing(CS)[1], which is a new sampling theory, points out that, the acquisition for sparse and compressible signals can be processed as a non-adaptive linear measuring at rates significantly lower than Nyquist rate. The CS sampling not only reduces the data collection requirement for hardware devices, but also executes data compression while signal sampling, which saves the storage space.

Compressed sensing theory mainly consists of three components: signal sparse representation, measuring matrix design and reconstruction algorithm. The reconstruction algorithms that have been proposed include convex algorithm, non-convex algorithm and greedy algorithm, etc. The recovery algorithms based on greedy algorithm have received considerable attention in the recent years and been widely applied in the related fields for their advantages such as the simple structure, easy implementation and low reconstruction complexity. The proposed greedy algorithms include matching pursuit(MP)[2], orthogonal matching pursuit(OMP)[3], stage-wise OMP(StOMP)[4], regularized orthogonal matching pursuit(ROMP)[5], compressive sampling matching pursuit(CoSaMP)[6], subspace pursuit(SP)[7], and

G. Lee (Ed.): Advances in Automation and Robotics, Vol. 1, LNEE 122, pp. 363–368.
springerlink.com

sparsity adaptive matching pursuit(SAMP)[8]. Among these algorithms, the SP algorithm has the higher reconstruction precision and low reconstruction complexity. However, SP needs the prior information of the signal sparsity, which constrains the application of SP in many fields such as image processing for the signal sparsity is unknown. SAMP is an improved algorithm based on SP and it can be executed without prior sparsity information while the real sparsity would be approached step by step. The SAMP is better than SP in terms of reconstruction precision but more complicate.

A fast sparsity adaptive matching pursuit algorithm is proposed in this paper. In the proposed algorithm the updating of support set performs only once when the support set expands so that the algorithm could be converged faster while the complexity is reduced greatly. Simulation results show that the FSAMP has the faster reconstruction speed and the higher reconstruction precision compared with SAMP.

2 Image Processing Based on Compressed Sensing

The main idea of compressed sensing lies in: suppose a signal(or image) **X**, if it is sparse in the time domain or has a sparse representation in the transform domain $\mathbf{\Psi}$, it can be processed by non-adaptive linear measuring with a M×N(M<<N) measuring matrix $\mathbf{\Phi}$ and the measuring samples **y**, which can be taken as the compressed Nyquist sampling data, is acquired

$$\mathbf{y}=\mathbf{\Phi X}=\mathbf{\Phi \Psi S}=\mathbf{\Theta S} \tag{1}$$

Where $\mathbf{\Theta}=\mathbf{\Phi\Psi}$ is an M×N matrix. **S** is a sparse representation of **X** in the transform domain $\mathbf{\Psi}$. If the measuring matrix $\mathbf{\Phi}$ is designed under a restricted isometry property(RIP) [9], the sparse representation **S** could be acquired by some certain reconstruction algorithms with the known information about measuring matrix $\mathbf{\Phi}$, transform matrix $\mathbf{\Psi}$ and the measuring samples **y**. And then the signal **X** is available, that is

$$\mathbf{X}=\mathbf{\Psi S} \tag{2}$$

The sparsity of image signal in the wavelet domain has been fully verified. Consider a signal X, which has a sparse representation S in the wavelet domain, by Eqn.(1), the **y** can be seen as a inner product of $\mathbf{\Theta}$ and **S**. With some certain reconstruction algorithms the **S**, which is a sparse representation of **y** in the transform domain $\mathbf{\Theta}$, can be recovered directly and then with wavelet inverse transform the original image signal **X** is acquired.

3 Fast Sparsity Adaptive Matching Pursuit Algorithm

The basic idea of the OMP family is to select out the support set of the original signal by greedy iterative, namely select out the columns that were involved in the composition of **y** from matrix $\mathbf{\Theta}$ and the sparse vector **S** is obtained. The OMP mainly employs forward pursuit which expands the support set step by step until the size of the support set is consistent with the size of the real one. As for the SP, the support set shall be updated many times by backward pursuit at a fixed size I=K. Obviously the SP algorithm calls for the signal sparsity as prior conditions. SAMP algorithm uses the advantages of the two algorithms, while the forward pursuit is processed by expanding the support set with a step size, the backward pursuit is executed to eliminate the wrongly selected elements in the support set. Because the backward pursuit should be processed for many times when the support set of SAMP expands one time, redundant update calculations are included.

The proposed fast sparsity adaptive matching pursuit algorithm(FSAMP) simplifies the reconstruction into two phase and takes variable P as a indicator of the reconstruction state, P=0 indicates phase one and P=1 phase two. In the phase one, the backward pursuit just performs only once after every time the support set expands until the condition is met to enter phase two, while in the phase two the backward pursuit is repeated many times at a fixed support set size until the halting condition is met. Though there may be wrongly selected atoms when the support set expends the wrong ones will be eliminated in the subsequent update. The atoms with big weights which are selected out earlier will be tested for many times while the ones which are selected out later get fewer tests. Compared with SAMP algorithm the proposed algorithm greatly improves the reconstruction efficiency and has the guarantee of reconstruction precision. The flow chart of the FSAMP algorithm is demonstrated in Fig.1 and the main steps of FASMP algorithm are summarized below.

Input: measuring sample vector $\mathbf{y}$, reconstruction matrix $\mathbf{\Theta}$, step size st
Output : $\Delta\mathbf{S}$ is the sparse approximation of the original signal
Initialization:
$\mathbf{r}_0=\mathbf{y}$ {initial residual}
$F_0=[]$ {index set of support set}
I=st {initial size of the F_0}
j=1 {iteration index}
P=0 {initial stage index}

1) $\mathbf{B}_j$ =Max ($|\mathbf{\Theta}^*\mathbf{r}_{j-1}|$, I) {find the best match I columns from $\mathbf{\Theta}$}
2) $\mathbf{C}_j=\mathbf{F}_{j-1}\cup\mathbf{B}_j$ {make the index set of candidate set}
3) $\mathbf{Ft}$=Max ($|\mathbf{\Theta}^{\dagger}{}_{Cj}{}^*\mathbf{y}|$, I) {find the best match I columns from $\mathbf{C}_j$}
where $\mathbf{\Theta}_{Cj}$ denotes a matrix which consists of the columns from $\mathbf{\Theta}$ according to $\mathbf{C}_j$, $\mathbf{\Theta}^{\dagger}=(\mathbf{\Theta}'\mathbf{\Theta})^{-1}\mathbf{\Theta}'$ denotes the pseudo-inverse of the matrix $\mathbf{\Theta}$, where $\mathbf{\Theta}'$ denotes the transposition of the matrix $\mathbf{\Theta}$.
4) $\mathbf{r}=\mathbf{y}-\mathbf{\Theta}_{Ft}\mathbf{\Theta}^{\dagger}{}_{Ft}{}^*\mathbf{y}$ {calculate residual}
5) If P=1, then shift to 7) otherwise shift to 6).
6) Update $\mathbf{F}_j=\mathbf{Ft}$, update residual $\mathbf{r}_j=\mathbf{r}$. If $\|r_{j-1}\|_2-\|r\|_2\leq\varepsilon$, then j=j+1, P=1 and shift to 2); otherwise expend the support set size I=I+st, j=j+1 and shift to 2). $\|\bullet\|$ denotes the 2 norm of vector (•), ε is a threshold.
7) If $\|r_{j-1}\|_2\leq\|r\|_2$, then quit iteration and calculate the sparse approximation $\Delta\mathbf{S}$, otherwise shift to 8).
8) Update $\mathbf{F}_j=\mathbf{Ft}$, update residual $\mathbf{r}_j=\mathbf{r}$, j=j+1 and shift to 2).

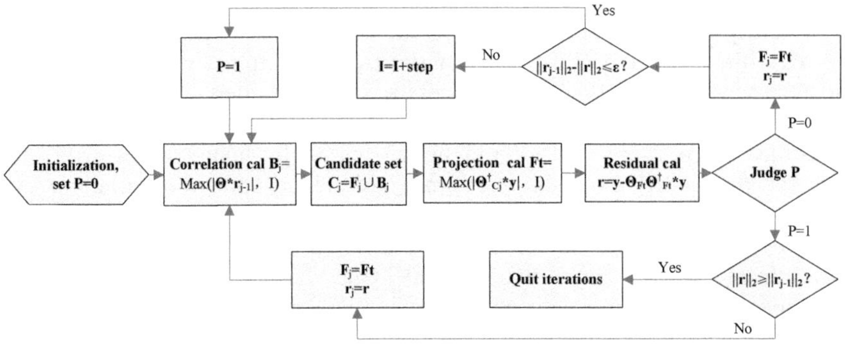

Fig. 1. FSAMP algorithm flow chart

4 Simulation Results

The 512*512 image Lena and 512*512 image Boat would be processed by compressed sensing. The sparsifying matrix is Daubechies wavelet 9/7 wavelet and the structurally random matrices are used as the measuring matrix [10]. The sampling rate of measuring matrix are 0.2,0.3,0.4,0.5 and 0.6.

The SP, SAMP and FSAMP are respectively used to realize the image reconstruction. Let the step size in the FSAMP st=M/(2*log2(N)) as well as in the SAMP . The input sparsity K of SP is estimated by setting it equal to the sparsity that SAMP detects. The experimental platform is a computer with AMD AthlonTM 64 X2 4000+ CPU and the software is Matlab 7.1. The reconstruction results of the two image by FSAMP at different sampling rate are respectively shown in Fig. 2 and Fig. 3. It is obviously seen from Fig. 2 and Fig. 3 that the proposed algorithm can reconstruct the original image well at different sampling rate.

Fig. 2. The reconstruction results of Lena512 by FSAMP at different sampling rate (a) M/N=0.2; (b) M/N=0.4; (c) M/N=0.6; (d)original image

Fig. 3. The reconstruction results of Boat512 by FSAMP at different sampling rate (a) M/N=0.2; (b) M/N=0.4; (c) M/N=0.6; (d)original image

In order to quantitatively compare the performance of SP, SAMP and FSAMP algorithms, their running time and reconstruction PSNR of the two images are given in Table 1 and Table 2.

Table 1. Three algorithms' running time and reconstruction PSNR of Lena512

M/N Sampling rate	Running time/(s)			PSNR/(dB)		
	SP	SAMP	FSAMP	SP	SAMP	FSAMP
0.2	164.23	166.26	115.32	28.19	28.21	28.21
0.3	159.52	132.31	103.59	30.60	30.67	30.72
0.4	179.98	168.51	104.88	32.44	32.69	32.70
0.5	205.29	242.79	105.80	33.84	34.38	34.45
0.6	193.72	286.87	117.87	34.92	35.74	35.88

Table 2. Three algorithms' running time and reconstruction PSNR of Boat512

M/N Sampling rate	Running time/(s)			PSNR/(dB)		
	SP	SAMP	FSAMP	SP	SAMP	FSAMP
0.2	270.60	198.45	135.16	28.36	28.76	28.77
0.3	225.68	237.11	118.82	30.88	31.39	31.35
0.4	288.59	269.75	125.55	32.59	33.37	33.42
0.5	288.71	325.45	121.64	33.86	35.22	35.30
0.6	219.61	402.50	140.37	36.51	36.89	36.97

As can be seen, the running time of SAMP is close to that of SP at low sampling rates but with the increasing of sampling rate the speed of SAMP declines rapidly. When the sampling rate M/N=0.6, the running time of SAMP for the reconstruction of that two images are 286.7s and 402.50s which are higher than that of SP. The running time of FSAMP is far below that of SP and SAMP at all of sampling rates and with the increasing of sampling rate the speed of FSAMP just declines slightly. At the same time the PSNR of FSAMP is higher than that of SP and SAMP.

5 Conclusion

To reduce the redundant calculation of SAMP, the fast sparsity adaptive matching pursuit algorithm (FSAMP) is proposed in this paper. The proposed algorithm greatly improved the reconstruction efficiency while with the guarantee of reconstruction precision by eliminating the unnecessary updating calculation of SAMP algorithm. Simulation results show that the FSAMP has the faster reconstruction speed and the higher reconstruction precision than SAMP.

References

1. Donoho, D.L.: Compressed sensing. IEEE Transactions on Information Theory 52(4), 1289–1306 (2006)
2. Mallat, S., Zhang, Z.: Matching pursuit with time-frequency dictionaries. IEEE Trans. on Signal Processing 41(12), 3397–3415 (1993)
3. Tropp, J., Gilbert, A.: Signal recovery from random measurements via orthogonal matching pursuit. Transactions on Information Theory 53(12), 4655–4666 (2007)
4. Donoho, D.L., Tsaig, Y., Starck, J.-L.: Sparse solution of underdetermined linear equations by stage-wise orthogonal matching pursuit. Technical Report (March 2006)
5. Needell, D., Vershynin, R.: Uniform uncertainty principle and signal recovery via regularized orthogonal matching pursuit. Found. Comput. Math. 9(3), 317–334 (2009)
6. Needell, D., Tropp, J.A.: CoSaMP: Iterative signal recovery from incomplete and inaccurate samples. Applied and Computational Harmonic Analysis 26(3), 301–321 (2009)
7. Dai, W., Milenkovic, O.: Subspace pursuit for compressive sensing signal reconstruction. In: 2008 5th International Symposium on Turbo Codes and Related Topics, pp. 402–407 (2008)
8. Do, T.T., Lu, G., Nam, N., Tran, T.D.: Sparsity adaptive matching pursuit algorithm for practical compressed sensing. In: Asilomar Conference on Signals, Systems, and Computers, Pacific Grove, California (2008)
9. Baraniuk, R.: A lecture on compressive sensing. IEEE Signal Processing Magazine 24(4), 118–121 (2007)
10. Do, T.T., Trany, T.D., Gan, L.: Fast compressive sampling with structurally random matrices. In: Proceedings of the IEEE International Conference on Acoustics, Speech and Signal Processing, Washington D.C., USA, pp. 3369–3372 (2008)

Modeling Research on Internal Manufacturing Supply Chain Based on SCOR Model

Hailili Maimaiti, Zailin Guan*, and Baoxi Wang

The School of Mechanical Science and Engineering, Huazhong University of Science and Technology, 430074 Wuhan, Hubei, China

Abstract. This paper introduces the supply chain management, comments the basic process and levels of supply chain operation reference model SCOR 10.0, and combining the characteristics of the internal manufacturing enterprises, this paper puts forward an internal manufacturing supply chain modeling method based on SCOR model, which can better give consideration to the performance of the whole supply chain and realize the enterprise process's optimization.

Keywords: supply chain management, supply chain operation reference model (SCOR), modeling.

1 Introduction

With the development of the times, the contemporary enterprise competition has changed from the competition between enterprises to the competition between supply chains. Therefore, when we analyze the enterprise internal problems and improve the weakness, we can't just limit to the enterprise interior, we should expand the scope of survey and look at the more inclusive supply chain network which is in close touch with the enterprise. Only from the Angle of the supply chain can we observe, analyze and improve the enterprise internal various processes and problems more comprehensively and systematically [1].

The concept of enterprise internal supply chain forms like this: each process of the enterprise interior is one of the supply chain links, the former link is a supplier of the latter link, and the latter link is a customer of the former link. In this light, the products flow from suppliers to the final customer, forms a complete "supply chain which is from the supplier's supplier to the customer's customer" [2].

In recent years, the SCOR model has achieved ever-growing importance, most importantly in the North America field of business, but also increasing in Asia and Europe. It is surprising that hardly any scientific studies are available with reference to the model and its application. In actual fact, its reference to realization and its

* Author introduction: Zailin Guan, male, professor of huazhong university of science and technology, doctoral supervisor, research direction: digital factory, workshop production operation and optimization, manufacturing system modeling simulation analysis; Hailili·Maimaiti, male, graduate student of huazhong university of science and technology; Baoxi Wang, female, doctoral student of huazhong university of science and technology.

G. Lee (Ed.): Advances in Automation and Robotics, Vol. 1, LNEE 122, pp. 369–378.
springerlink.com

efficiency are simply taken to be a given quantity. SCOR has already been obtained in many professions. The practice of many domestic and foreign companies proves that using SCOR model to manage the business logistics, the company can guarantee the delivery service level and reduce inventory by 50% as well. We may have a close look at the application of SCOR model from two aspects: On the one hand, SCOR believes that internal activities of any enterprises can be divided into the plan, source, make, delivery and return. On the other hand, any company activities mentioned above can be classified according to the characteristic of different products.

SCOR model is a form of standard symbol, which can analyze the characteristics of the supply chain from top to bottom and is an effective identification method to the supply chain and its management. Process optimization can be in some levels, but must be established in the language description which is unified and truly reflects the real situation. SCOR model gives the second and the third levels in different detailed degree, which can let the manager eliminate, combine, delete and change the process in different levels. In the second level, such adjustment is macro and can significantly change the mode of production, such as changing push type production to pull type production. While the third level is the detail of the second level, at this level, the process optimization is for the specific process within department, such as the change of the logistics mode and the establishment of the production unit, and so on.

2 The Supply Chain and SCOR Model

Supply chain is around the core enterprise, through the control of information flow, logistics and cash flow, from the beginning of purchasing raw materials, to producing intermediate products and final products, at last sending the products to the hands of consumers by the sales network, which make suppliers, manufacturers, distributors and retailers, until the end customers a whole function nets chain structure [3].

Supply Chain Operation Reference Model (SCOR) is the results of Supply Chain Council (SCC) [4]. SCOR model offers a unique system frame, combining the business process, index, best practices and technology to a combined structure to support the communication among supply chain partners, so as to improve the effectiveness of the supply chain and supply chain improvement activities.

SCOR model is a set of effective reference model to give the enterprise supply chain structure a graphical performance, with five abstract levels, five different management processes to describe the enterprise process, at the same time, through stressing the plan to strengthen the enterprise department's unified coordination [5].

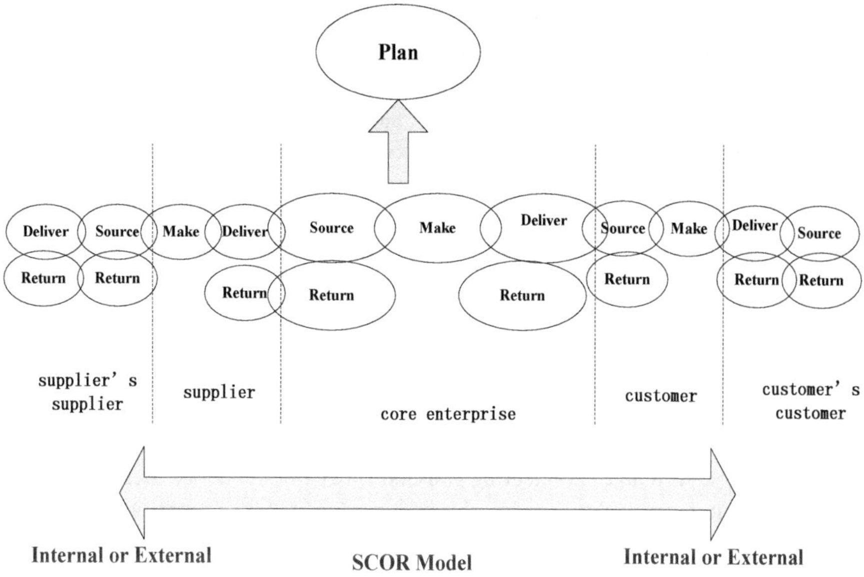

Fig. 1. SCOR model structure schematic picture

The first level of SCOR decomposes supply chain operation into five related management processes, namely Plan, Source, Make, Deliver and Return, their specific meanings as follows.

Plan (P): The Plan processes describe the planning activities associated with operating a supply chain.

Source (S): The Source processes describe the ordering (or scheduling) and receipt of goods and services.

Make (M): The Make processes describe the activities associated with the conversion of materials or creation of the content for services.

Deliver (D): The Deliver processes describe the activities associated with the creation, maintenance, and fulfillment of customer orders.

Return (R): The Return processes describe the activities associated with the reverse flow of goods back from the customer.

3 S Enterprise Secondary SCOR Model

The SCOR model's peculiar supply chain-oriented structure, clear description method, unified performance evaluation standard, can largely make up for the traditional process optimization methods' shortcomings. Although SCOR is a operation reference model developed for supply chain, considering the enterprise internal business departments and processes are similar with the operation characteristics among supply chain node enterprises, in the single enterprise's internal process optimization, also can use the SCOR's internal supply chain concept, adapt innovative modeling method, base on the performance assessment index's targeted selection, carry on the enterprise internal operation process optimization.

In this paper the research on the S enterprise industrial estate's layout with L workshop building as the center, extends to each park warehouse. L workshop covers an area of nearly 90000 m^2 and consists of a structural components area, a machining area and an assembly area of pumping, mixing equipments. L workshop's future outside warehouses mainly include D workshop, G workshop, E workshop, steel warehouse, rubber tubes center and F workshop (center warehouse), among them, D workshop is used to store arms frame and oil pipe; G workshop is used to store hopper, oil tank, turret, water tank and other big elements; E workshop is used to store outrigger, S pipe, turntable, connecting rod and other big elements; steel warehouse, rubber tubes center are used to store steels and hoses needed by pump truck, the vehicle pump and trailer pump; center warehouse is mainly used for storing outsourcing big elements (100 big elements, 200 big elements, 301 big elements, D01 big elements, D02 big elements) and some small materials.

Homemade raw materials are stored in B workshop, according to the production plan they are cut in this workshop, after cutting they are transported by forklift to C workshop (pumping machining), G workshop (pumping structural components), H workshop (pumping structural components) and E workshop (pumping outrigger and turntable) to machining and weld. The machining or welding finished products coated in D workshop or I workshop, after finishing coating the materials are sent into warehouse or deposited in the cache area and distributed online according to the assembly line production plan.

Semi-finished products storage and outside-factory distribution: F workshop includes high shelves and sorting area, high shelves' materials are pulled off the shelves by 8 sets, sorting area packs dish according to the size of materials the single sets, two sets or four sets, and send materials in accordance with 8 sets by flat car to L workshop's stereoscopic warehouse cache area or big elements transit zone. For the big elements from G workshop and E workshop, such as hopper, turret, outrigger, turntable, etc are distributed by flat car in accordance with 4 sets to air logistics online point; the arms frame from D workshop are distributed by fork car in accordance with single time two arms frame to arms frame air logistics online point.

SCOR's different elements are mainly used to describe the production and management processes of enterprises within the supply chain (including the supplier's supplier, suppliers, the core enterprise, the customer, the customer's customer), for a single enterprise, it's merely a node in the supply chain. But if seeing from the enterprise internal supply chain system, the enterprise internal operation process has many typical characteristics of the supply chain. So, it's feasible to apply SCOR to the enterprise internal supply chain, model the enterprise internal process, evaluate enterprise's key performance indicators and provide best practice analysis.

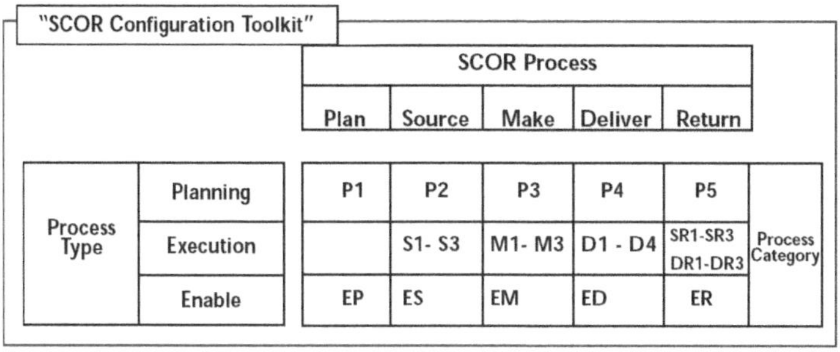

Fig. 2. SCOR structure toolbox

On the base of the above division of the supply chain structure, we can choose the right processes from the second level of the SCOR toolbox (as shown in fig. 2) to build SCOR internal process model [5].

Now taking S enterprise pumping internal supply chain for example, establish models by SCOR. The whole internal supply chain is divided into four level suppliers, core enterprise and customers, six levels in all. L workshop is as the core enterprise, its upstream work units are its suppliers, the parking and testing area is the final customer. The suppliers are divided into 4 levels according to the production logistics process: level 1 supplier is steel warehouse and F workshop; Level 2 suppliers is B workshop and rubber tube center; Level 3 suppliers is C, E, G, H workshop; Level 4 supplier is D, I workshop. Now choose the right process detailed elements from the SCOR secondary configuration layer to construct the internal process model. The following fig. 3 shows.

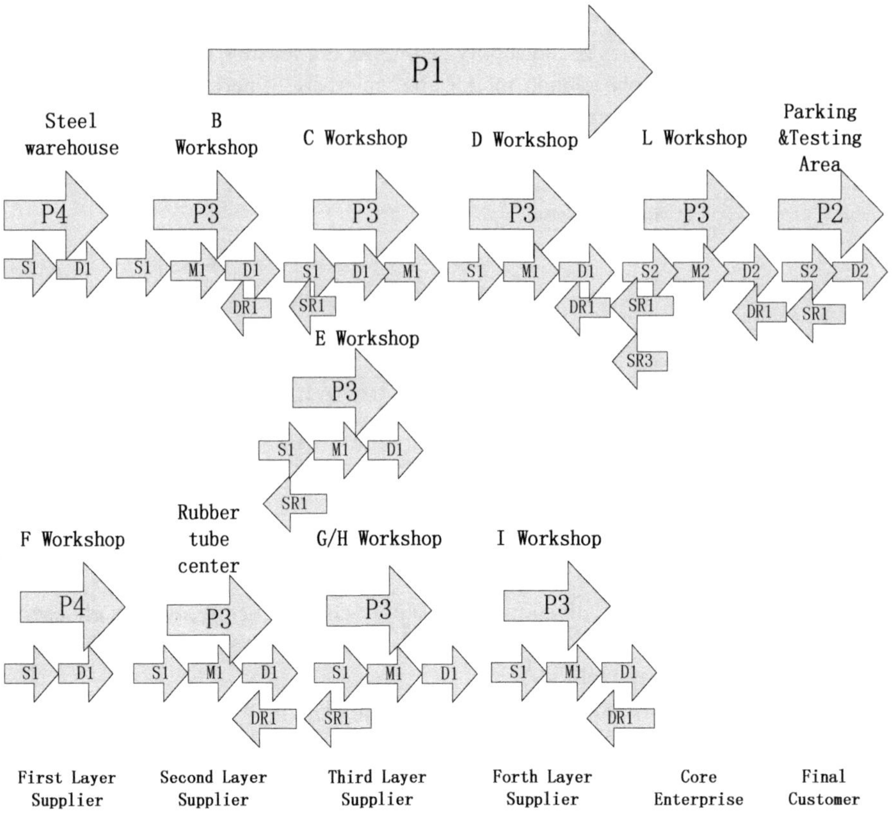

Fig. 3. S enterprise internal supply chain secondary layer model

Lever 1 suppliers:
Steel warehouse , F workshop
P4: plan deliver;
S1: source stocked product;
D1: deliver stocked product.

Lever 2 suppliers:
B workshop, Rubber tube center
P3: plan make;
S1: source stocked product;
M1: make-to-stock;
D1: deliver stocked product;
DR1: deliver return defective product.

Lever 3 suppliers:
C, E, G, H workshop
P3: plan make;
S1: source stocked product;
M1: make-to-stock;
D1: deliver stocked product;
SR1: source return defective product.

Lever 4 suppliers:
D, I workshop
P3: plan make;
S1: source stocked product;
M1: make-to-stock;
D1: deliver stocked product;
DR1: deliver return defective product.

Core enterprise:
L workshop
P3: plan make;
S2: source make-to-order product;
M2: make-to-order;
D2: deliver make-to-order product;
SR1: source return defective product;
SR3: source return excess product;

DR1: deliver return defective product.
Final customer:
Parking and Testing area
P2: plan source;
S2: source make-to-order product;
D2: deliver make-to-order product;
SR1: source return defective product.

SCOR provides three different material flow mode, they are specific in the process as S/ M/ D1, S/ M/ D2 and S/ M/ D3, on behalf of the three categories of production modes respectively: store-oriented, order-oriented and custom design. The operation process of the whole S enterprise pumping is according to the customer orders, but the enterprise's internal operation is not entirely the production mode of make-to-order, such as enterprise internal leverl, 2, 3, 4 suppliers' practical production mode is make-to-stock, while the L workshop is assembly-to-order.

The second layer model of SCOR doesn't describe each work unit's specific process, but the role and position the work unit occupied in the whole supply chain. Therefore, macroscopically speaking, if want to describe the whole supply chain structure and realize the whole supply chain structure configuration (between enterprise and enterprise), to this layer is enough; But if want to further optimize enterprise internal process, we must continue refine process to the third layer (process elements layer), and establish each work unit's specific operation process steps.

4 S Enterprise Internal Supply Chain Core Company's SCOR Third Level Refining Process

L workshop assembly area's basic layout is shown in fig. 4, which is mainly comprised of a, b, c, d, e, f, g seven logistics passages and assembling/department production line corresponding with each logistics passage. Assembly area logistics passage is 205 m long, 2.5 m wide.

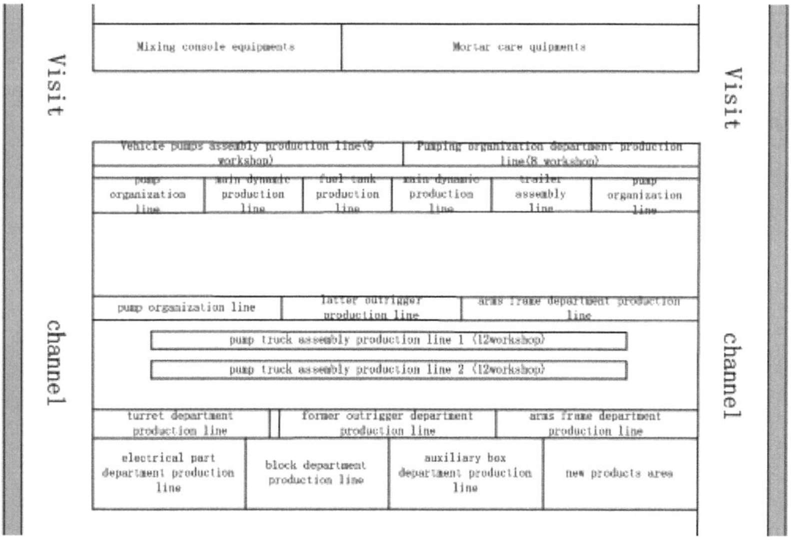

Fig. 4. L workshop assembly area basic layout picture

Logistics passage a has electrical part department production line, block department production line and auxiliary box department production line, on the right side the new products area temporarily stores new products, in the lower side is big elements cache area; b logistics channel has pump truck assembly production line 2 , turret department production line, former outrigger department production line and arms frame department production line; c logistics channel has pump truck assembly production line 1, pumping organization department production line, latter outrigger department production line and arms frame department production line; d logistics channel on the left side is vehicle pumps assembly production line, pumping organization department production line, main dynamic department production line and fuel tank department production line, on the right side is the trailer assembly production line, pumping organization department production line, car body department production line and main dynamic department production line; e logistics channel belongs to agitate equipment (mixing console, mortar car); f, g logistics channels belong to road machine business division. On both sides of the logistics channels are 8-meter-wide and 150-meter-long visit channels. The core enterprise L workshop's second layer SCOR model is shown as the following picture.

According to S enterprise supply chain's second model, we can make the third layer SCOR model of the core enterprise L workshop further. First we can specify the second layer processes.

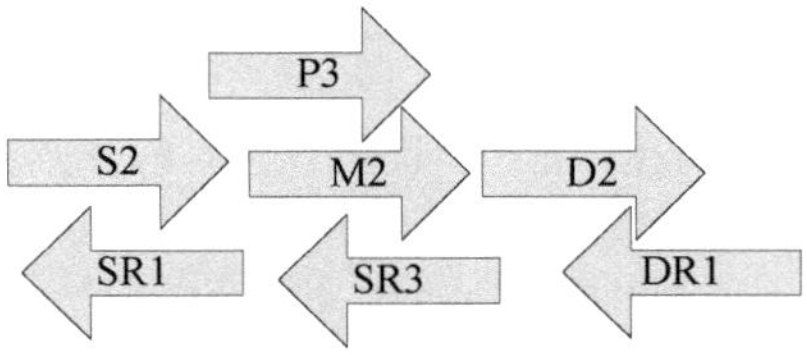

Fig. 5. L workshop second layer SCOR model picture

According the processes above, L workshop's third layer SCOR model is shown as the following picture.

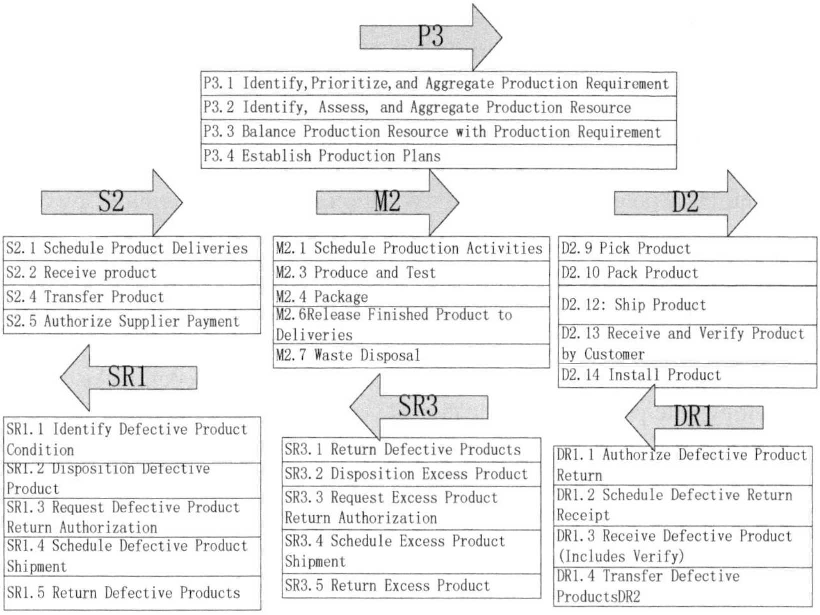

Fig. 6. L workshop third layer SCOR model schematic picture

After thinning the model includes each process's detailed work process, which will play an important role in process reorganization. Then, according to the input/output contact between each third layer process in SCOR, we can draw the L workshop third layer SCOR model structure picture, as shown in fig. 7.

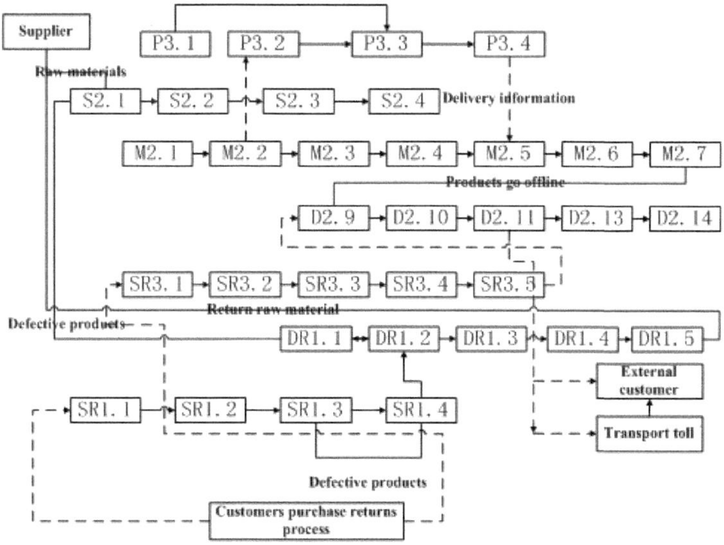

Fig. 7. L workshop third layer SCOR model structure picture

The core company is the heart of the whole enterprise internal supply chain, which is responsible to assemble all the semi-finished products, mechanical parts, raw materials and other materials provided by the suppliers mentioned, and make the final products needed by the end customers. This link is the core of a supply chain, because the whole supply chain materials demand and information flow demand come from here, suppliers of the supply chain adjust they own production plans for this link's requirement, the whole supply chain's production rhythm is determined by this link, too.

5 Conclusion

SCOR is the first standard supply chain operation reference model, make models for manufacturing enterprise by SCOR can make enterprises accurately exchange supply chain problems, evaluate the performance objectively make decisions based on the implementation, and supervise the competitor's performance evaluation, choose which improvement of the supply chain can produce the biggest strategic value, and measure the influence of performance improvement. The model helps the applied enterprise to make the way short, and regulate the software vendors and consulting institutions' behavior and quality standard, is the lean engineering which is beneficial to the whole manufacturing industry.

References

1. Jie, H., Huang, P., Zhang, C.: Supply chain modeling method based on SCOR model. Industrial Engineering and Management 2 (2004)
2. Pen, Z.: Research on manufacturing enterprise internal supply chain system– SCOR model's application and exploration in enterprise process optimization, vol. 3 (2008)
3. Supply Chain Council: Supply Chain Operations Reference Model, Version 10.0. Supply Chain Council Inc. U.S.A (2011)
4. Paul, J.: Supply Chain Practice–A guide to Supply Chain Operations Reference Model (SCOR)
5. Cohen, S., Rowe, J.: Strategic supply chain management– supply chain's best performance management principle. The People's Posts and Telecommunication Publishing House 1 (2006)

A Non Real-Time Registration Algorithm for Multi-radar System

Hongyan Li[1], Xinxi Feng[1], Zhaoying Wang[1], and Tao Liu[2]

[1] Institute of Telecommunication Engineering, Air Force Engineering University , 710077, Xi'an, China
[2] Air Force Military Communication Agency Radar and Electronic Division, 100000, Beijing, China
lihongyan71@sohu.com

Abstract. In order to improve the initiate fusion tracking performance of multi-radar system and keep the stability of fused tracks when the primary station is changed, a non real-time registration algorithm for multi-radar is presented in this paper. This algorithm use simplex method to obtain the best resolution, target function is the sum of the difference between the multiple radar measurements about the same target. Both measurement bias and location bias are considered simultaneously in this algorithm. To demonstrate its efficiency, the computer simulation results are presented.

Keywords: multi-radar system, error registration, simplex method.

1 Introduction

In modern air defense, multi-radar systems or netted radar systems are used to provide an accurate, complete and current air picture. In the systems, measured data from multiple radars will be sent to the processing center for fusion. If all the data are accurate, there would be only one track for one target in the center, and the fused track quality is higher than single radar track. But in practical engineering, it is often found that the fusion of radars without alignment may lead to system performance that is worse than any of the individual radar in the system. Even worse, large errors will result in two apparent aircrafts when only one real exists. This alignment problem is usually referred as the registration problem [1]. So, one of the prerequisites for successful multi-radar fusion is the removal of bias errors (also known as registration, systematic, or alignment errors) from radar measurement data.

Measurement errors include random error and systematic error. Random error can be reduced by filter methods, while systematic error are considered constant ,which can't be eliminated by filter. So registration must be done before multi-radar track fusion. Major source of registration errors includes: 1) sensor position and alignment errors; 2) sensor biases including the azimuth bias error with respect to a common reference and the range offset errors of each sensor; and, 3) coordinate conversion errors [1].

Various registration algorithms have been proposed in recent years, such as RTQC, LS, EML, KF and UKF and so on [2],[3],[4],[5],[6],[7]. However, most of the

G. Lee (Ed.): Advances in Automation and Robotics, Vol. 1, LNEE 122, pp. 379–384.
springerlink.com

algorithms only considered registration error 2) and 3) mentioned above. Because of the coupling between 1) and 2), it is difficult to resolve them simultaneously. So there are few of literature about this problem. But these two type registration errors exist in practical system simultaneously, and strongly affect the fusion performance.

In order to improve the initiate fused tracking performance of netted radar system and reduce the effect on the stability of track when the primary station is changed. A non real-time registration algorithm for multi-radar system is presented in this paper. It treats registration as a nonrestraint optimization problem and uses simplex method to estimate range, azimuth bias errors and radar position error simultaneously.

2 Description of Registration Problem

Assume that the position of radar is represented by (λ, ϕ, H), where λ is the geodetic longitude, ϕ is the latitude, and H denotes the altitude above the mean sea level. Target measurement from multiple radar is given by (r_i, θ_i, h_i), $i = 1, 2, \cdots, m$, where r is the slant range, θ is the azimuth, and h is the altitude. Then assume that there are two radar, radar1 is primary station, without bias error, radar2 have two types error. Measurement in local polar coordinate frame is given by

$$S_k^i = \left[r_k^i, \theta_k^i, h_k^i\right]^T \qquad i = 1,2 . \tag{1}$$

Let $X^i(k)$ be the Cartesian measurement of radar i,

$$X_k^i = \left[x_k^i \quad y_k^i \quad z_k^i\right]^T = \Phi_i\left(r_k^i, \theta_k^i, h_k^i\right) \qquad i = 1,2 . \tag{2}$$

$$\begin{cases} x_k^i = \sqrt{\left(r_k^i\right)^2 - \left(z_k^i\right)^2}\, sin\left(\theta_k^i\right) \\ y_k^i = \sqrt{\left(r_k^i\right)^2 - \left(z_k^i\right)^2}\, cos\left(\theta_k^i\right) . \\ z_k^i = h_k^i \end{cases}$$

Define the measurement model of sensor 1 and sensor 2 as follows:

$$r_k^1 = r_t^1(k) + \Delta r_n^1(k) . \tag{3}$$

$$\theta_k^1 = \theta_t^1(k) + \Delta\theta_n^1(k) . \tag{4}$$

$$h_k^1 = h_t^1(k) + \Delta h_n^1(k) . \tag{5}$$

$$r_k^2 = r_t^2(k) + \Delta r_b^2(k) + \Delta r_n^2(k) . \tag{6}$$

$$\theta_k^2 = \theta_t^2(k) + \Delta\theta_b^2(k) + \Delta\theta_n^2(k) . \tag{7}$$

$$h_k^2 = h_t^2(k) + \Delta h_n^2(k) . \quad (8)$$

The subscripts t, b and n represent true value, sensor bias error and measurement noise, respectively. Radar1 is located at (λ_1, ϕ_1, H_1), radar2 is located at $(\lambda_2 + \Delta\lambda, \phi_2 + \Delta\phi, H_2)$, where $\Delta\lambda$ is longitude bias, $\Delta\phi$ is latitude bias. The registration process is based on ECEF coordinate system, detailed transformation refers to literature [7].

3 A Non Real-Time Registration Algorithm Based on Simplex Method

3.1 Establish Target Function for Error Registration

If there are no errors, the track from two radars for one target should be the same in the central coordinate system. So, we use the sum of difference between the measurement from two radar for one target as the target function. These measurements must be synchronous. The function is as follows:

$$f = \sum_{i=1}^{N} (x_{1i} - x_{2i})^2 + (y_{1i} - y_{2i})^2 . \quad (9)$$

Error source considered here include range bias Δr, azimuth bias $\Delta\theta$, and position bias $(\Delta\lambda, \Delta\phi)$. When the target function reach minimum, it means that radar2 has been aligned to radar1. Because of the coupling between the errors, it is difficult to give a rigorous analytic equation for the target function as the errors. Therefore, simplex method is used to optimize the target function. According to the minimum target function value, systematic errors can be estimated.

3.2 Simplex Method

Simplex method is a directly nonrestraint optimization method. When a simplex shape is given in R^n, the function value of the $n+1$ vertexes can be obtained, the maximum and minimum can also be determined. Then, echoing, spreading and compressing etc. method (not always used at the same time) can be used to calculate a second maximum point. Next, this point is used to replace the maximum vertex, thus a new simplex shape is established. The new simplex shape can also be established by closing to the minimum vertex [8].

3.3 Registration Process

The two radars are assumed synchronous. Registration process is as follows:

1) Give five set initiate value of bias, the bias vector is $X_n = \left[\Delta r_b^2 \;\; \Delta\theta_b^2 \;\; \Delta\lambda^2 \;\; \Delta\phi^2\right]^T$.

2) Correct the measurements of radar2 by the initiate value, then transform them to the coordinate system of radar1, next calculate target function by equation (9). The target function can also be calculated based on more than one target. These targets must be overlapped between two radars.

3) Use simplex method to do iterative search, until reach the threshold. One of the five set value which is the minimum is the final result.

4) Use the bias estimated vector to correct radar2'measurement.

If there are other radars in the net, the same method can be used. In practical engineering, this method can be used across the different primary station to keep the stability of the track.

4 Simulation

Assume that radar1 is located at $(86^\circ, 37^\circ, 500)$, radar2 is located at $(89^\circ + 0.0112^\circ, 39^\circ - 0.0045^\circ, 500)$, the parameters of systematic errors and random error are assumed as follows:

$$\Delta r_b^2 = 1000m\ . \tag{10}$$

$$\Delta \theta_b^2 = 0.025rad\ . \tag{11}$$

$$\sigma_{rn} = 100m\ . \tag{12}$$

$$\sigma_{\theta n} = 0.001rad\ . \tag{13}$$

Then, produce three targets moving along a straight line with constant velocity. Simulations are run 30 times in DELL5100, configure parameters of the computer are as follows: CPU: Pentium(R)4 3.0GHz; RAM: 512M.

The simulation results are shown by table 1 and figure 1.

Table 1. The statistical result of 30 simulations

number	Range bias (m)	Azimuth bias (rad)	Longitude bias (°)	Latitude bias (°)	time (s)
1	1155.0138	0.025	0.014	-0.0043	22.7736
2	1048.4494	0.025	0.0126	-0.0044	30.8496
3	983.3277	0.025	0.0117	-0.005	21.3184
4	1068.7344	0.025	0.0131	-0.0046	40.4008
5	944.1566	0.025	0.0124	-0.0051	22.0419
6	963.2005	0.0251	0.0115	-0.006	23.0071
7	1105.3846	0.025	0.0127	-0.0046	24.4296
8	1090.4216	0.025	0.0134	-0.0046	23.8423
9	1164.0185	0.025	0.0129	-0.004	23.1588
10	1095.4702	0.025	0.0125	-0.0041	24.8896
11	1188.7622	0.0249	0.0129	-0.0035	22.6498

Table 1. (*continued*)

12	981.195	0.025	0.0129	-0.0048	32.1709
13	1252.2722	0.0249	0.0134	-0.0034	25.5391
14	1148.9228	0.0249	0.0133	-0.0033	30.5335
15	1056.9106	0.025	0.0127	-0.0047	26.8347
16	1079.7861	0.025	0.0129	-0.0044	25.3899
17	1056.2816	0.0249	0.012	-0.0041	24.8241
18	1089.6123	0.025	0.0126	-0.0049	26.1190
19	1068.2513	0.0249	0.0122	-0.0035	26.5862
20	1029.5808	0.0251	0.0125	-0.0048	23.8146
21	1131.9324	0.0249	0.0128	-0.0039	28.0028
22	1025.6845	0.0249	0.0115	-0.0038	34.9938
23	1058.4675	0.0251	0.013	-0.0046	21.2320
24	1237.2964	0.0248	0.0132	-0.0027	28.1941
25	1152.5591	0.025	0.0128	-0.0041	23.0623
26	1002.3206	0.025	0.0125	-0.0038	24.9071
27	1044.497	0.025	0.0124	-0.0048	29.5178
28	1327.1447	0.025	0.0141	-0.0039	29.9238
29	1144.8045	0.0249	0.0134	-0.0038	28.7536
30	1089.6547	0.025	0.0125	-0.0045	27.2281
mean	1092.8038	0.025	0.0128	-0.0043	26.5663
Standard deviation	126.2767	6.4363e-005	0.0017	6.8455e-004	4.2809

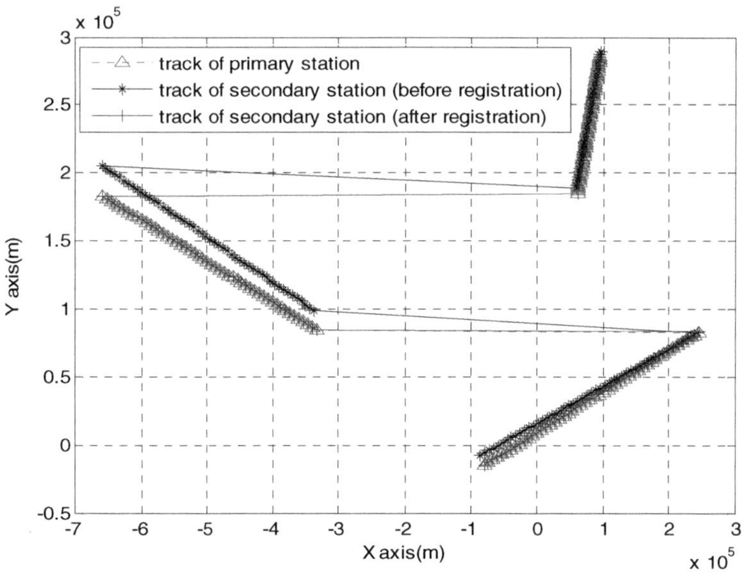

Fig. 1. The comparison of tracks before registration and after registration

5 Summary

This paper presents a non real-time registration algorithm for multi-radar system. It can estimate range bias, azimuth bias and position offset simultaneously. Simulation results show its good performance. The algorithm can be used in practical netted radar system. It can improve the initiate fused tracking performance and reduce the effect on the stability of the track when the primary station is changed. It is worthy to be researched.

References

1. Dana, M.: Registration: A prerequisite for multiple sensor tracking. In: Bar-Shalom (ed.) Multitarget-Multisensor Tracking: Advanced Applications, pp. 51–60. Artech House, Chicago (1990)
2. Leung, H., Gault, K.: Comparison of registration error correction techniques for air surveillance radar network. In: SPIE Proc. Signal and Data Processing of Small Target, San Diego, CA, pp. 498–508 (1995)
3. Zhou, Y., Henry, L.: An exact maximum likelihood registration algorithm for data fusion. J. IEEE Trans. on Signal Processing 45(6), 1560–1572 (1997)
4. He, X., Li, J., Jing, Z.: An overview for development of the techniques on multiple sensor registration. Journal of Air Force Engineering University (Natural Science Edition) 2(2), 11–14 (2001)
5. Wang, B., Wang, J., Zhang, B.: Algorithm for multi-platform multi-sensor registration based on unscented Kalman filter. Journal of Projectiles, Rockets, Missiles and Guidance 28(2), 245–247 (2008)
6. Hu, H., Jing, Z., Hu, S.: An unscented Kalman filter based multi-platform multi-sensor registration. Journal of ShangHai JiaoTong University 39(9), 1518–1521 (2005)
7. Li, H., Feng, X.: Least square registration algorithm based on the ECEF coordinate transformation. System Engineering and Electronics 24(1), 92–95 (2002)
8. Chen, B.: Optimization Theory and Algorithm. TsingHua University Press, Beijing (2005)

An Efficient Video Panorama Constructing Algorithm Based on Color Invariant Features

Ping Zhou and Xiling Luo

School of Electronic Information Engineering,
Beijing University of Aeronautics and Astronautics,
Beijing, China
zhouping880814@yahoo.com.cn, luoxiling@buaa.edu.cn

Abstract. Construct a panorama is an efficient way to handle the limitation of camera visual angle. The process of constructing a video panorama can be defined as image registration and fusion of video frames to a composite wide-view image. This paper proposes a video panorama constructing algorithm based on color invariant features. The panorama constructing process contains three steps. At the start, we build the camera motion model based on color invariant features in order to apply our algorithm in a scene where there are illumination, highlights and shadows. After that, for the purpose of reducing redundancy and accelerating speed, key frames are selected for the following image fusion process. At last, the input frames are transformed to a same coordinate to fuse a panorama. Experimental results show that the proposed algorithm is effective and accurate.

Keywords: video panorama, image registration, color invariant features.

1 Introduction

Construct a panorama is an efficient way to handle the limitation of camera visual angle. Panorama construction has been an active area of research in computer vision and it has various applications such as video surveillance, stabilization, compression, indexing, and virtual environments[1][2].

The process of constructing a video panorama can be defined as image registration and fusion of video frames to a composite wide view image. According to the method used in image registration step, the video panorama constructing algorithm can be classified broadly into two categories, direct methods and feature based methods.

Direct methods can make use of the whole image data and thus provide an accurate registration. However, the weakness of direct methods is they have a constraint on input images[3]. For example, FFT-based algorithm in [4] is a typical representative of direct methods in video panorama constructing algorithm, it has a precise result, but it require a close image initialization.

Feature based algorithms generally use local descriptors representing neighborhoods of feature points extracted from each video frame[5]. Recently, local SIFT (Scale Invariant Feature Transform) features have been widely used in image

G. Lee (Ed.): Advances in Automation and Robotics, Vol. 1, LNEE 122, pp. 385–392.
springerlink.com

registration[6]. SIFT has been proven to be the most robust among the local invariant feature descriptors with respect to different geometrical changes[7][8]. However, SIFT descriptors is calculated by gray intensities and sometimes this lead to error in image registration, since valuable information provided by color are ignored.

This paper proposes a video panorama constructing algorithm based on color invariant features, our algorithm contains three steps. At the start, we build the camera motion model based on color invariant features in order to apply our algorithm in a scene where there are illumination, highlights and shadows. After that, for the purpose of reducing redundancy and accelerating speed, key frames are selected for the following fusion. At last, the input frames are transformed to a same coordinate to fuse a panorama.

The rest of the paper is organized as follows. We describe camera motion model building algorithm in section 2. And in the following section3, key frames selection algorithm and image transforming process is described. In section 4, experimental results are presented. Experimental results show the proposed algorithm is effective and robust. Finally, conclusions and future work are drawn in section 5.

2 Camera Motion Model Building

The aim of camera motion model building is to obtain the homography matrix between video frames. With the matrix, we could transform frames to a same coordinate for constructing a panorama. In this section, we build the camera motion model based on color invariant features. Color Scale Invariant Feature Transform (CSIFT) features combines color information and Scale Invariant Feature Transform (SIFT) features, which makes them obtain inherently the robustness of SIFT to different geometrical transformations, as well as the robustness to photometric changes[9].

2.1 Color Invariance Calculation

According to Kubelka-Munk theory, reflected spectrum of most objects can be modeled, as in

$$E(\lambda,\vec{x}) = i(x)\{[1-\rho_f(\vec{x})]R_\infty(\lambda,\vec{x}) + \rho_f(\vec{x})\} \ . \tag{1}$$

Where λ is the wavelength and $\vec{x}$ is a two dimension vector which describes the image direction. $\rho_f(\vec{x})$ is the Fresnel reflectance at $\vec{x}$ and $R_\infty(\lambda,\vec{x})$ is the material reflectivity. $E(\lambda,\vec{x})$ describes the reflected spectrum in the observing direction.

Color invariance can be calculated as dividing the first derivative of λ by the second derivative of λ, as in

$$H = \frac{E_\lambda}{E_{\lambda\lambda}} = \frac{\partial R_\infty(\lambda, \vec{x})}{\partial \lambda} / \frac{\partial^2 R_\infty(\lambda, \vec{x})}{\partial \lambda^2} = f(R_\infty(\lambda, \vec{x})) \ . \tag{2}$$

In the formula, $H = \frac{E_\lambda}{E_{\lambda\lambda}}$ is defined as the reflectance property of objects which is independent of viewpoint, illumination direction, illumination intensity and surface orientation.

Based on the human vision system and CIE 1964 XYZ, $(E, E_\lambda, E_{\lambda\lambda})$ can be approximately calculated from the known RGB color space, as in

$$\begin{bmatrix} E \\ E_\lambda \\ E_{\lambda\lambda} \end{bmatrix} = \begin{bmatrix} 0.06 & 0.63 & 0.27 \\ 0.30 & 0.04 & -0.35 \\ 0.34 & -0.60 & 0.17 \end{bmatrix} \times \begin{bmatrix} R \\ G \\ B \end{bmatrix} \ . \tag{3}$$

2.2 Features Extraction and Descriptors Building

Suppose the reference image is $I_1(x, y)$ and the sensed image is $I_2(x, y)$, the corresponding color invariance are $H_1(x, y)$ and $H_2(x, y)$. The process of extracting color invariant features and building descriptors is mostly the same in SIFT algorithm, in another word, CSIFT descriptors can be obtained by employing SIFT algorithm in the color space[10].

The process can be described as follows. At first, the extremes in difference of Gaussian (DOG) space are selected as interest points. Meanwhile, to boost up the matching stability, interest points with low contrast and edge response are wiped off. Then, the orientation parameter of each interest point is calculated by the gradients of surrounding pixels. Therefore, each interest point has three parameters: orientation, scale and localization.

After localizing the interest points, feature descriptors are built to characterize them, the local gradient-orientation histograms for the same-scale neighboring pixels of an interest point are used as the key entries of descriptors. At the end of this section, an interest point can form its corresponding CSIFT feature descriptor of 128 dimensions vector.

The advantage of CSIFT descriptors is that, instead of using gray values, we use color invariance as the calculation space of each input image. Building CSIFT descriptors in this way makes them obtain inherently the robustness of SIFT to different geometrical transformations, as well as the robustness to photometric changes, with using of color invariance instead of gray values.

2.3 Features Matching

The matching process is performed by identifying the nearest neighbor feature pairs between two images. A rough feature matching can be defined as finding the feature pairs with minimum Euclidean distance. In a two-dimension space, Euclidean distance is defined as follows.

$$D = \sqrt{(x_j^{'} - x_i)^2 + (y_j^{'} - y_i)^2} \quad . \tag{4}$$

2.4 Camera Motion Model Building

After the feature matching section, we can get corresponding feature pairs. Suppose $\{(x_1, y_1)...(x_m, y_m)\}$ is the feature set of image $I_1(x, y)$, and $\{(x_1^{'}, y_1^{'})...(x_m^{'}, y_m^{'})\}$ is the corresponding feature set of image $I_2(x, y)$. The projective transform between $I_1(x, y)$ and $I_2(x, y)$ can be modeled as follows.

$$\begin{bmatrix} x^{'} \\ y^{'} \\ 1 \end{bmatrix} = \begin{bmatrix} h_1 & h_2 & h_3 \\ h_4 & h_5 & h_6 \\ h_7 & h_8 & 1 \end{bmatrix} \cdot \begin{bmatrix} x \\ y \\ 1 \end{bmatrix} \quad . \tag{5}$$

Where, $H = \begin{bmatrix} h_1 & h_2 & h_3 \\ h_4 & h_5 & h_6 \\ h_7 & h_8 & 1 \end{bmatrix}$ is defined as the Homography Matrix.

3 Key Frame Selection and Image Transforming

3.1 Key Frame Selection

A video sequence consists of hundred frames and consecutive frames do not provide much valuable information, therefore, not all frames are required to create a panorama. In this section, key frames are selected for the purpose of reducing redundancy and accelerating speed.

Considering accelerating speed, we set a video frames interval threshold to reduce the number of images for the following mosaic process, in this way, a frame interval threshold is set. Otherwise, the adjacent key frames must have an appropriate balanced overlap area, because a large overlap increases the computational cost without providing much data while a small overlap results in less accurate image registration[11]. The overlapping area between relative key frames should be large enough to avoid the lost of scene information. Thus, frame interval and overlap area parameters are designed to select key frames.

Overlap area between two video frames is defined as

$$overlap = \frac{F(m) \cap F(m+L)}{F(m)} \quad . \tag{6}$$

3.2 Image Transform Using RANSAC Estimation

Once features have been extracted from all key frames, they must be matched. Since there is always wrong matches among matching pairs because of noise, in order to find the right matches pairs to the fullest extent, we use RANSAC (Random Sample Consensus) algorithm for model parameters estimation. RANSAC estimation algorithm can remove the false matching features (outliers) and identity the correct matching features (inliers) from the candidate features[12]. For each pair of potentially matching images we have a set of feature matches that are geometrically consistent (RANSAC inliers) and a set of features that are inside the area of overlap but not consistent (RANSAC outliers).

We select four sets feature correspondences to compute the homography matrix with the direct linear transformation method, and then video frames are aligned together with a 3x3 homography matrix. All selected key frames are transformed and projected onto a common mosaic surface.

Image transform using RANSAC estimation increases the accuracy of the matching by eliminating the outlier corresponding points. The perspective transform for the image matching is then derived from the selected corresponding points as follows.

$$\begin{bmatrix} x_1 & y_1 & 1 & 0 & 0 & 0 & -x_1 * x_1^{'} & -y_1 * y_1^{'} \\ x_2 & y_2 & 1 & 0 & 0 & 0 & -x_2 * x_2^{'} & -y_2 * y_2^{'} \\ x_3 & y_3 & 1 & 0 & 0 & 0 & -x_3 * x_3^{'} & -y_3 * y_3^{'} \\ x_4 & y_4 & 1 & 0 & 0 & 0 & -x_4 * x_4^{'} & -y_4 * y_4^{'} \\ 0 & 0 & 0 & x_1 & y_1 & 1 & -x_1 * y_1^{'} & -y_1 * y_1^{'} \\ 0 & 0 & 0 & x_2 & y_2 & 1 & -x_2 * y_2^{'} & -y_2 * y_2^{'} \\ 0 & 0 & 0 & x_3 & y_3 & 1 & -x_3 * y_3^{'} & -y_3 * y_3^{'} \\ 0 & 0 & 0 & x_4 & y_4 & 1 & -x_4 * y_4^{'} & -y_4 * y_4^{'} \end{bmatrix} \cdot \begin{bmatrix} h_1 \\ h_2 \\ h_3 \\ h_4 \\ h_5 \\ h_6 \\ h_7 \\ h_8 \end{bmatrix} = \begin{bmatrix} x_1^{'} \\ x_2^{'} \\ x_3^{'} \\ x_4^{'} \\ y_1^{'} \\ y_2^{'} \\ y_3^{'} \\ y_4^{'} \end{bmatrix} \quad . \tag{7}$$

Once we get the homography matrix, we could project all selected key frames to a same coordinate to fuse a panorama.

4 Experiments

Our proposed algorithm contains three steps. At the first, we build the camera motion model based on color invariant features, after this section, we can obtain the homography matrix between video frames. Color invariant features obtain the robustness to different geometrical transformations and photometric changes.

Fig. 1. Input test images

Fig. 2. Building camera motion model based on color invariant features

On the second step, key frames are selected in order to reduce redundancy and accelerating speed. In this paper, we depend on frame interval and overlap area parameters to select key frames, the frame interval threshold is set to be 20 and overlap area is 0.6.

Fig. 3. Select key frames

On the third step, the selected frames are transformed to a same coordinate to fuse a panorama.

Fig. 4. Image transforming to fuse a panorama

From the experiments, we can see that, our proposed algorithm is accurate and effective.

5 Conclusion

In this paper, we introduced a video panorama constructing algorithm based on color invariant features, which combines both color and geometrical characteristics. Besides, for the purpose of reducing redundancy and accelerating speed, key frames are selected for the following image transforming process. Experiment results proved our proposed algorithm is accurate and effective.

Acknowledgments. This work was supported by the National Natural Science Foundation of China (NSFC) under Grant No.60879013.

References

1. Junlan, Y., Schonfeld, D.: Robust Video Stabilization Based on Particle Filter Tracking of Projected Camera Motion. Circuits and System for Video Technology 7(19), 945–954 (2009)
2. Glowacz, A., Grega, M.: Compression and Distribution of Panoramic Videos Utilising MPEG-7-based Image Registration. Multimedia Tools and Applications 40, 321–339 (2008)
3. Gonzalez, Woods, R.E.: Digital Image Processing, 2nd edn. Publishing House of Electronics Industry (2007)
4. Tzimiropoulos, G., Argyriou, V.: Robust FFT-Based Scale-Invariant Image Registration with Image Gradients. Pattern Analysis and Machine Intelligence 10(32), 1899–1906 (2010)

5. Brown, M., Lowe, D.G.: Automatic Panoramic Image Stitching using Invariant Features. Computer Vision 74(1), 59–73 (2007)
6. Lowe, D.G.: Distinctive Image Features from Scale-Invariant Keypoints. J. Computer Vision 60(2), 91–110 (2004)
7. Mikolajczyk, K., Schmid, C.: A Performance Evaluation of Local Descriptors. J. Pattern Analysis and Machine Intelligence 27(10), 1615–1630 (2005)
8. Huynh, D.Q., Sanini, A., Wei, L.: Evaluation of Three Local Descriptors on Low Resolution Images for Robot Navigation. In: 24th International Conference on Image and Vision Computing, New Zealand, vol. 11, pp. 23–25 (2009)
9. Alaa, E., Abdel-Hakim, Farag, A.A.: CSIFT: A SIFT Descriptor with Color Invariant Characteristics. In: Computer Vision and Pattern Recognition, vol. 2, pp. 1978–1983 (2006)
10. Koen, E.A., de van Sande, S., et al.: Evaluating Color Descriptors for Object and Scene Recognition. Pattern Analysis and Machine Intelligence 32(9), 1582–1596 (2010)
11. Li, J., Pan, Q.: Automated Feature Points Management for Video Mosaic Construction. In: Proceedings of the 3rd Information Technology and Applications, vol. 1, pp. 760–763 (2005)
12. Dianjian, Y., Songchen, C.: Two-View Motion Segmentation with Model Selection and Outlier Removal by RANSAC-Enhanced Dirichlet Process Mixture Models. Computer Vision 88, 489–501 (2010)

Approximate Output Regulation of Spherical Inverted Pendulum by Neural Network Control

Zhaowu Ping and Jie Huang

Department of Mechanical and Automation Engineering,
The Chinese University of Hong Kong, Shatin, N. T., Hong Kong, China
zwping@mae.cuhk.edu.hk, jhuang@mae.cuhk.edu.hk

Abstract. Recently, the output regulation problem of the spherical inverted pendulum was studied in [8]. It is known that the solvability of the output regulation problem depends on the solvability of the regulator equations which are a set of nonlinear partial differential equations. Since the exact solution of the regulator equations associated with the spherical inverted pendulum is not available due to the complexity of the equations, the paper [8] tried a polynomial approximation of the regulator equations. In this paper, we first show that the regulator equations associated with the spherical inverted pendulum exist and then propose a neural network approach to obtain an approximate solution to the output regulation problem of the spherical inverted pendulum. We also make some comparison with the method in [8] and show that, the neural network controller leads to a much better tracking performance for larger exogenous signals.

Keywords: Output regulation, multivariable nonlinear systems, spherical inverted pendulum, neural network control.

1 Introduction

Shown in Figure 1 is the spherical inverted pendulum [8]:

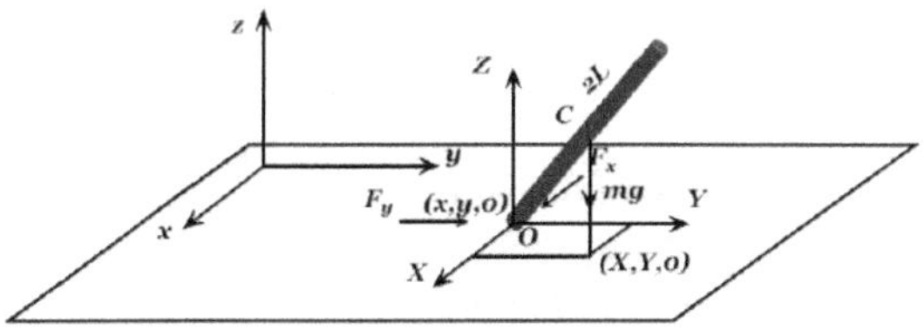

Fig. 1. Spherical inverted pendulum

where $x, y \in R$ represent the position of the base of the pendulum in the horizontal plane and $X, Y \in R$ represent the x and y positions of the vertical projection of the center of the pendulum onto the horizontal plane, $F_x, F_y \in R$ are the control forces being applied to the cart at the base of the pendulum, m is the mass of the uniform rod, L is the distance from the base of the pendulum to the center of mass, and g is the gravitational constant.

G. Lee (Ed.): Advances in Automation and Robotics, Vol. 1, LNEE 122, pp. 393–400.
springerlink.com

The motion equations of the spherical inverted pendulum is as follows [8]:

$$\dot{x} = f(x) + g(x)u, \; y = h(x) \tag{1}$$

where

$$x_1 = x, x_2 = \dot{x}, x_3 = y, x_4 = \dot{y}, z_1 = X, z_2 = \dot{X}, z_3 = Y, z_4 = \dot{Y}, u_1 = F_x, u_2 = F_y$$

$$u = [u_1 \quad u_2]^T, x = [x_1 \quad x_2 \quad x_3 \quad x_4 \quad z_1 \quad z_2 \quad z_3 \quad z_4]^T$$

$$f(x) = \begin{bmatrix} x_2 \\ \dfrac{-3(L^2 - z_1^2 + 3z_3^2)(-Bz_1 + Cz_1) + 12z_1z_3(-Bz_3 + Cz_3)}{A} \\ x_4 \\ \dfrac{12z_1z_3(-Bz_1 + Cz_1) - 3(L^2 + 3z_1^2 - z_3^2)(-Bz_3 + Cz_3)}{A} \\ z_2 \\ \dfrac{3(L^2 - z_1^2 + 3z_3^2)(-Bz_1 + Cz_1) - 12z_1z_3(-Bz_3 + Cz_3)}{A} \\ z_4 \\ \dfrac{-12z_1z_3(-Bz_1 + Cz_1) + 3(L^2 + 3z_1^2 - z_3^2)(-Bz_3 + Cz_3)}{A} \end{bmatrix}$$

$$g(x) = \begin{bmatrix} 0 & 0 \\ \dfrac{4(L^2 + 3z_3^2)}{A} & \dfrac{-12z_1z_3}{A} \\ 0 & 0 \\ \dfrac{-12z_1z_3}{A} & \dfrac{4(L^2 + 3z_1^2)}{A} \\ 0 & 0 \\ \dfrac{-3(L^2 - z_1^2 + 3z_3^2)}{A} & \dfrac{12z_1z_3}{A} \\ 0 & 0 \\ \dfrac{12z_1z_3}{A} & \dfrac{-3(L + 3z_1^2 - z_3^2)}{A} \end{bmatrix}, \; h(x) = \begin{bmatrix} x_1 \\ x_3 \end{bmatrix}$$

$$A = m(L^2 + 3(z_1^2 + z_3^2)), B = \frac{4m(L^2(z_2^2 + z_4^2) - (z_2z_3 - z_4z_1)^2)}{3(L^2 - z_1^2 - z_3^2)^2}, C = \frac{mg}{\sqrt{L^2 - z_1^2 - z_3^2}}.$$

The stabilization problem of spherical inverted pendulum was studied in [7]. More recently, an asymptotic tracking problem of the spherical inverted pendulum with sinusoidal reference signals was further studied in [8]. The problem was formulated as a nonlinear output regulation problem as studied in [4], [6]. It is known that the solvability of the output regulation problem depends on the solvability of the regulator equations which are a set of nonlinear partial differential equations. Typically, the exact solution of the regulator equations associated with a nonlinear system is not available. Therefore, finding various approximate solutions to the nonlinear output regulation problem has become the more practical approach. So far, there are mainly two approximate methods for the nonlinear output regulation problem. The first one is based on the polynomial approximation of the solution of the regulator equations [4], [5], and the second one is based on the neural network approximation of the solution of the regulator equations [9]. Both methods have been tested in some single-input, single output nonlinear systems such as the ball and beam system [2], [5] and the inverted pendulum on a cart system [9], and have shown quite satisfactory results. Nevertheless, the real challenge to any numerical approximation method is the complexity of the system. The spherical inverted pendulum system, which has eight state variables, two inputs and two outputs and is much more complicated than most benchmark systems such as the ball and beam system and the inverted pendulum on a cart system, may well serve as an interesting test-bed for the approximate approaches developed in [4], [5], and [9]. The work in [8] has shown the performance of the polynomial approach in some detail. In this paper, we will further investigate the applicability of the neural network approach to the approximate output regulation problem of the spherical inverted pendulum, and make some comparison with the polynomial approach. Our investigation shows that the neural network approach not only offers some significant computational advantage over the polynomial approach, but also results in a much smaller approximation error for larger exogenous signals.

The rest of the paper is organized as follows. In Section 2 we summarize the basic theory of the nonlinear output regulation problem. In Section 3, we synthesize an approximate control law for the output regulation problem of the spherical inverted pendulum based on the neural network approximate solution of the regulator equations of the spherical inverted pendulum and make some comparison with the result in [8]. Finally, we conclude this paper with some concluding remarks.

2 Preliminaries

In this section, we summarize some results on the nonlinear output regulation problem based on the treatment in [4], [6], [9]. Consider a nonlinear plant as follows:

$$\begin{aligned} \dot{x}(t) &= f(x(t), u(t), v(t)), x(0) = x_0, t \geq 0 \\ y(t) &= h(x(t), u(t), v(t)), t \geq 0 \end{aligned} \tag{2}$$

where $x(t)$ is the n-dimensional plant state, $u(t)$ the m-dimensional plant input, $y(t)$ the p-dimensional plant output representing the tracking error, $v(t)$ the q-dimensional exogenous signal representing both reference inputs and disturbances. The exogenous signal $v(t)$ is generated by a q-dimensional exosystem

$$\dot{v}(t) = a(v(t)), v(0) = v_0, t \geq 0. \tag{3}$$

We assume the functions f, h, and a are sufficiently smooth in a neighborhood of the origins of the respective Euclidian spaces and vanish at their origins.

We consider a state feedback control law of the following form

$$u = \psi(x, v) \tag{4}$$

where ψ is also sufficiently smooth in a neighborhood of the origin vanishing at the origin. The composition of the plant and the control law will lead to the following closed-loop system

$$\begin{aligned} \dot{x}(t) &= f_c(x, v) \triangleq f(x, \psi(x, v), v) \\ y(t) &= h_c(x, v) \triangleq h(x, \psi(x, v), v). \end{aligned} \tag{5}$$

By state feedback output regulation problem, we mean the design of a controller of the form (4) such that, for all sufficiently small x_0 and v_0, the trajectories of the closed loop system (5) exist and are bounded for all $t \geq 0$, and is such that $\lim_{t\to\infty} y(t) = 0$.

Some standard assumptions for ensuring the solvability of the above problem are listed as follows.

Assumption 2.1: The pair $\left(\frac{\partial f}{\partial x}(0,0,0), \frac{\partial f}{\partial u}(0,0,0)\right)$ is stabilizable.

Assumption 2.2: The equilibrium of the exosystem (3) at $v = 0$ is Lyapunov stable, and all the eigenvalues of $\frac{\partial a}{\partial v}(0)$ have zero real parts.

Assumption 2.3: There exist two sufficiently smooth functions $\boldsymbol{x}(v)$ and $\boldsymbol{u}(v)$ defined in a neighborhood of the origin of R^q such that $\boldsymbol{x}(0) = 0, \boldsymbol{u}(0) = 0$ and for all $v \in V$,

$$\begin{aligned} \frac{\partial \boldsymbol{x}(v)}{\partial v} a(v) &= f(\boldsymbol{x}(v), \boldsymbol{u}(v), v) \\ 0 &= h(\boldsymbol{x}(v), \boldsymbol{u}(v), v). \end{aligned} \tag{6}$$

Remark 2.1: (6) is known as regulator equations [6]. Assumption 2.3 is a necessary condition for the solvability of the nonlinear output regulation problem. Moreover, Assumption 2.3 together with Assumptions 2.1 and 2.2 also leads to a state feedback control law of the form

$$\psi(x, v) = \boldsymbol{u}(v) + K(x - \boldsymbol{x}(v)) \tag{7}$$

where K is such that the matrix $\frac{\partial f}{\partial x}(0,0,0) + \frac{\partial f}{\partial u}(0,0,0)K$ is Hurwitz.

However, for most practical nonlinear systems, the exact solution $\boldsymbol{x}(v), \boldsymbol{u}(v)$ of the regulator equations (6) is not available. Therefore, some efforts have been made on finding the approximate solution of the nonlinear output regulation problem based on polynomial [5] or neural networks [9]. Given some real number $\gamma > 0$ and some compact subset $\Gamma \subset R^q$ containing the origin of R^q, we call two functions $\hat{\boldsymbol{x}}(v), \hat{\boldsymbol{u}}(v)$ an approximation solution of the regulator equations up to the accuracy γ on Γ if

$$\begin{aligned} &\max_{v\in\Gamma} ||\boldsymbol{x}(v) - \hat{\boldsymbol{x}}(v)|| < \gamma, \max_{v\in\Gamma} ||\boldsymbol{u}(v) - \hat{\boldsymbol{u}}(v)|| < \gamma \\ &\max_{v\in\Gamma} \left\|\frac{\partial(\boldsymbol{x}(v) - \hat{\boldsymbol{x}}(v))}{\partial v}\right\| < \gamma. \end{aligned} \tag{8}$$

Here, we will summarize the result of [9] as follows.

Theorem 2.1: Under Assumptions 2.1, 2.2 and 2.3, for any $\gamma > 0$ and any compact subset $\Gamma \subset R^q$ containing the origin of R^q, there exist three layer feedforward neural network functions $\hat{\boldsymbol{x}}(v), \hat{\boldsymbol{u}}(v)$ satisfying (8). Moreover, let K be such that the matrix $\frac{\partial f}{\partial x}(0,0,0) + \frac{\partial f}{\partial u}(0,0,0)K$ is Hurwitz. Then, the following control law

$$\psi(x, v) = \hat{\boldsymbol{u}}(v) + K(x - \hat{\boldsymbol{x}}(v)) \tag{9}$$

solves the practical nonlinear output regulation problem of system (2) in the sense that the closed-loop system (5) has the property that, for any $\epsilon > 0$, there exists γ and Γ such that, for all sufficiently small initial conditions x_0 and v_0, the closed-loop system has a bounded solution for all $t \geq 0$, and $\lim_{t\to\infty} ||h_c(x(t), v(t))|| \leq \epsilon$.

3 Synthesis of Neural Network Based Approximate Control Law

In this section, we first show that a local solution of the regulator equations associated with spherical inverted pendulum exists. Then we will further design an approximate control law of the form (9) and evaluate the performance of the control law. Recall from [8] that both of the two reference signals for the two outputs of the spherical inverted pendulum are sinusoidal functions and can be generated by the following exosystem

$$\dot{v} = A_1 v = \begin{bmatrix} 0 & \omega_1 & 0 & 0 \\ -\omega_1 & 0 & 0 & 0 \\ 0 & 0 & 0 & \omega_2 \\ 0 & 0 & -\omega_2 & 0 \end{bmatrix} \begin{bmatrix} v_1 \\ v_2 \\ v_3 \\ v_4 \end{bmatrix} \tag{10}$$

whose output is

$$y_d(t) = \begin{bmatrix} v_1 & v_3 \end{bmatrix}^T. \tag{11}$$

Thus we can define the tracking error $e = [e_1 \; e_2]^T$ as follows:

$$e_1 = x_1 - v_1, e_2 = x_3 - v_3. \tag{12}$$

This exosystem clearly satisfies Assumption 2.2 as the eigenvalues of A_1 are on the imaginal axis and are semi-simple.

We will use the same data as those in [8] to conduct our design. That is, the plant parameters are $m = 0.050\text{Kg}, L = 0.650\text{m}$. The reference trajectory is $y_d(t) = [A_1 \sin(1.5t) \;\; A_2 \cos(1.5t)]^T$.

The regulator equations associated with (1), (10) and (12) are as follows

$$\begin{aligned} \frac{\partial \boldsymbol{x}(v)}{\partial v} A_1 v &= f(\boldsymbol{x}(v), \boldsymbol{u}_1(v), \boldsymbol{u}_2(v), v) \\ 0 &= \begin{bmatrix} \boldsymbol{x}_1(v) - v_1 \\ \boldsymbol{x}_3(v) - v_3 \end{bmatrix} \end{aligned} \tag{13}$$

where $\boldsymbol{x}: R^4 \mapsto R^8, \boldsymbol{u}_1: R^4 \mapsto R^1$ and $\boldsymbol{u}_2: R^4 \mapsto R^1$ are sufficiently smooth functions vanishing at the origin. Note that $\boldsymbol{x} = \mathrm{col}(\boldsymbol{x}_1, \cdots, \boldsymbol{x}_4, \boldsymbol{z}_1, \cdots, \boldsymbol{z}_4)$.

By [8], or by a mere inspection, it can be verified that the solution of the regulator equations (13) can be partially obtained as follows:

$$\begin{aligned} &\boldsymbol{x}_1(v) = v_1, \boldsymbol{x}_2(v) = \omega_1 v_2, \boldsymbol{x}_3(v) = v_3, \boldsymbol{x}_4(v) = \omega_2 v_4 \\ &\boldsymbol{u}_1(v) = \frac{\Gamma_1(\boldsymbol{z}(v), v)}{4L^2\left(L^2 + 3\boldsymbol{z}_1^2(v) + 3\boldsymbol{z}_3^2(v)\right)} + \frac{3}{4L^2} \boldsymbol{z}_1(v) \\ &\qquad \times \left(-B(\boldsymbol{z}(v)) + C(\boldsymbol{z}(v))\right)\left(L^2 - \boldsymbol{z}_1^2(v) - \boldsymbol{z}_3^2(v)\right) \\ &\boldsymbol{u}_2(v) = \frac{\Gamma_2(\boldsymbol{z}(v), v)}{4L^2\left(L^2 + 3\boldsymbol{z}_1^2(v) + 3\boldsymbol{z}_3^2(v)\right)} + \frac{3}{4L^2} \boldsymbol{z}_3(v) \\ &\qquad \times \left(-B(\boldsymbol{z}(v)) + C(\boldsymbol{z}(v))\right)\left(L^2 - \boldsymbol{z}_1^2(v) - \boldsymbol{z}_3^2(v)\right) \end{aligned} \tag{14}$$

where

$$\begin{aligned} \Gamma_1(z, v) &= -(L^2 + 3z_1^2) A \omega_1^2 v_1 - 3 z_1 z_3 A \omega_2^2 v_3 \\ \Gamma_2(z, v) &= -(L^2 + 3z_3^2) A \omega_2^2 v_3 - 3 z_1 z_3 A \omega_1^2 v_1 \end{aligned}$$

and $\mathbf{z}(v) = \text{col}(\mathbf{z}_1(v), \mathbf{z}_2(v), \mathbf{z}_3(v), \mathbf{z}_4(v))$ satisfies the following partial differential equations

$$\frac{\partial z(v)}{\partial v} A_1 v = \delta(\mathbf{z}(v), v). \tag{15}$$

The expression of $\delta(z, v)$ is omitted due to space limit.

The system $\dot{z} = \delta(z, v)$ is the zero dynamics of the system (1), (10) and (12) and equation (15) is called the center manifold equation. Since the eigenvalues of $\frac{\partial\delta(0,0)}{\partial z}$ are 6.7254, -6.7254, 6.7254, -6.7254, which have nonzero real parts, and the eigenvalues of A_1 are all on the imaginary axis, by the center manifold theorem [1] and [6], equations (15) admits a local solution $\mathbf{z}(v)$ in the neighborhood of the origin of R^4. Thus Assumption 2.3 is also satisfied. Nevertheless, there is no clue how to obtain the exact solution of (15). Therefore, here we will consider to use a three layer feedforward neural network to approximate the solutions of (15) based on the work in [9]. The main idea is to approximate the solution $\mathbf{z}_i(v), i = 1, \cdots, 4$ of (15) by the following functions

$$\mathbf{z}_i(W, v) = \textstyle\sum_{j=1}^{N} w_{ij}^O \phi\left(W_j^I v + b_j\right) + wd_i^O \tag{16}$$

where $i = 1, \cdots, 4$, N is some positive integer, $W \in R^{9N+4}$ is the weight with components given by w_{ij}^O, w_{ij}^I for $i = 1, \cdots, 4, j = 1, \cdots, N$, b_j for $j = 1, \cdots, N$ and wd_i^O for $i = 1, \cdots, 4$, $\phi(y) = \frac{1-e^{-y}}{1+e^{-y}}$ is the bipolar sigmoid nonlinear activation function, v_1, v_2, v_3, v_4 are the inputs, $\mathbf{z}_1(W, v), \mathbf{z}_2(W, v), \mathbf{z}_3(W, v), \mathbf{z}_4(W, v)$ are the outputs respectively, $W_j^I = [w_{1j}^I \ w_{2j}^I \ w_{3j}^I \ w_{4j}^I]$.

Functions of the form (16) are called three layer feedforward neural networks, the integer N is called the number of neurons in the hidden layer, and the parameter W is called the weights of the neural networks [3]. By the universal approximation theorem [3], since $\mathbf{z}(v)$ is sufficiently smooth, given any compact subset Γ of R^4 containing the origin and a real number ϵ, there exist N and W such that

$$||\mathbf{z}(W, v) - \mathbf{z}(v)|| \leq \epsilon, \forall v \in \Gamma.$$

Next, following the idea given in [9], we will devise a parameter optimization method to find a suitable N and weight vector W. For this purpose, let $e_{2k-1} = \frac{\partial z_{2k-1}(W,v)}{\partial v} A_1 v - \mathbf{z}_{2k}(W, v), e_{2k} = \frac{\partial z_{2k}(W,v)}{\partial v} A_1 v - \delta_k(W, v)$ where, for $k = 1,2$, $\delta_k(W, v) = \delta_k(\mathbf{z}(W, v), v)$. Also let $J(W, v) = e_1^2 + e_2^2 + e_3^2 + e_4^2$.

Our goal is to find N and W such that $\sup_{v\in\Gamma} J(W, v)$ is sufficiently small. For this purpose, we further define a function of W by $Q(W) = \sum_{v\in\Gamma_d} J(W, v)$ where Γ_d is a finite set. If Γ_d is sufficiently *dense* in Γ, then a W that makes $Q(W)$ sufficiently small may likely make $\sup_{v\in\Gamma} J(W, v)$ sufficiently small. Now, for each fixed N, $Q(W)$ only depends on weight W. Therefore, we can use any parameter optimization method to find W. In fact, let

$$\Gamma = \left\{v \in R^4 \middle| \| (v_1, v_2) \| \leq 2, \| (v_3, v_4) \| \leq 2\right\}$$

$$\Gamma_d = \{(\rho_1 \sin\theta_1, \rho_1 \cos\theta_1, \rho_2 \sin\theta_2, \rho_2 \cos\theta_2) \mid \rho_1 = 0.5, 1, 1.5, 2, \theta_1 = \frac{2\pi}{5},$$

$$\frac{4\pi}{5}, \frac{6\pi}{5}, \frac{8\pi}{5}, 2\pi, \rho_2 = 0.5, 1, 1.5, 2, \theta_2 = \frac{2\pi}{5}, \frac{4\pi}{5}, \frac{6\pi}{5}, \frac{8\pi}{5}, 2\pi\}.$$

We have found using the gradient descent algorithm that, when $N = 40$, there exists a weight $\widehat{W}$ such that $Q(\widehat{W}) \leq 0.0572$.

Based on the approximate solution $\mathbf{z}(\widehat{W}, v)$, we can obtain an approximate solution of the regulator equations denoted by $(\widehat{\mathbf{u}}(v), \widehat{\mathbf{x}}(v))$ which in turn leads to a controller of the form (9) where K is such that the eigenvalues of $\frac{\partial f(0)}{\partial x} + g(0)K$ are as follows: -316.33, -316.33, -3.1676, -3.1676, -2.8504+ 0.9505i, -2.8504-0.9505i, -2.8504+0.9505i, -2.8504-0.9505i.

The performance of the controller is evaluated by computer simulation with the same initial condition as given in [8], that is, $x(0) = [0,0,0,0,0,0,0,0]^T$. Table 1 shows the maximal percentage steady-state tracking errors with different amplitudes. Figure 2 shows the tracking performance of the neural network controller and the third order polynomial controller for $A_1 = 2, A_2 = 2$.

Table 1. Maximal percentage steady-state tracking errors

Amplitude		Neural network design (%)		Third order design (%)	
A_1	A_2	e_1	e_2	e_1	e_2
0.5	0.5	1.00	0.31	0.06	0.03
1.0	1.0	1.20	0.51	0.50	0.28
1.5	1.5	0.98	0.83	1.76	0.99
2.0	2.0	1.93	1.23	4.34	2.49

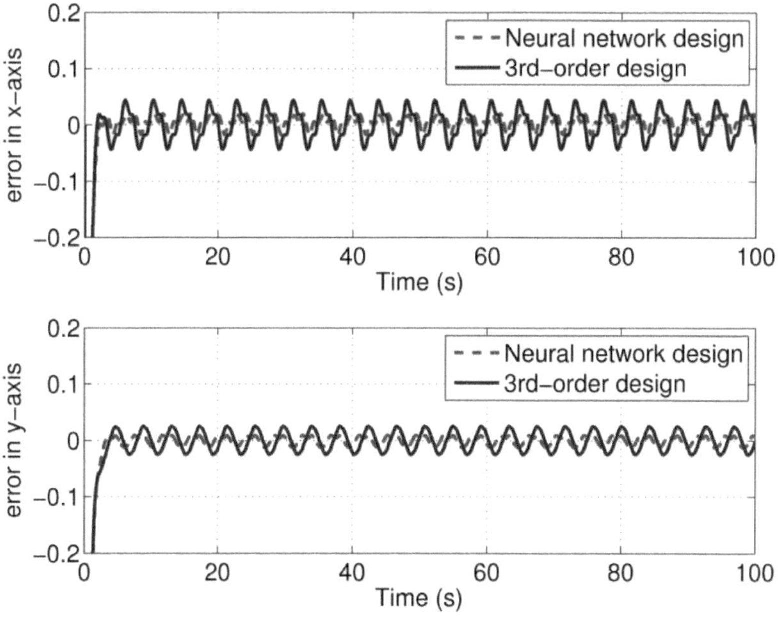

Fig. 2. Error in x-axis and y-axis with $A_1 = 2, A_2 = 2$

4 Conclusion

In this paper, we have considered an alternative approach to the approximate output regulation problem of the spherical inverted pendulum by a neural network approximation of the solution of the regulator equations. The performance of our control law has been compared with a third order polynomial based control law in [8], and shown that the steady-state tracking errors of the method in this paper maintains a more uniform performance over a range of amplitudes of the reference input signals.

Acknowledgments. The work of this paper was substantially supported by the Research Grants Council of the Hong Kong Special Administration Region under grant No. CUHK412609.

References

1. Carr, J.: Applications of the Center Manifold Theory. Springer, New York (1981)
2. Chu, Y.C., Huang, J.: A Neural Network Method for Nonlinear Servomechanism Problem. IEEE Transactions on Neural Networks 11, 1412–1423 (1999)
3. Hornik, K.: Approximation Capabilities of Multilayer Feedforward Networks. Neural Networks 4, 251–257 (1991)
4. Huang, J., Rugh, W.J.: Stabilization on Zero-error Manifolds and the Nonlinear Servomechanism Problem. IEEE Transactions on Automatic Control 37, 1009–1013 (1992)
5. Huang, J., Rugh, W.J.: An Approximation Method for the Nonlinear Servomechanism Problem. IEEE Transactions on Automatic Control 37, 1395–1398 (1992)
6. Isidori, A., Byrnes, C.I.: Output Regulation of Nonlinear Systems. IEEE Transactions on Automatic Control 35, 131–140 (1990)
7. Liu, G., Nešić, D., Mareels, I.: Non-local Stabilization of a Spherical Inverted Pendulum. International Journal of Control 81, 1035–1053 (2008)
8. Postelnik, L., Liu, G., Stol, K., Swain, A.: Approximate Output Regulation for a Spherical Inverted Pendulum. In: Proceedings of the 2011 American Control Conference, San Francisco, California, USA, pp. 539–544 (2011)
9. Wang, J., Huang, J., Yau, S.S.T.: Approximate Output Regulation Based on Universal Approximation Theorem. International Journal of Robust and Nonlinear Control 10, 439–456 (2000)

Discrete-Time Sliding-Mode Control of a Mobile Platform with Four Driving/Steering Wheels*

Adrian Filipescu[1], Viorel Minzu[1], Adriana Filipescu[1], and Eugenia Minca[2]

[1] Department of Automation and Electrical Engineering,
University "Dunarea de Jos" of Galati, Domneasca, 47, Galati, Romania
Adrian.Filipescu@ugal.ro, Viorel.Minzu@ugal.ro,
Adriana_Filipescu@yahoo.ca

[2] Department of Automation, Computer Science and Electrical Engineering,
University "Valahia" of Targoviste, Unirii, 18, Targoviste, Romania
Eugenia.Minca@gmail.com

Abstract. In this paper the problem of trajectory tracking of the four-driving/steering wheels (4DW/SW) mobile platform SEEKUR, by means of a discrete-time sliding mode controller is presented. SEEKUR is a mobile platform which may be assimilated as an autonomous vehicle. Commercially available, both holonomic and nonholomic, all-weather, outdoor autonomous vehicle platform suitable for a wide range of security, military inspection and research applications. The discrete-time sliding-mode controller has been designed using a discrete time nonholonomic model of SEEKUR. Simulation and real-time results are presented in trajectory tracking control.

Keywords: SEEKUR mobile platform, discrete-time, sliding-mode control.

1 Introduction

Autonomous vehicles are suitable for a wide range of applications, such as transporting components between workstations in factories, routine, repetitive 24 hrs work, etc. They can even be used in hazardous areas like contaminated zones, mine fields or in combat areas. For this reason, research on autonomous vehicles has been very active during the past decades. It is obvious that these applications require precise control, and the trajectory tracking problem is crucial for the overall performances of the vehicle. Therefore, a significant number of studies address this problem: [1], [2], [3], [4]. Variable structure control (VSC) proved to be a robust solution in various applications, and can be used for controlling wheeled mobile robots. One particular type of variable structure control is the sliding mode control (SMC) [5]. The theory of SMC has been applied in continuous time to many control systems, [6], [7], [8], [9], [10], [11], [12], [13], thanks to its robustness against large parameter variations and disturbances. This recommends the SMC methodology for the trajectory tracking of autonomous vehicles, which is subject to uncertain perturbations. Designing switch functions of state variables or output variables to

* This work was supported by CNCSIS –UEFISCSU, project number PNII – IDEI code 506/2008 and CNCS-UEFISCDI, project number PN-II-ID-PCE-2011-3-0641.

G. Lee (Ed.): Advances in Automation and Robotics, Vol. 1, LNEE 122, pp. 401–409.
springerlink.com

form sliding surfaces insures that the trajectories are maintained on the surfaces, leading to the desired system dynamics. However, the SMC is affected by the chattering phenomenon, caused by high frequency switching over discontinuity of the control signal. Sliding mode control has been mostly considered for continuous-time systems, but continuous-time control can not be implemented on digital computers or microprocessors. As a result, the research of sliding mode control focused more on discrete-time sliding mode control (DSMC) [14], [15], [16], [17] in an attempt to eliminate the problems caused by the discretization of continuous-time controllers.

This paper addresses the problem of trajectory tracking of the four-driving/steering wheels (4DW/SW) mobile platform SEEKUR (fig. 4), by means of a discrete-time sliding mode controller. SEEKUR is a commercially available both holonomic and non-holonomic, all-weather, outdoor autonomous vehicle platform suitable for a wide range of security, inspection and research applications. Having an unique design, combined with omni-directional steering, it allows holonomic and nonholonomic motion. The discrete-time sliding mode controller has been designed using a discrete-time nonholonomic model of SEEKUR, rather than the usual continuous-time model, in order to avoid the problems created by the discretization of the continuous-time controller. The simulations were conducted using the MobileSIM simulator, offered by the manufacturer of SEEKUR, Mobile Robots [18], and a dedicated control application.

Beyond this introduction, the structure of this paper is as follows: the second section presents the model of the vehicle, including lateral slips, the third section is a detailed presentation of the discrete-time sliding mode controller, the fourth section shows the simulation results and the fifth section presents the results of the experiments with the real vehicle. Some conclusions are given in the last section.

2 The Kinematics Model of the Autonomous Vehicle

The kinematic model of the model including lateral slips is shown in fig. 1. Each wheel has a certain steering angle δ_i and slip angle β_i. The slip angle defines how large the lateral slip is and it is calculated using the longitudinal and linear velocities of the wheel

$$\beta_i = \tan^{-1}\left(\frac{v_{y_{wi}}}{v_{x_{wi}}}\right). \tag{1}$$

The subscript i denotes each wheel ID as shown in fig.1, (x_{CG}, y_{CG}, ψ) defines the position and orientation of the center of gravity of the vehicle, (x_{wi}, y_{wi}) defines the position of the ith wheel, v and v_{wi} are the linear velocities of the vehicle and of each wheel. Also, β represents the sideslip of the vehicle, l_f and l_r are the distance from the center of gravity of the vehicle to the front and rear wheels, l_f, l_r, d are constants.

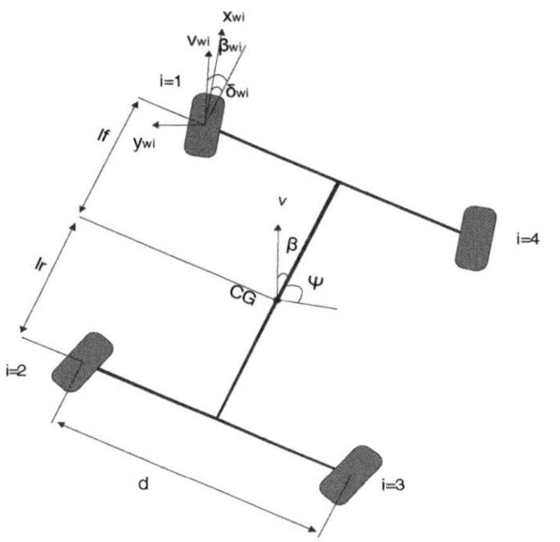

Fig. 1. Kinematics of the four driving/steering wheels

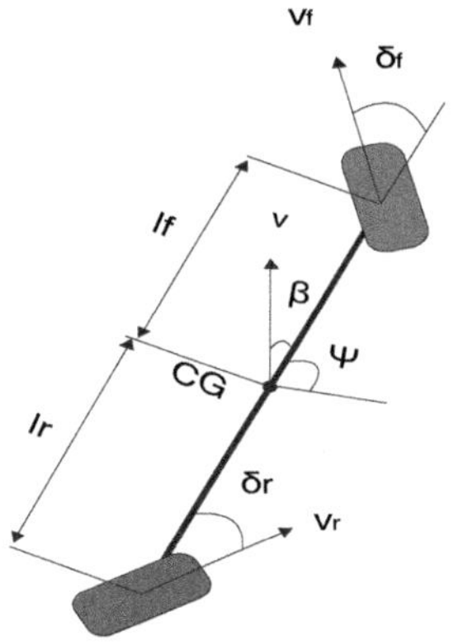

Fig. 2. Kinematic model of a bicycle

By taking into account the lateral slip, the nonholonomic constraints can be expressed as

$$\dot{x}_{CG} \cdot \sin(\beta + \psi) + \dot{y}_{CG} \cdot \cos(\beta + \psi) = 0 \tag{2}$$

$$\dot{x}_{wi} \cdot \sin(\beta_{wi} + \delta_{wi} + \psi) + \dot{y}_{wi} \cdot \cos(\beta_{wi} + \delta_{wi} + \psi) = 0 \ . \tag{3}$$

The geometric constraints between each wheel and the center of gravity can be written as

$$\begin{cases} x_{w1} = x_{CG} + l_f \cdot \cos\psi - d/2 \cdot \sin\psi \\ x_{w2} = x_{CG} - l_f \cdot \cos\psi - d/2 \cdot \sin\psi \\ x_{w3} = x_{CG} - l_r \cdot \cos\psi + d/2 \cdot \sin\psi \\ x_{w4} = x_{CG} + l_f \cdot \cos\psi + d/2 \cdot \sin\psi \end{cases} \cdot \tag{4}$$

$$\begin{cases} y_{w1} = x_{CG} + l_f \sin\psi - d/2 \cdot \cos\psi \\ y_{w2} = x_{CG} - l_f \cdot \sin\psi - d/2 \cdot \cos\psi \\ y_{w3} = x_{CG} - l_r \cdot \sin\psi + d/2 \cdot \cos\psi \\ y_{w4} = x_{CG} + l_f \cdot \sin\psi + d/2 \cdot \cos\psi \end{cases} \cdot \tag{5}$$

Considering wheels 1 and 2 and writing (3), (4) and (5) in matrix form, the resulting model is

$$A_{12} \cdot \dot{q}_0 = 0 \tag{6}$$

where

$$A_{12} = \begin{bmatrix} \sin\phi_{w1} & -\cos\phi_{w1} & -l_f \cdot \cos(\phi_{w1} - \psi) - \frac{d}{2}\sin(\phi_{w1} - \psi) \\ \sin\phi_{w2} & -\cos\phi_{w2} & l_r \cdot \cos(\phi_{w2} - \psi) + \frac{d}{2}\sin(\phi_{w2} - \psi) \\ \sin\phi_0 & -\cos\phi_0 & 0 \end{bmatrix} \quad (7)$$

$$\dot{q}_0 = \begin{bmatrix} \dot{x}_{CG} & \dot{y}_{CG} & \dot{\Psi}_{CG} \end{bmatrix}^T = 0 \quad (8)$$

$$\phi_{wi} = \beta_{wi} + \delta_{wi} + \psi, \quad \phi_0 = \beta + \delta \quad (9)$$

Assuming a planar motion, rigid body and no slippage of the tire, the model can be approximated by the bicycle model (fig. 2)

$$\begin{bmatrix} \sin(\delta_f + \psi) & -\cos(\delta_f + \psi) & -l_f \cdot \cos\delta_f \\ \sin(\delta_r + \psi) & -\cos(\delta_r + \psi) & l_r \cdot \cos(\delta_r \\ \sin(\beta + \psi) & -\cos(\beta + \psi) & 0 \end{bmatrix} \cdot \begin{bmatrix} \dot{x}_{CG} \\ \dot{y}_{CG} \\ \dot{\psi} \end{bmatrix} = 0 \; . \quad (10)$$

3 Discrete-Time Sliding-Mode Controller

The discrete-time sliding mode controller is a variable structure controller, which performs measurements and applies control signal at discrete, regular moments, and holds the control signal constant between sampling moments. This results in a discontinuous control. Discrete-time sliding mode control offers invariance to uncertain parameters, compensating the uncertainties that exist in real mobile robot applications, thus making it a good choice for trajectory tracking. To design a controller capable to track a desired trajectory, a virtual robot, moving along the target trajectory $q_d(t) = [x_d(t) \;\; y_d(t) \;\; \theta_d(t)]^T$, is considered (fig. 3). The purpose of the controller is to keep the trajectory tracking error of the real vehicle as close to 0 as possible. The block schema of the closed loop together with SEEKUR mobile platform is presented in fig. 4. The equation defining the tracking errors is

$$\begin{bmatrix} x_e \\ y_e \\ \theta_e \end{bmatrix} = \begin{bmatrix} \cos\theta_d & \sin\theta_d & 0 \\ -\sin\theta_d & \cos\theta_d & 0 \\ 0 & 0 & 1 \end{bmatrix} \begin{bmatrix} x_r - x_d \\ y_r - y_d \\ \omega_r - \omega_d \end{bmatrix} \quad (11)$$

In discrete-time, (14) becomes

$$\begin{cases} x_e[k] = x_{rd}[k] \cdot \cos\theta_e[k] + y_{rd}[k] \cdot \sin\theta_e[k] \\ y_e[k] = -x_{rd}[k] \cdot \sin\theta_e[k] + y_{rd}[k] \cdot \cos\theta_e[k] \\ \theta_e[k] = \theta_r[k] - \theta_d[k] \end{cases} \quad (12)$$

$$x_{rd}[k] = x_r[k] - x_d[k], \quad y_{rd}[k] = y_r[k] - y_d[k] \quad (13)$$

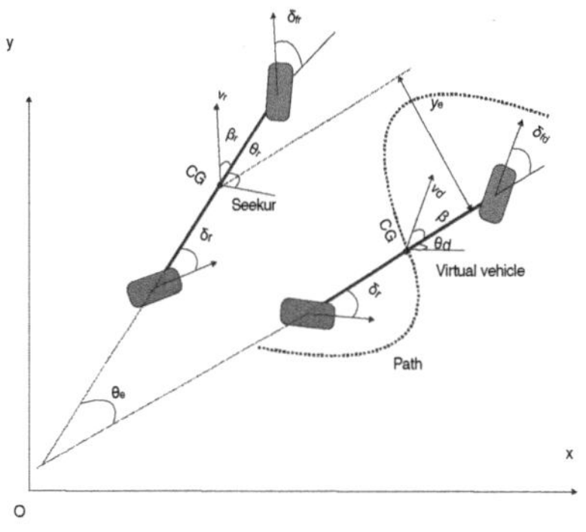

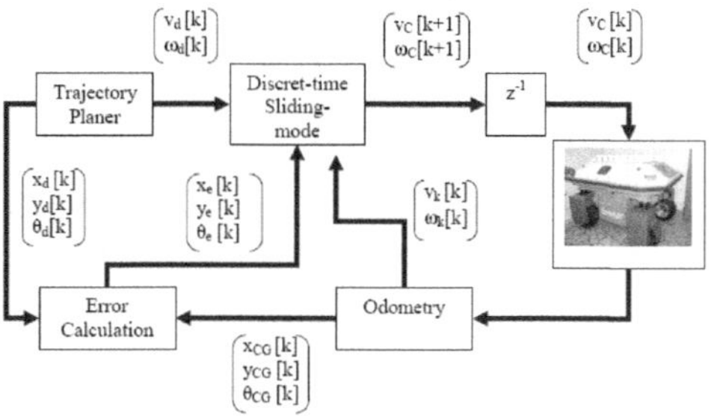

Fig. 3. Trajectory tracking errors of virtual vehicle **Fig. 4.** Closed loop structure

In continuous time, the dynamics errors for the trajectory tracking are defined as

$$\begin{cases} \dot{x}_e = -v_d + v_r \cdot \cos\theta_e + \omega_d \cdot y_e \\ \dot{y}_e = v_r \cdot \sin\theta_e - \omega_d \cdot x_e \\ \dot{\theta}_e = \dot{\theta}_r - \dot{\theta}_d \end{cases} \tag{14}$$

and in discrete-time

$$\begin{cases} x_e[k+1] = x_e[k] + (v1 - v_d[k] + y_e[k] \cdot \omega_d[k]) \cdot T_s \\ y_e[k+1] = y_e[k] + (v2 - x_e[k] \cdot \omega_d[k]) \cdot T_s \\ \theta_e[k+1] = \theta_e[k] + (\omega_r[k] - \omega_d[k]) \cdot T_s \end{cases} \tag{15}$$

where

$$v1 = v_r[k] \cdot \cos\theta_e[k], \quad v2 = v_r[k] \cdot \cos\theta_e[k] \tag{16}$$

The following sliding surface attractiveness condition, in discrete-time, has been used

$$s[k] \cdot (s[k+1] - s[k]) < 0 \tag{17}$$

The reaching law proposed in [14] is

$$s[k+1] = (1 - q \cdot T_s) \cdot s[k] - \varepsilon \cdot T_s \cdot \mathrm{sgn}(s[k]) \tag{18}$$

with

$$0 < 1 - q \cdot T_s < 1, \quad 0 < \varepsilon \cdot T_s < 1 \tag{19}$$

where $T_s \rangle 0$ is the sampling period, $\varepsilon \rangle 0$ is the reaching velocity, $q \rangle 0$ is the converging exponential. The sliding mode trajectory will move within the quasi-sliding mode band

$$|s[k]| < \frac{\varepsilon \cdot T_s}{1 - q \cdot T_s} \ . \tag{20}$$

The sliding surfaces are defined as

$$\begin{cases} s_1[k] = x_e[k+1] + k_1 \cdot x_e[k] \\ s_2[k] = y_e[k+1] + k_2 \cdot y_e[k] + k_0 \cdot \mathrm{sgn}(y_e[k])\theta_e[k] \end{cases} \tag{21}$$

where k_0, k_1, k_2 are positive constant parameters, x_e , y_e and θ_e are trajectory tracking errors. From (23) and (27) we get:

$$\begin{aligned} s_1[k+1] &= x_e[k+2] + k_1 \cdot x_e[k+1] \\ s_1[k+1] &= (1 - q \cdot T_s) \cdot s_1[k] - \varepsilon \cdot T_s \cdot \mathrm{sgn}(s_1[k]) \end{aligned} \tag{22}$$

$$\begin{aligned} s_2[k+1] &= y_e[k+2] + k_2 \cdot y_e[k+1] + k_0 \cdot \mathrm{sgn}(y_e[k]) \cdot \theta_e[k] \\ s_2[k+1] &= (1 - q \cdot T_s) \cdot s_1[k] - \varepsilon \cdot T_s \cdot \mathrm{sgn}(s_1[k]) \end{aligned} \tag{23}$$

The discrete-time sliding-mode controller is obtained from (12), (15), (22), (23):

$$\begin{aligned} v[k+1] = {} & \frac{1}{T_S \cos\theta_e[k]}[-(1 - q_1 \cdot T_s) \cdot s_1[k] + \varepsilon_1 T_s \,\mathrm{sgn}(s_1[k] - x_e[k+1] \cdot (1 + k_1) \\ & - (v_d[k+1] - v_r[k] \cdot \theta_e[k+1] \sin\theta_e[k] - w_d[k+1] y_e[k] - w_d[k] y_e[k+1]) \cdot T_s] \end{aligned} \tag{24}$$

$$\begin{aligned} \omega[k+1] = {} & w_d[k] + \frac{1}{v_r[k]\cos\theta_e + k_0 \,\mathrm{sgn}\, y_e[k+1]}[(q_2 T_s - 1) s_2[k] \\ & + \varepsilon_2 T_s \,\mathrm{sgn}\, s_2[k] - \theta_e[k] - y_e[k+1](k_2 + 1) \\ & - (v_r[k+1] \sin\theta_e[k] + w_d[k+1] x_e[k] - w_d[k] x_e[k+1]) T_s)] \end{aligned} \tag{25}$$

4 Simulation Results with Discrete-Time Sliding-Mode Controller

This paragraph presents the simulation results for the proposed discrete-time sliding mode controller. The actual control program is written in C++ running on a PC, with sampling time set to T_s=100 ms. The vehicle was simulated with MobileSIM, a software simulator offered by Mobile Robots Inc. – manufacturer of the SEEKUR platform. The trajectory of the simulated robot, recorded using MobileSim is presented in fig. 5. The real trajectory of the simulated SEEKUR is displayed in fig. 6 with a continuous red line, while the desired trajectory is plotted with an interrupted blue line. Fig. 7 shows the trajectory tracking error on the x axis. Fig. 8 shows the trajectory tracking error on the y axis. In fig. 9 the heading error of the vehicle is presented. Fig. 10 shows the sliding surface s_1. Fig. 11 shows the sliding surface s_2.

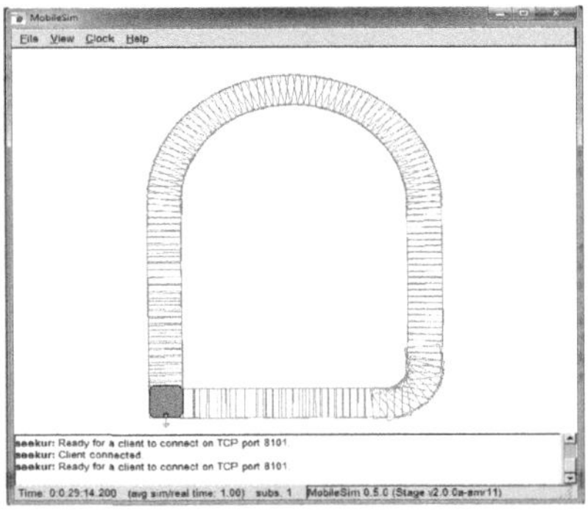

Fig. 5. Trajectory of the simulated vehicle

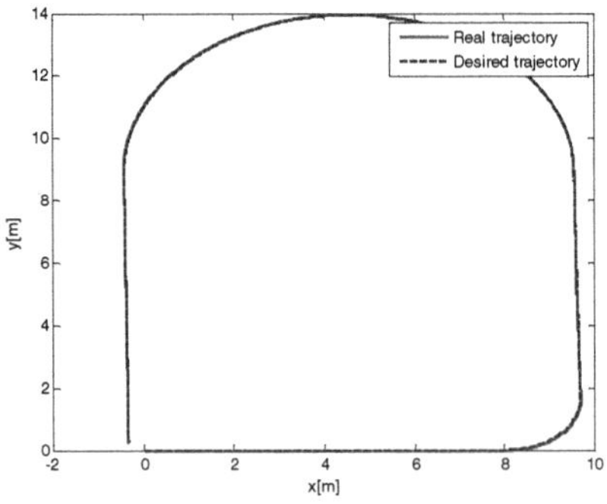

Fig. 6. Real and desired trajectories

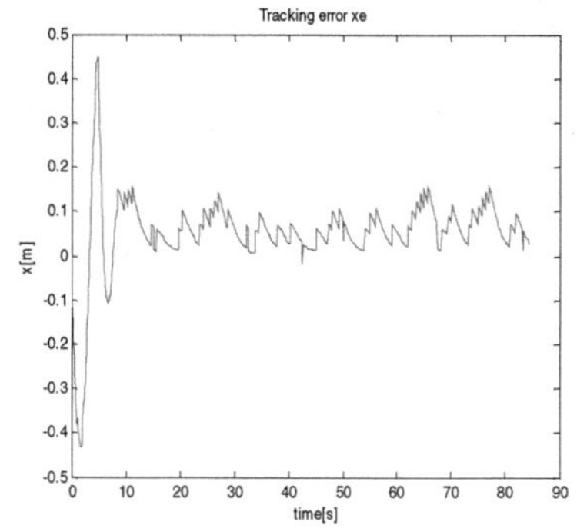

Fig. 7. Trajectory tracking error, x_e

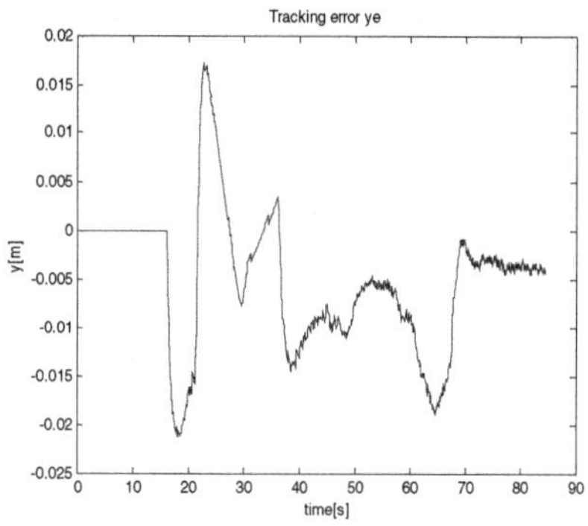

Fig. 8. Trajectory tracking error, y_e

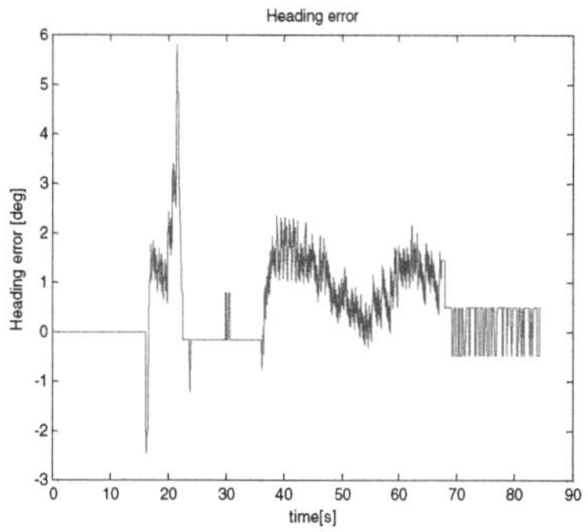

Fig. 9. Heading error, θ_e

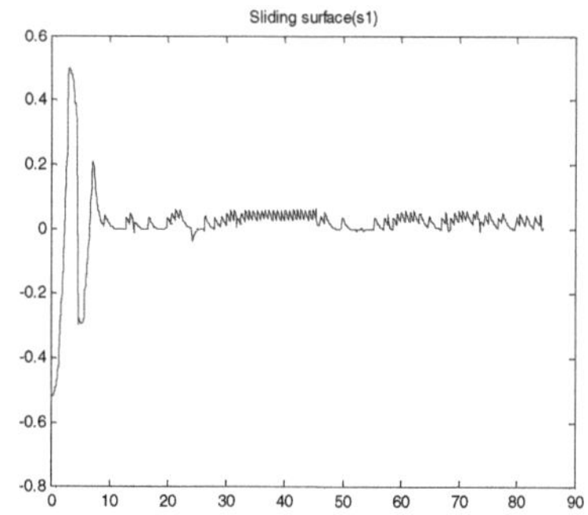

Fig. 10. Sliding surface, s_1

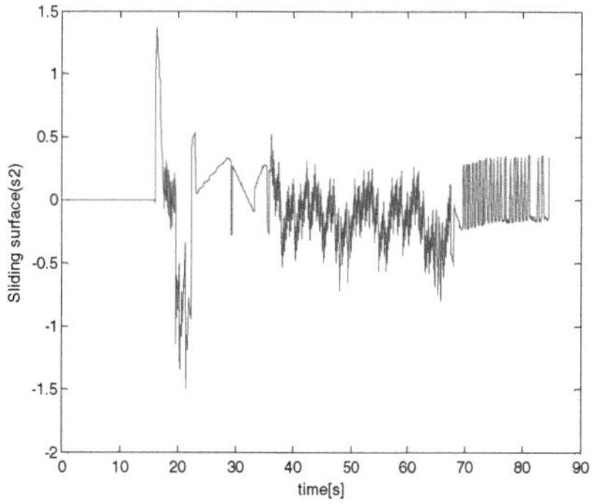

Fig. 11. Sliding surface, s_2 .

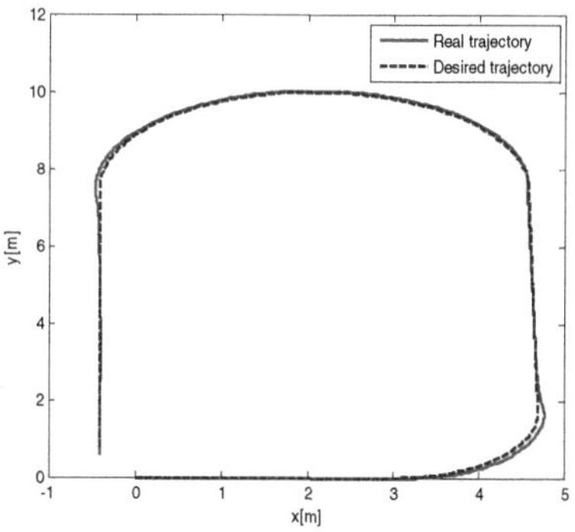

Fig. 12. Real and desired trajectories

5 Real-Time Control of the SEEKUR Mobile Platform

In order to validate the proposed controller, a second set of experiments were conducted, using a real SEEKUR platform. The control program was written in C++ and run on a PC. Sampling time was set to T_s=100 ms, and the commands to the vehicle were sent using a wireless connection through a wireless access point and a universal device server that adds Ethernet connectivity to the vehicle's serial connection. The robot followed the desired trajectory, presented in fig. 12 with an interrupted blue line. Fig. 12 also shows the real trajectory with a continuous red line. Fig.13. shows the tracking error on the x axis, fig. 14 shows the error on the y axis and the heading error is displayed in fig. 15. Fig. 16 shows the sliding mode surface s_1. Finally, the sliding surface s_2 is presented in fig. 17.

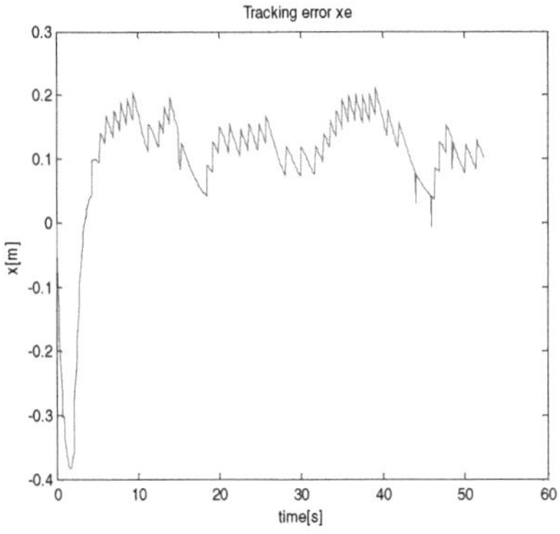

Fig. 13. Trajectory tracking error, x_e

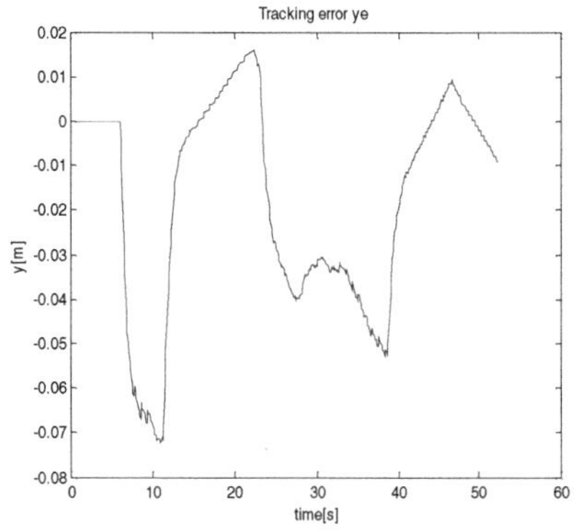

Fig. 14. Trajectory tracking error, y_e

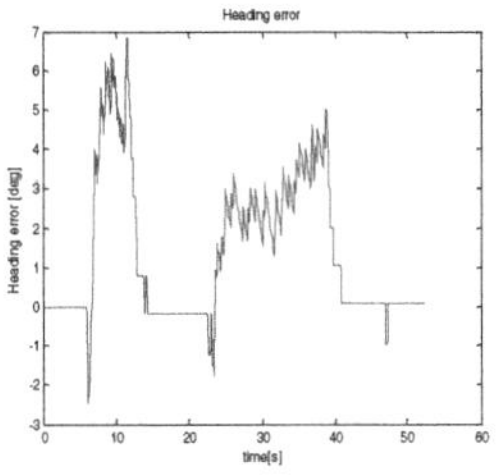

Fig. 15. Heading error, θ_e

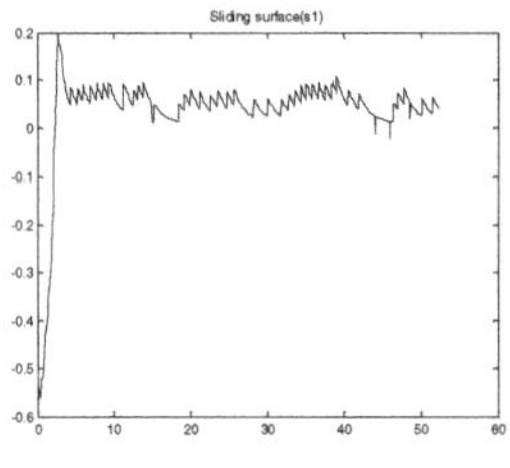

Fig. 16. Sliding surface, s_1

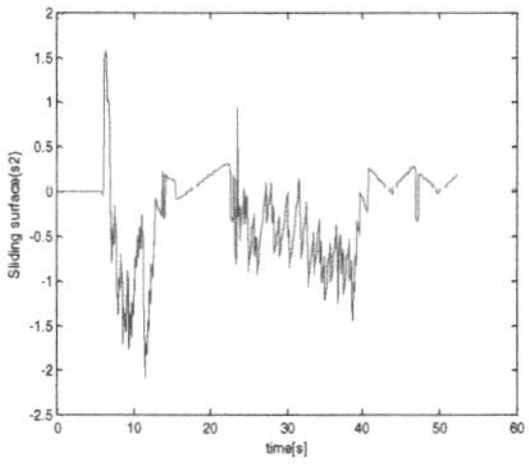

Fig. 17. Sliding surface, s_2

6 Conclusions

In this paper, a discrete-time sliding mode controller for trajectory tracking of four driving-steering wheels mobile platform SEEKUR is presented. The simulation and real time experimental results prove the effectiveness of the proposed controller. The desired performances for this controller are achieved. The controller is simple and easy to use. Considering the versatility of the target platform, this research might be of interest in many practical implementations.

References

1. Chu-Quing, C., Lian-Zheng, G., Rui-Feng, L.: Mobile robots target tracking using finite-time convergence sliding mode controller. In: Proceedings of the 8th IEEE International Conference on Control and Automation, Xiamen, China, pp. 460–464 (2010)
2. Jinag, Z.P., Nijmeijer, H.: Tracking control of mobile robots: a case study in backstepping. Automatica 33, 1393–1399 (1997)
3. Jinag, Z., Nijmeijer, H.: A recursive technique for tracking control of nonholonomic systems in chained form. IEEE Trans. Automatic Control 44, 265–279 (1999)
4. Yuan, Z.P., Wang, Z.P., Chen, Q.J.: Trajectory tracking control of a nonholonomic mobile robot. In: Proceedings of the 8th IEEE International Conference on Control and Automation, Xiamen, China, pp. 2207–2211 (2010)
5. Utkin, V.I.: Sliding modes in optimization and control. Springer, New York (1992)
6. Utkin, V.I., Guldner, J., Shi, J.: Sliding mode control in electromechanical systems. Taylor & Francis, London (1999)
7. Solea, R., Filipescu, A., Stamatescu, G.: Sliding-mode real-time mobile platform control in the presence of uncertainties. In: Proceedings of the 48th IEEE Conference on Decision and Control and 28th Chinese Control Conference, Shanghai, China, pp. 7747–7752 (2009)
8. Solea, R.: Sliding mode control applied in trajectory-tracking of WMRs and autonomous vehicles, PhD Thesis, University of Coimbra, Portugal (2009)
9. Solea, R., Filipescu, A., Nunes, U.: Sliding-mode control for trajectory tracking of a wheeled mobile robot in presence of uncertainties. In: Proceedings of the 7th Asian Control Conference, Hong Kong, China, pp. 1701–1706 (2009)
10. Solea, R., Filipescu, A., Filipescu, S., Dumitrascu, B.: Sliding-mode Controller for Four-Wheel-Steering Vehicle: Trajectory-Tracking Problem. In: Proceedings of the 8th World Congress on Intelligent Control and Automation, Jinan, China, pp. 1185–1190 (2010)
11. Solea, R., Filipescu, A., Manzu, V., Filipescu, S.: Sliding-mode Trajectory-tracking Control for a Four-Wheel-Steering Vehicle. In: Proceedings of the 8th IEEE International Conference on Control and Automation, Xiamen, China, pp. 382–387 (2010)
12. Solea, R., Filipescu, A., Cernega, D.: Lateral Motion Control of Four-Wheels Steering Vehicle Using a Sliding-Mode Controller. In: Proceedings of the 29th Chinese Control Conference, Beijing, China, pp. 3699–3703 (2010)
13. Gao, W., Hung, J.C.: Variable structure control of nonlinear systems: A new approach. IEEE Transactions on Industrial Electronics 40, 45–55 (1993)
14. Gao, W., Wang, Y., Homaifa, A.: Discrete-time variable structure control systems. IEEE Transaction on Industrial Electronics 42, 117–122 (1995)
15. Furuta, K., Pan, Y.: Discrete-time variable structure control. LNCIS, vol. 274, pp. 57–81. Springer, Berlin (2002)
16. Bandyopadhyay, B., Janardhanan, S.: Discrete-time sliding mode control a multirate output feedback approach. LNCIS, vol. 323. Springer, Berlin (2006)
17. Wang, B.: On discretization of sliding mode control systems. RMIT University, Melbourne (2008)
18. SEEKUR-Autonomous all-weather robot. Mobile Robots Inc., `http://mobilerobots.com/ResearchRobots/ResearchRobots/SeekurUGV.asp`

Tuning of Unscented Kalman Filter Based Biogeography-Based Optimization for UGVs Navigation

Kuifeng Su[1,2], Zhidong Deng[1], and Zhen Huang[1]

[1] Department of Computer Science, State Key Laboratory of Intelligent Technology and Systems, Tsinghua National Laboratory for Information Science and Technology, Tsinghua University, Beijing 100084, China
[2] Academy of Armored Force Engineering, Beijing, China 100072
skf08@mails.tsinghua.edu.cn, michael@mail.tsinghua.edu.cn, chinaheart2003@qq.com

Abstract. In the case of linear systems, corrupted by white Gaussian noise, the Kalman filter is proved to be an optimal filter in the mean least square sense. When the system model and measurements are non-linear, variation of Kalman filter like extended Kalman filter (EKF), Unscented Kalman filters (UKF) and Particle filter （PF） are used. However, the best performance of UKF is achieved when the random variables only are Gaussian, and affected seriously by the estimation precision of noise covariance. In this paper, we give a noise covariance estimation method based Biogeography-based Optimization (BBO). The experimental results obtained from real-world road tests validate the performance of our approach.

Keywords: Unmanned ground vehicle, Unscented Kalman filter, Noise Covariance, Biogeography-based Optimization.

1 Introduction

Real-time information about position is the basis of the autonomous driving of unmanned ground vehicle (UGV). No matter what kind of sensor configuration is used to establish navigation system, the estimation strategy of optimal online information fusion is the basis of integrated navigation system. For decades, a variety of information fusion and state estimation methods emerge, such as Kalman filter (KF), extended Kalman filter (EKF), unscented Kalman filter (UKF), particle filter (PF) and so on [1], [2]. Under the hypothesis of linear and Gaussian, KF can get the linear unbiased optimal estimation. However, most systems are nonlinear and non-Gaussian, therefore, non-linear filter has become a hot topic in the state estimation area; on this basis, the EKF based on local linearization is also developed, Benes and Daum filtering based on finite element approximation, UKF and PF based on probability statistical properties of similar random variables. Although EKF is widely used, due to its requirement for a strong differentiability of the system, it is apt to be impacted by errors. Therefore, UKF filter was developed on the basis of UT proposed by Julier [3], which can directly utilize the systemic nonlinear dynamic equation without linearization, so there is no need to calculate the Jacobian matrix [4].

G. Lee (Ed.): Advances in Automation and Robotics, Vol. 1, LNEE 122, pp. 411–419.
springerlink.com

However, KF filter is still regarded as the basic framework of UKF filter, so in order to obtain better estimated performance, the priori knowledge about the statistical characteristic of system and measurement noise is required, which cannot be implemented in the practical application. Inaccurate statistical characteristic of noise may lead to estimated error, even the filter diverges [3], [4]. To overcome this kind of problem, this paper applied the Biogeography-Based Optimization (BBO) algorithm to adjust UKF parameters and noise covariance matrices. The BBO-UKF adaptive state estimation algorithm proposed in this paper can not only dynamically estimate the statistical properties of noise but also effectively solve high-dimensional optimization problems, thus achieving the efficient adaptive state estimation of autonomous vehicle.

2 Problems

For intelligent Vehicle THIV-I of Tsinghua University, the installation position of various sensors is as shown in Figure 1. The location system can acquire the speed and steering angle of 4 wheels input by ABS sensor through the CAN bus of the vehicle itself. Meanwhile, the encoders on the rear transmission shafts can be used to obtain the accurate information about the travelled distance during sample period.

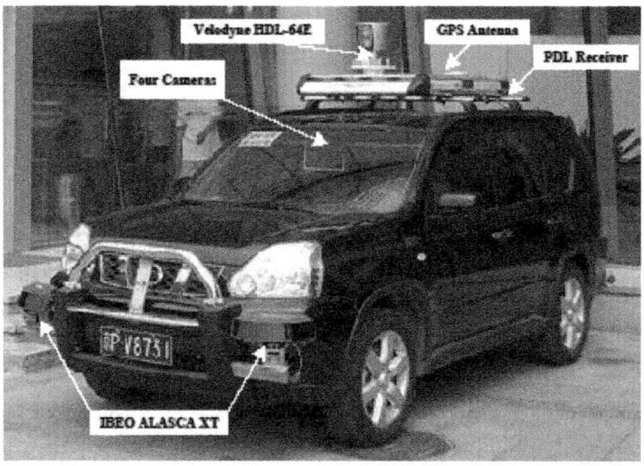

Fig. 1. Intelligent Vehicle of Tsinghua University

2.1 System Model

In this paper, the coordinate origin of the vehicle body locates in the center point of the rear axle, that is, the installation position of IMU, whose position is $P_k(x_k, y_k)$ at the time of t_k, and the heading angle is ϕ_k. In 2D environment, UGV's position can be expressed by the state vector $\chi_k = (x_k, y_k, \phi_k)$. The position relation of UGV is as shown in Figure 3. In this figure, H represents a half of distance between rear wheels, while L represents the distance between front axle and rear axle of the UGV.

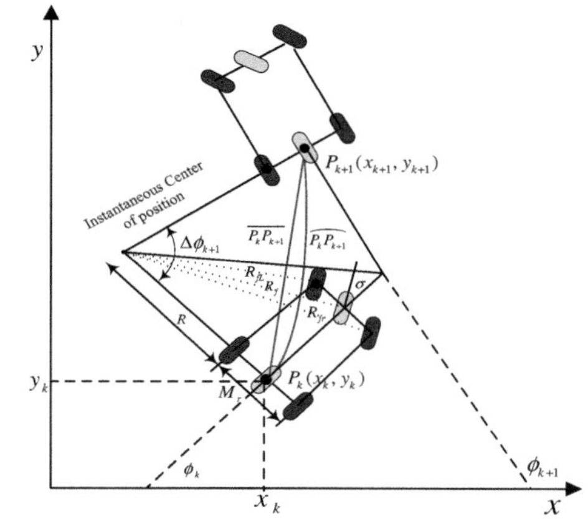

Fig. 2. Schematic diagram of the vehicle's kinematics parameter model

As shown in Figure 2, the vehicle moves from P_k to P_{k+1}. $\widehat{P_kP_{k+1}}$ refers to the arc length between P_k to P_{k+1}, $\overline{P_kP_{k+1}}$ refers to the linear distance between these two points and $\Delta\phi_{k+1}$ is the increment of heading angle. According to the geometric theorem, we can conclude that:

$$\frac{\widehat{P_kP_{k+1}}}{\overline{P_kP_{k+1}}}=\frac{(R+M_r/2)\Delta\phi_{k+1}}{2(R+M_r/2)\sin[\Delta\phi_{k+1}/2]}=\frac{\Delta\phi_{k+1}/2}{\sin[\Delta\phi_{k+1}/2]} \tag{1}$$

Therefore, the linear distance $\overline{P_kP_{k+1}}$ between P_k and P_{k+1} is:

$$\overline{P_kP_{k+1}}=\frac{\sin[\Delta\phi_{k+1}/2]}{\Delta\phi_{k+1}/2}\widehat{P_kP_{k+1}}=\frac{\sin[\Delta\phi_{k+1}/2]}{\Delta\phi_{k+1}/2}\Delta D_{k+1} \tag{2}$$

In this formula, ΔD_{k+1} refers to the travel distance of the back wheel center measured by encoder. Therefore, if the sampling period is sufficient, the vehicle's kinematics state equation can be described as:

$$\mathrm{X}(k)=\begin{bmatrix} x_{k+1} \\ y_{k+1} \\ \phi_{k+1} \end{bmatrix}=\begin{bmatrix} x_k+\Delta x_{k+1} \\ y_k+\Delta y_{k+1} \\ \phi_k+\Delta\phi_k \end{bmatrix}=\begin{bmatrix} x_k+\overline{P_kP_{k+1}}\cos\left[\phi_{k+1}+\Delta\phi_{k+1}/2\right] \\ y_k+\overline{P_kP_{k+1}}\sin\left[\phi_{k+1}+\Delta\phi_{k+1}/2\right] \\ \phi_k+\Delta\phi_{k+1} \end{bmatrix} \tag{3}$$

2.2 Observation Model

The system of autonomous vehicle utilizes odometer and IMU measurement unit to achieve the estimation on basic position and attitude. In order to overcome the error accumulation caused by the relative positioning system, the system applies 64-line HDL lidar, which is mounted in the anterior part of middle top of UGV, and the position relation of the relative vehicle is as shown in Figure 3.

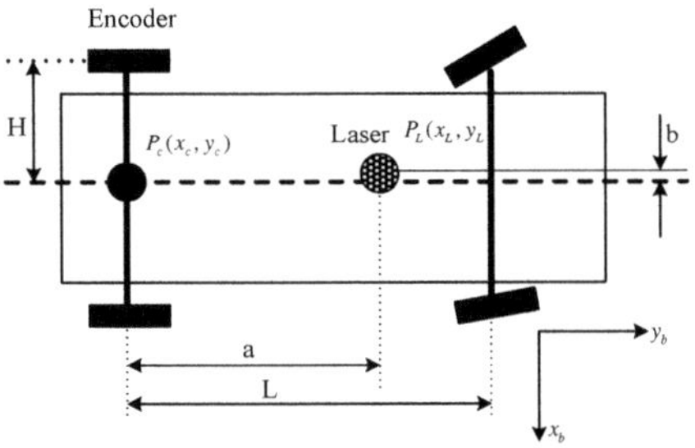

Fig. 3. Installation diagram of UGV sensor

Then the relationship that the rear axle center transforms to the lidar centre is shown as follows:

$$\begin{aligned} x_L &= x_c + a\cos(\phi) + b\cos(\phi + \pi/2) \\ y_L &= y_c + a\sin(\phi) + b\sin(\phi + \pi/2) \end{aligned} \tag{4}$$

The corresponding measurement equation is:

$$z = h(X, x_i, y_i) = \begin{bmatrix} z_\gamma^i \\ z_\beta^i \end{bmatrix} = \begin{bmatrix} \sqrt{(x_i - x_L)^2 + (y_i - y_L)^2} \\ atan\left(\dfrac{(y_i - y_L)}{(x_i - x_L)}\right) - \phi_L + \pi/2 \end{bmatrix} + v \tag{5}$$

In this formula, z refers to the measurement vector, that is, the coordinate of landmark; $\{x_i, y_i\}$ refers to the location of landmark identified by the 3D lidar sensors and v is the measurement noise.

3 BBO-UKF State Estimation Algorithm

3.1 Biogeography-Based Optimization

According to the island migration model of biogeography, Dan Simon proposed BBO in 2008[5]. BBO is also a species optimization algorithm, but compared with GA and evolutionary strategy optimization algorithm, it is not related to the problems of regeneration and producing next generation; There is also a marked difference between ACO and BBO. ACO will produce a new set of solutions in each iteration, while BBO will maintain its solution set to the next iteration in each iteration and adjust solution space according to the migration probability. BBO has more in common with PSO and DE. Compared with PSO and DE, BBO directly updates by the migration of the solution. Therefore, the solutions of BBO algorithm can share properties (suitability index variable, SIVs) with each other. BBO mainly consists of the two process of species migration and species variation[6], [7], [8]. BBO algorithm flow is as follows:

```
1) Initialize the BBO parameters.
2) Initialize the random parameter of habitat.
3) For each habitat, HIS is mapped as the number and
mobility (λ,μ) of species S.
     Calculate and determine whether HIS can satisfy the
     suspensive condition.
     If HIS dose not satisfy the suspensive condition
        4) Utilize the immigration rate and emigration
             rate to restore the habitat, and then
             recalculate each HIS.
        5) For each habitat, refresh the probability
             distribution of species, update the species
             according to mutation operator, and then
             recalculate the fitness.
      End if
6) End the algorithm cycle according to the termination
condition.
```

3.2 Adaptive UKF Algorithm Based on BBO

Standard UKF algorithm involves six parameters, including the initial state $\overline{X}_0$, the initial variance P_0, process noise covariance Q, the measurement noise covariance R and the transformation parameters α and β of UT. In the landmark-based navigation system, the impact of initial state and covariance may gradually be weakened as time goes by. Transformation parameters α and β of UT can only affect the higher-order term of nonlinear system. As the priori knowledge of the UKF algorithm, process and measurement noise covariance matrix Q and R have an impact on the estimated accuracy and stability of UKF algorithm [9], [10].

This paper uses the method of the minimum filter of a recursive algorithm estimating the deviation between the covariance and that of the actual information to estimate Q and R.The average covariance can be used to approximate the actual variance of the system [11], [12] as follows:

$$S_k = \frac{1}{N} \sum_{i=k-N+1}^{k} \upsilon_k \upsilon_k^T \tag{6}$$

Where N refers to the estimated window. The information sequence is shown as:

$$y_k = y_k - \hat{y}_{k|k-1} \tag{7}$$

y_k and $\hat{y}_{k|k-1}$ are the actual measurement and predicted value; based on UKF algorithm the prediction equation can be expressed as:

$$\hat{S}_k = \sum_{i=0}^{2n} w_i^c \left(\hat{y}_{i,k|k-1} - \hat{y}_{k|k-1}\right)\left(\hat{y}_{i,k|k-1} - \hat{y}_{k|k-1}\right)^T \tag{8}$$

The adaptation function of the adaptive BBO-UKF algorithm is minimized as:

$$FIT_k = tr\left(\Delta S_k^2\right) = tr\left[\left(S_k - \hat{S}_k\right)^2\right] \tag{9}$$

BBO-UKF algorithm achieves the real-time estimation of the noise statistical characteristics, so as to provide more accurate statistical information for filter. Figure 4 shows the flowchart of BBO-UKF algorithm.

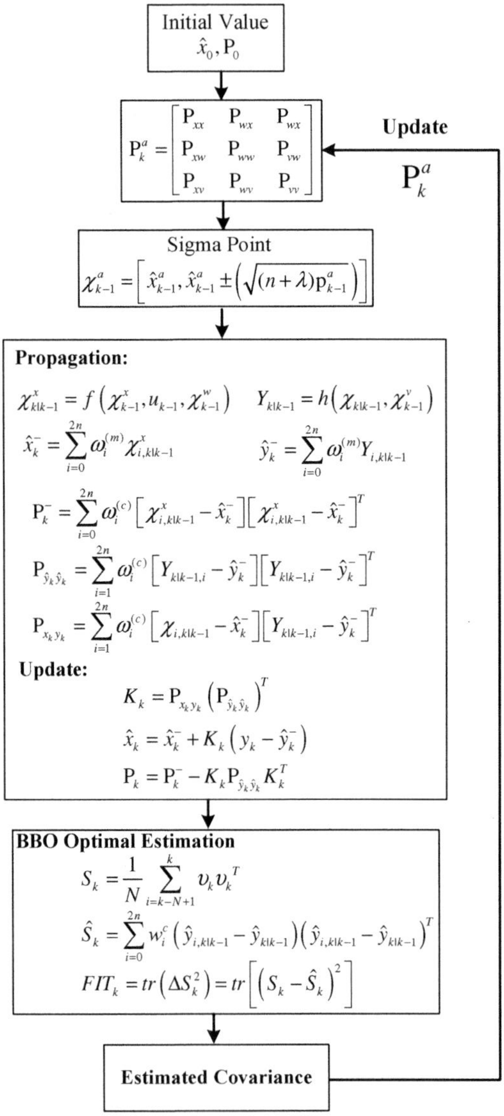

Fig. 4. Flow chart for BBO-UKF Algorithm

4 Experimental Result and Analysis

To test the validity of BBO-UKF algorithm, this paper has carried out experiment for several times on the road of school. The test sites are all ideal two-dimensional environment, in addition, the traffic signs are also placed as the landmark of lidar detection. We first utilize the high-precision Span/DGPS integrated navigation system from Novatel to record the actual position and attitude of autonomous vehicle, that is, $\chi\left(x_{ref}, y_{ref}, \theta_{ref}\right)$, thus assessing the state estimation result of the UKF algorithm.

To achieve the BBO-UKF algorithm, BBO is mainly used to estimate information variance, the process noise variance and measurement noise variance. To simplify the system design and reduce the complexity of BBO optimization algorithm, make $\mathrm{P}_k^a = \left(\mathrm{P}_k^a\right)^T$, and then a total of 8 parameters require to be estimated. According to the estimated noise range, the parameters to be estimated are mapped as 8D integer vector, and each integer vector in the solution-space is regarded as a habitat suitability index variable (SIV), therefore, the optimal solution is the vector with high habitat suitability index variable. The testing environment and basic trajectory of BBO-UKF algorithm is as shown in Figure 5, and figure 6 indicate the comparison of vehicle trajectory, the eastward and northward deviation. The estimation precision of BBO-UKF algorithm proposed in this paper has been greatly improved compared with that of UKF algorithm.

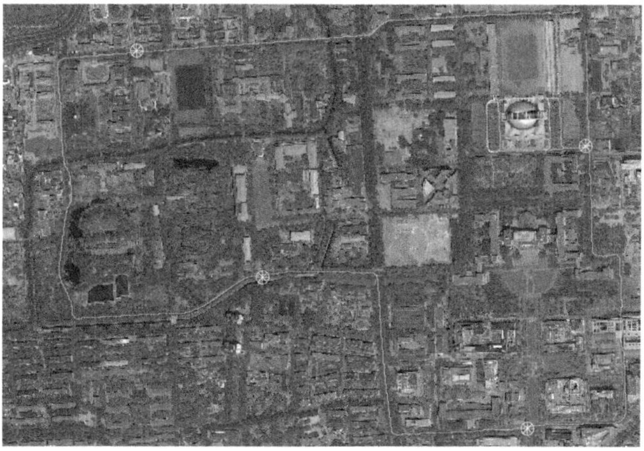

Fig. 5. Google earth map of testing environment in Tsinghua University

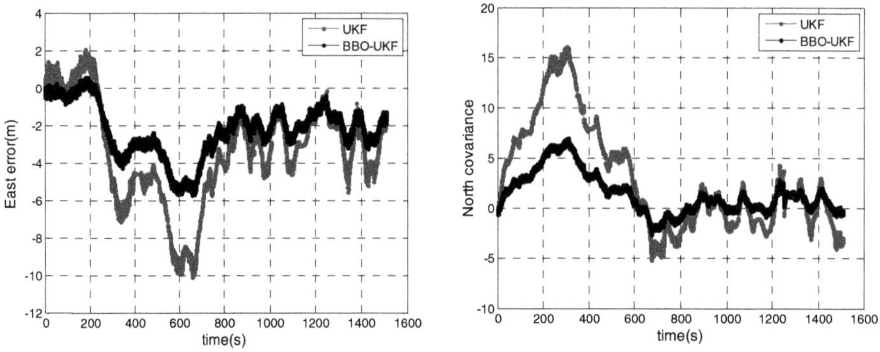

(a) eastward deviation and covariance (b) northward deviation and covariance

Fig. 6. Comparison between eastward and northward deviation and covariance

5 Conclusions

UKF filter utilizes UT transformation to achieve the non-linear mapping of random variables, which can effectively solve the estimation problems of nonlinear system. However, its estimation precision deeply relies on the system and statistical characteristic of measurement noise, so this paper proposes that we should use BBO optimization algorithm to dynamically estimate the statistical characteristic of noise regarding the minimization of information sequence variance as the target. Experimental results show that the BBO-UKF algorithm can effectively improve the location accuracy of autonomous vehicles. In addition, due to the inherent characteristics of BBO algorithm, its efficiency is higher than other intelligent optimization algorithms such as PSO in solving high-dimensional optimization problems, which has provided a guarantee for real-time application of BBO-UKF.

References

1. Smith, R., Self, M., Cheeseman, P.: Estimating uncertain spatial relationships in robotics. In: Cox, I.J., Wilfon, G.T. (eds.) Autonomous Robot Vehicles, pp. 167–193. Springer, New York (1990)
2. Julier, S.J., Uhlmann, J.K.: A new approach for filtering nonlinear systems. In: Proceedings of the American Control Conference, pp. 1628–1632 (1995)
3. Julier, S.J., Uhlmann, J.K.: New extension of the Kalman filter to nonlinear systems. In: The Robotics Research Group, Department of Engineering Science, The University of Oxford, Oxford, OX1 3PJ, UK
4. Julier, S.J., Uhlmann, J.K.: Unscented filtering and nonlinear estimation. Proc. IEEE, 401–422 (2004)
5. Simon, D.: Biogeography-based optimization. IEEE Transactions on Evolutionary Computation 6(12), 702–713 (2008)
6. Simon, D., Ergezer, M., Dawei, D.: Markov analysis of biogeography-based optimization [OL], `http://academic.csuohio.edu/simond/bbo`

7. Simon, D., Ergezer, M., Du, D.: Markov models for biogeography-based optimization and genetic algorithms with global uniform recombination [OL], http://academic.csuohio.edu/simond/bbo/markov/MarkovJournal.pdf
8. Simon, D.: A probabilistic analysis of a simplified biogeography-based optimization algorithm [OL], http://academic.csuohio.edu/simond/bbo/simplified/bbosimplified
9. Jwo, D.-J., Weng, T.-P.: An adaptive sensor fusion method with applications in integrated navigation. The Journal of Navigation 61, 705–721 (2008)
10. Yang, Y., Cui, X., Gao, W.: Adaptive integrated navigation for multi-sensor adjustment outputs. The Journal of Navigation 57(2), 287–295 (2004)
11. Jwo, D.-J., Weng, T.-P.: An Adaptive Sensor Fusion Method with Applications in Integrated Navigation. Journal of Navigation 61(04), 72–79 (2008)
12. Zhang, S., Yang, S.: A Novel Fusion Method for DR /GPS Integrated Navigation System. In: Proc. of IEEE Int. Conf. on Industrial Informatics, pp. 108–111. IEEE Press, Indin (2008)

Robust Real-Time Lane Marking Detection for Intelligent Vehicles in Urban Environment

Wentao Yao and Zhidong Deng*

State Key Laboratory of Intelligent Technology and Systems
Tsinghua National Laboratory for Information Science and Technology
Department of Computer Science and Technology, Tsinghua University
Beijing, 100084, China
ywt@mails.tsinghua.edu.cn, michael@tsinghua.edu.cn

Abstract. In this paper, we propose a framework of robust and real-time lane marking detection and tracking for the autonomous driving of intelligent vehicle under urban road environments. Our framework use hyperbola model as the lane marking model and several adaptive techniques to robustly extract lane marking pairs for the current lane. The framework provides a complete solution from environment perception of intelligent vehicle to vehicle control. The experimental results achieved on urban road scenarios show that our system performs well under various situation.

Keywords: lane marking detection, intelligent vehicle, hyperbola model.

1 Introduction

Environment perception is one of the fundamental challenges for the autonomous driving of intelligent vehicle. As the main tasks of environment perception, detection and tracking of painted lane markings on urban road scenarios are a tough problem because there exist different types of lane markings and the detection is often affected by illumination, weather, occlusion, shadow, defect or non-lane markings like arrow, crosswalk, and traffic symbol.

For decades, a lot of advanced lane marking detection and tracking systems have been developed. [9] gives a complete and detailed review of some representative lane marking detection and tracking systems. The commonly-used lane marking models include simple straight line model, quadratic curve model[7], complex clothoid model[4], and spline model[12], etc. Other systems employ implicit models, such as yaw angle and lateral offset[11], or generate vehicle control commands directly[1].

There are various method to detect lane markings in images and videos. Edge features are most widely used in the detection of lane markings[7][2][10]. Frequency-based techniques, such as the LANA system[8] and the use of Gabor filter[13], are shown to be effective in dealing with extraneous edges. Inverse projection mapping (IPM)[1][5] is another effective way for lane marking detection

* Corresponding author.

G. Lee (Ed.): Advances in Automation and Robotics, Vol. 1, LNEE 122, pp. 421–428.
springerlink.com

by performing the detection on top-view images. Some new techniques, such as probability based method[3] and SVM classifier[6] are becoming popular these years.

In this paper, we propose a robust real-time lane marking detection and tracking system for intelligent vehicles in urban road environments. Fig. 1(a) shows our intelligent vehicle THIV-I developed by Tsinghua University. A monochrome CCD camera is mounted near the rear-view mirror with a little tilt down to capture the images of the road. Fig. 1(b) shows a typical road image captured by THIV-I. It is usually hard to robustly detect lane markings because of shadow, occlusion, heavy traffic and some other road markings such as arrows and crosswalks.

(a) THIV-I developed by Tsinghua University (b) A typical urban road image captured by THIV-I

Fig. 1. THIV-I intelligent vehicle and typical urban road image

Our system can be divided into the following successive procedures: road image pre-processing, straight line extraction, lane marking pixel scanning, model fitting, lane marking tracking and vehicle controlling parameters calculation. The structure of our framework is shown in Fig.2.

2 Lane Marking Model

We adopt the hyperbola lane marking model proposed in [7]. The model assumes that in the world coordinate frame, a lane marking with curvature k can be approximated by a parabola of the form

$$x = \frac{1}{2}ky^2 + my + b \tag{1}$$

In the case that the camera is tilted down, suppose that H is the height of the camera focal point above the ground, rf and cf are the height and the width of a pixel divided by the focal length, and (cc, cr) is the center point of the

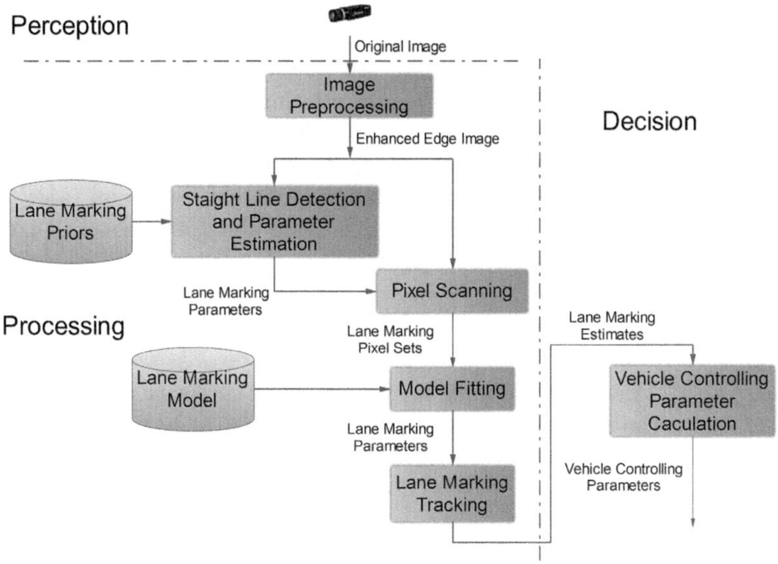

Fig. 2. The structure of our framework

image. According to Eq.(2), this curvature can be expressed by a hyperbola in the image coordinates of the following form

$$c = \frac{k'}{r-h} + m'(r-h) + b' \tag{2}$$

where (c, r) denotes the projection of the point (x, y) from the world coordinate frame to the image coordinate one, h is the height of the horizon in the image coordinate frame.

In Eq.(2), the parameters k', m' and b' are calculated below

$$\begin{aligned} k' &= \frac{0.5 * k * H * \sqrt{1 + cr^2 * rf^2}}{rf * cf} \\ m' &= \frac{rf * (b - m * H * cr * rf + 0.5 * k * H^2 * cr^2 * rf^2)}{cf * H * \sqrt{1 + cr^2 * rf^2}} \\ b' &= \frac{\sqrt{1 + cr^2 * rf^2} * (m - k * H * cr * rf)}{cf} + cc \end{aligned} \tag{3}$$

3 Lane Marking Detection and Tracking Algorithm

In this section, the whole processing procedure of our framework is introduced. The input of the system is the raw road images and the vehicle controlling parameters are generated on real-time and robustly as outputs after the processing.

3.1 Road Image Pre-processing

The first step of the detection is to calculate the edge of the input road image. We present an adaptive Canny algorithm with double threshold segmentation. To eliminate the influence of illumination and weather, the threshold in the Canny operator is chosen according to the average illumination on the road. We use a double threshold segmentation to eliminate most non-lane-marking pixels in the edge image. Let us denote $g(i)$ as the gradient value of pixel i, $b(i)$ as the intensity value of it, and $p(i)$ as the segmentation result of the same pixel. The double threshold segmentation can be express as Eq.(4).

$$p(i) = \begin{cases} 1, \text{ if } g(i) > T_g \text{ and } b(i) > T_b \\ 0, \text{ otherwise} \end{cases} \tag{4}$$

where T_g and T_b denote thresholds for gradient and intensity, respectively.

Fig.3 shows the result of the pre-processing. Fig.3(a) is the original image. Fig.3(b) gives the Canny edge image and Fig.3(c) is the results obtained after the double threshold segmentation is done. Compared to the Canny edge image, the double threshold segmentation has the removal of a crack on the road while keeping most of the lane marking pixels.

(a) The original image (b) The canny edge (c) The edge after double threshold

Fig. 3. Road Image Pre-processing

3.2 Straight Line Extraction and Parameter Estimation

It is observed that even for a winding road, the lane markings closely near the vehicle is approximately straight and easy to be extracted. Thus we can detect nearby straight lines using probabilistic Hough transform, and find out correct lane markings among them. Only lines that satisfy certain constrains, such as length and gradient, are remained. In the edge image, a lane marking usually corresponds to two adjacent lines that converge at the horizon. Thus the lines detected are partitioned into several equivalence classes. Eq.(5) is the equivalence relation. Each valid class must contain at least two lines.

$$\begin{cases} |A(i) - A(j)| < T_a \\ |D(i) - D(j)| < T_d \end{cases} \tag{5}$$

where $A(i)$ and $D(i)$ are the angle of line i and the horizontal distance in the image, respectively, and T_a and T_d are thresholds.

Lines in each equivalence class are then grouped into one, and the left and right lane markings of the current lane are specified as the nearest pair of lane markings that locate in the left and right part of the image, respectively. Fig.4(a) shows the straight line detection results, where the green rectangle indicates the region of interest (ROI). Note that some other lines are also detected besides the lane markings (on the arrow). Fig.4(b) shows the extraction of the left and right lane markings, together with the horizon estimation. Note that some false detections on the arrow have been eliminated since there is only one line in its equivalence class.

After the detection for the lane marking pair is done for current lane, the horizon can be regarded as the horizontal line that pass through the intersection of the lane marking pair. A triangle model of current lane width from bottom to horizon is established.

3.3 Lane Marking Pixel Scanning and Model Fitting

After the nearby lane markings of the current lane are detected, the pixels of the entire lane markings are scanned iteratively from the bottom of the image to the horizon. A mid-to-side strategy is used to search for the inner edge of the lane marking pair. We maintain the two sets of the edge points that correspond to the left and right lane markings of the current lane. The procedure is described as follows:

1. Set the beginning searching point to the middle of the current lane on the bottom of the image.
2. Scan the edge image in a small range around the expected position at both sides, according to the width model, and choose the lane marking pixels if they exist. Then add them to the corresponding sets.
3. Update the beginning searching point for the next row according to the detection in the current row.
4. Repeat Step 2 and 3 until the current scanning line is near the horizon.

After the scanning procedure is completed, the two sets of lane marking pixels are fitted using hyperbola model respectively. The RANSAC algorithm is utilized to remove outliers. After that, we use least square method (LSM) to fit the model in order to refine the result.

Fig.4 gives the result of the lane marking detection. Fig.4(c) shows the pixels that belong to the left and right lane markings, respectively, after the scanning procedure is done. Fig.4(d) is the result of the model fitting, where the green curve indicates the centerline of the current lane, which should be tracked by the vehicle.

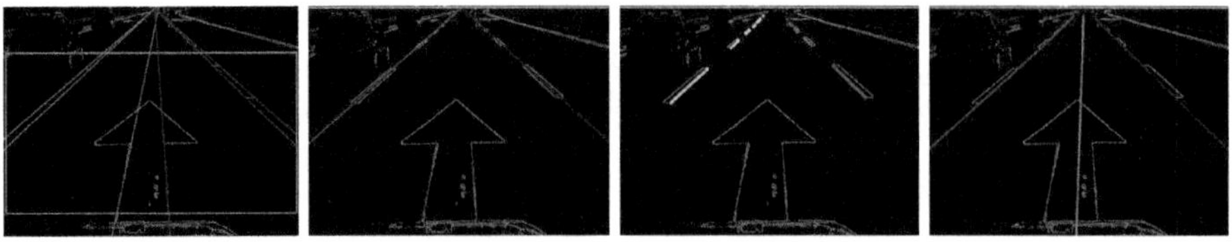

(a) Line detection (b) Lane marking extraction and horizon estimateion (c) Pixel scanning (d) Model fitting

Fig. 4. Lane marking detection

3.4 Lane Marking Tracking and Vehicle Controlling Parameter Calculation

We adopt a simple but effective linear filtering in our system to keep the current detection result for a short time when the lane markings is missing or false in successive frames. A new tracking process starts when a new lane marking pair is found in several continuous frames. For example, when the vehicle is changing lanes.

In order to get the vehicle controlling parameters, we set the centerline of the current lane as the target curve. Then we transform the target line into a parabola in the world coordinate frame. The transformation is done using camera calibration techniques. Then the vehicle controlling parameters, such as road curvature, lateral offset and yaw angles, can be obtained directly.

4 Experimental Results

The framework of lane marking detection and tracking was implemented on the intelligent vehicle THIV-I. Test videos were captured by the camera with a resolution of 768×576. We tested our framework under a variety of road scenarios. The average accuracy of our system is above 95% on urban road and above 98% on expressway. The average processing time for each frame is less than 30ms on a 2.66GHz CPU. Fig.5 shows some detection results on complex conditions.

Fig.6(a) and Fig.6(b) show respectively the lateral offset and yaw angle of THIV-I on the urban road of Fig.5, using fuzzy control. The lateral offset is mostly less than 0.4m and yaw angle less than 8°, which shows good performance in lane marking detection and vehicle controlling.

(a) Road with crack and crosswalk (b) Road with shadows (c) Heavy traffic (d) Heavy traffic with arrow

(e) Passing an overpass (f) Curve road with crosswalk (g) Occlusion and arrow (h) Other types of lane markings

Fig. 5. Some sample result in different situations

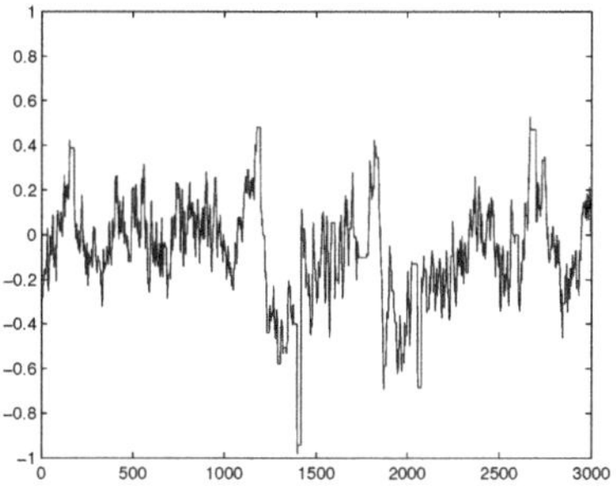

(a) lateral offset

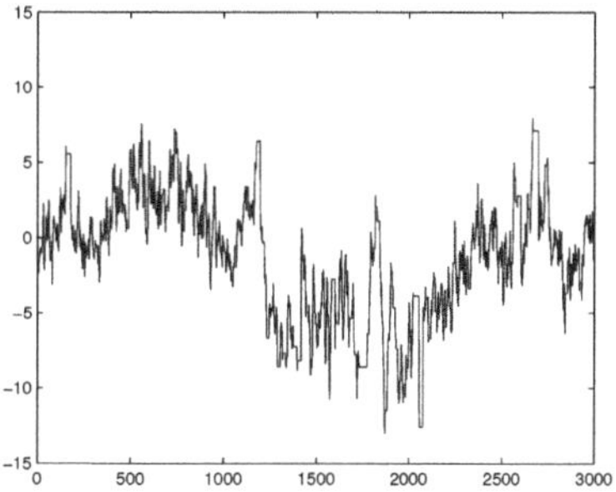

(b) yaw angle

Fig. 6. Lateral offset and yaw angle on urban road

5 Conclusion

In this paper, we propose a framework of robust and real-time lane marking detection and tracking for intelligent vehicles. The system is completely adaptive with only a few parameters pre-specified. It runs very well under different road scenarios in the presence of shadow, occlusion, and other non-lane markings. Experimental results show that our system can meet the need for intelligent vehicle applications.

Acknowledgement. This work was supported in part by National Natural Science Foundation of China (NSFC) under grant No.90820305 and No.60775040.

References

1. Bertozzi, M., Broggi, A.: Gold: a parallel real-time stereo vision system for generic obstacle and lane detection. IEEE Transactions on Image Processing 7(1), 62–81 (1998)
2. Broggi, A.: Robust real-time lane and road detection in critical shadow conditions. In: Proceedings, International Symposium on Computer Vision 1995, pp. 353–358 (1995)
3. Danescu, R., Nedevschi, S.: Probabilistic Lane Tracking in Difficult Road Scenarios Using Stereovision. IEEE Transactions on Intelligent Transportation Systems 10(2), 272–282 (2009)
4. Dickmanns, E.D., Mysliwetz, B.D.: Recursive 3-D road and relative ego-state recognition. IEEE Transactions on Pattern Analysis and Machine Intelligence 14(2), 199–213 (1992)
5. Felisa, M., Zani, P.: Robust monocular lane detection in urban environments. In: 2010 IEEE Intelligent Vehicles Symposium (IV), pp. 591–596 (2010)
6. Kim, Z.: Robust Lane Detection and Tracking in Challenging Scenarios. IEEE Transactions on Intelligent Transportation Systems 9(1), 16–26 (2008)
7. Kluge, K.: Extracting road curvature and orientation from image edge points without perceptual grouping into features. In: Proceedings of the Intelligent Vehicles 1994 Symposium, pp. 109–114 (1994)
8. Kreucher, C., Lakshmanan, S.: LANA: a lane extraction algorithm that uses frequency domain features. IEEE Transactions on Robotics and Automation 15(2), 343–350 (1999)
9. McCall, J.C., Trivedi, M.M.: Video-based lane estimation and tracking for driver assistance: survey, system, and evaluation. IEEE Transactions on Intelligent Transportation Systems 7(1), 20–37 (2006)
10. Paetzold, F., Franke, U., Von Seelen, W.: Lane recognition in urban environment using optimal control theory. In: Proceedings of the IEEE Intelligent Vehicles Symposium IV, pp. 221–226 (2000)
11. Pomerleau, D.: Ralph: rapidly adapting lateral position handler. In: Proceedings of the Intelligent Vehicles 1995 Symposium, September 1995, pp. 506–511 (1995)
12. Wang, Y.: Lane detection and tracking using B-Snake. Image and Vision Computing 22(4), 269–280 (2004)
13. Zhou, S., Jiang, Y., Xi, J., Gong, J., Xiong, G., Chen, H.: A novel lane detection based on geometrical model and Gabor filter. In: 2010 IEEE Intelligent Vehicles Symposium (IV), pp. 59–64 (2010)

Research and Application of Elevator Energy-Saving Devices with Super Capacitor to Store Energy

Zehua Yao[1], Jianru Wan[1], Bin Li[1], Jianxiong Qian[2], and Sheng Wang[1]

[1] School of Electrical Engineering and Automation, Tianjin University, Tianjin, 300072, China
[2] China Special Equipment Inspection and Research Institute, Beijing, 100013, China
libin6199@163.com

Abstract. For the problems of complex control and harmonic interference when elevator's regenerative braking energy feed back to the grid, The paper presents an energy saving program. Renewable energy is stored with super capacitors and used locally. The paper analyzes the basic operating principle of the super-capacitor energy storage device and power operation curves in different conditions. The elevator energy consumption experiments are completed in five typical working conditions. Experimental results show that super capacitor energy storage device of the elevator is stable and has a good energy saving effect.

Keywords: super capacitor, bi-directional DC-DC converter, energy feedback.

1 Introduction

When the elevator in the state of light loading uplink, heavy loading down or braking, it generates renewable power. The renewable power is transported to the filter capacitor in DC bus side of the inverter and generating pumping voltage. It seriously threats to the safety of the system. At present, energy feedback system is extensively researched. It can convert the energy stored in the capacitor of the inverter into current in the same phase with the grid voltage and feedback to the grid. But it is complex to control the system and when the elevator is running, the power changes in a large extent. So the system could cause great impact on the grid and great disturbance on the power distribution. This may endanger the safe operation of power system. Connecting a super-capacitor energy storage device to DC bus between the rectifier and the inverter through bi-directional DC-DC converter circuit can solve the above problems. They can efficiently absorb the energy of the elevator back. When required to power to the elevator and its auxiliary devices, they can do it, achieving the power of the local use of feedback and saving energy; they can balance the power to reduce the impact on power; avoid the negative impact on the grid when the energy back to the grid; in the case of elevator suddenly loses power, super capacitor groups as a UPS provide power leveling necessary.

G. Lee (Ed.): Advances in Automation and Robotics, Vol. 1, LNEE 122, pp. 429–436.
springerlink.com

2 Elevator Saving Device Architecture

The schematic of super capacitor energy storage device is shown in Figure 1. Dotted line is the super-capacitor energy storage device which is made up of the super-capacitor, emergency power supply EPS, bi-directional DC-DC converter and controller, etc, used in conjunction with the inverter. When the elevator at a light load up, overloaded downlink or brake down state, mechanical energy generates AC power (renewable energy),which become DC power through the freewheeling diode and DC voltage rises. Through the device direct current will be stored into super capacitor by bi-directional DC-DC converter. On the one hand through the emergency power supply EPS conversion lift assist system, the other they were used when the supply of tractor use through the bi-directional DC-DC converter.

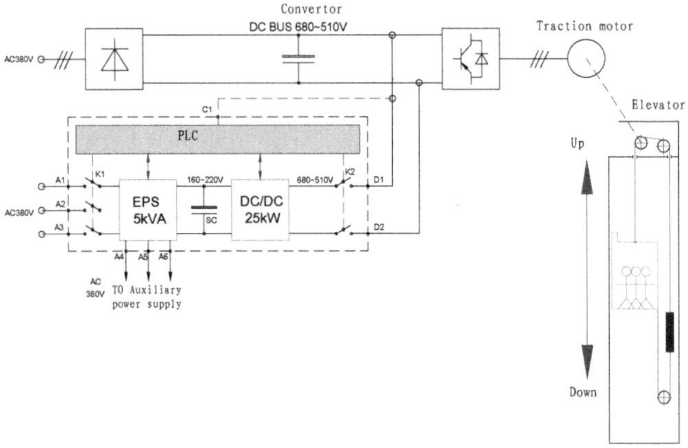

Fig. 1. Elevator energy-saving devices through super capacitor system schematic

3 Elevator Energy-Saving Devices Design Principles

3.1 Elevator Energy-Saving Devices Functional Design

According to design requirements, the elevator energy-saving devices through super capacitor should have the energy recovery, power compensation, power supply for assistance system and UPS function.

(1) energy recovery

Generally, energy feedback devices are based on the inverter DC link voltage V_d to decide whether to back the power. Feedback voltage uses a fixed value V_k. As the voltage fluctuations, When the value of V_k is too small, the high voltage grid will produce error feedback; When the value of V_k is too large, then the feedback effect is decreased.

The energy-saving device uses voltage adaptive control, that no matter how voltage fluctuations, and only when the DC bus voltage of the static and dynamic voltage difference Δv satisfy the set value, the energy-saving devices absorbed feedback energy into the capacitor timely, and stored, effectively avoiding the error feedback.

According to Figure 1, the DC bus voltage of the static and dynamic voltage difference Δv_1 was detected through C1 port. When Δv_1 meet the set value, the controller starts D-DC pass anti-wizard, storing the feedback energy in the capacitor, while stabilizing v_d in the DC510-680V. The energy flow diagram is shown in Figure 3.

(2) power compensation

Power compensation uses the same voltage adaptive control methods. In the figure 1, D1, D2 sides are connect to inverter DC BUS. The DC bus voltage of the static and dynamic voltage difference Δv_2 was detected through C1 port. When Δv_2 meet the set value, the controller starts DC-DC pass forward. According to the value of Δv_2, adjusting the DC power output to compensate for the power output of the grid. The maximum output power supply is in 20kW or less stably. The energy flow diagram is shown in Figure 4.

(3) Power supply for assistance system

The power of elevator assistance system is provided by output port A3, A4, A5 in figure 1. When the grid supply power normally, A3, A4, A5 connected with grid through the EPS. When the capacitor energy exceeds the set value or grid failure to work due to fault, A3, A4, A5 powered from the side of the capacitor through the EPS.

(4) UPS function

No matter in what state, super capacitor should set aside part of the energy in work. In the case of elevator suddenly loses power or grid failure to work, super capacitor groups work as a UPS, providing power to elevator and assistance system, providing the necessary power.

3.2 Super Capacitor Parameters Selection

We calculate the super-capacitor to standard that super-capacitors are able to absorb all the energy full load from the top of the elevator back down to the bottom.

$$W_g = \frac{1}{2} t_1 P_g(t_1) \times 2 + (t_2 - t_1) P_g(t_1) = \frac{v_m}{a_m}(m_1 - m_2) g v_m + \frac{(s - \frac{v_m^2}{a_m})}{v_m}(m_1 - m_2) g v_m \tag{1}$$

Where m_1 is the quality of the car and full load; $m_2$is quality of the elevator counterpoise; v is instantaneous velocity; a is the instantaneous acceleration; v_m is elevator rated speed; s is the building height.

The selected test parameters of the elevator are shown in Table 1. we can get the full energy feedback down in a full load when the parameters in Table 1 are taken into the energy feedback equation. w_g=410KJ=114wh.

Table 1. Main technical indicators of elevator

project	Technical parameters	explanation
rated load (kg)	1000	
elevator speed (m/s)	2.5	
number of floors (floor)	22	-1,1,2,4,6,8,10,12,14,16,18,19,20
floor height (m)	4	
Motor Power (KW)	16.3	
Motor voltage (V)	340	Three-phase AC
Drive type	AVs1022	Xiwei, 22KW
Auxiliary power(KVA)	5	lighting, ventilation, elevator doors and communications

Select the basic parameters of super-capacitor is shown in Table 2, in which the work of single super-capacitor voltage is 1-1.4V, the work of super-capacitor group voltage is 160-240V.

Table 2. Parameters of super capacitor

project	Technical parameters	project	Technical parameters
single capacitance	12000F	total capacitance	75F
work voltage V	160-240	Energy reserves WH	270
Maximum feedback current	100A	Life cycle time	≥500000
Charge and discharge efficiency %	≥90	Operating temperature range ℃	-25-55
Dimensions mm	800×600×1600	weight kg	2000

In the course of work, the super-capacitor's voltage limit is set to 175V. When the super-capacitor voltage is higher than 175V, EPS access into super capacitor, auxiliary system get power through super capacitor; When the elevator is in power state, the super capacitor provides power for the elevator as compensation; When the super capacitor is lower than 175V, EPS take electricity, super capacitors do not provide any energy for the elevator. The limit of super-capacitor voltage is 240V, when the super capacitor voltage exceeds this value, if the elevator is still in the energy feedback state, the super capacitor do not charge this part of power, open the discharge circuit to discharge, these energy consumed in the brake resistor. The part of the energy between 160V and 175V use as the elevator UPS, in the case of elevator suddenly loses power, super capacitor groups used as a UPS to provide necessary power.

3.3 Analysis Typical Conditions of Elevator

Conditions of elevator include load rise, load down, full rise, full down, half load rise, half load down, light load rise, light load down, overload rise, overload down. Focus on analysis of the power curve at the operating conditions of no-load and full load.

(1) Empty rise, full down

The condition of empty rise, full down is typical of the energy feedback process (Figure 2). Running continuously charge to the super capacitor. When the super-capacitor voltage $v_c \geq 175V$, the controller controls the output of EPS connect with the super capacitor, the store energy in super capacitor convert to three-phase AC power to supply to the lift assist system using off part of the recovery of energy within the super-capacitor. The capacitor can be continuously recovering renewable energy of elevator.

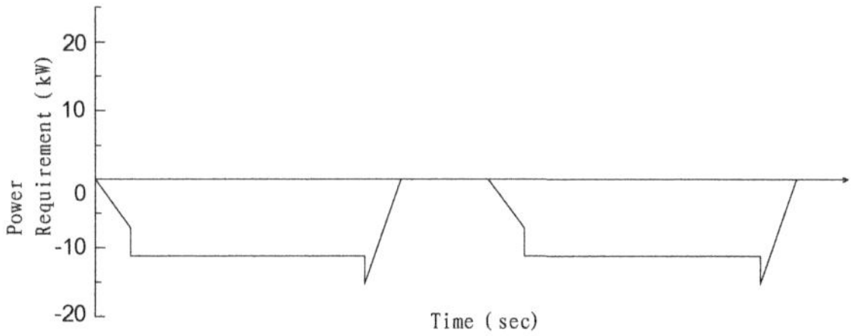

Fig. 2. Power curve at load rise, full down conditions

(2) Full rise, empty down

The condition of full rise, empty down is typical of the super-capacitor power compensation process (Figure 3). In the process of starting and accelerate the capacitor output 20kW power to compensate the output of grid. Capacitor energy consumption is added through the grid by the DC -DC when the elevator stop. Therefore, in the elevator full up empty down conditions the capacitor can be continuously output to compensate grid, stable at 20kW or less.

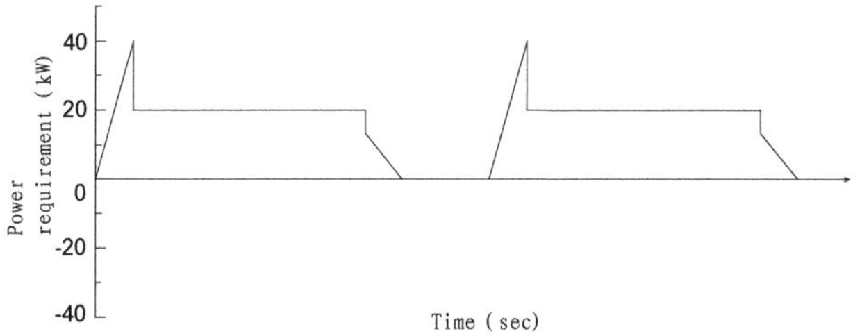

Fig. 3. Power curve at full rise, load down conditions

(3) Full rise, full down

In the process of full rise, the super-capacitor output 20kW power to compensate grid. In the process of full down, the capacitor saves renewable energy (Figure 4).

When the super-capacitor output compensation power is less than the recovery energy (monitoring the capacitor voltage changes, the super capacitor voltage $v_c \geq 175V$), the controller control output of EPS connects with super capacitor.

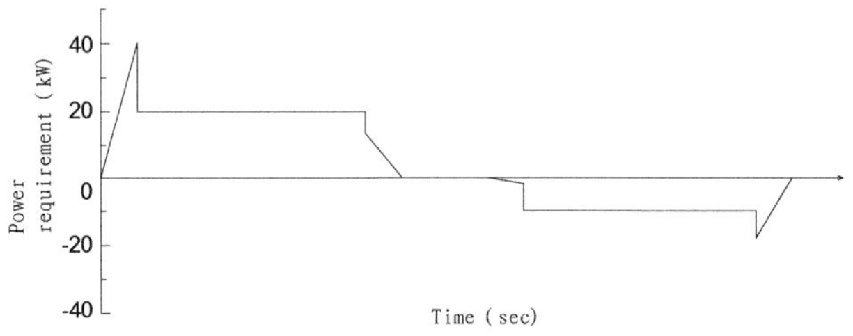

Fig. 4. Power curve at full rise, full down conditions

(4) Non-empty rise, non-empty down

In this condition, according to the different load, energy-saving devices choose to work between the feedback compensation function of energy and power.

4 Experimental Results

Elevator energy consumption measurement method use reference trip. At condition of load, 25% load, 50% load, 75% load, full load, the elevator run up and down 10 times and at the bottom and top end station open and close a door once, run 20 minutes. Respectively, without super capacitor energy storage and super capacitor energy storage device join the case of measuring energy consumption of the elevator, as shown in Table 3.

Table 3. The table of elevator energy consumption and energy saving

Condition	Super capacitor energy storage device is added?	Power consumption WH	Energy saving
no load	Y	800	36%
	N	1250	
25%load	Y	550	8.3%
	N	600	
50%load	Y	600	14.2%
	N	700	
75%load	Y	750	37.5%
	N	1200	
full load	Y	1450	21.6
	N	1850	
total	Y	4150	26%

From Table 3 we can see the elevator in the 75% load energy savings is ideal, reach 37.5%; 25% load only save 8.3%; the elevator with super capacitor energy storage device energy save up to 26% average.

The figure 5show the elevator with a super capacitor energy storage device run in five typical operating conditions, the inverter DC bus voltage, super-capacitor voltage, and ultra-capacitor discharge current impulse test waveforms.

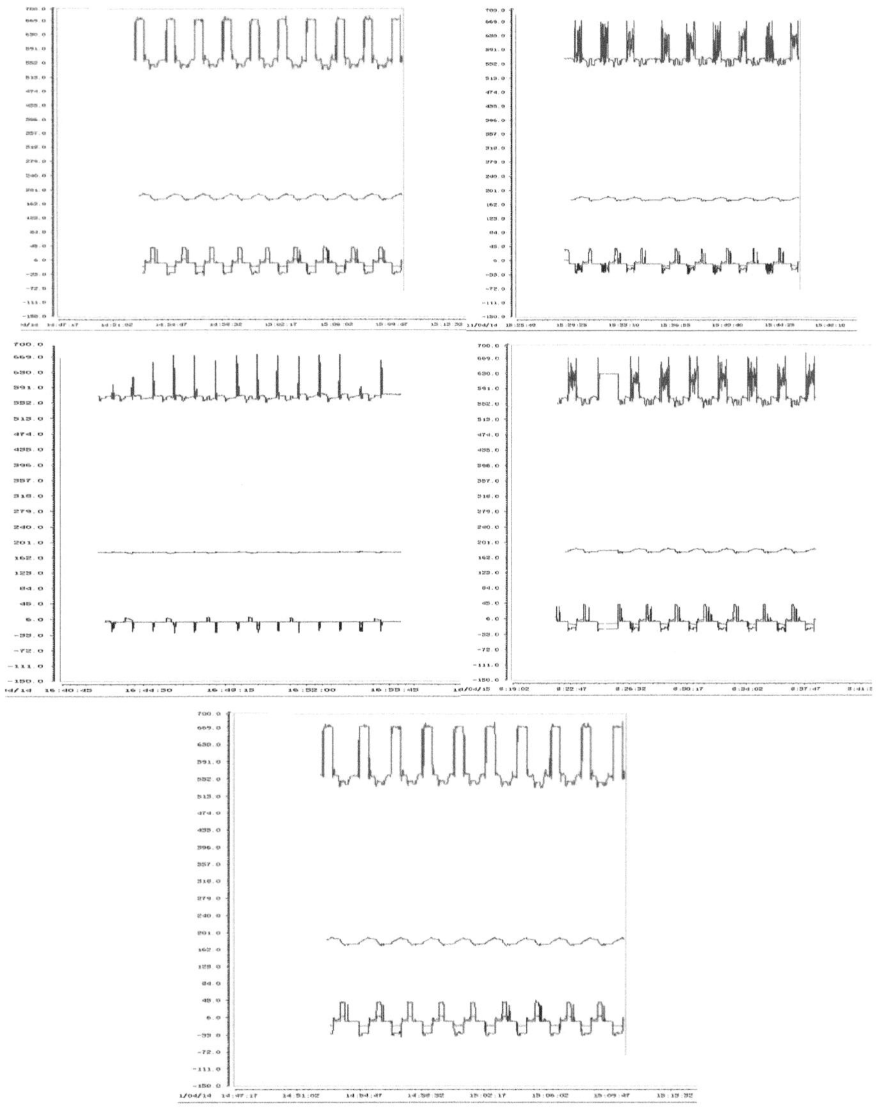

Fig. 5. DC bus voltage, super capacitor voltage, super capacitor charge and discharge current waveforms at the condition of empty, 25% load, 50% load, 75%load and full load

It can be seen from the figure that no matter in what condition the drive DC bus voltage are effectively controlled within the 510-680V, showing super capacitor energy storage device can effectively absorb the energy feedback of the elevator and the power compensation. Super capacitor voltage is always in the vicinity of its normal operating voltage 175V,the energy feedback be absorbed by super capacitor is effective used in place. Super capacitor charge and discharge current is controlled at less than 40A.

5 Conclusions

This paper analyzes the basic working principle and the work process of elevator energy-saving devices with super capacitor energy storage. The work of super capacitor energy storage device's design and test were completed. Experimental results show that the elevator with a super capacitor energy storage device can efficiently use the elevator feedback power, achieving using the renewable energy locally, avoiding the impact of feedback power on the grid. Actual measurement data show that energy savings can be more than 25%.This has a broad research and application value.

Acknowledgments. The authors gratefully acknowledge the financial support of the National Natural Science Foundation of China (No.60874077) and AQSIQ Nonprofit Industry Specialized Research Funding Projects (No.200910097 and 201010057).

References

1. Makuian:The Design of Bi-directional DC/DC Converter Applied in Super-capacitor Storage Systems. Zhejiang University (2010) (Chinese)
2. Xu, D.: Modeling and Control of Power Electronic System. Machinery Industriy Press, Beijing (2006) (Chinese)
3. Ribeiro, P.F., Johnson, B.K., Crow, M.L., Arsoy, A., Liu, Y.: Energystorage systems for advanced power applications. Proceedings of the IEEE 89(12), 1744–1756 (2001)

The Design of Elevator Failure Monitoring System

Zehua Yao[1], Jianru Wan[1], Xiangwei Li[1], Liguang Shi[1], and Jianxiong Qian[2]

[1] School of Electrical Engineering and Automation, Tianjin University, Tianjin, 300072, China
[2] China Special Equipment Inspection and Research Institute, Beijing, 100013, China
libin6199@163.com

Abstract. In view of inconformity of various elevator controller interface signals, a common elevator failure remote monitoring system is designed in this paper. In the monitoring system, a common elevator signal acquisition device is developed, and the collected signals are processed and made logical judgment by Mono-Chip Computer (MCU) as processor, then the results are transferred to the elevator failure monitoring interface by communication system. It can provide a convenient and visual solution for discovery, analysis, and elimination of the elevator failure in service.

Keywords: elevator failure remote monitoring, signal acquisition, monitoring interface, elevator failure.

1 Introduction

The elevators are widely used and play an important role in residential, hotels, shopping malls, office buildings and other high-rise buildings as a vertical transport in buildings. Thus, it is necessary to monitor the running status and failure conditions of elevator for the elevator management, maintenance and safe operation.

In recent years, elevator failure arose frequently, and has been gradually increasing, the main reasons as follow:

a. With the rapid economic development, the existing elevator components are very complex: all the products since 1950s are in service, the imported products have their share in the elevator market. Numerous types, different drive ways and control methods give more difficulty to the elevator safety management.

b. For the moment, the elevator maintenance services industry has not yet formed in our country. The industry standard and specification remain to be perfected.

c. The lack of elevator elimination mechanism. The elevators should be eliminated when operating for a long time due to the debasement of hardware security, increased energy consumption and lag in technology. However, currently there is no corresponding standard for elevator elimination mechanism.

In the current elevator market, elevator manufacturers develop their own monitors and controllers that are incompatible for common use. So the elevator controller interface signals can not be obtained directly. A common elevator failure remote monitoring system is developed in this paper. It can provide convenient and visual solutions for failure discovery, analysis and elimination of the running elevator. The elevator

G. Lee (Ed.): Advances in Automation and Robotics, Vol. 1, LNEE 122, pp. 437–442.
springerlink.com

failure is classified, the related elevator signals are collected by choosing proper signal acquisition devices, and MCU is used as CPU to process the failure signals, then the results are transferred to elevator failure monitoring interface via the use of network communication technology and database technology.

2 System Components

Elevator failure remote monitoring system is that after many elevators installed in a building area, we make remote monitoring, management, data maintenance, statistics, analysis, failure diagnosis for them. Its purpose is to make remote data maintenance, remote failure diagnosis, and statistics and analysis of the elevator running status (group control effects, frequency of utilization, failure number and failure type), and at last, transfer the failure information to the elevator maintenance service center, which can also initiatively track the operation condition of the elevator. This system can be installed on a building, a community, a district (county) or a city to achieve hierarchical management. Through the elevator failure remote monitoring, the passive failure repair could become the active failure maintenance. When in elevator failure conditions, the monitoring system can, on the one hand, timely inform maintenance staff and assist them in repairing the elevator immediately to reduce the downtime under failure type analysis, on the other hand, take emergency operations to appease the trapped passengers, preventing use of improper reaction.

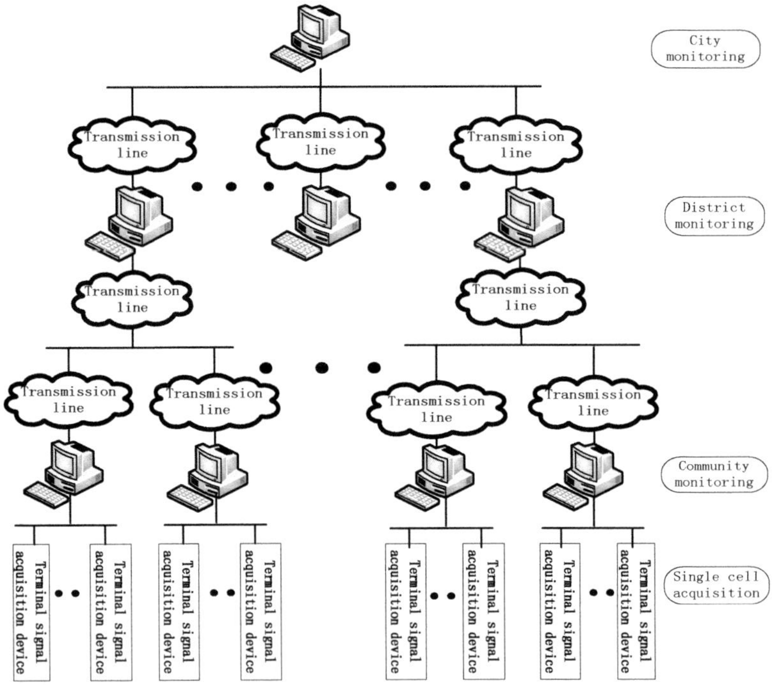

Fig. 1. Structural representation of city elevator failure monitoring system

Fig.1 is structural representation of city elevator failure monitoring system. Sensors are installed in reasonable position of the elevator, and form the terminal signal acquisition module with MCU. Related elevator signals are collected and transmitted through the transmission line to the monitoring interface. Lower-level monitoring information can also be uploaded to the higher-level monitoring interface.

3 Signal Acquisition

Signal acquisition is the most basic and most complex work to the elevator failure monitoring system. Failure judgment is based on signal acquisition, but there are various elevator signals: analog and digital, electric and non-electric. Therefore, the principle of signal acquisition is collecting as little as possible signals, which have been ultimately converted to electric signals, to complete determine of the failure type. Fig.2 is structure diagram of elevator signal acquisition. The elevator running state signal and some failure signals are collected by front-end data acquisition devices. The MCU processes the collected signals (failure judgment), stores the results to a designated area in a certain format, and completes the final output of the failure code.

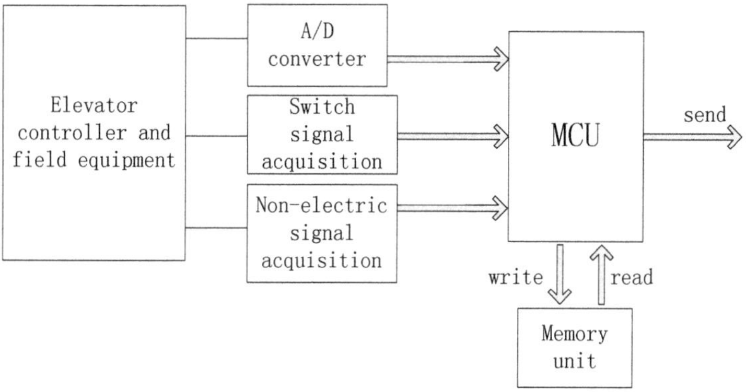

Fig. 2. Structure diagram of elevator signal acquisition

The design ideas of elevator location and speed signal acquisition: the rotary encoders are installed on the traction machine, the Pulse Generator card (PG card) convert the encoder outputs into pulse number, which is proportional to motor speed and location. The elevator speed and location can be determined by the cumulative number of pulses, the status signal, such as elevator leveling signal, speed changing signal, upper and lower limit signals, can be got according to the elevator location. Fig.3 shows the circuit diagram of the encoder pulse signal acquisition. Rotary encoder products rotation pulse signals when the elevator traction machine works, then the signals are processed through the PG card and sent to the pulse counting chip, and the resulting signal is transferred to the MCU.

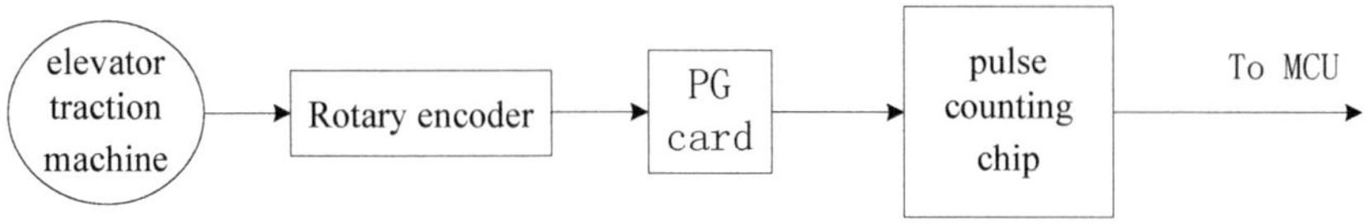

Fig. 3. The circuit diagram of the encoder pulse signal acquisition

The switching signals such as the calling elevator signals on each floor and the layer selection signals within the elevator car use Hall SS40A type switch current sensor. As shown in Fig.4, the SS40A is installed to the contact switch of the out calling device and inner layer select device, the secondary signal of the sensor is connected to MCU through the conditioning circuit.

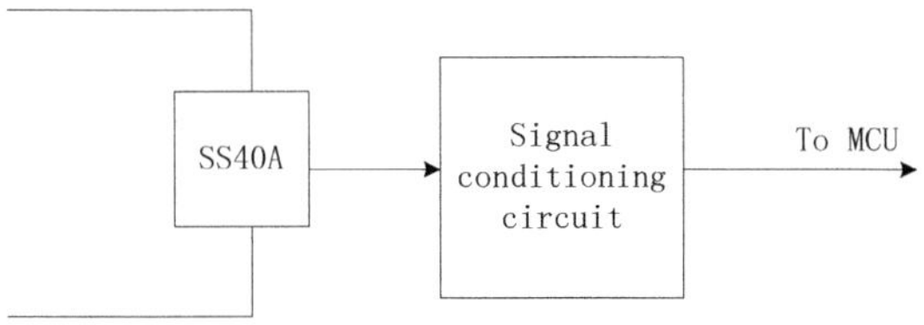

Fig. 4. The acquisition block diagram of switch signal sensor

To signals relating to the relay, such as the leveling signal, safety relay, door lock loop, the corresponding relay switch state can be collected. As shown in Fig.5, the signal lines lead from the contacts on the relay switch, conditioned by the signal conditioning circuit, then sent to MCU. Fig.6 is the safety relay switch state collector.

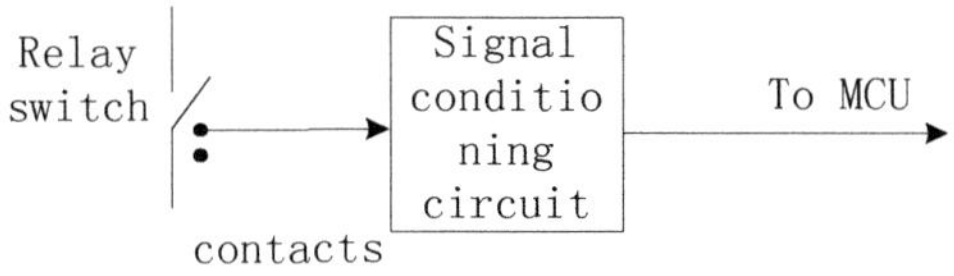

Fig. 5. The relay switch state collector

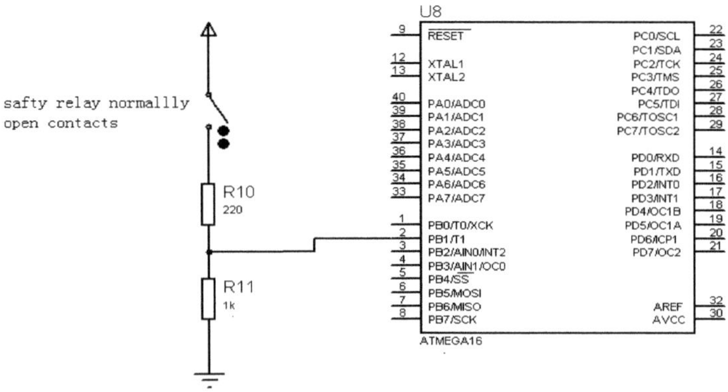

Fig. 6. The safety relay switch state collector

4 Failure Judgment and Monitoring Interface Development

4.1 Failure Judgment

The main failure of elevator are divided into 10 types: start failure, operation failure, stop failure, door failure, signal system failure, communication failure, inverter failure, power failure, relay failure, parameter setting failure, respectively. Among various elevator signals, the system designed in this paper only collects rotary encoder signal, the door switch signal, and the elevator out calling and inner layer select signal. The integrated logic judgment of these three types signals can lead to have failure or not and the failure type. Fig.7 is part of the failure judgment logic diagram, for example, the rotary encoder signal can determine the elevator location signal. The integrated logic judgment of elevator location signal, the elevator design parameters and the door switch signal can determine whether the elevator door open and shut normally.

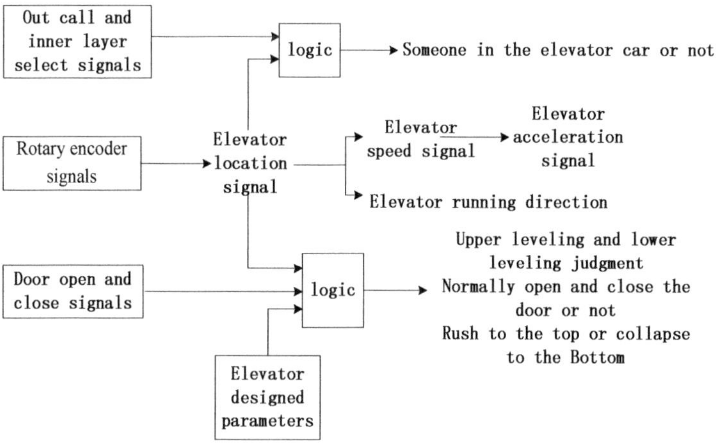

Fig. 7. Failure judgment system

4.2 Monitoring Interface Development

The elevator failure remote monitoring interface is based on VB programming language, with its visual, event-driven characteristics, the database uses Microsoft Access format database. Fig.8 is the total interface of the elevator failure remote monitoring system, with community selection, unit selection and elevator selection. The building height and elevator model will be shown after the elevator is chosen. Click the real-time monitoring of the elevator, the elevator specific operation states and failure conditions can be seen.

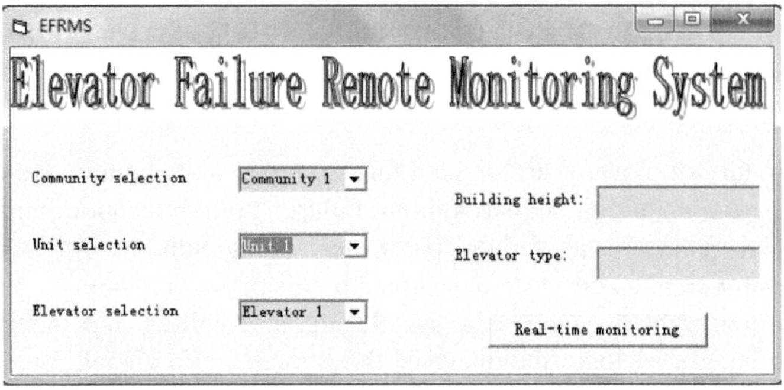

Fig. 8. The total interface of the elevator failure remote monitoring system

5 Conclusions

Considering inconformity of various elevator controller interface signals, a common elevator failure remote monitoring system is designed in this paper. In the monitoring system, a common elevator signal acquisition device is developed using added devices, and the collected signals are processed and made logical judgment by MCU, then the results are transferred to the elevator failure monitoring interface by communication system. It can provide a convenient and visual solution for discovery, analysis, and elimination of the elevator failure in service.

Acknowledgments. The authors gratefully acknowledge the financial support of the National Natural Science Foundation of China (No.60874077) and AQSIQ Nonprofit Industry Specialized Research Funding Projects (No.200910097 and 201010057).

References

1. Flores, A.Q., Carvalho, J.B., Cardoso, A.J.M.: Mechanical Fault Detection in an Elevator by Remote Monitoring. In: ICEM 2008, pp. 1–5 (2008)
2. Wan, J., Chen, C., Sun, Y.: Remote Monitoring System of Elevator Based on a Duai-serial-port Single-chip. Computer Measurement and Control 14, 900–901 (2006)
3. Qiu, C., Zhang, G., Wu, Z.: Design of intelligent elevator data acquisition system based on FPGA and DSP. Computer Engineering and Design 30, 1577–1579 (2009)
4. Zhang, H., Wang, X., Zhu, X.: Modern elevator control technology. Harbin Institute of Technology Press, Harbin (1994)

Reconfiguration of Aircraft Distributed Power Distribution System Based on Multi-Agent

Ling Zhang, Lin Cai, and Li Wang

Department of Electrical Engineering, Nanjing University of Aeronautics and Astronautics
Nanjing, Jiangsu Province, 210016
tbytmz@nuaa.edu.cn, lcolinqq@163.com, liwang@nuaa.edu.cn

Abstract. The reconfiguration strategy of aircraft distributed power distribution system using the Multi-Agent approach was explored. Firstly sequential storage structure was used for network topology coding, thus feeding branches between adjacent bus nodes were recorded. Secondly, a multi-agent system was established based on layered and federal structure. Structure of the MAS and the information exchange method were described in detail. Thirdly, an example was cited to illustrate the configuration method. Simulation results showed that the method proposed in this paper could trade and detect network topology and working status automatically and restore power supply to load points in the context of fault isolation.

Keywords: Multi-Agent, Distributed Power System, Reconfiguration.

1 Introduction

In the flight stage, various failures and abnormal operation will threaten the safety and reliability of the distribution system. When failures occur, reconfiguration is used to isolate failures from the system, restore the distribution system. Therefore, reconfiguration is an important tool to improve reliability and survivability of distribution system and it is essential to conduct research on reconfiguration towards aircraft distributed power distribution system.

In terrestrial and ship power distribution system, heuristics, modern stochastic optimization algorithm such as ant colony algorithm, partical swarm optimization, genetic algorithm, simulated annealing and hybrid approach combined with advantages of various algorithms[1][2][3] are used to solve power system reconfiguration.

Because distributed power system is used in modern aircraft, distributed control or centralized-distributed control is adopted correspondingly. Multi-Agent System(MAS) has the characteristics of autonomy, adaptability, interactivity and cooperation[4] and thus is appropriate for management and control of distributed control system. Currently, MAS is applied in reconfiguration in power system at home and aboard[5][6][7][8][9]. Paper [5] and [8] propose setting agents for power station, bus and line. Every agent communicates with its adjacent agents and completes reconfiguration in collaboration. This method is especially suited for reconfiguration of simple radial distribution system. When distribution system gets complicated or turns into ring structure, this method may

G. Lee (Ed.): Advances in Automation and Robotics, Vol. 1, LNEE 122, pp. 443–450.
springerlink.com

become invalid. Paper [7] proposes breaking the ring structure distribution system through setting labels for every agent and then conducting reconfiguration with the method suited for radial structure distribution system. Paper [8] proposes a multi-agent approach to reconfiguration. Agents of abnormal substation make decisions according to status of their neighbors or agents of normal substation make and launch reconfiguration strategies. Paper [6] proposes a multi-agent approach to reconfiguration for the More Electric Aircraft (MEA) power system. All agents are divided into two levels: negotiation level and operation level. Agents at negotiation level communicate with each other.

In this paper, a multi-agent system approach is applied to the distributed aircraft power distribution system reconfiguration. A multi-agent system was established based on layered and federal structure. The proposed reconfiguration method can trade and detect network topology and working status automatically and restore power supply to load points in the context of fault isolation.

2 Structure of Aircraft Power System

Our study is applied to the distributed aircraft power distribution system, seen as in fig.1. It is a redundant power system with AC and DC areas. G1 and G3 supply power in parallel. G2 and G4 supply power in parallel. APU works as an auxiliary power supply while BAT1 and BAT2 work as emergency power supply. When fault occurs among main powers, APU can provide extra power for loads and batteries BAT1 and BAT2 provide sensitive electronic loads with uninterrupted DC power.

The distribution system is divided into four main distribution centers. They are left AC power distribution center, right AC power distribution center, left DC power

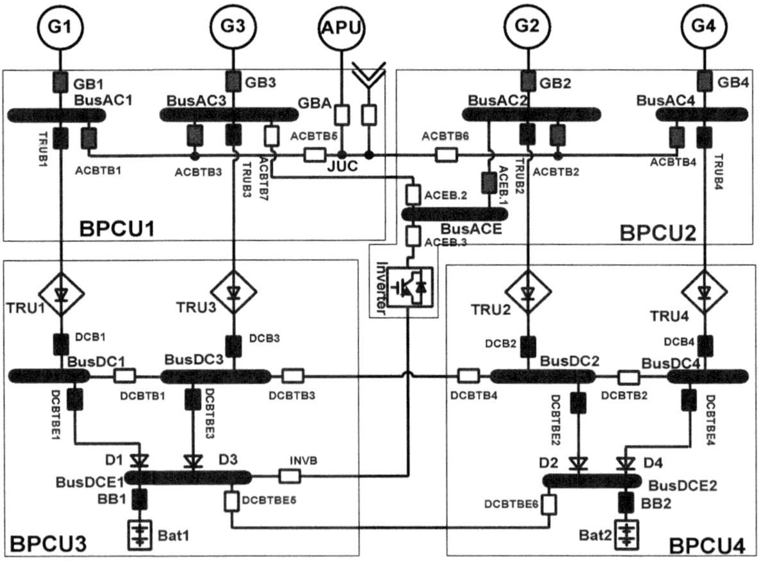

Fig. 1. Structure of aircraft power system

distribution center and right DC power distribution center and are separately monitored by corresponding Bus Power Control Unit BPCU1~ BPCU4.

3 Network Topology Coding and Tracking

Network topology coding and tracking is the first problem to be solved on conducting reconfiguration. Available power supply paths can be searched through network topology coding and tracking. And only in this context, the control system can realize the decision-making and implementation of the fault reconfiguration.

All buses in the distribution system are seen as nodes. One bus works as a power receiver which accepts energy from other buses. It also works as a power distributor which sends energy received from corresponding power supply and buses to other buses. This paper uses sequential storage structure to record every non-zero element in the adjacency matrixes of the system. Every element has three data areas which are used to store the row, column where it is in the adjacency matrix and its value. Every row vector of the sequential storage structure array records all components of the feeder from some power distributor to some power receiver.

The sequential storage structure array is shown in table 1. It can be shown in the table that every row vector records all components of the feeder from some power distributor to some power receiver. It means that array Adj_pre records all information of the feeders which starts from every bus node to its neighbor, its adjacent bus node.

For example, in fig 1, components on the feeder connecting BusAC1 and BusDC1 contain BusAC1, DCB1, TRUB1, TRU1 and BusDC1. Then in the 15th row of the Adj_pre, all these information is recorded.

Table 1. The sequential storage structure array of the distribution system (partial)

Adj_pre					
ID of power receiver	ID of power distributor	ID of switch close to receiver	ID of switch close to sender	ID of power supply on the feeder	ID of Adj_pre row vector
Idx_BusAC1	Idx_BusAC1	Idx_GB1	Idx_GB1	Idx_G1	1
Idx_BusAC1	Idx_BusAC3	Idx_ACBTB1	Idx_ACBTB3	0	2
Idx_BusAC1	Idx_JUC	Idx_ACBTB1	Idx_ACBTB5	0	3
……	……	……	……	……	……
Idx_BusAC3	Idx_BusAC1	Idx_ACBTB3	Idx_ACBTB1	0	8
……	……	……	……	……	……
Idx_BusDC1	Idx_BusAC1	Idx_DCB1	Idx_TRUB1	Idx_TRU1	15
……	……	……	……	……	……
Idx_JUC	Idx_BusAC1	Idx_ACBTB5	Idx_ACBTB1	0	36
……	……	……	……	……	……

4 Reconfiguration of Distribution System Based on Multi-Agent

Number of components in the system is relatively large and every component has a variety of normal and abnormal working states. If we analyze the influence of every component in the system with the permutations and combinations method and make fault configuration strategy, this may not only make workload large, but also ignore some cases. Actually, every bus and its adjacent bus can interact through the feeder connecting them, while buses which don't have adjacency relationship interact with each other through buses between them instead of interacting directly. This idea can not only simplify the analysis but also meet the basic idea of control and management of distributed control system. In order to put the idea into practice, the whole distribution system is divided into several subsystems. Every subsystem consists of a bus as a core and its adjacent switches. In fig 1, component BusAC1, GB1, ACBTB1, TRUB1 form a subsystem.

Every subsystem is only influenced by its adjacent subsystem directly and will only directly influence its adjacent subsystem. Any two subsystems which don't have adjacency relationship interact with each other through subsystems between them instead of interacting directly. For every subsystem, what it needs is whether its adjacent subsystem can supply or is supplying power to it, whether diffusion fault that can influence it occurs in its adjacent subsystem and whether it needs supplying power to its adjacent subsystem and so on. Subsystems do not need to know specific working status of other subsystems which do not have adjacency relationship with them.

4.1 Structure of MAS of Power Distribution Control System

This paper adopts a MAS structure based on layered and federal structure as the power distribution control system.

Each distributed controller of the power distribution control system is an entity agent of the MAS. In normal working status, BPCU1 monitors buses and switches located in its control area. It also works as coordination level agent in MAS. Other distributed controllers BPCU2~BPCU4 work as function level agents. They receive coordination control commands from the coordination level agent, manage their internal logic agents and finally complete management and control of distribution areas they monitor respectively. All distributed controllers coordinate and complete the whole control of the distribution system under the coordination of the coordination level agent.

Each bus and switch in the entity agents are assigned a logic agent. These logic agents are programs running in the entity agents.

In order to make all logic agents interact effectively and avoid conflict, these agents are classified as several logic agent federations according to topology of the distribution system. A logic agent federation is corresponding to a subsystem mentioned above. Bus agent is selected as the promoting agent in every logic agent federation. A bus agent monitors working status of corresponding bus and generates control variables that make a request to or forbid the closing operation of its adjacent switch agents. According to perceived status, a switch agent generates control signal of corresponding switch combined with request and lock information received from its adjacent bus agent and coordination control command. It also tells its adjacent bus agent its own status. As the promoting agent, bus agent is also responsible for the status synthesis of subsystem corresponding to federation where the bus agent allocates and distributes the status. Meanwhile, it reads the status distributed by agent federation of adjacent subsystem and takes it as its own foundation of making strategy.

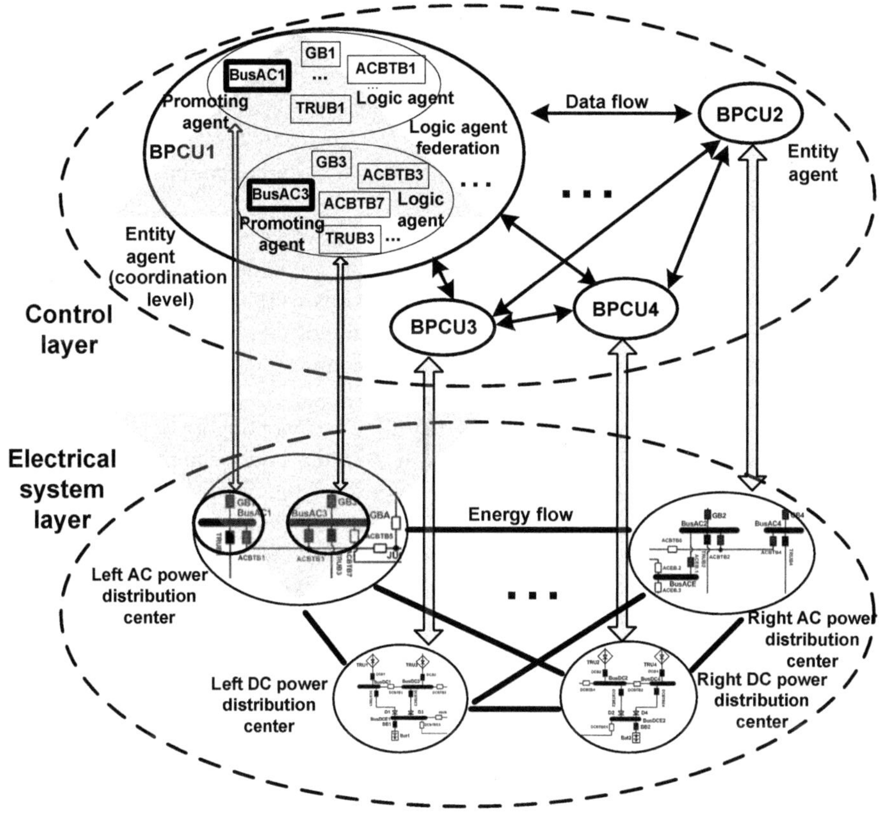

Fig. 2. MAS of the power distribution control system

Structure of MAS of power distribution control system is set up through the analysis above. It can be shown in fig 2. Each distributed controller which works as an entity agent includes several logic agents which are classified as several logic agent federations according to subsystems forming above. All agents interact with each other and finally complete the control task.

4.2 Synthesis of Working Status of Logic Agent Federation

In the process of reconfiguration, each logic agent federation makes strategies according to its own status and status distributed by adjacent logic agent federation.

On the basis of strategies obtained above and coordination command, control signal of switches is generated. Therefore, configuration method proposed in this paper is based on fault event driving. The main problem is status synthesis of every logic agent federation.

When bus works as power receiver, it may present fault diffusion state towards the corresponding bus acting as power distributor through the feeder connecting them. Similarly, when bus works as power distributor, it may present the state of being able to supply power towards the corresponding bus acting as power receiver through the feeder connecting them. Therefore, three column vectors are defined: pa_{path}, ps_{path} and $expd_{path}$. They respectively record the information whether the corresponding bus can supply power to its adjacent bus through the feeder connecting them, whether the corresponding bus is supplying power to its adjacent bus through the feeder connecting them, whether it presents fault diffusion state towards its adjacent bus through the feeder connecting them.

The basic idea of pa_{path} synthesis is that if the bus working as the power distributor itself meets two conditions at the same time: (a) it is not in the state of fault diffusion, (b) it can receive power from other bus except the bus working as the power receiver, it means that the bus working as the power distributor can supply power to the bus working as the power receiver through the feeder connecting them.

The basic idea of ps_{path} synthesis is similar with pa_{path}.

The basic idea of $expd_{path}$ synthesis is that if the bus working as the power receiver itself meets one of the three conditions: (a) it is in the state of fault diffusion, (b) it presents fault diffusion due to its adjacent bus except the bus working as the power distributer, (c) the converter on the feeder connecting power distributor and power receiver is faulty, it means that the bus working as the power receiver presents the state of fault diffusion towards its adjacent bus working as the power distributor through the feeder connecting them.

4.3 Simulation and Results

In order to verify the accuracy of the configuration method, we conduct many groups of simulation tests. The simulation is conducted in MATLAB. However, due to space limitations, we selected a group simulation test to illustrate the fault reconfiguration method proposed in this paper. Fault events and recovery results of simulation test can be shown in table 2.

When fault 1 occurs, subsystem which takes BusAC2 as the core is in the state of fault diffusion and it will affect its adjacent subsystems. Thus adjacent switches of BusAC2 are all disconnected. According to the status synthesis of the logic agent federation, BusAC2 can not supply power to BusACE and BusDC2, that is pa_ac2_ace=0, pa_ac2_dc2=0. Also according to the status information, BusAC3 and BusDC4 can supply power to BusACE and BusDC2 separately, that is pa_ac3_ace=1, pa_dc4_dc2=1, therefore switches DCBTB2, ACBTB7 and ACEB2 are connected.

Table 2. Fault events and recovery results of simulation test

No.	fault events	recovery results	
1	Short circuit of BusAC2	disconnect	GB2、ACBTB2、DCB2、TRUB2、ACEB1
		connect	DCBTB2、ACBTB7、ACEB2
2	Fault disconnection of DCB3	disconnect	TRUB3
		connect	DCBTB1
3	Short circuit of BusDC3, Fault connection of DCBTB1	disconnect	DCBTBE3、DCB1、DCBTBE1、TRUB1
		connect	DCBTBE5、DCBTBE6

When fault 2 occurs, according to the status synthesis of the logic agent federation, pa_ac3_dc3=0, it means BusAC3 can not supply power to BusDC3 and BusDC3 is interrupted. According to the status information, BusDC1 can supply power to BusDC3, that is pa_dc1_dc3=1, therefore switch TRUB3 is disconnected and switch DCBTB1 is connected.

When fault 3 occurs, subsystem which takes BusDC3 as the core is in the state of fault diffusion and it will affect its adjacent subsystems. Thus adjacent switches of BusDC3 are all disconnected. But since fault connection of DCBTB1 has occurred, subsystem which takes BusDC1 as the core is in the state of fault diffusion and it will also affect its adjacent subsystems. Adjacent switches of BusDC1 also need disconnecting. Therefore, switches DCBTBE3, DCB1, DCBTBE1 and TRUB1 are disconnected. Then, BusDCE1 is supplied only by Bat1. According to the reconfiguration strategies, BusDCE2 should supply power to BusDCE1. According to status synthesis of the logic agent federation, BusDCE2 can supply power to BusDCE1, that is pa_dce2_dce1=1, therefore switches DCBTBE5 and DCBTBE6 which are located on the feeder between BusDCE1 and BusDCE2 are connected.

At last, the configuration of the whole system is done.

5 Conclusion

Research on configuration method of aircraft distributed power distribution system based on MAS is conducted in this paper. This paper uses sequential storage structure for network topology coding, thus feeding branches between adjacent bus nodes are recorded. This makes the foundation of network reconfiguration. A multi-agent system was established based on layered and federal structure in this paper. Simulation results shown that the method proposed in this paper can trade and detect network topology and working status automatically and restore power supply to load points in the context of fault isolation.

Acknowledgments. The authors would like to express their sincere thanks for the financial support of the Aeronautical Science Foundation of China (20091952022) from Aviation science and Technology Committee.

References

1. Zhang, Y., Yang, X.: Review of shipboard power system network reconfiguration. Transactions of China Electrotechnical Society 20(8), 1–6 (2005)
2. Wang, X., Li, J., Xiao, J.: Network configuration of the shipboard power system based on gradient discretization method of partical swarm optimization. Transactions of China Electrotechnical Society 22(12), 140–145 (2007)
3. Li, Z., Xia, L.: Service reconfiguration of shipboard power system using improved ant colony algorithm. In: Proceedings of the CSU-EPSA, vol. 21(2), pp. 47–52 (2009)
4. Chen, J.: Research on multi-agent system's application in power system (2007)
5. Wang, P., Yu, K., Li, Z., Chen, X., Xu, W.: Application of multi-agent technology to enhance power supply reliability of high voltage distribution network. In: Proceedings of the CSEE, vol. 29(S1), pp. 50–54 (2009)
6. Bian, X., Ma, Q., Zhou, Y.: A Multi-Agent Approach to the More Electric Aircraft Power System Reconfiguration. In: IEEE International Conference on Mechatronics and Automation, ICMA 2007, pp. 978–982 (2007)
7. Huang, K., Cartes, D.A., Srivastava, S.K.: A Multiagent-Based Algorithm for Ring-Structured Shipboard Power System Reconfiguration. IEEE Transactions on Systems, Man, and Cybernetics, Part C: Applications and Reviews 37(5), 1016–1021 (2007)
8. Nagata, T., Tao, Y., Fujita, H.: An autonomous agent for power system restoration. In: IEEE 2004 Power Engineering Society General Meeting, 1069 – 1074 (2004)
9. Momoh, J.A., Xia, Y., Alfred, K.C.: Dynamic reconfiguration for shipboard power system using multi-agent system. In: IEEE Power and Energy Society General Meeting - Conversion and Delivery of Electrical Energy in the 21st Century (2008)

A Method of Delay Compensating for UPQC Compensations Detection without PLL

Jianru Wan[1], Hailiang Zhou[1], Zehua Yao[1], and Jianxiong Qian[2]

[1] School of Electrical Engineering and Automation, Tianjin University, 300072, China
[2] China Special Equipment Inspection and Research Institute, 100013, Beijing, China
gaoge2004@163.com

Abstract. UPQC compensation detecting has characteristics of high precision and fast dynamic response. The article gave the solution in order to mitigate the influences of hardwire and calculation delay problems which existed in compensations detection used synchronous coordinate transformation without PLL. The accuracy has been improved by correcting the reference rotation angle in the matrix of synchronous coordinate transformation to compensate the delay coursed by hardwire circuit such as sampling. Meanwhile, state feedback was introduced to make up the calculation defect of low-pass filter in to speed up dynamic response. Finally, the simulation results verified that this solution applied in UPQC compensation detection is feasible and effective.

Keywords: UPQC, compensations detection, delay compensation, without PLL.

1 Introduction

Distortion and unbalance are common for the grid voltage. Meanwhile, the increasing number of power electronic equipments and other nonlinear devices produce lots of harmonics and reactive current, which make the grid voltage more serious and lower Energy utilization because of the circuit impedance. In order to avoid that the power quality problems cause damage and loss for some important loads, the power quality improving devices have been applied and developed such as LC passive filter, static var generation(SVG), active power filter(APF), dynamic voltage resistor(DVR) and unified power quality conditioner(UPQC) and so on. Specially, UPQC has function of both APF and DVR, and because of this, it has been the focus in the research of power quality.

Detecting the voltage and current compensations accurately and in a timely manner is very important to implement the functions for UPQC. As is known, We can implement the compensations detection in frequency domain or time domain, such as FFT[1] and Wavelet transform[2] in frequency domain while $p-q$, $i_p - i_q$ and synchronous coordinate transformation[3] based on instantaneous reactive power theory[4] in time domain. Currently, many scholars committed to the improvement of synchronous coordinate transformation used to detect UPQC compensations. Reference [5] and [6] applied double synchronous coordinate transformation with PLL and a decoupling network to get positive sequence voltage component under

G. Lee (Ed.): Advances in Automation and Robotics, Vol. 1, LNEE 122, pp. 451–458.
springerlink.com

unbalance, but when the voltage contains harmonics, the fundamental component can not be isolated. Reference [7] analysis the basic principle of non-PLL and proposed a method of harmonics detection without PLL based on the average theory. However, these articles just focus on how to solve the calculation delay and rarely mention the hardwire delay caused by hardwire circuit such as sensors measurement, signal conditioning, ADC and so on. With the rapid development of microprocessor, solving the delay problem just from the point of reducing calculation already exist limitations.

This paper gave out a solution in which the accuracy has been improved by correcting the reference rotation angle in the matrix of synchronous coordinate transformation to compensate the delay coursed by hardwire circuit such as sampling, meanwhile, state feedback was introduced to make up the calculation defect of low-pass filter in to speed up dynamic response.

2 UPQC Hardwire Delay Compensation

According to Fourier analysis theory and symmetrical component method, three-phase three-wire system voltage can be decomposed into sinusoidal-shaped fundamental positive sequence, negative sequence fundamental and high harmonics. The sampled grid voltage is shown as:

$$\begin{bmatrix} v_a' \\ v_b' \\ v_c' \end{bmatrix} = \sqrt{2}\begin{bmatrix} V_1^+\sin(\omega t+\phi_1^+-\theta_1)+V_1^-\sin(\omega t+\phi_1^--\theta_1)+\sum_{n=2}^{\infty}V_n\sin(n\omega t+\phi_n-\theta_1) \\ V_1^+\sin(\omega t-\frac{2\pi}{3}+\phi_1^+-\theta_1)+V_1^-\sin(\omega t+\frac{2\pi}{3}+\phi_1^--\theta_1)+\sum_{n=2}^{\infty}V_n\sin(n(\omega t-\frac{2\pi}{3})+\phi_n-\theta_1) \\ V_1^+\sin(\omega t+\frac{2\pi}{3}+\phi_1^+-\theta_1)+V_1^-\sin(\omega t-\frac{2\pi}{3}+\phi_1^--\theta_1)+\sum_{n=2}^{\infty}V_n\sin(n(\omega t+\frac{2\pi}{3})+\phi_n-\theta_1) \end{bmatrix} \quad (1)$$

Where, the hardwire delay is representative by delay angle θ_1 while ϕ is the initial phase. On right-hand side, subscript of each variables indicates the order while superscript shows the phase sequence.

The grid voltage is transformed to the components under pq coordinate frame through synchronous coordinate transformation. Then, after LPF, we can get the DC components as follows:

$$\begin{bmatrix} \overline{v_p'} \\ \overline{v_q'} \end{bmatrix} = \overline{\mathbf{C}\cdot\begin{bmatrix} v_a' \\ v_b' \\ v_c' \end{bmatrix}} = \begin{bmatrix} \sqrt{3}V_1^+\cos(\omega t+\phi_1^+-\theta_1-\theta) \\ -\sqrt{3}V_1^+\sin(\omega t+\phi_1^+-\theta_1-\theta) \end{bmatrix} \quad (2)$$

Where, the transformation matrix is:

$$\mathbf{C}=\sqrt{2/3}\begin{bmatrix} \sin(\theta) & \sin(\theta-2\pi/3) & \sin(\theta+2\pi/3) \\ -\cos(\theta) & -\cos(\theta-2\pi/3) & -\cos(\theta+2\pi/3) \end{bmatrix} \quad (3)$$

Where, θ is the rotated angle derived from the virtual sine & cosine function table.

Then, conducting inverse transform of the DC component shown in (2), we can get the fundamental positive component of the grid voltage:

$$\begin{bmatrix} v_{1a}'^{+} \\ v_{1b}'^{+} \\ v_{1c}'^{+} \end{bmatrix} = \mathbf{C}^{-1} \cdot \begin{bmatrix} \overline{v_p'} \\ \overline{v_q'} \end{bmatrix} = \begin{bmatrix} \sqrt{2}V_1^+ \sin(\omega t + \phi_1^+ - \theta_1) \\ \sqrt{2}V_1^+ \sin(\omega t + \phi_1^+ - 2\pi/3 - \theta_1) \\ \sqrt{2}V_1^+ \sin(\omega t + \phi_1^+ + 2\pi/3 - \theta_1) \end{bmatrix} \tag{4}$$

Obviously, the hardwire delay angle has been transferred to the fundamental positive sequence component. In order to solve the influence of the delay angle θ_1, we can break through from the point of changing rotated angle in $\mathbf{C}^{-1}$.

Assuming the angle corrected is θ_{d1}, the rotated angle in $\mathbf{C}^{-1}$ will be $\theta'=\theta+\theta_{d1}$, that mains:

$$\mathbf{C}^{-1'} = \sqrt{2/3}\begin{bmatrix} \sin(\theta') & -\cos(\theta') \\ \sin(\theta'-2\pi/3) & -\cos(\theta'-2\pi/3) \\ \sin(\theta'+2\pi/3) & -\cos(\theta'+2\pi/3) \end{bmatrix} = \sqrt{2/3}\begin{bmatrix} \sin(\theta+\theta_{d1}) & -\cos(\theta+\theta_{d1}) \\ \sin(\theta+\theta_{d1}-2\pi/3) & -\cos(\theta+\theta_{d1}-2\pi/3) \\ \sin(\theta+\theta_{d1}+2\pi/3) & -\cos(\theta+\theta_{d1}+2\pi/3) \end{bmatrix} \tag{5}$$

Then, conducting inverse transform of the DC component in (2) we get the fundamental positive component of the grid voltage:

$$\begin{bmatrix} v_{1a}^{+} \\ v_{1b}^{+} \\ v_{1c}^{+} \end{bmatrix} = \sqrt{2/3}\begin{bmatrix} \sin(\theta+\theta_{d1}) & -\cos(\theta+\theta_{d1}) \\ \sin(\theta+\theta_{d1}-2\pi/3) & -\cos(\theta+\theta_{d1}-2\pi/3) \\ \sin(\theta+\theta_{d1}+2\pi/3) & -\cos(\theta+\theta_{d1}+2\pi/3) \end{bmatrix} \cdot \begin{bmatrix} \overline{v_p'} \\ \overline{v_q'} \end{bmatrix} = \begin{bmatrix} \sqrt{2}V_1^+ \sin(\omega t + \phi_1^+ - \theta_1 + \theta_{d1}) \\ \sqrt{2}V_1^+ \sin(\omega t + \phi_1^+ - 2\pi/3 - \theta_1 + \theta_{d1}) \\ \sqrt{2}V_1^+ \sin(\omega t + \phi_1^+ + 2\pi/3 - \theta_1 + \theta_{d1}) \end{bmatrix} \tag{6}$$

Let $\theta_{d1} = \theta_1$, the fundamental positive sequence component of voltage will be the standard form without delay. In other words, hardwire delay will be compensated by rectifying the synchronous coordinate inverse transformation matrix.

Similarly, the load current can be broke down into fundamental positive sequence, negative sequence, and high harmonics components. Assuming that the current sampling delay angle is θ_2, then the load current will be written as:

$$\begin{bmatrix} i_{la}' \\ i_{lb}' \\ i_{lc}' \end{bmatrix} = \sqrt{2}\begin{bmatrix} I_1^+ \sin(\omega t + \varphi_1^+ - \theta_2) + I_1^- \sin(\omega t + \varphi_1^- - \theta_2) + \sum_{n=2}^{\infty} I_n \sin(n\omega t + \varphi_n - \theta_2) \\ I_1^+ \sin(\omega t - \frac{2\pi}{3} + \varphi_1^+ - \theta_2) + I_1^- \sin(\omega t + \frac{2\pi}{3} + \varphi_1^- - \theta_2) + \sum_{n=2}^{\infty} I_n \sin(n(\omega t - \frac{2\pi}{3}) + \varphi_n - \theta_2) \\ I_1^+ \sin(\omega t + \frac{2\pi}{3} + \varphi_1^+ - \theta_2) + I_1^- \sin(\omega t - \frac{2\pi}{3} + \varphi_1^- - \theta_2) + \sum_{n=2}^{\infty} I_n \sin(n(\omega t + \frac{2\pi}{3}) + \varphi_n - \theta_2) \end{bmatrix} \tag{7}$$

To simplify the calculation and get active current, we use the unit fundamental positive sequence voltage component to calculate the p axis component directly replacing the synchronous coordinate transformation, and the process is as follows:

$$i_p' = \sqrt{2/3}[i_{la}' \times \sin(\omega t + \phi_1^+) + i_{lb}' \times \sin(\omega t - 2\pi/3 + \phi_1^+) + i_{lc}' \times \sin(\omega t + 2\pi/3 + \phi_1^+)] \tag{8}$$

We obtain the DC component by low-pass filtering:

$$\overline{i_p'} = \sqrt{3}I_1^+ \cos(\varphi_1^+ + \theta_2 - \phi_1^+)] = \sqrt{3}I_1^+ \cos(\phi_1^+ - \varphi_1^+ - \theta_2)] \tag{9}$$

From (7), we found that it has an error of θ_2 compared with the actual value. Let $\theta''=\theta+\theta_{d1}+\theta_{d2}$, and then conduct inverse transformation of the pq axis voltage components once more. Assuming the matrix is $\mathbf{C}_2^{-1}$, and:

$$\mathbf{C}_2^{-1}=\sqrt{2/3}\begin{bmatrix}\sin(\theta+\theta_{d1}+\theta_{d2}) & -\cos(\theta+\theta_{d1}+\theta_{d2})\\ \sin(\theta+\theta_{d1}-2\pi/3+\theta_{d2}) & -\cos(\theta+\theta_{d1}-2\pi/3+\theta_{d2})\\ \sin(\theta+\theta_{d1}+2\pi/3+\theta_{d2}) & -\cos(\theta+\theta_{d1}+2\pi/3+\theta_{d2})\end{bmatrix} \tag{10}$$

Then, we can get:

$$\begin{bmatrix}v_{1a}''^{+}\\ v_{1b}''^{+}\\ v_{1c}''^{+}\end{bmatrix}=\mathbf{C}_2^{-1'}\cdot\begin{bmatrix}\overline{v_p'}\\ \overline{v_q'}\end{bmatrix}=\begin{bmatrix}\sqrt{2}V_1^{+}\sin(\omega t+\phi_1^{+}-\theta_1+\theta_{d1}+\theta_{d2})\\ \sqrt{2}V_1^{+}\sin(\omega t+\phi_1^{+}-2\pi/3-\theta_1+\theta_{d1}+\theta_{d2})\\ \sqrt{2}V_1^{+}\sin(\omega t+\phi_1^{+}+2\pi/3-\theta_1+\theta_{d1}+\theta_{d2})\end{bmatrix} \tag{11}$$

Furthermore, the unit fundamental positive sequence component will be obtained:

$$\begin{bmatrix}\sin(\omega t+\phi_1^{+}-\theta_1+\theta_{d1}+\theta_{d2})\\ \sin(\omega t+\phi_1^{+}-2\pi/3-\theta_1+\theta_{d1}+\theta_{d2})\\ \sin(\omega t+\phi_1^{+}+2\pi/3-\theta_1+\theta_{d1}+\theta_{d2})\end{bmatrix}=\frac{1}{\sqrt{2}V_1^{+}}\begin{bmatrix}v_{1a}''^{+}\\ v_{1b}''^{+}\\ v_{1c}''^{+}\end{bmatrix}=\sqrt{\frac{3}{2(\overline{v_p'}^2+\overline{v_q'}^2)}}\begin{bmatrix}v_{1a}''^{+}\\ v_{1b}''^{+}\\ v_{1c}''^{+}\end{bmatrix} \tag{12}$$

Then calculating the p axis component like formula (6) and (7) again, we can get the DC component of active current:

$$\overline{i_p''}=\sqrt{3}I_1^{+}\cos(\varphi_1^{+}+\theta_2-\phi_1^{+}+\theta_1-\theta_{d1}-\theta_{d2}) \tag{13}$$

As what mentioned before, $\theta_{d1}=\theta_1$, if making $\theta_{d2}=\theta_2$, the delay of current detection would be compensated. Furthermore, the fundamental positive sequence active component could be represented as follows:

$$\begin{bmatrix}i_{1ap}^{+}\\ i_{1bp}^{+}\\ i_{1cp}^{+}\end{bmatrix}=\sqrt{\frac{2}{3}}\begin{bmatrix}\sin(\omega t+\phi_1^{+}-\theta_1+\theta_{d1})\\ \sin(\omega t-2\pi/3+\phi_1^{+}-\theta_1+\theta_{d1})\\ \sin(\omega t+2\pi/3+\phi_1^{+}-\theta_1+\theta_{d1})\end{bmatrix}\cdot\overline{i_p''}=\frac{1}{\sqrt{\overline{v_p'}^2+\overline{v_q'}^2}}\begin{bmatrix}v_{1a}^{+}\\ v_{1b}^{+}\\ v_{1c}^{+}\end{bmatrix}\cdot\overline{i_p''}=\begin{bmatrix}\sqrt{2}I_1^{+}\cos(\varphi_1^{+}-\phi_1^{+})\sin(\omega t+\phi_1^{+})\\ \sqrt{2}I_1^{+}\cos(\varphi_1^{+}-\phi_1^{+})\sin(\omega t-2\pi/3+\phi_1^{+})\\ \sqrt{2}I_1^{+}\cos(\varphi_1^{+}-\phi_1^{+})\sin(\omega t+2\pi/3+\phi_1^{+})\end{bmatrix} \tag{14}$$

So far, the delay caused by hardwire has been solved.

3 UPQC Calculation Delay Compensation

As is known, compared with the detecting method with PLL, non- PLL method still applies the low-pass filter though there is no PI calculation, and this is the fundamental reason for the delay. Simulation result indicated that the low-pass filtering process will take about 5 Industrial frequency cycles to be stable when voltage or current compensation changes. It makes the system can not response quickly and then lower the dynamic property. Therefore, we introduce the states feedback in the control in order to improve the filtering process of LPF.

Because the output of the filter is DC component, we should not use delay angle to discuss it, so begin from the amplitude of compensation.

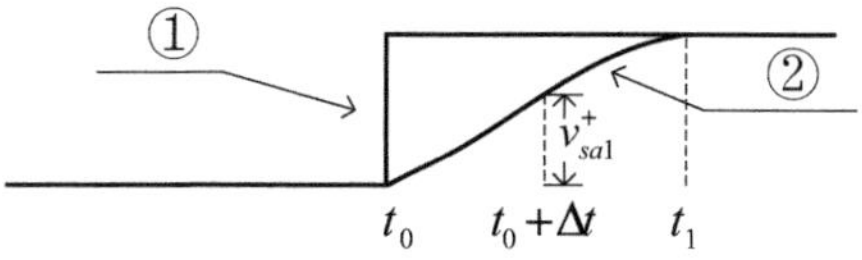

Fig. 1. Schematic plot of LPF calculation delay analysis

Take A-phase grid voltage for example. In Fig.1, firstly, assuming that the fundamental positive sequence component of grid voltage increases in time t_0, and curve ① shows the ideal vary curve of the p axis component, while ② is the calculated result curve by low-pass filter with the whole stable time $t_1 - t_0$. Analyzing every moment $t_0 + \Delta t$ in the calculating process, and assuming the result got from LPF is litter Δv^+_{sal} than the actual value, the harmonic of voltage must increase Δv^+_{sal} correspondingly in this time. In other words, the deviation caused by LPF has been increased into the final harmonic component of voltage, which also means the harmonic component v'^+_{sah} includes part of fundamental positive sequence component Δv^+_{sal} in time $t_0 + \Delta t$. Therefore, we provide feedback v'^+_{sah} to the front side as shown in Fig.2. As known, the harmonic component will be removed by synchronous coordinate transformation and LPF. As a result, it is equivalent to just make fundamental voltage get bigger, and thus make up the amplitude loss caused by LPF. The analyzing process is similar when the fundamental positive sequence component deceased or load current changed, so there no need to discuss.

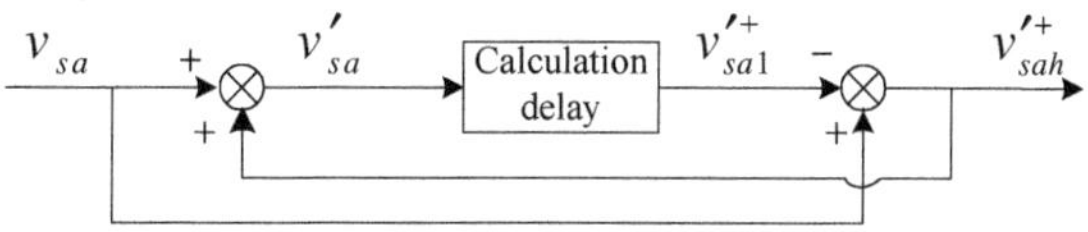

Fig. 2. Block diagram of state feedback

4 Simulation

Simulation tests have been carried to demonstrate the validity of the method proposed in MATLAB. The grid voltage sag and distorted occurred at 0.15s, of which wave was shown in Fig.3. Meanwhile, the load current wave is shown in Fig.4. Then, we will compare the differences between the simulation results with the method proposed and without the method.

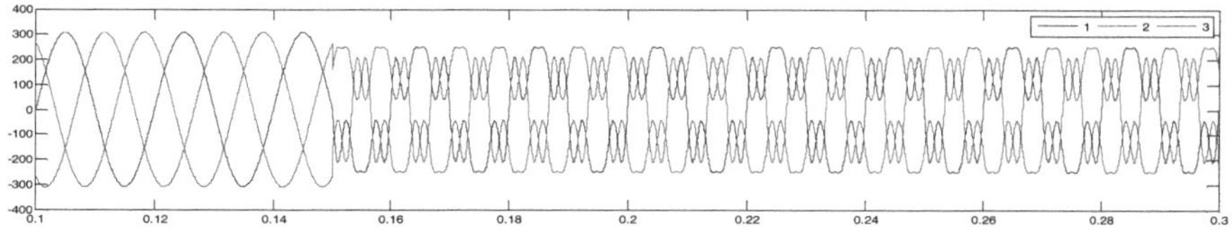

Fig. 3. The grid voltage curves

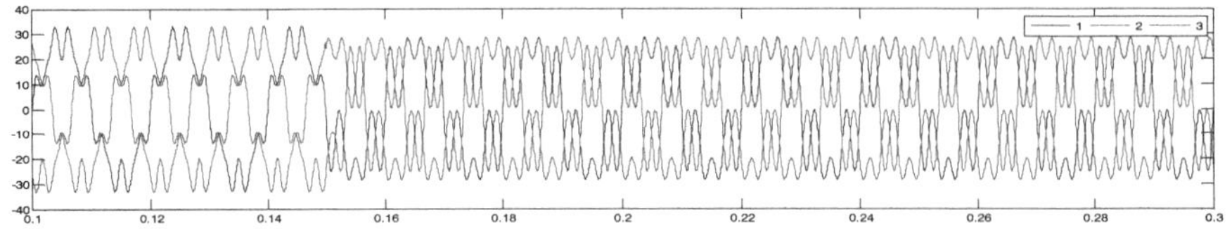

Fig. 4. The load current curve

Firstly, we will compare the difference in terms of the solution for hardwire delay. The voltage simulation results are shown in Fig.5 while the current waves in Fig.6.

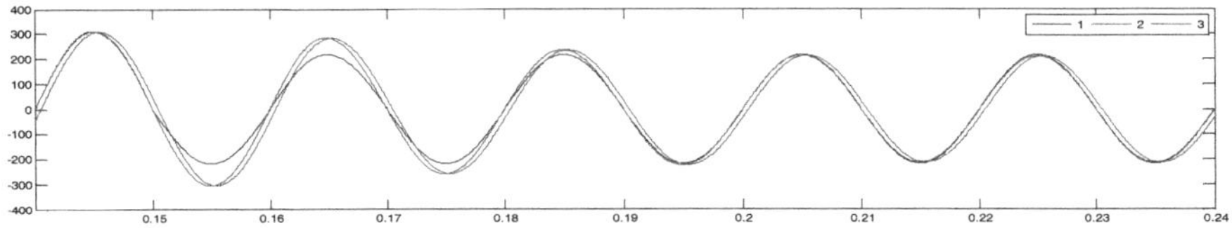

Fig. 5. Simulation waves of voltage compensation detection

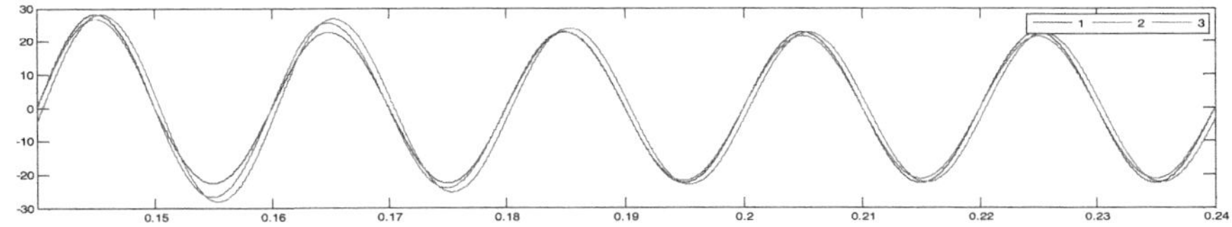

Fig. 6. Simulation waves of current compensation detection

In Fig.5, 1 curve 1is the actual fundamental positive sequence voltage component while curve 2 and 3 are the simulation result with and without hardwire delay compensation respectively. Obviously, the delay caused by hardwire sampling has been compensated and the fundamental positive sequence component detected keeps in-phase with the actual curve, but the result without hardwire delay compensation is delayed an angle. The current results that the simulation reaches in Fig.6 confirmed the validity of the solution proposed similarly.

Secondly, contrast waves about the solution for calculation delay is shown in Fig.7~9.

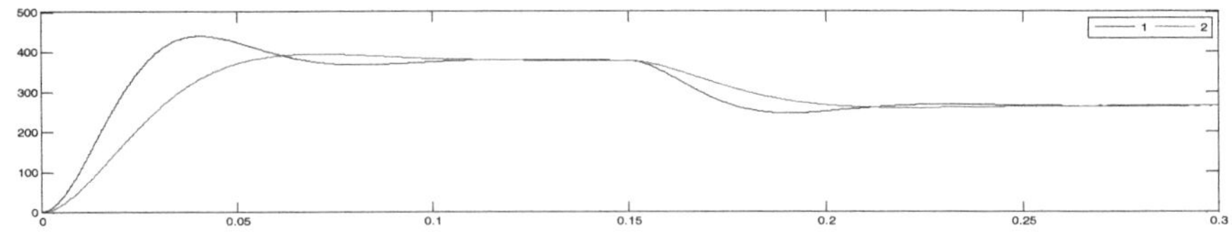

Fig. 7. $\overline{v_p'}$ waves with and without state feedback

In Fig.7, curve 1 and 2 are the calculation result of $\overline{v_p'}$ with and without state feedback respectively. Through comparative analysis, we can find that the dynamic response is improved, but it took more time to reach stable state. Besides, the simulation results agree well with the theoretical values when the compensations varying.

Fig.8 and Fig.9 indicated the compensation detecting results. In Fig.8, curve 1is the actual fundamental positive sequence voltage component while curve 2 and 3 are the simulation result with and without state feedback respectively. By contrasting, curve 2 is faster to track the variation of the grid voltage than curve 3. It is the same to the current compensation detection which showed in Fig.9.

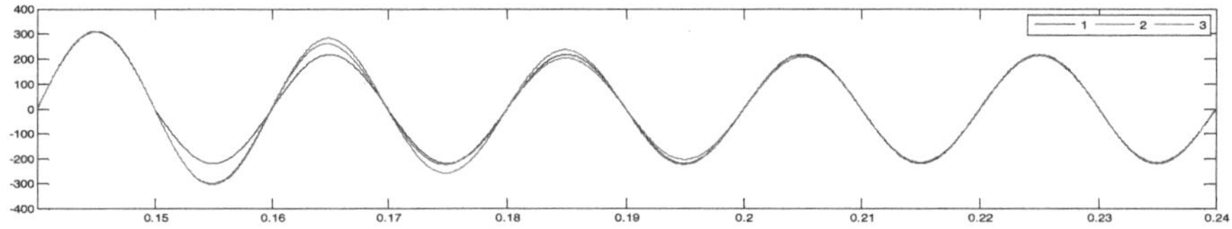

Fig. 8. Simulation waves of voltage compensation detection

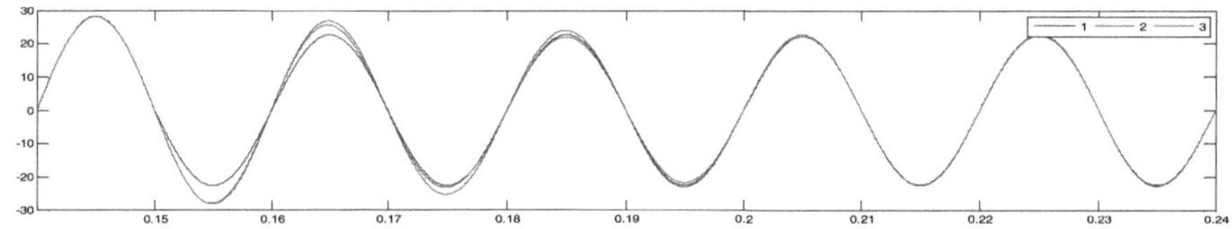

Fig. 9. Simulation waves of current compensation detection

Finally, Integrating hardwire delay compensation and state feedback together, curve 1is the actual value, while 2 and 3 are curves with and without hardwire delay compensation and state feedback respectively in Fig.10. We can see that both detecting precious and dynamic response has been improved. The same conclusion can be obtained for the current compensation detection which shows in Fig.11.

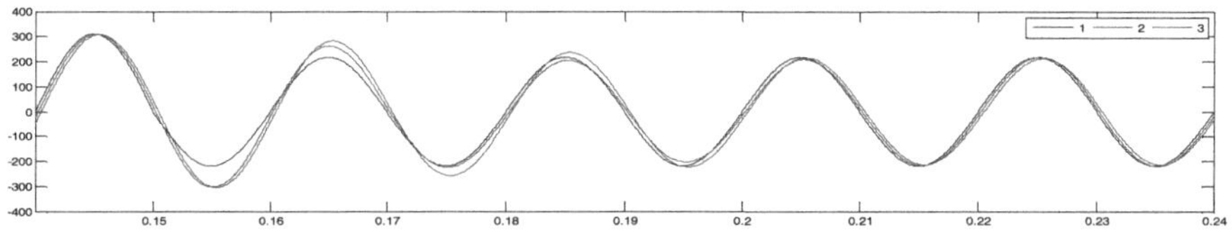

Fig. 10. Simulation waves of voltage compensation detection

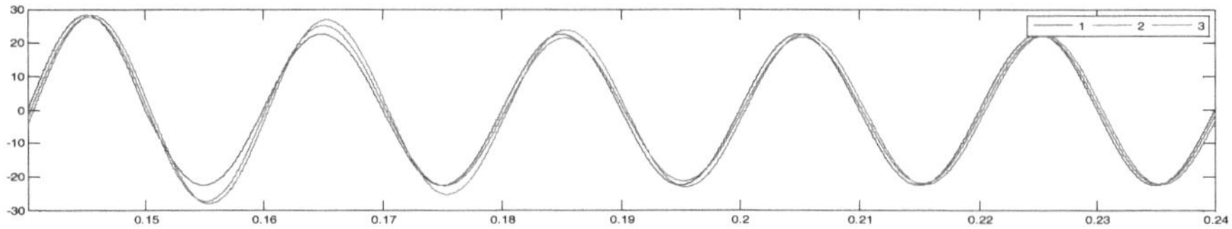

Fig. 11. Simulation waves of current compensation detection

5 Conclusion

This paper analyzed the influence that the delay exerted into UPQC compensation detection from the point of hardwire and calculation. Then it gave out corresponding resolution, and that is compensate the delay caused by the hardwire sampling by correcting the reference rotated angle of synchronous coordinate inverse transformation and speed up dynamic response of the low-pass filter by introducing state feedback into the control process.

Acknowledgments. This work is supported by the National Natural Science Funds (NO. 60874077) and AQSIQ Nonprofit Industry Specialized Research Funding Projects (NO.200910097 and 201010057), and thanks here.

References

1. Jiang, X., Xiao, X., Liu, H., Ma, Y.: The Output Spectrum Analysis of High-Power Multilevel Voltage Source Converters using Double Fourier Series. In: Transmission and Distribution Conference and Exhibition: Asia and Pacific, pp.1–5. IEEE/PES (2005)
2. Sangeethapriya, S., Prabhakar, M.: Harmonic Detection using Wavelet Transform. In: Proceedings of International Joint Conference on Neural Networks, Montreal, Canada, pp. 2228–2231 (2005)
3. Suh, Y., Tijeras, V., Lipo, T.A.: A Nonlinear Control of the Instantaneous Power in dq Synchronous Frame for PWM AC-DC Converter under Generalized Unbalanced Operating Conditions. In: Industry Applications Conference, pp. 1189–1196 (2002)
4. Akagi, H., Kanazawa, Y., Nabae, A.: Instantaneous Reactive Power Compensators Comprising Switching Devices without Energy Storage Components. IEEE Transactions on Industry Applications 20(3), 625–630 (1984)
5. Rodriguez, P., Pou, J., Ergas, J., et al.: Decoupled Double Synchronous Reference Frame PLL for Power Converters Control. IEEE Trans. on Power Electronics 22(2), 584–592 (2007)
6. Tian, G., Wang, S., Wang, Z.: Design of Phase Locked Loop Based on Double Synchronous Reference Frame under Unbalance Grid Voltage in Wind Power Systems. Electric Drive 40(7), 53–57 (2010)
7. Ding, J., Zhang, H., Zhang, X.: A Harmonics-detection Method without PLL and Based on Average Theory in Three-phase Circuit. Electrical Measurement & Instrumentation 47(553), 43–47 (2010)

Research on Optimum Energy Control Method for UPQC Based on Super Capacitor Energy Storage System

Jianru Wan[1], Zehua Yao[1], Shuchao Li[1], Jianxiong Qian[2], and Shuang Han[1]

[1] Shool of Electrical Engineering and Automation, Tianjin University, 300072, Tianjin, China
[2] China Special Equipment Inspection and Research Institute, Beijing, 100013, China
gaoge2004@163.com

Abstract. As UPQC based on super capacitor energy storage costs a long time to compensate the power, the optimum energy control strategy is proposed in this paper. This strategy obtains the injection angle of the compensation voltage by PSO by analyzing the optimal stable power of UPQC. This strategy uses bi-direction DC/DC circuit to reduce the capacity of UPQC devices, which can extends the compensation time without increasing energy storage capacity and enhances operation economy. The control method is verified possible and effective by theoretical analysis, simulation and experiment results.

Keywords: Unified Power Quality Conditioner, Super capacitor storage, optimum energy control, bi-direction DC/DC control, PSO.

1 Introduction

The growing of various load in power grid, and voltage problems are harming the safety of power system. The traditional power quality compensation device can't meet the consumers' requirements of power quality in [1]. UPQC based on super capacitor energy storage system can compensate not only harmonics and reactive current, but also supply voltage problems. And super capacitor can improve the load quality and play a role as short-time backup UPS. On the basis of analyzing energy flow, optimum energy control strategies are proposed which can reduce the need of active power and increase the maximum compensation time without increasing the energy storage capacity. This paper analyzes optimum energy control strategy quantitatively. Then the bi-direction DC/DC is introduced. At last, the control method is verified effective by the results of simulation.

2 Topology of UPQC Based on Super Capacitor Energy Storage System

UPQC based on super capacitor storage can not only compensate supply voltage and harmonics, but also compensate the reactive and harmonic component of the load

G. Lee (Ed.): Advances in Automation and Robotics, Vol. 1, LNEE 122, pp. 459–466.
springerlink.com

current. Topology of UPQC based on super capacitor energy storage is as Fig. 1, and it is composed of series voltage source inverter (SVSI), parallel voltage source inverter (PVSI) and the DC link. SVSI is in series with the load at the input of the circuit through the second side of the transformer, works as the controlled voltage source that compensating the voltage harmonic and the voltage fluctuation from the power system. PVSI is series with passive filter at output of the circuit through a transformer and then connected to the load in parallel, works as a controlled current source that compensating the harmonics and reactive load current. Super capacitors energy storage device, as DC link, can adjust power and can play a role as UPS.

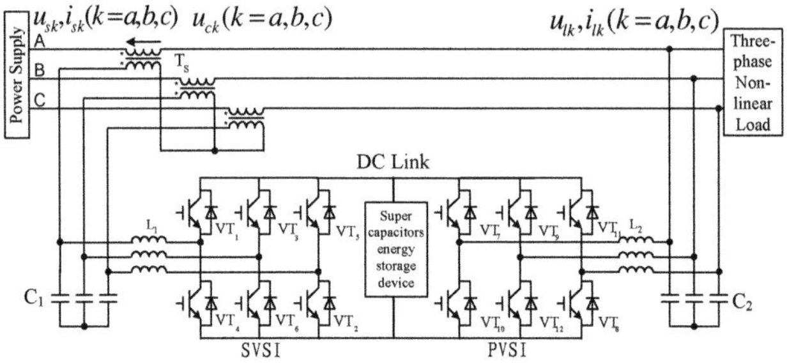

Fig. 1. The structure of UPQC based on super capacitor storage system

3 Optimum Energy Control Method for UPQC Based on Super Capacitor Energy Storage System

3.1 Power Analysis of Optimum Energy Control Method

The equivalent circuit diagram of UPQC when it comes to voltage sag is shown as Fig.2. On this basis, power in steady state is analyzed, and the phase graph in this condition is as Fig.3.

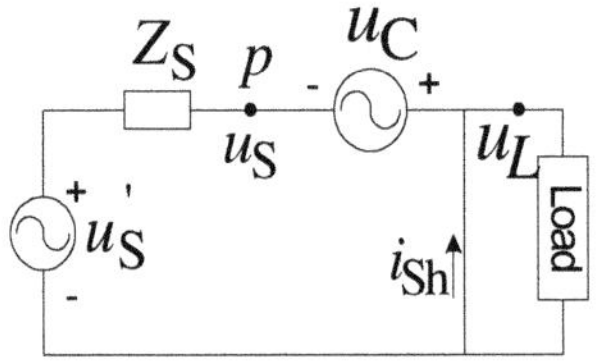

Fig. 2. Equivalent circuit

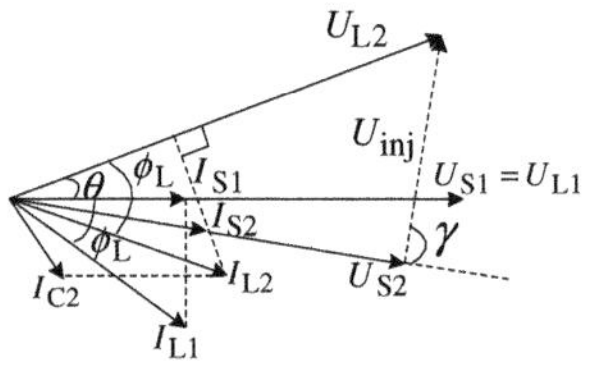

Fig. 3. Vector diagram of UPQC

In Fig.2, u_S'、 Z_S are respectively the equivalent voltage source and resistance of the input voltage. u_S is the actual grid supply voltage and u_C the compensation voltage of series side, u_L the load voltage, i_{Sh} the compensation voltage in the parallel side.

According to the theory of power balance, active power required of load is supplied by the point of p, and the power factor of UPQC is 1. The parameters that will be used later are given as follows: U_{L}, load voltage; U_{S}, input voltage amplitude; I_{S1}, virtual value of source current before the sag; I_{S2}, virtual value of source current after the sag; I_{L1}, virtual value of load current before the sag; I_{L2}, virtual value of load current after the sag; U_{L1}, virtual value of load voltage before the sag; U_{L2}, virtual value of load voltage after the sag; U_{inj}, virtual value of compensation voltage; K, degree of the input voltage sag; γ, angle of voltage injection in the series side; θ, angle variations of load voltage after compensation; ϕ_L, load power factor angle.

In order to analyze the problem, given: $U_{\mathrm{L1}} = U_{\mathrm{L2}} = U_{\mathrm{S1}} = U_{\mathrm{L}} = 1\,\mathrm{p.u.}$, and $I_{\mathrm{L1}} = I_{\mathrm{L2}} = I_{\mathrm{L}} = 1\,\mathrm{p.u.}$, related power calculation is as follows:

$$K = (U_{\mathrm{L}} - U_{\mathrm{S}})/U_{\mathrm{L}} \tag{1}$$

$$P_{\mathrm{S}} = U_{\mathrm{S1}} I_{\mathrm{S1}} = U_{\mathrm{S2}} I_{\mathrm{S2}} \tag{2}$$

According to the theory of power balance:

$$P_{\mathrm{S}} = P_{\mathrm{L}} \tag{3}$$

I_{S2} can be obtained with (2) and (3):

$$I_{\mathrm{S2}} = I_{\mathrm{L}} \cos\phi_{\mathrm{L}} / (1-K) \tag{4}$$

According to the phase angle relation in Fig.3, using relation of trigonometric function, I_{C2} can be obtained:

$$I_{\mathrm{C2}} = I_{\mathrm{L}} \sqrt{A} / (1-K) \tag{5}$$

In which $A = \cos^2\phi_{\mathrm{L}} + (1-K)^2 - 2(1-K)\cos\phi_{\mathrm{L}} \cos(\phi_{\mathrm{L}} - \theta)$

In Fig.3, U_{inj} can be obtained by cosine law:

$$U_{\mathrm{inj}} = U_{\mathrm{L}} \sqrt{(1-K)^2 + 1 - 2(1-K)\cos\theta} \tag{6}$$

The expression of power in parallel P_{Sh} can be obtained by (6) and formula for power in parallel:

$$P_{\mathrm{Sh}} = U_{\mathrm{L2}} I_{\mathrm{C2}} + I_{\mathrm{C2}}^2 Z_{\mathrm{shunt}} \tag{7}$$

Similarly, the expression of power in series P_{Ser} can be obtained by the relation among all the parameters in Fig.3:

$$P_{\text{Ser}} = U_{\text{inj}} I_{\text{S2}} = \frac{U_{\text{L}} I_{\text{L}} \sqrt{(1 - K)^2 + 1 - 2(1 - K)\cos\theta\cos\phi_{\text{L}}}}{1 - K} \tag{8}$$

The required capacitor of compensation process P_{UPCC} can be obtained by (7) and (8):

$$P_{\text{UPCC}} = \sqrt{\frac{\cos^2\phi_{\text{L}}}{(1-K)^2} - \frac{2\cos\phi_{\text{L}}\cos(\phi_{\text{L}}-\theta)}{1-K} + 1} + \left[\frac{\cos^2\phi_{\text{L}}}{(1-K)^2} - \frac{2\cos\phi_{\text{L}}\cos(\phi_{\text{L}}-\theta)}{1-K} + 1\right] Z_{\text{shunt}} + \frac{\sqrt{(1-K)^2 + 1 - 2(1-K)\cos\theta\cos\phi_{\text{L}}}}{1-K} \tag{9}$$

Equation (9) shows that the total capacity of compensation process of UPQC is a function about K、ϕ_{L}、Z_{shunt} and θ. In which, K、ϕ_{L} and Z_{shunt} are three independent parameters. When K take a special value as well as ϕ_{L} and Z_{shunt}, optimal energy control strategy can be realized on condition that θ is the optimal solution to ensure that P_{UPCC} can realize compensation at a minimum capacity.

3.2 Calculation of Optimal Angle θ_{opt}

Calculation of the injection angle is a core problem of optimal energy control strategy. Optimal angle θ_{opt} is determined by Particle Swarm Optimization (PSO) in [2]. Firstly, PSO initialize a group of particle whose position and speed are random, and then obtain the optimal solution through iteration. Suppose that at the t iteration time, p_{id} is the optimal solution of particle i, called individual extremum, and p_{gd} is the optimal solution of the hole particle swarm ,called global extremum. At the $t+1$ iteration time, particle i updates its speed according to equation (10) using the two extremums, so

$$v_{id}(t+1) = wv_{id}(t) + c_1 r_1 (p_{id}(t) - x_{id}(t)) + c_2 r_2 (p_{gd}(t) - x_{id}(t)) \tag{10}$$

And then particle i gets to the next position at the new speed,

$$x_{id}(t+1) = v_{id}(t) + v_{id}(t+1) \tag{11}$$

Where, x_{id} is the position of the particle at the t iteration time, $v_{id}(t)$ is the vector of at the t iteration time, w is inertia weight factor, c_1 and c_2 are learning factors, ranging from 0 to 2, r_1 and r_2 are uniformly distributed random numbers ranging from 0 to 1. Particles constantly trace individual extremum and global extremum in the space until it comes to the prescriptive iteration times or the error is permissible. After the t iteration time, $p_{id}(t+1)$ will be updated by (12), $p_{gd}(t+1)$ and by (13).

$$p_{id}(t+1)=\begin{cases} v_{id}(t+1) & f(v_{id}(t+1))\le f(p_{id}(t)) \\ p_{id} & f(v_{id}(t+1))> f(p_{id}(t)) \end{cases} \tag{12}$$

$$p_{gd}(t+1)=\begin{cases} p_{id}(t+1) & f(p_{id}(t+1))\le f(p_{gd}(t)) \\ p_{gd} & f(p_{id}(t+1))> f(p_{gd}(t)) \end{cases} \tag{13}$$

When it comes to the condition that the number of particles is 50, the times of iteration is 40, w=(0.4 ~ 0.8)and $c_1 = c_2 = 1.4$, the voltage optimal Angle θ_{opt} can be obtained through calculation, and then the amplitude and phase of the injected voltage can be obtained.

$$U_C = U_L\sqrt{1+(1-x)^2-2(1-x)\cos\theta_{opt}} \tag{14}$$

$$\alpha = \tan^{-1}\left(\frac{U_{S1}\sin\theta_{opt}}{U_{S1}\cos\theta_{opt}-U_{S2}}\right) \tag{15}$$

4 Control Strategy

Fig. 4 is the block diagram of UPQC optimum energy control strategy, made up by SVSI and PVSI. Instruction signal is formed respectively by measuring the voltage and current of the grid and load, and it then drives SVSI and PVSI to compensate the power quality comprehensively.

4.1 The Control Strategy for SVSI

SVSI mainly consists of measuring voltage sag, calculation module, load power factor detection module, optimal injection voltage calculation module and PWM. The grid voltage sag is measured by average method, the voltage sag degree x is calculated by the voltage virtual value of grid and load. The harmonic current of the load can be filtered out by Low Frequency Band Pass Filter, so the virtual value and phase angle of the load fundamental current can be calculated, then they will be analyzed together with phase angle of the rated voltage of the load. On this condition, the power factor of the load $\cos\theta_L$ can be calculated, and the amplitude and phase of the optimal injected voltage can be obtained by (14) and (15).

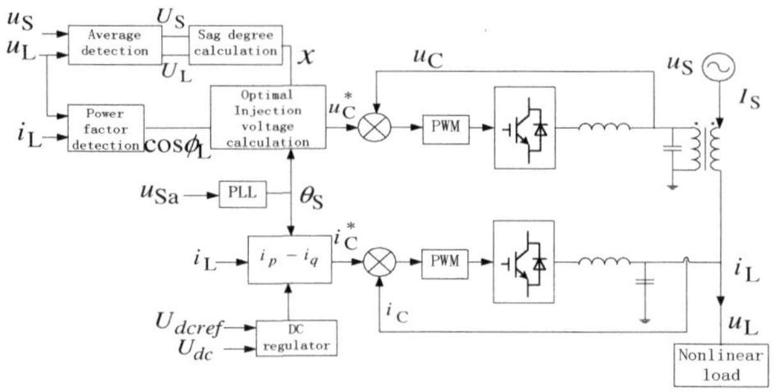

Fig. 4. The block diagram of UPQC optimum energy control strategy

4.2 The Control Strategy for PVSI

PVSI should detect the real time harmonic current of the load by instantaneous reactive power theory in this paper. Instantaneous reactive power theory is applied widely with its good steady state accuracy and dynamic performance. In the process of instantaneous reactive power calculation, fundamental active current can indirectly control the voltage and power flow of the DC side. In Fig.4, U_{dcref} is the fiducial value of DC capacitor, and U_{dc} is the feedback value of it. The output of PI adds to the DC component of active current, which can change the active power absorbed from the AC side, so the DC bus capacitor voltage can be controlled.

4.3 Control Strategy for the Energy Storage Unit at the DC Side

UPQC must stay the DC voltage steady to realize compensation. Super capacitor can realize high-power charging and discharging in a short time, so it can control the voltage sharing active power to keep the DC voltage stable.Fig.5 shows the DC power storage segment, made up by double DC/DC chopping circuit and a super capacitor, among them, the part with dotted wire frame is the super capacitor ,the others are the double DC/DC chopping circuit.

In Fig.5, u_{SC} is the terminal voltage of supper capacitor group, u_{PWM} is the modulating voltage of bi-direction DC/DC converter, u_{dc} is DC bus voltage, i_{SC} is the charging current of the super capacitor, i_{PWM} is the modulating current of bi-direction DC/DC converter, i_{dc} is DC bus current.

This converter can exchange energy between the super capacitor group in the low-voltage DC side and capacitor in the high-voltage DC side. Single-loop control of the DC current is that using difference value between DC bus voltage and reference voltage controls switch to obtain PWM signal. Control strategy for the energy storage unit at the DC side is as Fig.6.

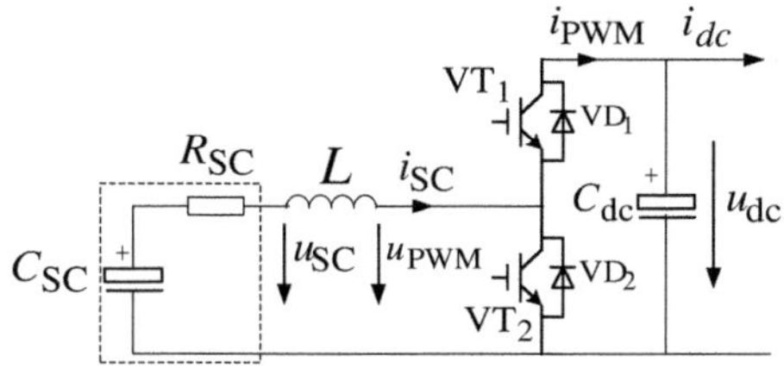

Fig. 5. Energy storage unit at the DC side

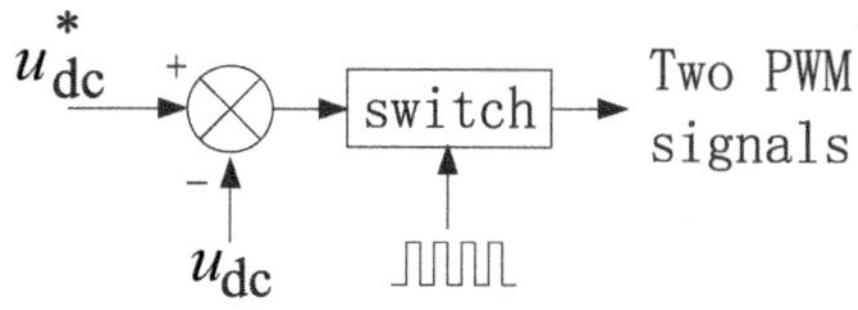

Fig. 6. Control strategy for the energy storage unit at the DC side

5 Simulation and Analysis

Power system dynamic simulation software adopted in computer simulation is MATLAB/Simulink, the main circuit is as Fig. 1. Optimum energy control strategy is adopted in UPQC. The model parameters are as follows: the supply voltage is 220V, supply frequency is 50Hz, the ratio of compensation transformer is 1:1, the compensation inductance of series is 1mH, compensation capacitor is 1μF, θ_{opt} is 29°, the compensation inductance of parallel is 5mH, the filter capacitor is 1μF, DC side is controlled by bi-direction DC/DC, the super capacitor is 1F, the filter capacitor is 2000μF, the stable voltage value of DC link is 800V,the load is three-phase full-controlled bridge concatenated 35Ω resistance and 2mH inductance. The simulation results are as Fig. 7.

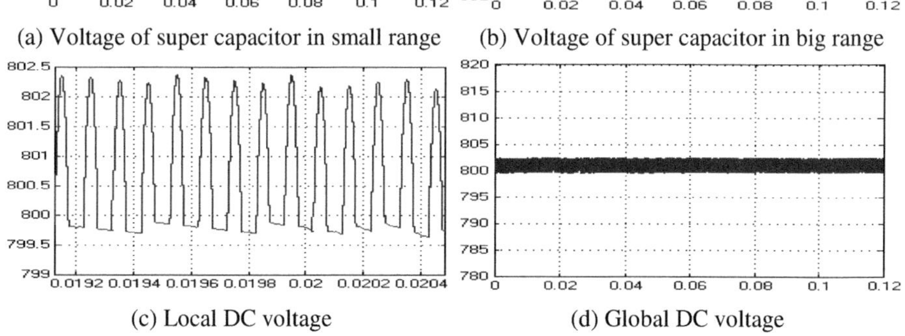

Fig. 7. Simulation results

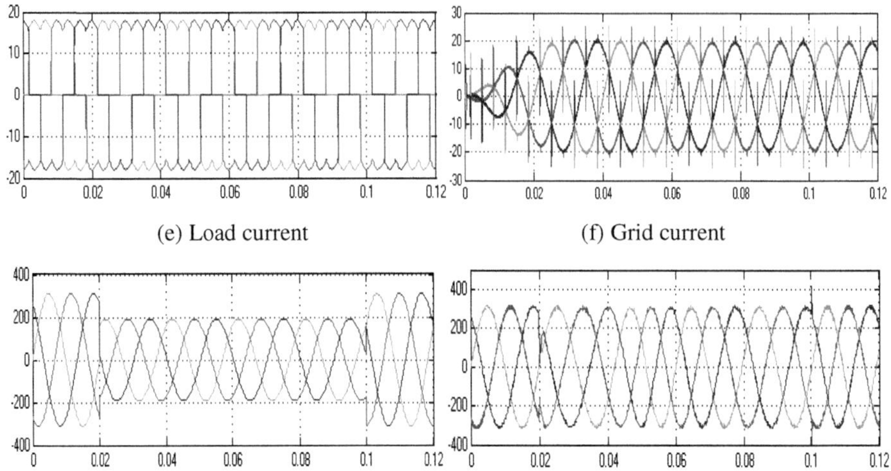

(e) Load current (f) Grid current

(g) Grid voltage (h) Load voltage

Fig. 7. (*continued*)

6 Conclusion

As the characteristics of UPQC, this paper proposed the optimal energy control strategy. Firstly, this strategy analyzes the Stable Power of UPQC, then derives out the functional relationship between optimal energy and the injection angle of compensation voltage, optimizes the energy and the injection angle by PSO. In this strategy, bi-direction DC/DC circuit makes the DC bus stable. The simulation results show that this control method can compensate not only the harmonic and reactive current caused by nonlinear load but also the supply voltage rise and drop, quality, reduce the capacity of UPQC devices, extend the compensation time without increasing energy storage capacity, enhance operation economy.

Acknowledgments. The authors gratefully acknowledge the financial support of the National Natural Science Foundation of China (No.60874077) and AQSIQ Nonprofit Industry Specialized Research Funding Projects (No.200910097 and 201010057).

References

1. Hong, S.: Research on Compensation Technique of Current Harmonic and Voltage Sag in UPQC. Tianjin University, China (2009)
2. Jin, X., Tang, Z.-D., Yu, X.-C., Wu, M.-P.: Optimal Energy Control Strategy of Unified Power Quality Controller and Simulation. Computer Simulation 28(3), 321–324 (2011)

Quality Control of Sensor Network Data

Davood Izadi[1], Jemal Abawajy[1], and Sara Ghanavati[2]

[1] School of Information Technology, Faculty of Science and Technology,
Pigdons Road, Waurn Ponds Campus, Deakin University,
Geelong, Victoria, 3220, Australia
{dizadi,jemal.abawajy}@deakin.edu.au
[2] School of Engineering, Faculty of Computer and Science,
Shahid Chamran University of Ahvaz, Iran
Sarah.ghanavati@yahoo.com

Abstract. A wireless sensor network (WSN) is a group of sensors that are geographically distributed and interconnected by wireless networks. Sensors gather information about the state of physical world. Then, after processing forward them to the main destination. To be able to provide this service, there are many aspects of communication techniques that need to be explored. Supporting quality of service (QoS) will be of critical importance for pervasive WSNs that serve as the network infrastructure of diverse applications. To illustrate new research and development interests in this field, this paper examines and discusses the requirements, critical challenges, and open research issues on QoS management in WSNs. A brief overview of recent progress is given.

Keywords: Wireless sensor network, Quality of service, Accuracy, Node failures, Data aggregation.

1 Introduction

Wireless sensor network (WSN) technology has been explored for about a decade. A WSN consists of a sink node, which collects data from other sensors and a large number of small and low cost sensors that can communicate wirelessly. The sensor devices that involve memory, processor, transceiver and power supply have the ability of converting physical phenomenon (e.g. heat, light, etc) into electrical signals. Nowadays, WSN is popularly used for gathering; processing and transmitting data for interested users. Such a service can be applied to many applications such as habitat monitoring, agriculture, environment sensing and health monitoring. Many of these applications require good quality of data to properly function. For instance, in environmental application, it is necessary to have reliable data about air and water pollution. Also, in health monitoring, it is important to have reliable data for tracking humans' heart functions.

In order to provide good quality data, apart from the intensive exploration of different aspects of WSN technology, such as hardware of sensor devices, architecture of networks in different environments and data security, quality of service (QoS) is still an interesting wide area for research.

G. Lee (Ed.): Advances in Automation and Robotics, Vol. 1, LNEE 122, pp. 467–480.
springerlink.com

QoS in WSN is used to control and enhance the quality of data in WSNs [1]. Then, the network can be more reliable for monitoring systems. In order to analyse and increase quality of service in a network there are many challenges, which can be categorized into hardware and software aspects of an ad hoc network, need to be considered. This research is more focused on software aspects of a WSN. In this case, there are many challenges such as real-time constraints, sensor node failures; memory limitation and the dynamic nature of the networks need to be addressed.

This paper provides a brief overview of QoS provisioning in the context of WSNs. Some critical challenges and possible research topics are discussed and then, related work is reviewed. The primary aim is to demonstrate new research and development interests in this field.

2 QoS Requirements

It is understandable that QoS requirements, based on conditions of WSNs, can be used in different meanings and prospectives. In fact, different techniques in WSNs may recognize QoS in different ways, which can be designed based on the application requirements. For instance, in a safe control system, delay and packet loss during transmission may not be allowed, while it might be acceptable in air conditioning systems in an office. Nevertheless, in general, methods of QoS in any WSN are applied to enhance the performance of WSNs for users [1]. However, there is not a common or formal definition of QoS in a WSN. Thus, in order to provide a desired quality of data in ad hoc networks, QoS in WSN can be analysed based on the following aspects. Although, there are many other QoS parameters worth mentioning, but these seven are the most fundamental.

2.1 Memory Limitation

Since a sensor node is a small device, the memory capacity in the device cannot be much. In fact, mostly, there is not enough space to run complicated algorithm and also storing data. Hence, overflow can be a significant issue in providing high quality data [2].

2.2 Delay

Multi-hop routing, network congestion, processing information and traffic are the reasons that can increase delay in communication. That can affect synchronization in the network and consequently make issues in data collection processing, especially in critical events. So, in order to ensure that data can be received on time to the destination in real time application, it is necessary to be aware of delay metrics in developing data collection mechanisms. Notice that, on-time does not mean fast communication or computation but it is a unique time for every network that needs to be delivered within the timing requirements [3].

2.3 Power Consumption

Energy consumption is the biggest constraint that needs to be concerned in WSN that can be categorized into three main aspects (i) energy for the sensor transducer, (ii) energy for communication among sensor nodes, and (iii) energy for microprocessor computation. In this case, it has been shown [4] that the required power for transmitting, which is the most consumption, for each bit in WSNs can be 800 to 1000 instructions. So, it is important to be aware of energy consumption in collecting data. Thus, in order to increase the life time of such a network and enhance the performance of the network, it is required to develop a routing protocol by taking energy constraint into account [5].

2.4 Complexity

In a WSN, mathematic functions, coding and encoding of data, etc can increase the complexity of processing data. In addition to that, the complexity of the developed mechanism can also be noticeably increased by integrating the QoS parameters. In this case, solving unpredictable problems in processing data, especially in random events, in such a large sized network could be difficult [6]. So, the performance of the data collection mechanism can be enhanced by considering the complexity of the applied protocol.

2.5 Flexibility

Node movement, removal or addition, system update or reconfiguration causes the network topology, routing, and node traffic load will change. This can then influence QoS attributes such as transmission delay, packet loss rate, and utilization [7]. In some situations, the QoS in WSNs may become unsatisfactory when delay and/or packet loss rate are too large. Therefore, QoS management paradigms are needed to enhance the flexibility and adaptability of WSNs with respect to the changing network conditions.

2.6 Accuracy

The accuracy of ad hoc networks is not only dependant on physical properties of the environment, e.g., the channel characteristics, and the performance of the integrated hardware components, but the localization algorithms and system protocols also have a significant role in providing an accurate network [8].

2.7 Location /Global Required Information

The requirement of global knowledge can increase traffic noticeably in a network. Also, due to keeping the nodes alive for each communication process, energy consumption cannot be avoidable. Consequently, by developing a mechanism in order to use local information instead of information of an entire network, the energy consumption can be decreased.

3 Challenges

In this section some of the major features that can influence QoS in a WSN will be discussed.

3.1 Data Collection in a Dynamic Network

Dynamic and static networks, with their advantages and disadvantages in different environment conditions are the two main classifications of WSNs. In dynamic networks, unlike the static network, sensor nodes are capable of moving accidently or in required directions with different speeds. For example, storm, earthquake, flood, etc are some of the factors that can move the sensor devices to the unpredicted location. Apart from the accident movements, the mobility of the sensor nodes could be categorized into many groups [9]. For example, random mobility patterns, which the speed and direction of any individual node are selected randomly from a predefined range. Next, the random value of speed and direction can be granted to a specific group of nodes. So, instead of one node, a number of nodes can be moved in a specified direction. The third pattern of node movement is defining the value for groups of nodes based on their geographic locations. Finally, complex mobility pattern, which mixes the mentioned patterns together [9].

WSN with the capability of handling the movements can provide the entire monitoring system with the following advantages. First, mobile sensors can increase the performance of the network by increasing the coverage area of the network [10]. Also, by using the mobile sink, even multiple mobile sinks[11], it has been proven that the life time of the network can be increased [12]. On the other hand, reliability and accuracy, life time of the network, delay and traffic, while they move, are the most challenging aspects of the dynamic networks. So far, there are many sophisticated protocols and frameworks that have been developed, but this research will prove that data collection is still an open research issue that needs to be investigated.

3.2 Fault-Tolerance Mechanisms

It is understandable that QoS, based on conditions of WSNs, can be used in different meanings and prospectives. In fact, different techniques in WSNs may recognize QoS in different ways. Nevertheless, in general, methods of QoS in any WSNs are applied to enhance the performance of WSNs for users [1]. Then, the network can be more reliable for systems that need to be significantly accurate such as health monitoring, asset tracking, automation and security and fire detection. In this case, a failure is the most common problem in WSN that can affect the performance of the network. Insufficient battery, hardware failure of the battery, memory, CPU, antenna and radio may also be damaged during a deployment process or even after installations. In fact, poor and defective hardware components also need to be considered as main reasons of failure [13]. Additionally, software integration of the components is another main reason that can destroy connectivity among senor nodes. For example, routing loops is an abnormality in an ad hoc network that can cause packets be transferred across many nodes and coming back to the origin. Failure can also happen in the low quality links that can be affected by the environment. Storm, fire, magnetic fields in the interested area are the factors that can produce interference in communication links [13]. As the consequence of the issues, obviously, the performance of the network can be influenced.

3.3 WSN Data Fusion

Data generated from neighbouring sensors is often redundant and highly correlated. In addition, the amount of data generated in large sensor networks is usually enormous for the base station to process. Hence, a mechanism is required to combine data into high-quality information at the sensors or intermediate nodes. Then, the transmitting process would be reduced by decreasing the number of packets for sending. As a result, energy and bandwidth in a network can be conserved. This can be accomplished by data aggregation. Data aggregation is defined as the process of aggregating data from multiple sensors to eliminate redundant transmission and provide fused information to the base station.

4 Open Issues

Over the years, in order to meet the desired QoS in a WSN, significant protocols have been developed. Although they improve QoS in a WSN, they still have weaknesses and are not able to provide desired end to end QoS. Thus, for real time application, reliable data delivery is still a real challenging issue in WSN. In this case, three open research question topics of interest will be identified.

4.1 How to Develop a Mechanism for Data Collection Based on QoS Obligations?

In order to collect data in an ad hoc network there are many protocols that have been developed. Direct diffusion [14] is a basic data collection protocol that is designed based on a data-centric method and determined by localized interactions. This is different from IP-style communication, where nodes are identified by end-points. In fact, in direct diffusion, hop to hop communication is used which requires less energy and helps to overcome failures. Direct diffusion protocol, in terms of energy consumption, delay, traffic and also overflow has some limitations. In fact, networks with such a protocol, due to broadcasting messages to find the source and also the best route in the network cannot have the desirable performance due to the high energy consumption of the system. Then after, Intanagonwiwat et al [14] discussed that by using semantic approaches to the direct diffusion protocol the performance of the network can be increased. In this case, in order to improve the process of the detecting sources, transferring requested data messages and also identifying the best route to the sink, the interest message can consist of some specific information, such as type of collected data, location, time and intensity. As a result of applying semantic approaches in broadcasting, energy-efficiency, traffic and time delay could be achieved. However, since the mentioned protocol did not consider priorities in urgent events, this cannot be a significant data collection in WSNs.

Jabeur and Graniero [15] discussed another technique, based on geographic locations and interested areas, to collect data in WSN. The proposed technique requires both virtual WSN (VWSN) and physical WSN to work in parallel and communicate well. VWSN is responsible for assigning multi agent systems to enhance the context awareness of the sensor nodes. The agents can identify the minimum number of the sensors that can cover the interested areas and also, by using some software, they can control the network. Consequently, the sensors can be improved in making their decisions, cope with the frequent changes in a WSN and communicate with each other. This approach [15] also used a semantic based routing approach to use the specific nodes based on the system requirements in the applied network. By using a virtual network, the ability of flexible processing data, with limited power sensor nodes, can be improved. Additionally, by semantic routing connections in the network, the process of data collection is more focused just on the interested area. However, due to the complexity of the method, the protocol cannot be a significant solution for collecting data in a WSN.

Software based structure is a communication framework that has been suggested by Zhang et al [16] to collect data in WSN. The framework consists of transport, network and MAC layers. Each layer has its own responsibilities. The framework has the ability of enhancing the reliability of the network by counting the successful transmissions. Also, the framework can compare them with the desired result by using balanced energy consumptions, latency and the trade off among nodes. Additionally, multi-application is the significance of this system. It means different sub areas or applications, which can be identified with a specific ID number, can be monitored by the system. Furthermore, since the framework has divided the entire network and components into different parts for setting up the system, then, finding and solving unpredictable issue would be much easier. However, the framework cannot be applied in WSNs with immobile sensors [16]. Also, since the successful transmissions are counted and compared by the desired ones, the reliability of the network with a huge number of data packet transmissions is decreased.

In order to increase the reliability of data collection in a WSN, Sun [17] proposed a Automatic Communication Framework (ACF) that is the combination of timeout based [18], streaming[19] and ExOR[20] protocols. The main contribution of the ACF is increasing the adaptively and reliability of data collection in WSNs with acceptable life time. Also energy efficiency and reliability of the system has been improved by using negative acknowledgements, multipath delivery protocols and also nodes' decision in finding their forwarding direction. However, due to applying different protocols, controlling them and switching on different protocols based on the result of monitoring performance of the network, the complexity of the network can be considerably high.

In order to control delay, accuracy and latency in collecting data in WSN, delay sensitive feedback control (DSFC) proposed by Shao-liang et al [21]. DSFC was designed to find the optimal point between delay and accuracy. In general, DSFC adaptively aggregate different data from different sources with the aim of decreasing delay while the accuracy is desirable. Peng Shao-liang et al [21] proved that DSFC can help the system to meet the desired QoS in a WSN by frequently checking the end to end delay and accuracy, while decreasing data transformation. However, DSFC, due to holding different data for aggregating and eliminating, decrease the latency and accuracy.

To sum up, all the mentioned protocols and frameworks have weaknesses and strengths. Table 1 demonstrates the comparison of the described protocols based on their fundamental challenges.

Table 1. A comparison of data collection protocols

Protocols	Delay	Static/Dynamic	Accuracy	Complexity
Direct diffusion and semantic approach [14]	Low	Static	Low	Low
Virtual WSN with Physical Network [15]	Low	Dynamic	High	High
Software based structure [16]	Low	Static	Low	Low
Automatic communication framework [17]	High	Dynamic	High	High
Delay sensitive feedback control [21]	High	Dynamic	Low	Low

4.2 How to Collect Data in a Reliable Manner in the Presence of Sensor Node Failures?

In order to prevent the affections of failures and increase the fault tolerance in WSN, there are many proposed methods. This section, by comparing the most popular of them, will prove that fault tolerance in a network is still an open research problem.

In order to decrease the energy consumption and also fault tolerance, PROC (Proactive Routing with Coordination) [22] has been proposed to find the most appropriate nodes for directing data to the destination. Routing packets has been improved by PROC due to the use of application-specific information that is broadcasting through the network. However, the mechanism can be used just for simple failure detections. Furthermore, the proposed technique is not technically sound due to the required long time and huge programming for detecting failures and recovering. Also, it does not have a control scheme on traffic flow patterns in the entire network [23].

Direct spanning tree (DST) [24] is another proposed technique that can increase fault tolerance. Spanning tree model connects all the sensors in WSNs to each other based on tree form connective paths. There are many routing protocols have been designed based on spanning tree. For example Minimum Energy Spanning Tree (MEST) that is sending data through the nodes to the sink based on the energy consumption of each node [24]. Since MEST does not work properly in failure situations, DST has been proposed. DST uses a mechanism that can make the network to establish paths from sources to sink by sender nodes and improves the capability of fail tolerance. However, DST cannot be a perfect technique in WSNs due to the following weaknesses. Firstly, DST is generating traffic because of its concentration on a specified Node. Next, high delay and memory overflow can be another result of congestion on the nodes, which are located near to sink.

For the purpose of detecting link and node failures, the highly resilient, energy efficient and multipath routing protocol [25], based on direct diffusion protocol [14], has been proposed. Ganesan et al [25] suggested constructing multiple paths between sources and sink, rather than just considering the best path as a primary route. Thus, there are a few routes that are held to be used in the case of failures in primary paths. Consequently, the sink has the opportunity to choose another route instead of searching and creating immediate routes when a failure is detected, which is the advantages of the solution. On the other hand the proposed protocol cannot be accepted as a desirable solution due to its high energy consumption for keeping the other routes alive and accessibility at any required time for the sink.

In addition to the described proposed methods, Reliable Energy Aware Routing (REAR) [26] is a routing protocol that is designed based on the energy consumption of the transaction. REAR uses a backup path for every single primary path; which is organized by the sink. This protocol is not efficiently solving the problem due to the high energy consumption. It is believed that if a node can decide to send data by using another neighbour, instead of changing the entire direction, obviously the energy consumption would be reduced. And also, the delay which is caused by changing the entire path cannot be avoided.

The combination of clustering and multi hop techniques [27] is another routing protocol mechanism that has been proposed to forward collected data from CH to CH on the way of sink. Since the protocol is dividing the network into different clusters dynamically, failure links and nodes in clusters can be detected by CHs faster. So, new CHs can be ensured of their nodes' life and communicated links [23]. Also, for intra cluster, CHs can figure out the connection to other CHs.

To sum up, the proposed protocols, which have been described above, cannot perfectly provide fault tolerance in WSN. Table 2 shows that none of the protocols are able to detect failures in a network with the respect of providing QoS.

Table 2. The comparison of fault tolerance protocols

Protocols	Delay	Location/global required info	Low/High consumption	Overflow possibility
PROC [22]	High delay	Global	High	Low
DST [24]	High delay	Global	Low	High
Multi hop hybrid routing protocol based	High delay	Local	High	High
Highly resilient, energy efficient multi path routing protocol	Low delay	Local	High	Low
REAR [26]	High delay	Local	High	Low

4.3 How Can We Perform Data Fusion in WSN with QoS Assurance?

Aggregating protocols are categorized into two main sections, which are structure and structure free based. Following that, the structure based is further classified into tree, grid and cluster architecture. Figure 1 illustrates the data aggregation structure.

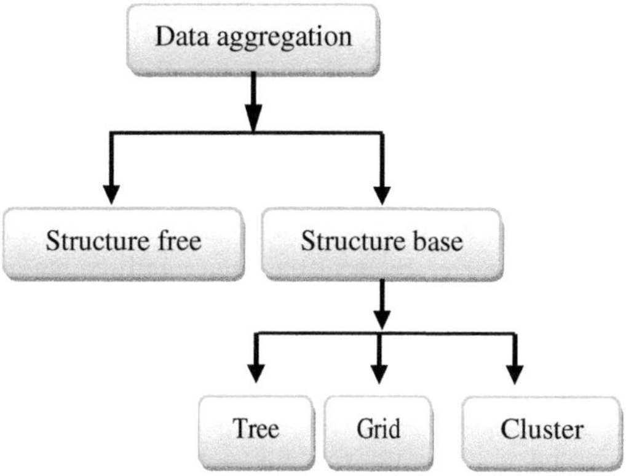

Fig. 1. Classification of aggregating data protocols in a WSN

4.3.1 Structure Based

Structure based can be divided into three of the most popular protocols, which are clustering, tree and grid protocols.

4.3.1.1 Clustering. In recent years, a huge number of hierarchical data aggregation based approaches have been proposed. In such protocols some special sensors are responsible for fusing data for the purpose of reducing the transmission data to the sink [28]. This approach can be useful in preventing every single energy-constrained node to send data.

LEACH (Low Energy Adaptive Clustering Hierarchy) [29] is the main and the first clustering protocol. Heinzelman [29] divided the protocol into two main sections.

Firstly, clustering section, which clusters dynamically and will be established based on self-adaptive mode. Then, all the CHs need to be introduced to the nodes. This can help the other sensors to select the best CH for themselves, based on some priorities. Then, the last step (steady step), which is longer than the first step, is for transferring packets in the WSN.

LEACH is more suitable for a WSN that is required to periodically monitor the covered area and also report the foundations. However, the performance of LEACH was improved later by LEACH-C [30]by 20 to 40 percentages [31]. But since CHs can be any node, it has been assumed that they need to have same conditions (e.g. battery power and capability of aggregating data) to be able to reach the main destination at any time that they are required [31]. This condition of the LEACH can increase the cost of energy consumption in the entire network and can also increase the probability of overflow into the network.

Apart from LEACH, clustered diffusion, with dynamic data aggregation (CLUDDA) [32], is a combination of clustering, with the diffusion mechanism that can aggregate the sensed data dynamically. CLUDDA has the ability to fuse data in an unfamiliar environment by using query definitions. In CLUDDA, unlike LEACH-C [30], the aggregation point is not assigned to any specific point or any specific group, which means the aggregation node can be changed when the source nodes are changed. In fact, following the changes, the aggregating point is identified by the query definition. Although CLUDDA can improve LEACH by sending data to the specific node instead of broadcasting to the network, this protocol cannot be a desired solution due to its high energy consumption because CHs need to be changed following any source change. In addition, since the new CH needs to inform the other nodes, the traffic and also overflow can be another shortage of the protocol.

Chain Based data aggregation [33] is a clustering based algorithm that is assuming every single sensor in WSNs has the ability of aggregating data and then, forwarding them to the next node. Lindsey et al [33] proposed a power efficient data gathering protocol for sensor information systems (PEGASIS), with the aim of simplifying the data fusion process in a WSN. This technique is a more energy efficient method than the previous described protocols. However, since all the nodes need to aggregate the collected data, the delay, especially for the end nodes, cannot be ignored [31]. Furthermore, PEGASIS cannot be a desirable solution for big WSNs or even for the very far distance sink. That is because, the leader nodes need to forward all the aggregated packet data to the sink [31].

4.3.1.2 Tree Based Protocols. Tree based protocols arrange all the sensor nodes into a tree structure and organize them to be able to aggregate the sensed data by immediate nodes in the network [28]. Then, a brief version of data is transferred to the root node along the tree.

This method can be analysed into two main aspects. Firstly, query routing based data aggregation algorithm, that is designed based on direct diffusion. The main thought behind this method is naming the data by using the information of the entire network. Firstly, the energy consumption is quite high and also the existed delay of the protocol are not avoidable especially in a large network [34].

Sub optimal aggregation tree is another protocol that is more focusing on constructing a possible minimum tree to the sink [34]. The main idea is designing a protocol to be able to use three different algorithms. Firstly, greedy internal tree (GIT) to create the tree in the network gradually and cover the entire network [34]. In this protocol, the closest node to the sink is finding a route to the main destination (sink) then, all other nodes are connecting themselves to the created route. Shortest past tree (SPT) [35] is another sub algorithm that is applied in the network to be able to identify the shortest data route to the sink. However, overlapping routes is a common issue of SPT. The last suboptimal algorithm is CNS (Centre at Nearest Source) [34]. This algorithm gives the responsibility of aggregating data to the very closest nodes to the sink. Then, all the other nodes forward their sensed data to the end node to be aggregated.

Since the conditions of different networks are changing, hence the performance of the mentioned sub optimal algorithms is changing. For example, if the distance of the closest node to the sink changed and became very far from the destination, the CNS cannot perform as it used to. Consequently, by changing the topology of the network, the way of the three algorithms application might be changed, which can influence the energy consumption and cause delay in the entire network.

4.3.1.3 Grid Based Data Aggregation. Vaidhyanathan et al. [16] have proposed two data aggregation schemes, which are based on dividing interested regions for monitoring by a sensor network into several grids. They are: grid based data aggregation, which nodes are not allowed to communicate with each other, and in-network data aggregation [36]. In grid-based data aggregation, the sensors in a particular grid transmit the data directly to the data aggregator, which has been assigned before in the grid [31]. In-network aggregation is similar to grid based data aggregation with two major differences. First, unlike the grid based, any sensor can be the aggregator and also, each sensor, within a grid, communicates with neighbouring sensors [31]. In other to increase the performance of the data aggregation, a hybrid scheme which combines the in-network and grid-based aggregation schemes has been proposed [36]. This protocol, in terms of overflow problems and also data latency can be accepted. However, apart from the energy consumption, complexity of the protocol, especially for a large number of nodes, the protocol cannot approach the desired performance. In addition, this protocol is not considering node and link failure.

Table 3 illustrates the comparison of the described protocols. The table clearly shows that data fusion in WSN with QoS assurance is still an open research issue that needs to be investigated.

Table 3. The comparison of data aggregation protocols

Protocols	Delay	Low/High consumption	Overflow possibility
CLUDDA [32]	High	High	Low
Chain Based [33]	High	Low	High
Sub-optimal Aggregation Tree Algorithms [34]	High	High	Low
Cluster Structure Data Aggregation Algorithms LEACH [29]	Low	High	High
Grid based data aggregation [16]	Low	High	Low

4.3.2 Structure Free Based

In order to not maintain or reconstruct any structure, especially in the environment with the frequent changes, for example node failure, structure-free-based has been proposed [28]. There are two main challenges in performing structure free data aggregation. Firstly, since there is no pre constructed structure, routing decisions for the efficient aggregation of packets needs to be made on-the-fly. Secondly, as nodes do not explicitly know their upstream nodes, they cannot explicitly wait on data from any particular node before forwarding their own data. Many researchers [37] have been investigated in this area for aggregating data and shows that the complexity of using structure free based algorithm in a WSN is very high. For example, randomized waiting impact, the cost of the routing process, some models in the MAC layer, etc., need to be studied.

5 Conclusion

Recent years have witnessed the intensive exploration of different aspects of wireless sensor networks such as architecture and protocol design, energy conservation and data management. Apart from that, QoS in WSNs is still an interesting and important field that has a wide area for research. So far, there are many protocols which have been proposed, with the aim of influencing the performance of ad hoc networks, but all of them have weaknesses.

This research has discussed the requirements and challenges for supporting QoS in WSNs. Then, in order to design a WSN for delivering data in a real time application, some interesting open research topics have been identified, through the spectrum of research in this field. Subsequently, many protocols and frameworks have been discussed and analysed based on fundamental QoS challenges in individual tables.

References

1. Tang, S., Li, W.: QoS supporting and optimal energy allocation for a cluster based wireless sensor network. Computer Communications 29, 2569–2577 (2006)
2. Masoum, A., Meratnia, N., Dilo, A., Taghikhaki, Z., Havinga, P.J.M.: Cross-Layer Analyses of QoS Parameters in Wireless Sensor Networks. In: Advances in Networks and Communications, pp. 595-605
3. Xia, F.: Qos challenges and opportunities in wireless sensor/actuator networks. Sensors 8, 1099–1110 (2008)
4. Sen, J.: A survey on wireless sensor network security, Arxiv preprint arXiv:1011.1529
5. Hill, J., Szewczyk, R., Woo, A., Hollar, S., Culler, D., Pister, K.: System architecture directions for networked sensors. Acm Sigplan Notices 35, 93–104 (2000)
6. Lu, Q., Luo, W., Wang, J., Chen, B.: Low-complexity and energy efficient image compression scheme for wireless sensor networks. Computer Networks 52, 2594–2603 (2008)
7. Xia, F., Zhao, W., Sun, Y., Tian, Y.C.: Fuzzy logic control based QoS management in wireless sensor/actuator networks. Sensors 7, 3179–3191 (2007)
8. Boukerche, A.: Algorithms and protocols for wireless sensor networks. Wiley-IEEE press (2009)
9. Li, W., Han, J.: Dynamic Wireless Sensor Network parameters optimization adapting different node mobility, pp. 1–7
10. Liu, B., Brass, P., Dousse, O., Nain, P., Towsley, D.: Mobility improves coverage of sensor networks, pp. 300–308 (2005)
11. Kim, H.S., Abdelzaher, T.F., Kwon, W.H.: Minimum-energy asynchronous dissemination to mobile sinks in wireless sensor networks, pp. 193–204 (2003)
12. Wang, Z.M., Basagni, S., Melachrinoudis, E., Petrioli, C.: Exploiting sink mobility for maximizing sensor networks lifetime (2005)
13. Jurdak, R., Wang, X.R., Obst, O., Valencia, P.: Wireless Sensor Network Anomalies: Diagnosis and Detection Strategies. Intelligence-Based Systems Engineering, 309–325
14. Intanagonwiwat, C., Govindan, R., Estrin, D.: Directed diffusion: A scalable and robust communication paradigm for sensor networks, pp. 56–67 (2000)
15. Jabeur, N., Graniero, P.: A Hybrid Location-Semantic Approach to Routing Assisted by Agents in a Virtual Network, pp. 523–533 (2007)
16. Zhang, S.Q., Cui, Y., Dong, Z.Y., Dong, Y.J.: A Communication Framework for WSN Based Data Collection Applications. Applied Mechanics and Materials 40, 482–487 (2011)
17. Sun, J.: An autonomic communication framework for wireless sensor networks (2009)
18. Gnawali, O., Yarvis, M., Heidemann, J., Govindan, R.: Interaction of retransmission, blacklisting, and routing metrics for reliability in sensor network routing, pp. 34–43 (2004)
19. Cao, Q., He, T., Fang, L., Abdelzaher, T., Stankovic, J., Son, S.: Efficiency centric communication model for wireless sensor networks, pp. 1–12 (2006)
20. Biswas, S., Morris, R.: ExOR: opportunistic multi-hop routing for wireless networks. ACM SIGCOMM Computer Communication Review 35, 133–144 (2005)
21. Shao-Liang, P., Shan-Shan, L., Yu-Xing, P., Pei-Dong, Z., Nong, X.: A delay Sensitive Feedback control Data aggregation Approach in wireless sensor network. In: Computational Science–ICCS 2007, pp. 393–400 (2007)
22. Macedo, D.F., Correia, L.H.A., dos Santos, A.L., Loureiro, A.A.F., Nogueira, J.M.S.: A rule-based adaptive routing protocol for continuous data dissemination in WSNs. Journal of Parallel and Distributed Computing 66, 542–555 (2006)

23. Al-Fares, M.S., Sun, Z., Cruickshank, H.: A Hierarchical Routing Protocol for Survivability in Wireless Sensor Network (WSN). In: Proceedings of the International MultiConference of Engineers and Computer Scientists, vol. 1 (2009)
24. Ji, P., Wu, C., Zhang, Y., Jia, Z.: Research of directed spanning tree routing protocol for wireless sensor networks, pp. 1406–1410 (2007)
25. Ganesan, D., Govindan, R., Shenker, S., Estrin, D.: Highly-resilient, energy-efficient multipath routing in wireless sensor networks. ACM SIGMOBILE Mobile Computing and Communications Review 5, 11–25 (2001)
26. Hassanein, H., Luo, J.: Reliable energy aware routing in wireless sensor networks (2006)
27. Zhao, J., Erdogan, A.T., Arslan, T.: A novel application specific network protocol for wireless sensor networks, vol. 6, pp. 5894–5897 (2005)
28. Pandey, V., Kaur, A., Chand, N.: A review on data aggregation techniques in wireless sensor network (2010)
29. Heinzelman, W.R., Chandrakasan, A., Balakrishnan, H.: Energy-efficient communication protocol for wireless microsensor networks, vol. 2 , p. 10 (2002)
30. Heinzelman, W.B., Chandrakasan, A.P., Balakrishnan, H.: An application-specific protocol architecture for wireless microsensor networks. IEEE Transactions on Wireless Communications 1, 660–670 (2002)
31. Rajagopalan, R., Varshney, P.K.: Data-aggregation techniques in sensor networks: a survey. IEEE Communications Surveys & Tutorials 8, 48–63 (2006)
32. Chatterjea, S., Havinga, P.: A dynamic data aggregation scheme for wireless sensor networks (2003)
33. Lindsey, S., Raghavendra, C., Sivalingam, K.M.: Data gathering algorithms in sensor networks using energy metrics. IEEE Transactions on Parallel and Distributed Systems, 924–935 (2002)
34. Chen, Y., Shu, J., Zhang, S., Liu, L., Sun, L.: Data fusion in wireless sensor networks, pp. 504–509 (2009)
35. Mikami, S., Aonishi, T., Yoshino, H., Ohta, C., Kawaguchi, H., Yoshimoto, M.: Aggregation efficiency-aware greedy incremental tree routing for wireless sensor networks. IEICE Transactions on Communications 89, 2741–2751 (2006)
36. Vaidyanathan, K., Sur, S., Narravula, S., Sinha, P.: Data aggregation techniques in sensor networks. Osu-cisrc-11/04-tr60, The Ohio State University (2004)
37. Fan, K.W., Liu, S., Sinha, P.: Structure-free data aggregation in sensor networks. IEEE Transactions on Mobile Computing, 929–942 (2007)

The Design on Intelligent Logging Unblocking Device

Yang Yu and Ming Chi

College of Mechanical Engineering Xi'an University of Science and Technology
Xi'an Shannxi Province, China
yuy@xust.edu.cn, chiming_hz@163.com

Abstract. Based on the actual situation that the well logging instrument is blocked on its way down to the well, by using electrical integration technology, we put forward an intelligent logging unblocking device which integrated with detection technology, control technology and mechanical technology. We designed the overall program, chose the model of acceleration sensor and DC motor, designed a mechanical system structure of the device which includes a logging instrument, a shell and a drilling, etc, developed an acceleration collection system of the intelligent logging unblocking device based on single chip and a DC motor control system. When the well logging instrument can not go down to the oil well or is blocked, the system can dredge the way of the well and then enable the well logging instrument go down to the oil well successfully.

Keywords: logging, unblocking device, mechanical structure, control detection system.

1 Introduction

In oil exploration and development processes, After the drilling work is done, a series work of cementing well, completing well and logging should be done to get parameters including porosity and permeability, etc. At present, the way of logging is putting the logging instrument down to the welll with cable. With the help of its own gravity, the logging instrument could complete the logging task. If the slope of the well is happened to be far away from the plumbline or the wall of the well is in a bad situation like collapsing or clogging, then the well logging instrument won't reach the bottom of the well. So it is difficult to get well logging data. Usually in this situation, engineers will have to pull up logging instrument from the well, start a second drilling, expanding and dredging well to make sure the logging instrument can go down of the well. The second drilling is time-consuming and the cost is high.

This thesis aims to invent an intelligent logging unblocking device. Installing it in the front of the logging instrument. When the logging instrument gets stuck, by using the front drill, the device can help correct the slope of the wall and dredge the well automatically under the effect of the control system. So the logging instrument can go down to the well smoothly. This device integrated with detection technology, control technology and mechanical technology. It can not only avoid the repeating drilling work, but the repeating measurement. In this way, the logging speed and precision can be improved effectively.

G. Lee (Ed.): Advances in Automation and Robotics, Vol. 1, LNEE 122, pp. 481–488.
springerlink.com

2 Overall System Design

The composition of intelligent logging unblocking device includes power supply, acceleration sensor, control system with STC89C52 as core, DC servo motor and its drive system, mechanical transmission system and the executive organs(Fig.1).

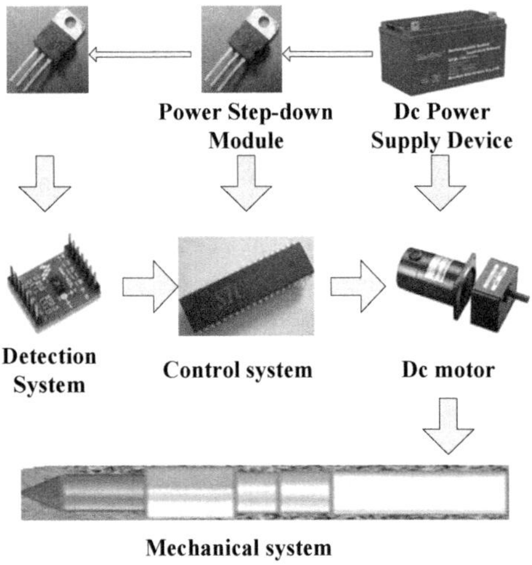

Fig. 1. System composition of intelligent logging unblocking device

The working process of intelligent logging unblocking device is: when the system with uniform speed in the process of down to the well, if it does not run into obstacles, the vertical acceleration measured by the acceleration sensor should be +g. If the system meets obstrutions, it'll slow down. The vertical acceleration measured by acceleration sensor will have great changes in the opposite direction.(Even the intelligent logging unblocking device can't work. In other words, the descent velocity will be 0). In this case, control system will drive DC motor. Through mechanical transmission system, DC motor is connected to roller cone bit to make bit work, remove obstacles and enable the well logging instrument to go down to the oil well successfully. The DC motor won't stop working until the acceleration measured by acceleration sensor return to +g.

3 Design of Mechanical Body

The main function of the mechanical body is: the DC motor and controller will rely on DC power to realize the power supply. The DC motor will drive the bit rotation through the reducer, transmission gears and coupling. Then it will remove obstacles due to the tilt wall or well clogged. So the intelligent logging unblocking device can go down successfully, finish the operation and realize logging function.

The intelligent logging unblocking device is shown in Fig.2. the mechanical structure contains Roller cone bit, transmission shaft, transmission gears, motor, battery, Logging tubular column and so on.

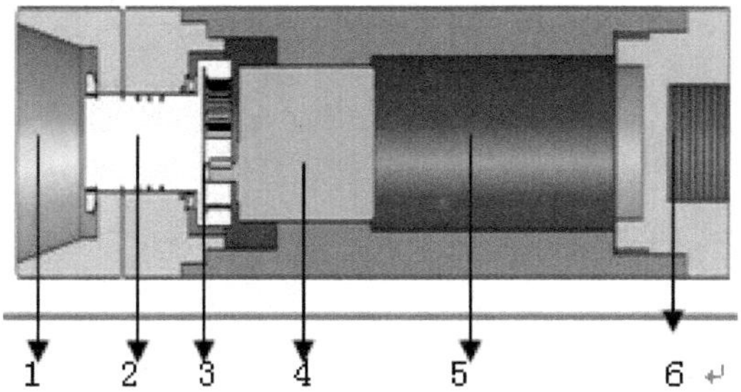

Fig. 2. The structure of intelligent logging unblocking device

1—Roller cone bit 2—Transmission shaft 3—Transmission gears
4—DC motor 5—Battery 6—Logging tubular column

4 Detection and Control System Design

The detection and control system plays the role of real-time detecting the downward speed and acceleration of the intelligent logging unblocking device. When its acceleration is less than the threshold, we can start the DC motor, drive the front drill of the device and remove obstacles. When the acceleration sensor is restored, we can stop the DC motor.

4.1 System Hardware Block Diagram

The detection system is mainly composed by MMA7455L acceleration sensor modules produced by Freescale Company. Control system as STC89C52 be the core includes the motor drive module, data storage module, power supply module and reset circuit etc(Fig.3). It uses I²C Communication mode Between STC89C52 and MMA7455L. The acceleration sensor will get acceleration value. Then the value will be transmitted to the MCU for judgment. If the value reachs threshold, the DC servo motor will work driven by the single-chip microcomputer. The acceleration value will be saved in SD memory card at the same time.

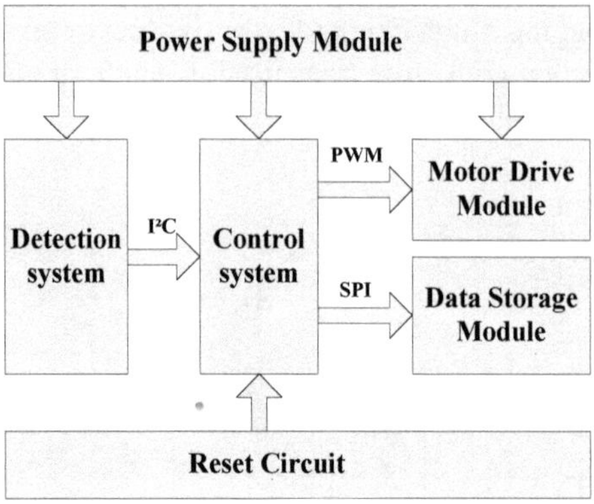

Fig. 3. The block diagram of the detection and control hardware system

4.2 Acceleration Sensor

The detection system by using small range acceleration sensor made by MMA7455L3 axis has three measurement ranges from 2g,4g to 8g. It also supports digital interface. User can read registers to get the digital quantity output directly which will help data processing.

The MMA7455L acceleration sensor will change its output signal based on object motion and direction changing. If the device moves along a certain direction or is affected by gravity, the the output signal will change based on the moving direction of the device and sensitivity of the sensor. Fig.4 shows the tube feet connection diagram of the acceleration sensor. It uses I²C Communication mode Between the single-chip microcomputer and the sensor. In which the SCL line(signal of clock)is connected with the MCU p1.7 port, SDA data line is connected with MCU p1.6. Acceleration sensor is used in measurement model. It can output the value of XYZ acceleration in three directions.

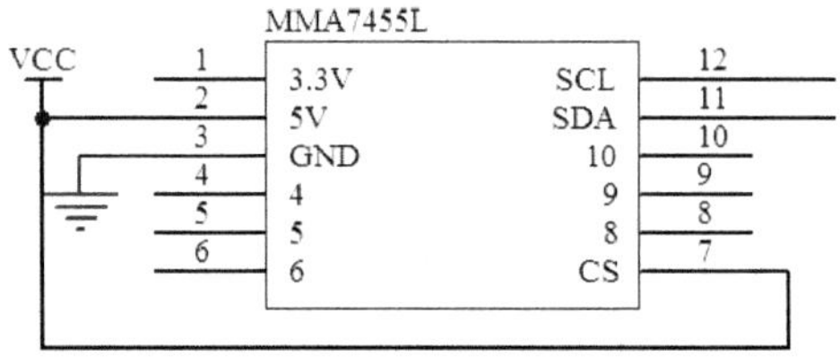

Fig. 4. Acceleration sensor hardware connection diagram

4.3 The Motor Drive Circuit and PMW Speed Control

The Motor drive circuit adopts mosfet drive mode(fig.5). The type of the mosfet is IRF9540N. It is an unipolar type of voltage control device. Following shows some of its features. Not only it has the ability of self-shut off, but also it's driven by less power.

The Switch speed is high. It has the secondary breakdown protection function. The safe operation area is wide, etc.

Control for the motor speed can be achieved by controlling the voltage located both sides of the DC motor. The Single-chip microcomputer can output certain frequency pulse width modulation signal to control the connection of the base electrode in IRF9540N. Through bridge circuits constituted by 4 mosfets, we can control the speed and direction of the DC motor.

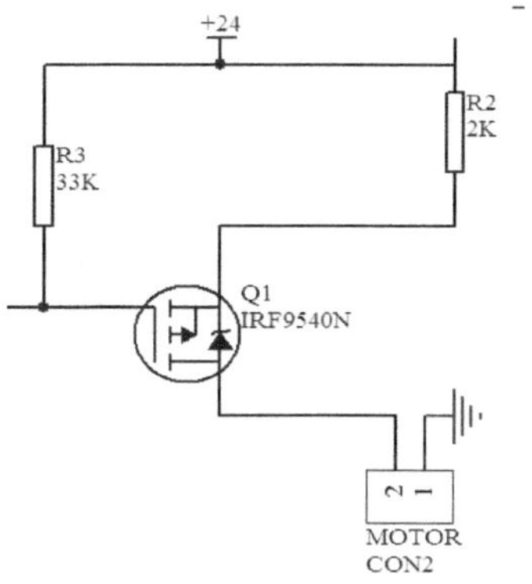

Fig. 5. The circuit driven by motor principle diagram

4.4 Power Supply Module Circuit

The power supply module circuit can change its 24V battery voltage into 5V MCU voltage. By using the way of two levels reducing voltage realized by the type of 7809 and 7805 three-terminal integrated stabilivolt voltage circuits, the conversion efficiency of power will be improved, the power consumption and heat productivity will be reduced. The connection diagram of circuit 7809 is shown in Fig.6.

4.5 Data Storage Module

The AT45DB161D is a serial-interface Flash Memory chip(Fig.7). It can work between 2.5V to 2.7V, support high speed operations, adopt SPI communication mode. Its frequency can be up to 66MHz. Its 17,301,504 bits of memory are organized as 4,096 pages with 512 bytes or 528 bytes in each of them. In addition to its main memory, the AT45DB161D also contains two SRAM buffers with 512/528 bytes in each of them. The buffers will allow receiving data and support data stream is written while the main Memory is programming. Then it will have data communication via three-wire interface(SI, SO, SCK).

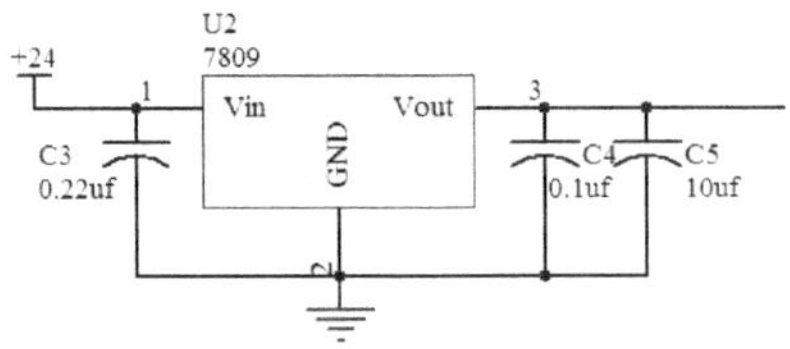

Fig. 6. 7809 circuit connection diagram

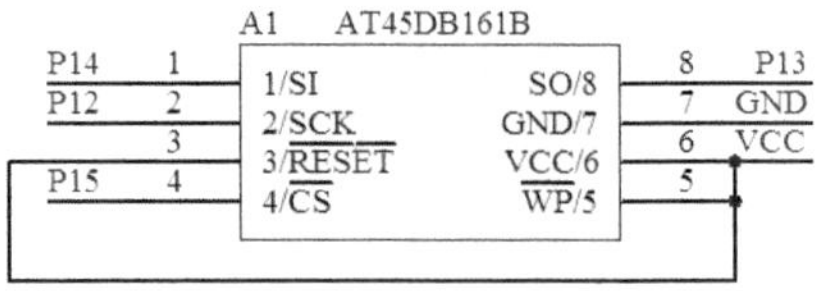

Fig. 7. Data storage circuit

4.6 Software Data Processing and Flow Chart

The acquisition value of acceleration should adopt appropriate digital processing. Here we applied both the extremum filter and media filtering algorithm. Extreme value filter means removing the maximum and minimum sample value in sampling time interval. By using extremum filter, the extreme value caused by interference can be eliminated. The impulse interference caused by occasional reasons will be overcome. The media filter means calculating the arithmetic average in sampling time interval. It mainly be used in system which aims to resisting acute and mutations interference. Working process of control system in intelligent logging unblocking device is shown in Fig.8.

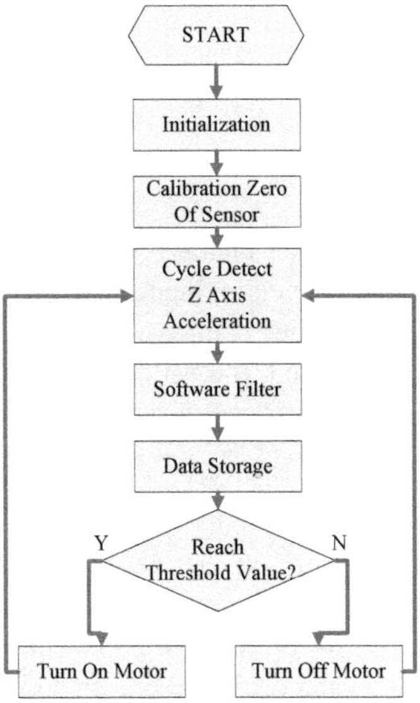

Fig. 8. System software flow chart

5 Simulation Experiment

The simulation of the system by using software c++ Builder mainly include of serial port communication design and acceleration waveform display etc. The simulation experiment includes three cases. The first case is when the system is on its way down to the well with uniform speed without deviating the plumbline, then it'll reach there directly. The second case is when the slope of well deviates certain angles from plumbline, the device will revise the well angle automatically. The third one is when the well is totally blocked, the device will dredge the well automatically.

When the system is on its way down to the well with uniform speed without deviating the plumbline, the acceleration value will be +g detected by the acceleration detect system(Fig.9).

When the slope of well deviates certain angles from plumbline, or encounter certain obstacles, the acceleration value will decrease from +g. When it reaches the threshold setting by the system, the drill will work to help remove obstacles. After awhile, the device will keep working and the acceleration value will return to +g(Fig.10).

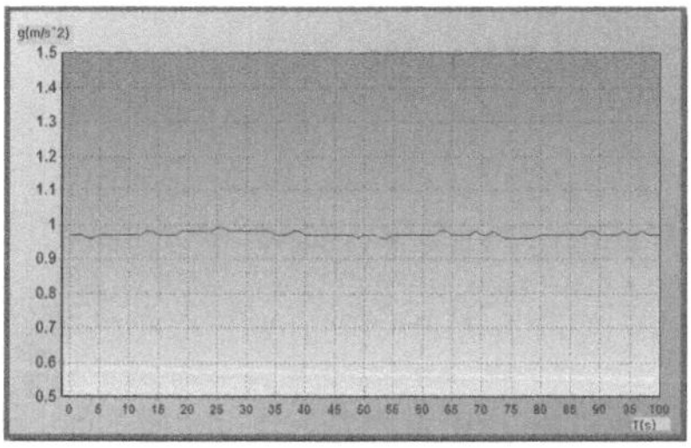

Fig. 9. Without deviating the plumbline

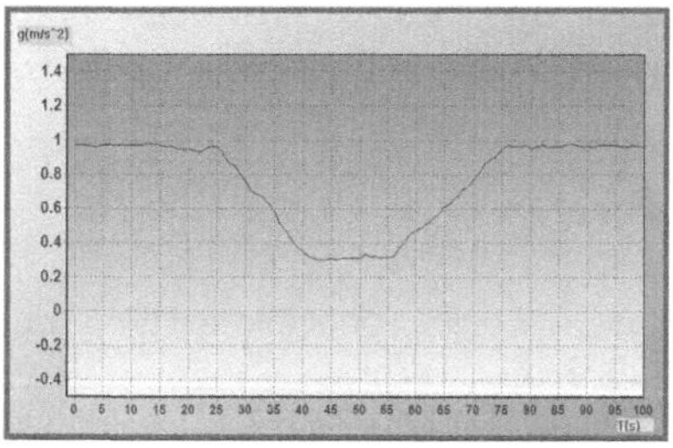

Fig. 10. Deviating from the the plumbline

When the well is totally blocked while the system is going down, the device will revise the well angle and dredge the well automatically. The acceleration value will have a great change which will decrease rapidly from +g to near 0. Then the device will start to work and dredge the well. After awhile, the acceleration will return to +g. This means the system operates well and it achieves the expecting effect(Fig.11).

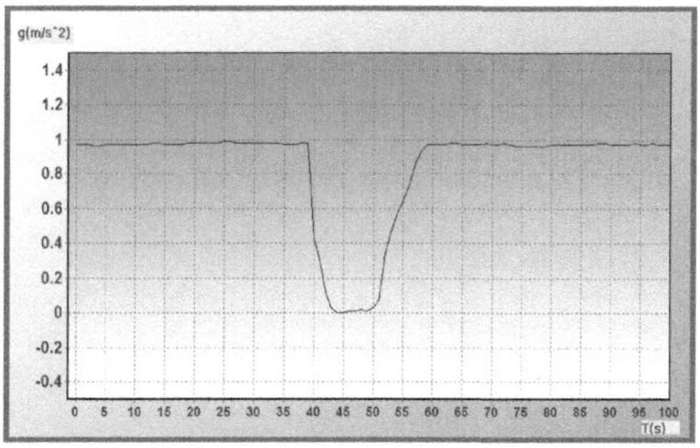

Fig. 11. The well is totally blocked

6 Conclusion

An intelligent logging unblocking device is designed according to the common phenomenons and shortages of logging equipments happened in current logging. This device integrated with mechanical technology, control technology and detection technology. When came across situations like sidewall tilt, narrow or jam in the process of logging, the intelligent logging unblocking device can use its front drill dredging the way of well automatically. It can avoid secondary drilling and repeating logging effectively and lie a foundation of logging automatically. This device has a deep sense of engineering practice and a practical engineering application value.

Acknowledgements. This research was supported by Scientific Research Program Funded by Shaanxi Provincial Education Department (2010JC11) and Science Research Program Funded by Beilin District Xi'an (GX1112).

References

1. Liu, X.: Logging principle and the engineering application. Petroleum industry press, Beijing (2006)
2. Hou, W., Dai, J., Zhen, X.: Forearm movement attitude detection based on the acceleration sensor. The Sensor and the Micro System 28(1) (2009)
3. Li, X.: Wireless sensor network technology, vol. 9. Beijing institute of technology press, Beijing (2007)
4. Tian, X., Lu, Q., Xiong, C.: Design for inclinometer based on the acceleration sensor. The Sensor Technology Journal 19(2) (2006)
5. Wang, C.: Design for intelligent surveying slant sensor. Southwest jiaotong university, Cheng du, 4-6, 8–9 (2007)
6. Ma, B., Najafi, M.: Development and applications of trenchless technology in China. Tunnelling and Underground Space Technology 3(8) (2007)

Generation, Testing and Calibration of Periodic Signal Distortion

Huixin Zhang[1], Sai Qi[2], and Yong Zhou[1]

[1] Electrical Engineering Department, Harbin Institute of Technology, 150000 Harbin, China
[2] Heilongjiang Electric Power Staff University, 150000 Harbin, China
qichao@hit.edu.cn, qisaiqisai@163.com, zhouyonghit@163.com

Abstract. The module of uniform transmission line is established. Using complex frequency analysis, the generation of signal distortion is indicated by the attenuation coefficient and phase coefficient. According to the different effect of each component on the total signal, the level of signal distortion is judged and the filter situation is decided, by the introduction of Fourier Transformation and the recording of the signal input at the terminal of the transmission line. In the case of parallel resonance, the resonance point is determined when the signal, after one component input superimposes on it, matches the signal input. Both correction methods in sine and non-sine transmission are demonstrated aiming at the burr, by a standard sine wave recorded at the terminal of the transmission line.

Keywords: Signal Distortion, Fourier transform, Burr, Calibration.

1 Introduction

Since lumped parameter does not match the uniform transmission lines, a micro-segment dx is set, which has x long apart from the starting point. The part of dx must be small enough, and then the segment dx can be set as the lumped parameter circuit system approximately. Assume the resistor and the inductor of the micro-segment (for both sides) are R_0dx and L_0dx respectively, also the conductance and the capacitor between two lines are C_0dx and C_0dx. Among them, R_0 (the SI unit is Ω/m), L_0(the SI unit is H/m)、G_0 (the SI unit is S/m) and C_0 (the SI unit is F/m) are unit length parameters of even distribution along the line. The circuit model of microsegment uniform line is showed as below.

G. Lee (Ed.): Advances in Automation and Robotics, Vol. 1, LNEE 122, pp. 489–496.
springerlink.com

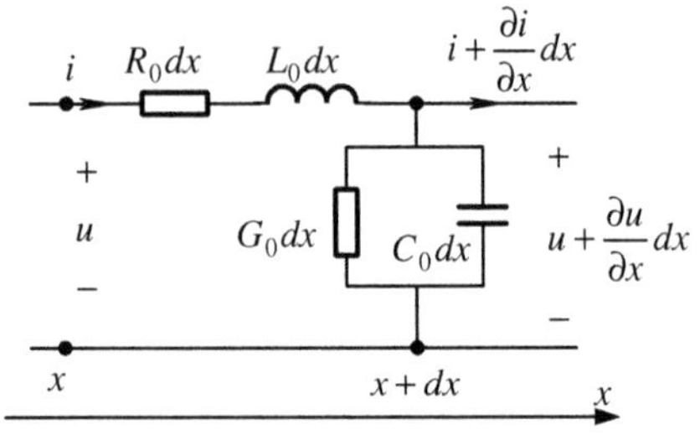

Fig. 1. The Circuit Model of Micro-segment Uniform Line

2 The Production of Distortion

From Figure.1., according to the Kirchhoff's current law (KCL) and Kirchhoff's voltage law (KVL), we can get

$$u-(u+\frac{\partial u}{\partial x}\mathrm{d}x)=(R_0\mathrm{d}x)i+(L_0\mathrm{d}x)\frac{\partial i}{\partial t}$$

$$i-(i+\frac{\partial i}{\partial x}\mathrm{d}x)=G_0\mathrm{d}x(u+\frac{\partial u}{\partial x}\mathrm{d}x)+C_0\mathrm{d}x\frac{\partial}{\partial t}(u+\frac{\partial}{\partial x}\mathrm{d}x)$$

Rearrange and omit the opponent of the second order differential

$$-\frac{\partial u}{\partial x}=R_0 i+L_0\frac{\partial i}{\partial t} \tag{1a}$$

$$-\frac{\partial i}{\partial x}=G_0 u+C_0\frac{\partial u}{\partial t} \tag{1b}$$

The equation of the uniform line above is the partial differential equation for the distance x and the time t, including u and i. Set x as the parametric variable, and turn the equation into the phasor formula in frequency domain.

$$-\frac{\mathrm{d}\dot{U}(x)}{\mathrm{d}x}=(R_0+\mathrm{j}\omega L_0)\dot{I}(x) \tag{2a}$$

$$-\frac{\mathrm{d}\dot{I}(x)}{\mathrm{d}x}=(G_0+\mathrm{j}\omega C_0)\dot{U}(x) \tag{2b}$$

Solve, get

$$\dot{U}(x)=\dot{U}'\mathrm{e}^{-\gamma x}+\dot{U}''\mathrm{e}^{\gamma x} \tag{3}$$

where attenuation coefficient

$$\gamma=\sqrt{(R_0+\mathrm{j}\omega L_0)(G_0+\mathrm{j}\omega C_0)}=\alpha+\mathrm{j}\beta \tag{4}$$

Since that, (3) can be presented as

$$\dot{U}(x)=\dot{U}'\mathrm{e}^{-\alpha x}\mathrm{e}^{\mathrm{j}(-\beta x)}+\dot{U}''\mathrm{e}^{\alpha x}\mathrm{e}^{\mathrm{j}\beta x}$$

α , defined as attenuation coefficient, is related with the wave amplitude; β , defined as phase coefficient, is related with wave phase.

$$\alpha=\sqrt{\frac{1}{2}\left(|Z_0|\cdot|Y_0|+R_0G_0-\omega^2L_0C_0\right)}$$

$$\beta=\sqrt{\frac{1}{2}\left(|Z_0|\cdot|Y_0|-R_0G_0+\omega^2L_0C_0\right)} \tag{5}$$

where $|Z_0|=\sqrt{R_0^{\ 2}+(\omega L_0)^2}$, $|Y_0|=\sqrt{G_0^{\ 2}+(\omega C_0)^2}$.

From the analyzation above, it is clear that attenuation coefficient and phase coefficient are the complex functions of frequency. When non-sinusoidal signal is transmitting, since the attenuation coefficient is changed along with the frequency, the distortion of amplitude engenders. And since the phase speed of each harmonic is changed along with the frequency, the phase shift of each harmonic is not proportional to frequency, which changes the relative position of each harmonic at the axis of time. That's how the distortion of phase engenders.

3 Testing of Distortionless

Signal without distortion refers to the output and input signal compared to the same waveform, but the signal strength and the emergence of time can be different.

3.1 Testing on the Level of Distortion and the Judge of Band-Pass

After the Fourier transform of the periodic signal, we set that the fundamental frequency is ω , the k-th harmonic frequency is $k\omega$, the amplitude of the input singnal of the k-th harmonic is A_k and the phase is θ_k , the amplitude of output signal of the k-th harmonic is B_k and the phase is φ_k . According to the actual need, the first n components are taken as the test object for both signals input and output, the signal input remarked as a_k ($a_k = A_k\cos(k\omega+\theta_k)$), the signal output remarked as b_k ($b_k = B_k\cos(k\omega+\phi_k)$). Assume the signal input and output are presented in "Periodic Signal Input" and "Periodic Signal Output" in Fig.2. respectively. And the amplitude frequency characteristic analyses of them are presented in "Spectrum of Periodic Signal Input" and "Spectrum of Periodic Signal Output" in Fig.2. respectively.

Because as k increases the amplitude A_k , B_k decreases and the k-th component reduces its impact on the overall waveform, so it can be set $\sigma_k = s_k * |A_k - B_k|$, where s_k is the weight factor , which gets smaller as k increases according to the actual need. The standard to judge the level of distortion is $\sum \sigma_k \cdot |A_k - B_k|$ is presented in "The Absolute Value of Deviation along with the Frequency" in Fig.2.

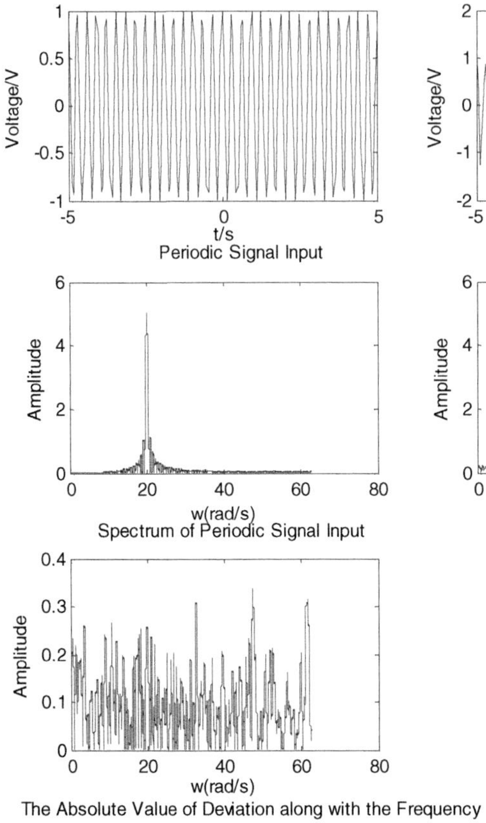

Fig. 2. Frequency Analysis Chart

From "The Absolute Value of Deviation along with the Frequency" in Fig.2, it is clear to see the level of distortion along with the frequency. Meanwhile, if the deviation of one rang is zero, we can judge the frequency corresponded to the rang is the frequency of band-pass. And if the deviation of one rang is large, we can judge the frequency corresponded to the rang is the frequency of band-stop. Also the low-pass and the high-pass are easy to distinguish. Thus, find out the generation of the distortion. But as to the parallel resonant, since the resonant frequency is just one point, instead of one range, it is hard to figure out intuitively. So the method following is needed.

3.2 The Test of Resonance Phenomenon

For parallel resonance, the periodic signal output after FFT will lost certain resonant frequency components, namely the amplitude corresponding to the frequency is zero.

The first n components of the signal input after FFT are taken as the test object, according to actual need. Then they are stored in an array with n components, which are marked as $a_k = A_k \cos(k\omega + \theta_k)$. The first n component respectively overlaps with the signal output to get the new curve. If the curve coincides with the original signal input, then the component a_k is the lost one because of the resonance , while the frequency $k\omega$ is the corresponding resonant frequency.

For series resonance, the amplitude corresponding to the resonance point will be very high, it is easy to judge directly from the map.

4 Calibration Method of Distortion

According to the definition of distortionless signal, which allows the change of the signal strength or the delay of signal time, the calibration method of amplitude distortion of the periodic signal has been designed. Under the ideal condition in this calibration system, "burr phenomena" will be avoided completely.

In the Fourier transform, minimum unit is the sinusoidal component, any periodic signal can be seen as the result of the superposition of n components. For clearer discussion, the situation can be divided into sinusoidal communication system and non-sinusoidal communication system.

4.1 Sinusoidal Communication System

For two distortionless sines in-phase, dividing one cyclic time into n parts, two values of every corresponding point from two sines respectively are in the same multiple relationship. Since that, we only need to store $A_i\left(i = 1,...,n\right)$ - values of n points from one standard sine wave, as presented in "Standard Sine Wave" in Fig.3. At the terminal of the line, what only need to be checked is the amplitude of fundamental wave- B_0 and the frequency ω . The signal output is presented in "Distorted Signal" in Fig.3. Finally, send out one point every $2\pi / (n\times\omega)$ second, which value is $B_i\left(B_i = B_0 * A_i\right)$, as is presented in "Corrected Signal Output" in Fig.3. The reason why we should use B_0 is because that burr, the frequency of which is much higher than the fundamental wave, has little effect on the fundamental wave due to its small amplitude.

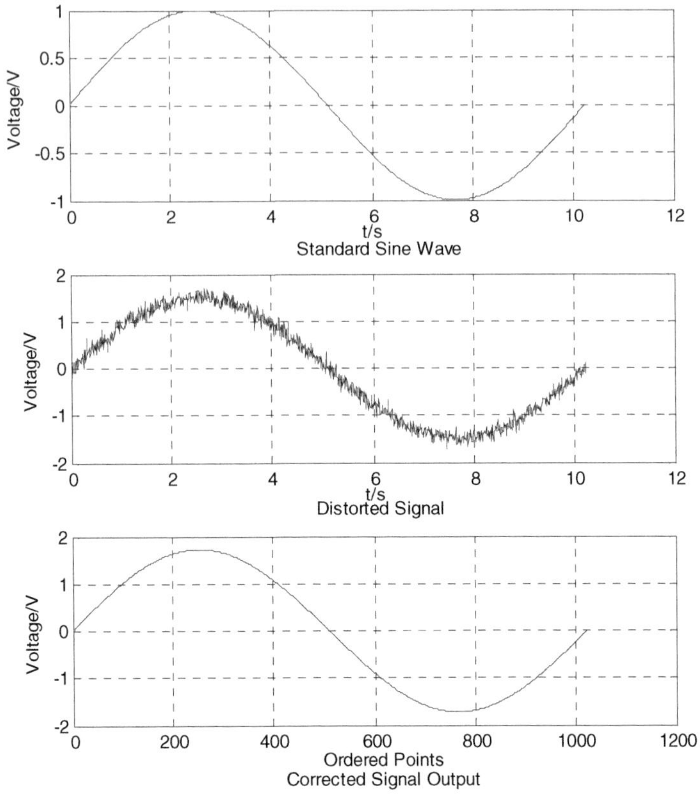

Fig. 3. Sinusoidal Communication System

4.2 Nonsinusoidal Communication System

According to the theory of Fourier transform, each periodic signal can be decomposed to innumerable sinusoidal components. The periodic signal output is analyzed according to the frequency, using Fourier Transfer. First set the standards, deciding whether the component should be filtered out. The standards are different according to the frequency, since the higher harmonic component has relatively less effect on the total waveform. Next, n sinusoidal components are sent out according to the calibration method mentioned in 4.1 Periodic Sinusoidal Communication System. Finally, we get the corrected signal while n components stacked.

But in this method, we may lose some components away with the burr, especially higher harmonics. In addition, the degree that the burr is filter out is concerned with the precision deciding the range of frequency. One reasonable way to solve this problem is to find out the range of frequency where the biggest amplitudes are presented by FFT. And then these frequencies which are the integral multiple ranges of above are preserved. Except them, the frequencies whose corresponding amplitudes are more than 1% (1% is assumed, it should be decided by the actual situation) of the biggest value are rejected. For example, if the range of biggest amplitudes lie in (50.0, 51.0), then the integral multiple ranges are (100.0, 102.0), (150.0, 153.0), (200.0, 204.0)and so on.

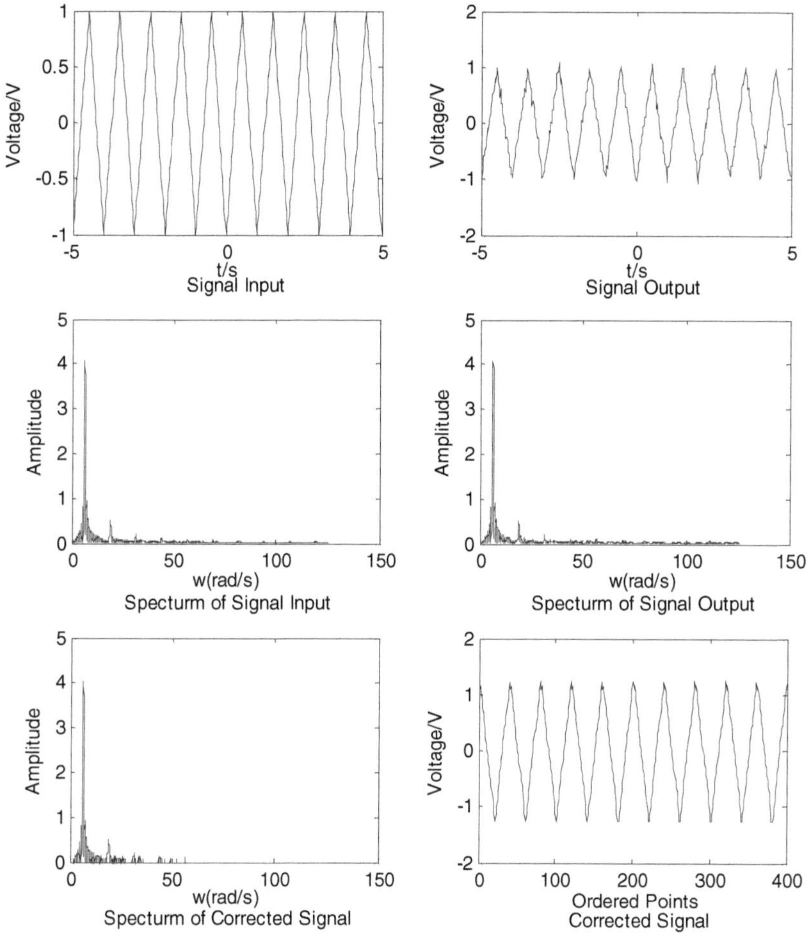

Fig. 4. Nonsinusoidal Communication System

The calibration method above is aimed at the distortion of amplitude, unrelated to the distortion of phase. But from the calibration-process of the distortion of amplitude, we can get the attenuation coefficient of each harmonic, then we can calculate $Z_0, Y_0, R_0, G_0, L_0, C_0$.

By inserting these dates above in formula (5), we can get the phase coefficient of each harmonic. As a result, the corresponding phase corrected is obtained.

5 Concluding Remarks

In the paper, the verification and the calibration method is put forward for the transmission without aberration. However, these methods remain to be perfected and simplified, since the premise of the correction method is that burr is in small value and with high frequency, while the signal input is relatively in low frequency. Moreover, we have to give up some higher harmonics. In addition, aiming at the phase coefficient, we need to do further experimental test.

References

1. Chen, X.: Fundamentals of Electrical Circuit Theory, 3rd edn. Higher Education Press (2004)
2. Oppenheim: Signal and System, 2 edn. Electronic Industry Press (2009)

Properties of the Fifth Type of Covering-Based Rough Sets

Jiaqing Zhou[1,2] and Hongmei Nie[1]

[1] Math, Physics and Information Engineering College of Zhejiang Normal University, Jinhua, China, 321004
[2] Department of Information, Southwest Jiao Tong University, Cheng Du, China, 610031
jhzjq@zjnu.cn, nhm@zjnu.cn

Abstract. Rough set theory has been proposed by Pawlak as a useful tool for dealing with the vagueness and granularity in information systems. Classical rough set theory is based on equivalence relation. The covering rough sets is an improvement of Pawlak rough set to deal with complex practical problems which the latter one can not handle. This paper studies covering-based generalized rough sets. In this setting, a covering can also generate a lower approximation operation and an upper approximation operation, but some of common properties of classical lower and upper approximation operations are no longer satisfied. We investigate conditions for a covering under which these properties hold for the covering-based lower and upper approximation operations.

Keywords: Rough set, Granular computing, Covering, Upper approximation, Lower approximation.

1 Introduction

Rough set theory has been proposed by Pawlak [1] in 1982. Classical rough set theory is based on equivalence relation. However, in modern various fields of science and technology, the complex relations are extended to tolerance relation [3], general binary relation [4], similarity [4] and coverings [4, 5, 10, 12, 13]. For investigating them, some researchers even proposed fuzzy set theory [8], rough set theory [1, 2], computing with words [6, 9], granular computing[7].

Papers [15-18] have proposed four type of covering generalized rough set. These authors started from the essential properties of the classical lower and upper approximation operations, for example, normality, extension, and idempotency, to explore the relations satisfy the specific properties of classical rough sets. In paper[19], the author proposed the fifth type of covering generalized rough set. In this paper, we address similar issues in the fifth type of covering-based rough sets.

The other parts of this paper are organized as follows: In Section 2, we present the fundamental concepts and properties of the Pawlak's rough set theory. Section 3 gives basic concepts and properties of the fifth type of covering-based rough sets. Section 4 is the major contribution of this paper. For every property of classical rough sets listed in Section 2, we investigate the essential properties for a covering such that the lower and

G. Lee (Ed.): Advances in Automation and Robotics, Vol. 1, LNEE 122, pp. 497–503.
springerlink.com

upper approximation operations generated by such a covering satisfy this specific property of classical rough sets. This paper concludes in section 5.

2 Fundamentals of the Pawlak's Rough Sets

Let U be a finite set, the domain of discourse, and R an equivalence relation on U . R is generally called an indiscernability relation in rough set theory [2]. R will generate a partition $U / R = \{Y_1, Y_2, ..., Y_m\}$ on U where $Y_1, Y_2, ..., Y_m$ are the equivalence classes generated by the equivalence relation R , and, in the rough set theory, they are also called elementary sets of R . For any $X \subseteq U$ we can describe X by the elementary sets of R and the two sets $R_*(X) = \bigcup\{Y_i \in U / R \mid Y_i \subseteq X\}$ $R^*(X) = \bigcup\{Y_i \in U / R \mid Y_i \cap X \neq \emptyset\}$ are called the lower and the upper approximation of X, respectively.

Let $\emptyset$ be the empty set, $-X$ the complement of X in U , from the definition of approximation sets, we have the following conclusions about them.

Proposition 1 *(The properties of the Pawlak's rough sets)*

(1L) $R_*(U) = U$ *(Co-normality)*

(1H) $R^*(U) = U$ *(Co-normality)*

(2L) $R_*(\emptyset) = \emptyset$ *(Normality)*

(2H) $R^*(\emptyset) = \emptyset$ *(Normality)*

(3L) $R_*(X) \subseteq X$ *(Contraction)*

(3H) $X \subseteq R^*(X)$ *(Extension)*

(4L) $R_*(X \cap Y) = R_*(X) \cap R_*(Y)$ *(Multiplication)*

(4H) $R^*(X \cup Y) = R^*(X) \cup R^*(Y)$ *(Addition)*

(5L) $R_*(R_*(X)) = R_*(X)$ *(Idempotency)*

(5H) $R^*(R^*(X)) = R^*(X)$ *(Idempotency)*

(6L) $X \subseteq Y \Rightarrow R_*(X) \subseteq R_*(Y)$ *(Monotone)*

(6H) $X \subseteq Y \Rightarrow R^*(X) \subseteq R^*(Y)$ *(Monotone)*

(7L) $R_*(-R_*(X)) = -R_*(X)$ *(Lower complement)*

(7H) $R^*(-R^*(X)) = -R^*(X)$ *(Upper complement)*

(8LH) $R_*(-X) = -R^*(X)$ *(Duality)*

(9L) $\forall K \in U / R$, $R_*(K) = K$ *(Granularity)*

(9H) $\forall K \in U / R$, $R^*(K) = K$ *(Granularity)*

The (3L), (4L), and (7L) are characteristic properties for the lower approximation operations [21-23], i.e., all other properties of the lower approximation operation can be deduced from these three properties. Correspondingly, (3H), (4H), and (7H) are characteristic properties for the upper approximation operation.

3 Fifth Type of Covering Generalized Rough Sets

Definition 1*[19]. Let $<U,C>$ be a covering approximation space. For $\forall x \in U$, $\bigcap\{K \in C \mid x \in K\}$ is called the neighborhood of x and denoted as $Neighbor(x)$.*

Definition 2 *[19]. For a set $X \subseteq U$, the set $X_* = \bigcup\{K \in C \mid K \subseteq X\}$ is called the fifth type of covering lower approximation of X and the set $X^* = X_* \bigcup \{Neighbor(x) \mid x \in (X - X_*)\}$ is called the fifth type of covering upper approximation of X .*

Definition 3. *Let C be a covering of U , $P(U)$ the power set of U . The operations FIL_C and FIH_C : $P(U) \to P(U)$ are defined as follows: $\forall X \in P(U)$, $FIL_C(X) = X_*$, $FIH_C(X) = X^*$.*

We call them the fifth type of covering lower approximation operation and the fifth type of covering upper approximation operation, coupled with the covering C , respectively. When the covering is clear, we omit the lowercase C for the two operations.

Definition 4 *(Reducible element [11]). Let C be a covering of a domain U and $K \in C$. If K is a union of some sets in $C - \{K\}$, we say K is reducible in C , otherwise K is irreducible.*

Definition 5 *(Reduct [11]). For a covering C of a domain U , the new irreducible covering by deleting all reducible elements is called the reduct of C and denoted by $reduct(C)$.*

Proposition 2*[19].* Let *$<U,C>$ be a covering approximation space, $X \subseteq U$ $FIH(X) = \bigcup\{\bigcap_{K \in Md(x)} K \mid x \in X\}$.*

Proposition 3. *$X_* = X$ if and only if X is a union of some elements of C .*

Proposition 4. *$X^* = X$ if X is a union of some elements of C .*

The following example 1 shows that the contrary of proposition 4 do not hold.

Example 1. Let $U = \{a,b,c\}$, $K_1 = \{a,b\}$, $K_2 = \{a,c\}$, $K_3 = \{b,c\}$ .Then $\forall X \subseteq U$,we have $X^* = X$,let $X = \{a\}$, *X isn't a union of some elements of C .*

Corollary 1. *FIH satisfies $FIH(X) = X$ if FIL satisfies $FIL(X) = X$.*

Corresponding to the properties of the Pawlak's rough sets listed in Section 2, we have the following results:

Proposition 5. *The fifth type of covering lower and upper approximations have the following properties:*

$(1L)\ U_* = U$ *(Co-normality)*

$(1H)\ U^* = U$ *(Co-normality)*

$(2L)\ \varnothing_* = \varnothing$ *(Normality)*

$(2H)\ \varnothing^* = \varnothing$ *(Normality)*

$(3L)\ X^* \subseteq X$ *(Contraction)*

$(3H)\ X \subseteq X^*$ *(Extension)*

$(4H)\ (X \cup Y)^* = X^* \cup Y^*$ *(Addition)*

$(5L)\ (X_*)_* = X_*$ *(Idempotency)*

$(5H)\ (X^*)^* = X^*$ *(Idempotency)*

$(6L)\ X \subseteq Y \Rightarrow X_* = Y_*$ *(Monotone)*

$(6H)\ X \subseteq Y \Rightarrow X^* = Y^*$ *(Monotone)*

$(9L)\ \forall K \in U/R,\ R_*(K) = K$ *(Granularity)*

$(9H)\ \forall K \in U/R,\ R^*(K) = K$ *(Granularity)*

From the example 2-4,we have

Proposition 6. *The following six properties of the Pawlak's lower and upper approximations do not hold for the fifth type of covering lower and upper approximations.*

$(4L)\ (X \cap Y)_* = X_* \cap Y_*$ *(Multiplication)*

$(7L)\ (-(X_*))_* = -X_*$ *(Lower Complement)*

$(7H)\ (-(X^*))^* = -X^*$ *(Upper Complement)*

$(8LH)\ R_*(-X) = -R^*(X)$ *(Duality)*

Example 2*(Multiplication)*. Let $U = \{a,b,c,d\}$, $K_1 = \{a,b\}$, $K_2 = \{a,b,c\}$, $K_3 = \{c,d\}$. $C = \{K_1, K_2, K_3\}$ .Clearly, C is a covering of U .Let $X = \{a,b,c\}$ and $Y = \{c,d\}$ .Then we have $X \cap Y = \{c\}$, $(X \cap Y)_* = \varnothing$,but $X_* \cap Y_* = \{c\} \neq \varnothing$,so $(X \cap Y)_* \neq X_* \cap Y_*$.

Example 3*(Lower Complement and Upper Complement)*. Let $U = \{a,b,c,d\}$, $K_1 = \{a,b\}$, $K_2 = \{a,b,c\}$, $K_3 = \{c,d\}$. $C = \{K_1, K_2, K_3\}$ .Clearly, C is a covering of U . Let $X = \{a,b,c\}$,we have, $(-X_*)_* = (-\{a,b,c\})_* = \{d\}_* = \varnothing$, $-X_* = \{d\}$, $(-X^*)^* = (-\{a,b,c\})^* = \{d\}^* = \{c,d\}$, $-X^* = \{a,b\}$ .So, $(-X_*)_* = -X_*$, $(-X^*)^* = -X^*$ do not hold.

Example 4 *(Duality)*. Let $U = \{a,b,c,d\}$, $K_1 = \{a,b\}$, $K_2 = \{a,b,c\}$, $K_3 = \{c,d\}$. $C = \{K_1, K_2, K_3\}$ .Clearly, C is a covering of U .Let $X = \{a,b,c\}$. Then we have $R_*(-X) = R_*(\{d\}) = \varnothing$, $R^*(X) = \{a,b,c\}$, $R_*(-X) = -R^*(X)$ do not hold.

4 Conditions under Which Covering-Based Rough Sets Satisfy Certain Classical Properties

As we can see from the previous section, some of properties of classical rough sets are no longer valid for covering-based rough sets. Now we consider the issue when they hold. Firstly, we investigate what conditions a covering should hold so that the lower approximation operation generated by such a covering will satisfy property (4L) in Proposition 1.

Theorem 1. *FIL satisfies (4L)* $FIL(X \cap Y) = FIL(X) \cap FIL(Y)$ *if and only if* C *satisfies the following properties:* $\forall K_1, K_2 \in C$, $K_1 \cap K_2$ *is a union of finite elements in* C.

Proof: $\Rightarrow$:

Since $K_1 \cap K_2 = FIL(K_1) \cap FIL(K_2) = FIL(K_1 \cap K_2)$,by Proposition 3, $K_1 \cap K_2$ is a union of finite elements in C .

$\Leftarrow$: By (6L) in Proposition 5, it is easy to see that $FIL(X \cap Y) \subseteq FIL(X) \cap FIL(Y)$.

On the other hand, by Proposition 2, let $FIL(X) = K_1 \cup ... \cup K_m$ and $FIL(Y) = K_1' \cup ... \cup K_n'$ where $K_i, K_j' \in C$, $1 \le i \le m$, $1 \le j \le n$.

For any $1 \le i \le m$ and $1 \le j \le n$, $K_i \cap K_j' \subseteq X \cap Y$ .By the assumption in this Theorem, $K_i \cap K_j'$ is a union of finite elements in C , let us say $K_i \cap K_j' = W_1 \cup ... \cup W_l$ where $W_h \in C$, $1 \le h \le l$, so $W_h \subseteq FIL(X \cap Y)$ for $1 \le h \le l$. Thus $K_i \cap K_j' \subseteq FIL(X \cap Y)$ for $1 \le i \le m$ and $1 \le j \le n$.

From $FIL(X) \cap FIL(Y) = \bigcup_{i=1}^{m} \bigcup_{j=1}^{n} (K_i \cap K_j')$,we prove that $FIL(X) \cap FIL(Y) \subseteq FIL(X \cap Y)$.

Therefore, $FIL(X \cap Y) = FIL(X) \cap FIL(Y)$.

Now we start to consider property Lower Complement Relation and Upper Complement Relation.

Theorem 2. *FIL satisfies (7L)* $FIL(-FIL(X)) = -FIL(X)$ *if and only if* $\forall K_1, ..., K_m \in C$, $-(K_1 \cup ... \cup K_m)$ *is a union of finite elements in* C .

Proof: $\Rightarrow$:

$$\forall K_1, ..., K_m \in C, \quad FIL(-(K_1 \cup ... \cup K_m)) = FIL(-FIL(K_1 \cup ... \cup K_m)) = -FIL(K_1 \cup ... \cup K_m) = (-K_1 \cup ... \cup K_m) .$$

So $-(K_1 \cup ... \cup K_m)$ is a union of finite elements in C .

$\Leftarrow$: By Proposition 3, $FIL(X)$ is a union of finite elements in C , say $FIL(X) = K_1 \cup ... \cup K_m$ where $K_1, ..., K_m \in C$. $FIL(-FIL(X)) = FIL(-(K_1 \cup ... \cup K_m))$. Again by Proposition 3 and the assumption of this Theorem, $FIL(-(K_1 \cup ... \cup K_m)) = -(K_1 \cup ... \cup K_m)$.

So $FIL(-FIL(X)) = -FIL(X)$.

Theorem 3. *FIH satisfies that* $\forall X \subseteq U$ *,(7H)* $FIH(-FIH(X)) = -FIH(X)$ *,if* $\forall x, y \in U$ *,* $y \notin Neighbor(x)$ *,and .* $K \in Md(y)$ *,then* $K \cap Neighbor(x) = \varnothing$ *.*

Proof: $\Leftarrow$:

$\forall y \in -FIH(X)$,since $FIH(X) = \bigcup_{x \in X} Neighbor(x)$, $y \notin Neighbor(x)$ for any $\forall x \in X$ .By the assumption, $Neighbor(y) \cap Neighbor(x) = \varnothing$ .Now we have $\bigcup_{y \in -FIH(X)} Neighbor(y) \cap FIH(X) = \varnothing$,i.e., $FIH(-FIH(X)) \subseteq -FIH(X)$ .Combining $-FIH(X) \subseteq FIH(-FIH(X))$,we conclude that $FIH(-FIH(X)) = -FIH(X)$.

The following example 5 shows that the contrary of theorem 3 do not hold.

Example 5. Let $U = \{a,b,c\}$, $K_1 = \{a,b\}$, $K_2 = \{a,c\}$, $K_3 = \{b,c\}$ .Then $a \subseteq U$, we have $b \notin Neighbor(a)$ *and* $K_1 \in Md(b)$,but $K_1 \cap Neighbor(a) = \{a\} \neq \varnothing$.

For property(8LH),we have the following conclusion.

Theorem 4. *FIL and FIH satisfies (8LH)* $FIH(-X) = -FIL(X)$ *if* $reduct(C)$ *is a partition.*

Proof: By definition 2.

5 Conclusions

Rough set is one of the useful tools for data mining. The covering rough sets model is an extension to the classical rough sets. It is more flexible in dealing with uncertainty and granularity information systems. In this paper we start from the properties of classical rough sets to find out the conditions of a covering under which covering-based rough sets produced by such a covering will satisfy these specific properties. There are research topics to apply the theory to data mining.

References

1. Pawlak, Z.: Rough sets. Internat. J. Comput. Inform. Sci. 11, 341–356 (1982)
2. Pawlak, Z.: Rough sets: Theoretical aspects of reasoning about dat. Kluwer Academic Publishers, Boston (1991)
3. Skowron, A., Stepaniuk, J.: Tolerance approximation spaces. Fundamenta Informaticae 27, 245–253 (1996)
4. Slowinski, R., Vanderpooten, D.: A generalized definition of rough approximations based on similarity. IEEE Trans. on Knowledge and Data Engineering 12(2), 331–336 (2000)
5. Tsang, E., Cheng, D., Lee, J., Yeung, D.: On the upper approximations of covering generalized rough sets. In: Proc. 3rd International Conf. Machine Learning and Cybernetics, pp. 4200–4203 (2004)
6. Wang, F.-Y.: Outline of a computational theory for linguistic dynamic systems: Toward computing with words. International Journal of Intelligent Control and Systems 2(2), 211–224 (1998)

7. Yao, J.T., Yao, Y.Y.: Induction of classification rules by granular computing. In: Rough Sets and Current Trends in Computing, pp. 331–338 (2002)
8. Zadeh, L.A.: Fuzzy sets. Information and Control 8, 338–353 (1965)
9. Zadeh, L.A.: The concept of a linguistic variable and its application to approximate reasoning – I. Information Sciences 8, 199–249 (1975)
10. Zhu, W.: Topological approaches to covering rough sets. Information Sciences 177(6), 1499–1508 (2007)
11. Zhu, W., Wang, F.-Y.: Reduction and axiomization of covering generalized rough sets. Information Sciences 152, 217–230 (2003)
12. Zhu, W., Wang, F.-Y.: A new type of covering rough sets. In: IEEE IS 2006, London, September 4-6, pp. 444–449 (2006)
13. Zhu, W., Wang, F.-Y.: Relationships among three types of covering rough sets. In: IEEE GrC, pp. 43–48 (2006)
14. Zhu, W., Wang, F.-Y.: Three types of covering rough sets. To appear in Transactions on Knowledge and Data Engineering 19, 1131–1144 (2007)
15. Zhu, W., Wang, F.-Y.: Properties of the First Type of Covering-Based Rough Sets. In: Workshop on Foundations of Data Mining and Novel Techniques in High Dimensional Structural and Unstructured Data, ICDM 2006, Hongkong, China, pp. 407–411 (2006)
16. Zhu, W.: Properties of the Second Type of Covering-Based Rough Sets. In: ICWI 2006 Workshop on GrC & BI, Hongkong, China, December 18-22, pp. 494–497 (2006)
17. Zhu, W., Wang, F.-Y.: William Zhu, Fei-Yue Wang:Properties of the Third Type of Covering-Based Rough Sets. In: ICMLC 2007, Hong Kong, China, August 19-22, pp. 3746–3751 (2007)
18. Zhu, W.: Properties of the Fourth Type of Covering-Based Rough Sets. In: HIS 2006, AUT Technology Park, Auckland,New Zealand, December 13-15, pp. 43–43 (2006)
19. Wu, M., Wu, X., Shen, T., Cao, C.: A New Type of Covering Approximation Operators. In: 2009 International Conf. Electronic Computer Technology, pp. 334–338 (2009)
20. Zhu, W., Wang, F.-Y.: On Three Types of Covering-Based Rough Sets. IEEE Transactions on Knowledge and Data Engineering 19(8), 1131–1144 (2007)
21. Lin, T.Y., Liu, Q.: Rough approximate operators: axiomatic rough set theory. In: Ziarko, W. (ed.) Rough Sets, Fuzzy Sets and Knowledge Discovery, pp. 256–260. Springer, Heidelberg (1994)
22. Zhu, F., He, H.-C.: The axiomization of the rough set. Chinese Journal of Computers 23(3), 330–333 (2000)
23. Zhu, F., He, H.-C.: Logical properties of rough sets. In: Proc. of The Fourth International Conference on High Performance Computing in the Asia-Pacific Region, pp. 670–671. IEEE Press (2000)

Research on Strain Measurement of Abyssal Riser with FBG Sensors

Baojun Li[1], Haiyan Wang[1], Xiaohong Shen[1], Dexing Yang[2], Shiquan Jiang[3], and Liangbing Xu[3]

[1] College of Marine Engineering, Northwestern Polytechnical University, Xi'an, 710072, China
[2] Shaanxi Key Laboratory of Optical Information Technology, School of Science, Northwestern Polytechnical University, Xi'an 710072, China
[3] Research Center of China National Offshore Oil Corporation, Beijing, 100027, China

Abstract. Marine riser fatigue damage in the hostile ocean environments is the fatal issue for deep and ultra deep waters petroleum and gas drilling and productions during. Strain detecting provides a direct measure of the riser fatigue. One riser fatigue monitoring formula based on FBG sensor with orthogonal distribution was researched which takes into account temperature compensation function. Mathematical model for calculating stress and bending moment of riser with the relative change in the Bragg wavelength was presented. One riser stress-strain measuring device based on FBG sensor was designed. Strain measurement tests of a riser scale model were performed. The study shows that the riser fatigue monitoring formula based on FBG sensor presented in this paper is reasonable, and the research results provide the technical support for field practice.

Keywords: Marine riser, FBG sensor, fatigue, monitoring, strain.

1 Introduction

It well known that the fatigue damage of marine riser in the hostile ocean environments is the fatal issue for petroleum and gas drilling and productions during deep and ultra deep waters [1, 2, 3]. With respect to fatigue damage, riser curvature is the most important parameter as it is directly related to wall stresses. It is significant to monitor and detect the bending moment and stress that characterize the riser curvature. Different from the atmospheric environment, it presents severe challenges to monitor and detect riser fatigue parameters in the hostile abyssal environments.

FBG (fiber Bragg grating) sensors have many advantages for strain sensing in such applications including their small size, the potential to multiplex hundreds of sensors with a single ingress/egress fiber, their immunity to electromagnetic interference, and their corrosion resistance[4,5,6]. Recently, FBG has become the most challenging strain sensor and is progressively being utilized for structure fatigue monitoring, such as bridges, tunnel aspects and others.

G. Lee (Ed.): Advances in Automation and Robotics, Vol. 1, LNEE 122, pp. 505–512.
springerlink.com

The fatigue monitoring of marine riser said above is far more complex than challenging than the common civil structure, for the harsh abyssal ocean environments and complex riser structural curvature and oscillation response. As a result, it provides one arena for FBG to display its talents in strain measurement subjected to a complex strain field.

In this paper, the primary strain measuring principle of FBG sensor was introduced and a riser fatigue monitoring formula based on FBG sensor with orthogonal distribution was designed which takes into account temperature compensation function. Mathematical model for calculating stress and bending moment of riser with the relative change in the Bragg wavelength was presented. Strain measurement tests of a riser scale model were performed. One purpose of the experiment was to verify the practicability and rationality of riser fatigue monitoring method presented in this paper.

2 Fundamental Principle of FBG Sensing

The Bragg wavelength which reflects from the grating is a function of effective refractive index and its period. Bragg wavelength is changed by temperature and strain variation. During strain measurement, a separated, strain-free FBG as the temperature sensor to measure the temperature of the strain field directly. The reference FBG is located in the same thermal environment as the strain sensor. The strain error caused by the temperature variation can be compensated to the wavelength shift induced by temperature variation from the total wavelength shift obtained with the strain sensor. The structural strain is expressed as [7, 8].

$$\varepsilon = [\Delta\lambda_B - K_T\Delta T] / K\tau\lambda_B \tag{1}$$

where, $K\tau$ is coefficient of sensitivity of fiber Bragg grating strain sensors, K_T is temperature sensitivity coefficient, $\Delta\lambda_B$ is wavelength shift, ΔT is temperature change.

3 Riser Strain Detection Method

There is no proper formula to character directly riser structural response subject to particularly harsh loading conditions in the abyssal ocean environments at present. However, riser structural curvature can be illustrated by Fig.1 [9].

One riser stress-strain measuring device based on FBG sensor was designed which possess temperature compensation function corresponding with riser structural curvature conditions under abyssal ocean environments. FBG strain sensors were arranged along the riser direction of sensitive axis. The distribution of FBG strain sensors is orthogonal shown in Fig.2. Demodulation devices are occupied in wavelength scanning. After sampling, signal processing and data analysis on reflective spectrum, the respective reflected wavelength can be gained and further the Bragg wavelength shift of each FBG sensor is figured out.

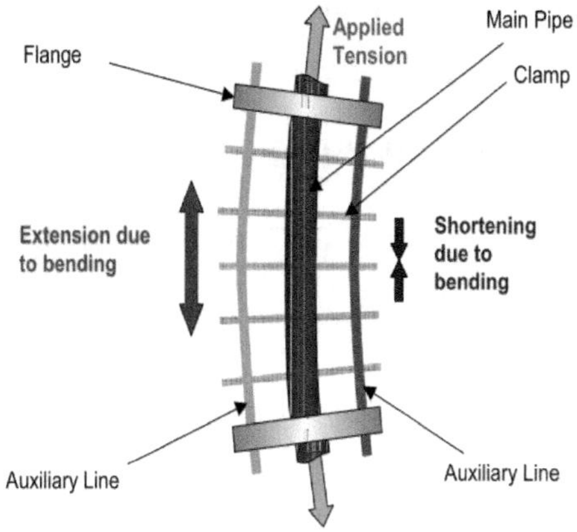

Fig. 1. Schematic of riser elastic curvature

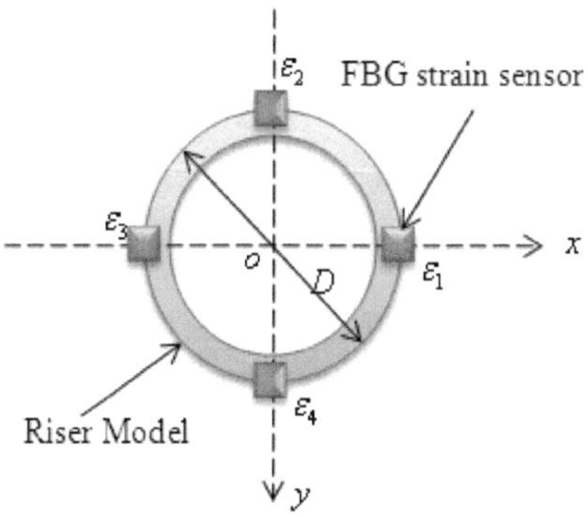

Fig. 2. Principle of FBG sensor arrangement

The theoretical calculation method of the riser maximum stress and bending moment according to the Bragg wavelength shift is deduced. The stress and bending moment of elastic beam is expressed as follows:

$$\delta = E\varepsilon \tag{2}$$

$$M = 2EI\varepsilon/D \tag{3}$$

Where, E is modulus of elasticity, I is inertia moment of section, D is external diameter of riser.

Assuming the riser with pure bending, the strain extent is $\varepsilon_1$equal to ε_3, and ε_2 is equal to ε_4. In order to reduce measurement error, we get:

$$\varepsilon_x = (\varepsilon_1 + \varepsilon_3)/2, \varepsilon_y = (\varepsilon_2 + \varepsilon_4)/2 \tag{4}$$

And then the bending moment and stress of the riser curvature strain are given respectively:

$$M(z,t) = EI(D^2 - d^2)(\Delta\lambda_x^2 + \Delta\lambda_x^2)^{0.5}/64DK_\tau\lambda_B \tag{5}$$

$$\delta(z,t) = E(\Delta\lambda_x^2 + \Delta\lambda_x^2)^{0.5}/2K_\tau\lambda_B \tag{6}$$

Where, d is riser's inner diameter, $\Delta\lambda_x=\Delta\lambda_{B1}(z, t)+\Delta\lambda_{B3}(z, t)-2K_T\Delta T$, $\Delta\lambda_y=\Delta\lambda_{B2}(z, t)+\Delta\lambda_{B4}(z, t)-2K_T\Delta T$.

4 Experiments on Dynamic Strain Measurement

In order to validate the performance of FBG strain monitoring equipment, a dozen of standard metal foil resistances train gauge, whose measured result is as comparison value, is adhered to riser scale model. A set of test is figure out in Fig.3 which is field experiment photograph.

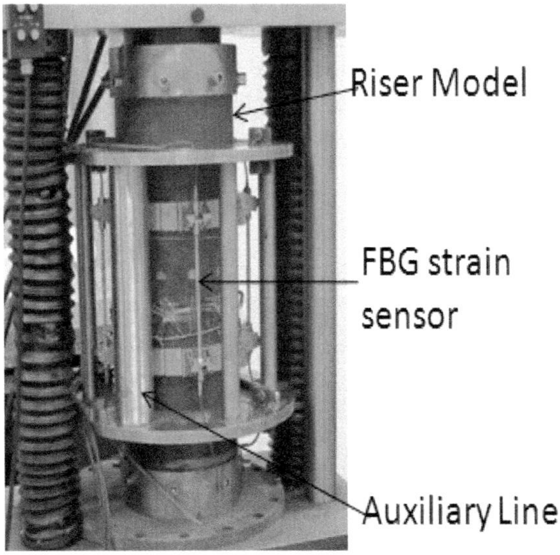

Fig. 3. Test equipment schematic

In this section, one bend test will be performed. Various bending loads are put on the riser scale model. The measured results of the riser model strain and the Bragg wavelength shifts are figured out in Fig.4. The line indicates strain of riser model measured by standard metal foil resistances train gauge. The dots indicate the reflected wavelength of FBG strain sensor. The reflected wavelength changes are quite accord with the strain measured by standard metal foil resistances train gauge according to the Fig.8. The bending moment and stress are calculated respectively. They are objective illustrated in the Figure 5. One conclusion can be observed that the FBG measured results approximate totally that from standard resistances train gauge.

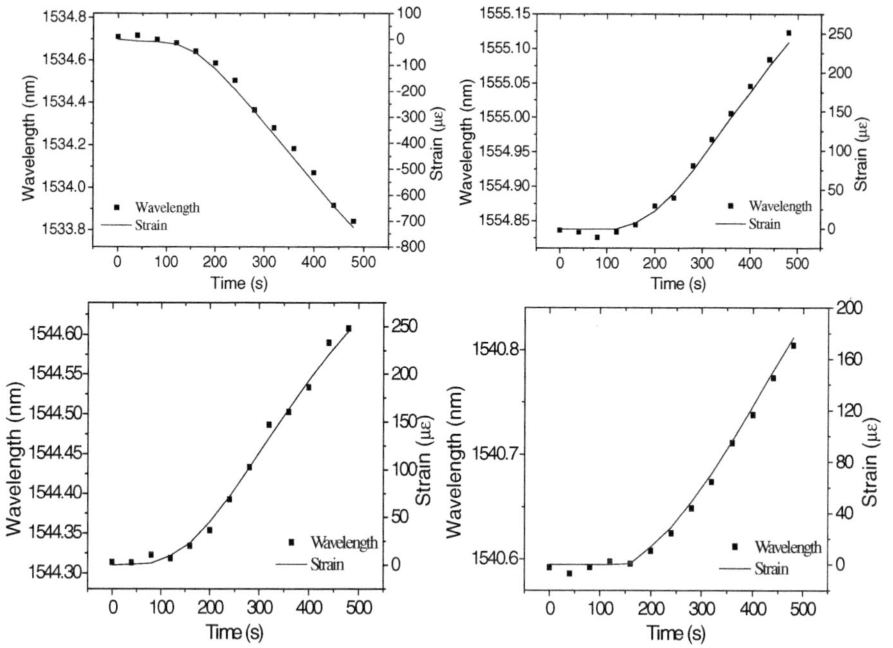

Fig. 4. Graph of test results of the riser model strain

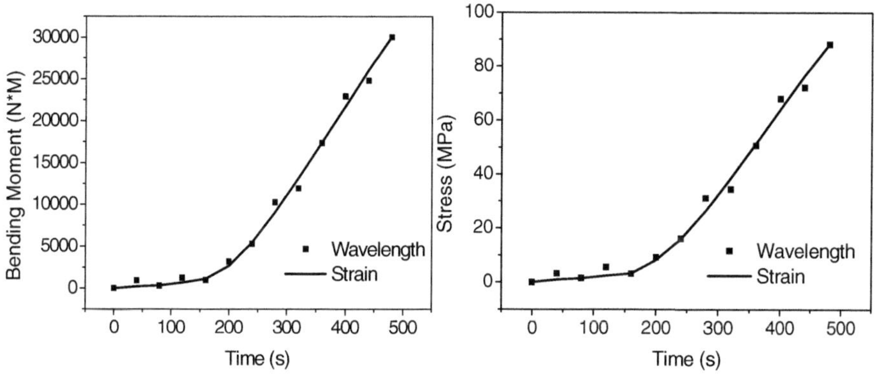

Fig. 5. Graphic measured values of bending moment and stress

A good deal of field experiment has been carried out. The measured value of bending moment and stress, their absolute measurement error and relative error were listed in Table 1 and Table 2 separately.

Table 1. The result and error of bending moment on bend test

Case	Moment /N*m	Abs Error/ N*m	Rel Error /%
1	30123.34	-1021.93	3.39
2	29689.17	1083.39	3.65
3	28392.36	1001.31	3.53
4	27934.60	849.93	3.04
5	31510.93	-740.23	2.35
6	34976.89	-1606.52	4.59
7	35445.67	1008.66	2.85
8	30021.86	1229.08	4.09
9	29227.07	1045.94	3.49

Table 2. The result and error of stress during bend test

Case	Stress /MPa	Abs Error/MPa	Relative Error /%
1	87.12	-3.24	3.72
2	85.71	3.74	4.36
3	92.43	3.33	3.60
4	90.74	-3.48	3.84
5	100.09	2.39	2.39
6	105.54	-5.90	5.59
7	105.26	3.51	3.33
8	88.19	4.93	5.59
9	85.22	4.51	5.29

One elongation test is carried out in this section. The calculated results of bending moment and stress according to elongation test data are shown in Fig.6. The dots indicate the bending moment and stress values calculated by the strain measurement of FBG strain sensor, with the solid curves representing the actual results obtained through the strain value measured by standard metal foil resistances train gauge. Clearly, there is a very good agreement between the FBG measured results and the approximate standard metal foil resistances train gauge solutions.

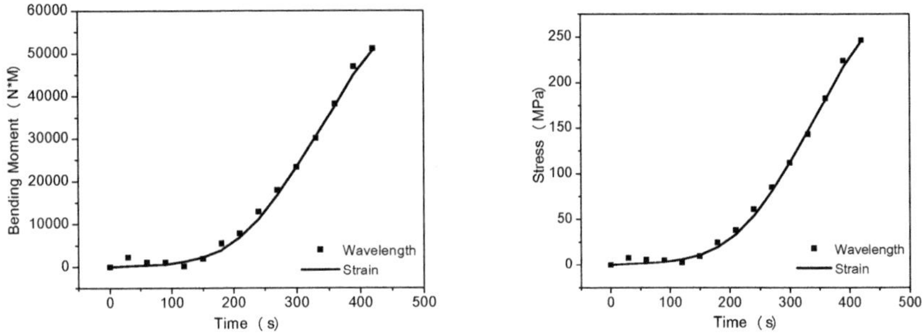

Fig. 6. Graphic measured values of bending moment and stress

The statistical results of bending moment and stress and their error according to several elongation tests are summarized in Table 3 and Table 4 respectively.

Table 3. Results and error of bending moment during elongation test

Case	Moment/N*m	Abs Error/ N*m	Relative Error /%
1	40985.42	2005.39	4.89
2	46665.10	1216.08	2.61
3	46215.06	-1163.50	2.52
4	51138.30	1969.20	3.85
5	39232.20	-1237.30	3.15
6	49837.80	2768.76	5.56

Table 4. Results and error of stress during elongation test

Case	Stress/MPa	Abs Error/ MPa	Relative Error/%
1	98.95	-2.94	2.97
2	202.36	10.01	4.95
3	228.57	5.94	2.60
4	228.06	-5.07	2.22
5	246.26	7.03	2.85
6	182.30	-5.17	2.84

The relative error of FBG strain detection result for bending moment and stress is less than 6% depend on statistical experiment result, which meet the error detection requirements. In addition, the resistance to pressure of FBG strain sensor has been tested. It can endure the pressure more than 30MPa during underwater environment. Its excellent watertight performance and resistance to pressure has been verified, which satisfy deep-water monitoring conditions.

5 Conclusions

This paper offers an insight into measuring deepwater riser bending moment and stress.

Firstly, the primary strain measuring principle of FBG sensor was introduced and a riser fatigue monitoring formula based on FBG sensor with orthogonal distribution was designed which takes into account temperature compensation function. Furthermore, Mathematical model for calculating stress and bending moment of riser with the relative change in the Bragg wavelength was presented. At last, one strain measurement tests of a riser scale model were performed. the relative error of FBG strain detection result for bending moment and stress is less than 6% depend on statistical experiment result, which meet the error detection requirements.

This research results provide the technical support for field practice. Of course, it needs to further optimize the monitoring system to meet the field riser fatigue measurement.

References

1. Iranpour, M., Taheri, F., Kim Vandiver, J.: Structural life assessment of oil and gas risers under vortex-induced vibration. Marine Structures 21, 353–373 (2008)
2. Mukundan, H., Modarres-Sadeghi, Y., Dahl, J.M., Hover, F.S., Triantafyllou, M.S.: Monitoring VIV fatigue damage on marine risers. Journal of Fluids and Structures 25, 617–628 (2009)
3. Chezhian, M., Mørk, K., Meling, T.S., Makrygiannis, C., Lepinasse, P.: NDP Review of State-of-the-Art in Riser Monitoring: Lessons Learned and Experiences Gained. In: Offshore Technology Conference, Houston, Texas, USA, OTC 17810 (2006)
4. Ma, G.-M., Li, C.-R., Jiang, J., Quan, J.-T., Meng, C.-P., Cheng, Y.-C.: A Novel Fiber Bragg Grating Tension Sensor for Ground Conductor. International Conference on e-Education, e-Business, e-Management and e-Learning, 240–243 (2010)
5. Fu, H.Y., Liu, H.L., Chung, W.H., Tam, H.Y.: A Novel Fiber Bragg Grating Sensor Configuration for Long-Distance Quasi-Distributed Measurement. IEEE Sensors Journal 8(9), 1598–1602 (2008)
6. Jia, Z., Ren, L., Li, D., Li, H.: Fiber Bragg grating strain sensor for measurement of vortex-induced vibration. In: Proceedings of the 12th International Conference on Engineering, Science, Construction, and Operations in Challenging Environments - Earth and Space, pp. 1601–1608 (2010)
7. Du, Y., Li, J., Liu, C.: A Novel Fiber Bragg Grating Temperature Compensated Strain Sensor. In: First International Conference on Intelligent Networks and Intelligent Systems, wuhan, china, pp. 569–572 (2008)
8. Kreuzer, M.: Strain measurement with Fiber Bragg Grating Sensors, `http://micronoptics.com/uploads/library/documents/FBGS_StrainMeasurement_mo.pdf`
9. Hariharan, M., Thethi, R.: Drilling Riser Management in Deepwater Environments. In: The 7th International Oil & Gas Conference and Exhibition, New Delhi, India, January 15-19 (2007)

Comparison Evidential Fusion Network with Decision Tree for Reliable Contextual Information

Hyun Lee, Byung Rak Son, Jung Eun Kim, and Sang Cheol Lee

CPS Research Lab, Division of Robotics System, DGIST,
50-1 Hyunpung, Dalsung, Daegu, 711-873, South Korea
{hlee,brson,jekim,sclee}@dgist.ac.kr

Abstract. In intelligent environments, detecting possible errors and completing missing values then deciding about the quality and validity of the sensed data are roles of context reasoning in order to improve the reliability of information. Particularly, learning to associate operations and roles with dynamic contexts is necessary for adapting user needs and making optimal classification to context-aware computing. However, a top-down based inference has difficulty to deal with various kinds of uncertainty such as unknown, ambiguous, imprecise, and erroneous related problems. Since misclassification by high-level classification to choice for learning the role acceptance may lead to wrong decision. Thus, we propose Evidential Fusion Network (EFN) that uses a belief or confidence level of information as a bottom-up based inference. In order to analyze advantage and disadvantage of the proposed EFN, we compare our approach (bottom-up) with decision tree approach (top-down). Finally, we suggest the combination of top-down based analysis and bottom-up based inference as a context reasoning method for obtaining reliable information.

Keywords: Evidential Fusion Network (EFN), ID3, Multi-Agent System, Context-Aware, Situation Network, Inference.

1 Introduction

A multi-agent system in smart environment perceives user activities and identities user needs correctly in order to react in an appropriate way. However, user needs and system services evolve in the course of time and space. In particular, different users may have different types of needs for the same activities. Thus, [1] introduces an abstract context model, which consists of situation, roles and relations, in the form of a situation network based on the supervised learning that involves learning a function from examples of its inputs and outputs to adapt the system services to the changing needs of the user. However, the abstract context model in the form of the situation network is a top-down model. In this case, misclassification by high-level classifier or wrong decision which entity or entities (i.e., physical object or person) to choice for learning the role acceptance test may not be corrected. In addition, decision three learning may not determine how many different decision trees are consistent with the available training examples. This means that the supervisor feedback is not always consistent in reality. Even though machine learning (i.e., supervised [2], unsupervised

G. Lee (Ed.): Advances in Automation and Robotics, Vol. 1, LNEE 122, pp. 513–520.
springerlink.com

[3], or reinforcement [4]) is necessary for adapting user needs to the context-aware computing and for making optimal classification in smart environments, each context or situation network has ambiguous factors depending on the circumstances. Thus, we consider the confidence value of context or situation network that will be integrated to the learning algorithm so as to obtain more correct information. Finally, we compare our approach with decision tree for analyzing the advantages and disadvantages of the two approaches then suggest the combination of the two approaches.

The rest of the paper is organized as follows. As a top-down model, decision tree learning (i.e., ID3 [5]) is introduced in section 2. We introduced Evidential Fusion Network (EFN) as a bottom-up model in section 3. In section 4, we perform a case study to distinguish the ID3 and EFN. We compare and analyze the two approaches to show the advantage and disadvantage. We then conclude this work in section 6.

2 Decision Tree Learning (ID3)

In this section, we define decision tree learning (ID3) formulation [5] as a supervised learning technique (a top-down model). Decision tree are commonly used for gaining information for the purpose of decision-making. Decision tree starts with a root node on which it is for users to take actions. From this node, the users split each node recursively according to decision tree learning algorithm. The final result is a decision tree in which each branch represents a possible scenario of decision and its outcome. Decision tree learning is a method for approximating discrete-valued target functions, in which the learned function is represented by a decision tree. Decision tree learning is one of the most widely used and practical methods for inductive inference [6].

In particular, we introduce ID3 that is a simple decision tree learning algorithm [5]. The basic idea of ID3 algorithm is to construct the decision tree by employing a top-down, greedy search through the given sets to test each attribute at every tree node. In order to select the attribute that is most useful for classifying a given sets, we need to do is to minimize the questions asked (i.e. minimizing the depth of the tree). Thus, we need some function which can measure which questions provide the most balanced splitting. In this case, the information gain metric is such a function. In order to define information gain precisely, we need to discuss entropy, which means that measuring homogeneity of the learning set. Given a set S, containing these positive and negative targets, the entropy of S related to this boolean classification is:

$$\text{Entropy}(S) = \text{-P (positive)} \log_2 \text{P (positive)} - \text{P (negative)} \log_2 \text{P (negative)}. \tag{1}$$

where P (positive): proportion of positive examples in S and P (negative): proportion of negative examples in S. In the case that the target attribute can take N values:

$$Entropy(S) \equiv \sum_{i=1}^{N} - p_i \log_2 p_i \tag{2}$$

and then we measure the expected reduction in entropy related to specified attribute when splitting a decision tree node (i. e., Information Gain) to minimize the decision tree depth. The information gain, Gain (S, A) of an attribute A,

$$Gain(S, A) \equiv Entropy(S) - \sum_{v \in Values(A)} \frac{|S_v|}{|S|} Entropy(S_v) \tag{3}$$

where values (A) are all possible values for attribute A, $S_v \in S$ contains all examples for which attribute A has the value v, and $|S_v|$ is the cardinality of set S_v.

3 Evidential Fusion Network (EFN)

We introduce a probabilistic reasoning method that utilizes a belief, confidence level of evidence in order to deal with different types of uncertainty related problems such as incompleteness, inexactness, or ignorance. For instance, there exist certain aspects of the world that is not captured by employing physical sensor (i.e., incompleteness). Sometimes, the reports of multiple sensors which monitor one and the same situation can make less concern (i.e., inexactness). It is difficult to judge how much we should trust a report of a given sensor because there is no technical specification to tell us about the accuracy or resolution. In addition, it is difficult to determine at design time what aspects of a situation of interest we might be able to capture (i.e., ignorance).

Hence, we propose Evidential Fusion Network (EFN) as a probabilistic reasoning method (a bottom-up model). Particularly, we apply the Dezert-Smarandache Theory (DSmT) of evidence [7] so as to deal with conflict and ignorance [8] characteristics. Among various kinds of probabilistic reasoning methods, we focus on the reasoning method based on DSmT, since other methods such as classical inference, Fuzzy Logic, Bayesian inference, Dempster-Shafer theory, or voting have disadvantages [9], [10].

As shown in Figure 1, EFN consists of six steps is performed as a fusion process to make a higher confidence level of information. First, the threshold of each sensor is calculated. In addition, each sensor, associated context state and relevant situation can be represented by the evidential form as the frame of discernment, which represents all possible sensors' value and their combination values.

Second, we apply a discounting rate (error rate) that reflects sensor's credibility to each sensor. Since sensors has default error rate when they are made or environmental variations can make error factors. We calculate a discounting rate using equation (4). In this case, Θ is the frame of discernment and X is a subset of Θ.

$$m^r(X) = \begin{cases} (1-r)m(X) & X \subset \Theta \\ r + (1-r)m(\Theta) & X = \Theta \end{cases} \tag{4}$$

Where the source is absolutely reliable (r=0), the source is reliable with a discount rate r ($0 < r < 1$), and the source is completely unreliable (r=1) (e.g., a discounting rate of "pressure sensor" is 10%, a discounting rate of "environmental sensors" is 20% and a discounting rate of "medical body sensor is 5% in Figure 1).

Third, we apply a multi-valued mapping that represents a relationship between two frames of discernment. It represents evidence to same problems with different views. So, a multi-valued mapping Γ describes a mapping function Γ: $\Theta_A \leftarrow 2^{\Theta_B}$ by assigning a subset $\Gamma(ei)$ of Θ_B to each element *ei* of Θ_A. Based on the multi-valued mapping, a *translation is* utilized to determine the impact of evidence originally appearing on a frame of discernment on elements of a compatibly related frame of discernment. For example, suppose that Θ_A carries a mass function *m*, the translated mass function over the compatibly related Θ is defined as equation (5):

$$m'(B_j) = \sum_{\Gamma(e_i)=B_j} m(e_i) \tag{5}$$

After multi-valued mappings are applied to context states in Figure 1, context states are summed up by adapting different quality of sensed data. Each context attribute has different weights for clustering the context attributes within context state. We assume that the weight of "Pressure sensor" is 60%, that of "Location sensor" is 20% and that of "Motion sensor" is 20% with in context state (CS1) in Figure 1. Particularly, we show the case that "Location sensor" is not operated.

Forth, several independent sources of evidence combine belief mass distribution on the same frame in order to achieve conjunctive consensus with conflicting masses. In this case, we apply the Proportional Conflict Redistribution no. 5 (PCR5) rule [11] as a combination rule using equations (6) and (7). PCR5 is operated by following three steps. 1) PCR5 rule calculates the conjunctive rules of the belief masses of the sources. 2) It calculates the total or partial conflicting masses. 3) It redistributes the conflicting masses proportionally to non-empty sets involved in the model according to all integrity constraints. PCR5 rule for two sources is defined by:

$m_{PCR5}(\phi) = 0$ and $\forall(X \neq \phi) \in G\Theta$

$$m_{12}(X) = \sum_{\substack{X_1,X_2 \in G^{\Theta} \\ X_1 \cap X_2 = X}} m_1(X_1) m_2(X_2) \tag{6}$$

$$m_{PCR5}(X) = m_{12}(X) + \sum_{\substack{Y \in G^{\Theta} \setminus \{X\} \\ c(X \cap Y) = \emptyset}} \left[\frac{m_1(X)^2 m_2(Y)}{m_1(X) + m_2(Y)} + \frac{m_2(X)^2 m_1(Y)}{m_2(X) + m_1(Y)}\right] \tag{7}$$

Fifth, degrees of belief such as lower (Belief) and upper (Plausibility) bounds on probability are calculated. Then the uncertainty levels (i.e., ignorance = Plausibility – Belief) of evidence are measured by using equation (8) after applying PCR5 rule into EFN. Belief (*Bel*) and Plausibility (*Pl*) for PCR5 rule are defined by:

$$\begin{aligned} Bel(\{X\}) &= m_{PCR5}(\{X\}) \\ Pl(\{X\}) &= m_{PCR5}(\{X\}) + m_{PCR5}(\{X, \neg X\}) \\ Pl(\{X\}) - Bel(\{X\}) &= m_{PCR5}(\{X, \neg X\}) \end{aligned} \tag{8}$$

Finally, we use the expected utility and the maximum of the pignisitic probability such as Generalized Pignistic Transformation (GPT) [12] as a decision criterion. The result of context reasoning (e.g., the situation of the user) is inferred by calculating the belief, uncertainty, and confidence (i.e., GPT) levels of information. GPT is defined by: $\forall(A) \in D^{\Theta}$,

$$P\{A\} = \sum_{X \in D^{\Theta}} \frac{C_M(X \cap A)}{C_M(X)} m(X) \tag{9}$$

where $C_M(X)$ denotes the DSm cardinal of a proposition X for the DSm model M of the problem under consideration. For example, we get a basic belief assignment with non-null masses only on X_1, X_2 and $X_1 \cup X_2$. After applying GPT, we get:

$$\begin{aligned} &P_{PCR5}(\{X\}) \\ &= m_{PCR5}(\{X\}) + \tfrac{1}{2} m_{PCR5}(\{X, \neg X\}) \end{aligned} \tag{10}$$

4 A Case Study

In order to compare the operation of ID3 algorithm with the evidential fusion process, we assume a simple scenario as shown in Figure 1. In this case, the symbolic context attribution is similar to the frame of discernment. We represent the status of sensors as active (1) or inactive (0). In addition, we apply the weight (i.e., degree of importance) to each sensor. For instance, Pressure Sensor is 0.35, Location Sensor is 0.10, Motion Sensor is 0.05, Blood Pressure Sensor is 0.1, Body Temperature Sensor is 0.15, and Respiratory Rate Sensor is 0.25. Then, we make the decision (negative (N) or positive (P)) based on the operation of each sensor. After calculating the cases of learning set, we can obtain the following aspects. Total learning set is 63, negative case is 34, and positive case is 29. Based on these assumptions and pre-defined conditions, we obtain

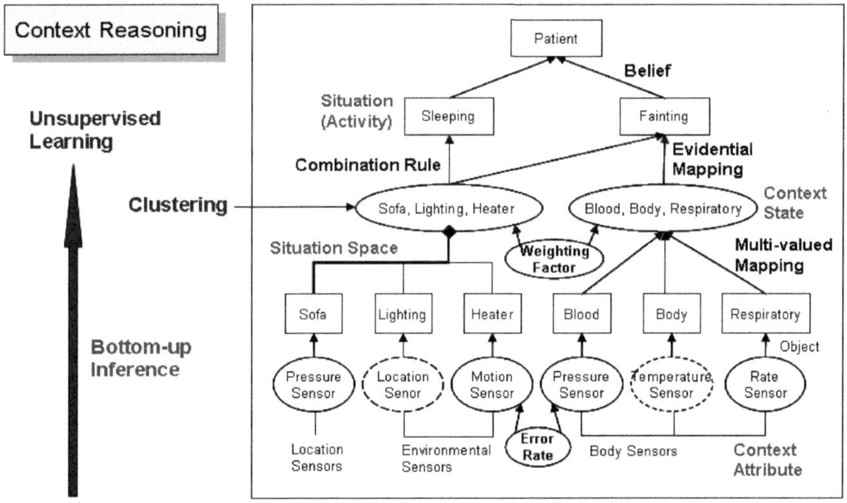

Fig. 1. A simple bottom-up based inference scenario so as to figure out the situation of a patient (e.g., sleeping or fainting). Probabilistic context reasoning is performed by following an evidential fusion process within EFN then we obtain the degree of belief of the situation.

the root node of Entropy (S) and information gains (Gain(S, A)) for each attribute. For instance, we recursively calculate each case to obtain the maximum information gain based on the defined Entropy (S) then split the maximum information gain (Gain(S, A)) until the Entropy (S) is equal to zero. Finally, the recursive calculation of ID3 is finished.

After we split the context attribute descriptions applied to the example scenario of Figure 1, the context classification is constructed by the splitting context attribute as shown in Figure 2. Based on the Figure 2, we know that the learning process of ID3 is similar to the top-down based inductive inference (i.e., one of case-based reasoning). By [13], the case-based reasoning is the classification of new sensed context based on known similar context. Since, the top-down based inductive inference has advantage and disadvantage compared to our proposed approach (a bottom-up based inference), we compare and analyze the two approaches in the next section.

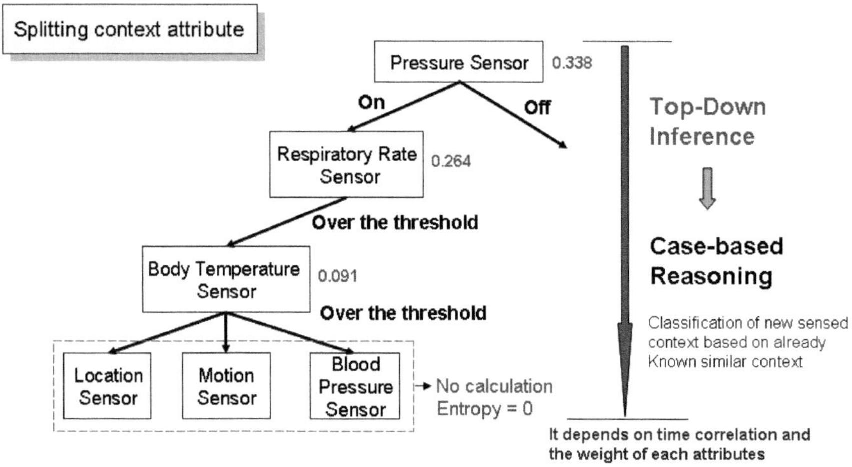

Fig. 2. A Context Classification using ID3 algorithm based on the example scenario of Figure 1. The results show the top-down based inference (i.e., one of case-based reasoning) and we can make a classification depending on the importance of each context attribute.

5 Comparison and Analysis

The top-down based inference can be used to predict the utility of a particular sensory action with respect to a goal at top. In the case of a given example, the situation space of "fainting" or "sleeping" of the user will fuse the most relevant context state and context attribute [14]. The selection of an information source or the activation of a process to compute new information is regarded as a set of actions (i.e., context state) available to the decision maker in decision theory.

However, the top-down based inference has disadvantages as mentioned in section 1. Thus, we also consider the bottom-up based inference, which allows the integration of the context attributes from a sensory action. By [15], the higher level context (i.e., meeting) will be obtained sensed data based on time and space correlation.

Therefore, we suggest the combination of the top-down based inference and the bottom-up based inference to make a context classification and a context reasoning as shown in Figure 3. In particular, we suggest that context reasoning will be followed by the bottom-up based inference then the context analysis will be performed by the top-down based inference for adapting user's needs.

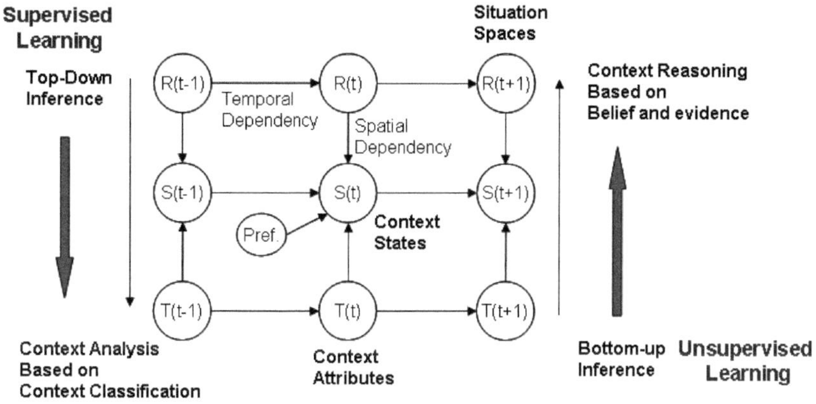

Fig. 3. The combination of the top-down based inference and the bottom-up based inference. In this fig 3, T(t) represents current context attribute description and T(t-1) and T(t+1) represent previous time-indexed context attribute and next time-indexed context attribute, respectively to show temporal dependency of consecutive time-indexed context attribute. S(t) and R(t), which have spatial dependency, represent context states and situation spaces of the user, respectively.

6 Conclusion

Until now, we compared our proposed approach as a bottom-up based inference (i.e., EFN) with decision tree based approach as a top-down based inference to discuss the advantage and disadvantage of the two approaches for obtaining reliable information. Even though the decision tree based approach can solve problems in domains that are only partially understood and it can easily transforms raw sensed data into meaningful information to make decisions that may lead to actions, there exist some uncertainty factors about the quality and the validity of the sensed data in real situation. Thus, we proposed the EFN approach that can deal with these kinds of uncertainty. Moreover, we suggested the combination of the top-down model and the bottom-up model. The combination of the two approaches is more helpful to understand the situation of the user in context-aware computing applications. Since the bottom-up inference reduces the uncertainty level of selected information then the top-down analysis classifies the context of the user more clearly.

Acknowledgments. This work was supported by the DGIST R&D Program of the Ministry of Education, Science and Technology of Korea (11-BD-01).

References

1. Brdiczka, O., Reignier, P., Crowley, J.L.: Supervised Learning of an Abstract Context Model for an Intelligent Environment. In: The Proc. of Joint sOc-EUSAI Conf., Grenoble, Switzerland, pp. 259–264 (2005)
2. Raya, M., Papadimitratos, P., Gligor, V., Hubaux, J.: On Data-Centric Trust Establishment in Ephemeral Ad Hoc Network. In: IEEE INFOCOM 2008, pp. 1238–1246 (2008)
3. Dargie, W.: The Role of Probabilistic Schemes in Multisensor Context-Awareness. In: PerCom Workshops, pp. 27–32 (2007)
4. Wu, H.: Sensor Data Fusion for Context-Aware Computing Using Dempster-Shafer Theory. CMU-RI-TR-03-52, Thesis (2003)
5. Ross Quinlan, J.: Induction of Decision Trees. Machine Learning 1(1), 81–106 (1986)
6. Mitchell, T.M.: Machine Learning, Intl. edition. McGraw-Hill, New York (1997)
7. Smarandache, F., Dezert, J.: Advances and Applications of DSmT for Information Fusion, vol. 2. American Research Press, Rehoboth (2006)
8. Maskell, S.: A Bayesian approach to fusing uncertain, imprecise and conflicting information. Information Fusion 9, 259–277 (2008)
9. Hong, X., Nugent, C., Mulvenna, M., McClean, S., Scotney, B., Devlin, S.: Evidential Fusion of sensor data for activity recognition in smart homes. J. of Pervasive and Mobile Computing 5(3), 236–252 (2008)
10. Rogova, G.L.: Reliability in information fusion: literature survey. In: The Proc. of the 7th Intl. Conf. on Information Fusion, pp. 1158–1165 (2004)
11. Smarandache, F., Dezert, J.: Information fusion based on new proportional conflict redistribution rules. In: The Proc. of the 8th Intl. Conf. on Information Fusion, pp. 907–914 (2005)
12. Dezert, J., Smarandache, F., Daniel, M.: The Generalized Pignistic Transformation. In: The Proc. of the 7th Intl. Conf. on Information Fusion, pp. 384–391 (2004)
13. Kofod-Petersen, A., Cassens, J.: Explanations and Context in Ambient Intelligent Systems. In: Kokinov, B., Richardson, D.C., Roth-Berghofer, T.R., Vieu, L. (eds.) CONTEXT 2007. LNCS (LNAI), vol. 4635, pp. 303–316. Springer, Heidelberg (2007)
14. Roy, N., Pallapa, G., Das, S.: A Middleware Framework for Ambiguous Context Mediation in Smart Healthcare Application. In: IEEE Intl. Conf. on Wireless and Mobile Computing, Networking and Communication, pp. 72–80 (2007)
15. Schmidt, A.: Resource, `http://www.hcilab.org/teaching/rom2006/`

Research on Key Technology of LXI Bus Instrument Interface System

Li Ming[1], Liang Siyang[2], and Fu Wei[1]

[1] Ordnance Engineering College, Shijiazhuang 050003, P. R. China
[2] Beijing Institute of Technology, Beijing 100081, P. R. China

Abstract. The research and development on LXI bus technology are still in its infancy at present, this paper in-depth study on LXI bus specification standard and track the development of international auto-test technology. Mastering LXI bus core technologies and developing LXI bus interface system equipment can lay the foundation of develop LXI bus instruments.

Keywords: LXI bus, interface system, key technology.

1 Introduction

LXI bus is the new generation of instrument platforms, it is the trend of information and network technology development, and it will become mainstream bus technology in the future. The research and development on LXI bus technology is still in its infancy at present, this project track the development of international auto-test technology, especially on frontier research of LXI bus technology. In-depth study on LXI bus specification standard, master LXI bus core technologies, develop LXI bus interface system equipment, and vigorously strengthen the independent innovation ability, research the advanced level in develop the LXI instruments.

2 Key Technology of LXI Bus

Most of the LXI bus specifications content is based on the existing mature industry standards, and its main task is to describe how to use these mature specifications prescribe what functions the equipment should provided. In order to test and measurement, LXI bus specification describes a control interface which based on Ethernet interface and web access, the key technologies of LXI bus specification is mainly focus on Ethernet interface design. The key technologies involved: Ethernet technology, Web interface technology, equipment discovery and identification technology, driver program technology, precision time synchronization, hardware trigger bus and distributed applications technologies.

3 Function Analysis on LXI Bus Instrument System

During LXI bus instrument design, in order to maximize improve the reusability of develop different LXI bus instruments, the LXI instrument is divided into functional

G. Lee (Ed.): Advances in Automation and Robotics, Vol. 1, LNEE 122, pp. 521–524.
springerlink.com

units and the bus interface unit, functional units only provide measurement function associated with specific equipment, the interface unit need to achieve all the network services based on the Ethernet interface which are asked by LXI bus specification, and also need to define instrument control standard interface between the functional unit. Interface unit is used to pass information between the various test equipment, let all information between different test equipment pass successfully with different bus specifications prescribes the code system, speed, level and methods. In the test system, the collectivity including software and hardware interfaces is called interface system; the interface system actually is a liaison officer that each equipment can communicate effectively. In other words, the difference between the various bus devices is mainly reflected in the interface system, and therefore research on LXI bus interface system and replace different functional circuits, can achieve a variety of testing capabilities.

Through the LXI bus interface systems, LXI instruments can follow LXI bus standard to achieve LAN communication, IEEE 1588 time synchronization, hardware trigger; and users can also operate and control LXI instrument by LXI bus. From the structure diagram, we can see that LXI bus interface circuit is mainly consist of power supply module, host processor, network interface module, clock module and hardware triggered module. The interfaces include power interface, network interface, functional board circuit interface, hardware trigger interface, ARM debug interface, FPGA debug interface and virtual terminal debug serial port and so on.

4 Hardware Design of LXI Bus Instrument Interface System

As LXI bus interface hardware structure determines the instrument category, the same type but different functions of the LXI bus instrument has the same LAN interface hardware architecture, therefore, in the design of LXI instrument, the hardware unit is divided into interface unit and functional unit. The connection between functional unit and LXI bus interface unit is standardized design, it is propitious to achieve LXI bus instrument modularization and standardization, and accelerate the development of the LXI instrument bus schedule, while reduce development difficulty and improve the reliability.

The basic functions of various parts of LXI bus interface embedded system are described below.

(1) Serial debug interface: terminal interface of embedded system, it can be used to print debugging information and receive user input information, mainly used in device development phase;

(2) JTAG interface: used to download boot code and debug the underlying process by ARM emulator;

(3) IIC Interface EEPROM: 256kbit erasable EEPROM is used to save the device configuration information;

(4) 4MB FLASH: save the device's startup code, the operating system kernel and file system image;

(5) 16MB SDRAM: the operating space of system and applications;

(6) 10MHz active crystal: through on-chip PLL circuit become 50MHz and used as microprocessor clock;

(7) External bus interface: the main interface unit of LXI common interface module and functional modules. It can face can achieve read and write operations to the equipment functional unit and response interrupt request submitted by equipment functional unit;

(8) 10M/100M Ethernet interface: Through this interface, the device can access network with a rate of 10Mbps or 100Mbps, using the DP83640 can not only accomplish basic physical layer chip transceiver function ,but also provide capture IEEE 1588 event message time stamp and LXI application functions based on IEEE 1588.

5 Software Design of LXI Bus Instrument Interface System

The goal of software design is to develop a highly real-time, flexibility and versatility software platform of measure data communications. Therefore, the design of specific software platform uses the modular and multi-process management thinking. Accordance to the LXI bus requirements, LXI instrument need to achieve LAN communication and Web interface design. The main task of LXI bus interface circuit is to transmit instrument information and measurement data to the host computer via Ethernet, or sent commands from host computer to LXI instrument via Ethernet and controls the instrument. LXI bus interface circuit need to send and receive and process large amounts of information and data while running multiple tasks, such as instrument control, communications, information detection, which requires reasonable scheduling and interrupt handling to multi-tasking.

The software design of LXI bus interface system includes building platform software and develops network service program and instrument control program which run in the platform software. Through the analysis of LXI bus specification, LXI bus interface system software should provide the following functions.

(1) Network protocol support. In order to achieve the LXI bus network interface function, platform software should be able to support the following protocols.

① TCP, UDP, ICMP protocols;

② IPv4 - had best support both IPv4 and IPv6;

③ determine the IP address protocol - DHCP, Auto-IP (Dynamic Link-Local Addressing);

④ ARP protocol, used for IP conflict detection;

⑤ DNS client support;

⑥ Multicast support for sending IEEE 1588, direct inter-module communication, mDNS packets;

⑦ ONC-RPC, for instruments found based on VXI-11.

(2) Web server and back-end dynamic web processor support. LXI bus instruments need to include the Web server to provide access to the Web page service. The web pages of equipment will include a number of dynamic information such as IP address, DNS server address and so on, so it is need some mechanism to update the dynamic data on a Web page, not just contain static pages. We can use ASP, JSP, CGI and other technologies to generate dynamic web pages.

6 Conclusions

Based on the module hardware and software integration, debugging, we established test methods and procedures, analyzed and verified the test results of LXI bus interface system. The test results show that the interface system met the design specifications. LXI bus technology will provide better performance, compatibility and ease of use with lower cost.

References

1. Bai, Y.: LAN configuration of LXI instrument. Electronic Test (7), 87–90 (2009)
2. Zhang, Y., Xue, D., Wu, C., Ji, P.: Research of portable information terminal based on MIPS processor and windows CE. In: Internationnal Conference on Advanced Computer Control, ICACC 2009, pp. 528–531 (2009)
3. Wang, L., Wang, C., Li, M.: Design of digital multimeter based on ARM. In: International Conference on Computer, Mechatronics, Control and Electronic Engineering, Changchun, pp. 110–112 (2010)
4. Xu, X., et al.: LXI bus interface design and testing systems in military applications. Computer Measurement and Control 10 (2009)
5. Li, G., et al.: LXI-based distributed network bus automatic test system design. Electronic Measurement and Instrument, Supplement (2009)
6. Zhu, S., et al.: LXI bus test systems and in military applications. Electronic Measurement and Instrument, Supplement (2008)
7. LXI: beyond GPIB, PXI and VXI test response to major challenges in the application guide, Agilent Technologies
8. Li, X.: LAN-based automated test equipment set up technology computer measurement and control, 14 (2006)
9. LXI - the next generation of automated test system peak Symposium, 07 (2011)

A Study about the Prediction of University Library Lending Based on Multiple Regression Analysis

Bao Sun

Library, North China Institute of Science and Technology
065201 Langfang, China
sunbaoadam@163.com

Abstract. The change of library lending depends on the change of the number of readers and the books that a library possesses. Library lending has a strong association with the number of readers and the library collection. The author has investigated library lending, the number of readers and the collection of a university library in China for 18 school years. The prediction model of library lending with the number of readers and the collection as the independent variables is constructed. The result of F-test and t-test proves that a strong linear relationship exists between library lending and the number of readers, the collection. The prediction value of new observations of library lending is obtained in accordance with the prediction equation. The author gives the methods of calculating confidence interval of the mean library lending and prediction interval of a single library lending in the school year 2010-2011.

Keywords: Multiple linear regression, Library lending, The number of readers, The library collection, Test, The adjusted coefficient of multiple determination, Prediction, Confidence interval, Prediction interval.

1 Introduction

It is necessary for a university library administrator to predict library lending in terms of increase of librarian and device or in terms of decrease of them. So it needs to construct the prediction model of library lending. There are many methods for predicting a variable such as time series analysis [1] and average prediction method [2].These two kinds of prediction theory merely involve the change of the variable itself, whereas the prediction of library lending should involve the influence of other important factors such as the readers and the books belonging to a library.

The theory of multiple linear regression analysis is an important prediction method that has been used in many fields of nature science and of social science such as business management, market policy decision, financial analysis [3]. It is mainly used to discuss the relationship of some variables and to predict one variable through the other ones.

It is found by the author that the change of library lending depends on the change of the number of readers and the library collection. Then the author wishes to

G. Lee (Ed.): Advances in Automation and Robotics, Vol. 1, LNEE 122, pp. 525–532.
springerlink.com

construct a multiple linear regression model to relate library lending to the number of readers and the collection, so the library lending can be predicted through the number of readers and the collection based on the multiple linear regression analysis.

2 Construction of the Prediction Model of Library Lending

2.1 Investigation of Library Lending, the Number of Readers and the Library Collection for Eighteen School Years

The prediction model is mainly used in university in China and its data come from the university library. The library discussed in this article belongs to North China Institute of Science and Technology. In the university the freshmen register in October and the seniors graduate in June. Every school year consists of the first semester and the second semester. All students' first semester except the freshmen starts in September of one year and ends in January of the next year. The first semester of the freshmen starts in October and ends at the same time as the ordinary students. The second semester of all ordinary students except the seniors starts in February and ends in July. The second semester of the seniors starts at the same time as the ordinary students but ends in June. So the number of readers is fixed within a school year. The author has investigated the library lending from August of one year to July of the next year, the library collection at the end of the every school year and the number of readers for eighteen school years, i.e. from August 1993 to September 2010.In China August is summer vocation and there is a few students borrowing books from library. So the library lending in August should be included in its school year's library lending.

Table 1. Library lending, number of readers and library collection of eighteen school years

School year	Number of Readers	Library Collection	Library Lending
1993-1994	1,869	68,094	53,677
1994-1995	1,484	77,432	40,737
1995-1996	1,842	82,873	45,587
1996-1997	2,024	94,064	45,920
1997-1998	2,605	108,953	44,343
1998-1999	3,527	20,164	62,520
1999-2000	5,709	99,563	149,608
2000-2001	6,850	137,109	168,494
2001-2002	8,728	171,600	191,353
2002-2003	9,873	198,374	231,200
2003-2004	10,525	236,079	254,829
2004-2005	11,021	263,603	285,733
2005-2006	12,315	387,815	299,971
2006-2007	13,706	461,573	372,197
2007-2008	15,692	535,918	369,425
2008-2009	16,678	805,298	342,177
2009-2010	17,493	902,451	316,390
2010-2011	21,249	982,451	

2.2 Acquisition of Linear Regression Parameters

The linear relationship between library lending and the number of readers, the library collection can be expressed as the following equation [4].

$$L_i = \beta_0 + \beta_1 R_i + \beta_2 C_i + \varepsilon_i (note : i = 1,2,\cdots,n) \tag{1}$$

This model is a system of n equations that can be expressed in matrix notation as

$$\mathbf{L} = \mathbf{X}\beta + \varepsilon \tag{2}$$

Therefor, the least squares estimate of $\boldsymbol{\beta}$ is [5]

$$\hat{\beta} = (\mathbf{X'X})^{-1}\mathbf{X'L} \tag{3}$$

2.3 Evaluation of the Significance of Prediction Model Using Method of F-test

The appropriate hypotheses are

$$\begin{aligned} &H_0 : \beta_1 = \beta_2 = 0 \\ &H_1 : \beta_j \neq 0 \quad for \quad at \quad least \quad one \quad j \end{aligned} \tag{4}$$

Rejection of $H_0{:}\beta_1{=}\beta_2{=}0$ implies that at least one of the regressor variables contributes significantly to the model [6].The test statistic for $H_0{:}\beta_1{=}\beta_2{=}0$ is

$$F_0 = \frac{SS_R/2}{SS_E/(n-3)} = \frac{MS_R}{MS_E} \tag{5}$$

We should reject H_0 if the computed value of the test statistic in Equation (5),f_0,is greater than $f_{\alpha,\ 2,\ n-3}$[7]. The procedure is usually summarized in an analysis of variance table such as Table 2.

Table 2. Analysis of variance for testing significance of regression in multiple regression

Source of variation	Sum of squares	Degrees of freedom	Mean square	F_0
Regression	SS_R	2	MS_R	
Residual	SS_E	n-3	MS_E	$F_0 = \frac{MS_R}{MS_E}$
Total	SS_T	n-1		

2.4 R^2 and Adjusted R^2

We may use the coefficient of multiple determination R^2 as a global statistic to assess the fit of the prediction model. Computationally,

$$R^2 = \frac{SS_R}{SS_T} = 1 - \frac{SS_E}{SS_T} \tag{6}$$

Many regression users prefer to use an adjusted R^2 statistic:

$$R^2_{adj} = 1 - \frac{SS_E/(n-3)}{SS_T/(n-1)} \tag{7}$$

The greater the adjusted R^2 is, the more significantly the number of readers and the library collection contribute to the model [8].

2.5 Tests on Individual Regression Coefficients of the Number of Readers

The null and alternative hypotheses to test are

$$\begin{aligned} H_0 &: \beta_1 = 0 \\ H_1 &: \beta_1 \neq 0 \end{aligned} \tag{8}$$

The test statistic for this hypothesis is

$$T_0 = \frac{\hat{\beta}_1}{\sqrt{\sigma^2 C_{11}}} = \frac{\hat{\beta}_1}{se(\hat{\beta}_1)} \tag{9}$$

The null hypothesis H_0:β_1=0 is rejected if $|t_0| > t_{\alpha/2,n-3}$. If H_0:β_1=0 is not rejected, this indicates that the regressor R, that is, the number of readers can be deleted from the model [9].

2.6 Tests on Individual Regression Coefficients of the Library Collection

The test of the library collection is very similar to that of the number of readers. The hypothesis to test on individual regression coefficients of the collection is

$$\begin{aligned} H_0 &: \beta_2 = 0 \\ H_1 &: \beta_2 \neq 0 \end{aligned} \tag{10}$$

The test statistic for this hypothesis is [10]

$$T_0 = \frac{\hat{\beta}_2}{\sqrt{\sigma^2 C_{22}}} = \frac{\hat{\beta}_2}{se(\hat{\beta}_2)} \tag{11}$$

The null hypothesis H_0:β_2=0 is rejected if $|t_0| > t_{\alpha/2,n-3}$ .If H_0:β_2=0 is not rejected, this indicates that the regressor C, that is, the collection can be deleted from the model.

2.7 Prediction Value of New Observations of Library Lending

The following regression equation between library lending and the number of readers, the library collection can be used to predict values of library lending for various values of the number of readers (R_i) and the library collection (C_i).

$$\hat{L}_i = \beta_0 + \beta_1 R_i + \beta_2 C_i \tag{12}$$

We can obtain the fitted or prediction value by substituting new observation (R_0,C_0) into the foregoing regression equation [11].

$$\hat{L}_0 = \beta_0 + \beta_1 R_0 + \beta_2 C_0 \tag{13}$$

2.8 Confidence Interval on the Mean Response of Library Lending

A 100(1-α)% confidence interval on the mean response of library lending at the point R_0,C_0 is[12]

$$\hat{\mu}_{L|\mathbf{x_0}} - t_{\alpha/2,n-3}\sqrt{\hat{\sigma}^2 \mathbf{x'_0}(\mathbf{X'X})^{-1}\mathbf{x_0}} \le \mu_{L|\mathbf{x_0}} \le \hat{\mu}_{L|\mathbf{x_0}} + t_{\alpha/2,n-3}\sqrt{\hat{\sigma}^2 \mathbf{x'_0}(\mathbf{X'X})^{-1}\mathbf{x_0}} \tag{14}$$

2.9 Prediction Interval for the Future Observation of Library Lending

A 100(1-α)% prediction interval for the future observation of library lending is[13]

$$\hat{L}_0 - t_{\alpha/2,n-3}\sqrt{\sigma^2\left(1 + \mathbf{x'_0}(\mathbf{X'X})^{-1}\mathbf{x_0}\right)} \le L_0 \le \hat{L}_0 + t_{\alpha/2,n-3}\sqrt{\sigma^2\left(1 + \mathbf{x'_0}(\mathbf{X'X})^{-1}\mathbf{x_0}\right)} \tag{15}$$

3 Procedure of Computation Practice

3.1 Parameters and Equation of Multiple Linear Regression Model

The 17 observations are in Table 1.We will now use the matrix approach to fit the regression model between library lending and the number of readers, the library collection to these data. The least squares estimates are found from Equation (3) as

$$\begin{bmatrix} \hat{\beta}_0 \\ \hat{\beta}_1 \\ \hat{\beta}_2 \end{bmatrix} = \begin{bmatrix} -1{,}480.930393 \\ 29.128650 \\ -0.186850 \end{bmatrix}$$

Therefore, the fitted regression model with the regression coefficients rounded to six decimal places is

$$\hat{L}_i = -1{,}480.930393 + 29.128650\mathrm{R}_\mathrm{i} - 0.186850\mathrm{C}_\mathrm{i} \tag{16}$$

3.2 Evaluation of the Significance of Prediction Model Using Method of F-test

The analysis of variance is shown in Table 3 attained in Microsoft Excel by running the Regression command of Data Analysis menu item.

Table 3. Test for significance of regression for library lending

Source of variation	Sum of squares	Degrees of freedom	Mean square	F_0	P-value
Regression	2.450394E+11	2	1.225197E+11		
Residual	8,566,401,112	14	6.118858E+08	200.232976	5.017348E-11
Total	2.536058E+11	16			

Since $f_0 > f_{0.01,2,14} = 6.514884$(or since the P-value is considerably smaller than $\alpha = 0.01$),we reject the null hypothesis and conclude that library lending is linearly related to either the number of readers or the library collection, or both.

3.3 R^2 and Adjusted R^2

Based on the computed result above in 3.2, the coefficient of multiple determination R^2 is calculated as follows:

$$R^2 = \frac{SS_R}{SS_T} = \frac{2.450394\text{E}+11}{2.536058\text{E}+11} = 0.966222$$

Thus the model accounts for about 96.6222% of the variability in library lending. Similarly, the adjusted coefficient of multiple determination R^2_{adj} is computed as follows:

$$R^2_{adj} = 1 - \frac{SS_E/(n-3)}{SS_T/(n-1)} = 1 - \frac{8{,}566{,}401{,}112/(17-3)}{2.536058\text{E}+11/(17-1)} = 0.961396$$

3.4 Tests on Individual Regression Coefficients of the Number of Readers

The t-statistic in Equation (9) is

$$t_0 = \frac{\hat{\beta}_1}{\sqrt{\sigma^2 C_{11}}} = \frac{29.128650}{\sqrt{(6.118858\text{E}+08)(1.100845\text{E}-08)}} = 11.223337$$

Since $t_{0.01/2,14} = 2.97684$,we reject $H_0{:}\beta_1 = 0$ and conclude that the variable R (the number of readers) contributes significantly to the model.

3.5 Tests on Individual Regression Coefficients of the Library Collection

The t-statistic in Equation (11) is

$$t_0 = \frac{\hat{\beta}_2}{\sqrt{\sigma^2 C_{22}}} = \frac{-0.186850}{\sqrt{(6.118858\text{E}+08)(5.880630\text{E}-12)}} = -3.114909$$

Since $t_{0.01/2,14}$=2.97684,we reject H_0:β_2=0 and conclude that the variable C (the library collection) contributes significantly to the model.

3.6 Prediction Value of Library Lending in School Year 2010-2011

From Table 1, the new observation has R_0=21,249 and C_0=982,451, and the prediction value is

$$\hat{L}_0 = -1{,}480.930393 + 29.128650\,R_0 - 0.186850\,C_0 = 433{,}903.042469$$

3.7 Confidence Interval of Mean Library Lending in School Year 2010-2011

We would like to construct a 99% confidence interval on the mean library lending with R_0=21,249,C_0=982,451.Therefore, a 99% confidence interval on the mean library lending at this point is found from Equation (14) as

$$374{,}077.558934 \le \mu_{L|\mathbf{x}_0} \le 493{,}728.526004$$

3.8 Prediction Interval of Library Lending in School Year 2010-2011

Suppose that we wish to construct a 99% prediction interval on library lending at the point R_0=21,249, C_0=982,451.Therefore, from Equation (15) the 99% prediction interval is

$$339027.538886 \le L_0 \le 528778.546052$$

4 Conclusions

The author has observed the change of library lending and that of the number of readers and the library collection for eighteen school years in the library of North China Institute of Science and Technology and has found that there is a linear relationship between library lending and the number of readers, the collection. Four kinds of methods, such as F-test, the coefficient of multiple determination, the adjusted coefficient of multiple determination and t-test on individual regression coefficients also prove that library lending is linearly related to either the number of readers or the collection, or both. Based on the theory of multiple linear regression analysis, the paper discusses how to construct the prediction model of library lending of university library. The prediction value, the confidence interval and the prediction interval of library lending in the school year 2010-2011 has been obtained. Then library lending in the school year 2010-2011 may be 433,903.0 books, between 339,027.5 books and 528,778.5 books and the mean library lending in the same school year may be between 374,077.6 books and 493,728.5 books.

The prediction of library lending will play a key role in the library administration. The administrators of most libraries are paying more and more attention to library lending which is often mentioned in their annual reports. The prediction of library lending may also have great effects on the annual plan of most libraries.

Acknowledgments. This work is supported by the National Natural Science Foundation of China (No. 60872055), Postdoctor Science Foundation of China (No. 20090460320) , Natural Science Foundation of Hebei Province(No.F2010001745) and State Administration of Work Safety Key Projects(No. 10-120).

References

1. http://www.statsoft.com/textbook/time-series-analysis/
2. http://www.crowdcast.com/blog/2010/04/02/not-your-average-prediction-market/
3. http://www.stat.yale.edu/Courses/1997-98/101/linreg.htm
4. http://www.jerrydallal.com/LHSP/slr.htm
5. Obremski, T.: Pricing Models Using Real Data. Teaching Statistics 30(2), 44–48 (2008)
6. Korkut, I., Acir, A., Boy, M.: Application of regression and artificial neural network analysis in modelling of tool-chip interface temperature in machining. Expert Systems with Applications 38(9), 11651–11656 (2011)
7. Lorenzo, C., Maggio Ruben, M., Carlo, B., et al.: Application of partial least square regression to differential scanning calorimetry data for fatty acid quantitation in olive oil. Food Chemistry 127(4), 1899–1904 (2011)
8. http://stattrek.com/AP-Statistics-4/Test-Slope.aspx
9. Suarez Sanchez, A., Garcia Nieto, P.J., Riesgo Fernandez, P., et al.: Application of an SVM-based regression model to the air quality study at local scale in the Aviles urban area (Spain). Mathematical and Computer Modelling 54(5-6), 1453–1466 (2011)
10. Freire, R.Z., Oliveira, G.H.C., Mendes, N.: Development of regression equations for predicting energy and hygrothermal performance of buildings. Energy and Buildings 40, 810–820 (2008)
11. Devore, J.L.: Probability and Statistics for Engineering and the Science, 6th edn., pp. 496–554. Brooks/Cole (2004)
12. http://www.weibull.com/DOEWeb/confidence_intervals_in_simple_linear_regression.htm
13. Mann, P.S.: Introductory statistics, 5th edn., pp. 580–642. John Wiley & Sons inc. (2004)

Feature Fusion and Prototype Learning Algorithm Based Smoking Shot Recognition

Wuping Zhang, Pengfei Zhu, Xinzhe Wang, and Wentao Tian

Radio, Film and Television Design and Research Institute,
Beijing, China
{zhangwp,zhupf,wangxz,twt}@drft.com.cn

Abstract. This paper present a novel human smoking shot recognition method based on feature fusion and the prototype learning algorithm. We first extract the color feature and the texture of the skin patch and the cigarette patch. Then we use the subspace learning algorithm to fuse the features. The prototype learning algorithm based on metric learning is used to classify the fusion features. The experimental results show the effectiveness of our method.

Keywords: smoking shot recognition, feature fusion, prototype learning, subspace learning, metric learning.

1 Introduction

Smoking is from Columbus discovered the New World, after several hundred years of history so far. Smoking not only harmful to your body, will cause harm to others, also have adverse social impacts.The state Administration of Radio Film and Television of China instructed that the quantities of smoking shots should be controlled in the movies in recent years.

Many action recognition methods are proposed in recent years. Bobick and Davis [1] present a viewed-based approach to the representation and recognition of human movement, motion-energy images (MEI) and motion-history images (MHI) are used here. Babu and Ramakrishnan [2] propose a technique for building coarse Motion History Image (MHI) and Motion Flow History (MFH) from the compressed video and extract features from these static motion history information for characterizing human actions. Ke et al [3] introduce a volumetric feature framework for analyzing video by extending Viola and Jones' work for static-scene object detection to the spatio-temporal domain. Laptev and Linderberg [4] propose to extend the notion of spatial interest points into the spatio-temporal domain and show how the resulting features often reflect interesting events that can be used for a compact representation of video data as well as for its inter presentation. Gorelick et al [5] represent actions as space-time shapes and show that such a representation contains rich and descriptive information about the action performed. In [6] , a color based ration histogram analysis is introduced to extract the visual clues from appearance interactions between lighted cigarette and its human holder. HMM framework is used in this paper.

G. Lee (Ed.): Advances in Automation and Robotics, Vol. 1, LNEE 122, pp. 533–540.
springerlink.com

Many classifiers can be used to recognize human behavior, such as nearest neighbors, Neural networks, SVM and Bayes, etc. Prototype learning [6] is effective in improving the classification performance of nearest-neighbor classifier and in reducing the storage and computation requirements. There are many prototype learning algorithms, such as Minimum Classification Error (MCE) [7], Generalized Learning Vector Quantization (GLVQ) [8][9], Soft Nearest Prototype Classifier (SNPC) [10], Log-Likelihood of Margin (LOGM) [13],etc. We used a kind of prototype learning algorithm based on metric learning in [15], that promoted the effectiveness of the prototype learning.

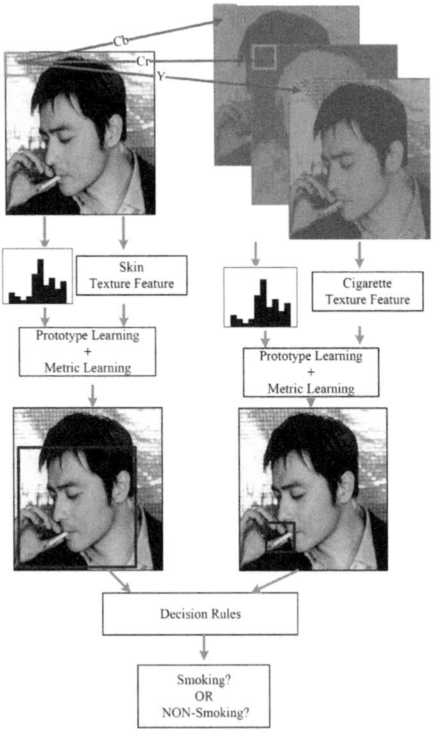

Fig. 1. The framework of our method

Figure 1 shows the framework of our method. The rest of this paper is organized as follows. Section 2 shows the region of interest detection. The formulation of prototype learning, metric learning algorithm, and our prototype learning algorithm using metric learning are introduced in section 3. To evaluate our method, the experiments are shown in section 4. Section 5 shows the conclusion of our paper.

2 Region of Interest Detection

In this paper, we proposed a kind of region of interest detection method based on feature fusion using subspace learning.

2.1 Block Based Skin Feature Extraction

We first convert the original image from RGB color space to orthogonal YCbCr color space. [14] The YCbCr space is one of the most popular choices for skin tone detection and has been used by many researchers such as Hsu et al. Y is the luminance (or intensity) component and Cb and Cr are the blue-difference and red-difference chrominance components. The image is split into small patches, and we extract the color feature and the texture feature separately. The color feature is represented by the histogram. The texture feature is represented by the covariance descriptor. We extract image intensity and norm of first and second order derivatives of intensity in both x and y directions. Each pixel is mapped to a 5 dimensional feature space, and each patch can be represented as a 5×5 dimensional symmetric covariance matrix:

$$C_R = \frac{1}{S-1}\sum_{i=1}^{S}(z_i - u)(z_i - u)^T \tag{1}$$

2.2 Block Based Cigarette Feature Extraction

Here we use the detection method like the skin region detection. We first split the image into small patches, then extract the color feature and the texture feature separately.

2.3 Feature Fusion Based on Subspace Learning

We used a kind of subspace learning algorithm [17] to get the fusion feature of skin color feature and skin texture feature.

Given a data set $X = \{x_i\}_{i=1}^n$, each datum is labeled by $l_i \in L$, where $L = \{l_1, l_2, \cdots, l_c\}$ for the c classes. After applying the above two feature extraction operations on each original datum, we obtain n feature sets $F_i^{(1)} = \{f_i^{(k)}\}_{k=1}^m$ for $i = 1, 2, \cdots, n$. On the other hand, we can also consider the feature sets as $F_k^{(2)} = \{f_i^{(k)}\}_{i=1}^n$ for $k = 1, 2$, each of which contains the same features for different data vectors. Here, when $k = 1$, it denotes the color feature, and when $k = 2$, it denotes the texture feature.

Table 1. Feature Matrix

	$F_1^{(1)}$	$F_2^{(1)}$	$\cdots$	$F_n^{(1)}$
$F_1^{(2)}$	$f_1^{(1)}$	$f_2^{(1)}$	$\cdots$	$f_n^{(1)}$
$F_2^{(2)}$	$f_1^{(2)}$	$f_2^{(2)}$	$\cdots$	$f_n^{(2)}$

We use the Canonical Correlation Analysis to measure the similarity of two feature sets. Suppose F_1 and F_2 represent two feature matrix of two feature sets, which are represented by feature vectors in the columns of the two feature matrices.

Given two orthogonal basis matrices $P_1 \in \mathbb{R}^{D_0 \times d_1}$ and $P_2 \in \mathbb{R}^{D_0 \times d_2}$, we will get $F_1 F_1^T = P_1 \Lambda_1 P_1^T$ and $F_2 F_2^T = P_2 \Lambda_2 P_2^T$, here Λ_1 and Λ_2 denote the diagonal matrices of corresponding eigenvalues. Choose the same dimension, denoted as d_0, of the two subspaces. $P_1^T P_2 \in \mathbb{R}^{d_0 \times d_0}$ is $Q_{12} \Lambda_0 Q_{21}^T$, and $Q_{12}^T Q_{12} = Q_{21}^T Q_{21} = Q_{12} Q_{12}^T = Q_{21} Q_{21}^T = I$. The similarity can be written as

$$S(F_1, F_2) = \max Tr(Q_{12}^T P_1^T P_2 Q_{21}) \tag{2}$$

Where Tr denotes the trace operation.

The subspace matrix P is found by solving the optimization problem.

$$\begin{aligned} P &= \arg\max_P (J_1 + J_2) \\ &= \arg\max_P \Big(\sum_{k_1=1}^{m}\sum_{k_2=1}^{m} S_P(F_{k_1}^{(2)}, F_{k_2}^{(2)}) + \sum_{i_1=1}^{n}\sum_{l_{i_2}=l_{i_1}} S_P(F_{i_1}^{(1)}, F_{i_2}^{(1)}, L)\Big) \end{aligned} \tag{3}$$

Where

$$S_P(F_{k_1}^{(2)}, F_{k_2}^{(2)}) = \max Tr(Q_{k_1k_2}^T P_{k_1}^T P P^T P_{k_2} Q_{k_2k_1}) \tag{4}$$

$$S_P(F_{i_1}^{(1)}, F_{i_2}^{(1)}, L) = \max Tr(Q_{i_1i_2}^T P_{i_1}^T P P^T P_{i_2} Q_{i_2i_1}) \tag{5}$$

J_1 and J_2 are rewritten as

$$J_1 = Tr(P^T A P), J_2 = Tr(P^T B P), \tag{6}$$

here

$$A = \sum_{k_1=1}^{m}\sum_{k_2=1}^{m} (P_{k_1} Q_{k_1k_2} - P_{k_2} Q_{k_2k_1})(P_{k_1} Q_{k_1k_2} - P_{k_2} Q_{k_2k_1})^T$$

$$B = \sum_{i_1=1}^{m}\sum_{l_{i_2}=l_{i_1}} (P_{i_1} Q_{i_1i_2} - P_{i_2} Q_{i_2i_1})(P_{i_1} Q_{i_1i_2} - P_{i_2} Q_{i_2i_1})^T$$

The matrix $P = [p_1\ p_2 \cdots p_d]$ is obtained by solving the eigen-decomposition problem

$$(A + B)p = \lambda p \tag{7}$$

Here $\{p_i\}_{i=1}^d$ are engenvectors corresponding to the d largest eigenvalues.

3 The Prototype Learning Algorithm Based on Metric Learning

3.1 The Formulation of Prototype Learning[6]

For N -class classification problem, prototype learning is to design a set of prototype vectors $\{p_{ls}, s=1,2,\cdots,S\}$ for each class l. An input pattern $x \in R^d$ is classified to the class of the nearest prototype.

For learning the prototypes, consider a labeled training dataset $T=\{(x_n, y_n)$, where is the class label of the training pattern x_n. The objective of learning is to minimize the expected risk based on loss function ϕ:

$$E_{\phi} = \int \phi(f(x) \mid x) p(x) dx \tag{8}$$

where $\phi(f(x) \mid x)$ is the loss that x is classified by the decision $f(x)$, and $p(x)$ is the probability density function at the sample point x. In practice, the expected loss if replaced by the empirical average loss on a training set:

$$\hat{E}_{\phi} = \frac{1}{N} \sum \phi(f(x_n) \mid x_n) \tag{9}$$

Generally, the loss function depends on the parameters of prototypes $\Theta = \{p_{ls}, l = 1,2,\cdots,L, s=1,2,\cdots,S\}$. At the stationary, the loss function R_{ϕ} satisfies $\nabla \hat{R}_{\phi} = 0$.

For minimizing the empirical loss, the prototypes are updated on each training pattern by stochastic gradient descent

$$\Theta(t+1) = \Theta(t) - \eta(t) \nabla \phi(f(x) \mid x) \mid_{x=x_n} \tag{10}$$

where $\eta(t)$ is the learning rate at the t-th iteration.

3.2 Distance Metric Learning Algorithm

There are many metric learning algorithm. In this paper, Eric P.Xing et al.' [18] method is used here.

Suppose we have some set of points x_i, and are given information that certain pairs of them are similar:

$$S: (x_i, x_j) \in S \text{ if } x_i \text{ and } x_j \text{ are similar} \tag{11}$$

Consider learning a distance metric of the form

$$d(x,y) = d_A(x,y) = \| x - y \|_A = \sqrt{(x-y)^T A (x-y)} \tag{12}$$

This problem can convert to an optimization problem

$$\min \sum_{(x_i, x_j) \in S} \| x_i - x_j \|_A^2 \tag{13}$$

$$s.t. \sum_{(x_i, x_j) \in D} \| x_i - x_j \|_A \geq 1 \tag{14}$$

$$A \succeq 0 \tag{15}$$

Here D can be a set of points known to be dissimilar if such information is explicitly available; otherwise, we may take it to be all pairs not is S .

3.3 Prototype Learning Algorithm Using Metric Learning

In prototype learning algorithm, Euclidean distance is directly used in classification. And the distance metric learning is necessary here. It can advance the performance of the prototype learning. The algorithm is showed here.

Prototype learning using metric learning

●1. Initialize the prototype with K-means algorithm.
●2.Select randomly the pairs from the cluster and construct the optimization problems.
●3. Call the Eric algorithms for the distance learning.
●4. Choose a pattern x_n from the pattern pool.
●5. Find the prototype vector m_{ki} and m_{rj} are the nearest prototype to x_n from the positive class and from the negative class.
●6. Update the prototype vector m_{ki} and m_{rj} :

$$m_{ki} = m_{ki} - 2\eta(t)\frac{\partial \phi(x_n)}{\partial g_{ki}}(x_n - m_{ki})$$

$$m_{rj} = m_{rj} - 2\eta(t)\frac{\partial \phi(x_n)}{\partial g_{rj}}(x_n - m_{rj})$$

●7. Repeat 4-6 until convergence.

Output: assign x to the class of the nearest prototype.

4 The Experimental Results

In the experiment, we use the relation of the cigarette region (CR) and skin color region (SCR) to recognize the action.

We use the rule as:

1. No skin color region : Non-smoking.
2. No Cigarette color region : Non-smoking.
3. Two regions are detected \& the distance of the CR and the SCR is far: Non-smoking.
4. Two regions are detected \& the distance of the CR and the SCR is near: Smoking.

To analyze the performance of our proposed method, we used 8100 images in the experiment. These images included 5640 smoking shots and 2460 non-smoking shots. Figure 2 show the sample frames of our data sets.

Fig. 2. The sample frames of our data sets

In the experiment, we use three prototype learning algorithms which are GLVQ, SNPC, LOGM. For the parameters of prototype learning, the initial learning rate $\eta(0)$ was set to $0.1\tau * \text{cov}$, where τ was drawn from $\{0.1, 0.5, 1, 1.5, 2\}$ and cov is the average distance of all training patterns to the nearest cluster center. The prototype number S was selected from $\{1,2,3,4,5\}$; the regularization coefficient α was selected from $\{0, 0.001, 0.0005, 0.01, 0.05\}$. The training parameters and model parameters were optimized in the space of the cubic grid (α, β, τ) by cross-validation on the training data.

Table 2. The experimental Results

Algorithm	Recall	Precision	F_1
GLVQ	94.4%	86.7%	90.4%
GLVQ+ML	95.7%	91.2%	93.4%
SNPC	93.8%	87.1%	90.3%
SNPC+ML	94.5%	90.4%	92.4%
LOGM	94.9%	86.9%	90.7%
LOGM+ML	95.9%	90.1%	92.9%

From the table 2, we can see that the method using prototype learning method based on metric learning is better than using prototype learning method directly. The experimental results show the effectiveness of our method.

5 The Conclusions

In this paper, we proposed a novel smoking shot recognition method. We first extract the color feature and the texture of the skin patch and the cigarette patch. Then we use the subspace learning algorithm to fuse the features. The prototype learning algorithm based on metric learning is used to classify the fusion features. The experimental results show the effectiveness of our method. Our method can be used to recognize other actions in the future.

References

1. Bobick, A., Davis, J.: The recognition of Human Movement Using Temporal Templates. IEEE Trans. on Pattern Analysis and Machine Intelligence 23, 257–267 (2001)
2. Babu, R., Ramakrishnan, K.: Recognition of human actions using motion history information extracted from the compressed video. Image and Vision computing 22, 597–607 (2004)
3. Ke, Y., Sukthankar, R., Hebert, M.: Efficient Visual Event Detection using Volumetric Features. In: Proceedings of International Conference on Computer Vision, vol. 1, pp. 166–173 (2005)
4. Laptev, I., Lindeberg, T.: Space-time Interest Points. In: Proceedings of the Ninth IEEE International Conference on Computer Vision, vol. 1, pp. 432–439 (2003)
5. Gorelick, L., Blank, M., Shechtman, E., Irani, M., Basri, R.: Actions as Space-Time Shapes. IEEE Trans. on Patten Analysis and Machine Intelligence 29, 2247–2253 (2007)
6. Liu, C., Nakagawa, M.: Evaluation of prototype learning algorithms for nearest-neighbor classifier in application to handwritten character recognition. Pattern Recognition 34, 601–615 (2001)
7. Juang, B., Katagiri, S.: Discriminative learning for minimization error classification. IEEE Trans. Signal Process 40, 3043–3054 (1992)
8. Sato, A., Yamada, K.: Generalized learning vector quantization. Advances in Neural Information Processing Systems 8, 423–429 (1998)
9. Sato, A., Yamada, K.: A formulation of learning vector quantization using a new misclassification measure. In: Proceedings of the 14th ICPR, vol. 1, pp. 322–325 (1998)
10. Seo, S., Obermayer, K.: Soft learning vectory quantization. Neural Computation 15, 1589–1604 (2003)
11. Schuldt, C., Laptev, I., Caputo, B.: Recognition human actions: A local svm approach. In: Proceedings of the 17th International Conference on Pattern Recognition, vol. 3, pp. 32–36 (2004)
12. Meng, H., Pears, N., Bailey, C.: Human action classification using SVM 2K classifier on motion features. In: Proceedings of International Workshop on Multimedia, Content Representation, Classification and Security, vol. 1, pp. 458–465 (2006)
13. Jin, X., Liu, C., Hou, X.: Prototype learning with margin-based conditional loglikelihood loss. In: International Conference on Pattern Recognition, vol. 1, pp. 1–4 (2008)
14. Zuo, H., Hu, M., Wu, O.: Patch-Based Skin Color Detection and Its Application to Pornography Image Filtering. In: www, vol. 1, pp. 1227–1228 (2010)
15. Zhu, P., Hu, M., Yuan, C., Li, L.: Prototype Learning using Metric Learning based Behavior Recognition. In: International Conference on Pattern Recognition, vol. 1, pp. 2604–2607 (2010)
16. Wu, P., And Hsieh, J., And Cheng, S., And Tseng, S.: Human Smoking Event Detection Using Visual Interaction Clues. In: International Conference on Pattern Recognition, vol. 1, pp. 4344–4347 (2010)
17. Fu, Y., Cao, L., Guo, G., Huang, T.: Multiple Feature Fusion by Subspace Learning. In: CIVR, vol. 1, pp. 127–134 (2008)
18. Xing, E., Ng, A., Jordan, M.: Distance metric learning with application to clustering with side information. In: NIPS (2003)

Information Visualization of Water Supply and Groundwater

Sang Gon Woo, JinLong Piao, Hyung Chul Kim, Cheol Min Kim, and Seong Baeg Kim

Dept. of Computer Education, College of Education
Jeju National University
ggoong4u@gmail.com, herodragon107@hotmail.com,
k2youngc@naver.com, cmkim@jejunu.ac.kr, sbkim@jejunu.ac.kr

Abstract. The demand of high-quality water for drinking is growing as the interest of well-being increases and the lack of water resource is concerned. So, We propose how to visualize the information of water supply and groundwater, which are hidden under the ground, using IT fusion approaches. First, we developed a visualization system of the water supply information using IT tools like Flex and MAP API supported by a search portal. Second, we implemented a graphical view of the groundwater information with 3D Max. As a result, we can provide the information visualization for people who are interested in the flow of the fresh water to their home through water supply. Also, we can show how the groundwater level changes, depending on the precipitation per region. Information visualization for water supply can clearly give a sound trust that tap water is safe to drink because the whole process that the fresh water flows from the underground to the home can be vividly seen. In case of the groundwater, information visualization would be helpful in managing the groundwater and predicting the groundwater deposits.

Keywords: IT fusion, Information visualization, Water supply, Groundwater.

1 Introduction

Compared with inland region, Jeju island, which lies in the Korea Strait, southwest of Korea, relies heavily on the groundwater by getting most of the necessary water including tap water from the groundwater because there exists no river or lake due to geological features in Jeju. Since people couldn't visually get reliable information about the inner system of the water supply and the underground structure of the groundwater, there has been a concern about the safety of the tap water, which is provided by the water supply. As a good example for enhancing the reliability of the water supply, there is Seoul Water-Now System, which is developed by the office of the water supply, Arisu, Seoul Metropolitan government in Korea. As shown in Figure 1, the system manages the running water in 72 branches and represents its quality and flow visually using the Flash tool on the web[1]. People can understand the process delivered to their homes and check out the water quality. The system helps to reduce a vague concern that tap water is not proper to drink.

G. Lee (Ed.): Advances in Automation and Robotics, Vol. 1, LNEE 122, pp. 541–547.
springerlink.com

Recently, as 3D technology evolves, 3D graphical representation is popular in various fields. For example, 3D movies or animations has been developed.

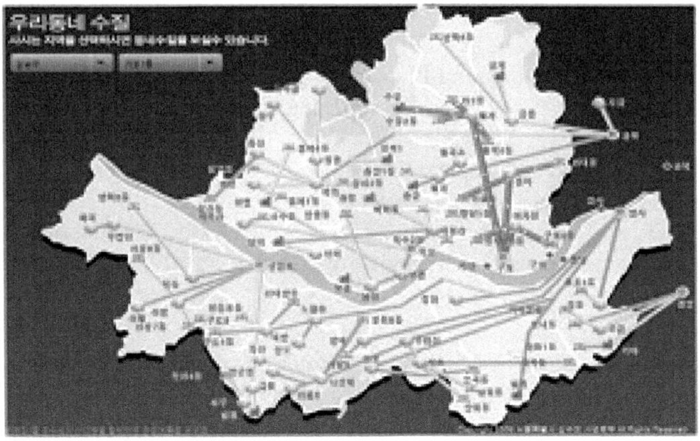

Fig. 1. Seoul Water-Now System

By displaying graphically the process delivered to the home through the water supply from the invisible Jeju groundwater using IT fusion technology, the reliability of Jeju groundwater and the water supply can be increased. To achieve this, using OPEN MAP API, we represent a Jeju map, which visualizes the delivery processes to the home through the water supply from each source and the water quality of each source on the map. Also, we represent 3D map on the groundwater information using 3D Max[5]. As a result, the reliability of Jeju groundwater and the water supply can be higher and the clean image of Jeju can be highlighted.

2 Purpose of the Research

To enhance the international compctitiveness and create new industries, in all kinds of industry fields, more than all, fusing the various related industries into one is very important in the future. To meet this situation, recently, the interest of the studies on the convergence technology of the related industry fields has been raised[2].

In order to boost the competitiveness of the water industry with high potentiality and to achieve a high growth by creating new fields, it is required to apply scientific and advanced technology. In this research, we will study the fusion of the related disciplines and technologies in the field of Jeju water industry. Especially, regarding the groundwater, which is a key element in Jeju water industry, we propose the information visualization methods using the state-of-the-art information technology. Also, we aim to show the process from the underground, which draws out the groundwater, to the home that is the ultimate destination of the groundwater by providing information visualization using graphical tools. In the end, information visualization of water supply can enhance the safety of tap water. The groundwater information visualization can show how the groundwater level and precipitation change.

3 Information Visualization of Water Supply

3.1 Map API (NAVER) and Flash

Using Naver Map API, which is provided by Naver search portal, we can display a map on the web and represent the necessary locations and information on the map [3].

Using Flash Map API, we complete the API map with the packages such as 'Map control registration or use', 'Marking on the map and information window usage', 'Poly-line', '3D map display', and so on. At this point, after finding out the coordinates of each water source using the coordinate transformation API, we mark the point of each water source with the specific image symbol on the map using the transformed coordinates and also represent with arrows on the map how the tap water flows from each water source to the water supply using polylines, as shown in Fig. 2.

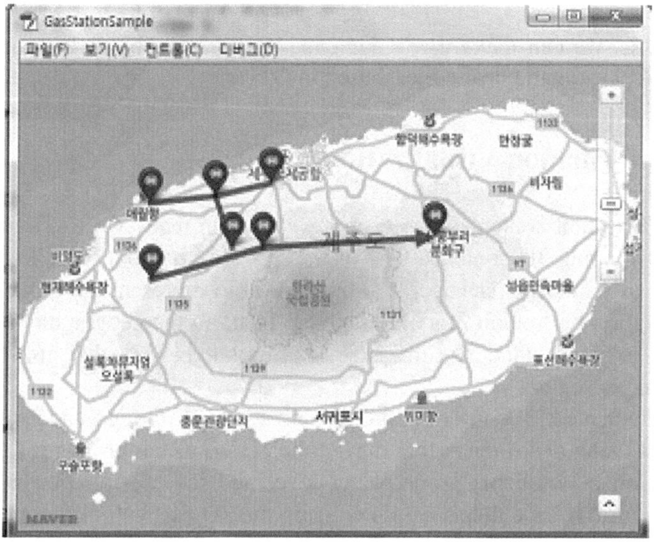

Fig. 2. Visualization of the water supply information using Map API

3.2 Flex

Flex is a kind of software development tool for the development of rich internet applications, which can write the XML tag-based script and generate a Flash module after compiling it[4]. Also, by providing various user interface designs and multimedia replay tools, Flex supports the creation of an executable SWF file not only in the web pages also in PC without browser[4].

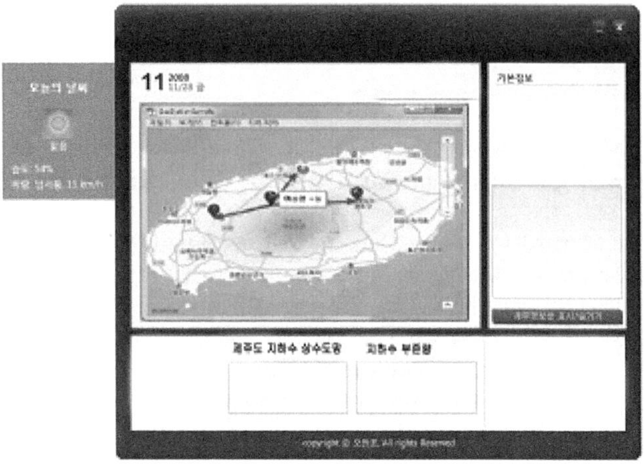

Fig. 3. Interface snapshot using Flex

To use this, we can make user's contents easily accessible, which are developed with MAP API through porting Flex-based interface.

4 3D Information Visualization

Recently, 3D digital contents have been developed in many ways. There are several tools for developing 3D contents. One of them is 3D Max tool, which is a modeling, animation and rendering package. It's meaningful to represent the groundwater level and precipitation per region in a 3D graphical form on the geographic shape because it is easy to understand the massive data related with the groundwater and precipitation per region.

We implemented 3D information visualization of the groundwater level and precipitation using the 3D max tool during the period of our project [6]. To represent the correlation between the groundwater level and the precipitation per region, as shown in Figure 4-7, we displayed them using the 3D bar with different colors. Also, we provided animation to represent their changes according to time. So, it's possible to detect the correlation between them by simply watching the animation of a specific period.

Figure 4 shows the groundwater level and precipitation when there is little precipitation during the winter. To the contrary, Fig. 5 shows the groundwater level and precipitation when there is much precipitation during the summer. Fig. 6 and 7 show the effect of cyclone causing much precipitation in bursts. Fig. 6 represents the groundwater level and precipitation before cyclone is coming while Fig. 7 shows the groundwater and precipitation right after cyclone is gone. We can see from these two figures that both the groundwater level and precipitation become very high after cyclone is gone.

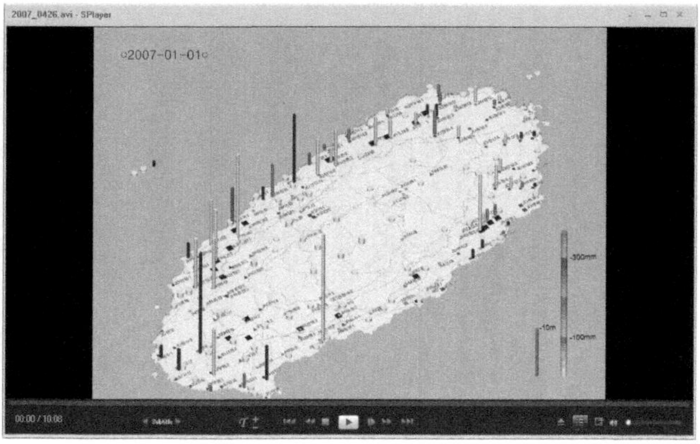

Fig. 4. Groundwater level and precipitation (less rain)

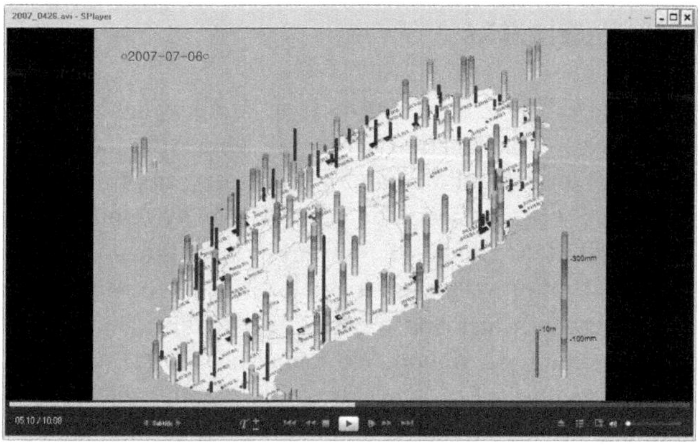

Fig. 5. Groundwater and precipitation (more rain)

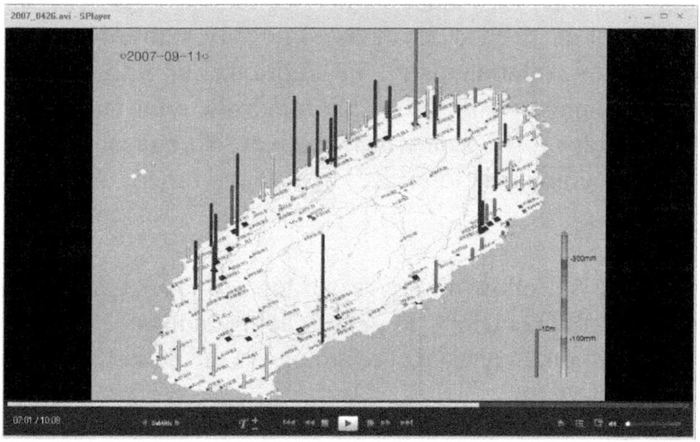

Fig. 6. Groundwater level and precipitation (before cyclone)

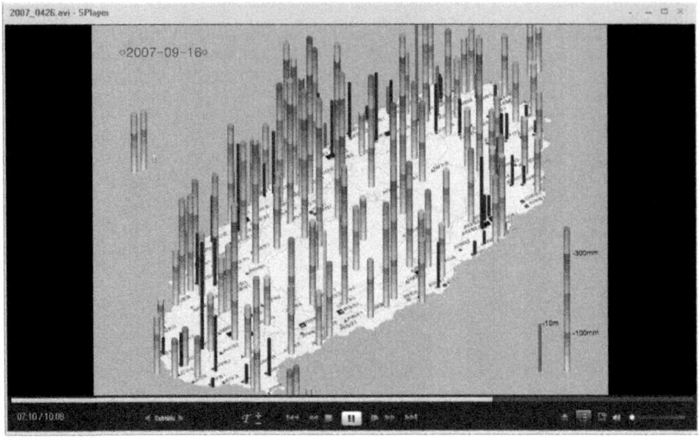

Fig. 7. Groundwater level and precipitation (after cyclone)

5 Concluding Remarks

This paper has proposed how to visualize Jeju groundwater and the water supply information using Flex 3, Flash Action Script 3.0, Map API, and 3D Max. By getting the knowledge of the hierarchical diagram of Jeju water supply, Jeju people can build up the trust of Jeju tap water provided by water supply system from the groundwater and also recognize the clean image of Jeju through information visualization. Also, the information visualization of groundwater level and precipitation per region will help to manage the groundwater and predict the groundwater deposits according to the changes of precipitation per region. We found that the 3D representation can give an intuitive understanding of the massive data collected. If there is no 3D representation, it's so difficult to catch up the meaning of the data collected because the amount of the data is massive as the time period and the spots for collection increase.

Furthermore, in order to make this research a practical and successful development case, it is needed to devise the systematized methodology of IT fusion in the field of water industry. In addition, our information visualization results require the sophisticating process through the continuous discussions with related experts in order to be a practical product. To enhance interaction service with users, human computer interface like touch-based interaction should be examined.

Acknowledgements. A Following are results of a study on the "Human Resource Development Center for Economic Region Leading Industry" Project, supported by the Ministry of Education, Science & Technology (MEST) and the National Research Foundation of Korea (NRF).

References

1. Seoul Water-Now System, `http://arisu.seoul.go.kr/`
2. Ministry of Education, Science and Technology, National Research Foundation of Korea. 2009 Future Convergence Technology Forum - Our Choice of the Competitive Age through Technology fusion (2009)
3. Oh, C.: Mashup guide using Open API. Acorn Inc. (2009)
4. Ok, S.: ADOBE Flex learned by Examples. Acorn Inc. (2008)
5. 3D Max, `http://en.wikipedia.org/`
6. Kim, S.B.: The Development of an Analysis Tool for Analyzing Relations between Precipitation and Groundwater Level by Regional Groups and An Efficient Management Plan through the Fusion of Groundwater and Advanced Information Technology in Jeju-do. Technical Report, Jeju Water Industry Center (2010)

IT Fusion-Based Correlation Analysis between Jeju Regional Rainfall and Groundwater Level

JinLong Piao, Seong Baeg Kim, and Hanil Kim

Dept. of Computer Education, Jeju National University
66 Jejudaehakno, Jeju-si, 690-756, Korea
herodragon107@hotmail.com, {sbkim,hikm}@jejunu.ac.kr

Abstract. The shortage of groundwater deposits and a pollution problem unexpected can occur if Jeju water industry grows fast and the amount of use of groundwater increases. Therefore, it is required to prepare more sophisticated and systematic management plan for maintaining Jeju high-quality groundwater using advanced information technology. So, we propose how to find out the correlation analysis between Jeju regional rainfall and groundwater level. For the correlation analysis, we used data analysis tools like SPSS and Weka based on data mining techniques. We introduced the concept of the sensitivity and devised an approach for finding out the strong positive correlation between them. Our analysis scheme will be helpful to predict the changes of the groundwater deposits according to rainfall per region on Jeju island.

Keywords: Correlation analysis, rainfall, groundwater, data mining, IT fusion.

1 Introduction

In order to bring high growth by creating a new area and boosting international competitiveness of water industry, which is high potential in Jeju island, it is essential to advance Jeju water industry and make it scientific through the systematic convergence of the related discipline and technology[1]. Currently, there isn't yet a problem according to the use of Jeju groundwater considering the current amount of use due to the establishment of groundwater management system for years in Jeju island[2]. However, in the future, if Jeju water industry achieves high growth and the amount of use of groundwater increases rapidly, it is possible to experience the shortage of the potential groundwater reserve amount and some pollution problem[3]. Therefore, it is necessary to devise a new scheme to measure the change amount of groundwater for the management of Jeju high-quality groundwater[4]. The water budget of Jeju groundwater is determined by precipitation, evaporation loss, and efflux. So, we should draw a method of correlation analysis in order to know the degree of the positive correlation between regional rainfall and groundwater level[5].

G. Lee (Ed.): Advances in Automation and Robotics, Vol. 1, LNEE 122, pp. 549–556.
springerlink.com

2 The Overview of SPSS & Weka

The Predictive Analytics software of the package SPSS supports to draw an efficient decision using data in important decision-making. It's possible to reflect on work process by providing prediction, based on analysis of the necessary information collected.

Weka is a suite of machine learning software written in Java used for data mining task. It includes data mining tools such as data preprocessing, classification, regression, clustering, association rules, and visualization. It's popular as a proper tool for data analysis and predictive model.

In this study, we draw a correlation analysis scheme between precipitation and groundwater level using a pair of tools : SPSS and Weka. These tools help to analyze the massive data that is difficult to catch up the correlation between data.

3 Obtaining the Correlation Degree Using SPSS

3.1 Definition of Correlation

When there are two variables x, y, the relation that there is a change of y depending on a change of x, is called correlation. There are two types of correlation : positive and negative. The positive correlation indicates the case that if x increases, then y also increases. On the contrary, the negative correlation implies the case that x and y are in inverse proportion. The correlation coefficient is used for identifying whether the correlation between two variables exists or not. The correlation coefficient r ranges normally from -1 to 1. The sign of r represent positive or negative correlation. The strength of the correlation can be represented by the absolute value | r |. As the value is closer to 1, the strength of the correlation is stronger. When there is little correlation, the correlation coefficient is close to 0.

3.2 Correlation Formula

The correlation coefficient of variables x and y can be calculated in the following steps. There are four steps for calculating the correlation coefficient r that indicates the degree of the correlation. The correlation coefficient r is based on the deviation of variables x and y that have a correlation.

① S(xx) : deviation sum of squares of x is calculated.

$$S(xx) = \sum_{i=1}^{n} (x_i - \bar{x})^2 = \sum_{i=1}^{n} x^2 - \left(\sum_{i=1}^{n} x_i\right)^2 / n$$

② S (yy) : sum of squared deviations of y is calculated.

$$S(yy) = \sum_{i=1}^{n} (y_i - \bar{y})^2 = \sum_{i=1}^{n} y_i^2 - \left(\sum_{i=1}^{n} y_i\right)^2 / n$$

③ S (xy) : multiplied by the sum of the deviation of x and y is calculated.

$$S(xy) = \sum_{i=1}^{n}(x - \bar{x})(y_i - \bar{y}) = \sum_{i=1}^{n} x_i y_i - \left(\sum_{i=1}^{n} y_i\right)/n$$

④ The correlation coefficient r is calculated.

$$r = \frac{S(xy)}{\sqrt{S(xx)S(yy)}}$$

3.3 Correlation Analysis

From the correlation coefficient r calculated, the criteria to identify the strength or weakness of the correlation are as follows.

$|r| > 0.8$: a very strong correlation
$0.6 < |r| < 0.8$: a strong correlation
$0.4 < |r| < 0.6$: a weak correlation
$|r| < 0.4$: little correlation

The correlation is stronger as the absolute value of the correlation coefficient r is bigger.

3.4 Sensitivity

We introduced *sensitivity* that indicates the degree of how fast groundwater level is up. In our study, we found that the sensibility is lower as precipitation is higher. The formula to calculate the sensitivity is as follows. As shown in the formula, sensitivity to precipitation Sp is determined by three factors : groundwater level, precipitation, time lag. The time lag represents the time elapsed after rain. The high value of *sensitivity* indicates that the correlation between groundwater and precipitation exists.

Sensitivity to precipitation $Sp=(100*(L-L_b))/\{P*(T+1)\}$
where L : groundwater level (m)
L_b: base groundwater level (m)
P : precipitation (mm)
T : time lag between rain and water level rise (day)

Fig. 1 shows a snapshot that lists up the *Sp* values calculated using the formula above. In the figure, we can see that *sensitivity Sp* varies according to region and day.

DL3 =(100*(DH3-DI3))/($B3*(1+1))

	A	B	C	D	E	DF	DG	DH	DI	DJ	DK	DL	DM	DN	DC
1	일시	제주평균	서귀평균	고산평균	성산평균	JM도남2	도남2수위차	JM신엄	기준수위	신엄수위차이	base_level	민감도	기준수위3	JM안성	안성수
2	2009-01-01	2	2	1	3	6.9		5.5	1.3		4.2	120		13.8	
3	2009-01-02	1	0	0	1	6.89	-0.01	5.46	1.26	-0.04	4.2	189.6774194	-4.24	13.77	-
4	2009-01-03	1	0	0	0	6.87	-0.02	5.45	1.25	-0.01	4.2	367.5	-4.21	13.76	-
5	2009-01-04	1	0	0	0	6.84	-0.03	5.44	1.24	-0.01	4.2	420	-4.21	13.73	-
10	2009-01-09	3	2	5	2	6.74	-0.01	5.35	1.15	-0.01	4.2	71.70731707	-4.21	13.65	
11	2009-01-10	2	3	1	4	6.71	-0.03	5.33	1.13	-0.02	4.2	96.39344262	-4.22	13.63	-
12	2009-01-11	3	2	2	2	6.69	-0.02	5.32	1.12	-0.01	4.2	77.36842105	-4.21	13.62	-
13	2009-01-12	1	2	1	2	6.67	-0.02	5.28	1.08	-0.04	4.2	150.7692308	-4.24	13.22	
14	2009-01-13	1	0	0	1	6.67	0	5.26	1.06	-0.02	4.2	226.1538462	-4.22	13.49	
15	2009-01-14	1	0	0	1	6.65	-0.02	5.24	1.04	-0.02	4.2	392	-4.22	13.55	
17	2009-01-16	2	0	0	0	6.6	-0.02	5.2	1	-0.02	4.2	140	-4.22	13.53	-
18	2009-01-17	1	0	0	0	6.57	-0.03	5.19	0.99	-0.01	4.2	267.2727273	-4.21	13.53	
19	2009-01-18	3	4	1	2	6.55	-0.02	5.19	0.99	0	4.2	66.81818182	-4.2	13.53	
20	2009-01-19	0	0	0	0	6.51	-0.04	5.15	0.95	-0.04	4.2	#DIV/0!	-4.24	13.48	-
21	2009-01-20	0	0	0	0	6.5	-0.01	5.14	0.94	-0.01	4.2	#DIV/0!	-4.21	13.46	-
22	2009-01-21	6	5	3	5	6.48	-0.02	5.13	0.93	-0.01	4.2	37.21518987	-4.21	13.45	-
23	2009-01-22	1	0	0	0	6.46	-0.02	5.13	0.93	0	4.2	420	-4.2	13.43	-
24	2009-01-23	0	0	0	2	6.43	-0.03	5.06	0.86	-0.07	4.2	#DIV/0!	-4.27	13.39	-
25	2009-01-24	2	2	2	4	6.41	-0.02	5.06	0.86	0	4.2	98	-4.2	13.38	-
26	2009-01-25	3	1	1	3	6.39	-0.02	5.04	0.84	-0.02	4.2	62.55319149	-4.22	13.36	-
27	2009-01-26	0	0	0	0	6.36	-0.03	5.01	0.81	-0.03	4.2	#DIV/0!	-4.23	13.32	-
28	2009-01-27	1	0	0	0	6.36	0	4.99	0.79	-0.02	4.2	420	-4.22	13.24	-
29	2009-01-28	1	0	0	0	6.34	-0.02	4.99	0.79	0	4.2	367.5	-4.2	13.24	
30	2009-01-29	10	7	5	12	6.33	-0.01	4.99	0.79	0	4.2	21.77777778	-4.2	13.28	
31	2009-01-30	20	10	7	25	6.32	-0.01	4.99	0.79	0	4.2	10.61371841	-4.2	13.32	
32	2009-01-31	0	0	0	0	6.34	0.02	4.96	0.76	-0.03	4.2	#DIV/0!	-4.23	13.29	-
33	2009-02-01	0	0	0	0	6.35	0.01	4.93	0.73	-0.03	4.2	#DIV/0!	-4.23	13.24	-
34	2009-02-02	2	2	2	2	6.36	0.01	4.92	0.72	-0.01	4.2	117.6	-4.21	13.23	-
35	2009-02-03	13	6	6	11	6.35	-0.01	4.92	0.72	0	4.2	15.72192513	-4.2	13.27	
36	2009-02-04	0	0	0	0	6.36	0.01	4.91	0.71	-0.01	4.2	#DIV/0!	-4.21	13.26	-
37	2009-02-05	0	0	0	0	6.36	0	4.88	0.68	-0.03	4.2	#DIV/0!	-4.23	13.23	-
38	2009-02-06	0	0	0	0	6.36	0	4.86	0.66	-0.02	4.2	#DIV/0!	-4.22	13.19	-
39	2009-02-07	0	0	0	0	6.35	-0.01	4.85	0.65	-0.01	4.2	#DIV/0!	-4.21	13.16	-
40	2009-02-08	0	0	0	0	6.34	-0.01	4.84	0.64	-0.01	4.2	#DIV/0!	-4.21	13.17	
41	2009-02-09	3	1	1	1	6.33	-0.01	4.84	0.64	0	4.2	72.59259259	-4.2	13.19	
42	2009-02-10	0	0	0	0	6.34	0.01	4.82	0.62	-0.02	4.2	#DIV/0!	-4.22	13.18	-

Fig. 1. Sensitivity Calculation

3.5 Making the Correlation Clear

In order to make the correlation clear, we calculated the correlation by classifying precipitation into the units of 10mm, 20mm, and 30 mm above. In the approach, the cases that show high correlation are at the following.

- More than 10mm: 0.628
- More than 20mm: 0.708
- More than 30mm: 0.654

In case the precipitation is more than 20 mm, the degree of the correlation is high. So, it's possible to make the correlation clear by filtering the data between 20 mm and 30 mm. Fig 2, 3, and 4 show how the correlation coefficient differs according to the unit of precipitation.

The correlation coefficient

Correlation coefficient

		Jeju	Sineom	Jocheon
Jeju Average	Pearson	1	.628**	.245
	p-value		.000	.200
	N	29	29	29
Sineom water Level difference	Pearson	.628**	1	.372*
	p-value	.000		.047
	N	29	29	29
Jocheon water Level difference	Pearson	.245	.372*	1
	p-value	.200	.047	
	N	29	29	29

Fig. 2. Rainfall with more than 10mm

The correlation coefficient

Correlation coefficient

		Jeju	Sineom	Jocheon
Jeju Average	Pearson	1	.708**	.162
	p-value		.000	.494
	N	20	20	20
Sineom water Level difference	Pearson	.708**	1	.301
	p-value	.000		.197
	N	20	20	20
Jocheon water Level difference	Pearson	.162	.301	1
	p-value	.494	.197	
	N	20	20	20

Fig. 3. Rainfall with more than 20 mm

The correlation coefficient

		Jeju	Sineom
Jeju Average	Pearson	1	.654*
	p-value		.011
	N	14	14
Sineom water Level difference	Pearson	.654*	1
	p-value	.011	
	N	14	14

Fig. 4. Rainfall with more than 30 mm

4 Association Analysis Using Weka

We used the data mining method provided by Weka3.7.0 program to do association analysis. The association analysis was applied for the purpose of drawing out the association rules. In the process of the association analysis, we gave the following two conditions. First, Aprioir algorithm applicable for only discrete data used 'NumericToNominal' filter to converts data type to Nominal. Second, the Aprioir algorithm's minMetric(reliability critical value) is set to 0.7 and numRules(extraction rules number) is set to 30. If there was no rain on Jeju island, Samyang groundwater level was 0 and the reliability was 80%; the same condition Hagwi ground water level was -0.02 and the reliability was 76%. If the precipitation, which was highest, was 244, the difference value of Samyang, Shindo, Taeheuang, and Hagwi are 0.93, 1.48, 0.59, and 1.06, respectively.

Fig. 6, 7, and 8 show the results of the association analysis. The degree of correlation is represented on the screen using dots. Fig. 9 shows the correlation between total rainfall and water level.

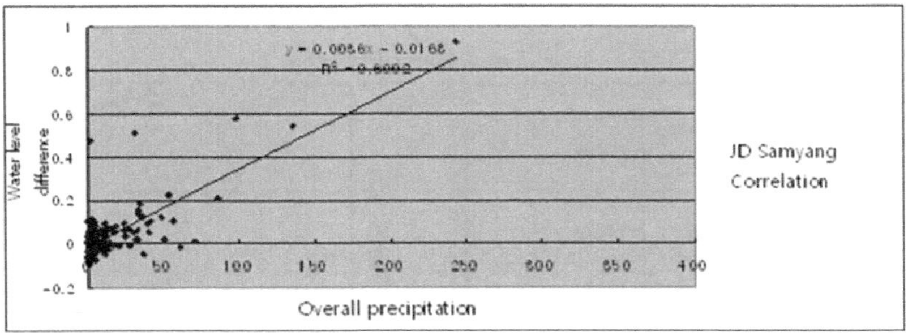

Fig. 6. Association analysis 1

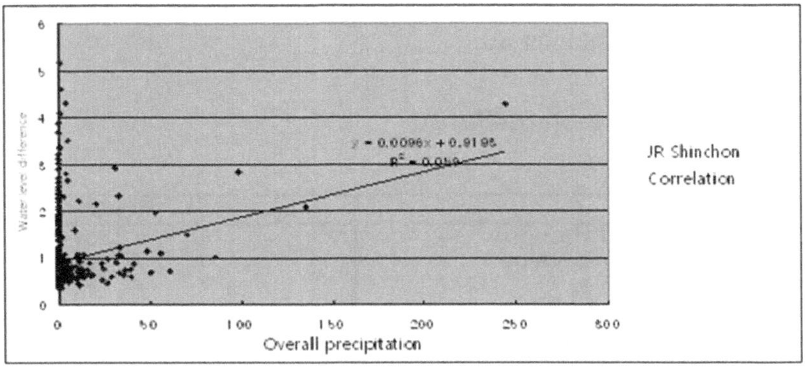

Fig. 7. Association analysis 2

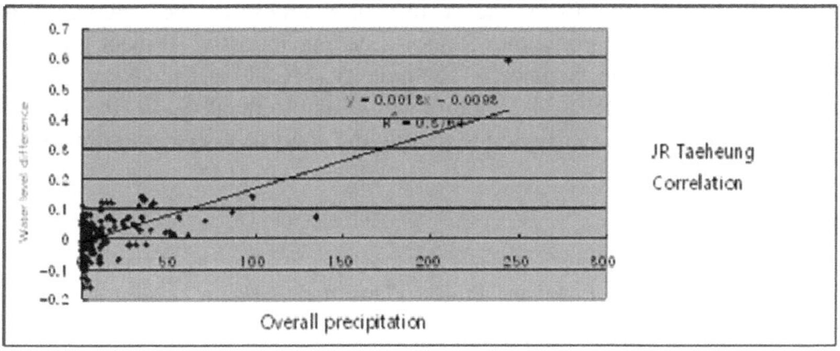

Fig. 8. Association analysis 3

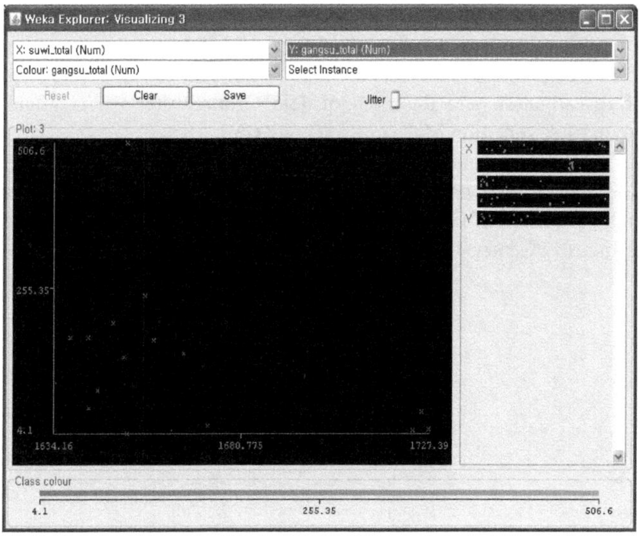

Fig. 9. Correlation between total rainfall and water level

5 Conclusion

This paper has drawn a correlation analysis scheme associated with precipitation and change of groundwater level, which have an effect on Jeju groundwater, using statistical software such as SPSS and Weka. With this scheme, Jeju groundwater can be managed more efficiently and scientifically because the prediction of groundwater change can be achieved. It is required to develop an analysis tool that can manage simply the correlation-related formula proposed in this paper. Further, a method to make the correlation more clear between them should be examined continuously by applying advance information technology.

Acknowledgements. Following are results of a study on the "Human Resource Development Center for Economic Region Leading Industry" Project, supported by the Ministry of Education, Science & Technology (MEST) and the National Research Foundation of Korea (NRF). The authors would like to thank Soon Hee Im and Prof. Ik-Chan Pang who have contributed to getting the analysis results with this work.

References

1. Ministry of Knowledge Economy, Project Plan for Fostering a Leading Industry in Jeju Metropolitan Economy Area (2009)
2. Chang, W.I., Kim, S.H., Park, S.J., Park, S.H.: IT-based convergence technology (medical, IT-oriented convergence) Commercialization Trends. E-Communications Trend Analysis 23(5) (2009)
3. Ministry of Knowledge Economy, The 5th Future Planning Committee Report: IT KOREA Five Future Strategies (2009), `http://itstartup.go.kr/upload/bbs/20090902.hwp`
4. Ko, K.W.: Characteristic of Jeju Island Groundwater and Water Resource Management (I)- History of Investigation and Research of Jeju Groundwater - Water and Future. Korea Water Resources Association Magazine 39(7) (2006)
5. Kim, S.B.: The Development of an Analysis Tool for Analyzing Relations between Precipitation and Groundwater Level by Regional Groups and An Efficient Management Plan through the Fusion of Groundwater and Advanced Information Technology in Jeju-do, Jeju Water Industry Center Technical Report (2010)

Mobile Digital Storytelling Development for Energy Drink's Promotion and Education

Yong-Seok Kim[1], Min-Ji Han[1], Chang-Jun Hong[1], Seong Bo Ko[2], and Seong Baeg Kim[1]

[1] Dept. of Computer Education, College of Education, Jeju National University
[2] Dept. of Industrial and Applied Economics, College of Applied Life Sciences, Jeju National University
sbkim@jejunu.ac.kr, rlapstmd@nate.com, juneoovv@nate.com
minji3661@gmail.com, ksb5263@jejunu.ac.kr

Abstract. Digital storytelling is the process in which stories are expressed in digital media, which is invaluable due to the limitless potential in the field of advertisement and education. An intriguing digital storytelling is to have engaging and high sensual stimulating animations that contain the narration required for the desired result. For this study, we deal with energy drinks which have recently had a high attention and rapidly growing markets. In particular, we focus on Jeju specific energy drinks along with a narration for hand-held devices such as mobile phones by implementing interactive animations which include a touch interface and a narration involved.

Keyword: Digital storytelling, Energy drinks, Jeju specific energy drinks, Interaction, Mobile-Web, Touch interface.

1 Introduction

Recently, commercials with in depth stories have been invaluable to the industry as a key component and thus are the focus of attention in the world of today and tomorrow. Digital storytelling is a technique in which messages are expressed through the means of digital technology like using the media for websites such as Youtube. In our culture, the fad of energy drinks haven't been dominant in today's society so government focus have not shifted towards the energy drink industry like the western culture have. In this regard, this study aims to develop the animation about Jeju specific energy drinks, which uses digital storytelling techniques. Worldwide energy drink markets are growing, but there is a lack of interest in energy drinks in Korea. However, the general public needs to be aware about energy drinks because energy drinks markets will increase in Korea.

The main ingredient of Jeju specific energy drinks is Jeju Jingyul, which is a type of orange, designated as a natural monument with a large historical value. But the Jeju tangerine has suffered from tariffs elimination and introduction of seasonal tariffs by the Korea-US FTA. In this situation, Jeju Jingyul, which has

G. Lee (Ed.): Advances in Automation and Robotics, Vol. 1, LNEE 122, pp. 557–566.
springerlink.com

more than 1,000 year history, can be a core of digital storytelling and be developed as an animation that introduces Jeju history, allowing people to become aware of the Jeju tangerine protection. This study focuses on aspects of interaction. The animation of enhancing interactivity will become a novel animation for users rather than providers. In addition, this animation using storytelling techniques is expected to bring the effect of Jeju-specific energy drinks' promotion and education.

2 Background

2.1 Energy Drinks

Energy drinks are energy supplement drinks, with existing beverages focusing on thirst, fatigue, and relieving hangovers but Jeju brand drinks are designed to drink before you start work or an activity. More than 14 billion dollars in the industry have already formed the energy drink market and has matured as an industry with a stable annual growth. The need for energy drinks have already been integrated into the social as people demand such drinks for sports, clubs, and parties, making new energy drinks very easy to introduce and grow as the demand increases. Back in 1997, *Red Bull* was launched from the energy drink market in the United States. However, it has become *Red Ocean* already. In the early days, only small and medium-sized companies produced energy drinks, but now even *Coca-Cola, Pepsi* and *Starbucks* have joined the energy drinks market. In Nigeria, as well as the United States, more than 30 brands compete in the production of energy drinks. In Austria, energy drinks increased by nearly 20 percent from 2007 to 2008, demonstrating just how in demand the energy drink markets are.

In Korea, energy drinks were introduced in August 2007. A New York-based *Turnon* of 'Turn-Berber Ridges' was known as the first importer to sell. Then in March 2010, *Lotte* released a 'Hotsix' and Dong-A Otsukas 'X cocas', Bokaenbokui Dash, Kwang-Dong Pharmaceutical for 'PowerShot' , Myung-Mun Pharm 'Power Ten' and other competitive products to the market, as the market began to be formed.

2.2 Jeju-Specific Energy Drinks

Jeju Island has also developed an energy drink referred to as 'Gemurosa'. Many people drink Gemurosa to be more focused, as well as to mix with alcohol, relieving hangovers and fatigue. Gemurosa is made of Jeju native bark and Dangyuja, Jingyului, Mugwort, Cassia, licorice, and various herbs. When it was tested, the degree of hepatotoxicity was greater than other products. Gemurosa also includes 'dried orange peel' that is completely dried before the orange ripens. The dried orange peel is not only recommended as fragrant bath salts, but also has effects that make skin smooth.

2.3 Digital Media and Storytelling

Stories expressed on paper are considered literature. If they are expressed in visual media, they are referred to as film. Finally, if expressed in digital media, they become digital storytelling.

Digital storytelling include narrative activities on your computer and web-based interactive multimedia narrative creations, which are made up of texts, images, songs, voices, videos, and animations[4].

Storytelling allows children to not only improve their vocabulary, but also stimulates boundless imagination. It satisfies people emotionally and is one of the effective ways to do it creatively. Moreover, storytelling is one of the essential factors to develop edutainment contents[5]. A company makes a good story out of their products and uses the story as a marketing tool in promoting and advertising, which is brand communications so-called "Marketing Story". A story made collaboratively by a business and customer together is an interactive method of communication to stimulate the consumer's emotion and gives them a lot of pleasure. Above all, storytelling can be used to emphasize a company's image or the relevance of products, adding value for their customers, which is important for making the customers interested or creating a good impression.

3 Interactive Mobile Web Animation

3.1 Definition of Interaction

Interaction consists of personal or social exchanges of information and knowledge. It can also be invoked by the media in order to exchange information, in which interaction is a necessary catalyst to achieve the bi-directional, dynamic, self-directed. Interactions that appear in our daily lives are examples of important communication activities.

Studies about interaction and interactivity have flourished since with the advent of the Internet and the emergence of new media. Through interaction, today's mobile web animation aids in the effective upgrade of knowledge easily and the animation helps to convey information effectively.

3.2 Interactive Mobile Web Animation Effect

3.2.1 Education Effect

Mobile web animation interaction provides a wealth of information and open webs and integrated environments not to experience traditional education. Educational media found on the web are highly interactive unlike the traditional one-way communication. Learners using the computer communications and online discussions with other learners and other experts can exchange ideas and information to interact creatively and actively. Also, it's possible to provide a lesson by developing an educational fairy tale for children.

3.2.2 Promotion Effect

Ads produced by animation techniques are clearly different from the real picture. Animation is a unique narrative, which is simple and imaginative. Fiction image and exaggerating techniques can maximize the effect that traditional ads don't have. Also, a real picture is limited by physical phenomena, but animation ads are realistic and less restrictive [2].

4 Mobile Digital Story 'Devoted Daughter Rosa'

4.1 Character

A character isn't simply a tool for the production of the narration but the heart and soul of the story due to the fact that the characters sets the mood and creates the feelings for the story. Through that, it shall be to show at it in order to express the general inside the world of a work. Screening time is short on this animation, it is decided concretely on inner and external setting of a character[3].

Table 1. Main characters' characteristics of 'Devoted Daughter Rosa'

	Name	Characteristics
Main character	**Myeong-su**	A main character that lives in the modern society. A young man who is 20 years old. Myeong-su lives with his aging mother, however, he is a childish university student and likes to play with his friends. One day when Myeong-Su is suffering from a hangover, he overhears someone talking about energy drinks and becomes interested.
	Rosa	A main character at the medieval times. A girl of 15 years old. She is living with sick widows. As filial piety is deep, in order to heal her mother's disease, she has tried everything but nothing works. She is growing a Jeju Jingyul tree, but gets a small number of profits only. Although she is poor, she has a cheerful character.
Helper	**Gemu**	A neighborhood friend of Rosa from the past. He is a special friend for Rosa because he helps her whenever Rosa has difficulties. He passed a government official test when he was 15 years old and he is a subordinate of the king.

Table 1. (*continued*)

Supporting character	**Rosa\`s mother**	Rosa's mother raised Rosa brightly in a situation to have been hard. Although she has a life-threatening disease, she tries not to be a burden to Rosa.
	King	The King has the same as disease as Rosa's mother. He tries to take medicine but the disease cannot be easily cured.

4.2 Period and Space Background

There are current and past background in it. In the current points in time, an entertainment area of the city center is the background. In this place, the main character Myeong-su is drunk of liquor every night. In the past points in time, a king appears in it. The main character is Rosa who is living with widows that are in pain. Through comparing current and past times, it can be an educational element.

4.3 Story Line

Table 2 shows the outline of the story that includes three elements. The three elements are public information effect, educational effect, and interest effect. For public information effect, the superiority of Jeju-specific energy drink is emphasized naturally as the story flows. The storytelling provides an educational effect because it gives a lesson to children, based on a fairy tale. For interest effect, we provide a mobile game and interactive animation.

Table 2. Story Line

Story summary	Myeong-su is suffering from a hangover because of his excessive drinking yesterday. He hears that energy drinks are good for hangovers and tries to search for more information, and finds some animation on the Internet. Rosa that who is living with painful widows is main character in the animation. She is growing a Jeju Jingyul tree, but she gets a small number of profits only. Rosa\`s mother is suffering from a continuous cough and digestion problems. One day, Rosa runs out of money and cannot buy medicine for her mother anymore. Rosa makes tea out of Jingyul\`s peel and lets her mother drink it. Unexpectedly, the state of the disease improves gradually and her mother is recovered completely after having drunk Jingyul\`s peel tea. On the other hand, Rosa\`s friend Gemu is a subordinate of the king. He is under enormous pressure and

Table 2. (*continued*)

	stain in order to heal the king's disease. Gemu hears of Rosa`s mother news and gives the Jingyul`s peel tea for the king to drink. The king's health becomes all better. The king gives a large compensation to Rosa, and Rosa lives happily with her mother.
Public information effect	- Inform the record which a king eats Jeju Jingyul through stories. - Introduce the energy drink's effect through stories. - Main ingredient introduction.
Educational Element	- History repeats itself. - Filial duty emphasis. - Deliver message of hope through Rosa who lives hard in difficult situations.
Interest element	- Story itself - Interactive animation

4.4 Interactive Animation

Possibility to optionally accommodate the content that users want at any time and the expansion of interactivity which allows users to participate actively gives an active choice in their consumption. This changes the one-way communication occurs between the mass media and users.

Users expect to participate more actively than one-way web-animation for education or promotion. In addition, main portal sites are focusing on mobile business strategy while the popularity of smart phone and Tablet PC is growing. Due to the rapid growth of smart phones, this affects every aspect of society as well as the IT field. This especially changes the web at a rapid pace. The web is not only viewed simply on PC but also viewed on mobile.

If digital animation is made for mobile, it is easy to modify existing data and the design. Furthermore, although there are disadvantages, such as the need to spend a lot of loading time and should be downloaded to view when existing 2D/3D digital animations and videos are expressed on web, it becomes data with low capacities and a variety of animation effects can be represented by using flash animation.

Accordingly, the digital story based on Jeju Jingyul's history makes users interact with their mobile devices.

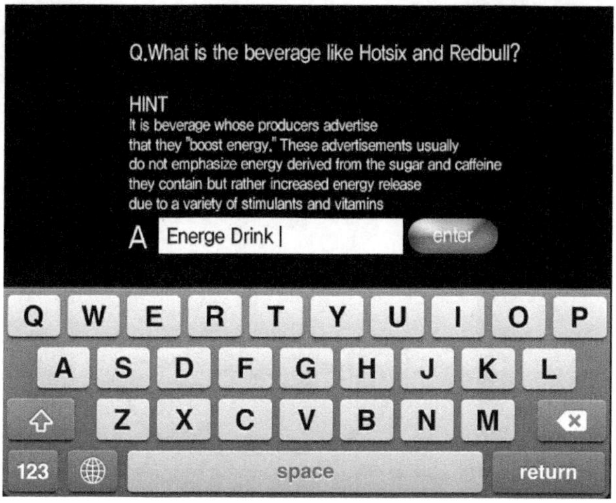
Q.What is the beverage like Hotsix and Redbull?
HINT
It is beverage whose producers advertise
that they "boost energy." These advertisements usually
do not emphasize energy derived from the sugar and caffeine
they contain but rather increased energy release
due to a variety of stimulants and vitamins
A
Energe Drink
enter
Q W E R T Y U I O P
A S D F G H J K L
Z X C V B N M
123
space
return

Fig. 1.

harvest
There's a big piece of fruit growing on that tree up there!
6
148
05

Fig. 2.

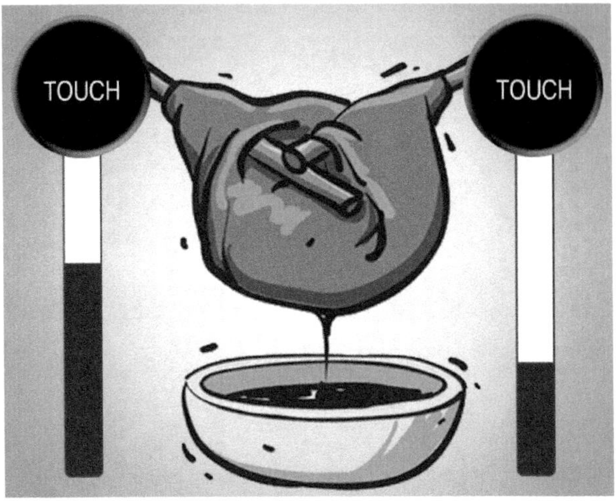

Fig. 3.

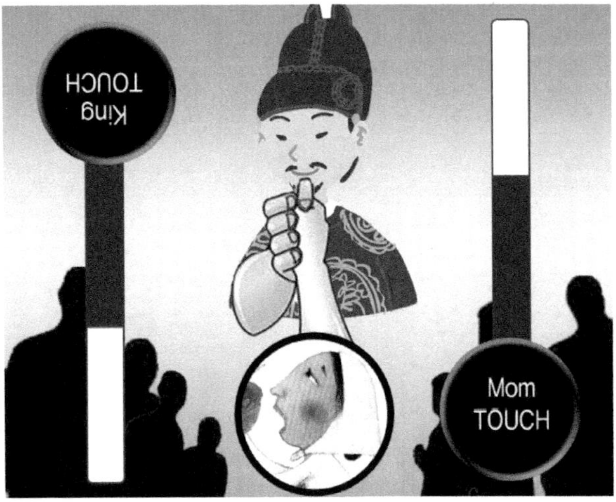

Fig. 4.

<Mobile digital animation ‘Devoted daughter Rosa’>

Fig.1. is an introduction to the mobile digital animation ‘devoted daughter Rosa’, which is the screen before this animation is started. When the main character Myeong-Su in the current time becomes interested in energy drinks, he is supposed to find this animation called ‘Devoted daughter Rosa’. If users want to watch this

animation, they should solve the question shown in Fig.1. That question is: 'What are energy drinks?' Therefore, through this game, this animation leads to see great results from promoting and stimulate their curiosity.

Fig.2. is one of this animation's scenes. The main character Rosa, taking place in the historic times, is supposed to plant Jeju Jingyul trees. In this scene, users play the game like 'Planting a tree' by using touch interface. It's familiar to users because there are many games like this on the internet. This game can be good in that it forces users to concentrate on this animation.

Fig.3. is the scene that Rosa makes the tea from Jingyul`s peel. This part is similar to the flash game like 'Making a juice'. It gives users virtual experience to make Jeju specific energy drinks on a computer. Consequently, users become the hero of this animation and solve some mission with mobile touch skills such as simple drag and touch operation, which makes users interact with their mobile devices

Fig.4. is expressed to appeal Rosa's mother and King who get their health and have a strong power after this story is finished. They are supposed to life a stone in this short game. Therefore, this scene emphasizes the effect of the tea which is made of Jingyul's peel.

Through the virtual experience like short flash games, users don't simply watch the animation but it makes users interact with their mobile devices using mobile touch skills. It also must solve a mission to continue watching the animation, which makes users draw their interest and gives them some information about energy drinks.

5 Conclusion

In this study, the application of how a interactive story can be use in order to sell the product as well as engage the user in more detailed and in depth content shows how the usage of digital storytelling is more advantageous than the ordinary commercials. It breaks from the existing formal of single direction caused superficial activity. This fact has meaning rather than simply viewing. It can experience to interaction with users. Through the animation, users can experience public information and educational function. Also, it isn't to provide simply information through 'digital storytelling' techniques, but also uses storytelling technique to stimulate the consumer's emotions and gives them a lot of pleasure.

Acknowledgements. Following are results of a study on the "Human Resource Development Center for Economic Region Leading Industry" Project, supported by the Ministry of Education, Science & Technology (MEST) and the National Research Foundation of Korea (NRF).

References

1. Next economy, Energy drinks and emerged as a new Blue Ocean (September 1, 2011), http://www.nexteconomy.co.kr/news/articleView.html?idxno=6208
2. Kim, M., Boo, K.-M., Im, K.-D., Ko, S.B., Kim, S.B.: An Interactive UCC Creation and the Effect Analysis. Korean Journal of the Science of Emotion & Sensibility (2010)
3. Kim, H.-J.: A Study on the Stop Motion Animation for Television Advertising. Master's thesis, In-Cheon Graduate School Fine Arts (2010)
4. An, J.: The Easiest Short Animation Free-Production. JungMun, Seoul (2005)
5. Choi, H.-S.: Digital Storytelling, The Korean Institute of Information Scientists and Engineers (2003)
6. Baekbompul: Effective Storytelling Strategy of Children's Edutainment Animation. Master's Thesis, GyeongSeong University (2009)

Agile Automation Systems Based on Cyber-Physical Systems and Service-Oriented Architectures

Detlef Zühlke[1] and Lisa Ollinger[2]

[1] Innovative Factory Systems, German Research Center for Artificial Intelligence, Trippstadter Str. 122, 67663 Kaiserslautern, Germany
zuehlke@dfki.de
[2] Institute of Production Automation, University of Kaiserslautern, Gottlieb-Daimler-Str. 42, 67663 Kaiserslautern, Germany
lisa.ollinger@mv.uni-kl.de

Abstract. Nowadays, manufacturing companies have to face increasing challenges to keep pace in our globalized world. The key enabler to fulfill the rising demands is an industrial automation system that supports dynamically changes of the manufacturing equipment and the automation software with low effort. Since current automation systems aren't able to meet these requirements, we need significant changes. Therefore, we present our vision of the automation network that is based on the concepts of Cyber-Physical Systems and Service-oriented Architectures. For realizing this vision we point out which aspects and influencing factors have to be taken into account and propose approaches for a gradual implementation.

Keywords: Industrial automation, SOA, Cyber-Physical Systems.

1 Introduction

Production is facing dramatic changes over the next decade. The worldwide markets demand for new goods in ever shorter time leading to shorter product life cycles mostly combined with an increasing product complexity. Additionally, the customer is asking for a high product quality at lower and lower prices. To keep pace with global competition, manufacturing companies have to adapt their production planning and their manufacturing plants in increasingly shorter time cycles to a wide-range of influencing factors. Therefore, highly adaptive, flexible, and reconfigurable production plants are needed. The key enabler for realizing organizational and technical adaptions dynamically and with low effort is a suitable automation system of the production plant. However, current control structures in industrial automation don't meet the requirements for realizing such agile automation systems.

This paper describes a new control paradigm based on service-orientation for future factory automation solutions and points out what is needed to establish such control architectures in future.

G. Lee (Ed.): Advances in Automation and Robotics, Vol. 1, LNEE 122, pp. 567–574.
springerlink.com

2 Control Architectures Today

A well-known representation of industrial control architectures is the automation pyramid (Fig. 1). It displays various layers that represent different automation levels. Each layer has its typical automation systems that realize the automation tasks, like PLC, MES, or ERP systems. The lowest level contains the field devices that execute the production process and are operated via their input and output channels (I/Os). The layer above, the process control level, contains the control devices that control and monitor the actions of the field devices, wherefore typically programmable logic controllers (PLC) are used. For implementing the control programs the process logic reads the current output signals and sets the input signals of the field devices depending on the ongoing process. The connection between the field devices and the process control devices is realized through field bus systems or old-fashioned wiring by using a wide variety of communication standards.

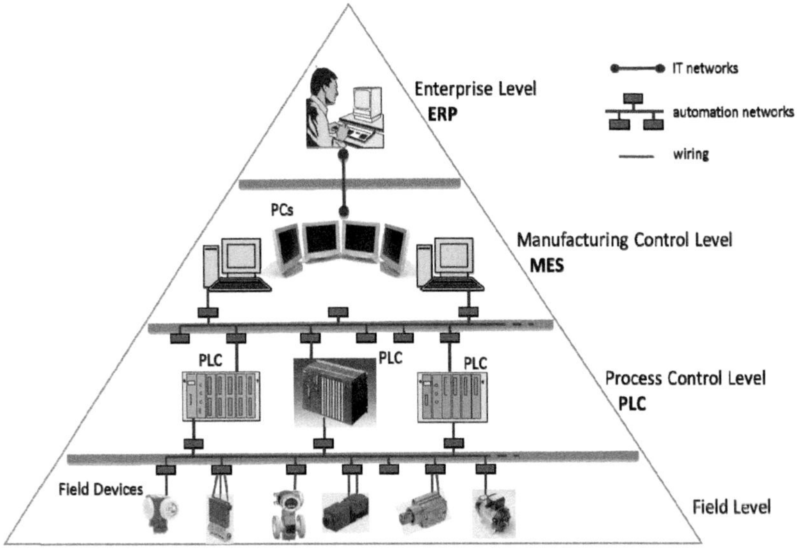

Fig. 1. The pyramid of automation

On the higher layers of the pyramid, the more high-level management and business applications of the production control takes place. Whereas these applications already work with standard IT systems and standards like ordinary PCs, Ethernet, TCP/IP in the lower automation layers special hardware and automation standards are still used. This situation is due to the strict requirements concerning reliability, availability, real-time capability, etc., so that the integration of new IT concepts in lower automation levels is rather restrained.

The drawback of these traditional structures in automation is the overall lack of integration capacity and a high complexity. Since the programming takes place on signal level the control programs are hardware-dependent and complex. For a layer-spanning connection of different automation devices special middleware is needed which is tailored to the respective application. Thus, today's control architectures don't meet the requirements concerning horizontal and vertical integration and an advanced software design and implementation. To cope with these challenges new concepts for industrial automation are required.

3 The Vision of the Automation Network

Nowadays, the biggest drivers for the technical progress are advances in micro electronics and information and communication technology (ICT). Applying new concepts from theses domains to industrial automation can help us to improve our traditional control structures. In the following two concepts are used to develop a new control paradigm for industrial automation; namely Cyber-Physical Systems (CPS) and Service-oriented Architectures (SOA).

In 1991 Mark Weiser described the vision of future computing: "Ubiquitous computing names the third wave in computing, just now beginning. First were mainframes, each shared by lots of people. Now we are in the personal computing era, person and machine staring uneasily at each other across the desktop. Next comes ubiquitous computing, or the age of calm technology, when technology recedes into the background of our lives" [1]. Since then this new paradigm of ubiquitous (or pervasive) computing has more and more become reality. Each of us owns many computers mostly invisible in every day devices like smartphones, cameras, or heart pace makers. The second accompanying revolution can be seen in the communication area. The embedded computers will be even more powerful when they are linked in a network and can exchange information. In the meantime, we even talk of an Internet-of-Things characterized by a situation where each network node is becoming smart. This means that each node can act as an independent addressable computer in a worldwide standardized network. Combining the ideas of ubiquitous computing and the Internet-of-Things will lead us to a new paradigm: Cyber-Physical Systems (CPS). A CPS comprises a network of physically distributed embedded sensors and actuators equipped with computing and communication capabilities [2]. They constitute integrations of computation with physical processes by monitoring and controlling the physical process by embedded computers [3].

Applying this vision to industrial automation structures we will achieve a new control architecture, the automation network (Fig. 2). This vision comprises the automation systems as a CPS so that every device is able to interact with other devices and has some computational power. This enables a strong horizontal and vertical integration within automation systems. For realizing such an automation network, the question arises how the communication is realized with high interoperability and how the tasks can be organized in a transparent, hardware-independent, and flexible way. To avoid the problems due to the signal-based communication we have to move to more abstract descriptions of control processes. Then hardware and software can be linked at a later stage so that processes within the complete life cycle from planning to design, operation, and maintenance can be realized more dynamically and with lower effort. Furthermore, abstract component and task models can improve the interoperability between communication partners so that the basis for high horizontal and vertical integration is provided [4].

A concept that enables such a higher degree of abstraction by encapsulating reusable functions in an open and standardized way is the paradigm of Service-oriented Architectures (SOA).

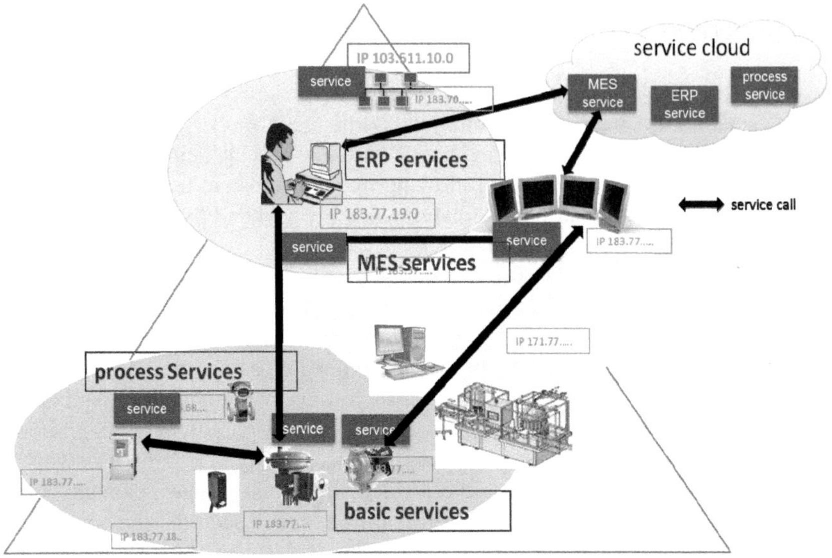

Fig. 2. The vision of the automation network based on SOA and CPS

4 Service-Oriented Automation

Applying the paradigm of service-oriented architectures to industrial automation is a promising approach to decrease significantly the integration and programming efforts within automated production systems. Therefore, we developed the approach SOA-AT to describe the application of service-orientation in the context of industrial automation. Service-oriented architectures have been used for IT applications for a while, particularly in the field of business processes. However, SOA itself is defined as a general concept for software architectures that represents versatile, different methods and applications as reusable and openly accessible services for enabling a platform and implementation independent usage [5]. These services have a standardized and openly accessible interface, whereby the implementation of the functionality is hidden. Thus, SOA is technology independent, high-level concept that provides architectural blueprints [6].

In terms of industrial control we will receive a SOA-AT when all automation functions within the automation network are encapsulated as services. Therefore, the mechatronic functions on field level that execute and monitor the production process are represented as basic services. They constitute the link between the physical and computational processes within the automated production facility. For implementing higher automation functions like control programs, theses basic services can be aggregated to composed services [7]. The process logic for building the high-level functions of the composed services by using other existing services is called service orchestration. The foundation for orchestrating services is the abstract description and representation of mechatronic and computational functionality, and the process logic. This abstraction offers the possibility that automation processes can be planned independently of the production equipment so that the hardware can be selected and assigned at later stages of the planning process. Thus, dependencies during the planning phase can be reduced so that planning processes can be parallelized and the complexity of planning and implementation processes can be reduces [8].

Altogether, our vision of the automation network based on the concept of SOA-AT ensures a holistic automation architecture with far-reaching positive effects on the whole life cycle of automated production facilities.

5 A Roadmap for Service-Oriented Automation Systems

The benefits that could be leveraged by applying service-orientation to industrial automation have been presented. The question arises how such advanced automation systems can be realized based on the current state of the art. We need a suitable migration path to realize the technological and organizational change in industrial automation. In the following we will discuss the most important issues that have to be considered on the way to a service-oriented automation network and some first methods of resolution will be proposed.

5.1 SOA Technology

Regarding the aspect of horizontal and vertical integration the used SOA technology plays a major role. This applies to the requirements on the service specification, the supporting service methods, and on the properties of the communication protocols. To specify a service it has to be decided whether a service should be part of a group of services, like the operations within a web service, which parameters, address, input and output messages, etc., the service has. Additionally, the service technology can comprise special supporting methods of a service, for example a discovery function or the subscription of events. Furthermore, the communication protocol that exchanges the data between service client and provider has to be specified.

Today, the most widely used SOA technology is the web service standard. However, web services were developed for IT applications, so they don't meet the special requirements we have in the automation domain. In the last years, other SOA technologies with the focus on realizing services for technical devices emerged, notably OPC-UA and the device profile for web services (DPWS) [9, 10]. But today, mature, freely accessible, and plain development tools are still missing so that we need further developments concerning tool support.

5.2 Cyber-Physical Automation Systems

Since the capability to provide services isn't supported by field devices today, the idea of CPS supports the realization of a field device bringing its own service functionality. Therefore, current research concerning CPS has to be continued so that each field device could contain an embedded device with the respective service technology implemented for an acceptable price. Furthermore, vendors of automation devices have to equip their products with such embedded devices with service functionality directly.

Of course, this development won't take place promptly but it can be realized step by step. A medium-term solution would be to do the service encapsulation of non SOA-ready devices on a higher computational level so that the device itself hasn't to be changed. However, the far-reaching benefits cannot be achieved until a cyber-physical automation system can be realized so that physical changes to the production equipment can be done easily.

5.3 Process Implementation and Execution

Applying service-orientation to the automation domain enables completely new possibilities for implementing and executing processes. Today, the PLC standard IEC 61131 defines the prevailing way for representing processes that most devices on control level are working with. Since it doesn't support any kind of service orchestration or high-level programming, we need new methods and technologies. However, to create a continuous development, we shouldn't exclude the PLC standards completely, because they are extremely established and accepted. In a first step the standard could be enhanced for using services and generating service orchestrations.

Since ordinary SOA applications comprising pure software systems differ in many ways from automation processes, we have to deal with additional constrains and dependencies when it comes to the execution of a physical process [7]. In particular, services representing mechatronic functionality are heavily dependent on the location of the service and the mechatronic design of the hardware. Thus, we need additional information about the physical structure of the plant and the hardware itself for developing service orchestrations. A characteristic principle of SOA that enables a high degree of flexibility is late binding. This means that the respective service is dynamically chosen and assigned within the service orchestration during runtime. However, high requirements on reliability, repeatability, and predictability have to be considered so that we need also information about the time performance and failure probability of the services. As a start, the ability of late binding could be limited so that we assign the services during the last planning steps before implementation. This makes it much simpler and current engineering processes can be already improved by planning in a hardware-independent way.

5.4 Engineering Concepts

In the last sections, we discussed mainly implementation issues for realizing service-oriented control architectures. However, the far-reaching benefits of SOA based on the higher abstraction degree have to be achieved with the help of suitable engineering concepts and methods. Today, planning processes are too sequential, too comprehensive in content, too much hardware-oriented, and too product-specific [4].

Additionally, the complexity of planning processes grows more and more with the rising complexity of the production facilities.

By means of encapsulating the functionality of automation components in a more abstract way, the gap between planning and implementation phases can be closed, so that seamless engineering processes can be realized. Additionally, the degree of dependence between different engineering domains can be lowered and concurrently, the complexity of planning processes can be handled in more sophisticated way. For setting up or reconfigure automation systems, this affects not only the process control but also other supporting applications, like maintenance, human-machine interaction, monitoring, and diagnosis. Therefore, we need new engineering methods for planning and implementing automation systems that make use of the advantages provided by the paradigm of service-orientation.

5.5 Standardization and Service Definitions

To enable the full degree of flexibility and interoperability a vendor- and customer-comprehensive way to describe and use services is needed. This comprises a standardized service technology that defines how the services communicate with each other and the way the service interface is structured and described. Concerning W3C web services the use of HTTP and SOAP for the communication and WSDL for the interface representation is defined. The much tougher question is how the services themselves are defined in a standardized way so that the services become vendor-independent. Therefore, we would need an overall standard that specifies how automation functionality has to be encapsulated to services and how the services have to be named. However, establishing such an overall standard seems to be quite unrealistic.

A much more sophisticated approach constitutes the semantic description of services [11]. Since the broadly use of semantic technologies may be realistic in the long term but not in the next years, we need a temporary solution. For introducing the use of service-orientation company-wide automation standards constitute a good foundation for at least company-wide standardization of services.

6 Conclusion and Outlook

To accomplish the growing demands on manufacturing systems new concepts for industrial automation are needed. By applying the paradigms of Cyber-Physical Systems and Service-oriented Architectures we can develop agile automation architectures that meet the requirements we have to cope with in future. In order to gain the potential benefits, we have to develop a multi-dimensional approach that comprises a wide range of influencing research topics. Besides the described topics concerning technological issues, planning concepts, and implementation methods there are many more aspects that have to be considered like safety and security aspects, educational and organizational matters, and acceptance problems. Since these changes can't be realized overnight, a suitable migration path and integration methods can help us to do changes in gradual way.

In future, we will continue our work concerning the SOA-AT framework where we focus the development of engineering processes for designing and implementing service-oriented control software. For evaluating the practical suitability of our concepts we will keep on with setting up real demonstration scenarios in the SmartFactory^{KL}. There we have already built a first demonstration system running with a service-oriented control architecture [12] that will constitute the foundation for further applications.

References

1. Weiser, M.: The Computer for the 21st Century. In: Scientific American, Special Issue on Communications, Computers, and Networks (September 1991)
2. Tabuada, P.: Paulo: Cyber-Physical Systems: Position Paper, `http://varma.ece.cmu.edu/CPS/Position-Papers/Tabuada.pdf` (retrieved August 31, 2011)
3. Lee, E.: Cyber Physical Systems: Design Challenges. University of California, Berkeley Technical Report No. UCB/EECS-2008-8 (2008)
4. Zühlke, D.: SmartFactory – Towards a Factory-of-Things. Annual Reviews in Control 34(1), 129–138 (2010)
5. Melzer, I.: Service-orientierte Architekturen mit Web Services: Konzepte - Standards - Praxis, 3rd Aufl. Spektrum Akademischer Verlag, Heidelberg (2008)
6. Krafzig, D., Banke, K., Slama, D.: Enterprise SOA: Service-Oriented Architecture Best Practices. Prentice-Hall, Upper Saddle River (2004)
7. Ollinger, L., Schlick, J., Hodek, S.: Leveraging the Agility of Manufacturing Chains by Combining Process-Oriented Production Planning and Service-Oriented Manufacturing Automation. In: Proceedings of the 18th IFAC World Congress, Milan, Italy (2011)
8. Pohlmann, E.G.: Prozessorientierte Planung von serviceorientierten Fabriksteuerungssystemen, Fortschritt-Berichte pak, Band 17. Technische Universität Kaiserslautern, Kaiserslautern (2008)
9. Souza, L., Spiess, P., Guinard, D., Köhler, M., Karnouskos, S., Savio, D.: A Web Service based Shop Floor Infrastructure. In: Proceedings of the Conference Internet of Things, Zurich, Switzerland, pp. 50–67 (2008)
10. Stopper, M., Katalinic, B.: Service-oriented Architecture Design Aspects of OPC UA for Industrial Applications. In: Proceedings of the International MultiConference of Engineers and Computer Scientists, IMECS 2009, Hong Kong, vol. II (2009)
11. Loskyll, M., Schlick, J., Hodek, S., Ollinger, L., Gerber, T., Pirvu, B.: Semantic Service Discovery and Orchestration for Manufacturing Processes. In: Proceedings of the Confence on Emerging Technologies on Factory Automation, ETFA 2011, Toulouse, France (2011)
12. Ollinger, L., Schlick, J., Hodek, S.: Konzeption und praktische Anwendung serviceorientierter Architekturen in der Automatisierungstechnik. In: Proceedings of VDI Congress AUTOMATION 2011, Baden-Baden, Germany (2011)

A Straightforward Algorithm to Transform Robot Control State Machines into Verifiable Lotos Specifications

Reza Babaee and Seyed Morteza Babamir

University of Kashan, Ravand Blvd., Kashan, Iran
rbabaeecar@grad.kashanu.ac.ir,
babamir@kashanu.ac.ir

Abstract. Formal methods have exhibited as a promising technique that provides verifiable specifications for highly reliable software. However from the practical robotics perspective, formal methods remain unaccustomed among real-world robotic control software projects. The underlying reason is the unreasonable fear in working with purely theoretical foundations of formal methods and as a consequence many of their remarkable abilities are often ignored.

We try to establish a particular relationship between a particular form of State Machines as the prevalent modeling language applied in many robotic control software industries on one side, and the formal Lotos specification language as the general specifying language for open distributed systems, on the other side. We present a quite straightforward algorithm to transform refined state machines into Basic Lotos Specifications so that it can be further implemented as a program to mechanize the transformation process.

1 Introduction

Formal methods have been proved to be an essential technique for specifying safety-critical systems in which desired properties have to be satisfied by the software. Formal specification languages provide the capability of constraints proof by means of mathematical foundations. However from the practical point of view, developing mathematical models is supposedly an onerous task disregard of considerable overall benefits obtained by using these models.

On the other hand, visual semi-formal diagrams (such as illustrative diagram sets in Unified Modeling Language (UML) [1]), support practitioners by means of visual conceptual illustrations while they usually suffer from lack of formality required for specifying sufficiently reliable systems.

We strongly believe that instead of making these methods strictly formal by introducing new tools, exploiting current conventional formal methods is rather more acceptable in many industrial situations. However the conceptual visuality earned by such diagrams should also be taken into account in the software development processes which are more actively engaged by human.

G. Lee (Ed.): Advances in Automation and Robotics, Vol. 1, LNEE 122, pp. 575–583.
springerlink.com

The solution to this problem is trivial: depute automatic programs to work with formal methods while human-centric approaches (such as UML) provide inputs for these programs. In fact we need to develop specific programs that receives visual models as inputs and produce corresponding formal specification as output.

Achieving this goal requires specific algorithms to translate prevalent visual illustrations, into desired formal specification languages regarding system application domains. In this paper we have chosen *Lotos* formal specification language as one of the predominant formal methods in order to specify Open Systems Interconnection (OSI) architecture [2]. On the other side, we use UML 2.0 State Machines (SM) [3] as the modeling charts to specify a robot system behavior, and then translate them into a *Basic* form of Lotos specification language in a relatively straightforward manner.

In the next Section we briefly introduce the Basic Lotos specification language, then we proceed with presenting the core elements of SM in the Section 3. We explain our algorithm on a simple robot control system case study described in the Section 4. In the Section 5 we provide our main approach to transform SMs into Basic Lotos specifications. Needed refinements on SM to generate a better Lotos specification is presented in the Section 6 followed by the related works and our conclusion in the Section 7 and Section 8, respectively.

2 Basic Lotos

Lotos (Language Of Temporal Ordering Specification) is mainly developed within International Standards Organization (ISO) for specifying Open Systems Interconnection or OSI computer network architecture [2]. The considerable strength of Lotos in expressing agents parallelism and synchronization, makes it an appropriate choice for specifying robots control software which remarkably require to be synchronized with other agents.

Lotos's main features have been inspired by *process algebra* in which the process synchronized communications are described in terms of Calculus of Communicating Systems (CCS). Lotos has two basic components: one for control aspects of the system which is based on CCS, and one for data structures description which is based on abstract data type technique ACT-ONE [4]. In this paper we focus on the control aspect of the system, which is also called *Basic Lotos*, and leave the abstract data type description for future researches.

The constitute element of a Basic Lotos specification is *process*. A process itself might involve other processes called subprocess; thus a Lotos specification describes a system by means of hierarchy of process descriptions [2]. Considerable strength of Lotos stems from its outstanding process *parallelization* features as well as process *synchronization* [5]. The elementary atomic parts in Lotos Specifications are events, or generally *actions*. Actions are the elements by which processes can be synchronized with each other. In the case that the data is not exchanged between processes (our focus), the actions and the *gates* (process input parameteres) are identical. All possible actions in a system specification shape the *action alphabet* of that specification.

A combination of actions assembled by Lotos operators builds *behavior expressions*. Processes are described in terms of behavior expressions which represent states of processes. Behavior expressions might be combined together with actions by Lotos operators to form larger behavior expressions. To be fitted into the convention standards applied in Lotos context, the actions are given lower-case names while behavior expressions are appeared in upper-case names. For the space sake we are not going to explain each operator individually, but we will clearly explicate them when we want to describe the mapping function between them and the corresponding SM notations.

3 UML 2.0 State Machines

We use behavior state machines as our modeling language since it has wider basic elements to describe the behavior of a robot control system. Thankfully SM possesses extremely advantageous properties and features that make it an eminently rational modeling language choice for Lotos specifications:

- As we mentioned at Section 2, Lotos describes processes in terms of their states (i.e. behavior expressions) thus we need to a specific modeling language to show states of processes precisely.
- The main power of Lotos arises from process parallelization and synchronization, therefore its correspondent modeling language requires appropriate instruments for representing parallelization and synchronization. The former is satisfied in UML 2.0 SM by using *composite states* which allows to define several parallel state machines, and the latter is represented by special *choice* or *signal* pseudostates (See Section 5).

4 Robot Controller Case Study

For understandability we illustratively present our approach step-by-step on a simple robot controller case study [6]. Before explaining the transformation process into Lotos specification (Section 5), let disclose more information on the robot controller case study. An overall representation of the production cell plant is depicted in Figure 1.

We are going to construct a state machine model of the simple robot used in the plant according to the following description:

> The robot comprises two orthogonal arms. Once a metal blank is received at feed belt, the first arm takes it and puts it on the press plate. The robot waits until finishing pressing, then the second arm takes the pressed metal and transports it on the deposit belt. The robot may continues taking new metal blanks, or becomes switched off. It is also unreliable, means that it may crash at every possible state of the system and if it is repaired, it comes back to an initial state.

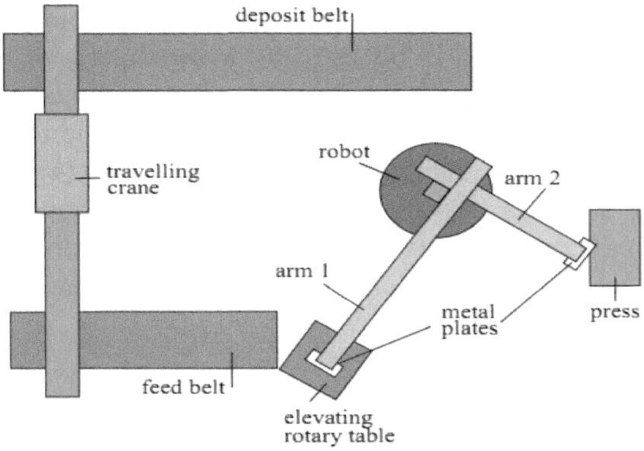

Fig. 1. Top view of production cell plant

Since there are two *waiting* points in the robot description, the robot has to synchronize itself with two other agents, i.e. the feed belt and the press. An equivalent state machine is shown in Figure 2.

As it is clear in the Figure 2, the robot has two waiting states which is represented by *send* and *receive* signal pseudostates. It also determines two points of robot controller termination: when it is switched off and when encounters an unrepairable crash. The former is a successful termination and the latter is an unsuccessful termination. A repair action will take robot controller into the ready state which is the initial state of the robot.

5 SM-to-Lotos Transformation

There are several significant steps towards transforming a system model represented by SM into the Lotos specification.

5.1 Process Extraction

As the first step it is vital to extract all possible modeled processes from SM. There may exist broad indications in the SM that point to process entities in the system. Generally a set of states, which might form a composite state, represents the condition of a *process* at a certain time.

In our example shown in Figure 2, obviously there are several main processes: *Robot Controller*, *Transport Belts* and the *Press*. Our main focus is the robot controller agent but in order to make our model more complete and precise we need to model two other agents which are interacting with the robot controller system. This indicates that the Lotos specification must have at least three processes which have to be synchronized with each other.

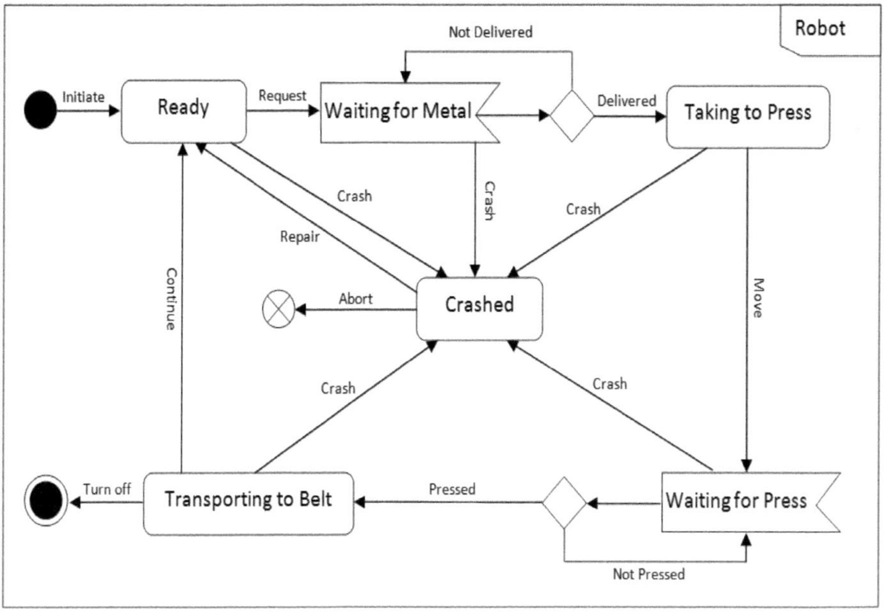

Fig. 2. Robot Control State Machine

5.2 Individual Process Description

Before we proceed with the whole system specification, we take a divide-and-conquer approach. In fact we do not concern about processes composition but rather we initially offer an individual description for each process. We try to create a fairly comprehensive description for each extracted process using SM models.

Actions and Behavior Expressions. Generally the states within a behavior state machine represent an activity or *behavior* of a process [7]. This properly signifies that a state in SM is simply mapped to a behavior expression in Lotos descriptions. Additionally the transition between two states is taken place when the specified written *event* on the transition occurs and that means the events on transitions are equivalent to actions in Lotos specifications. Consequently the composite states which includes other behavior expressions or actions are transformed into appropriate behavior expressions. If a state is a simple state, i.e. it is undividable to other expressions or actions, then a simple action with similar name is added to the actions alphabet. For example in Figure 2 *Ready* simple state is a behavior of the Robot Controller process and the *initiate* is the corresponding event or action. Please note that there might be also some internal behaviors for a single state. In this case a sub-process would be created to represent the internal behaviors of equivalent state; thus the remaining portion of *process extraction* step may be followed when the system becomes more expanded.

The Action Prefix Operator. The action prefix operator shown as ';' in Lotos, expresses the sequential combination of actions before a behavior expression. This operator is represented in SM by the sequential transitions (actions) before states (behavior expressions). In Figure 2 the *initiate* event is before *Ready* state so a possible behavior expression which shows *initiate* prefixes *READY* state is: `initiate;READY`

Choice Operator. One of the principal features of Basic Lotos is the choice operator shown as '[]' which indicates a non-deterministic selection between its two operands. Two leaving transitions with the same source state as well as identical *guards* or conditions on the transitions, can be transformed into a choice operator in Lotos with two target states as two choice operands. For instance two outgoing transitions *request* and *crash* leaving the *Ready* state and get into *Waiting for Metal* and *Crashed* states respectively, are translated into Lotos as follows:

```
initiate;READY;WAITING4METAL
[]
initiate;READY;CRASHED
```

5.3 Process Termination

There are two pseudostate symbols within a state machine indicating a termination: *Final* pseudostate and *Abort* pseudostate. Thankfully Lotos supports two kinds of process termination as well: `exit` and `stop`; the former implies a successful process termination and the latter denotes an unsuccessful or a *deadlock* termination. There can be also a recursive process invocation in order to express *live loops* in SM or *no exit* process termination. In our sample the robot can be whether turned off and successfully terminates or it can stop working if an unrepairable crash occurs. We can simply translate SM into the correspondent Lotos expression as below:

```
initiate;ready;CRASH;stop
[]
move;waiting4press;TRANSPORTING2BELT;exit
```

5.4 Process Composition

After finding how to describe an individual process, we should compose these processes (or behavior expressions) to form larger process specifications. There are principal operators in Lotos for process composition.

Enabling Composition. The *enable operator*, displayed by '>>' in Lotos has fairly the same semantic as prefix operator (Section 5.2); except that it is applied to illustrate two behavior expressions (or processes) sequential composition. For example Robot Controller can be enabled by process Feed Belt; thus we have in Lotos: `CONTROLLER >> FEED_BELT`

Disabling Composition. Lotos has a special operator to specify interrupting one behavior by another behavior called *Disable* operator and shown as '[>'. To find the comparable notations in the SM, we should semantically find a situation in which a state machine might be interrupted by another process. If for every state of a state machine (process) there is an outgoing transition to a single state (whether simple or composite), it can be mentioned that the process is interrupted by that state showing the behavior of an intervening process. In Figure 2 *crash* state is the state whose incoming transitions are originated from all other states; therefore the corresponding Lotos expression might be as:

```
initiate;READY;WAITING4METAL;TAKING2PRESS;
WAITING4PRESS;TRANSPORTING2BELT;exit
[>
crash;stop
```

Interleaving Operator. The interleaving operator displayed by '|||' allows the first operand to alternate its actions or behavior expressions with the second one. SM composite states provide a matching visual concept with interleaving operator: the *substate* machines within a composite state in several different *regions* can be executed simultaneously and independently. In other words states within various regions can be *interleaved* while the composite state machine is executing. This approach can be followed for two or more different processes, as well. In our example assuming two individual instances of the robots with the same state machines as in Figure 2, the interleaved execution of them leads to the following Lotos program:

```
CONTROLLER1 ||| CONTROLLER2
```

Synchronization. There is not only the process parallelism of importance but also it is crucial to synchronize processes with shared events or actions. Synchronization operator displayed by '||' synchronizes its two operands with the action appeared between two vertical lines. In other words two operand processes are interleaved but the process reaches to the synchronized action earlier than the other one, has to wait until the second process also reaches to the synchronization state. If there is no any actions between two synchronization symbol lines it means that all events occurred in two behavior expression operands have to be synchronized, otherwise they have to be synchronized with only determined actions (In this case the synchronization operator is shown as '|[...]|' within the shared actions between two brackets).

A synchronization between two processes is appeared in the SM when firstly there is at least one shared event between two processes and secondly a transition in one process may not be enabled unless the same event in the other process has occurred. This situation is modeled in SM by a combination of notations. There are two pseudostates in SM demonstrating signal exchanges between processes: *send* and *receive* signal pseudostates. They are often followed by conditional transitions that are regularly represented by *choice* pseudostates. For instance in the Figure 2, the robot has to be synchronized with the press in the *Press* event. A receive pseudostate followed by the choice pseudostate represents this synchronization in the robot state machine model while a send pseudostate represents the correspondent synchronization event in the press state machine. The equivalent Lotos expression is as follows:

```
taking2press;move;press;TRANSPORT2BELT;exit
|[press]|
close;forging;move2top;press;close;exit
```

Sometimes a synchronization is modeled with a conditional transition and a *waiting* state; although the waiting state can be eliminated.

6 State Machine Refinement

There may not be possible to transform any UML 2.0 State Machines into a Lotos specification, we still require some particular refinements in order to make SMs directly interpretable in terms of Lotos. According to above transformation algorithm we strongly recommend to undergo the following refinements on SM to have a much more validated specification:

1. Try to create a separate SM for each process within your system
2. Find specific events within your system that are the share points of your processes, i.e. they synchronize with each other on those events. Represent these points in every involved state machine with send/receive signal notations and identical names as well as a choice pseudostate. A *waiting* state is an alternative notation for signal pseudostates.
3. Specify successful/unsuccessful points of processes termination. Use final and abort pseudostate notations respectively to model them.
4. Determine the prohibitive behaviors for each process and model them by inserting an outgoing transition for every state of the process to a single destination state.

7 Related Works

The idea of transforming SM into Lotos is relatively new but not original as within [8] authors present a model transformation applied in the automotive industries. Lotos is used in [9] to give a semantic model for compositional UML state charts. Compare to our work, in some aspects it seems more inclusive, because we try to cover the essential elements of Lotos specification like process compositions. Moreover our presented algorithm is elaborately designed in such a way that could be straightforwardly developed as a program in order to mechanize transforming SM to Lotos specifications.

8 Conclusion and Future Works

A quite straightforward algorithm is proposed in this paper so that the robot control system specification is translated from UML 2.0 State Machines into Basic Lotos formal specifications. Lotos specification then can be comprehensively verified, validated and evaluated using favored toolboxes like CADP[1].

We hope to experiment the transformation algorithm with a practical real-world problem called Flight Computer Warning (FCW) whose corresponding Lotos specification has been already produced in [10]. This allows us to compare our generated Lotos specification with the original one; but before, we have to seek for a broadly similar approach to embed the data abstract specification into modeling language. Therefore it might be essential to deliberate on other UML 2.0 commercially desired diagrams in order to support the state machine models.

References

1. Booch, G., Rumbaugh, J., Jacobson, I.: Unified Modeling Language User Guide. The (Addison-Wesley Object Technology Series). Addison-Wesley Professiona (2005)
2. Bolognesi, T., Brinksma, E.: Introduction to the iso specification language lotos. Computer Networks and ISDN systems 14, 25–59 (1987)
3. (OMG), OMG: Unified modeling language: Superstructure (2005)
4. Ehrig, H., Fey, W., Hansen, H.: Act one an algebraic specification language with two levels of semantics (1983)
5. Logrippo, L., Faci, M., Haj-Hussein, M.: An introduction to lotos: learning by examples. Computer Networks and ISDN systems 23, 325–342 (1992)
6. Lewerentz, C., Lindner, T.: Case study production cell: A comparative study in formal specification and verification. In: KORSO: Methods, Languages, and Tools for the Construction of Correct Software, pp. 388–416 (1995)
7. Miles, R., Hamilton, K.: Learning UML 2.0. O'Reilly Media, Inc. (2006)
8. Chimisliu, V., Schwarzl, C., Peischl, B.: From uml statecharts to lotos: A semantics preserving model transformation (2009)
9. Mrowka, R., Szmuc, T.: Uml statecharts compositional semantics in lotos. In: ISPDC 2008, pp. 459–463 (2008)
10. Garavel, H., Hautbois, R.: An experiment with the lotos formal description technique on the flight warning computer of airbus 330/340 aircrafts. In: First AMAST International Workshop on Real-Time Systems, Iowa City, Iowa, USA, p. 20 (1993)

[1] `http://www.inrialpes.fr/vasy/cadp/`

A Multitasking System Based on RTX51

Guoan Zhang, Longmei Tang, and Jishi Zheng

Department of Computer and Information Science, Fujian University of Technology, Fuzhou, Fujian province, China
jvtczga@163.com

Abstract. In this paper, main task processing strategies of multi-tasking operating system RTX51 are discussed, with summary analysis of its internal kernel. On the basis of above results, a multi-task processing example is designed and implemented, which take a MCU as core with commonly used input and output components of intelligent instrument. The example can be easily transplanted into other applications, so it is of typical significance to the design of complicated multi-tasking system.

Keywords: RTX51, Multi-tasking, Intelligent Instrument, Transplant.

1 Introduction

RTX51 is a real-time operating system developed by KEIL company for MCS-51 series microcontroller missions [1] [2] [4]. It is of two versions, RTX51 FULL and an RTX51 TINY. RTX51 TINY is an subset of RTX51 FULL, only supports a time circulating for task scheduling and signals transferring between tasks as well. The biggest task number is 16, and parallel interruption can be used. But it cannot deal with information processing, neither the storage allocation and release nor preoccupied scheduling are supported. In RTX51, there is only one small kernel, fully integrated in KEIL C51 compiler [1] [3]. More important is, it occupies only about 800 bytes of program storage space, can run in a 8051 system without external data memory, but application can still access external memory. The technical parameters of RTX51 can be seen in table 1.

Table 1. RTX51 Specification

Description	RTX51 TINY
Number of tasks	16
RAM demand	7 byte DATA space 3 times of task IDATA space
Code capacity	900bytes
Hardware	Counter 0
System clock	1000~65535 cycles
Interrupt request interval	Less than 20 cycles
Task switch time	100~700 cycles Depend on workload of stack

G. Lee (Ed.): Advances in Automation and Robotics, Vol. 1, LNEE 122, pp. 585–590.
springerlink.com

2 Analysis of RTX51 TINY Kernel

2.1 Task Status

In RTX51, an user task is similar to the process in usual operating system [5], it is of the following status:

RUNNING: The task is in operation, at the same time only one task can be in "RUNNING" state.

READY: The task is waiting for a run, when the current task time slice in operation is used up, RTX51 run next task in "READY" state.

WAITING: The task is waiting for an event. If the waited event happen, the task goes into "READY" state.

DELETED : The task is not in executive queue.

TIME OUT: As the time slice of the task has been used up and the task will be in "TIME OUT" condition, and wait for running again. The status is similar to "READY" state, while internal operating process enable a cycling task to switch, the sign is marked.

2.2 Synchronous Mechanism

In order to ensure that the task execution sequence in coordination, synchronous mechanism must be used. In the kernel, following events **guarantee** synchronization and communication between tasks:

(1) SIGNAL: a bit reserved for communication between tasks, can be set or cleared by system function. If a task call os_wait function waiting for SIGNAL while SIGNAL is not set, the task will be hung up until SIGNAL is set, then it returns to "READY" state, and can be performed again.

(2) TIMEOUT: the delay is from the starting time of os_wait function, its lasting time can be determined by the number of timer pulse. The task calling os_wait function with TIMEOUT value will be suspended until the end of time delay, therefore return to the "READY" state, and can be performed again.

(3) INTERVAL : the time interval from starting time of the os_wait function, which can be determined by the number of pulse. The task calling os_wait function with INTERVAL value will be suspended until the end of time interval, then return to the "READY" state, and can be run again. Different from TIMEOUT, pulse counter of the task is no longer reset.

2.3 Scheduling Rules

In RTX51 TINY, an internal timer T0 8051 is used to produce timing pulse, each task can only be in execution in its respective distributed timing pulses (time slice). When previous time slice is used up, the operating system will switch to the next task to run, therefore, the task is executed in parallel.

Scheduling rules are as follows: (1) if specific event has not happened, (2) the task is in execution longer than stipulated time between switching, then the task will be broken. On the other hand, (1) if no other task is running, (2) some task is in "READY" or "TIMEOUT" state waiting to run, then another mission start.

3 Relevant Library Functions for Task Handling

In RTX51, several tasks are allowed in an "quasi parallel" performing at the same time. Each task will be implemented within a predefined time slice. Time over makes the executing task hung, and another task begins to carry out.

Os_wait function provides a more effective way to allocate processor time that task can use. The most simple event responsing os_wait is time over, then the current executing task is interrupted by the specified clock cycle. Os_wait function breaks the currently running task, and wait for a particular event. During the time the task is waiting for events, other tasks can be performed.

Another event affecting os_wait function is the signal. Signal is used to coordinate tasks, until another task sends signal, where wait state is over under the control of os_wait function. If the signal is sent out in advance, then the task will continue to carry out immediately.

Function isr_send_signal can be used to send signal in a interrupting function call, while os_send_signal function is used to signal to other tasks in ordinary tasks. The task received the signal should delete the signal with os_clear_signal function in real-time.

4 A Multitasking System

4.1 Hardware

Taking AT89S51 as the core, a general programmable interface chip 8255 A and eight LED digital tubes are expanded as the output devices, with eight keys for input device, a multitasking system is constructed, as shown in figure 1:

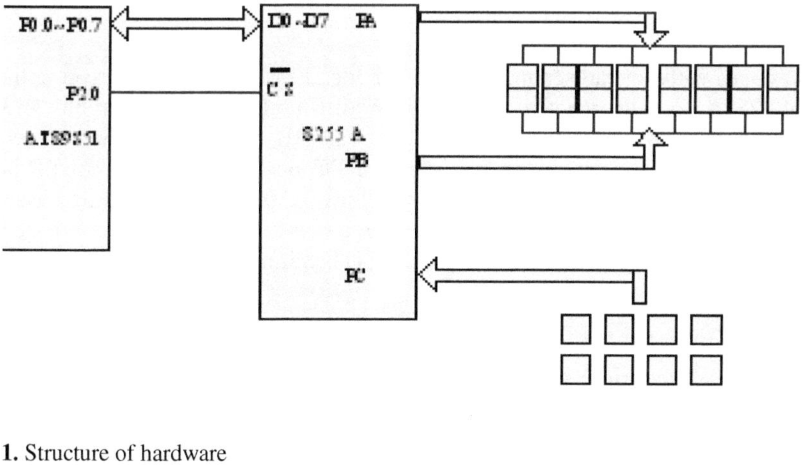

Fig. 1. Structure of hardware

4.2 Division of System Tasks

System is divided into five tasks including: task is 0 (INIT: initialization), task 1 (KEY_SCAN: keyboard scan), task 2 (DISP: digits display), task 3 (KEY_DISPOSE: key value processing), task 4 (TIME: the clock counting). Using timer T1, the initial counting value 0x3cb0 T1 interrupt service function can count 10 times 100 ms, namely one second, when isr_send_signal (TIME) function of RTX51 send signal to the task 4, and the clock counting is realized. The tasks of the system can be seen from table 2:

Table 2. Task structure

Name of task function	Comment
init	Task 0, counter T1 initialization create other tasks, and take other tasks into ready status
key_scan	Task 1, for keyboard scanning
disp	Task 2, for LED display
key_dispose	Task 3, for key pressing
timer	Task 4, for clock counting

5 Multitasking

In RTX51 TINY, after task 0 create other tasks, either each task can use its own CPU time slice or CPU switches between tasks by signal transmission. The realization of the tasks are explained as the following.

5.1 Task 0

Task 0 (init) mainly complete the following job: timer T1 is initialized, create other tasks, take other tasks into ready condition.

5.2 Task 1

Task 1 copes with keypad scanning, its design idea is as following, (1) read in the PC port of 8255A, to judge whether a key is pressed; If no key is pressed, return 0x30; If there is a key pressed, os_wait (K_TMO, 6 calls 0) function is called to remove pressing dither. (2) read in the value of the PC port again, and secondly judge, whether there is a key pressed; If not, also return 0x30 to tell misreading caused by interference, ignore the action; If there is, then a real key operation is confirmed. (3) then scan keys to judge which key is pressed. (4) handling key release.

5.3 Task 2

Task 2 (disp) description: with 8255A driving, time information of hour, minute and second are displayed on digital tubes on, each digital display delay is about one millisecond.

5.4 Task 3

Task 3 waits for a signal from task 2 (key_scan), which sends a to key value after keyboard scanning, value of the key pressed is processed.

When key_value = 0x30, indicates that no key is pressed; When key_value = 0 x01, said the key K0 is pressed, the timer T1 interruption shut down, begin to adjust clock; When key_value = 0x02, says the key K1 is pressed, moves to the previous clock adjusting position; When key_value = 0x03, says K2 key is pressed, moves to the next clock adjusting position; When key_value = 0x04, said key K3 is pressed, adjusted clock upward; When key_value = 0x08, said K7 key, adjust clock downward; When key_value = 0x05, said key K5 is pressed, put clock adjustment to the end, the clock goes back to normal counting.

5.5 Task 4

In task 4, os_wait (K_TMO + K_SIG calls, 1000, 0) is called to wait for a signal from the timer T1 interrupting service function and wait for one second overtime. When one second time arrives and signal from the timer T1 interrupting service function is received, the clock counts.

5.6 Timer 1 Interruption

In T1 timer interrupting service function, the initial counting value of the count is assigned again, function isr_send_signal (TIME) is called to send a signal to the task 4 (Timer) from the timer T1 interruption.

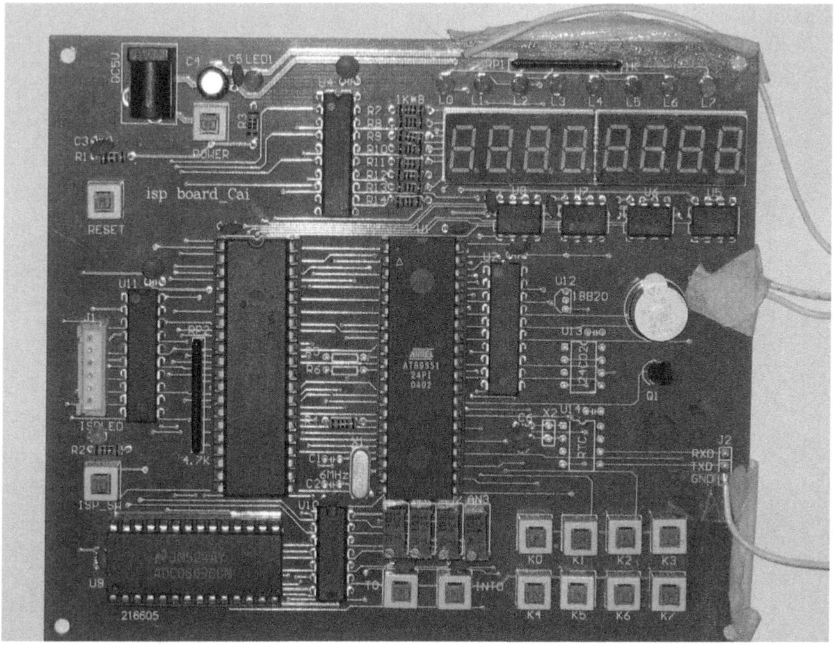

Fig. 2. The actual board

6 Conclusion

A multitasking system based on RTX51 is realized successfully, as shown in figure 2, and it is used to accomplish a real time clock display. It can display the current time and the time value can be modified at any time.

Multitasking method of realizing RTX51 involves: transplanting RTX51 is needed, will it take namely configuration files (Conf_tny.A51) should be added to the project file. The project file, together with written programs are compiled and connected.

Owing to the internal RAM of chip AT89S51 is 128 bytes, so will the RAMTOP value in Conf_tny.A51 file changed to 07FH.

Acknowledgment. This paper is supported by Technology Key Project of Fujian Province of China (2009H01010435), Science and Technical Project of Fujian Department of Education (JA10216).

References

1. Xu, A.: Single-chip Microcomputer High Level Language C51 Application Design. Electronic Industry Press, Beijing (2002) (in Chinese)
2. Tu, G., Yang, F., Hu, G.: Review of Embedded Operating System. Computer Application Research 11, 4–9 (2000) (in Chinese)
3. Zhang, P.: MCS-51 Single-chip Microcomputer Principle and Application Based on the C Programming Language. Tsinghua University Press, Beijing (2003) (in Chinese)
4. Curtis, K.E.: Embedded Multitasking (With Small Microcontrollers) Newnes Elsevier Inc. (2006), Copyright 2006
5. Liu, T., Luo, Z.: Computer operating system. Tsinghua University Press, Beijing (2008) (in Chinese)

A Small Tracked Robot for Cable Tunnel Inspection

Shufeng Tang, Shihui Chen, Qiong Liu, Biao Wang, and Xiaohong Guo

Inner Mongolia Electric Power Science Research Institute,
South Xilin Road, NO.21, Yuquan District, Hohhot City, Inner Mongolia 010020. China
tangshufeng@sohu.com,
{chenshihui,liuqiong,wangbiao,guoxiaohong}@dky.impc.com.cn

Abstract. A small tracked robot for cable tunnel inspection has been proposed and designed, which is made up of two side track movement mechanisms driven independently and posture adjusting mechanism, so the robot owns better movement flexibility and obstacle-climbing capability. Using posture adjusting mechanism between two side tracks, the robot could not only adapt to the tunnel with smooth pavement, but also adapt to pipelines with different diameters. The robot owns camera, CO, CO_2, SO_2, CH_4, smog and humiture sensors, and the information of tunnel collected by these sensors could be transmitted to in-hand controller on the ground and displayed with imaging pattern. So operators could find problems and maintain the cables in time to avoid accident. The inspection working of robot could enhance safe reliability of power transmission network effectively and improve security detection automation level of power transmission lines underground.

Keywords: Cable tunnel, Cable Pipe, Inspection, Track Robot.

1 Introduction

For reducing natural calamities' impact to power transmission and communication network and avoiding the interference between over-ground cables and on-ground buildings, more and more power transmission uses underground tunnel transmission pattern to improve the conveniences of adding and maintaining cables. The phenomenon of cable ageing, corrosion and so on would happen along with running time, at the same time tunnel water seepage could happen in humid zone. If cables always run in non-health status for long-time, then power transmission breakdown even fire disaster are avoidless. Once fire disaster happens, power-network scheduling would be impacted to affect manufacture and life power utilization seriously, at the same time abundant dust and poisonous harmful gas would be let out during cable burning which could cause enormous loss of social economy and environment. For this reason, the frequency of cable tunnel inspection should be increased urgently, finding the defect of cable and maintaining in time, accident could be avoided effectively.

Considering important significance of cable tunnel inspection job, relevant researches have been developed at home and abroad. Washington University's Bing Jiang [1-4] developed underground cable inspection robot named "Rovers" with 1.2 meters length in 2005, which used multi-section modularization configuration,

G. Lee (Ed.): Advances in Automation and Robotics, Vol. 1, LNEE 122, pp. 591–598.
springerlink.com

integrated with infrared sensor, medium testing sensor and sound sensor. The robot could run along with cable line owning 4 to 8 centimeters diameters, cross obstacle on the way and find position of breakdown happening. During working, the robot goes forward depending on infundibulate wheel and using "arm of stability" to maintain balance. Shanghai Jiao Tong University's Fu Zhuang [5-7] proposed a tracked cable inspection robot in 2007, which used the structure of track and front swing arm and used three motors to drive two side tracks and swing arm independently. The robot is 420mm length, 320mm width, 300mm height (not including antenna), the swing arm length is 200mm, and the height of vehicle bottom to ground is 45mm. The robot owns infrared camera with holder, ultrasonic sensor and gas sensor, which could collect image in tunnel and detection density of all kinds of toxic gas.

Above robots are suitable for cable tunnel, but biggish circle pipelines are used for power transmission in underground cable projects in many inland cities. So it is necessary to develop an inspection robot suitable for cable tunnel and cable duct for improving power network security.

2 Design Rules of Robot

For realizing cable tunnel or duct inspection job, considering movement flexibility and obstacle-climbing capability at the same time, the design of robot should satisfy following conditions:

1) Small contour dimension, light quality, entering into cable tunnel or duct through the manhole with 700mm diameter;

2) Not only being suitable for cable tunnel inspection, but also being suitable for cable duct inspection job in duct with more than 1 meter diameter, having better movement flexibility and obstacle-climbing capability, owning the maximum movement speed of 12 m/min and ability of climbing over 100mm height obstacle;

3) Having real-time image-collection equipment with own light source and all kinds of toxic harmful gas sensors;

4) Using wireless communication method with 250m communication distance, so operator could control robot on the ground;

5) Owning independent power supply which could work continuously more than 1 hour.

3 Mechanical System Design of Robot

Ground mobile robot could be divided into wheel type, tracked type and foot type according to movement pattern. Wheel type robot owns simple structure and better swerve ability, but worse obstacle-climbing capability and ground adaptability; tracked type robot owns better obstacle-climbing capability and ground adaptability, but more energy loss; foot type robot owns better ground adaptability but worse swerve ability, at the same time drive motor is needed for each joint so as to control being very complex. So tracked type is adopted during robot design, the main part of robot is made up of two side track movement mechanisms driven independently and posture adjusting mechanism in the middle, at the same time the camera with holder and all kinds of environment detection sensors are allocated on robot body. The intersection angle between two side tracks could be adjusted to fit with cable tunnel and cable duct with different diameter by middle posture adjusting mechanism, shown as Fig. 1. The robot is designed using modularization unit to avoid interference among different function modules which including track movement module, posture adjusting module and camera module. Structure size of robot is 390mm(length) ×260mm(width) ×260mm(height), robot quality is 15Kg.

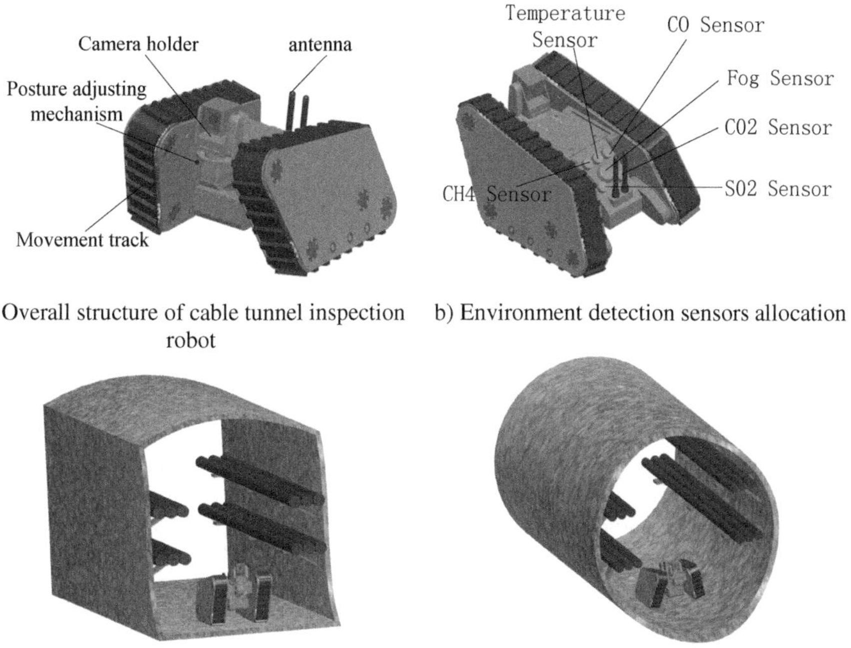

a) Overall structure of cable tunnel inspection robot b) Environment detection sensors allocation

c) Inspection working in cable tunnel d) Inspection working in cable duct

Fig. 1. The small track cable inspection robot

3.1 Track Movement Mechanism

Two side track movement mechanisms of robot are driven independently by one DC deceleration motor for each side, left and right track movement mechanism are mirror image for each other. The schematic of left side track movement mechanism is shown as Fig. 2, in which the motor, transmission mechanism and driver are integrated into a closed movement track unit.

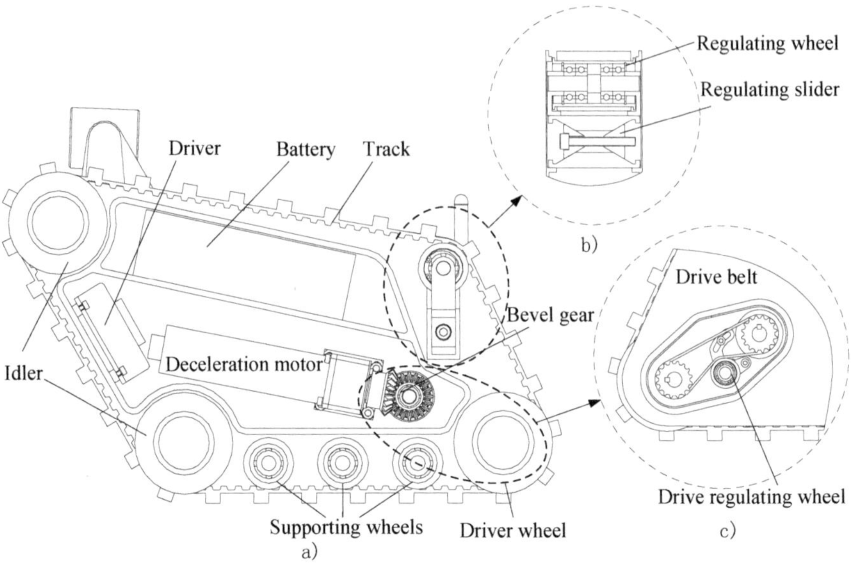

a) Structure diagram of left side movement track; b) Diagram of track strainer; c) Structure diagram of drive belt

Fig. 2. The schematic of left side track movement mechanism

The track is supported by driver wheel, idler and regulating wheel to form a quadrilateral structure, front-slant track structure is helpful to improve obstacle-climbing capability. Three supporting wheels are equipped on the bottom to improve movement ability of robot. The driver wheel is driven by deceleration motor through bevel gear and drive belt, so the whole movement track could be driven to go forward or draw back. Considering structure compact and operation conveniences, slider mechanism is used to tension track, in which distance between regulating sliders is adjusted by regulating screws to push regulating wheel seat moving up to realize track tensioning. Storage battery is allocated in each movement track unit which could be used in parallel to supply energy for the whole robot.

3.2 Posture Adjusting Mechanism

The linkage is used in posture adjusting mechanism. Steering engine drives linkage to rotate by driving shaft; driving linkage drives two side track unit to rotate around the robot body, thus the intersection angle θ could be adjusted to fit with different ground, shown as Fig. 3. The brake is allocated on driving shaft. When posture is being adjusted by steering engine, power should be on to make break loose; after posture adjusting power should be off to make break work, thus robot could maintains adjusted posture. Simultaneously outward and inward motion of two side tracks could be realized by linkage mechanism, which owns simple structure and avoids return difference of intersection angle between two side tracks generated by using gear drive. The maximum intersection angle θ is 32° adjusted by linkage mechanism to fit with cable duct with the minimum diameter 1m.

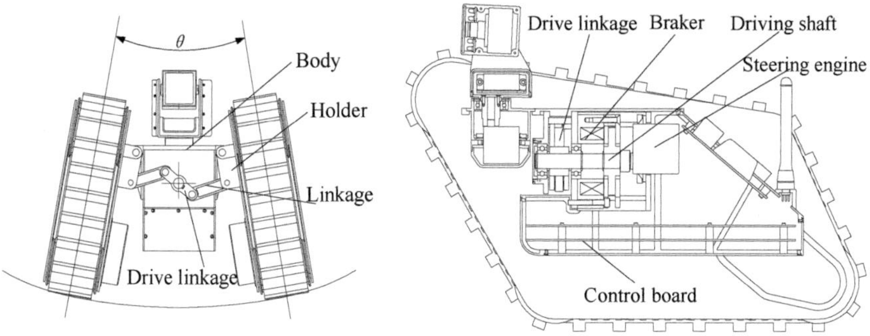

a) Front view of posture adjusting mechanism

b) Side cutaway view of posture adjusting mechanism

Fig. 3. Schematic of posture adjusting mechanism

3.3 Camera Holder

Camera holder owns two degrees of sheer and pitching to realize overall situation observation in the front of robot, whose sheer angle zone is ±90° and pitching angle zone is ±60°, shown as Fig. 4. The sheer degree is realized by sheer steering engine driving U type frame directly through sheer shaft to drive camera rotation; the pitching degree is realized by pitching steering engine driving camera shell to rotate around pitching shaft through timing belt. Led light source is integrated into camera package for video collection illumination in cable tunnel or duct.

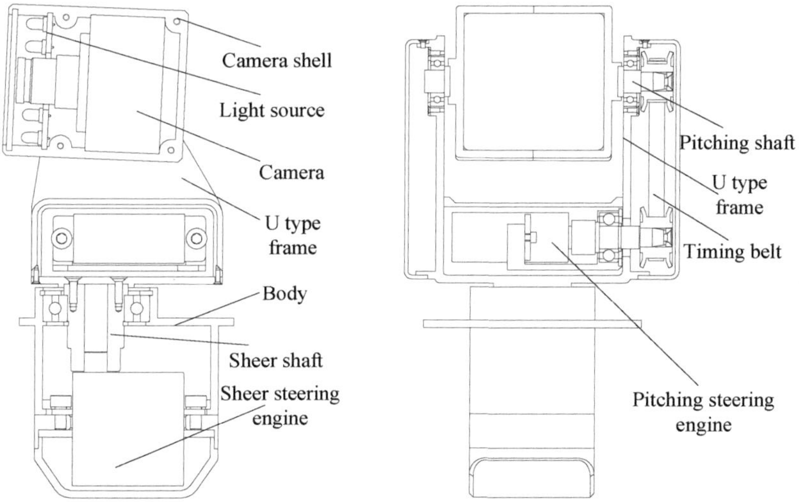

a) Front profile view of camera holder

b) Side profile view of camera holder

Fig. 4. Structure diagram of camera holder

4 Control System Design of Robot

Control system of cable tunnel inspection robot is made up of upper control and information display system and inferior drive and information collection system, the communication between two systems are realized by wireless communication pattern, shown as Fig. 5.

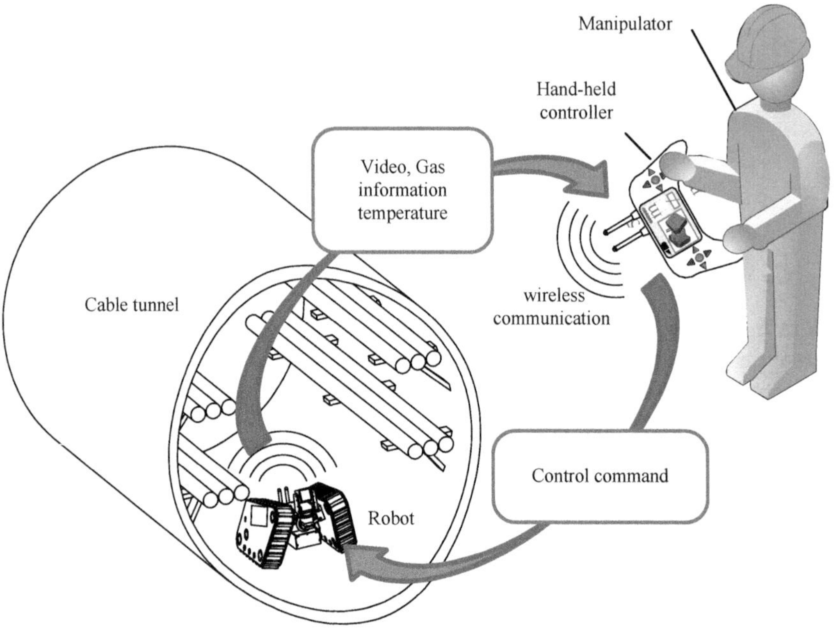

Fig. 5. Schematic of robot control system

When robot finishes inspection working, operator sends control command by hand-held controller on the ground to robot working in the tunnel through wireless communication, the image, gas and environment temperature information collected by camera and all kinds of sensor are send back to hand-held controller through wireless communication and displayed with imaging pattern for operator.

Inferior drive and information collection system could be used to control DC deceleration motor of track unit, steering engine of posture adjusting mechanism and steering engine of camera holder, and to collect information coming from CO, CO2, SO2, CH4, humiture and smog sensors and camera. Considering the maintainability and expandability of hard-system, modularization design idea is used; the whole system is divided into video collection module, data collection module, motor drive module and master control transmission module. Considering huge video data quantity, independent wireless module is used to send video back to upper controller, other modules realize communication by CAN bus and finally wireless module in master control module are used to realize communication with upper controller. The inferior system is shown as Fig. 6.

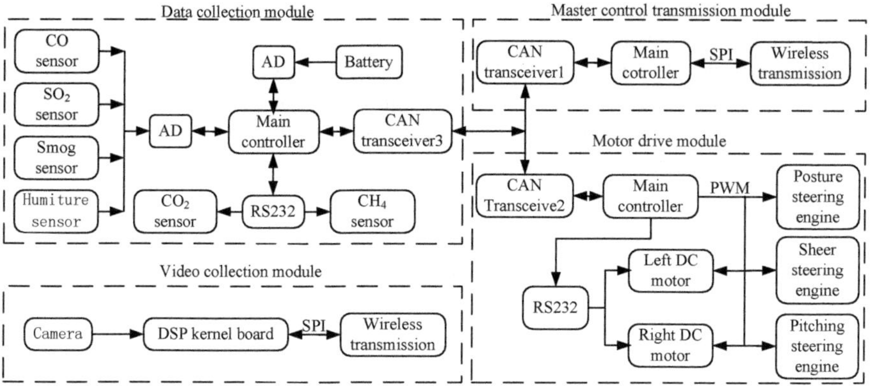

Fig. 6. Schematic of inferior drive and information collection system

Video collection module uses video encoder to convert simulate video signal collected by high-performance camera into digital signal, then digital signal is packaged and transmitted into upper controller through wireless module. Data collection module uses AD chips and RS232 serial interface to collect sensor information and battery electric quantity information, these information is packaged by main controller and transmitted into master control transmission module through CAN bus. Motor drive module receives the command coming from master control transmission module through CAN bus, the configuration of DC servo system is realized through RS232 and DC motors and steering engines are driven through PWM signal by main controller. Master control transmission module receives motion control command and transmits these commands into motor drive module by CAN bus, at the same time receives the environment information coming from data collection module and transmitted these information into upper controller through wireless module.

Upper control and information display system is used to receive all kinds of information coming from inferior system and display these information on touch screen of hand-held controller, hand shank on the hand-held controlled is used to realized robot motion control, schematic of system is shown as Fig. 7.

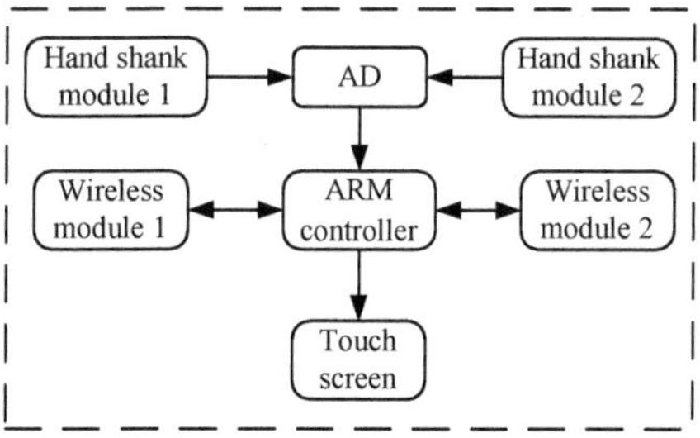

Fig. 7. Schematic of upper control and information display system

In upper system hand shank is used to finish DC motor and steering engine motion control operation, wireless module 1 is used to receive video information, wireless module 2 is used to receive sensor information and transmit motor configuration and control command. Touch screen is used to display all information coming from inferior system. Operation system is used to make sure performance of robot system.

5 Conclusion

For improving automatic level of cable tunnel or duct inspection and reliability of power transmission, a small track inspection robot is developed, which could be used not only in cable tunnel inspection but also in cable duct with different diameter. The robot is made up of two side track movement mechanisms driven independently and posture adjusting mechanism, owns better movement flexibility and obstacle-climbing capability. The camera and all kinds of sensors are integrated into robot, so the image in tunnel and environment information could be transmitted back to hand-held controller through wireless communication, these information would be displayed on touch screen and used to provide data support for inspection and maintain of underground power transmission line.

Acknowledgments. This work was supported by the 2011 application technology development fund of Inner Mongolia Autonomous Region.

References

1. Jiang, B., Sample, A.P., Mamishev, A.V.: Mobile Monitoring for Distributed Infrastructures. In: Proceedings of the IEEE International Conference on Mechatronics & Automation, Canada, pp. 138–143 (2005)
2. Jiang, B., Sample, A.P., Wistort, R.M., Mamishev, A.V.: Autonomous Robotic Monitoring of Underground Cable Systems. In: 12th Proceedings of International Conference on Advanced Robotics (ICAR 2005), pp. 673–679 (2005)
3. Jiang, B., Mamishev, A.: Robotic Monitoring of Power Systems. IEEE Transactions on Power Delivery 19, 912–918 (2004)
4. Jiang, B., Stuart, P., Raymond, M., Villa, D., Mamishev, A.V.: Robotic Platform for Monitoring Underground Cable Systems. In: Transmission and Distribution Conference and Exhibition 2002: Asia Pacific, pp. 1105–1109. IEEE/PES (2002)
5. Xie, Z., Fu, Z., Song, G., Chen, Z.: Control System Research of Behavior2based Cable Tunnel Inspection Robot. Machinery & Electronics (4), 47–51 (2008)
6. Xie, Z.: Research Of The Control System Of Cable Tunnel-Inspecting Robot. Master Degree Thesis, Shanghai Jiao Tong University (2008)
7. Yun, J., Zhuang, F.: Design of pint-size cable tunnel inspecting robots. East China Electric Power 37(1), 95–97 (2009)

A Novel RFID Authentication Protocol with Ownership Transfer

Han Jia and Jun Wen

School of Computer Science and Technology, University of Electronic Science and Technology of China, Chengdu, China
depart@126.com, wenjun@uestc.edu.cn

Abstract. RFID technology has a wide application in many fields. However, there are many security and privacy issues. The paper presents a RFID security proposal to enhance security levels. It is established as following steps. It first builds a security communication channel, then implements tags and corresponding reader authentication, finally solves the ownership transfer issue. This protocol involves minimal interaction between tags and corresponding reader, which can efficiently lower the computational burden on the tag. Its security is verified by BAN logic.

Keywords: Radio Frequency Identification, authentication, BAN, security and privacy.

1 Introduction

RFID is one of the rapidly developing techniques in recent years. It is widely used in many fields, such as retail trade, libraries, car tracking, product identification and passport. It would play an important role in the future. RFID system consist of tags, readers and a database server. In the life of tag, it may transfer ownership on many occasions, for example, it occurs when a manufacture delivers it to a retailer. Therefore, the seamless ownership transfer of tag is required in RFID system. It may suffer from attacks when both of previous owner and new owner accessed the information of the tag during the processing of ownership transfer. After the operation, the previous owner cannot access the information of tag. Typically, the tag have 5~10K logic gates, can store only hundreds bits. With the limit logic gates, about between 300 and 3000 gates can be devoted to security function. There is no such security transfer mechanism in RFID system due to limited computation capabilities and storage on tags. RFID system may suffer from some security threats. They are listed as follows:

Replay attack: An attacker transmits the information he got and spoofs legitimate tag. This attack may leak out the information of tag.

Impersonation: An attacker forges a tag or a reader as an authenticated one to steal the information in the database server.

G. Lee (Ed.): Advances in Automation and Robotics, Vol. 1, LNEE 122, pp. 599–606.
springerlink.com

Eavesdropping: It is easy for eavesdroppers to get the signal from the open wireless circumstance, which lead to leak the business information.

Dos attack: An attacker transmits some messages to interrupt the communication among tags, readers and database servers.

De-synchronization attack: The difference between the key in tags and the one in database results in the authenticated tag cannot be recognized.

Windowing problem: During the process of ownership transfer, both the old and new owners possess the information to authenticate the tag.

This paper proposes a RFID protocol that can resist the above attacks. It adopts random numbers to make sure every round of access is fresh. The message transmitted in the channel is cipher text to preventing leakage of the tag's information. If the bad occasion of de-synchronization happens, this protocol provides the corresponding mechanism to recover it.

The major contribution of this paper is to present a novel security and privacy RFID method with group ownership transfer. This protocol involves minimal interaction between reader and tags. The proposed method provides not only security but also efficiency.

This paper is organized as follows. Section 2 describes related work of RFID. Section 3 presents a new mutual authentication method. Section 4 verifies this protocol's security by BAN Logic. Section 5 draws a conclusion.

2 Related Work

Previous papers have done some research in RFID security. [1] pointed out there are several practical scenarios of group transfer. In addition, group transfer can substantially expand the application of RFID system. [2] is one of the earlier ownership transfer protocol. Unfortunately, there is a flaw in their solution that allows killing of the tag. [3] improved the protocol [2], [3]added some message to the last message form the database to the reader. However, this would lead to de-synchronization, and it cannot resist Dos attacks. [4] proposed a protocol to achieve group ownership transfer with the help of database server, which played a role of trusted third party. However, it may leak out the owner's privacy. [5] adopted a dynamic ID to avoid a replay attack, but it may suffer form a Dos attack. [6] adopted one-time secret to prevent attacks in his protocol, the secrets shared between tags and servers are changed once ownership transfer occurs. It may lead to de-synchronization even if a trusted third party is used. [7] achieved ownership transfer without TTP. This protocol vulnerable to suffer from eavesdropping attacks by the previous owner, and it cannot resist Dos attacks.

The above solutions can be divided into two kinds according to the involvement of a Trusted Third Party (TTP). One not involving a TTP requires many rounds to complete authentication. The other involving a TTP relies on the security and robustness of server, which required the TTP to be online anytime.

3 Paper Preparation a Novel RFID Security Protocol

An RFID security method that achieves all requirements based on xor and public key Infrastructure is proposed. The notations using throughout this paper are listed in table 1.

Table 1.

$E_k()$	Encryption function (under key k); It maybe xor , symmetrical encryption or asymmetric encryption , which depends the capability of computation about entity.
D ()	Decryption function ;
$K\&K_1$	The key for encryption ; Different group owner have different key , so it can identify the owner of tag. In this paper, K_1 represent a new owner.
ID	The unique identifier of tag ;
R_t	The random number generate by tag ;
R_r	The random number generate by reader;
R_s	The random number generate by database server;
$\oplus$	XOR operation;
Info(ID)	The specific information of tag which has this ID .

Assume public key infrastructure has been constructed between a reader and a server. The process of protocol is depicted as follows:

A section to establish a secure communication channel is as follows:

1. A reader generates a random number R_r, and transmits a ClientHello(contains R_r) which is encrypted using the public key of the server to the database server.
2. A server generates a random number R_s after receiving the random number R_r from the reader. ServerHello(contains R_r and R_s) is encrypted using the reader's public key, then sends ServerHello to the reader.
3. The Reader checks whether the random number received from server equal to R_r. If true, stores the random number R_s and sends ClientHelloDone to the server. Else it disposes this message and goes to step 1.
4. The server sends ServerHelloDone to the reader when the server receives ClientHelloDone from the reader. In this case, a secure communications channel has been established.

A section of mutual authentication is as follows:

5. The Reader sends a request (contains R_r) to a tag.
6. The tag stores R_r and generates a random number R_t. The tag sends $E_k(R_t)$, $E_k(ID)$ to the reader.

7. The reader encrypts $E_k(ID) \oplus R_s$, $E_k(R_t)$, R_s using public key of server and sends them to the server.

8. The server checks whether the random number received from reader equals to R_s. If false, the protocol is terminated, else the server gets Ek(ID) by computing $E_k(ID) \oplus R_s \oplus R_s$. The real ID is gotten by decryption function $D_k(E_k(ID))$. The server searches ID in the database. If it is found, authentication process is successful. The server sends Info(ID) encrypted using the public key of the reader to the reader. If it fails, decrypt $E_k(ID)$ using the last successful authenticated key. If ID is found, the authentication process is successful. The server sends info(ID) encrypted using public key of the reader to the reader and goes to step 9. Else the protocol is terminated.

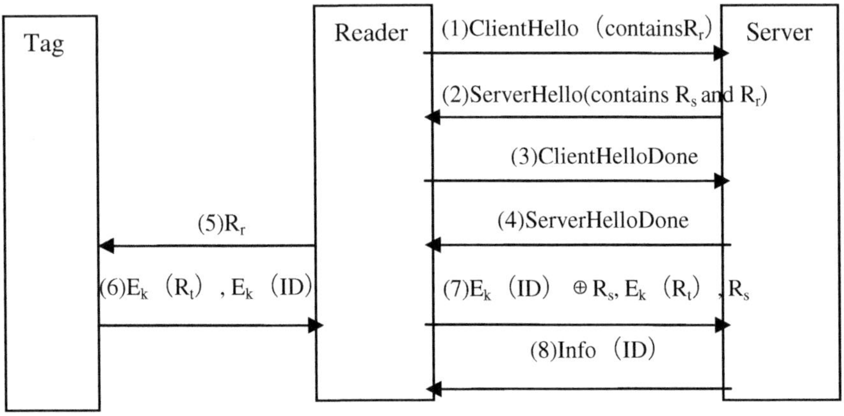

Fig. 1. Authentication Process

A section of ownership transfer is as follows:

9. The server gets the public key K_1 of new group owner. The server sends $E_k(ID \oplus R_t)$, $R_t \oplus K_1$, $E_k(R_t)$and R_r encrypted using the public key of the reader to the reader. Then updates the public key of owner to K_1 and stores K as the last successful authentication kcy.

10. The reader checks whether the R_r is equal to the random number the reader stored. If them equal, goes to the next step, else terminate the protocol.

11. The Reader sends $E_k(ID \oplus R_t)$, $E_k(R_t) \oplus R_r$ and $R_t \oplus K_1$ to the tag.

12. The tag gets R_r by computing $E_k(R_t) \oplus R_r \oplus E_k(R_t)$ and checks whether R_r equals to the random number the tag stored. If not equal, terminates the protocol. Else the reader is authenticated.

13. The tag checks whether $E_k(ID \oplus R_t)$ equals to what it stored. If true, goes to the next step, else terminates the protocol.

14. The tag gets the public key K_1 of new group owner by computing $R_t \oplus K_1 \oplus R_t$, and then replaces K with K_1. The process of group owner transfer is finished.

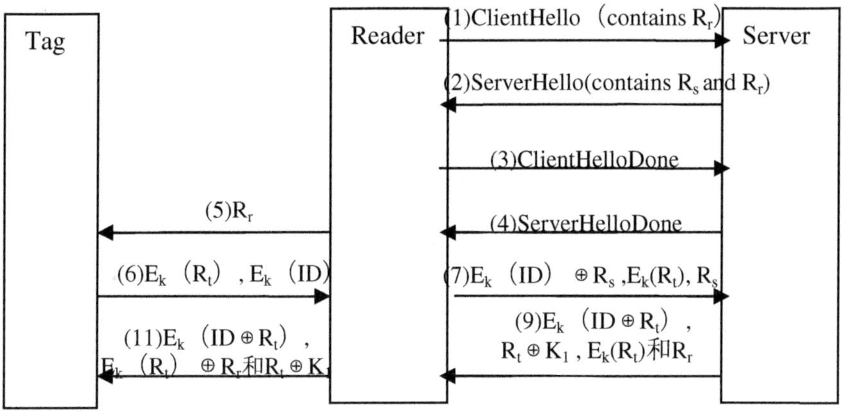

Fig. 2. Ownership Transfer Process

Figure 1 shows the process of mutual authentication. Figure 2 shows the process of group ownership transfer. This protocol can help resist the attack mentioned above. This protocol can adapt the capability of tag computation. If tag has strong computing capability, it can adopt hash or PKI Infrastructure. If tag has weak computing capability, it can adopt xor instead of $E_k()$, but the premise is public key large enough.

4 Analyzing Protocol with BAN Logic

Whether this protocol can achieve the security goal to expect can be proved by formal methods. BAN logic is a well-known authentication logic. Protocol security can be verified by BAN logic to decide whether a protocol can reach expected target and some flaws can thus be found. Syntax and Semantics of BAN logic is shown as follows [8]:

$P|\equiv X$: P trusts the message X is true , P believes X .
$P \lhd X$: P received a message contains X , P sees X .
$P|\sim X$: P has transmitted a message contains X . P said X .
$P| \Rightarrow X$: P controls X
$\#(X)$: X is fresh . X has not been transmitted in any message before.
$P \overset{K}{\longleftrightarrow} Q$: P and Q communicate to each other with the shared key K . No one discovered K except P , Q or a third party trusted by P or Q .
$\{X\}_K$: It means that X is encrypted under K.

Rules of BAN Logic

1. message-meaning rule

   Rule 1 : $P|\equiv P \overset{K}{\longleftrightarrow} Q$, $P \lhd \{X\}K \quad \vdash P|\equiv Q|\sim X$

2. nonce-verification rule

   Rule 2 : $P|\equiv\#\{X\}$, $P|\equiv Q|\sim X \ \vdash P|\equiv Q|\equiv X$

3. jurisdiction rule

$$\text{Rule 3}: P|\equiv Q|\Rightarrow X,\ P|\equiv Q|\equiv X \ \vdash P|\equiv X$$

4. seeing rules

$$\text{Rule 4}: p \lhd (X,Y) \ \vdash P \lhd X$$

$$\text{Rule 5}: P \lhd \langle X \rangle K \ \vdash P \lhd X$$

$$\text{Rule 6}: P|\equiv P \stackrel{K}{\longleftrightarrow} Q,\ P \lhd \{X\}K \ \vdash P \lhd X$$

5. freshness rule

$$\text{Rule 7}: P|\equiv \#\{X\} \ \vdash P|\equiv \#\{X,Y\}$$

6. belief rules

$$\text{Rule 8}: P|\equiv X, P|\equiv Y \ \vdash P|\equiv (X,Y)$$

$$\text{Rule 9}: P|\equiv (X,Y) \ \vdash P|\equiv X$$

$$\text{Rule 10}: P|\equiv Q|\equiv (X,Y) \ \vdash P|\equiv Q|\equiv X$$

$$\text{Rule 11}: P|\equiv Q|\sim (X,Y) \ \vdash P|\equiv Q|\sim X$$

Assume A represents a tag, B represents a reader, and S represents a database server. K_{BS} represents the shared key between B and S. The initial assumptions are as follows:

$$B|\equiv B \stackrel{K_{BS}}{\longleftrightarrow} S \tag{1}$$

$$S|\equiv B \stackrel{K_{BS}}{\longleftrightarrow} S \tag{2}$$

$$A|\equiv A \stackrel{K}{\longleftrightarrow} S \tag{3}$$

$$S|\equiv A \stackrel{K}{\longleftrightarrow} S \tag{4}$$

$$A|\equiv S| \Rightarrow A \stackrel{K}{\longleftrightarrow} S \tag{5}$$

$$A|\equiv \#(R_t) \tag{6}$$

$$B|\equiv \#(Info(ID)) \tag{7}$$

$$B|\equiv S| \Rightarrow Info(ID) \tag{8}$$

$$A|\equiv S| \Rightarrow K_1 \tag{9}$$

The idealization of the RFID protocol is listed as follows:

$$B \rightarrow S: \{R_r\}_{K_{BS}}$$

$$S \rightarrow B: \{R_s, R_r\}_{K_{BS}}$$

$$B \rightarrow A: R_r$$

$$A \rightarrow B: \{R_t, ID\}_K$$

$$B \rightarrow S: \{R_s, \{R_t, ID\}_k\}_{K_{BS}}$$

$$S \rightarrow B: \{Info(ID)\}_{K_{BS}}$$

$$S \rightarrow B: \{R_r, \{R_t, ID, K_1\}_k\}_{K_{BS}}$$

$$B \rightarrow A: \{R_t, ID, K_1\}_k,\ R_r$$

According to BAN Logic, the interpretation of the RFID protocol is as follows:

$$S \triangleleft \{R_r\}_{KBS} \tag{10}$$

$$B \triangleleft \{R_s, R_r\}_{KBS} \tag{11}$$

$$A \triangleleft R_r \tag{12}$$

$$B \triangleleft \{R_t, ID\}_K \tag{13}$$

$$S \triangleleft \{R_s, \{R_t, ID\}_k\}_{KBS} \tag{14}$$

$$B \triangleleft \{Info(ID)\}_{KBS} \tag{15}$$

$$B \triangleleft \{R_r, \{R_t, ID, A \xrightarrow{K_1} S\}_k\}_{KBS} \tag{16}$$

$$A \triangleleft \{R_t, ID, A \xrightarrow{K_1} S\}_k, R_r \tag{17}$$

The goals expected to achieve:

$$B|\equiv Info(ID),\ A|\equiv S \xleftrightarrow{K} A$$

Under the Rule 1, formula(15) and the assumption(1) is to obtain the following:

$$B|\equiv S|\sim Info(ID) \tag{18}$$

Under the Rule 2, formula(18) and the assumption(7) is to obtain the following:

$$B|\equiv S|\equiv Info(ID) \tag{19}$$

Under the Rule 3, formula(19) and the assumption(8) is to obtain the following:

$$B|\equiv Info(ID)$$

So the goal of B|≡Info(ID) has been proved.

Under the message-meaning rule, formula(17) and the assumption (3) is to obtain the following:

$$A|\equiv S|\sim (R_t, ID, A \xrightarrow{K_1} S) \tag{20}$$

Under the Rule 5 and assumption(6) is to obtain the following :

$$A|\equiv \#(R_t, ID, A \xrightarrow{K_1} S) \tag{21}$$

Under the Rule 2, formula(20) and (21) is to obtain the following:

$$A|\equiv S|\equiv (R_t, ID, A \xrightarrow{K_1} S) \tag{22}$$

Under the Rule 9 and formula(22) is to obtain the following :

$$A|\equiv S|\equiv A \xrightarrow{K_1} S \tag{23}$$

Under the jurisdiction rule, formula(23) and assumption(9) is to obtain the following:

$$A|\equiv S \xleftrightarrow{K} A$$

From the initial assumption, the goals are deduced by applying logic rules. By analyzing the RFID protocol with BAN Logic, it can verify that the RFID protocol is secure and flawless.

The protocol security of above issues is analyzed as follows:

This paper adopts random numbers to make sure that every round of communication is fresh. The random numbers are produced every round to prevent the replay attack. The message transmitted in the channel is cipher text so that attacker cannot figure out the original message. If an attacker forge a tag to involve the process of exchanged information, the database server can detect the identification does not exists in database, which will deny its next operations. When de-synchronization happens, the database server can identify the tag by using old owner's secret key. Then the database server sends the new secret key once again to recover it. This protocol transfers ownership by one-step operation. If this operation success, the old owner cannot access the tag because it does not know the new secret key; otherwise, the tag cannot identify the new owner, it will deny its access. By the way, it can resist windowing problems.

5 Conclusions

This paper proposes a RFID protocol, which can be implemented in either high-cost tags or low-cost tags. This protocol's security has been proved by BAN Logic. In this protocol, three random numbers are involved. How to reduce the random numbers in the protocol and simplify operations is what will be done in the future.

Acknowledgements. This research was financially supported by the Chinese postdoctoral science fund and Sichuan Jiuzhou Electronic Technology Co., Ltd.

References

1. Juels, A.: Yoking-proofs for RFID Tags. In: Second IEEE Annual Conference on Pervasive Computing and Communications Workshops, Washington DC, USA, pp. 138–142 (2004)
2. Osaka, K., Takagi, T., Yamazaki, K., Takahash, O.: An Efficient and Secure RFID Method with Ownership Transfer. Computational Intelligence and Security 2, 1090–1095 (2006)
3. Jappinen, P., Hamalainen, H.: Enhanced RFID Security Method with Ownership Transfer. In: Proceedings of the International Conference on Computational Intelligence and Security, pp. 382–385 (2008)
4. Lei, H., Cao, T.: RFID Protocol enabling Ownership Transfer to protect against Tractability and Dos attacks. In: The First International Symposium on Data, Privacy and E-Commerce (ISDPE 2007), November 1-3, pp. 508–510 (2007)
5. Tripathy, S., Nandi, S.: Robust Mutual Authentication for Low cost RFID Systems. In: 2006 IEEE International Conference on Industrial Informatics, August 2006, pp. 949–954 (2006)
6. Kulseng, L.: Lightweight mutual authentication, owner transfer, and secure search protocols for RFID systems. Master of Science thesis, Electrical & Computer Engineering Department, Iowa State University (2009)
7. Dimitrious, T.: RFIDDOT:RFID delegation and ownership transfer made simple. In: Proc.International Conference on Computational Intelligence and Security, pp. 382–385 (2008)
8. Bicakci, K., Baykal, N.: One-Time Passwords: Security Analysis Using BAN Logic and Integrating with Smartcard Authentication. In: Yazıcı, A., Şener, C. (eds.) ISCIS 2003. LNCS, vol. 2869, pp. 794–801. Springer, Heidelberg (2003)

Research of Trenchless Detection Technology and Equipment of Small-Caliber Buried Oil and Gas Pipeline

Zhen Zhang, Zhuxin Li, Yi Su, and Jun Chen

Dept. of Petroleum Supply Engineering, LEU, 401311, Chongqing, China
zh165zh@sina.com

Abstract. Pipeline transportation is a security, economical and effective transportation mode of fluid, and is widely used in petroleum, gas and other fields. The detection of small-caliber buried oil and gas pipeline under trenchless is an important means to master the situation of small diameter buried pipeline and to ensure the safe operation of pipeline. To ensure the straightway and safe operation and disaster prevention of oil and gas pipelines, the paper analyzes the status of trenchless detection technology of existing small-diameter underground oil and gas pipeline, proposed new technologies and countermeasures of detection of small-caliber buried oil and gas pipeline, constructed technology and equipment of detection of small-caliber buried pipeline.

Keywords: Trenchless Detection Technology and Equipment. Small-Caliber Buried Oil and Gas Pipeline.

1 Introduction

As a large number of pipeline transportation of oil, natural gas and other energy efficient means of transport, has been widely applied in the world, in order to protect the safe operation of oil and gas pipeline, to extend the service life, should the regular detecting, so that detect problems, and take timely measures. The detection technology and capability of buried pipeline cannot meet the needs of underground pipeline safety at present, especially in small-caliber buried pipeline detection and there is a big gap between the developed countries. Pipeline detection is an important means to master the situation and take preventive measures to ensure the safe operation of pipelines. The current small-caliber line detection of buried pipelines are used outside pipe sampling, Pipe will be measured by a certain sampling interval after excavation, hand-contact method to detect point by point. This method is time-consuming work fees, high cost, long cycle, cannot conduct a comprehensive detection and detection is not suitable for road, rail, marine and other areas under the pipeline, it is difficult to ensure the accuracy and reliability of detection. The development of new pipeline detection equipment, to enhance pipeline detection

G. Lee (Ed.): Advances in Automation and Robotics, Vol. 1, LNEE 122, pp. 607–614.
springerlink.com

technology and means, it has important economic and social benefits to study and solve the pipeline detection theoretical and technical issues.

2 Buried Pipeline Leak Detection Technology

Pipeline Leak Detection is essential to guarantee buried pipeline safety that delivers oil, natural gas and hazardous liquid medium for industrial. Leak detection equipment commonly used mobile devices, one is regular detection to detect leakage in the area and possible leakage of the site, the other is the pipes leak, discovered leak points by pipeline leak detection equipment, to take accurate excavation and sealing measures. According to the location of detection devices, leak detection technology can be divided into two types of external detection and internal detection.

2.1 Pipelines External Detection Technology

Pipelines External Detection Technology mainly includes olfactory acuity device method, acoustic method, airborne infrared, laser scanning leak detection technology. "Olfactory acuity device method" is the use of combustible gas sensor made of olfactory acuity meter trench in the valve shaft and other parts of direct detection, or detected by flame or plasma optical detection principles of creating portable or vehicle-mounted detector for road testing. This method is widely used gas leakage current detection in China's gas industry, generally found the gas leak by this method, and further using other methods to find the leak, and then plugging repair.

Acoustic method works, the fluid leakage through the walls leak, it will stimulate the transfer along the pipe wall of sound signals, mounted on the outer wall of the acoustic sensors to monitor pipeline leak size and location of the acoustic signal. No leakage occurs, the acoustic sensors to obtain the background noise signal, when a leak occurs, the resulting low-frequency acoustic signal leakage from the storage easy to distinguish from background noise and handling. The advantage is low cost, environmental adaptability, and can detection of buried pipelines at any time. The disadvantage is the limited detection range is not applicable long-distance pipeline.

In recent years, airborne infrared method was developed by the United States OILTON, this method is the use of helicopters lifting of a sophisticated infrared camera flying along the pipeline, through the infrared camera, and the results for spectral analysis, can determine the location of pipeline leaks. This method can be used for long pipeline leak detection, but test results influenced by the surrounding environment, require higher on the environment.

Laser scanning leak detection method is a leak detection method developed by gas automation design department of the former Soviet Union developed, the helicopter with the laser gas analyzer methane along pipeline flight, a laser beam to the gas pipeline along the gas pipeline scanning, laser beam reflected by the ground to the laser methane gas analyzer on the helicopter, when the gas pipeline leaks and the formation of methane in the pipe above the clouds, methane clouds will absorb the laser beam of a frequency domain spectrum , resulting in echo, the laser methane gas analyzer according to the attenuation of the frequency-domain spectroscopy to determine whether a gas leak, when the frequency spectrum of the attenuation exceeds a certain threshold, the laser methane analyzer issued a leak alarm signal indicator . This method can detect smaller leak, but subject to environmental and climate, the method used is very restricted.

2.2 Pipeline Internal Detection Technology

Detection pipeline mainly include magnetic flux leakage detection, ultrasonic detection method, the purpose is to prevent the detection of pipeline corrosion leaks.

Magnetic flux leakage detection equipment part is permanent magnet excitation, the N / S poles and the pipe wall contact, the magnetic circuit is generated in the wall. Sensor mounted between the two poles, Wall defects detected by the receiver coil due to magnetic flux leakage, wall to reaction conditions. For 20-30% of wall thickness of the corrosion depth of the corrosion detection, a description of the wall through the pipe can be used to determine whether the leak. This method requires close contact sensor and the wall, due to factors such as weld, uneven wall, so that these requirements are sometimes difficult to achieve. Such expensive detection equipment, detection costs are also higher.

Ultrasonic detection method using ultrasonic probe transmitting ultrasonic waves, to detect wall thickness and corrosion by the time difference of reflected waves between the pipe outer walls. As from the transmitter to the wall the need for homogeneous liquid as a sound wave media, so for natural gas pipeline, need to segment in a liquid to run the two ends of the two conventional pigging, Ultrasonic detector to run into the liquid segment. Japan Steel Corporation developed ultrasonic detection device can reproduce the pipeline pigging and pipe wall thickness of the surface of the image, detecting the weld corrosion, detection of the corrosion depth of the wall thickness of 10%.

Low-frequency ultrasonic guided wave method by automatically launch, receive and analyze low-frequency ultrasonic propagation along the pipe measured the reflected signal, to determine the extent of the damage pipe section. For urban buried gas pipeline, because the detection equipment import and export of settings, the probe contact with the wall, coupling and other issues, detection of the ongoing feasibility of the small. Therefore, low frequency ultrasonic guided wave method of application is even more necessary and urgent, and should be applied research.

3 The Constitute and General Layout of Trenchless Detection Equipment

Trenchless pipe detection equipment is the carrier of a variety of detection devices, which includes power and drive, control and positioning, support and walk, join with the regulation, devices and systems. These devices and systems research and development and rational allocation, is the key to implement pipeline detection, improve the quality and reliability of detection, especially develop the drive and walk installations for long-range small-caliber buried pipeline, with great challenges.

3.1 Performance Requirements for Pipeline Detection Equipment

As pipe detection equipment, the main task is to complete the pipeline detection, its performance must meet the testing requirements of the system. The magnetic memory / flux leakage testing systems, pipe detection equipment performance should reach the following basic requirements:

① walking steadily, attitude stability;
② driving force, high transmission efficiency;
③ controlled reliable, adaptive ability;
④ accurate positioning, control of travel associated with the detection.

3.2 Basic Composition of Pipeline Detection Equipment

Pipeline detection equipment is an autonomous walking in the electrical and mechanical installations inside the pipeline, it has the drive, walk, detection, control and other basic functions, it usually drives, running gear, controls and organization, detection devices and systems, supply system and devices and other components. Figure 1.

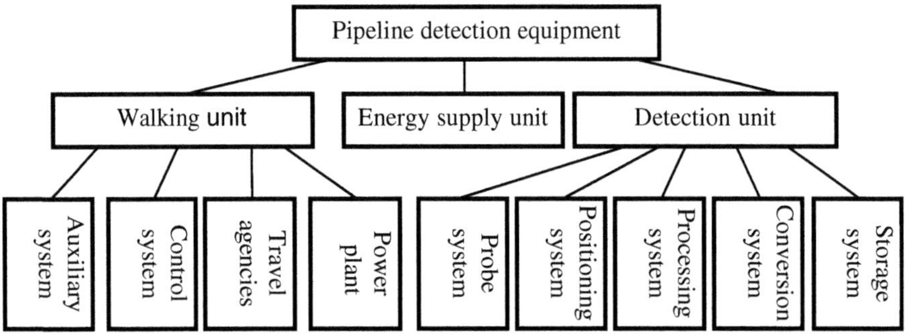

Fig. 1. The basic composition of detection equipment

In order to meet the pipeline's operation, according to mandate detection equipment and operating characteristics of the body, divided into multiple units, according to a certain degree of connection to link up to form multi-cell-type pipe detection equipment.

3.3 The Overall Layout of Pipeline Detection Equipment

Detection equipment system consists of walking drive, walking support organization, detection probe system, the detection system, positioning devices, energy devices and other components. These systems and devices according to their function, structure and connection between composition and structure of its relatively independent units, namely, drive unit, the probe and positioning unit, testing the system unit and the energy supply unit, and its overall layout and mutual relations between the Figure 2.

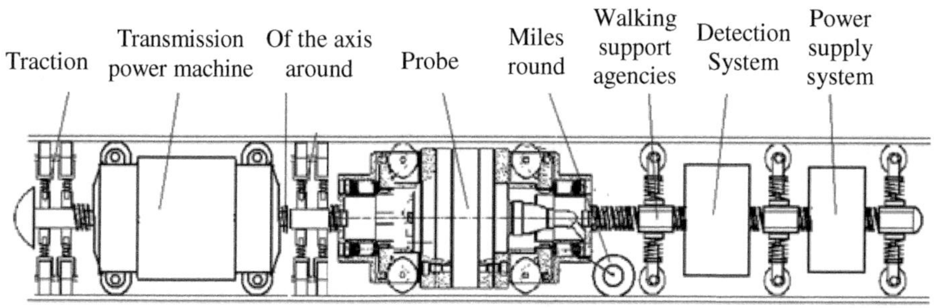

Fig. 2. The overall layout of pipeline inspection equipment

4 Key Technology of Pipeline Detection Equipment

4.1 Miniaturization Technology of Pipeline Detection Equipment

Combination of detection multi-channel detection equipment, operating heavy load, the required driving force, which makes the drive unit and the structure of the travel agencies larger, for small diameter pipes, the work is very conspicuous contradiction between the narrow space, this serious problems restricted the small diameter pipe detection equipment development and application. the key to solve the problem that research high performance, miniaturization of the detection equipment systems and detection systems, to study from this principle of the system, configuration, optimization, structural layout, circuit design and component selection other aspects, so that detection equipment is to ensure essential features and performance, but also compact and lightweight[1].

4.2 Drive and Control Technology of Detection Equipment

Efficient, reliable and compact drive and the device, and a large energy supply system, is technical and material assurance for autonomous walking distance pipeline detection equipment, the key to solve the driven technical problem is to research small size, Major driving force, high efficiency drive running gear and high-energy energy system; In addition, the pipeline detection equipment and communication links outside the pipe is very difficult, not in real time and control of manual intervention, so need to study the rational self-control system, to implement adaptive control.

A new type of spiral pipe wheel drive have been proposed, this device uses single-axis dual-driven approach, the so-called single-axis refers to the movement and power only by the motor shaft coaxial output, and dual-drive refers to a motor mounted before and after helical traction wheel, to achieve before and after the combination of traction, the traction drive principle shown in Figure 3. Devices powered by DC motor as the source, in order to obtain a low output speed, the motor is equipped with front and rear ends of the planet gear; traction with traction wheel axis double helix, two spiral traction wheel mounted between the two groups, respectively, before and after the ends of the motor reducer output shaft, in order to prevent structural interference of sync error caused resistance group set a reasonable fine-tune the front and rear automatic clearance compensation; motor housing with the battery pack, the former has a curved front wheel traction guide wheel, detection equipment and after the traction wheel bearings to achieve movement, through the connection between the motor is equipped with central support wheels, each part of the connection shafts are made of flexible spring steel shaft, all the wheels are elastic support, there is also universal joints after the end of traction wheels in order to connect to other devices.

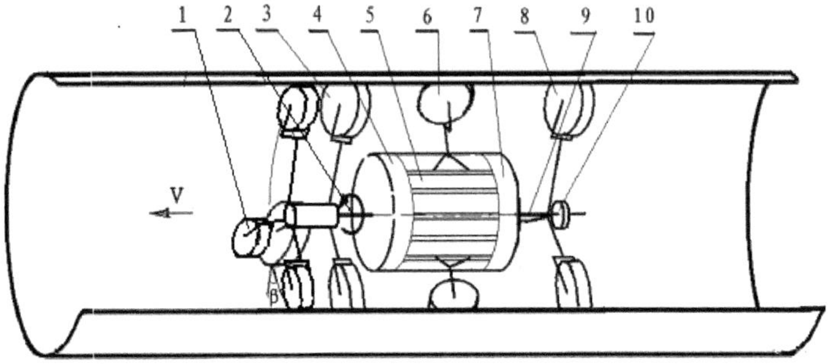

1. Guide the wheel 2. Around the axis of the wire 3. Before the traction wheels 4. Planetary gear 5. Motor housing and battery pack 6. Support wheels 7. Planetary gear 8. After the traction wheel 9. Flexible wire shaft 10. Rolling

Fig. 3. New spiral traction drives

4.3 Positioning Technology of Detection Equipment

Pipeline detection equipment detects in the pipeline, not only to detect defects in the pipe and its characters, but also to determine the exact location of the defect, which involves the positioning of pipe detection equipment. The positioning is quite difficult for buried pipeline detection equipment, mainly the low accuracy in the detection of long, is difficult to accurately determine the location of defects.

Detection information relative to the exact location of the pipeline, in fact, described by the positioning of the detection equipment, and its internal relations is to detection equipment clock. For detection equipment, usually with internal clock and external clock, the so-called internal clock is the electronic clock in self detection equipment, as long as the power to run, regardless of whether the detection equipment to walk, it runs according to its own natural frequency, free from the impact of external conditions, it can be called the absolute clock for detection equipment to run. Outer clock is set by a man-made contract, which start and stop according to the needs of the task, such as detection equipment to stop the clock stopped running, detection information and location information collection also stop that controlled by the external clock and realize not go without inspection. This will enable the detection information is accurate and reliable description of the location of tags, but also prevents too much redundant information occupy a lot of storage space, to maximize solve data storage problems in the long-range detection, the effective detection distance increased detection equipment.

Therefore, the structure of the process of detection location of the pipeline detection equipment shown in Figure 4, when the detection equipment, standby power systems, that the clock was started, the mileage wheel positioning system and acceleration positioning system is in working condition, detection of motion parameters; When the system determine the speed parameters of movement, indicating that the detection equipment has begun to walk, which starts external clock, while the detection system and position location system started detection and location, when the testing equipment for some reason or other need to stop walking, the external clock is stopped, this time location information and detection information extraction and processing also stop, wait outside the clock recovery Start and into work.

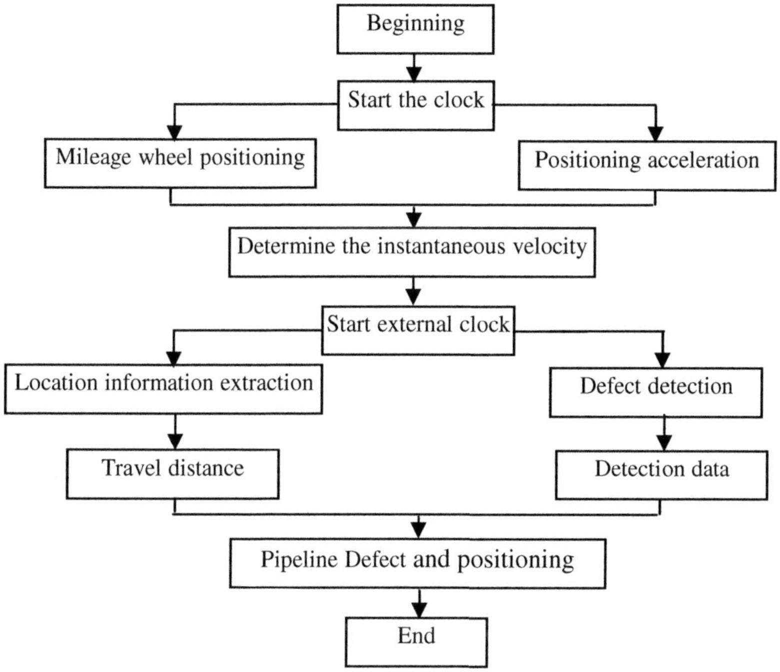

Fig. 4. The structure and localization process Detection equipment

4.4 Technical Crossing Obstacle of Detection Equipment

Self-detection equipment walks in the pipeline, it must overcome various obstacles encountered, such as the connection structure of pipe, valves, tees, elbows, and other access barriers to the local deformation, which requires pipeline detection equipment, drive and running gear with a strong obstacle ability to walk and maintain good posture to meet the needs of testing operations. Research a reasonable walking posture and related equipment size, from a modulation of superior performance, adaptive ability of the drive and running gear and guide mechanism to enhance the more impaired capacity of pipeline detection equipment.

Pipeline detection equipment is special detection equipment, walking in the pipeline, they often have to carry a variety of testing instruments and operating equipment. In special circumstances where a small operation, first, to ensure walk with the smooth by require, but in the actual project, the pipeline has some special pipe section would appear to walk barriers, these obstacles in place, structure, size, layout, etc of the detection equipment is unreasonable, even if the detection equipment is with sufficient power drive, cannot go beyond these obstacles, and even jamming cannot retreat, ranging from the work cannot be, while the pipeline system or testing equipment has been damaged, causing serious consequences. For the normal buried pipeline, these special pipe barriers mainly reflected in the elbow, tee and the local boss, and depression[2-5].

5 Conclusion

This paper describes the internal and external detection technology and the principle of buried pipeline; proposed the basic component and the overall layout of the technology program of pipeline detection equipment; discusses the key technology of pipeline detection equipment, Methods and principles of the driver is given more decision-making fusion technology information source localization. Detection system will need micro-depth technology research, develop more sophisticated detection system, to further expand detection applications. And the development of small size, high capacity energy systems, enhance self-driven testing capabilities and increase the robot's driving distance of continuous work. The high-performance signal acquisition and processing system to be researched, to further explore the signal to noise ratio to improve detection approaches.

References

1. Su, Y.: Electromagnetic detection of buried pipeline Thinking Research. J. Logistical Engineering University Learned Journal, 2 (2007)
2. Ong, J.K., Kerr, D.: Design of a semi-autonomous modular robotic vehicle for gas pipeline inspection. Systems and Contro l Engineering, 109–122 (2003)
3. Nguyen, T.T., Kim, D.K.: Dynamic, Modeling and Its Analysis PIG Flow through Curved Section in Natural Gas Pipeline. In: IEEE International Symposium on Computational Intelligence in Robotics & Automation, pp. 492–497 (2001)
4. Ong, J.K., Kerr, D., et al.: Design of a semi-autonomous modular robotic vehicle for gas pipeline inspection. In: Proc. Instn Mech. Engrs: J. Systems and Control Engineering, pp. 109–122 (2003)
5. Xu, F.: Pipeline robot in the corners of the study by Robot, J., 155–160 (2004)

Research of Path Planning in Virtual Scenes Based on Ant Colony Algorithm and Axis Aligned Bounding Boxes Collision Detection Technology

Jian-min Wang, Hong Yin, Wen-bin Xie, Guo-qing Qiu, Ji-heng Xu, and Ying Huang

Engineering Institute of Engineering Crops, PLA Univ. of Sci. & Tech., Nanjing 210007, China
wangjianmin1114@126.com

Abstract. A kind of path planning method based on ant colony algorithm is proposed for the path planning problem in Complex virtual scenes. The improved axis aligned bounding boxes collision detection technology is introduced to the algorithm, The constraint conditions of the path planning are combined with the searching of the following path points to reduce the search space and enhance the efficiency. Simulation examples show that we could finally acquire the optimum path by the method which is practicable and applicable.

Keywords: Ant colony algorithm, Axis aligned bounding boxes , Optimization path planning , Virtual scenes.

1 Introduction

Path planning problem is an important field of research on virtual reality technology, it requires to find an optimal path from the initial state to target state in the virtual scene avoiding obstacles based on some principle. In recent years, researchers have made some deep studies on the path planning problem in the virtual scenes and put forward a number of fast algorithms. In the literature[1] mesh is introduced into the path planning problem ,which has got some good results, but it simplifies three-dimensional scenes to two-dimensional scene, it is very difficult for us to effectively take into account the actual topography and constraints. In the literature[2] a robot path planning algorithm is proposed which is applied to the automatic roaming virtual scene path planning, it has achieved good result, but it is little practical application.

Recently a number of intelligent algorithm are increasingly used in the study of path planning and have achieved some results. And ant colony algorithm was firstly proposed by Dorigo, which is enlightened by the behavior of the ants finding path during the food searching[3-4]. Ant colony algorithm has some character such as global optimization, positive feedback of information, heuristic search and distributed computing and presents good capability in the traveling salesman problem[5],

G. Lee (Ed.): Advances in Automation and Robotics, Vol. 1, LNEE 122, pp. 615–622.
springerlink.com

Job-Shop Scheduling [6] and the secondary distribution of resources[7] and other combinatorial optimization problems. In this article we combine ant colony, algorithm with collision detection technology of axis aligned bounding boxes, and introduce the constraints of path planning into algorithm, which has achieved a good path planning results.

2 Path Planning

2.1 Grid Planning

Here we set S as the starting point which is also as the origin of coordinates, G as the target point, SG as the X-axis, Y axis is perpendicular to SG, Plot out SG as m equal parts and plot perpendicular whose perpendicular point is equal points of SG by which we could get a line $L_1, L_2, \dots L_{m-1}$ with the spacing $|SG|/m$, then set the X-axis as the center, plot n parallel lines on both sides respectively, with interval of $\gamma \cdot |SG|/m$ where γ is the constant coefficient, to get $H_1, H_2, \dots H_{2n+1}$. In the new coordinates system S-XY, there are (m-1) × (2n +1) points, ants transfer in the points.

At the same time the object is divided into a grid-based small objects ,the point covered by obstacles is set with the value -1 .

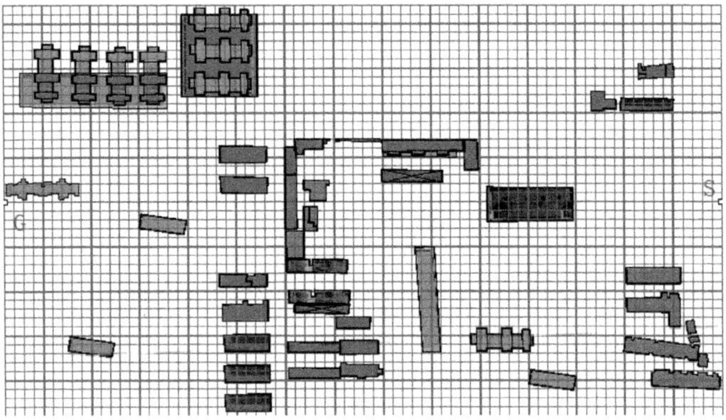

Fig. 1. Three-dimensional virtual scene meshing

2.2 Path Planning Objectives

Here we need to achieve two goals of path planning, The first is 100% survival rate on the path, the second is to achieved the shortest path on the basis of the survival rate.

That is $\min \sum_{i=1}^{n} V_i$ and the collision rate of any section of the path P_i is 0.

2.3 Constraint Handling

(1) Maximum turning angle θ, which is the angle of paragraph i and i +1is less than or equal to θ.
(2) The maximum travel positioning distance 2L. Set the straight line distance from the starting point to the target point as L, Provides the biggest travel path is 2L, that

$$\sum_{i=1}^{n} l_i < 2L \tag{1}$$

3 Improved Axis Aligned Bounding Boxes Collision Detection Technology[8]

As path planning for high real-time requirements in the virtual environment, most of the research simplify three-dimensional scenes into two-dimensional scenes, by losing accuracy for real-time. To solve this problem, we propose an improved Axis aligned bounding boxes collision detection algorithm to improve immediacy.

This method uses projection techniques to map the object plane, do collision and interference detection processing in two levels of a rough one and a precise one. The algorithm is to form a triangular surface for the detection of objects in the basic unit, while take the application of the triangular projection surface in the direction of motion instead of the swept surface contours, the application of differential linear path instead of the curve. As the straight lines in path planning are far more than the curves, and the efficiency of obtaining triangles is much higher than to construct contour surface, the algorithm makes the efficiency of a continuous collision detection in the path planning improved.

Collision algorithm description:

Step 1: Set point A as start point and the following point O as end point. Set O as the center, the projection plane XOY perpendicular to AO is determined .If AO is the mesh edge, the objects in the two grid adjacent AO are built with axis aligned bounding boxes and if the AO is the Diagonal of a grid, it only need to build axis aligned bounding boxes for the objects in this grid.

Step 2: Projected bounding box onto the XOY plane.

Step 3: If the projection of moving objects bounding box does not intersect the projection of obstacles object bounding box, O can be determined as the sequent point of A.

Step 4:Otherwise, it should be accurate collision detection. All collided triangles of the objects are projected onto the projection plane. if all triangles didn't not intersect, O can be determined as the sequent point of A. Otherwise, O is not the sequent point of A.

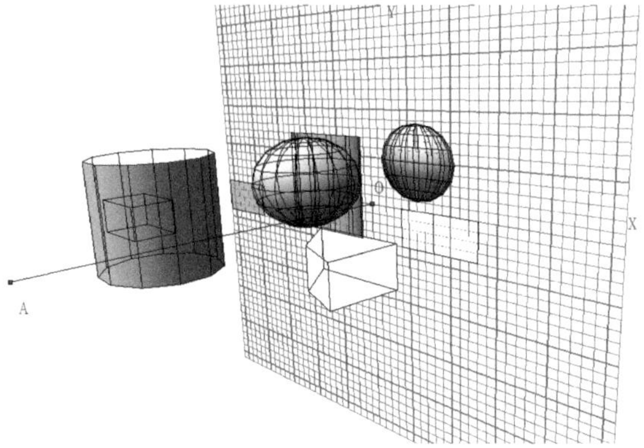

Fig. 2. The projection plane XOY

4 Planning Algorithm

4.1 Allow Sequent Points

Sequent point must meet the three conditions: Firstly, the path point can't be -1. Secondly, meet the constraints. Thirdly, Axis aligned bounding boxes collision detection, to ensure the P_i from the current point to the sequent point to be 0.

4.2 Set Pheromone Target

In order to induce the ants move fast to target, a pheromone source of which the strength is F is placed in the target, assuming that the target point distribution of pheromone intensity and distance is inversely proportional to the target point D with the distance R at the point W, the pheromone strength from the target is

$$\tau = F / R \tag{2}$$

4.3 Transfer Rules

Ants select the next path point based on certain transition probability function , which is the function of the point probability and the amount of pheromone in the connection edge .So the probability of ant k from point i to j is

$$p_{ij}^{k} = \begin{cases} \dfrac{\tau_{ij}^{a}\eta_{ij}^{\beta}}{\sum_{s\in allow_k} \tau_{ij}^{\alpha}\eta_{ij}^{\beta}}, j \in allow_k \\ 0, otherwise \end{cases} \tag{3}$$

Here α and β represent the parameters of the ants in the regulation of pheromone and heuristic information , τ_{ij} is the amount of pheromone in the edge ij. η_{ij} express the desired degree from the point i to j .

4.4 Pheromone Update

When all ants complete one iteration, the pheromone on each path updates by (4) to (5).

$$\tau_{ij}(t+N) = (1-\rho)\bullet\tau_{ij}(t) + \Delta\tau_{ij}$$
$$\Delta\tau_{ij} = \sum_{k=1}^{M}\Delta\tau_{ij}^{k} \quad (4)$$

Where: ρ (0 <ρ <1) is the evaporation coefficient of pheromone on the path wang,1-ρ indicates the increment of pheromone; $\Delta\tau_{ij}$ is the increment of pheromone on edge ij in this iteration; $\Delta\tau_{ij}^{k}$ is the pheromone left on edge ij in this iteration by the first K ant.

$$\Delta\tau_{ij}^{k} = \begin{cases} \frac{Q}{L_k}, \text{the path traversed by ant k in this iteration} \\ 0, otherwise \end{cases} \quad (5)$$

Where: Q is constant; L_k is the length of the path traversed by ant k in this iteration.

4.5 Algorithm Description

Step1:Set initial pheromone and the starting point, Calculate the broad cost of the edges , set the parameters in the ACA ,empty the ant path table, at the same time build Axis aligned bounding boxes of ants based is on need.

Step 2:Set the M ants on the starting point.

Step 3:Ant k to find the optimal path.

Step4: Search the ant sequent points which meets the constraints in point i ($allow_i$)..

Step 5:If ($allow_i$) is empty, set the path point i as -1 and discard the ant.

Step 6: Otherwise, move to the next point j according to the transition probability.

Step 7:The current point is not the target point and the search steps <maximum number of search steps, the path length <2Lmin , turn to step 4, otherwise stop the search.

Step 8:Calculate the path length of the ants, if the path length of ant k is less than the length of the optimal path, update the optimal path length.

Step 9:If the ants pass through the edge ij, then update the pheromone of the edge ij.

Step 10:reach the target point i, output the optimal path.

5 Path Optimization

As the ants are moving in the grid point, we set θ as 90 degrees, then only five points could be the sequent points, which shows that the resulting path has some limitations. we propose an optimized algorithm ,set three parameters i, j = i +1, k = i +2. Set the angle between i and j as α, and the one between j and k as β. If α is not equal to β, do collision detection on the path ij, if P is 0, then:

$$L_{ik} >= L_{ij} + L_{jk} \tag{6}$$

We can remove the point j in the path

Algorithm is as follows:

Step 1: Set the starting point as i = 0.
Step 2 : Calculate α and β.
Step 3: If equal, we can delete point i +1, turn to step 2.
Step 4: Otherwise, we assume a path from i to i +2, do collision detection.
Step 5: If P = 0, we can delete point i +1, turn to step 2.
Step 6: On the contrary, the point i +1 can't be deleted, Set i = i +1, turn to step 2.
Step 7: Delete all the point which could be deleted to result the final path.

6 Simulation Result

We use the algorithm in the path planning of fire in the virtual scene, the tests prove that we can get planned path rapidly and do further optimizing by using this algorithm.

Fig. 3. Search sequent point

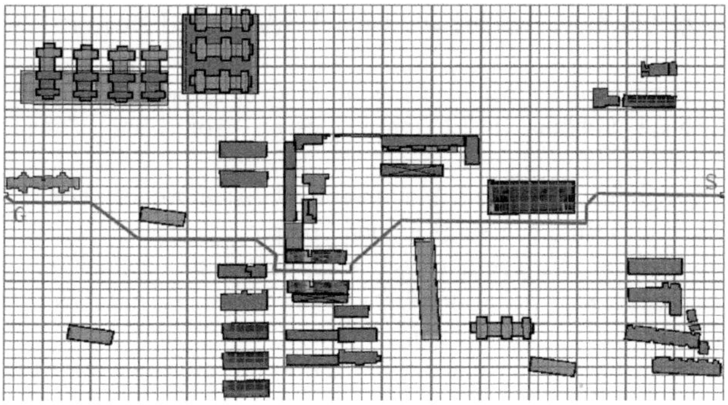

Fig. 4. The initial Path

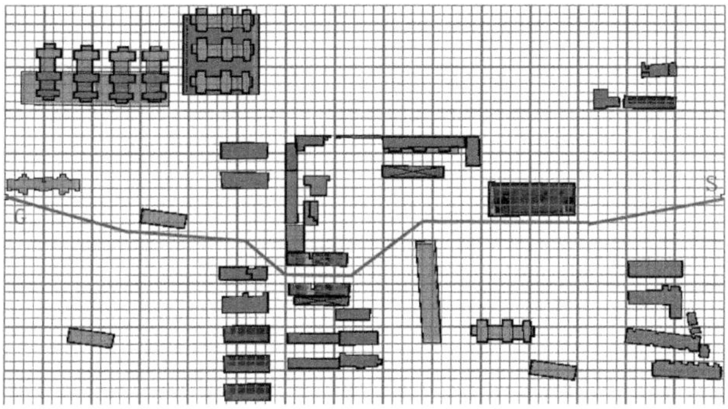

Fig. 5. The optimized path

7 Conclusion

In this paper we propose a path planning algorithm for the complexity in the virtual scene based on ant colony algorithm. The Practicality and efficiency of the algorithm have been improved by introducing axis aligned bounding boxes and constraint conditions. After getting the initial path, the path can be further optimized by the algorithm. Compared to previous algorithms, the path is more actual.

References

1. Li, Z.-L., Li, H., Zhan, C.-H., Peng, F.-Y.: Path-planning Algorithm Based on Mesh Partitioning in Virtual Environment. Journal of System Simulation 19(2), 214–217 (2007)
2. Shi, H.-B., Zhang, Y.-B., Tong, R.-F., Dong, J.-X.: Path Planning for Automated Navigation in Virtual Environment. Journal of Computer-aided Design & Computer Graphics 18(4), 592–595 (2006)
3. Colorni, A., Dorigo, M., Maniezzo, V.: Distributed optimization by ant colonies. In: Proceedings of the 1st European Conference on Artificial Life, 134-142. Elsevier Publishing, Amsterdam (1991)

4. Dorigo, M.: Optimization, learning and natural algorithms. Politecnico di Milano, Milano, Italy (1992)
5. Dorigo, M., Gambardella, L.M.: Ant colony system: a cooperative learning approach to traveling salesman problem. IEEE Transactions on Evolutionary Computing 1(1), 53–56 (1997)
6. Colorni, A., Dorigo, M., Maniezzo, V.: Ant colony system for job-shop scheduling. Belgian Journal of Operations Research Statistics and Computer Science 34(1), 39–53 (1994)
7. Maniezzo, V., Colorni, A.: The ant system applied to the quadratic assignment problem. IEEE Transactions on Knowledge and Data Engineering 1(5), 769–778 (1999)
8. Wang, L.-W., Liu, B.Y., Han, J.W.: Improvement AABB surrounds examination calculate way of collision box. Computer Engineering and Applications 43(33), 234–236 (2007)

An Effective and Efficient Perceptual Organization Method for Image Segmentation

Yingjian Qi[1] and Xu Yang[2]

[1] Science School, Communication University of China, Beijing, China
qiyingjian@cuc.edu.cn
[2] Computer and Information Technology School,
Beijing Jiaotong University, Beijing, China
yangxubj@126.com

Abstract. Existing approaches of perceptual organization are either less effective or less efficient. Here, we propose an effective and efficient perceptual organization method for image segmentation. The method is characterized by following: Firstly, Region Adjacency Graph (RAG) representation is adopted in the method, which reduces the computation complexity suffered by previous organization methods. Secondly, region-wise global weights are defined in RAG by utilizing global optimization criterion, which increases the effectiveness of our proposed organization method. Segmentation results validate the efficiency and effectiveness of the proposed organization method for image segmentation.

Keywords: Image Segmentation, Perceptual Organization Method, Region Adjacency Graph.

1 Introduction

Perceptual Organization (PO) is an important ability in human vision system (HVS) which helps us segment image into meaningful structures or even semantic objects [1]. Technically, PO refers to the ability of HVS to impose organization on sensory data, so as to group image primitives which arise from underlying common cause [2,3]. Since 1980's, computer vision community has been trying to simulate human's PO mechanism in order to improve the performance of image segmentation. Previous approaches of PO for image segmentation can be characterized by the following two factors [4]:

a) Image representation that serves as input of PO
b) The optimization criterion that is adopted by PO

Graph as image representation is the most commonly used one by existing PO approaches [5, 6]. An image is represented as a weighted undirected graph where each node in graph corresponds to a pixel in the image. Each edge is formed by connecting neighboring pair of pixels, and associated with a weight that reflects similarity or dissimilarity of pixels. However, computation under graph representation increases

G. Lee (Ed.): Advances in Automation and Robotics, Vol. 1, LNEE 122, pp. 623–630.
springerlink.com

greatly with the image size growing. This is because pixels are basic primitives, which leads to higher density of graph nodes.

Optimization criterion can be classified into two categories: global optimization criterion such as Normalized Cut criterion [5], and local optimization criterion, such as Minimum and Maximum criterion [6]. Organization based on global criterion can reliably segment image meaningful structures, but leads to heavy computation problem. Local criterion, however, can be implemented efficiently, but often falsely segment image structures.

In this paper, we propose an effective and efficient perceptual organization method for image segmentation. We adopt RAG as image representation, and integrating both optimization criteria to achieve maximum effectiveness and efficiency. This paper is organized as follows. Previous works are presented and analyzed in Section 2 and 3. An effective and efficient perceptual organization method is proposed in Section 4. Section 5 demonstrates the effectiveness and efficiency of the proposed organization method for image segmentation. Section 6 is our concluding remark.

2 Previous Perceptual Organization Method

Many PO methods were presented before, such as graph based optimization method [5, 6], and tensor voting method [7], etc. In this paper, however, we only concern with graph based optimization method for its popularity. Let $G = (V, E)$ be a graph with vertices $V_j \in V$ corresponding to a pixel in the image, and edge $(V_i, V_j) \in E$ representing the link between neighboring pixels. Weight W_{ij} is an element of matrix $\mathbf{W}$. When W_{ij} measures dissimilarity between V_i and V_j, $\mathbf{W}$ refers to dissimilarity matrix; and when similarity, it refers to affinity matrix. According to optimization criterion, two organization methods are compared here (cf. Fig.1.).

Firstly, Normalized Cut method (Ncut) is based on global criterion. Let $G = (V, E)$ be a graph and $\mathbf{W}$ be the affinity matrix. Let A and B be a bipartition of the graph, then the Ncut criterion is defined as [5]:

$$Ncut(A,B) = \frac{cut(A,B)}{assoc(A,V)} + \frac{cut(B,A)}{assoc(B,V)} \tag{1}$$

where $cut(A,B) = \sum_{V_i \in A, V_j \in B} W_{ij}$, $assoc(A,V) = \sum_{V_i \in A, V_k \in V} W_{ik}$. Minimization of Ncut criterion can bipartition the image and segment an image. However, computation of Ncut minimization increases greatly with the image size growing, limiting Ncut method to relatively small images.

Secondly, Minimum and Maximum method (MM) is based on local criterion. Let $G = (V, E)$ be a graph and $\mathbf{W}$ be the dissimilarity matrix of the graph. Let two subsets of pixels X and Y represent two neighboring regions of image. Then two regions are merged if they satisfy MM criterion [6]:

$$Ext(X,Y) \le \min\{Int(X), Int(Y)\} \tag{2}$$

Here $Ext(X,Y)$ means external difference, as the minimum weight between region X and Y. $Int(X)$ or $Int(Y)$ means internal difference of region X or Y, as the maximum weight within X or Y. Contrary to Ncut method, MM method is efficient. However, MM criterion leads to inappropriate merges of regions and we will analyze this problem in next section.

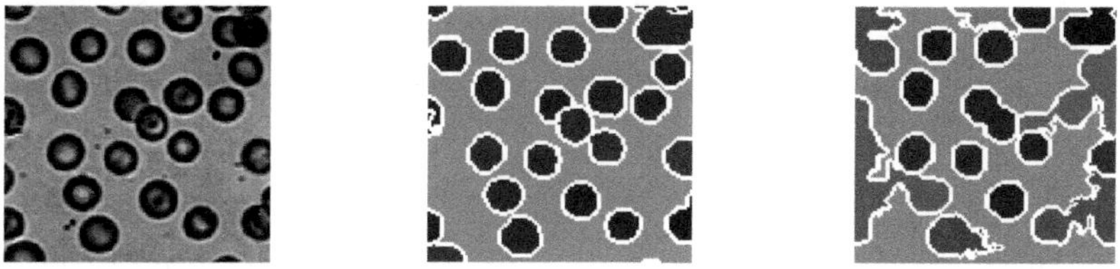

(a) Original Image (b) Ncut method (c) MM method

Fig. 1. Image segmentation: Ncut *vs.* MM, note that inappropriate merges of regions in (c)

3 Problem Analysis of Previous Methods

We claim that the both factors, namely image representation and optimization criterion, can influence the effectiveness and efficiency of perceptual organization. While previous methods focus on optimization criterion, we show that it's the graph defined on pixels that leads to inappropriate merges of regions and higher computation complexity. Inappropriate merges of regions will decrease effectiveness, yet higher computation complexity will decrease efficiency of perceptual organization.

Inappropriate merges of regions is mainly caused by pixel-wise local weights definition. Figure 2 illustrates the problem. Suppose there are two objects in an image, and they are represented as two clusters A and B in graph. A and B are connected with a 'bridge' of noise points (circle-shaped). Thus local weights of two neighboring pixels will suffer high possibility of noise transition from A to B. Therefore, the bridge points weaken the distinction between A and B. MM method would thus falsely merge clusters A and B into one cluster because MM criterion does not globally optimize organization results.

Ncut method can overcome the problem above compared with MM method, because it is based on global optimization criterion. However, global optimization will lead to unacceptable computation cost when the density of vertices in graph is high. Therefore, the computation complexity increases greatly with the image size growing.

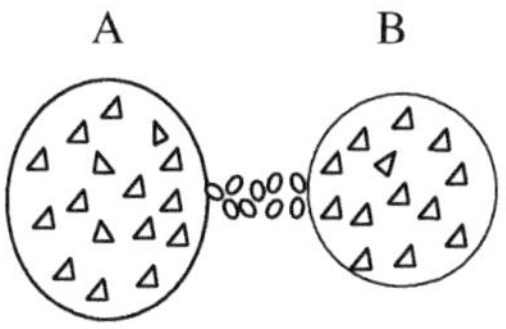

Fig. 2. Pixel-wise local weight definition weakens distinction between objects A and B because of the bridge of noise points

4 Proposed Organization Method

Contrary to traditional graph that is defined on pixels, we build Region Adjacency Graph (RAG) that is defined on regions. The proposed perceptual organization method contains following steps: Firstly, The image is segmented into preliminary regions and RAG is constructed based on local weight. Secondly, global weight replaces local weights with global weights by adopting Ncut organization criterion. Finally, MM organization is applied for RAG representation to get final image segmentation results.

4.1 Region Adjacency Graph Construction

Regions are the primitives in RAG construction. We segment an image into homogeneous color regions using a region segmentation technique called EDISON (Edge Detection and Image Segmentation System) [8]. Then RAG is constructed with weighted undirected graph $G_0 = (V^0, E^0)$, the vertices $V^0 = \{V_1^0, V_2^0, \ldots, V_N^0\}$ of this graph represent N image regions, and the edges E^0 are associated to the links between neighboring regions. The weight W_{ij}^0 reflects distance between two neighboring regions V_i^0 and V_j^0. We calculate weight W_{ij}^0 as Fisher distance between regions in HSV color space [9]. This weights definition is insensitive to noisy pixels, because the mean feature of regions and Fisher distance are both statistically stable. Moreover, the density of vertices in RAG is considerably low. For example, a typical 256×384 image has 98304 pixels, however, only approximately 1000 homogeneous color regions are contained in image. See Fig.3 (a) for homogeneous regions and Fig.3 (b) for its corresponding RAG representation.

(a) Homogeneous regions

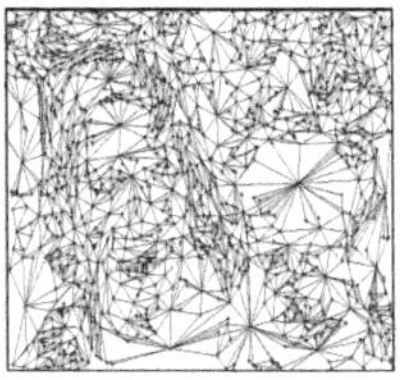

(b) Constructed RAG

Fig. 3. Region Adjacency Graph Representation

4.2 Global Weight Definition for RAG

In this section, we globally estimate the dissimilarity between any two vertices of RAG, so as to sharpen the distinction between objects in RAG (see Fig.2). The goal of this step is to replace local weight with global weight for each weight by applying Ncut optimization over RAG. Following is the detail.

Let $\mathbf{W}$ be the affinity matrix of RAG, here $W_{ij} = \exp\left(\frac{-W_{ij}^0}{\alpha}\right)$. Let $\mathbf{D}$ be the diagonal matrix such that $D_{ii} = \sum_j W_{ij}^0$. It's shown that minimization of Ncut criterion equals to solve generalized eigenvalue system [5]:

$$(\mathbf{D} - \mathbf{W})\mathbf{y} = \lambda \mathbf{D}\mathbf{y} \tag{3}$$

where $\mathbf{y}$ is a real vector. Note that there are N vertices in RAG, so that there are N elements in $\mathbf{y}$ and at most N eigenvectors exist. Therefore, solution to this system transforms RAG into a new feature vector space [10]. That is, each vertex $V_k^0 (1 \leq k \leq N)$ is represented as a global feature vector $F\left(V_k^0\right)$, which is closely related to some object. Therefore weight W_{ij}^0 can be redefined on this global vector and sharpen the distinction between objects. We simply use Euclidean feature distance between vertices V_i^0 and V_j^0, so the global weigh is calculated as:

$$W_{ij}^0 = \left\| F\left(V_i^0\right) - F\left(V_j^0\right) \right\| \tag{4}$$

4.3 RAG-Based MM Organization

In this section, MM organization applies to RAG derived from above. As illustrated in Fig.1, MM method is prone to cause inappropriate merges of regions because of local weight defined in graph. However, MM organization here can avoid inappropriate merges since RAG has global weight. Similar to MM method [6], an image is represented as multiple levels of resolution and any ith level is a graph G_i. Recursively, a vertex in $G_{k+1} (k \geq 0)$ is a merge of neighboring vertices of G_k according to MM criterion, until final level G_f that satisfies stopping rule:

$$Ext\left(V_i^f, V_j^f\right) > \min\left\{Int\left(V_i^f\right), Int\left(V_j^f\right)\right\} \tag{5}$$

V_i^f and V_j^f are any neighboring vertices in G_f. The output of RAG-based MM organization is a hierarchy composed by graph $G_0, G_1, \ldots, G_f$, and object of interest can exist at some intermediate level G_i. We manually choose the best G_i from $G_0, G_1, \ldots, G_f$, to get the final segmentation result.

To summarize, our proposed method can achieve the effectiveness and efficiency:

Effectiveness: We define region-wise weight that can reduce the noise sensitivity problem illustrated in Fig.2. Statistical distance between regions is adopted and hence more reliable than pixel-wise distance. **More important**, we introduce global weight definition, which replaces local weight with global weight calculated by Ncut optimization.

Efficiency: While graph defined on pixels leads to high computation, RAG representation can improve efficiency of PO, because number of regions is much less than number of pixels in an image. Particularly, RAG can considerably reduce the computation of Ncut optimization, which makes it possible to introduce Ncut criterion for globally defining weight of RAG.

Table 1. Efficiency comparison between Ncut, MM, and proposed method for image segmentation

Heading level	Execution Time (sec)		
Image from COREL	Ncut Method	MM Method	Proposed Method
Human	180.9	35.2	25.0
Flowers	140.5	33.6	26.7
Horses	142.4	32.8	24.6

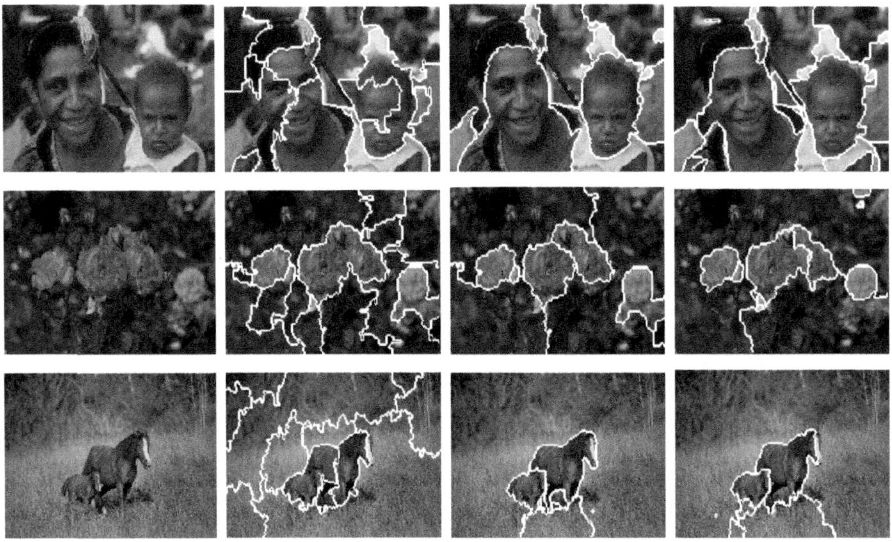

(a) Original image (b) MM Method (c) Ncut Method (d) Proposed Method

Fig. 4. Comparison between MM, Ncut, and the proposed method for image segmentation

5 Experimental Results

The proposed method has been implanted under the environment of MATLAB on a 2.5GHz Pentium IV machine, main memory of 512MB. In the experiment we set fixed value for parameter $\alpha = 0.0017$. Images from COREL image database serve as examples, with image size 256×384.

To test efficiency, we analyze the running time of the method into three parts: RAG construction, global weight definition, and RAG-based MM organization.

Firstly, EDISON for RAG construction requires approximately 4.5 sec for each 256×384 image. This part produces about 1000 regions from 98304 pixels. Secondly, the Lanczos algorithm [11] can solve eigenvectors problem in global weights definition. Since the density of vertices in RAG is quite low, this part only needs about 1.0 sec. The last part, RAG-based MM organization is the most computational, which takes about 20-30 sec. Table 1 shows efficiency comparison between Ncut method, MM method and proposed organization method. The execution time of proposed method is more efficient than MM method and Ncut method.

Fig.4 demonstrates the effectiveness of proposed organization method for image segmentation. Each row corresponds one experiment group. The first row shows performance comparison for the image 'Human'. We see that MM method can't segregate two human from background. The reason behind this problem is that under the local weight, MM method is prone to generate inappropriate mNE.Referges between foreground and background. After global weight definition, the proposed organization method achieves almost equivalent performance with Ncut method, but remarkably improves the performance of MM method. Limitation is also found in the experiment about parameters setting. When changing the size of image, we must carefully tune the parameters α to get the best segmentation results. Automatic parameters selection is important and it is the focus of our future work.

6 Conclusion

In this paper, we propose an effective and efficient perceptual organization method for image segmentation. Firstly, we construct RAG which greatly reduces computational complexity perceptual organization. Secondly, under RAG representation, we present our perceptual organization method by integrating two existing organization approaches in an effective and efficient manner. With image segmentation experiment, the proposed method is evaluated and proved that it has better performance than existing methods. In the future, we would like to process videos with perceptual organization for effective auto-parametric video objects segmentation.

Acknowledgment. This paper is supported by the follow projects. They are: Research of the image semantic based on scale invariant features (No.XNG0949,CUC). Visual Perception theory based image semantic and retrieve algorithm research (No. G2009382316,CUC). Super-resolution image restoration theory and technology in spatial microtherm environment(61071148,NSFC).

References

1. Randrianarisoa, V., Bernier, J.F., Bergevin, R.: Detection of Multi-part Objects by Top-down Perceptual Grouping. In: The 2nd Canadian Conf. on Computer and Robot Vision, pp. 536–543 (2005)
2. Vickery, T.J., Jiang, Y.V.: Associative grouping: Perceptual grouping of shapes by association. Attention, Perception, & Psychophysics 71(4), 896–909 (2009)
3. Demeyer, M., De Graef, P., Verfaillie, K., Wagemans, J.: Perceptual Grouping of Object Contours Survives Saccades. PLoS ONE 6(6), e21257 (2011)

4. Heuel, S., Förstner, W.: A Dual, Scalable and Hierarchical Representation for Perceptual Organization of Binary Images.In: IEEE Workshop on Perceptual Organization in Computer Vision, Santa Barbara (1998)
5. Shi, J., Malik, J.: Normalized Cuts and Image Segmentation. IEEE Transactions on Pattern Analysis and Machine Intelligence 22(8), 888–905 (2000)
6. Kropatsch, W.G., Haxhimusa, Y.: Grouping and Segmentation in a Hierarchy of Graphs. In: Bouman, C.A., Miller, E.L. (eds.) Computational Imaging, vol. 5299(2), pp. 193–204. SPIE, Bellingham, WA (2004)
7. Jia, J.Y., Tang, C.K.: Inference of segmented color and texture description by tensor voting. IEEE Transactions on Pattern Analysis and Machine Intelligence 26(6), 771–786 (2004)
8. Comaniciu, D., Meer, P.: Mean Shift: A Robust Approach Toward Feature Space Analysis. IEEE Transactions on Pattern Analysis and Machine Intelligence 24(5), 603–619 (2002)
9. Schettini, R.: A Segmentation Algorithm for Color Images. Pattern Recognition Letters 14(6), 499–506 (1993)
10. Malik, J., Belongie, S., Leung, T., Shi, J.: Contour and Texture Analysis for Image Segmentation. International Journal of Computer Vision 43(1), 7–27 (2001)
11. Golub, G.H., Loan, C.F.: Matrix Computations, 3rd edn. Johns Hopkins Univ. Press, Baltimore (1996)

Research on an Autonomous Traffic Information System Based on Inter-vehicle Communication Ad Hoc

Qi Zhang

School of Automation & Electrical Engineering,
University of Science & Technology Beijing, 100083, China
zhqi@ustb.edu.cn

Abstract. This paper studies and presents a pattern for an autonomous traffic information system based on Ad Hoc. In this pattern, the traffic information about travelling time of road section and average travelling speed is collected automatically by moving vehicles in the road net, with no need for fixed facilities for traffic information collection and publication. With the use of the vehicle flow in the road net and the change of Ad Hoc overlapping, the collected traffic information is transmitted, received and integrated through inter-vehicle communication and thus real-time traffic information of the whole road net is formed. Besides, this paper also studies and builds the models of automatic collecting, transmitting and integrating ,based on Ad Hoc, of traffic information, and the model of automatic updating of road net traffic information.

Keywords: Inter-vehicle communication, Ad Hoc, Traffic information, Information collection, Information transmission.

1 Introduction

The traffic congestion in the city has become a big traffic problem which is encountered by many countries and is needed to be solved as soon as possible. The real-time and effective traffic information system is an important guiding means for reducing traffic congestion, facilitating vehicle traveling, improving operating efficiency of road network. At present, the existing traffic information system mostly gets the real-time traffic information through many fixed collection equipment of traffic information. The collected traffic information is processed and integrated in the traffic information center in order to form the real-time traffic information which is needed in the road network traffic. Then the information is transmitted through all kinds of fixed display equipment and navigation equipment in the vehicle. However, in this system, the precision of information collection and the effect of information service are directly limited by the density of collection and publication equipment. In order to get collection data with high precision and dynamic road traffic information with high quality, we often need to increase the density of all kinds of equipment. Thus the building and realizing of this traffic information system is largely dependent on fund investment.

G. Lee (Ed.): Advances in Automation and Robotics, Vol. 1, LNEE 122, pp. 631–638.
springerlink.com

In order to realize real-time and accurate traffic information service and reduce fund investment which is used for building traffic information system, many experts and scholars in the world are devoting themselves to studying all kinds of new traffic information system. With the rapid development of communication technology and vehicle-to-internet technology, the traffic information system based on communication technology in vehicle is becoming a new research direction of intelligent transportation technology now and in the future.

2 Research Situation

At present, there are mainly two kinds of research about traffic information system based on inter-vehicle communication. For one kind of research , early warning information of traffic safety driving is transmitted through inter-vehicle communication and this research focuses on early warning of vehicle collision avoidance. For the other kind of research which is rare , traffic congestion information is transmitted through inter-vehicle communication and this research focuses on providing face-to-face vehicles with traffic congestion information of their frontal roads [1,2]. Because this traffic information system can only provide moving vehicles with traffic congestion information of their frontal roads and can not provide more traffic information about surrounding roads and road network, it is hard for the driver to know traffic congestion conditions of relative routes which lead to destination. Besides, this traffic information system can not provide vehicles with traffic congestion information of their frontal roads when the vehicles travel on the single lane where there are no face-to-face vehicles or the face-to-face vehicle doesn't pass through the frontal road which the current vehicle is interested in. Therefore, there is some limit in the real application of this traffic information system.

In addition, the present research about automatic traffic information collection, publication and transmission based on inter-vehicle communication is rare. In Japan, some experts and scholars put forward a scheme idea of collecting road traffic flow through inter-vehicle communication. In this idea, the floating vehicles system based on inter-vehicle communication is built, the floating vehicles communication platform which doesn't need fixed data processing center is built, and some relative algorithm of road traffic congestion information are studied [3]. Some Japanese experts and scholars also put forward that other vehicles information which is around the collection vehicle can be collected through inter-vehicle communication and the congestion information of surrounding vehicles can be shown on the electronic guiding map in the vehicle [4]. Besides, the author of this paper and some relative research fellows also present algorithm of traffic information collection based on inter-vehicle communication and carry out some corresponding simulation evaluation [5]. But the pattern of this traffic information system is still on the primary stage of research. The principle of the system pattern, the algorithm of congestion information generating, and the adaptability of the application will need to be discussed and complemented.

Because of many advantages of the traffic information system based on inter-vehicle communication such as low public investment, high information accuracy and good real-time service, this paper will study and present a new pattern of autonomous traffic information system which realizes the function of automatic collection, automatic transmission and dynamic updating of traffic information through moving vehicle flow and Ad Hoc. At the same time, math models of corresponding system will be built.

3 The Pattern for Autonomous Traffic Information System Based on Ad Hoc

This paper tries to present a pattern of autonomous traffic information system based on Ad Hoc. The automatic collection of traffic information is realized through the GPS which is installed in the vehicle and gets the traveling time and average traveling speed, which show that the road is blocked or unblocked, of certain road section. The automatic transmission of traffic information is realized through Ad Hoc which transmits the traffic information, which is owned by every vehicle and includes the information collected by the vehicle itself and the information integrated after receiving other vehicles information and the simplified information of road traffic information in the road network, to other vehicles. Thus every vehicle in the Ad Hoc network integrates the latest road traffic information collected by the current vehicle and the traffic information transmitted by other vehicles in the same Ad Hoc network to form the latest road traffic information, which is owned by the current vehicle, of certain road section. The generating and updating of the traffic information is realized through moving vehicles. When every vehicle transmits the integrated information of current collection road section and the simplified congestion information to other vehicles of other road sections in the road network , every vehicle probably owns congestion information of other road sections and dynamically updates the road traffic information , which is owned by the vehicle itself , of the whole road network according to the time effectiveness of congestion information , which the current vehicle receives , of other road sections.

Because of the moving of vehicles on the road, the constant change of relative VANETs, and the overlapping of several Ad Hoc networks, the road traffic information of certain road section can automatically spread to neighbouring road networks and the transmission speed and distance increase with the increase of density of vehicle flow in the road network.

According to this system pattern, this paper will build models of automatic collection, transmission, generating and updating of traffic information based on Ad Hoc.

4 Model for Automatic Collection of Traffic Information

As the collection base of traffic information, the road section is a basic unit and has the corresponding rule for the collection interval. For the long road, it can be divided according to the structure of road network and the characteristic of vehicle flow.

4.1 The Traffic Information Collected by the Vehicle Itself

In order to collect the traffic information of certain road section, every vehicle can use GPS and road electronic map to get the traveling time of the current road section through determining the starting point and terminal point of the current road section and to figure out the average traveling speed of the current road section. In order to simplify the description, this paper gives the following defining: the traffic information , which is collected by the current vehicle within the time interval m , of certain road section is the average traveling speed (or the traveling time) which is gotten by the current vehicle when it passes through the current road section within the time interval m.

The average traveling speed , which is collected by the vehicle itself , of certain road section is

$$V_{ik}(mT) = \frac{L_i}{t_{ik}(mT)}, \qquad k \in D_i(mT) \tag{1}$$

$V_{ik}(mT)$: average traveling speed collected by collection vehicle k on the road section i within the time interval m ;

$t_{ik}(mT)$: traveling time collected by collection vehicle k on the road section i within the time interval m ;

L_i : length of road section i ;

i : road section i in the road network , $i \in \{1,2,3,\ldots,N\}$;

N : the sum of road sections in the road network ;

T : the time interval of collection of road traffic information ;

m : the time interval sequence of collection of road traffic information ;

$D_i(mT)$: the collection vehicle set on the road section i within the time interval m.

4.2 Integrating and Processing of Information after the Vehicle's Receiving the Traffic Information of Other Vehicles

According to this research's premise that every vehicle transmits traffic information through Ad Hoc, every vehicle transmits the traffic information collected by the vehicle itself through information broadcast as well as receives traffic information transmitted by other vehicles. But because of the constant moving of vehicles and the constant change of VANETs, every vehicle can not necessarily receive the traffic information transmitted by all the vehicles that collect certain road section. Therefore, this paper presents that the model for integrating and processing of traffic information of certain road section can be gotten only through receiving average traveling speed collected by partial vehicles.

$$\overline{V}_{ik}(mT) = \frac{1}{n_{ik}(mT)}[\sum_{p \in D_{ik}(mT)} V_{ip}(mT)]\ \} \tag{2}$$

$\overline{V}_{ik}(mT)$: average traveling speed, which is figured out after information integrating and processing by the collection vehicle k on the road section i within the time interval m , of road section i ;

$D_{ik}(mT)$: vehicle set which is integrated and processed by the collection vehicle k on the road section i within the time interval m ;

$n_{ik}(mT)$: the vehicle sum of the vehicle set $D_{ik}(mT)$.

Because the number of vehicles which are integrated and processed by every vehicle is different, the average traveling speed which is figured out by every vehicle is a little different. The more vehicles the vehicle set includes, the more accurate the traffic information which is gotten through information integrating and processing is. And the average speed which is figured out is more likely to approach the real average speed of the current road section.

5 Model for Automatic Transmission of Traffic Information

The model for automatic transmission of traffic information mainly describe a process that the vehicle transmits the road traffic information , which is owned by the vehicle itself , of every road section to other vehicles in the road network through Ad Hoc. The vehicles in the Ad Hoc network include vehicles which are in the same collection road section and vehicles which are in the different collection road section. The target of information transmission and the content of information transmission include two situations.

5.1 Traffic Information Transmitted by the Vehicle Which Is on the Same Collection Road Section

The traffic information which is transmitted by the vehicle k to other vehicles on the same collection road section includes the original road traffic information which is collected by the vehicle k and the original road traffic information which is collected by other vehicles on the same collection road and is received by the vehicle k through inter-vehicle communication. The original road traffic information which is collected by other vehicles can be integrated and processed.

$$\{ V_{ip}(mT) \mid p \in D_{ik}(mT)] \} \tag{3}$$

5.2 Traffic Information Transmitted by the Vehicle Which Is on the Different Collection Road Section

Every vehicle collects traffic information not only to provide other vehicles on the same collection road section with original traffic information which is used for integrating and processing but also to transmit its integrated traffic information which is owned by the current vehicle to other vehicles on the different collection road section. In order to reduce information which is transmitted, this paper simplifies the traffic information which is transmitted according to the conditions of road congestion. The vehicle will only transmit the traffic information of congested road section.

$$V'_{ik}(mT) = \begin{cases} 0 & \overline{V}_{ik}(mT) \geq 40 \text{ km/h} \quad \text{(clear)} \\ 1 & 20 \leq \overline{V}_{ik}(mT) < 40 \text{ km/h} \quad \text{(slow travelling)} \\ 2 & 0 \leq \overline{V}_{ik}(mT) < 20 \text{ km/h} \quad \text{(congested)} \end{cases} \tag{4}$$

$V'_{ik}(mT)$: simplified road traffic information , which is transmitted by collection vehicle k on the road section i within the time interval m to other vehicles on the different collection road section , of the road section i.

6 Model for Automatic Generating and Dynamic Updating of Traffic Information

6.1 Automatic Generating of Traffic Information Which Is Owned by the Vehicle

Because every vehicle not only collects and integrates the traffic information of certain road section, but also receives the simplified traffic information of different road section in the road network through inter-vehicle communication, the traffic information, which is owned by the vehicle, of every road section in the whole road network can be formed quickly and can be shown on the guiding electronic map in the vehicle to give the driver some reference.

The traffic information which is owned by the collection vehicle k on the road section i within the time interval m includes the traffic information of the vehicle's collection road section and the traffic congestion information of different collection road sections. The two kinds of information will be transmitted to vehicles on the same collection road section and vehicles on the different collection road section. The composition of the whole road traffic information is as following:

$$\{ V_{ip}(mT) \mid p \in D_{ik}(mT), i \in \{1,2,\ldots,N\}\} \cup \{ V'_{jk}(m'T) \mid j \in \{1,2,\ldots,N\}, j \neq i \} \quad (5)$$

$\{ V_{ip}(mT) \mid p \in D_{ik}(mT), i \in \{1,2,\ldots,N\}\}$: average traveling speed , which is owned by vehicle k , of all the collection vehicles on the road section i within the time interval m ;

$\{ V'_{jk}(m'T) \mid j \in \{1,2,\ldots,N\}, j \neq i \}$: simplified information , which is owned by the vehicle k within the time interval m' , of congested road section j ;

m': the time interval within which the vehicle k owns the simplified information of congested road section j $(m' \leq m)$.

Among the traffic information owned by every vehicle, the traffic information which is collected by the current vehicle itself is the latest information, but the traffic information which is transmitted by other vehicles and is not collected by the current vehicle may include the collected congestion information which isn't or is unable to be updated within the past time interval, hence $m' \leq m$.

6.2 Dynamic Updating of Traffic Information Which Is Owned by the Vehicle

For the moving vehicle on certain road section in the road network, in order to get the latest congested road traffic information of other road sections in the road network , it needs to select the traffic information which is received by the current vehicle and determine whether to update the road traffic information , which is owned by the current vehicle , of every road section.

For the vehicle k , it will determine whether to update information according to the following situations when it receives the road traffic information of the vehicle h.

When the vehicle h and the vehicle k are on the same collection road section. the collection vehicle set $D_{ik}(mT)$ which is owned by the vehicle k will be updated according to the fact that there are collection vehicles which are not owned by the vehicle k in the collection vehicle set , which is owned by the vehicle h , of the original road traffic information. The updating of road traffic information will be carried out according to the formula (2) and the formula (4).

$$\begin{aligned} &\text{when} \quad D_{ik}(mT) \neq D_{ih}(mT) \\ &\text{then} \quad D_{ik}(mT) = D_{ik}(mT) \cup D_{ih}(mT) \end{aligned} \tag{6}$$

When the vehicle h and the vehicle k are on the different collection road sections or they transmit congested road traffic information which doesn't belong to the current collection road section, though they are on the same collection road section. there will be three situations which determine whether to update congested road traffic information.

If the vehicle k doesn't own the congested road traffic information. which is owned by the vehicle h , the congested road traffic information which is owned by the vehicle h will be added to the road traffic information of the vehicle k.

$$\begin{aligned} &\text{when} \quad V'_{jk}(m_{jk}T) = 0 \quad \text{but} \quad V'_{jh}(m_{jh}T) \neq 0, \quad k \neq h \\ &\text{then} \quad V'_{jk}(m_{jk}T) = V'_{jh}(m_{jh}T) \\ &\qquad m_{jk} = m_{jh} \end{aligned} \tag{7}$$

$V'_{jk}(m_{jk}T)$: congested road traffic information , which is owned by the vehicle k within the time interval m_{jk} , of the road section j which is not the collection road section of the vehicle k, $j \in \{1,2,\ldots,N\}$;

m_{jk} : the time interval within which the vehicle k owns the congested information of the road section j

If both the vehicle k and the vehicle h own the congested road traffic information of certain road section. it needs to compare the collection time interval of the congested road traffic information which is owned by the two vehicles .When the vehicle h has the latest collection time interval , the congested road traffic information of the vehicle k will be updated. When the vehicle h doesn't have the latest collection time interval, the congested road traffic information of the vehicle k will not be updated.

$$\begin{aligned} &\text{When} \quad m_{jk} < m_{jh} \quad k \neq h \\ &\text{then} \quad V'_{jk}(m_{jk}T) = V'_{jh}(m_{jh}T) \\ &\qquad m_{jk} = m_{jh} \qquad \text{(the same as the formula (7))} \end{aligned}$$

If the congested road traffic information is not updated within the limited time , it will be cleared. According to the time effectiveness of the traffic congestion information , when the congested road traffic information , which is owned by the vehicle k , of certain road section is not updated within the limited time , the information will be cleared.

$$\begin{array}{ll} \text{when} & mT - m_{jk}T \geq T_0 \\ \text{then} & V'_{jk}(m_{jk}T) = 0 \\ & m_{jk} = m_{jh} \end{array} \tag{8}$$

T_0: the time within which the congested road traffic information of the road network exists.

7 Conclusion

This paper presents a new pattern of autonomous traffic information system in which the function for automatic collection and transmission of the traffic information is realized by using moving vehicles and ad hoc and with no need for using fixed facilities for traffic information collection and publication. The corresponding models for automatic collection, automatic transmission, automatic generating and dynamic updating of traffic information are built. In addition, a question to be studied in the future is the transmission speed of traffic information in the road network and the covering range of the information. And it also needs some means and ways such as traffic simulation to simulate and evaluate the whole effect.

References

1. Narzt, W., Wilflingseder, U., Pomberger, G., Kolb, D., Hörtner, H.: Self-organising congestion evasion strategies using ant-based pheromones. J. IET Intelligent Transport Systems 4(1), 93–102 (2010)
2. Lu, X., Xie, K.L.: A Traffic Information System Based on Ad-Hoc Wireless Network. J. Computer Applications and Software 25(4), 122–123 (2008) (in Chinese)
3. Sato, M., Ishida, T., Horiguchi, R., Shimizu, K., Haruta, H., Wada, K., Uehara, K., Murai, J.: Evaluation of Road Traffic Information Generation Algorithm for Decentralized Probe Vehicle System. Transactions of Information Processing Society of Japan 49(1), 253–264 (2008) (in Japanese)
4. Nishimura, A., Koizumi, H.: A Real-time Traffic Flow Measurement Method Based on Inter-Vehicle Communication Considering the Volume of Vehicle Information. IPSJ SIG Technical Report, 2006-ITS-24(1) (2006) (in Japanese)
5. Wang, B.-L., Zhang, Q.: Method of Traffic Information Acquisition Based on Inter-vehicle Communication. Modern Electronics Technique, 34(7), 202–204 (2011) (in Chinese)

Development of a Virtual Humanoid Model Using the Denavit-Hartenberg Parameters as a Base for Visual Feedback Applications

Davinia Font, Tomàs Pallejà, Mercè Teixidó,
Marcel Tresanchez, and Jordi Palacín

Department of Computer Science and Industrial Engineering
University of Lleida, Jaume II, 69, 25001 Lleida, Spain
{dfont,tpalleja,mteixido,mtresanchez,palacin}@diei.udl.cat

Abstract. In this paper a virtual humanoid model based on the Denavit-Hantenberg (DH) parameters is described and proposed as a base for the development of future visual feedback applications to estimate human position, pose or motion. The virtual humanoid can replicate human body poses, has the center of the coordinate system located on the pelvis, and the stand up is the reference or zero pose. The position and orientation of the humanoid body links are determined according the DH frame conventions and the image of the model can be used in additional visual feedback applications.

Keywords: Denavit-Hartenberg parameters, humanoid robot, virtual model.

1 Introduction

The estimation of the human pose and motion [1-5] has many applications such as gaming, security, human-computer, interaction, telepresence, and health-care. The recent availability of depth cameras has spurred further progress in this field [6-8]. In this work we propose the development of a virtual humanoid model as a base for the development of visual feedback applications that compares humanoid and human pose. In such applications, the skeleton of the humanoid's model has to be similar to the human body which is integrated by bones, cartilage, ligaments and joints. In this work, the most important part is the definition of the links and the control of the joints because they define the poses achieved by the humanoid. There are three different types of joints which are defined as fixed joints, partially-movable joints and freely-movable joints. Fixed joints are the ones which don not allow any movement such the skull; partially-movable are those that only allow a slightly movement such as the vertebrae; and the free-movable joints allow the movement along different axes, such as translations or rotations. In a humanoid skeleton the rotation joints such as the ellipsoidal (Fig. 1-a) and hinged joints (Fig. 1-b) are the most important. Ellipsoidal joints have two degrees of freedom represented as a combination of two revolute joints, perpendicular to each other without separation. Hinged joints have one degree of freedom as a revolute joint.

G. Lee (Ed.): Advances in Automation and Robotics, Vol. 1, LNEE 122, pp. 639–646.
springerlink.com

2 Related Work

The development of the skeleton of a virtual humanoid model can be addressed computing the complete kinematics [9, 10] or using the simplified parametrical model proposed by Denavit & Hartenberg (DH) [11]. This model allows a simplified interpretation of the relative motion between articulations through the use of four basic transformations which are two revolutions and two translations along the x and z axes that are modified depending on the rotation parameters.

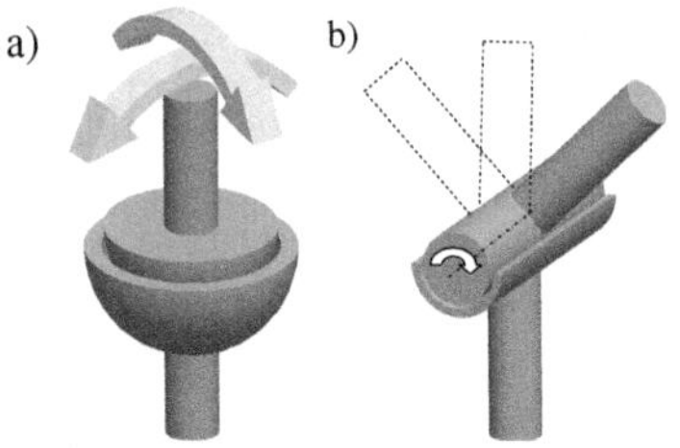

Fig. 1. Example joints: a) ellipsoidal b) hinged joint

There are a huge number of works based on the use of the DH parametric model. In [12], the geometry of the humanoid WABIAN-RIV was simulated using the simplified DH parameters. In [13] a consistent methodology using the DH parameters for a closed-form inverse kinematic joint in a general humanoid robot was explained. In [14] the walking pattern of a bipedal robot was generated using the simplified DH model. In [15], the DH parameters were used to define an exo-skeleton type master arm integrated in a KIST humanoid robot.

The new contribution of this work is the development of a complete humanoid model using the DH parameters that accepts individual and relative control of the joints as inputs and generates different image pose as outputs. Then, this model can be used in specialized human pose estimate applications such as walking analysis [16].

3 Denavit-Hartenberg Parameters

The DH parameters [11] create a reference frame for each articulated link used respect to the previous link frame based on its position and orientation. The DH parameters are represented as [a, d, α, θ], where "a" is the length of the translation along x axis between the initial and the end point; "d" is the offset along z axis between the initial and the end point; "α" is the rotation's angle about x axis, from old z axis to new z axis and, "θ" is the rotation's angle about z axis, from old x to new x being variable when a joint n is a revolution or constant when it is translational. The z-axis is always in the direction of revolution of the joint axes and the x axis is defined parallel to the common normal.

4 Humanoid Model

The structure of the basic skeleton of the humanoid model is as follows. There are five different paths which are defined to control the limbs' motion. These paths are left leg; right leg; left arm; right arm and the head. Each path starts at the center of the coordinate system located on the pelvis (Fig. 2 and Fig 3-a: P) and has different number of joints depending on the movement required.

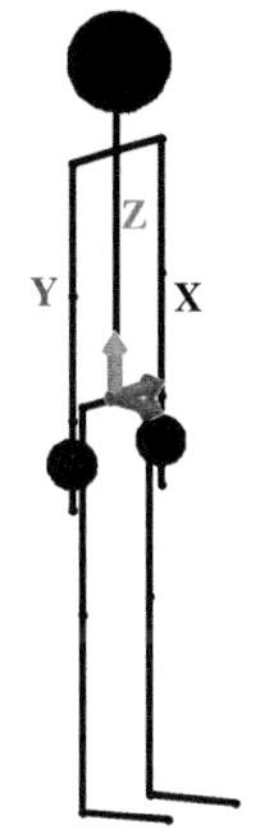

Fig. 2. Initial position of the humanoid's model specifying the the coordinate axis at the pelvis link

4.1 Links

The nomenclature used to define the humanoid's links is the following: L and R indicate the left and right part of the body respectively; A means ankle; K means knee; H means hip; P means pelvis; V means vertebra; E means elbow; S means shoulder and N means neck (Fig. 3-a). On the other hand, the assignment of the limb's length are L that means length; L_R, L_L means right and left part of the body; L_{LAn} means the length of n arm's limbs; L_{LLn} means the length of n leg's limbs and L_{LT} means torso (Fig. 3-b).

The limbs' lengths used in the humanoid model were obtained by graphically measuring a member of our team. The lengths measured were: 135 mm for the right (L_{LL1}) and left pelvis (L_{RL1}) link; 420 mm for the thighs (L_{RL2}, L_{LL2}); 400 mm for the calf (L_{RL3}, L_{LL3}); 200 mm for the feed (L_{RL4}, L_{LL4}); 500 mm for the torso (L_{LT}); 180 mm for the right (L_{RA1}) and left shoulder (L_{LA1}); 270 mm for the humerus (L_{LA2}, L_{RA2}); 250 mm for the forearms (L_{LA3}, L_{RA3}); 180 mm for the hands (L_{LA4}, L_{RA4}) and 180 mm for the neck (L_N).

The angular orientation and motion range of each articulation was determined by graphically measuring different extreme poses of the same member of our team. Table 1 shows the range values for the articulation which rotate around x axis (α parameter) whereas Table 2 has the same information for the rotation along z axis (θ parameter). The numbers of the articulations are defined in Fig. 4 to 6. For instance, the α10 parameter (Fig. 4-a: 10, knee) has an initial angular orientation of 0º and only a positive offset because the knee can only rotate in one direction whereas the θ10 parameter is always 0º as the rotation in the local z-axis is not allowed for this articulation, the same happens to the elbow.

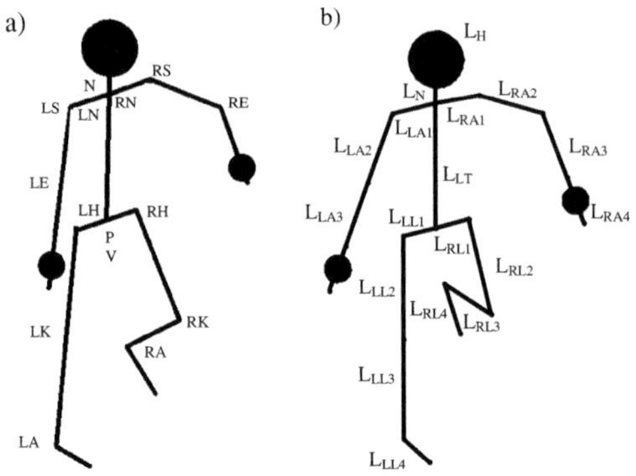

Fig. 3. Nomenclature used to define the humanoid's links/joints (a); nomenclature used to define the length of the limbs (b)

Table 1. Motion's range of alpha parameter (α)

Parameter	Initial position (degrees)	Offset range (degrees)	Range (degrees)
α 3	90	-70	70
α 5	0	0	160
α 6	0	-20	20
α 8	90	-70	70
α 10	0	0	160
α 11	0	-20	20
α 13	0	-20	20
α 14	0	-20	20
α 15	-90	-15	15
α 16	-90	-30	190
α 18	0	0	160
α 23	-90	-15	15
α 24	-90	-190	30
α 26	0	0	160
α 30	0	-60	60

4.2 Leg's Joints

Fig. 4 represents the articulations involved in human leg movement. The proposed DH model for both legs is symmetrical and composed by an ellipsoidal pelvis, an ellipsoidal hip, hinged knee, and a hinge ankle. The right leg has two joints (Fig. 4-a:1-2), located on the pelvis link (Fig. 3-a: P), providing two revolutions at this point through the x and z axes; joints 8-9 (Fig. 4-a) are located on the left hip (Fig. 3-a: LH) to provide movement to the whole leg through the composition of two revolutions; number 10 joint (Fig. 4-a) is the knee link, which generates movement to the tibia using one revolution whereas number 11 (Fig. 4-a), the ankle link, just provides motion to the feet having one free motion. There is another articulation, 12 (Fig. 4-a) that is created in order to draw the feet but it does not have any movement, for this reason it is not specified in Fig. 4.

The same standard is used for the left leg which has the same links working in the same way, represented as 1, 2 and from 3 to 7 (Fig. 4-b). Table 3 represents the DH parameters for the left and right path of the legs respectively

Table 2. Motion's range of theta parameter (θ)

Parameter	Initial position (degrees)	Offset range (degrees)	Range (degrees)
θ 1	0	-20	20
θ 2	0	-20	20
θ 4	0	-45	20
θ 9	0	45	-20
θ 15	0	-15	15
θ 16	0	-90	180
θ 17	90	-20	155
θ 23	0	-15	15
θ 24	0	-180	90
θ 25	90	-20	155
θ 31	-90	-20	70

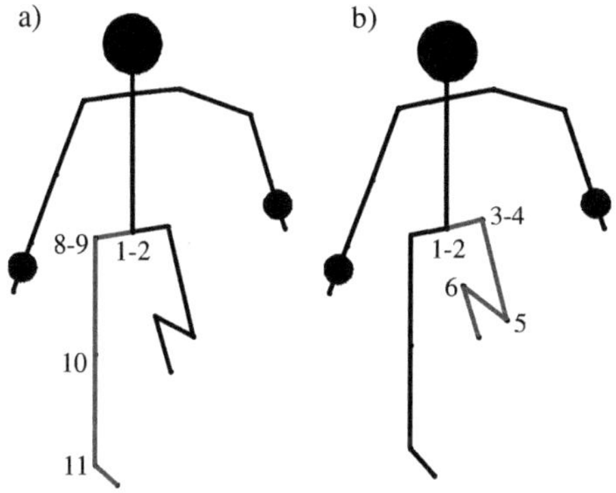

Fig. 4. Joint's nomenclature used to define right leg's motion (a) and left leg's motion (b)

Table 3. D-H Parameters for the leg's path

Left leg's path					Right leg's path				
Joint	d	a	α°	θ°	Joint	d	a	α°	θ°
1	0	0	-90	θ1	1	0	0	-90	θ1
2	0	0	0	θ2	2	0	0	0	θ2
3	0	$(L_{LL1}+L_{RL1})/2$	- α3	0	8	0	$-(L_{LL1}+L_{RL1})/2$	- α8	0
4	0	0	-90	-θ4	9	0	0	-90	-θ9
5	L_{RL2}	0	α5	0	10	L_{LL2}	0	α10	0
6	L_{RL3}	0	-α6	0	11	L_{LL3}	0	-α11	0
7	0	L_{RL4}	0	90	12	0	L_{LL4}	0	90

4.3 Arm's Joints

The arm's geometry has also two different paths, right (Fig. 5-a) and left arm (Fig. 5-b). Human arms are composed by an ellipsoidal pelvis, an ellipsoidal vertebra, ellipsoidal neck, ellipsoidal shoulder, and an ellipsoidal elbow. The right arm of the humanoid model is composed by joints 1, 2, 13, 14 and from 15 to 22 (Fig. 5-a). The joints 1-2 (Fig. 5-a) are the pelvis (Fig. 3-a:) link which has two free revolutions; 13-14 (Fig. 5-a) is a link located at the same point adding one more motion, this joint is used to move the torso part (Fig. 3-a: V) of the humanoid. The joint number 15 (Fig. 5-a) is the link responsible to give movement to the clavicle (Fig. 3-a: LN) whereas number 16 (Fig. 5-a) provides the movement to the shoulder (Fig. 3-a: LS). The last link that has motion along the arm is a two revolute joint, 17-18 (Fig. 5-a) represented as LE articulation. Links 19-22 are used to drawing the hand of the humanoid without generating any movement on it.

The same standard is used for the left arm representing the links as 1, 2, 13, 14 and from 23 to 28 (Fig. 5-b). Table 4 represents the DH parameters for the left and right path of the arm respectively.

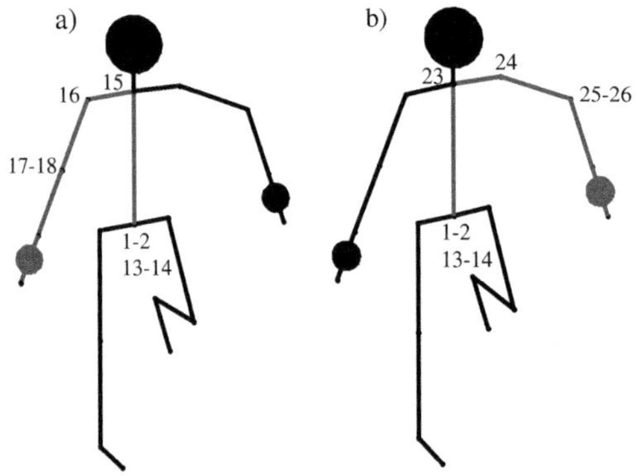

Fig. 5. Joint's nomenclature used to define right arm's motion (a) and left arm's motion (b)

Table 4. D-H Parameters for the arm's path

Left arm's path					Right arm's path				
Joint	d	a	α°	θ°	Joint	d	a	α°	θ°
1	0	0	-90	θ1	1	0	0	-90	θ1
2	0	0	0	θ2	2	0	0	0	θ2
13	0	0	α13	0	13	0	0	α13	0
14	0	0	α14	-90	14	0	0	α14	-90
23	L_{LT}	0	α23	θ23	15	L_{LT}	0	α15	θ15
24	$-(L_{LA1}+L_{RA1})/2$	0	α24	θ24	16	$(L_{LA1}+L_{RA1})/2$	0	α16	θ16
25	L_{RA2}	0	0	θ25	17	L_{LA2}	0	0	θ17
26	0	0	α26	0	18	0	0	α18	0
27	L_{RA3}	0	0	0	19	L_{LA3}	0	0	0
28	L_{RA4}	0	0	0	20	L_{LA4}	0	0	0

4.4 Neck's Joints

The neck's path motion of the humanoid is composed by an ellipsoidal pelvis, an ellipsoidal vertebra, and an ellipsoidal neck. The articulations used are 1, 2, 13, 14 and from 29 to 33 (Fig. 6). The joints 1-2 and 13-14 were described before and the 29 performs a vertical translation until the intersection between the neck and the shoulders where 30-31 links are located (Fig. 6). These links provide two revolutions to complete the motion of the head composed by a horizontal and vertical movement. On the other hand, the joints 32-33 are just used to define the center of two circles that represent the face of the humanoid and a short hair. The color of both circles can be configured to generate a more realistic virtual humanoid representation. Finally, Table 5 shows the DH parameters for the neck's path.

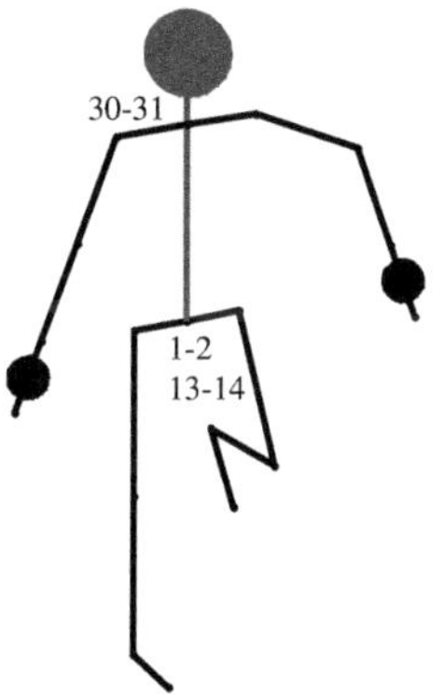

Fig. 6. Definition used to define the articulations along the neck's motion

Table 5. D-H Parameters for the neck's path

Joint	d	a	α°	θ°
1	0	0	-90	-θ1
2	0	0	0	-θ2
13	0	0	α13	0
14	0	0	α14	-90
29	L_{LT}	0	-90	0
30	0	0	α30	0
31	0	L_H	0	θ31
32	0	0	-90	0
33	10	0	0	0

5 Conclusions

This works presents the development of a virtual humanoid model using the Denavit & Hartenberg parameters. The links and joints used in the model have been described and the complete model presented. The virtual humanoid can replicate human body poses, has the center of the coordinate system located on the pelvis, and the stand up is the reference or zero pose. The virtual humanoid model has 18 links and 15 joints with a total of 106 fixed parameters and 26 parameters that can be modified or adjusted. The virtual model can generate different output parameters and images that can be used in the future development of specific visual feedback applications.

References

1. Moeslund, T.B., Hilton, A., Krüger, V.: A survey of advances in vision-based human motion capture and analysis. Computer Vision and Image Understanding 104, 90–126 (2006)
2. Poppe, R.: Vision-based human motion analysis: An overview. Computer Vision and Image Understanding 108, 4–18 (2007)
3. Ferrari, V., Marín-Jimenez, M., Zisserman, A.: Progressive Search Space Reduction for Human Pose Estimation. In: IEEE Conference on Computer Vision and Pattern Recognition, Anchorage, pp. 1–8 (2008)
4. Ferrari, V., Marín-Jimenez, M., Zisserman, A.: Pose Search: Retrieving People using Their Pose. In: IEEE Conference on Computer Vision and Pattern Recognition, Miami, pp. 1–8 (2009)
5. Kian, C., Lim, C.K., Luo, Z., Chen, I.-M., Yeo, S.H.: Wearable wireless sensing system for capturing human arm motion. Sensors and Actuators 166, 125–132 (2011)
6. Grest, D., Woetzel, J., Koch, R.: Nonlinear body pose estimation from depth images. In: Kropatsch, W.G., Sablatnig, R., Hanbury, A. (eds.) DAGM 2005. LNCS, vol. 3663, pp. 285–292. Springer, Heidelberg (2005)
7. Knoop, S., Vacek, S., Dillmann, R.: Sensor fusion for 3D human body tracking with an articulated 3D body model. In: Proceedings of the IEEE International Conference on Robotics and Automation, Orlando, pp. 1686–1691 (2006)
8. Plagemann, C., Ganapathi, V., Koller, D., Thrun, S.: Real-time identification and localization of body parts from depth images. In: IEEE International Conference on Robotics and Automation, Anchorage, pp. 3108–3113 (2010)
9. Hilton, A., Beresford, D., Gentils, T., Smith, R., Sun, W.: Virtual People: Capturing human models to populate virtual worlds. In: Proceedings of the Computer Animation, Washington, pp. 174–185 (1999)
10. Shin, S.Y., Kim, C.: On-line Human Motion Transition and Control for Humanoid Upper Body Manipulation. In: IEEE International Conference on Intelligent Robots and Systems, Taiwan, pp. 477–482 (2010)
11. Denavit, J., Hartenberg, R.S.: A Kinematic Notation for Lower Pair Mechanisms based on Matrices. Trans ASME J. Appl. Mech. 23, 215–221 (1955)
12. Carbone, G., Lim, H., Takanishi, A., Ceccarelli, M.: Stiffness Analysis of the Humanoid Robot WABIAN-RIV: Modelling. In: IEEE International Conference on Robotics & Automation, Taiwan, pp. 3654–3659 (2003)
13. Ali, M.A., Park, H.A., George Lee, C.S.: Closed-Form Inverse Kinematic Joint Solution for Humanoid Robots. In: International Conference on Intelligent Robots and Systems, Taiwan, pp. 704–709 (2010)
14. Choi, Y., Kim, D., Oh, Y., You, B.-J.: Posture/Walking Control for Humanoid Robot Based on Kinematic Resolution of CoM Jacobian With Embedded Motion. IEEE Transactions on Robotics 6, 1285–1293 (2007)
15. Lee, S., Park, S., Kim, M., Lee, C.-W.: Design of a Force Reflecting Master Arm and Master Hand using Pneumatic Actuators. In: IEEE International Conference on Robotics & Automation, Belgium, pp. 2574–2579 (1998)
16. Palleja, T., Teixido, M., Tresanchez, M., Palacin, J.: Measuring Gait Using a Ground Laser Range Sensor. Sensors 9, 9133–9146 (2009)

Knowledge Map and Knowledge Management Tools to Support Distributed Product Design

Jiang Zuhua, Su Hai, and Huang Yongwen

School of Mechanical Engineering, Shanghai Jiao Tong University, Shanghai 200030
P.R. China
zhjiang@sjtu.edu.cn

Abstract. In order to support distributed design work, it is necessary to use all kinds of geographically dispersed design expertise and knowledge as soon as possible. At the same time, all the knowledge base is asked to accumulate and update, as well as to retrieve and share. This paper focus on the knowledge map and knowledge management tools to support distributed design. Design activity in distributed product design includes calculation and reasoning, as well as simulation and evaluation. Design knowledge includes formula and rules, as well as constraints. The object of product design knowledge management includes the knowledge of the product domain, the design process knowledge and the expert with design knowledge. Knowledge management architecture can be described as four layers, which are management tool layer and basic management layer, as well as high management layer and knowledge integration layer. Knowledge map is used to connect all knowledge nodes with different kinds of knowledge base. To increase the transparency and agility of knowledge map, the knowledge indexing is a very important work. Knowledge expert can move along six paths that are central for effective knowledge visualization in product design contexts. The XML based knowledge index support building knowledge map automatically, and provide user information about the design knowledge. The map is constructed in the text based SVG format, so user can search the needed knowledge by key words. The basic element of product design knowledge is represented as BNF format. Some technologies of management tools had been analyses.

Keywords: Distributed design, Knowledge map, Knowledge management tool.

1 Introduction

Current marketplace is undergoing fast changes, and most great research and development centers in the world are asked to integrate their geographically dispersed design resource to response to market changes [1]. Zhou, S [2] presents a design support system for collaborative work based upon the composite knowledge representation. It can meet the main challenges of distributed environment. To overcome the limitations of commercial CAD and handbook form of knowledge expression, the distributed Knowledge Information System (DKIS) for product design is proposed [3]. Zhou also implement the mechanism on knowledge represented, stored and acquired quickly in

G. Lee (Ed.): Advances in Automation and Robotics, Vol. 1, LNEE 122, pp. 647–654.
springerlink.com

distributed intelligent information system (DIIS), which is based on the Internet [4]. The research on knowledge management for distributed product design is more important than before.

In order to support distributed design work, it is necessary to use all kind of design expertise and regulation as soon as possible. Designer like to use geographically dispersed design resource convenience. At the same time, all these knowledge base is asked to accumulate and update, as well as to retrieve and share. Most research works had been done to manage design knowledge for some special engineering domain, and this knowledge base is difficult to share and integrate to others. Knowledge map is useful to connect all kinds of knowledge node with knowledge base. With some knowledge management tools that meet designer cognition and habit, the design knowledge can be shared and integrated well.

2 Knowledge Management Content for Distributed Product Design

By the Internet, design knowledge can be used for many geographically dispersed design activities. Design activity includes calculation and reasoning, as well as simulation and evaluation. Respectively, design knowledge includes formula and rules, as well as constraints. When the product design scheme evaluation and redesign is done using fuzzy sets, the membership function will be used for fuzzy decision. At the meantime, the patent and successful design case is also the important knowledge for enterprise.

The goal of knowledge management is not only managing the knowledge, but also managing the people. Because some important knowledge is stored in individual's heads. For enterprises, how to manage their core knowledge efficiently is a very important problem. Currently researchers mainly concentrate on two aspects, one is building the knowledge base for intelligence design system (for example, expert system), and another is studying the general method that manage the distributed text knowledge or information by IT tool set. Here we focus on knowledge map and knowledge management tools. It's necessary to study the management technique of product design knowledge to collect and manage product design knowledge stored in different places and in different styles.

3 The Architecture of the Knowledge Management System for Product Design

Designer can manage and integrate structural knowledge, semi-structure knowledge and unstructured knowledge using different tools and different methods. Because of different utilization format, knowledge management can be divided by four layers, which are management tool layer and basic management layer, as well as high management layer and knowledge integration layer (Fig. 1).

Firstly, designer can obtain and accumulate design knowledge with tools in management tool layer. According to the designer's cognition custom, the design knowledge can be presented respectively as formula, constraint, fuzzy rule, membership function, design principle, patent, design manual, and design case. These types cover most design knowledge in the application.

In basic management layer, beside the basic function just like adding, deleting, browsing, modifying and searching design knowledge, management tool should also have the other function, for instance, the unique check and the consistency check, as well as validity check and conflict check to design knowledge.

In high management layer, the design knowledge management system obtains more knowledge by searching engine, enterprise knowledge portal, and knowledge map etc. It can help the user to discover implicit knowledge from product database and technical reports, and to refine meta-knowledge to support decision-making. Knowledge locking is a method of classifying and storing for all kinds of knowledge. Knowledge sharing aims at supporting knowledge transfer and communication, knowledge-exploiting is to externalize tacit knowledge, internalize explicit knowledge and use them to support design work. Knowledge evaluating is to evaluate the value of knowledge bases, then to help the user maintain knowledge.

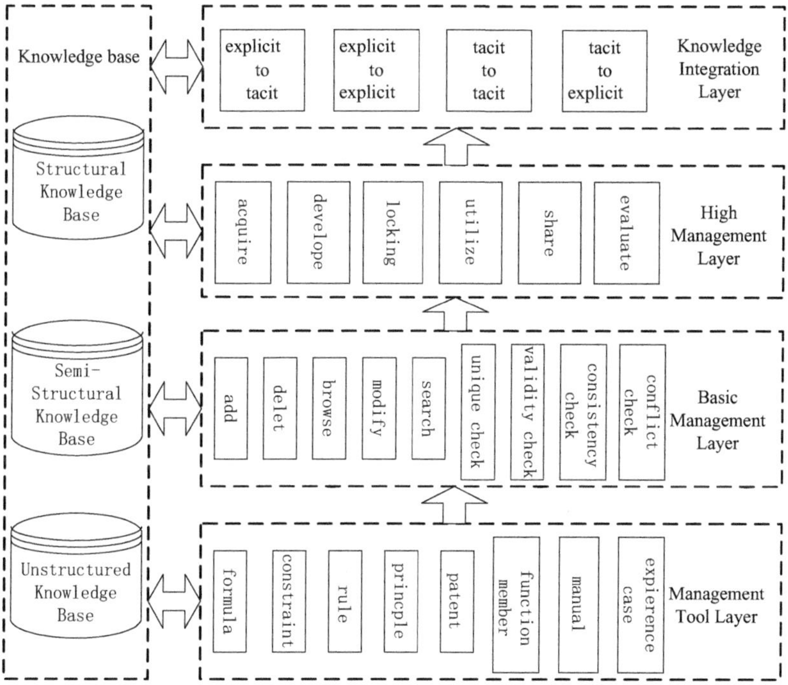

Fig. 1. The architecture of knowledge management system for product design

In knowledge integration layer, some experience and know-how can be concluded using background knowledge and data mining method. Some coding knowledge can be translated into another kind of knowledge using knowledge representation change, and it can be translated into tacit knowledge after understanding their relation and characteristic. By the translation from tacit to explicit, and from explicit to tacit, the design knowledge can be extended and spread.

4 Knowledge Map to Support Product Design

Knowledge map is a toolkit for visualizing and exploring contexts and relationships in distributed knowledge collections. To increase the transparency and agility of knowledge map, the knowledge indexing is a very important work. Some information about the knowledge and its context should be added. Knowledge expert can move along six paths that are central for effective knowledge visualization in product design contexts (see Table 1). These vectors will be useful guides for knowledge expert to create knowledge map for tailored to the specific context of different department in product development.

The index of knowledge nodes is based on above six vectors of knowledge attribute. The Backus-Naur Form of knowledge index is:

<knowledge index>:: = <context>|<classification>|<implement method>|<evaluation>|<relation>

<context > ::=<title> <key word> <description> <creator> <time of creation>

<classification >::=< classification system> <purpose> <sort>

<implement method>::=<format of the resource> <size> <location> <implement environment > < condition> <purview>

<evaluation >::=<expert> < evaluation of the knowledge>

<relation>::=<inherit>|<successively>|<reference>|<conflict>|< complementary>

Table 1. Six Vectors of Knowledge Attribute

Dimensions.	Details Captured	Example
Who	Details of Agents	The individuals involved, their skills
What	Details of Actions	Overall goals, outcome of application
Where	Details of Context	URL of distributed sources, the background of application
Why	Details of Purpose	Why was a certain decision made, what business logic dictated designs – capturing of reasoning
When	Details of Time	When did the knowledge was created
How	Details of Disposal	Tools and route involved to the implement

The XML based knowledge index support building knowledge map automatically, and provide user information about the knowledge. Knowledge engineer investigates knowledge resources at first, and then organizes the knowledge based on design task, add context information when indexing. The index of knowledge will be described in XML. Domain expert add relationship between two knowledge units. Finally, the drawing engine will out put a visualize knowledge to the user. The map is text based SVG format, so user can search the needed knowledge by key words. Fig. 2 shows the knowledge map of product design and its implement process. The complexity degree of the graphic layer is related to the number of design task. If the graphic layer contains too much knowledge units, its readability will decrease. So some complicated knowledge units should be showed as a sub-map to keep the readability of the map.

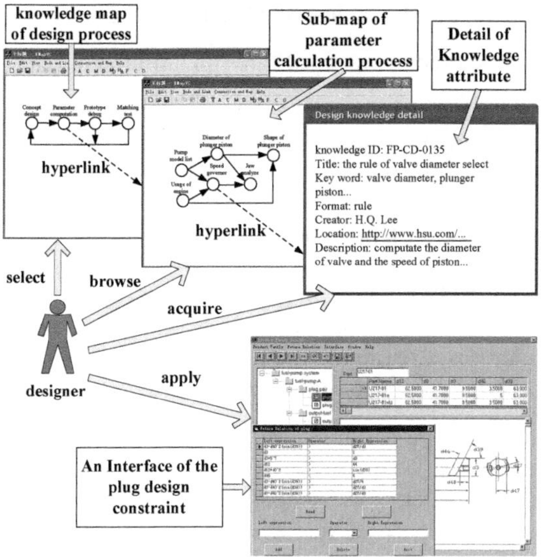

Fig. 2. The product design knowledge map and its implement process

5 The Description of Product Design Knowledge Elements

From the classification of the design knowledge, we found that the design knowledge element is algebra expression and proposition. The formula, constraint and membership function knowledge are mainly composed by algebra expression. The rule and principle, as well as patent and experience are mainly composed by propositions. The manual knowledge has both types of component.

5.1 Algebra Expression and Propositions

In algebra expression, the first work is managing the parameter and the expression. It can be described as BNF form as following:

<Expression>::=< Symbol>[{< Operator>[< Symbol>< Number>]} ...]
<Symbol>::=< Letter>{{< Letter or Number>} ...}
<Letter or Number>::= <Letter>|<Number>
<Letter>::= A| B| C| D| ...|X| Y| Z| a| b| c| d| ...|x|y|z
<Number>::= 0|1|2|3|4|5|6|7|8|9
<Operator>::= +|-|*|/

The parameter management includes parameter addition, edit, browse, deletion, unique check, usefulness check, unique check and the consistency check. The parameter validity checking is to check whether the new parameter name satisfy to the naming rule or not. The parameter unique checking is to check whether the new parameter duplicate the name of old parameter in knowledge management system. The parameter consistency checking is ask user not to modify parameter ID and parameter name, to make sure the parameter semantic is not changed when be modified.

The BNF of proposition can be described as following:

< Proposition >::= < Predication > < Status quantifier >

Here the predication can be expressed for material intensity check or life-span check for parts, homologous status quantifier can be expressed for "over strong", "qualified"," over weak", or " over high".

5.2 Description of Design Knowledge

(1) Formula type knowledge

Any formula can be assumed equality here. Considering expression-type formula, left section of equality is a parameter, and right section of equality is an expression. Formula type knowledge can be described as following according to BNF form:

<Algebra expression type formula>::=< Output parameter> < Mark of equality> <Algebra expression >

< Output parameter>::= <Symbol>

<Algebra expression>::= <Expression>

(2) Design constraint type knowledge

The design constraint type knowledge is making up of two expressions and relation operator. It can be described with BNF as follows.

< Design Constraint>::= < Expression> < Relation Operator> < Expression>

< Relation Operator>::= <|>|=| ≤ | ≥

(3) Fuzzy rule

The multidimensional fuzzy rule base can be described using BNF form:

<Fuzzy rule base >::= < Fuzzy rule>,< Fuzzy rule >, ……,< Fuzzy rule >

<Fuzzy rule>::= < premise> → < conclusion>

<Premise>:: = < Proposition > < Proposition > ……< Proposition >

<Conclusion>:: = < Proposition > < Proposition > ……< Proposition >

(4) Membership function

Membership function is a plain curve that links a series of coordinate point. Its domain, target, and fuzzy grade are also defined inside. Its BNF is described as following:

< Membership f unction >::=< Domain> < Target> < Fuzzy grade> < Point coordinates sets >

<Domain>::= < Symbol>

<Target>::=< Predication >

<Fuzzy grade>::=< Status quantifier>

< Point coordinates sets >::=< Point coordinates >,< Point coordinates >, …,< Point coordinates >

< Point coordinates >::=< Value of X coordinate >,< Value of Y coordinate >

<Value of X coordinate>::= <Number>

<Value of Y coordinate>::= <Number>

6 Knowledge Management Tools to Support Product Design

6.1 The Formula Type Knowledge Management Tool

The main function of the formula type knowledge management tool includes basic operation such as edit and addition, as well as browse and deletion. It has the function of construction validity analyze, consistency check and integrity check, as well as conflict check and syntactic analysis on the expression. And it has the function of searching knowledge in knowledge base by its parameter.

The unique check of the formula is to check whether the new formula has same name with the existing formula. The consistency constraint of the formula restricts user to modify the formula ID and its name when editing. It insures the engineering semantics of formula can not be changed. This tool has been applied to the formula-type knowledge management for air-conditioner compressor design.

6.2 Membership Function Type Knowledge Management Tool

Membership function is basic and important part for fuzzy set theory and its application. Here we describe membership function as a plain curve that links a series of coordinate point. It relate to domain and reference target, as well as some fuzzy grade. Membership function management tool control the attribute of variable, such as variable name and range, as well as unit. Its operation to membership function includes adding and deleting, as well as editing the polygonal line. It also store user's preference and control the grid of physics value. The interface of membership function management tool is composed of toolbar, fuzzy grade list, coordinate of selected point and editor window for the curve of membership function.

6.3 Fuzzy Rule Type Knowledge Management Tool

Using fuzzy rule type knowledge management tool, we can define the attribute of language variable and its value, including the language variable's range and unit as well as its name. We can also add, edit and delete some fuzzy rule. Encapsulating the inner data to C++ class, we can define rule class, rule list class and rule match degree list class. Making use of the C++, we let internal data exchange information with the outsider application procedure efficiently and safely. That tool has been used to resolve the redesign for inner bus decoration configuration.

6.4 Design Constraint Type Knowledge Management Tool

Design constraint type knowledge has very important effect on the design of complicated product, especially in concurrent development process. Making use of design constraint type knowledge management tool, we can express and manage the product structure constraint and assemble constraint, as well as recourse constraint. Constraint management tool can add, delete and modify design constraint type knowledge. It can also perform lexical analysis, syntax check and result check on every constraint. This tool includes the constraint editor and the constraint checker on relation among different parameters inside one part or inside one assembly. The constraint editor is responsible for inputting and outputting of constraint, the constraint checker is responsible for checking the constraint relation.

7 Conclusion

Many distributed system is used to allow geographically dispersed entities to work cooperatively towards product development. Knowledge map and knowledge management tool is useful to support distributed product design. Knowledge map can connect all kinds of design resource, and knowledge management tool can maintenance different kinds of design knowledge. The description of four types design knowledge had been introduced and their management tools had been developed. All these tools had been varied by fuel-pump design and air compressor design, as well as the configuration for inner bus.

Acknowledgement. The author is most grateful to National Nature Science Foundation of China (No. 70971085) and the Research Fund for the Doctoral Program of Higher Education of China (No. 20100073110035), for financial supports that made this research possible.

References

1. Salhieh, S., Monplaisir, L.: Collaboration planning framework (CPF) to support distributed product development. Journal of Intelligent Manufacturing 14(6), 581–597 (2003)
2. Borkowski, A., Branki, C., Grabska, E., Palacz, W.: Towards collaborative creative design. Automation in Construction 10(5), 607–616 (2001)
3. Zhou, S., Sang, C.K., Ling, W., Xie, Y.: Study on distributed knowledge information system for product design. Journal of Computer Science and Technology 16(5), 465–473 (2001)
4. Zhou, S.Q., Shin, K.S., Xie, Y.B., Yarlagadda Prasad, K.D.V.: The structure and implementation of a distributed integrated information system for product design. International Journal of Computer Applications in Technology 17(3), 129–138 (2003)

The Inverse Kinematics Solutions of Robot Manipulators with Offset Wrist Using the Offset Modification Method

Haihong Pan, Bing Fu, Lin Chen, and Junjie Feng

Guangxi Key Laboratory of Manufacturing System & Advanced Manufacturing Technology, School of Mechanical Engineering, Guangxi University, Nanning, Guangxi, 530004, China
hustphh@163.com

Abstract. It is no closed-form solutions for 6R robot manipulators without three consecutive joint axes parallel or three adjacent joint axes intersected a common point. In this paper, the offset modification (OM) method was applied to solve the inverse kinematics problem of these types of manipulators which didn't have closed-form solutions. A robot manipulator model of TA1400 with offset wrist was built by the OM method. And then three nonlinear equations were derived from the difference between the forward kinematics equations of the model robot manipulator and the TA1400 manipulator. Eight sets of inverse kinematics solutions of TA1400 were obtained by the three nonlinear equations with the fixed-point and the Steffensen iterative method. The results show that the orientation errors of the eight sets of solutions are all less than 10^{-15} and the position errors are less than 10^{-4} mm, which meet the requirements of the industrial robot's positions and orientations precision. The assembly experiments of TA1400 in ADAMS show that curve of each joint angle are smooth and continuous using the OM method. These results and experiments proved that the offset modification method can be applied to solve the inverse kinematics of industrial robot manipulators with front offset wrists.

Keywords: inverse kinematics, offset wrist, the offset modification (OM) method, ADAMS.

1 Introduction

The robot manipulators with offset wrist are forced to handle high payloads and to get long horizontal reach an appropriate angle [1], but they cannot get closed-form solutions for without three adjacent joint axes intersected or parallel[2, 3]. So the inverse kinematics of robot manipulators with offset wrist is an important problem that needs to be solved. Kucuk *et al.* [1] used the random value of joint one as the value of the variable to represent other joint values then used one-dimensional iterative search method to get the numerical solutions. Pashkevich [4] put forwarded two iterative algorithms to solve the problem in real time. Bu *et al.* [5] used the relationship between the robot joints' variables, transformed the high dimensional equations which contained six variables into a nonlinear equation that contained one variable and then used one dimensional iterative search to find the numerical solutions. Bingul *et al.*[6] put

G. Lee (Ed.): Advances in Automation and Robotics, Vol. 1, LNEE 122, pp. 655–663.
springerlink.com

forwarded an artificial neural network using backpropagation algorithm to find the numerical solutions.

These studies remained in the analysis of algorithms, and those algorithms are difficult to achieve in engineering practice for their complexity algorithm, substantial computation, and they required an iterative searching for the solutions. Here the OM method was adopted to solve the problem of inverse kinematics for TA1400 with offset wrist, and the application of the OM method was verified with TA1400 in ADAMS.

2 The Solutions of Forward and Inverse Kinematics

6R robot manipulators are widely used in industrial production and they produced huge technical and economic benefits. Coordinate systems were attached to TA1400 by means of the Denavit-Hartenberg (D-H) method as shown in Fig. 1. Table 1 was the D-H parameters of TA1400.

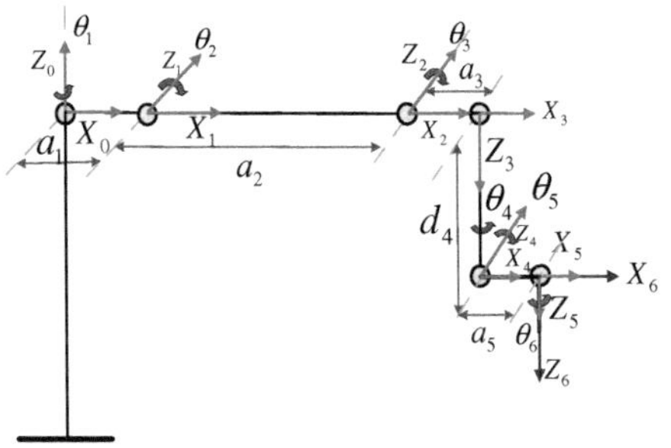

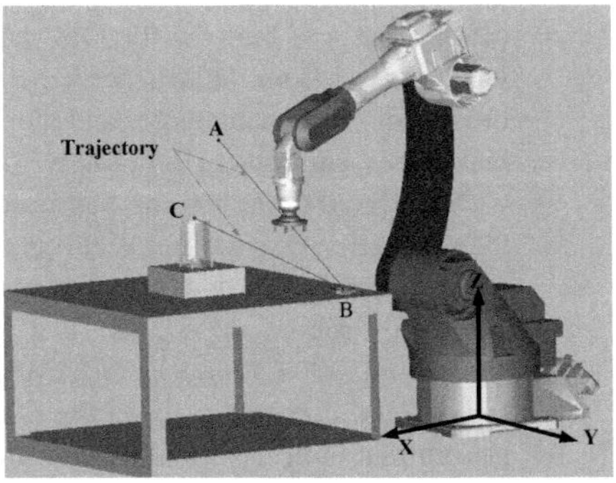

Fig. 1. The D-H parameters of the robot TA1400 **Fig. 2.** Robot trajectory simulation in ADAMS

Table 1. D-H parameters for TA1400

Joint	θ_i	α_i	a_i	d_i
1	θ_1	-90°	150	0
2	θ_2	0°	560	0
3	θ_3	-90°	130	0
4	θ_4	90°	0	600
5	θ_5	-90°	40	0
6	θ_6	0°	0	0

Table 2. D-H parameters for model robot

Joint	θ_i	α_i	a_i	d_i
1	θ_1	-90°	150	0
2	θ_2	0°	560	0
3	θ_3	-90°	130	0
4	θ_4	90°	0	600
5	θ_5	-90°	0	0
6	θ_6	0°	0	0

According to the D-H parameters in Table.1.The kinematics equation is described as follows:

$$T_6 = \begin{bmatrix} \vec{n} & \vec{o} & \vec{a} & \vec{p} \\ 0 & 0 & 0 & 1 \end{bmatrix} = A_1^0 A_2^1 A_3^2 A_4^3 A_5^4 A_6^5 \cdot \tag{1}$$

Where: $\vec{n} = \begin{bmatrix} n_x & n_y & n_z \end{bmatrix}^T$ is the normal vector of the robot manipulator, $\vec{o} = \begin{bmatrix} o_x & o_y & o_z \end{bmatrix}^T$ is the sliding vector of the robot manipulator, $\vec{a} = \begin{bmatrix} a_x & a_y & a_z \end{bmatrix}^T$ is the approaching vector of the robot manipulator, and $\vec{p} = \begin{bmatrix} p_x & p_y & p_z \end{bmatrix}^T$ is the position coordinates of the end-effector of the robot manipulator.

The pose of coordinate frame i with respect to the coordinate frame i-1 is described with homogeneous transformation matrix as Eq. 2.

$$A_i^{i-1} = \begin{bmatrix} \cos\theta_i & -\sin\theta_i\cos\alpha_i & \sin\theta_i\sin\alpha_i & a_i\cos\theta_i \\ \sin\theta_i & \cos\theta_i\cos\alpha_i & -\cos\theta_i\sin\alpha_i & a_i\sin\theta_i \\ 0 & \sin\alpha_i & \cos\alpha_i & d_i \\ 0 & 0 & 0 & 1 \end{bmatrix} \cdot \tag{2}$$

Substituting values of joint angles into the Eq. 2, and then took these results matrixes into Eq. 1. Thus, the position and orientation matrix of the end-effector and the forward kinematics solution can be obtained.

Because robot manipulators of TA1400 with offset wrist doesn't exist any closed-form inverse kinematics solutions and only numerical methods can be used to solve the problem. According to the offset modification method proposed by Gregory [7], the procedures of deriving the inverse kinematics algorithm of TA1400 are as follows:

(1) Build a model robot manipulator of TA1400

Modified the value of parameter a_5 (in Table 1) and made it equals to zero as shown in Table 2. Now the model robot has an Euler wrist, so it has closed-form inverse kinematics solutions.

Hence, the end-effector the model robot and real robot had the same position and orientation when they have the same joint values [7].

(2) Obtain three nonlinear equations

When the model robot and the robot TA1400 have the same joint angles ($\theta_r = \theta_m$), we can acquire Eq.3 and Eq. 4.

$$T_{ereal} - T_{emodel} = \begin{bmatrix} \vec{0} & \vec{0} & \vec{0} & \vec{p}_m - \vec{p}_r \\ 0 & 0 & 0 & 1 \end{bmatrix}; \tag{3}$$

$$\Delta\vec{p} = \vec{P}_m - \vec{P}_r = \begin{bmatrix} -((c_1c_2c_3 - c_1s_2s_3)c_4 + s_1s_4)a_5c_5 - (-c_1c_2s_3 - c_1s_2c_3)a_5s_5 \\ -((s_1c_2c_3 - s_1s_2s_3)c_4 + c_1s_4)a_5c_5 - (-s_1c_2s_3 - s_1s_2c_3)a_5s_5 \\ -(-s_2c_3 - c_2s_3)c_4a_5c_5 - (s_2s_3 - c_2c_3)a_5s_5 \end{bmatrix} \cdot \tag{4}$$

Where: T_{ereal} is the positions and orientations of the end-effector coordinate system of the robot; T_{emodel} is the positions and orientations of the end-effector coordinate system of the model robot; $\vec{P}_r = \begin{bmatrix} p_{rx} & p_{ry} & p_{rz} \end{bmatrix}^T$ is the position of the end-effector coordinate system respect to the robot's base coordinate system; p_{rx}, p_{ry}, p_{rz} stands for the projection of the end-effector position in the base coordinate system respectively; $\vec{P}_m = \begin{bmatrix} p_{mx} & p_{my} & p_{mz} \end{bmatrix}^T$ is the positions of the end-effector coordinate system respect to the model robot's base coordinate system, the p_{mx}, p_{my}, p_{mz} stands for the projection of the end-effector position in the base coordinate system respectively.

$\vec{P}_m$ is given by Eq. 5,

$$\vec{P}_m = \vec{P}_r + \vec{\Delta p}(\theta_m) \cdot \tag{5}$$

(3) Solve the inverse kinematics solutions of the model robot

$$T_6 = \begin{bmatrix} \vec{n} & \vec{o} & \vec{a} & \vec{p}_m \\ 0 & 0 & 0 & 1 \end{bmatrix} = A_1^0 A_2^1 A_3^2 A_4^3 A_{5m}^4 A_6^5 \cdot \tag{6}$$

Where: A_{5m}^4 is the new matrix of A_5^4 by D-H parameter in table 2.

Transformed the Eq. 6 then obtained Eq. 7:

$$A_3^2 A_4^3 A_{5m}^4 = (A_2^1)^{-1}(A_1^0)^{-1} T_6 (A_6^5)^{-1} . \tag{7}$$

Next, the element (3,4) in this matrixes of both sides of the Eq. 7 corresponds to equal. The value of joint one was solved as Eq. 8.

$$0 = -s_1 p_x + c_1 p_y \text{;} \quad \theta_1 = a\tan 2(p_y, p_x) - a\tan 2(0, \pm\sqrt{p_x^2 + p_y^2}) \cdot \tag{8}$$

And the elements (1,4) and (2,4) in the matrixes of both sides of the Eq. 7 corresponds to equal, the angle value of joint two and joint three were solved by the following equations:

$$-s_3 d_4 + a_3 c_3 = c_1 c_2 p_x + s_1 c_2 p_y - s_2 p_z - c_2 a_1 - a_2 ; \tag{9}$$

$$c_3 d_4 + a_3 s_3 = -c_1 s_2 p_x - s_1 s_2 p_y - c_2 p_z + s_2 a_1 . \tag{10}$$

Squared both sides of Eq. 9 and Eq.10, then computed the sum of the equations' same side and obtained the value of joint two. And then s_2 and c_2 can be solved.

$$\theta_2 = a\tan 2(c_1 p_x + s_1 p_y - a_1, p_z) - a\tan 2(t, \pm\sqrt{(c_1 p_x + s_1 p_y - a_1)^2 + p_z^2 - t^2}) .$$

Where, $t = [(c_1 p_x + s_1 p_y - a_1)^2 + p_z^2 + a_2^2 - d_4^2 - a_3^2] / 2a_2$.

The s_3, c_3 and θ_3 can be obtained by considering Eq. 9 and Eq. 10.

$$s_3 = \frac{a_3(-c_1 s_2 p_x - s_1 s_2 p_y - c_2 p_z + s_2 a_1) - d_4(c_1 c_2 p_x + s_1 c_2 p_y - s_2 p_z - c_2 a_1 - a_2)}{d_4^2 + a_3^2} ;$$

$$c_3 = \frac{a_3(c_1 c_2 p_x + s_1 c_2 p_y - s_2 p_z - c_2 a_1 - a_2) + d_4(-c_1 s_2 p_x - s_1 s_2 p_y - c_2 p_z + s_2 a_1)}{d_4^2 + a_3^2} ;$$

$$\theta_3 = a\tan 2(s_3, c_3).$$

Transformed the Eq. 6 then obtained Eq. 11:

$$A_4^3 A_{5m}^4 A_6^5 = (A_3^2)^{-1}(A_2^1)^{-1}(A_1^0)^{-1}T_6.$$

Next, the matrixes elements (3, 3) in both sides of the Eq. 11 correspond to equal. The value of joint five was solved as follows:

$$c_5 = (-s_3c_2 - c_3s_2)c_1a_x + (-s_3c_2 - c_3s_2)s_1a_y + (-c_3c_2 + s_3s_2)a_z;\ s_5 = \pm\sqrt{1-c_5^2};$$

$$\theta_5 = a\tan 2(s_5, c_5).$$

Use elements (1, 3) and (2, 3), when $s_5 \neq 0$, joint four value is solved as follows:

$$-c_4s_5 = (c_2c_3 - s_2s_3)c_1a_x + (c_2c_3 - s_2s_3)s_1a_y + (-s_3c_2 - c_3s_2)a_z;$$

$$-s_4s_5 = s_1a_x - c_1a_y;\ \theta_4 = a\tan 2(s_4, c_4).$$

The Eq.12 and Eq.13 are obtained by equated both the elements (3,1) and (3,2) in the matrixes two sides of the Eq. 11.

$$-s_4c_5 = (-s_1n_x + c_1n_y)c_6 - (-s_1o_x + c_1o_y)s_6; \tag{12}$$

$$-c_4 = (-s_1n_x + c_1n_y)s_6 + (-s_1o_x + c_1o_y)c_6. \tag{13}$$

s_6, c_6 and θ_6 solved by considering the Eq. 12 and Eq. 13:

$$s_6 = \frac{-(-s_1n_x + c_1n_y)c_4 + (-s_1o_x + c_1o_y)s_4s_5}{(-s_1n_x + c_1n_y)^2 + (-s_1o_x + c_1o_y)^2};\ c_6 = \frac{-(-s_1n_x + c_1n_y)s_4s_5 - (-s_1o_x + c_1o_y)c_4}{(-s_1n_x + c_1n_y)^2 + (-s_1o_x + c_1o_y)^2};$$

$$\theta_6 = a\tan 2(s_6, c_6).$$

When $s_5 = 0$, the 4th and 6th joint axes of the model robot are in the same line. In this case, only the sum or the difference of the angle of joint four and six can be calculated. In addition, the robot manipulator was now in singularities. Then the angle of joint four can be set any value, and the corresponding value of the joint six can be computed.

(4) Compute the inverse kinematics solutions with iterative methods

Fixed-point iterative algorithm is as follows:

$$\vec{P}_m^{k+1} = \vec{P}_{rd} + \Delta\vec{P}(\theta^k); \tag{14}$$

$$\vec{P}_r^{k+1} = \vec{P}_m^k - \Delta\vec{P}(\theta^k); \tag{15}$$

$$\theta^{k+1} = ikine(\begin{bmatrix} \vec{n} & \vec{o} & \vec{a} & \vec{p}_m^{k+1} \\ 0 & 0 & 0 & 1 \end{bmatrix}, c). \tag{16}$$

Where: $\vec{P}_{rd}$ is the desired endpoint position of the robot manipulator; ikine(T6, c) stands for the function to solve the inverse kinematics solutions of the model robot, the c stands for combination of symbols of the manipulator among the set of 8 solutions for the model robot, its value come from the combination of±1, ±1, ±1. The iterative algorithm stopped as Eq. 17, and then the effective inverse kinematics solutions were solved.

$$\left\| \vec{P}_r^{k+1} - \vec{P}_{rd} \right\| \le E_{\max} . \tag{17}$$

Where: $E_{\max}$ is the allowed position error of the end-effector.

Here if the Eq.17 does not hold by used the fixed-point iterative method, then the Steffensen iterative method was used to compute solution of inverse kinematics.

$$\vec{y}_m^k = \vec{p}_{rd} + \Delta \vec{p}(ikine\left(\begin{bmatrix} \vec{n} & \vec{o} & \vec{a} & \vec{p}_m^k \\ 0 & 0 & 0 & 1 \end{bmatrix}, c\right)); \tag{18}$$

$$\vec{z}_m^k = \vec{p}_{rd} + \Delta \vec{p}(ikine\left(\begin{bmatrix} \vec{n} & \vec{o} & \vec{a} & \vec{y}_m^k \\ 0 & 0 & 0 & 1 \end{bmatrix}, c\right)); \tag{19}$$

$$\vec{p}_m^{k+1} = \vec{p}_m^k - (\vec{y}_m^k - \vec{p}_m^k)^2 / (\vec{z}_m^k - 2\vec{y}_m^k + \vec{p}_m^k) . \tag{20}$$

Where: $\vec{y}_m^k$ 、 $\vec{z}_m^k$ are two intermediate variables in the Steffensen iterative method, they indicated the position of the model robot end-effector.

3 Experimental Results and Discussion

Here, we assumed that the angle values of joint one through six are 30°, -60°, 90°, 60°, 120° and 60°. Substituting them into Eq. 1 and Eq. 21 of the position and orientation matrix of the end-effector was obtained. Then the algorithm in Section 2 was used to calculate the inverse kinematics solutions of the robot. Eight sets of real numerical solutions were obtained and shown in Table 3.

$$T_6 = \begin{bmatrix} -0.7354968245269 & 0.4748797632096 & -0.4832531754731 & 187.5833024919770 \\ -0.6746392896287 & -0.4475158773653 & 0.5870190528383 & 128.3012701892220 \\ 0.0625000000000 & 0.7577722283114 & 0.6495190528383 & -124.6410161513776 \\ 0 & 0 & 0 & 1 \end{bmatrix} . \tag{21}$$

Table 3. Eight sets of solutions

Sets	Joint One	Joint Two	Joint Three
1	30.887728797904	-173.244873171616	112.603504792389
2	34.655359337414	150.426301647223	1.172383531573
3	30.000008374997	-59.999993457207	89.99999345723
4	36.617407017265	-28.477180848592	0.895233259568
5	-152.174164153989	-116.136968092029	139.923968362875
6	-139.743734586488	-125.376233730744	145.0049717999627
7	-140.985514687611	79.304509094676	64.052775959873
8	-150.067582246175	90.636273131246	60.198917695852
Sets	**JointFour**	**JointFive**	**JointSix**
1	124.177346464112	114.660970314742	166.214331699813
2	-49.305386160388	-92.131693942497	9.615503284586
3	60.000006044121	120.000003626468	59.999997582323
4	-127.684286242143	-106.344633857590	-150.892490633195
5	-80.723352993124	131.010721348346	126.284256823173
6	106.701376306486	-127.4620420202494	-60.2502809087677
7	-62.700890231282	58.827963070004	-129.512902501834
8	119.627473330126	-59.613698206703	60.857582373221

In the Table 3, the 3rd solution is very approaching to the former assume joint values. In order to verify the quality of these solutions, each set of solutions was substituted into Eq. 1 to compute results of corresponding positions and orientations matrixes of the end-effector. The maximum error of the eight sets solutions was computed by the Eq. 22.

$$T_{errorMax} = Max(T_6 - A_{1i}^0 A_{2i}^1 A_{3i}^2 A_{4i}^3 A_{5i}^4 A_{6i}^5)\,. \tag{22}$$

Where, $i = 1 \ldots 8$;

The maximum error matrix is as follows:

$$T_{errorMax} = \begin{bmatrix} -0.00000000000222 & 0.00000000000111 & -0.00000000000111 & -0.860897894767731 \\ 0.00000000000111 & -0.00000000000222 & 0.00000000000111 & -0.274830444482177 \\ 0 & -0.00000000000111 & 0 & 0.392137740874432 \\ 0 & 0 & 0 & 0 \end{bmatrix} \times 10^{-4}.$$

The 5th solution is the maximum error. But the position error of the 5th was less than 10^{-4} *mm*, and the orientation error was less then 10^{-15} which meet the requirements of the position and orientations precision of industrial robots.

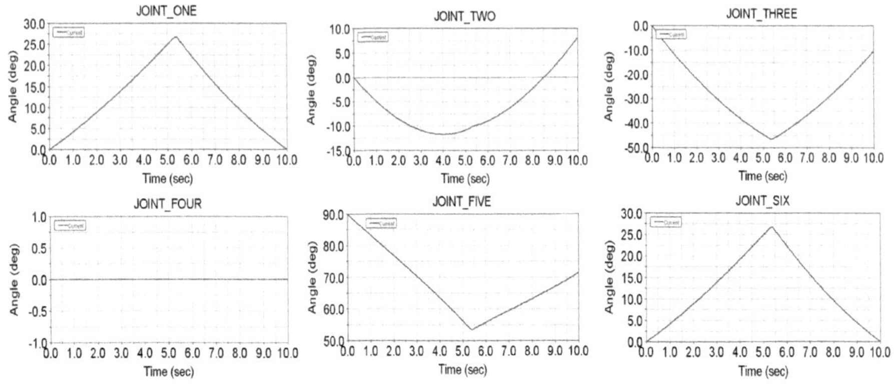

Fig. 3. Six curves of each joint angle in ADAMS

The virtual prototyping of robot manipulator TA1400 is built in ADAMS and do assembly experiments. The assembly task is that the end-effector of robot started from point A(800, 0, 690), moved to the point B(590, 300, 298), and reached to the point C(890,0,487). The coordinates of middle points in line segment's AB and BC were calculated by linear interpolation. The algorithm in Section 2 was used to calculate the corresponding solutions of inverse kinematics. Then, selecting a set of optimal solutions for each interpolation point as the joint trajectory, and import these trajectories into ADAMS for assembly experiments. In Fig.2The blue line is the particular trajectory of the robot. The robot TA1400 was moving along this trajectory in assembly experiments. Six curves of each joint angle were shown in Fig.3. These curves were smooth and continuous because the positions and orientations precision are very high by using the OM method.

4 Conclusion

The inverse kinematics solutions of TA1400 with offset wrist were solved by the OM method. Eight sets of solutions were obtained with the three nonlinear equations through the fixed-point and the Stenfensen iterative algorithm. The results show that the orientation errors of the eight sets of solutions are all less than 10^{-15} and the position errors are less than 10^{-4} mm, they all meet the requirements of the industrial robot's positions and orientations precision. The experiment indicates that the OM method can solve the inverse kinematics of the robot TA1400 and get the feasible solutions.

Acknowledgments. This work was supported by the National Natural Science Foundation of China (Grant No. 51065005), the Guangxi Key Laboratory of Manufacturing System & Advanced Manufacturing Technology (Grant No. 09-007-05S017) and the Science and High-Tech Research of Nanning (Grant No. 200802026A).

References

1. Kucuk, S., Bingul, Z.: Inverse Kinematics Solutions of Fundamental Robot Manipulators with Offset Wrist. Mechatronics. pp. 197–201 (2005)
2. Pieper, D.L.: The kinematics of Manipulators under computer control. PhD dissertation, Comput. Sci. Dept. Stanford Univ. (1968)
3. Duffy, J.: Analysis of Mechanisms and Robot Manipulators. Edward Arnold in press, London (1980)
4. Pashkevich, A.: Real-Time Inverse Kinematics for Robots with Offset and Reduced Wrist. Control Engineering Practice 5, 1443–1450 (1997)
5. Bu, W., Liu, Z., Tan, J.: Inverse displacement analysis of 6R robots with offset wrists based on decoupling degrees of freedom at the cutoff points. Chinese Journal of Mechanical Engineering 46, 1–5 (2010)
6. Bingul, Z., Ertunc, H.M., Oysu, C.: Comparison of Inverse Kinematics Solutions Using Neural Network for 6R Robot Manipulator with Offset. Computational Intelligence Methods and Applications, 1–5 (2005)
7. Grudic, G.Z., Lawrence, P.D.: Iterative Inverse Kinematics with Manipulator Configuration Control. Transactions on Robotics and Automation 9, 476–483 (1993)

Author Index

GPSR Compliance

The European Union's (EU) General Product Safety Regulation (GPSR) is a set of rules that requires consumer products to be safe and our obligations to ensure this.

If you have any concerns about our products, you can contact us on ProductSafety@springernature.com

In case Publisher is established outside the EU, the EU authorized representative is:

Springer Nature Customer Service Center GmbH
Europaplatz 3
69115 Heidelberg, Germany

Batch number: 10704025

Printed by Printforce, the Netherlands